ALLE · ZEIT · WACH · 1842

The Sea Urchin Embryo

Biochemistry and Morphogenesis

Edited by G. Czihak
With the Assistance of R. Peter

Contributors
F. Baltzer B. Brandriff P. S. Chen T. Gustafson R. Hinegardner
S. Hörstadius M. Ishikawa N. Isono Y. Isono R. Lallier
M. von Ledebur-Villiger G. Millonig K. Okazaki K. Osanai
J. Piatigorsky R. Rappaport J. Runnström R. C. Rustad
L. Schmekel T. Yanagisawa

Foreword by J. Brachet

Springer-Verlag Berlin Heidelberg New York 1975

Professor Dr. G. CZIHAK, Lehrkanzel für Genetik und Entwicklungsbiologie der Universität Salzburg, Porschestr. 8/V, A–5020 Salzburg/Austria

With 352 Figures

ISBN-13: 978-3-642-65966-9   e-ISBN-13: 978-3-642-65964-5
DOI: 10.1007/978-3-642-65964-5

Library of Congress Cataloging in Publication Data. Czihak, G. The sea urchin embryo--biochemistry and morphogenesis. 1. Sea urchin embryo. I. Peter, Roland, joint author. II. Baltzer, Friedrich, 1884– III. Title. [DNLM: 1. Sea urchins--Embryology. QL384.E2 C998s] QL958.C94 593'.95 74-16245

Offsetprinting: Julius Beltz, Hemsbach/Bergstr. Bookbinding: Brühlsche Universitätsdruckerei, Gießen.

Dedicated to the

STAZIONE ZOOLOGICA DI NAPOLI

where most of the classic work on the sea urchin embryo

had been performed

Foreword

Sea urchin eggs are objects of wonder for the student who sees them for the first time under the microscope. The formation of the fertilization membrane after insemination, the beauty of mitotic cleavage, the elegant swimming of embryos, remain an esthetic pleasure even for the eyes of seasoned investigators. But sea urchin eggs have other, more practical, advantages: they lend themselves to surgical operation without difficulty and they heal perfectly; they can be obtained in very large amounts and represent thus an extremely favorable material for biochemists and molecular embryologists.

It is not surprising that, in view of these exceptional advantages, sea urchin eggs have attracted the interest of innumerable biologists since O. HERTWIG discovered the fusion of the pronuclei (amphimixy), in *Paracentrotus lividus*, almost a century ago. The purpose of the present book is to present, in a complete and orderly fashion, the enormous amount of information which has been gathered, in the course of a hundred years of sea urchin embryology. JOSEPH NEEDHAM, in 1930, was still able to present all that was known, at that time, on the biochemistry of all possible species of developing eggs and embryos in his famous "Chemical Embryology" (Cambridge University Press). It would no longer be possible for one man to write a modern version of what was a "Bible" for the young embryologists of forty years ago. Science has made such progress that, even when only sea urchin eggs are considered, many authors would have to contribute to a treatise which could today play the eminent role filled by "Chemical Embryology" almost half a century ago. The present book, which was written by many specialists, will be, for years to come, the *opus magnum* of sea urchin embryology.

The book, which has about 700 pages, is not easy to read from the first to the last line; it is rather meant as a reference book; a tremendous wealth of factual material and a very extensive bibliography are provided after each chapter.

The first chapter is an excellent *historical introduction* to the subject by S. HÖRSTADIUS, the undisputed leader in sea urchin experimental embryology. The fundamental contributions of the German school of *Entwicklungsmechanik* (O. and R. HERTWIG, H. DRIESCH, T. BOVERI, C. HERBST, W. ROUX, etc.) are presented with authority and due respect for these great names. This chapter is particularly useful for molecular biologists wishing to shift to molecular embryology: they often know a good deal about gene regulation in Prokaryotes, less about the same problem in Eukaryotes, but very little or nothing about embryonic regulation - a still unsolved problem - or the pioneer work of H. DRIESCH.

The next chapter, by R. HINEGARDNER and by K. OSAKAI and others, gives a wealth of extremely valuable practical information: Where can one obtain sea urchins in North America, Japan or Europe? When are they ripe? How can one collect eggs and sperm, fertilize eggs, culture embryos and adults, etc.? Very precise answers are given to all these questions.

Chapter 3, by J. PIATIGORSKY, deals with *gametogenesis* (formation of the germ cells, spermatogenesis, oogenesis): structure and ultrastructure of the gonads are described in detail; but the cytochemical and biochemical aspects of the problem are not forgotten: oxygen consumption, DNA, RNA and protein synthesis are presented in a clear and detailed way. The chapter ends with a long list of still unanswered questions;they are important and it is to be hoped that these last pages will induce molecular biologists to try to answer them.

The subjects of the next two chapters are *fertilization* and *parthenogenesis*; these two important, closely linked, events are discussed, as is right, by a single author, M. ISHIKAWA. As in the whole book, the chronology of the embryological events is followed as a guide: attraction of spermatozoa,acrosome reaction of the spermatozoon, fusion of sperm and egg after the spermatozoon has crossed the jelly coat and the vitelline membrane, migration of the male pronucleus in the oocyte, sperm aster formation, fusion of the pronuclei, and cortical reaction are described in detail at the level of both light and electron microscopy. Many experiments on partial fertilization, sperm injection, effects of narcotics, refertilization and polyspermy, as well as the methods used in order to induce parthenogenesis, are presented and discussed. The author also discusses the classical theory of J. LOEB and the important biochemical changes (energy production, effects of sulfhydril reagents, protein and DNA synthesis) in fertilized and parthenogenetically activated eggs. As a kind of appendix to this section, M. VON LEDEBUR-VILLIGER, in Chapter 5, gives more details about the cytology of parthenogenetically activated eggs (formation of monasters and cytasters; polyploid and haploid mitoses, etc.).

Fertilization is followed by cleavage, hatching of the blastula, gastrulation,formation of prisms, plutei,late larval stages, and finally morphogenesis: all these stages of *sea urchin development* are masterly described in a single chapter by K. OKAZAKI. More detail is given about cell movements and cell interactions during morphogenetic movements by T. GUSTAFSON in Chapter 8: time-lapse studies, electron microscopy, cytochemistry, and effects of neuropharmacological agents such as serotonin have given important information about cellular behavior during cleavage and morphogenesis in ectoderm and mesendoderm. In the first part of the next chapter (Chapter 9, by L. SCHMEKEL), which deals with the ultrastructure of egg and embryo, there are unavoidable repetitions in the section dealing with the egg, spermatozoon and fertilization, which are, however, fully compensated by the excellent electron micrographs. The second part of the chapter, which describes the ultrastructure from the 2-cell stage to the gastrula, is more original and equally well illustrated.

In the next three chapters, we go back to *cleavage*. R. RAPPAPORT, in Chapter 10, discusses the various cleavage theories and the biophysics (viscosity, resistance to deformation, birefringence, etc.) of cleavage, as well as the alterations of cleavage pattern which follow compression or displacement of the mitotic apparatus in the fertilized egg. The suggestion is made that an alteration of the equatorial surface induced by the asters is responsible for furrow formation; the experiments by HIRAMOTO on cell division without mitotic apparatus and those by SCHROEDER on the contractile ring filaments are perhaps not discussed with enough detail. The following chapter, by B. BRANDRIFF and R. HINEGARDNER, is devoted to the many *centrifugation* experiments which have been done in order to analyze cleavage, formation of the micromeres, development of the dorso-ventral axis etc. in sea urchin eggs. The well-known experiments of E.B. HARVEY on the separation of nucleate and anucleate halves by centrifugation are presented and the biochemical data about respiration and protein synthesis in such halves are briefly given. Finally, in Chapter 12, R. RUSTAD discusses, on the basis of many experiments (irradiation, effects of mercaptoethanol, etc.) the regulation of the cell cycle during sea urchin egg cleavage. Some results concerning macromolecule synthesis are also given.

Chapter 13 is a particularly important one:it has been written by S. HÖRSTADIUS, to whom we owe so much of our knowledge about *experimental embryology* of sea urchin eggs. The author describes with great clarity his own classical experiments on determination of cleavage, on animal-vegetal polarity and on dorso-ventral axis. These experiments have led to the well-known double-gradient theory; the chapter ends by describing more recent experiments by JOSEFSSON and HÖRSTADIUS showing that unfertilized sea urchin eggs contain animalizing and vegetalizing substances.

Chapter 14, by G. MILLONIG, deals with a very interesting subject: *dissociation* and *reaggregation* of sea urchin embryos. The ultrastructure of reaggregating blastomeres is described in detail, but the biochemical work of G. GIUDICE on isolated blastomeres also receives the attention it deserves.

The subject of the following chapter, by P.S. CHEN and the much regretted F. BALTZER, is no less important and interesting: it deals with morphology (in particularly chromosome analysis) and biochemistry (respiration, DNA, RNA and protein synthesis) in *sea urchin hybrids*. Both diploid and haploid (androgenetic merogones) hybrids are discussed with F. BALTZER'S authority and P.S. CHEN'S biochemical competence. Having worked on the subject, I can express due admiration for this excellent presentation.

With Chapter 16, by R. LALLIER, we go back to morphogenesis: LALLIER again discusses the important question of *animalization* and *vegetalization*, but his outlook is different. While HÖRSTADIUS analyzed the properties of the animal and vegetal blastomeres (in particular the micromeres) by micro-surgery, LALLIER deals with the great variety of substances which can exert animalizing or vegetalizing effects when they are added to intact fertilized eggs. The classical vegetalizing agent, lithium, has been known since the days of HERBST; but other substances, such as isoniazid, chloramphenicol, etc. are also vegetalizing. The list of animalizing agents, on the other hand, is extremely long and varied (ions, heavy metals, polyanionic and polycationic substances, sulfhydril groups, proteases, etc.) The effects of carbohydrate and nucleic acid metabolism of these very useful substances, which also modify morphogenesis are also examined. But one must admit that, today, we have no simple and convincing molecular explanation of the mode of action either of animalizing or vegetalizing agents; an addendum to the chapter indicates, however, that we might now be moving on the right track.

The next three chapters are an incursion in the field of sea urchin egg *biochemistry*. T. YANAGISAWA gives an extensive, accurate, and complete account of *energy production* and *carbohydrate metabolism* in Chapters 17 and 18. One of the main problems goes back to Warburg's discovery, in 1908, that fertilization increases several-fold the oxygen consumption of sea urchin eggs; we now know that this is not a general phenomenon and that unfertilized sea urchin eggs are exceptional in having an exceedingly low *respiration*. But the control mechanisms which allow the very fast increase in oxygen consumption after insemination or parthenogenetic stimulation remain elusive despite the efforts made by many workers, in particular J. RUNNSTRÖM and his school. Many hypotheses have been proposed in order to explain the low oxygen consumption of unfertilized sea urchin eggs (lack of oxidizable substrates, etc.);

none of them has so far been convincingly demonstrated. T. YANAGISAWA'S excellent presentation might lead to new ideas and to experiments which will solve the mystery. It is of course very fitting to have chapters on *carbohydrate* (T. YANAGISAWA) and *lipid* (Y. and N. ISONO) *metabolisms* following the one dealing with respiration and energy production. The chapter on lipids clearly shows how much these important biochemical constituents have been neglected: the authors give extensive tables which are reminiscent of J. NEEDHAM'S "Chemical Embryology", but less than one page about lipid metabolism. Now that we know the essential role played by the lipid bilayer in all cell membranes, we must hope for a more dynamic approach to lipid metabolism in sea urchin eggs.

Chapter 20, by M. VON LEDEBUR-VILLIGER, is a first step in sea urchin *molecular embryology*; it deals with DNA in gametes and early development. The subject is handled in less than 15 pages; this brevity has the advantage of reducing repetition of what was already said about DNA in gametes to a minimum; but a more detailed treatment of such a fundamental problem would have been desirable (the same subject was treated, almost 10 years ago, in more than 50 pages by E. SCARANO and G. AUGUSTI-TOCCO). The reasons why only the essentials have been given in this chapter are clear: Prof. CZIHAK explains, in his Preface, that the molecular aspects of sea urchin embryology will be presented in a second volume. It is true that research on macromolecule synthesis during sea urchin development is very intensive today and connot be adequately discussed in a few pages; it is indeed moving forward so fast at the present time that it was probably a sound decision to postpone publication of chapters dealing with nucleic acid and protein synthesis, regulation of gene activity, etc. This is a field where substantial progress can no longer be made by spending 2 or 3 months working in a marine biological laboratory, but where full-time work on sea urchins in a well-equipped laboratory has become an absolute necessity.

The last chapter is deeply moving for all those who knew its author, J. RUNNSTRÖM, who started working on sea urchin eggs in 1914 and continued to work on them until his recent death. This chapter, entitled *Integrating Factors*, is the result of about 60 years of scientific research on sea urchin eggs: it is the scientific testament of a great scientist. The chapter ends with a theoretical explanation of sea urchin development which takes into account the existence of animalizing and vegetalizing substances (called morphogenes), their distribution along a double, antagonistic gradient, their possible regulatory effects on gene activity. The scheme presented on the very last page of the book (p. 664) attempts to explain both animal-vegetal and dorsal-ventral differentiation: it is a valid model, which summarizes years of thinking and experimenting by the "great old man" of sea urchin embryology.

This model and the huge amount of factual evidence gathered should be an incentive for further research. I am convinced that this book will indeed serve its purpose and that it will induce many young biologists (molecular or not) to work on that fascinating object: the sea urchin embryo.

Brussels, July 1975 J. BRACHET

Preface

Sea urchin eggs belong to the most favored research material in modern zoology. To the study of sea urchin eggs, indubitably the best known eggs, we owe some milestone discoveries such as the foundation of the chromosome theory of heredity, the development of isolated embryonic halves, the fertilization of eggs by spermatozoa, cell respiration, and important concepts such as the existence of gradients.

When I was invited by Prof. REVERBERI to write an article on experimental embryology of the sea urchin embryo in his handbook entitled: "Experimental Embryology of Marine and Fresh-water Invertebrates", I willingly accepted, not realizing how difficult it would be to condense the accumulated knowledge of a century's intense work into some 80 pages. Later becoming convinced that this brief survey could never mirror all the thougths and results in this field, I decided to edit a comprehensive handbook collecting all the important data with as few comments as possible by the reviewing authors.

Many generations of students have been fascinated by the transparency of the eggs in which for instance the penetration of sperm or appearance and breakdown of mitotic figures can easily be followed; as also by the synchrony of cleavage, which allows experimentation according to strictly set up time schedules. Some sea urchins are moreover unique in that they are ripe throughout almost the whole year. One animal may contain up to several millions of eggs or innumerable spermatozoa, being almost pure DNA preparations. The eggs are relatively poor in yolk, an adventage for the biochemist, who does not then need to seek for a needle in a hay-stack when isolating certain molecules.

The reader will miss chapters on molecular biology. The great progress in this field in our days has, however, not been supported through use of embryonic material. Our knowledge of transcription and translation in sea urchin eggs in most cases is only superficial and inconsistent and the field remains wide open for further studies. I finally became convinced that it would be premature to summarize the findings of molecular biology of the sea urchin embryo and therefore I suggest the possibility of an additional volume after some years. The reader will moreover miss a chapter on genetics of development. As a consequence of the difficulties in raising sea urchins to maturity, no research in genetics was carried out until now. In this respect the sea urchin egg has obvious disadvantages compared with other biological material.

Any handbook like this present one needs editorial treatment. A less severe one was carried out by Dr. R. PETER, P. CHANG and myself, a more severe by the copy editors of the publisher, which considerably condensed the information. We hope, however, to have produced a work which is satisfactory compromise between the widely differing demands.

My thanks go first of all to the authors for their patience and furthermore to Mrs. U. BUJARD at Springer-Verlag for her excellent final supervision of printing and to A. ASCHOFF for compiling the subject index.

Montpellier/Salzburg, July 1975 G. CZIHAK

Contents

List of Contributors

Professor Dr. F. BALTZER†, formerly: Finkenhubelweg 6, CH-3000 Bern/Switzerland

Dr. Brigitte BRANDRIFF, Division of Natural Sciences I, University of California, Santa Cruz, CA 95060/USA

Professor Dr. P.S. CHEN, Zoologisch-Vergl. Anatomisch. Institut der Universität, CH-8006 Zürich/Switzerland

Dr. T. GUSTAFSON, Wenner Grens Institute, Norrtullsgatan 16, S-11345 Stockholm Va/Sweden

Professor Dr. R. HINEGARDNER, Div. Natural Sciences I, University of California, Santa Cruz, CA 95060/USA

Professor Dr. S. HÖRSTADIUS, Box 561, S-75122 Uppsala/Sweden

Professor Dr. M. ISHIKAWA, Biological Institute, Faculty of Science, Nagoya University, Nagoya 464/Japan

Dr. N. ISONO, Dept. of Biology, Tokyo Metropolitan University, Setagaya-ku, Tokyo/Japan.

Dr. Yuhko ISONO, Dept. of Biology, Tokyo Metropolitan University, Setagaya-ku, Tokyo/Japan

Dr. R. LALLIER, Université de Paris VI, Station zoologique, F-06230 Villefranche-sur-mer (A.M.)/France

Dr. Marianne von LEDEBUR-VILLIGER, Boldistraße, CH-5415 Rieden/Switzerland

Professor Dr. G. MILLONIG, Centro Oncologico, Osp. Gen. Prov. S'Andrea, I-13100 Vercelli/Italy

Dr. K. OKAZAKI, Dept. of Biology, Tokyo Metropolitan University, 2-1-1, Fukazawa, Setagaya-ku, Tokyo/Japan

Professor Dr. K. OSANAI, Asamushi Marine Biological Station, Tohoku University, Asamushi, Aomori 039-34/Japan

Dr. J. PIATIGORSKY, National Institute of Health, Building 35, Bethesda, MD 20014/USA

Professor Dr. R. RAPPAPORT, Dept. of Biological Sciences, Union College, Schenectady, NY 12308/USA

Professor Dr. J. RUNNSTRÖM†, formerly: Wenner Grens Institute, Norrtullsgatan 16, S-11345 Stockholm/Sweden

Professor Dr. R.C. RUSTAD, Dept. of Radiology, Division Radiation Biology, Case Western Res. University, Cleveland, OH 44106/USA

Dr. Luise SCHMEKEL, Zoologisches Institut der Westfälischen Wilhelms Universität, D-4400 Münster/Federal Republic of Germany

Professor Dr. T. YANAGISAWA, Dept. of Biology, Tokyo Metropolitan University, Setagaya-ku, Tokyo/Japan

1. Introductory History

S. Hörstadius

Sea urchin eggs and larvae are well known experimental objects for studies in different branches of developmental physiology, where THEODOR BOVERI, HANS DRIESCH and CURT HERBST figure as the great pioneers. Already before GREGOR MENDEL's laws of inheritance became widely known, the cellular basis of genetics was brought into light by observations which to a great extent were made on sea urchin eggs.

OSCAR HERTWIG recognized the fusion of egg and sperm nuclei in eggs of *Paracentrotus lividus* as the essential factor in fertilization (1876). Later, in 1890, he noted the significance of meiosis and the formation of polar bodies in Ascaris. In the eighties several findings were presented pointing towards the nuclei and particularly the chromosomes as carriers of heredity. But as the chromosomes seemed to vanish in interphase they were not considered capable of continuity, which is a prerequisite of material carrying genetic information. This doubt was invalidated when BOVERI (1887) demonstrated in Ascaris that the chromosomes persist during interphase in the diffuse state.

The claim that egg and spermatozoon must contain the same amount of genetic material was answered by BOVERI's (1890) discoveries in sea urchins that the egg and spermatozoon contain the same number of chromosomes. This statement strengthened but did not prove the assumption that chromosomes are hereditary carriers. The other question was whether each chromosome would contain the whole mass of genetic factors or whether they would be qualitatively different. The answers were given by BOVERI's (1902, 1908) ingenious interpretations of sea urchin eggs which had been fertilized by 2 spermatozoa. After penetration of one spermatozoon the egg as a rule has the power of instantly blocking the entry of other spermatozoa. But with great excess of sperm the block against polyspermy does not always function. Entrance of 2 spermatozoa means 2 centrioles forming radiating centers resulting in 3 or 4 spindles, triaster and tetraster stages respectively. As a consequence the chromosomes of the 3 sets of haploid chromosomes in most cases must be atypically distributed to the 3 or 4 daughter cells. Only a few triaster eggs gave typical plutei. Many larvae showed mosaics of 3 or 4 regions with varying degrees of differentiation, evidently corresponding to the 3 or 4 initial cells. The conclusions were not only that the chromosomes are carriers of heredity but also that they are qualitatively different.

In 1885 WILHELM ROUX published a paper in the series, Beiträge zur Entwicklungsmechanik des Embryos. He had pricked amphibian eggs in different stages with hot steel needles in order to test whether parts of the egg were able of self-differentiation or if they were dependent on mutual interactions for their differentiation. 3 years later the famous paper appeared in which he described left, right, and anterior half embryos obtained after having killed 1 of the first 2 blastomeres. During cleavage they formed half morulae and blastulae, later corresponding half neural plates. Each half thus developed in accordance with its fate in normal development, as self-differentiation. But later the larvae could gradually grow to complete larvae, a process which ROUX called "Postgeneration", since the new structures had not existed before. The discovery that a half-egg gave rise to a half-larva was sensational.

The brothers OSCAR and RICHARD HERTWIG (1887) found that heavy shaking could break up sea urchin eggs into pieces, and that such fragments could be fertilized and start cleavage. Soon afterwards BOVERI (1889) reared such fragments into plutei, proving that larvae of typical shape could develop although the egg had been deprived of part of its material. A striking contrast to the results of ROUX was attained in the experiments of DRIESCH. In his first "Entwicklungsmechanische Studien" (1891) he used the shaking method of HERTWIG. He reproduced drawings of half-blastomeres cleaving as a half of an egg, with 4 meso-, 2 macro- and 2 micromeres, but on the other hand such isolated halves differentiated into plutei of half size. His results tallied with those of ROUX regarding cleavage, but the discovery that fragments developed into whole larvae opened a new line of research.

DRIESCH found that the separation of blastomeres was easier when the temperature was raised to 31°C (1892, 1893a). The same advantage was obtained by HERBST (1892) by adding substances as KCl, KBr, KJ, NaBr, and $NaHNO_3$ to the seawater, and the effect was still better by preparing artificial seawater devoid of some ions. Particularly efficient in loosening the adhesion between cells was seawater without calcium ions (1900). Brief treatment with Ca-free seawater facilitates the separation of blastomeres during operations, while longer treatment makes the blastomeres fall apart. This method has been widely used and renders great service to experimental embryology.

In a number of papers (1891-1910) DRIESCH studied the developmental potencies of fragments of different composition. He found that not only half-blastomeres but also quarter-blastomeres give dwarf plutei. Isolated blastomeres of 2- and 4-cell stages represent meridional fragments containing the same regions of cytoplasmic material along the egg axis. The next step was to obtain animal and vegetal fragments by isolation at 8- and 16-cell stages. DRIESCH judged the nature of the fragments obtained after shaking of 8-cell stages by observing whether they formed macro- and micromeres or not. He described dwarf plutei, gastrulae and ovoid larvae of vegetal type and even exogastrulae, as well as permanent blastulae with an enlarged tuft of long stereocilia.

Although the occurrence of permanent blastulae should have cautioned DRIESCH, his notion of the sea urchin egg as an "harmonious equipotential system" was based on the observation that at least some of the animal fragments, i.e. those with equal cleavage after isolation, developed into "whole" larvae. Another support was the differentiation of purely vegetal fragments, consisting of macro- and micromeres, into more or less harmonious plutei. BOVERI (1901a) thought that the whole vegetal half invaginates as mesenchyme and archenteron. The ectoderm in the vegetal fragments therefore had to be considered as a regulation to a harmonious individual. But this example is not valid. It was later proven that the half of the macromeres next to the equator normally gives ectoderm, and that this presumptive ectoderm in isolated vegetal halves may be partly endodermized (HÖRSTADIUS, 1931, 1935, 1936a, 1973). However, DRIESCH considered the blastomeres as totipotent and, in service of "the whole" (das Ganze), capable of forming a more or less typical organism.

DRIESCH was aware of the fact that potencies gradually become restricted and studied this problem by sectioning gastrulae. This could only be done at random (1895) and he found that the ectoderm had now lost the capacity to form endoderm, and vice versa. But the endoderm of both the animal and the vegetal fragments of gastrulae were said to be able to differentiate into esophagus, stomach, and intestine.

That parts could give rise to a whole led DRIESCH to further speculations. He speaks of autonomy of life processes and that this is governed by a mysterious force, the "entelechy" (1901a). As proof of this assumption he presents 2 concepts (1901b). If a piece were removed from a machine, the machine could not repair itself. An outside intervention is unavoidable. Similarly, a restitution of a living organism is ascribed to the entelechy. He finds a second proof in stating, that, when all cells in the ovary descend from a primary germ cell, the eggs, which are compared to threedimensional machines, cannot arise from 1 cell through countless divisions and yet remain "whole".

The results and ideas of DRIESCH soon became challenged by other observations. ZOJA (1895) separated animal and vegetal halves of 16-cell stages and found that all the animal halves developed to permanent blastulae. But he did not dare to draw the right conclusion against the authority of DRIESCH. He considered less favorable circumstances responsible for the lack of gastrulation. BOVERI (1901b, 1902) attained the same result with animal (unpigmented) fragments of unfertilized eggs. This significant difference between animal and vegetal regions has been confirmed in thousands of cases. JENKINSON (1911) was able to divide gastrulae individually by transverse sections at exact levels and to rear the 2 halves together. According to him archenteron and ectoderm are not equipotential systems. Only those parts of the archenteron greater than one half can form a tripartite digestive tract, and there are polar differences as to the faculty of forming coelomic sacs and intestine.

Instead of relying on a hypothetical force many scientists looked for differences in structure and also in metabolism along the egg axis. SELENKA (1883) described a reddish pigment band below the equator in the ripe eggs of *Paracentrotus lividus*. MORGAN (1894) observed that the pigment in the *Arbacia* egg at the 4-cell stage moved away from the micromere-forming pole, the micromeres becoming highly transparent. BOVERI (1901b) published his famous paper on the polarity of the sea urchin egg. He speaks of a stratification along the animal-vegetal axis, with layers of different physiological qualities. He later (1910) introduced the term gradient (Gefälle) into embryology. He also assumed that the most vegetal part would exert a special, determining influence upon the rest of the egg. If the most vegetal material were removed, the material next to it would act as its substitute, providing it was still vegetal enough to do so. A number of experiments mentioned in Chapter 13 will demonstrate the correctness of this farsighted idea. MORGAN (1905) considered polarity as a phenomenon of gradation of material, while CHILD (1907) defined polarity as axial physiological differentiation. HEFFNER (1908) and RUNNSTRÖM (1914) assumed gradients of concentration. CHILD and his school (1915-1936) showed the existence of layers of different physiological susceptibility in the eggs of starfishes, sea urchins, and other groups. As a result of susceptibility experiments RUNNSTRÖM (1928a, b, 1929) advanced the idea of 2 opposite gradients: animal and vegetal, and these interact but are also partially hostile to each other. This concept has been important towards understanding many phenomena encountered in experimental work.

It has not been possible to maintain the vitalistic philosophy of DRIESCH against the impact of new facts. Its main presumption has been found to be due to an experimental error. Shaking of early cleavage stages often results in inhibition of micromere formation, which, however, does not influence the further development. Many of the "animal" cells must in fact have been meridional or vegetal fragments, unable to form micromeres but able to gastrulate (HÖRSTADIUS, 1928, p. 124).

DRIESCH made many other classical observations and experiments. He found (1898, 1900) that in small whole larvae the cell size is the same as in normal larvae but the number of cells was reduced in direct proportion to the size of the fragment. This rule of constant cell size and variable number of cells is valid when comparing eggs of different size from the same species. DRIESCH further pointed out that the cells during cleavage decrease in size and asked how the final size becomes determined. BOVERI drew attention to the fact that the rule of fixed cell size only holds when the eggs have the same amount of nuclear material. Parthenogenetic eggs have nuclei of half size but the double number of cells. Fertilized eggs starting the development with a monaster have nuclei and cells of double size but only half the number compared to normal larvae (BOVERI, 1905). Giant eggs formed in the ovary with both nucleus and cytoplasm of double size also have doubled cell volume in the body (HÖRSTADIUS, 1971). Cell size after cleavage was found to be determined by fixed nucleus-cytoplasm relationship (RICHARD HERTWIG, 1903).

Dozens of papers, beginning with those by DRIESCH, 1891 and onwards, dealt with the development of more or less fused twins, obtained by an incomplete separation of the first 2 blastomeres: by displacement of later groups of blastomeres in relation to each other, or by fusion of 2 eggs, or cleavage stages, or blastulae. The results comprise all kinds of twins: double twins with 2 sets of organs but joined by the ectoderm, incomplete twins with organs partially doubled and partially unified, and harmonious giant larvae. If they originate from 2 whole eggs then all organs are of double size. To reach this state BOVERI (1901b) assumed that the egg axes must point in the same direction. If the vegetal poles are not too far apart the 2 invaginations fuse into a single archenteron. The results of the early random fusion studies have been confirmed by exact transplantations (HÖRSTADIUS, 1957).

WEISMANN (1892) thought that the nuclei distributed to different regions were qualitatively different and therefore the cause of larval differentiation. DRIESCH attacked this basic problem of development by his ingenious pressure experiments (1892 IV, 1893b). Fertilized sea urchin eggs were pressed under a coverslip. In accordance with the rule of OSCAR HERTWIG (1885) that spindles follow the longest diameter of the cytoplasm, spindles were formed parallel to the glass surface and the blastomeres became spread out in 1 layer instead of in rings on top of each other. Micromeres could appear in different positions, at the edge or in the middle of the plate. Plutei developed irrespectively of these variations. In the first paper DRIESCH saw this as a token of equivalency of the cells. But in the second paper he hits the crucial point by saying that he has the nuclei in view. It is obvious that the nuclei in a flat egg cannot reach the cytoplasmic regions in the same order of sequence as in the normal egg with its special cleavage pattern. This experiment, therefore, was the first proof against the hypothesis of WEISMANN, and led to the assumption that the cause of differentiation might be found in the cytoplasm.

DRIESCH and T.H. MORGAN (1895) gave a proof in this direction by removing parts of the cytoplasm from fertilized ctenophore eggs. Although the nucleus had not been touched the larvae showed well defined defects corresponding to certain parts of the body.

In a number of papers (1892-1904) CURT HERBST investigated anorganic substances in seawater that are necessary for development, as well as the effects of foreign substances added to the seawater. He stated that Na, K, Mg, and Ca as well as Cl, SO_4 and HCO_3 together with a slight alkalinity are indispensable. Na, K, Ca and Cl have to be present in the medium from the beginning of the development while Mg, SO_4 and HCO_3 are necessary only after the blastula stage. HERBST also found that some ions, at least to a certain degree, could be exchanged with others, e.g. K substituted with Rb or Cs, Cl with Br (not with J) and sulphates with thiosulphates.

A discovery of particular importance was that some combinations change the differentiation in a vegetal or animal direction. Addition of isotonic LiCl to the seawater (HERBST, 1892, 1893) results in larvae with the endoderm increased at expense of the ectoderm; the degree of vegetalization depending upon the concentration of the Li-ions in the solution and the time of exposure. At moderate concentrations the endoderm cannot invaginate, and an exogastrula is formed. With increasing concentrations the ectoderm vesicle becomes smaller. HERBST describes larvae completely transformed into endoderm. In SO_4-free seawater with increased amounts of Ca-ions HERBST (1897, 1904) obtained developmental changes in an animal direction, characterized by enlarged animal tuft and reduced archenteron. The importance of these discoveries is twofold. Firstly, a way was opened to study the effect of substances interfering with the 2 metabolic systems. Secondly, measurements as those of respiration are difficult to make on isolated animal and vegetal fragments. They are easier to determine on larger quantities of vegetalized or animalized eggs.

JACQUES LOEB succeeded in producing parthenogenetic larvae from unfertilized *Arbacia* eggs in 1899. He created hypertonic seawater by adding NaCl or $MgCl_2$ and obtained typical plutei. In a number of papers he reported that alkalinity, time of exposure, access to oxygen, and the presence of other substances were factors which improved the results (1909).

Centrifugation of sea urchin eggs leads to unexpected results. It was found that a stratification of the cytoplasm by centrifugal force does not prevent a normal development. The eggs differentiate in conformity with the original egg axis independently of the angle of the force to the axis. Therefore, the displaced substances may occupy any position in relation to the polarity of the egg. Yet a normal development ensues (LYON, 1906; MORGAN and LYON, 1907; MORGAN and SPOONER, 1909). MCCLENDON (1909) did not find great chemical differences between centrifugal and centripetal parts. Later investigations, particularly by ETHEL BROWN HARVEY (1933-1956) have shown that fragments are also able to develop although they contain different layers, i.e. not having all layers represented. It is not yet clear what gives the egg the power to reestablish the old polarity in spite of stratification at other angles, if, for example, the resistance is due to the cortical layer or to an internal structure unaffected by the centrifugal force.

Although BOVERI had proven that the nucleus carries hereditary factors he did not exclude the possibility that the cytoplasm also functions genetically. To attack this problem BOVERI (1889, 1914, 1918) isolated fragments of unfertilized eggs by violent shaking and fertilized them with sperm of the same or different species (homosperm, heterosperm merogony). As hybrids yield plutei with intermediate skeletal characters and as he found some merogones which were intermediate and others which were paternal, he considered the former as arising from fragments with, the latter from fragments without the egg nucleus. The latter cases should point towards the nucleus as the sole carrier of genetic information. A number of authors opposed the results of BOVERI and in his posthumous paper (1918) he admitted that real heterosperm merogones of the combination used, *Sphaerechinus granularis* cpl and *Psammechinus microtuberculatus* nucleus, did not develop further than to the onset of gastrulation. Shaking can make the whole nucleus invisible probably by breaking it into invisible pieces. The "merogones" developing to plutei therefore must have contained female nucleus elements. It has, however, later been possible to produce larvae with cytoplasm of one species and nucleus of another yielding paternal characteristics (HÖRSTADIUS, 1936b).

DRIESCH was not blind to the possibility of interactions between parts of the embryo. He saw that primary mesenchyme cells displaced by shaking in the early gastrula moved back to their original positions under obvious influence from the ectoderm (1892, 1896). HERBST arrived at the same conclusion with lithium larvae where the ecto-endoderm limit is moved towards the animal pole. In connection herewith the primary mesenchyme cells form the ring at a corresponding level. This was one of the observations leading HERBST (1892, 1893, 1896) to formulate his rule of formative stimuli (formative Reize). An example of this was the need of interaction between the ectoderm and the points of the skeletal rods in order to form the pluteus arms (now known to be the mesenchyme cells rather than the spicules that are involved in the process).

A new branch of developmental physiological studies was opened by OTTO WARBURG (1908) by his respiration measurements. He found a low oxygen consumption in unfertilized eggs. In comparison to this resting stage, the consumption raised to a level 6-7 times higher after fertilization, followed by a slow increase during early development.

SPEMANN recommended fine glass needles for operations on sea urchin embryos in 1906, but such needles were not used on echinoderms until much later (VON UBISCH, 1925b, HÖRSTADIUS, 1925). VOGT's (1925) local vital staining with small pieces of agar were also first employed by VON UBISCH on sea urchin larvae (1925a). With these methods a more exact analysis of the morphological events could be undertaken. The increasingly refined techniques in cytology, physiology, and biochemistry have produced an immense amount of knowledge which will be presented in the ensuing chapters of this book.

Before finishing this short historical sketch of the early decades, it may be allowed to do something as strange as paying tribute to a laboratory. The majority of early investigations on sea urchins have been performed at the Stazione Zoologica of Naples. It has, for many years, been the workshop of the HERTWIGS, DRIESCH, BOVERI, HERBST, MORGAN and WARBURG. The bold idea of ANTON DOHRN of founding an international institute for marine research, the first of its kind, has certainly richly awarded many fields, and this is the reason why this handbook is dedicated to this unique institution.

References

BOVERI, T., 1887. Über Differenzierung der Zellkerne während der Furchung des Eies von Ascaris megalocephala. Anat. Anz. 2, 688.
BOVERI, T., 1889. Ein geschlechtlich erzeugter Organismus ohne mütterliche Eigenschaften. S. Ber. Ges. Morph. Physiol. München 5, 73.
BOVERI, T., 1890. Zellenstudien. III. Über das Verhalten der chromatischen Kernsubstanz bei der Bildung der Richtungskörper und bei der Befruchtung. Jena. Z. Naturwiss. 24, 314.
BOVERI, T., 1901a. Die Polarität von Ovocyte, Ei und Larve des Strongylocentrotus lividus. Zool. Jahrb. Abt. Anat. Ontog. 14, 630-653.
BOVERI, T., 1901b. Über die Polarität des Seeigeleies. Verh. phys.-med. Ges. Würzburg (N.F.) 34, 145-175.
BOVERI, T., 1902. Über mehrpolige Mitosen als Mittel zur Analyse des Zellkerns. Verh. phys.-med. Ges. Würzburg (N.F.) 35, 67.
BOVERI, T., 1905. Zellenstudien V. Jena. Z. Naturwiss. 39, 445.
BOVERI, T., 1908. Zellenstudien VI. Die Entwicklung dispermer Seeigeleier. Ein Beitrag zur Befruchtungslehre und zur Theorie des Kerns. Jena. Z. Naturwiss. 43, 1.
BOVERI, T., 1910. Die Potenzen der Ascaris-Blastomeren bei abgeänderter Furchung. Festschrift f. Richard Hertwig 3, 131-214.
BOVERI, T., 1914. Über die Charaktere von Echiniden-Bastardlarven bei verschiedenem Mengenverhältnis mütterlicher und väterlicher Substanzen. Verh. phys.-med. Ges. Würzburg (N.F.) 43, 117.
BOVERI, T., 1918. Zwei Fehlerquellen bei Merogonieversuchen und die Entwicklungsfähigkeit merogonischer und partiell-merogonischer Seeigelbastarde. Arch. f. Entw. Mech. 44, 417.
CHILD, C.M., 1907. An analysis of form-regulation in Tubularia. VI. Arch. f. Entw. Mech. 24, 315.
CHILD, C.M., 1915. Axial gradients in the early development of the starfish. Amer. J. Physiol. 37, 203.
CHILD, C.M., 1936. Differential reduction of vital dyes in the early development of echinoderms. Wilhelm Roux' Arch. Entwicklungsmech. Organismen 135, 426-456.
DRIESCH, H., 1891. Entwicklungsmechanische Studien. I. Der Wert der beiden ersten Furchungszellen in der Echinodermenentwicklung. Experimentelle Erzeugung von Teil- und Doppelbildungen. Z. wiss. Zool. 53, 160-184.
DRIESCH, H., 1892. Entwicklungsmechanische Studien III-VI. Z. wiss. Zool. 53, 1-62.
DRIESCH, H., 1893a. Entwicklungsmechanische Studien VII-X. Mitth. Zool. Stat. Neapel 11, 221-253.
DRIESCH, H., 1893b. Zur Verlagerung der Blastomeren des Seeigeleies. Anat. Anz. 8, 348-357.
DRIESCH, H., 1895. Zur Analysis der Potenzen embryonaler Organzellen. Arch. f. Entw. Mech. 2, 169-203.
DRIESCH, H., 1896. Die taktische Reizbarkeit der Mesenchymzellen von Echinus microtuberculatus. Arch. f. Entw. Mech. 3, 362-380.
DRIESCH, H., 1898. Von der Beendigung morphogener Elementarprozesse. Arch. f. Entw. Mech. 6, 198-227.
DRIESCH, H., 1900. Die isolierten Blastomeren des Echinidenkeimes. Arch. f. Entw. Mech. 10, 361-410.
DRIESCH, H., 1901a. Die organischen Regulationen. Leipzig.
DRIESCH, H., 1901b. Zwei Beweise für die Autonomie der Lebensvorgänge. Verh. V. Int. Zool. Congress. Berlin.
DRIESCH, H., 1910. Neue Versuche über die Entwicklung verschmolzener Echinidenkeime. Arch. f. Entw. Mech. 30:1, 8-23.
DRIESCH, H., MORGAN, T.H., 1895. Zur Analysis der ersten Entwicklungsstadien des Ctenophoreneies. Arch. f. Entw. Mech. 2, 204.

HARVEY, E.B., 1933. Development of the parts of sea urchin eggs separated by centrifugal force. Biol. Bull. 64, 125-148.
HARVEY, E.B., 1956. The American Arbacia and other sea urchins. Princeton Univ. Press, 1-298.
HEFFNER, B., 1908. Über experimentell erzeugte Mehrfachbildungen des Skeletts bei Echinodermenlarven. Arch. f. Entw. Mech. 26, 1.
HERBST, C., 1892. Experimentelle Untersuchungen über den Einfluß der veränderten chemischen Zusammensetzung des umgebenden Mediums auf die Entwicklung der Tiere. I. Z. wiss. Zool. 55, 446-518.
HERBST, C., 1893. Experimentelle Untersuchungen über den Einfluß der veränderten chemischen Zusammensetzung des umgebenden Mediums auf die Entwicklung der Tiere. II. Weiteres über die Wirkung der Lithiumsalze und ihre theoretische Bedeutung. Mitth. Zool. Stat. Neapel 11, 136.
HERBST, C., 1896. D:o III-VI. Arch. f. Entw. Mech. 2, 455-516.
HERBST, C., 1897. Über die zur Entwicklung der Seeigellarven notwendigen anorganischen Stoffe, ihre Rolle und ihre Vertretbarkeit. I. Arch. f. Entw. Mech. 5, 649-793.
HERBST, C., 1900. Über das Auseinandergehen von Furchungs- und Gewebezellen in kalkfreiem Medium.Arch.f.Entw.Mech.9, 424-463.
HERBST, C., 1904. Über die zur Entwicklung der Seeigellarven notwendigen anorganischen Stoffe, ihre Rolle und ihre Vertretbarkeit. III. Arch. f. Entw. Mech. 17, 306.
HERTWIG, O., 1876. Beiträge zur Kenntnis der Bildung, Befruchtung und Teilung des tierischen Eies. Gegenbaurs morphol.Jahrb. 1, 374-434.
HERTWIG, O., 1885. Welchen Einfluß übt die Schwerkraft auf die Teilung der Zellen? Jena. Z. Naturwiss. 18, 175-205.
HERTWIG, O., 1890. Vergleich der Ei- und Samenbildung bei Nematoden. Eine Grundlage für celluläre Streitfragen. Arch. mikr. Anat. 36, 1-138.
HERTWIG, O., HERTWIG, R., 1887. Über den Befruchtungs- und Teilungsvorgang des tierischen Eies unter dem Einfluß äußerer Agentien. Untersuchungen zur Morphologie und Physiologie der Zelle. Jena. Z. Naturwiss. 20, 120.
HERTWIG, R., 1903. Über Korrelation von Zell- und Kerngröße und ihre Bedeutung für die geschlechtliche Differenzierung und die Teilung der Zelle. Biol. Cbl. 23, 49-52.
HÖRSTADIUS, S., 1925. Entwicklungsmechanische Studien an Astropecten aurancius. Arkiv Zool. 17B, 1-6.
HÖRSTADIUS, S., 1928. Über die Determination des Keimes bei Echinodermen. Acta Zool. (Stockholm) 9, 1-191.
HÖRSTADIUS, S., 1931. Über die Potenzverteilung im Verlaufe der Eiachse. Arkiv Zool. 23, 1-6.
HÖRSTADIUS, S., 1935. Über die Determination im Verlaufe der Eiachse bei Seeigeln. Pubbl. Sta. Zool. Napoli 14, 251-479.
HÖRSTADIUS, S., 1936a. Weitere Studien über die Determination im Verlaufe der Eiachse bei Seeigeln. Wilhelm Roux' Arch. Entwicklungsmech. Organismen 135, 40-68.
HÖRSTADIUS, S., 1936b. Studien über heterosperme Seeigelmerogone nebst Bemerkungen über einige Keimblattchimären. Mém. Mus. Hist. nat. Belg. 2me sér. fasc. 3, 801-880.
HÖRSTADIUS, S., 1957. On the regulation of bilateral symmetry in plutei with exchanged meridional halves and in giant plutei. J. Embryol. Exp. Morphol. 5, 60-73.
HÖRSTADIUS, S., 1971. Giant larvae of Paracentrotus lividus. Arkiv Zool. Ser. 2, 23, 417-422.
HÖRSTADIUS, S., 1973. Experimental Embryology of Echinoderms. Oxford: Clarendon Press, 1-192.
JENKINSON, J.W., 1911. On the development of isolated pieces of the gastrulae of the sea urchin Strongylocentrotus lividus. Arch. f. Entw. Mech. 32, 269.

LOEB, J., 1899. On the nature of the process of fertilization and the artificial production of normal larvae (plutei) from the unfertilized eggs of the sea urchin. Amer. J. Physiol. 3, 135-138.
LOEB, J., 1909. Die chemische Entwicklungserregung des tierischen Eies. Berlin: Springer-Verlag.
LYON, E.P., 1906a. Some results of centrifugalizing the eggs of Arbacia. Amer. J. Physiol. 15, 21.
LYON, E.P., 1906b. Results of centrifugalizing eggs. Arch. f. Entw. Mech. 23, 151.
MCCLENDON, J.F., 1909. Chemical studies on the effects of centrifugal force on the eggs of the sea urchin (Arbacia punctulata). Amer. J. Physiol. 23, 460-466.
MORGAN, Th.H., 1894. Experimental studies on echinoderm eggs. Anat. Anz. 9, 141-152.
MORGAN, Th.H., 1905. Polarity considered as a phenomenon of gradation of materials. J. Exp. Zool. 2, 495.
MORGAN, Th.H., LYON, E.P., 1907. The relation of the substances of the egg, separated by a strong centrifugal force, to the location of the embryo. Arch. f. Entw. Mech. 24, 147.
MORGAN, Th.H., SPOONER, G.B., 1909. The polarity of the centrifuged egg. Arch. f. Entw. Mech. 28, 104.
ROUX, W., 1885. Beiträge zur Entwicklungsmechanik des Embryo. III. Über die Bestimmung der Hauptrichtungen des Froschembryo im Ei und über die erste Teilung des Froscheies. Breslau. ärztl. Z.
ROUX, W., 1888. Über die künstliche Hervorbringung halber Embryonen durch Zerstörung einer der beiden ersten Furchungskugeln, sowie über die Nachentwicklung (Postgeneration) der fehlenden Körperhälfte. Virchows Arch. Anat. Phys. 114, 113 and 246.
RUNNSTRÖM, J., 1914. Analytische Studien über die Seeigelentwicklung. I. Arch. f. Entw. Mech. 40, 526-564.
RUNNSTRÖM, J., 1928a. Plasmabau und Determination bei dem Ei von Paracentrotus lividus Lk. Wilhelm Roux' Arch. Entwicklungsmech. Organismen 113, 556-581.
RUNNSTRÖM, J., 1928b. Zur experimentellen Analyse der Wirkung des Lithiums auf den Seeigelkeim. Acta Zool. (Stockholm) 9, 365-424.
RUNNSTRÖM, J., 1929. Über Selbstdifferenzierung und Induktion bei dem Seeigelkeim. Wilhelm Roux' Arch. Entwicklungsmech. Organismen 117, 123-145.
SELENKA, E., 1883. Die Keimblätter der Echinodermen. Studien über die Entwicklungsgeschichte der Tiere. 1:2. Wiesbaden.
SPEMANN, H., 1906. Über eine neue Methode der embryonalen Transplantation. Verh. Deutsch. Zool. Ges., 195-202.
VON UBISCH, L., 1925a. Entwicklungsphysiologische Studien an Seeigelkeimen. I. Über die Beziehungen der ersten Furchungsebene zur Larvensymmetrie, und die prospektive Bedeutung der Eibezirke. Z. wiss. Zool. 124, 361-381.
VON UBISCH, L., 1925b. D:O II. Die Entstehung von Einheitslarven aus verschmolzenen Keimen. Z. wiss. Zool. 124, 457-468.
VOGT, W., 1925. Gestaltungsanalyse am Amphibienkeim mit örtlicher Vitalfärbung. Wilhelm Roux' Arch. Entwicklungsmech. Organismen 106, 542-610.
WARBURG, O., 1908. Beobachtungen über die Oxydationsprozesse im Seeigelei. Hoppe-Seyler's Z. physiol. Chemie 75, 1-16.
WEISMANN, A., 1892. Das Keimplasma. Eine Theorie der Vererbung. Jena.
ZOJA, R., 1895. Sullo sviluppo dei blastomeri isolati dalle uova di alcune meduse (e di altri organismi). II. Arch. f. Entw. Mech. 2, 1-37.

2A. Care and Handling of Sea Urchin Eggs, Embryos, and Adults (Principally North American Species)

R. Hinegardner

Sea urchins are found along almost all coastlines, usually on shallow, rocky bottom, although some species live in deep water or in sand or silt (Table 2A.1). Seen from above, sea urchins are round or somewhat heart-shaped; from the side, they are more or less flat. *Scutillidae* - here referred to as sand dollars - are very flat. All sea urchins have spines: very long and sometimes poisonous as on the *Centrechinidae*, short and blunt as on the limpet-like *Colobocentrotus*, or very small and numerous as on the sand dollar. Sexes are separate - except for rare hermaphrodites - but can seldom be distinguished by external characteristics. Exceptions are *Lytechinus* and *Psammechinus*, the male and female of which have slightly different gonopores.

Table 2A.1. Sea-urchin habitats in North America and Hawaii

Species	Habitat
Arbacia punctulata	East coast of North America; subtidal
Dendraster excentricus	West coast; sand; intertidal and subtidal
Echinarachnius parma	East and northwest coasts; sand; below low tide
Echinometra mathaei	Hawaii; coral reefs
Lytechinus pictus	Southern California and south; just below low tide to over 25 m; variable habitat
Lytechinus variegatus	Southeast; sand; subtidal
Strongylocentrotus droebachiensis	Northeast and southeast coasts
Strongylocentrotus franciscanus	West coast; habitat similar to *S. purpuratus*; primarily subtidal
Strongylocentrotus purpuratus	West coast; rocky shore and bottom; from intertidal to at least 25 m
Tripneustes gratilla	Hawaii; subtidal; sand

Among the useful references on sea-urchin eggs are HARVEY (1956), COSTELLO et al. (1957), TYLER and TYLER (1966), HINEGARDNER (1967), and HÖRSTADIUS (1973). HARVEY's book deals primarily with *Arbacia punctulata*, but it includes other species as well as a thorough coverage of the sea urchin literature up to the early 1950s.

2A.1 Collecting and Keeping

Sea urchins may be collected from the shore at low tide or by diving or dredging. In cities where they are sold as food, the fish markets should not be overlooked as a ready source. They ship well by air freight, and several collectors in the United States guarantee live delivery.[1]

Sea urchins are easy to keep in a marine laboratory. Most of them adapt well to running seawater tanks and require little more than food such as algae. Many species can also be kept in closed systems providing there is ample clean water and aeration. They should not be crowded; for long-term maintenance, 1 ml of sea urchin requires about 400 ml of seawater. Aeration is essential and up to a point, increased aeration improves the urchins' general health. Sea urchins seem quite content amid swirling bubbles. Large porous air stones work well; they maintain adequate oxygen tension and probably help to remove excess ammonia from the water. Commercially available refrigerated seawater systems generally need additional aeration.

As a rule, North American sea urchins are easier to keep at relatively low temperatures. *Strongylocentrotus droebachiensis*, for instance, requires 10°C or less, but *Lytechinus pictus* must be kept above 8°C. A good compromise for many North American species is 15°C.

A good closed system is a 100-liter tank made of plastic or some other noncorroding material, with a filter powered by a motor-driven centrifugal pump. Controlled heating, or an external refrigeration unit and pump to circulate coolant through stainless steel tubing in the tank, may be used for temperature control. Each month 20 liters of seawater are replaced with fresh seawater and distilled water - known to lack toxicity - added to make up for evaporation. The pH of a closed tank usually decreases and may be controlled by crushed shell or dolomite gravel - usually in the filter (SPOTTE, 1970).

At times it is necessary to identify individual sea urchins so that a second sample of gametes may be obtained from the same specimen. Although there is no way to tag sea urchins without harming them, individuals may be isolated as in the chamber in Fig. 2A.1. Here the sea urchin is easily fed and remains as healthy as its nonisolated tankmates. The chamber shown is basically a large plastic tube through which water is circulated by an airlift pump attached to the side.

Feeding sea urchins in a closed system without immediate access to the ocean need not be difficult. Sea urchins are scavengers and eat many different foods including: large algae from the ocean (if available), lettuce, frozen shrimp, hard-boiled egg yolk, trout food and dried dog food.

Of the North American species, *L. pictus* adapts most easily to captivity, and specimens have been kept for more than 5 years and spawned many times. Its transparent egg is well suited to developmental studies, although adults are rarely larger than 6 cm in diameter and egg yield per female seldom exceeds 1 ml. A larger species such as *Strongylocentrotus purpuratus* is better suited for biochemical studies.

---

[1]Pacific Bio-Marine Supply Company, P.O. Box 536, Venice, Calif. 90291.
Stanley Becker, P.O. Box 62, Big Pine Key, Fla. 33043.
Gulf Specimen Company, P.O. Box 237, Panacea, Fla. 32346.

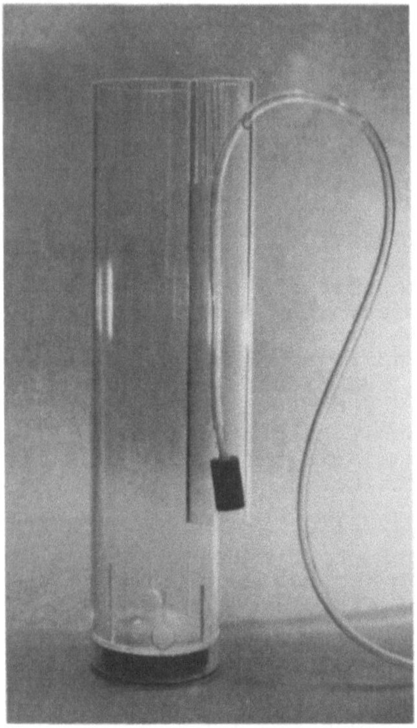

Fig. 2A.1. Chamber for isolating individual sea urchins (*Lytechinus pictus*) in tank. Made of plexiglass tubing (inside diameter: 7 cm) it extends about 1 cm above tank water level. Bubbles from air stone lift water up outside tube (inside diameter: 3.1 cm) and through slots into larger tube. Water then flows down and out through bottom holes and slots, flushing out fecal matter and debris. Iron piece sealed in base gives stability

2A.2 Spawning Season

Gamete production in captive sea urchins seems to depend on food. Even animals taken from the sea when they are normally infertile ripen within a month or two if they are well fed and they continue to produce gametes. It is possible that food supply directly or indirectly controls the spawning season in the wild (PEARSE, unpublished).

When a species ripens in the wild it can vary form year to year by more than a month and depends on geographical location. Those in northern waters tend to have a shorter season - beginning later in the year - than those in the south, while populations in tropical and semitropical waters often contain ripe individuals all year. Table 2A.2 lists the commonly used species and the seasons when they usually produce ripe eggs and sperm.

Table 2A.2. Spawning season and egg size of sea urchins commonly used in experiments. (From HARVEY, 1956; HYMAN, 1955; and BOOLOOTIAN, 1966)

Species	Spawning Season	Egg Diameter (μm)
Arbacia punctulata	Summer	79
Arbacia lixula	Fall-spring	79
Dendraster	Summer	114
Echinocardium cordatum	Summer (Atlantic) Winter (Mediterranean)	125
Echinometra mathaei	Winter (primarily)	80
Echinarachnius parma	Summer	145
Echinus esculentus	Spring-summer	180
Lytechinus pictus	Summer	111
Lytechinus variegatus	Winter	103
Paracentrotus lividus	Fall-spring	90
Psammechinus miliaris	Spring-summer	100
Strongylocentrotus droebachiensis	Late winter	160
Strongylocentrotus purpuratus	Fall-spring	80
Tripneustes gratilla	Summer-fall (primarily)	90

2A.3 Determining Sex

Sea urchins have separate sexes, but it is generally impossible to tell them apart by external characteristics. An exception is *L. pictus* in which the gonophores of the female look larger than those of the male because the female's are surrounded by a black pigment ring (Fig. 2A.2) of variable size. In some cases the sex is still hard to identify, although a large ring always indicates a female while no ring usually indicates a male.

In the European species of *Psammechinus* and *Paracentrotus* in the male the gonopore sits at the top of a low papilla in the female at the bottom of a shallow grove. The sex of any sea urchin may be determined by electrical stimulation or by biopsy. In the latter, a hypodermic needle is passed through the persistomatous membrane and into one of the gonads. The syringe contains seawater, a small amount of which is injected and then drawn up as the needle is slowly pulled out. If the specimen is ripe, gametes will probably be in the syringe, particularly if the specimen is a male.

2A.4 Obtaining Gametes

Spawning is commonly induced by coelomic injection of 0.5 *M* KCl in distilled water (TYLER, 1949). A hypodermic needle (about 24-gauge) is passed through the persistomatous membrane and aimed at the middle of the side where the gonads lie between the rows of tube feet. The

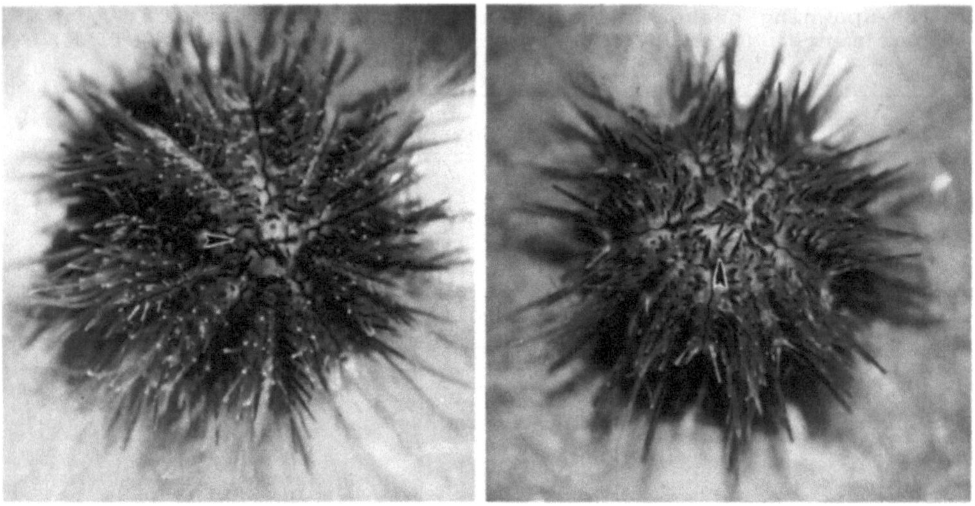

Fig. 2A.2. Male (left) and female (right) *Lytechinus pictus*. Arrows indicate large pigmental ring surrounding female gonopore and almost nonexistent male ring

amount of KCl injected depends on the size of the specimen and on the number of gametes desired. If 0.1 - 0.2 ml is carefully injected near one of the gonads, it is often possible to spawn that gonad only, leaving the others for future use. Larger amounts of KCl injected at several points induce complete spawning. Used in moderation, KCl seems not to harm eggs or specimen.

After injection, the sea urchin is inverted on a small diameter beaker filled with seawater and the gametes collected in the beaker. If survival of the specimen is not required, nothing further is necessary. If the animal is to be used again, it must be kept in water while spawning. For many species, exposure to air for even a few minutes allows the small amount of seawater adhering to the animal's epithelium to evaporate to a lethal hypertonic solution. Such a specimen may look well for a day or so; then its spines begin to fall off and it soon dies. To avoid this, invert it on a beaker within a larger beaker. If the water level in the larger beaker just covers the urchin and if that beaker is not much larger in diameter than the sea urchin, the sea urchin tends to hold on to the inside of the larger beaker and is less likely to right itself. If well fed, spawned animals regenerate new gametes in a month or so.

To induce spawning in *Arbacia*, HARVEY (1956) used electrical shock. Line voltage is stepped down by means of a transformer to about 10 V, and lead electrodes applied the current across the specimen. This method allows the animal to be spawned until the gonads are exhausted; sometimes spawning stops when the current is turned off. Though effective for a number of species, electrical shock is more cumbersome than KCl injection and harms the animals at the point of electrode contact.

Some species respond irregularly to KCl injection, and some do not respond readily to electrical shock. Another method is to inject acetylcholine or one of its analogues. A 0.01 *M* solution made up in seawater no more than 10-20 minutes before use, works well. When injected the same way as KCl, acetylcholine in moderation does not harm eggs or the urchin.

The oldest way to obtain gametes - removing the gonads - is used in special situations such as collecting oocytes. If the gonads are taken out carefully and allowed to shed in a dish of seawater, contamination from cellular debris is minimized and a great many eggs may be obtained. For sperm it is not necessary to remove the entire testes, as a small piece of gonad contains enough sperm to fertilize many eggs.

2A.5 Preparing Eggs

When shedding is complete, eggs should be filtered to remove such debris as broken spines and fecal pellets. Pouring the eggs through several layers of cheese cloth traps eggs and wastes them. Nylon mesh with openings 100-150 μm wide is far superior. A piece attached to a 28 x 100 mm plastic centrifuge tube with the bottom cut off, makes a good filter. The egg suspension is poured through into a beaker of clean seawater and the debris discarded. The filtered eggs are allowed to settle, the seawater poured off, fresh seawater added, and the eggs allowed to settle once more. Eggs from a number of species are difficult to fertilize unless they are washed well. Apparently a secretion from the adult urchins inhibits fertilization.

2A.6 Counting Eggs

A Coulter Counter with a 280 μm aperture gives the most accurate count of sea-urchin eggs. Because they settle rapidly, eggs should be stirred before each count. Two less expensive counters are reasonably accurate. One - devised by SHAPIRO (1935) - is a thin-walled capillary tube of known inside diameter, such as a 1-mm blood capillary or a disposable lambda pipet. This is dipped in a homogenous egg suspension and filled by capillary action. The tube is then placed on its side under a dissecting microscope and the eggs allowed to settle. The eggs are counted in a measured length of tube and from the tube diameter, the volume containing the counted eggs is calculated. The concentration of the original suspension is then determined. For easy counting, the eggs should not be too concentrated.

A simpler and more accurate way is to pipet a known volume of egg suspension into a petri dish such as the Falcon Plastic 60 x 15-mm "Integrid" tissue culture dish (No. 3030) which has a 2-mm ruling on the underside of the bottom surface. Egg concentration may be determined by counting the number of eggs per square. Accuracy depends on the uniformity of egg distribution in the dish, accurate pipetting, and the total number of eggs counted. The statistical counting error is the square root of the total number of eggs counted. Because eggs do not flow uniformly in pipets but tend to concentrate at the front of the water column, the entire contents of the pipet should be emptied into the dish.

Sea-urchin eggs are too large to be counted accurately in a hemocytometer. The accuracy of even a deep chamber is little better than an order of magnitude.

2A.7 Storing Gametes

A number of factors determine how long eggs and sperm can be kept without losing their ability to fertilize and develop normally. The eggs of *S. purpuratus* can be kept at 5°C for about 24 hours without significant loss of fertilizability. At higher temperatures they deteriorate more rapidly. The eggs of *L. pictus* are adversely affected by very low temperatures and cannot be stored at much less than 10°C. At 10°C they can be stored overnight. Survival is improved if eggs are kept oxygenated by slow stirring or by storage as a monolayer. Sand-dollar eggs do not keep well and should be fertilized shortly after spawning. Some attempts have been made to keep shed eggs for long periods (HARVEY, 1956), but it is far easier to store the eggs inside the female rather than outside.

Unlike eggs, spermatozoa may be stored as long as a week and still be usable. Because they are inactive without oxygen, sperm may be stored on ice if they are first centrifuged at several hundred times gravity and the supernatant removed. Once diluted, most sperm are active for less than an hour. A small amount of glucose increases their active life span.

2A.8 Agents Affecting Development

The eggs of many sea urchins - particularly those from temperate and tropical waters - may be raised at room temperature. The eggs of species living in colder climates usually require some temperature control. *Strongylocentrotus purpuratus* from central California and farther north begins to show developmental abnormalities above 15°C, and *S. droebachiensis*, which lives in even colder waters, develops abnormally above 10°C. *Echinarachnius parma* from Maine has an upper temperature limit of 18°C (RAPPAPORT, unpublished). Some warm-water species also have a lower temperature limit. *Lytechinus pictus* does not develop normally below 8°C.

Eggs are sensitive to many different substances, and it is important to keep all containers clean and free from extraneous chemicals. Many organic compounds and ions affect development. Synthetic seawater should always be made of the purest water and chemicals possible until it is known that less expensive products are not harmful. One should be particularly wary of the contaminants in magnesium chloride and distilled water. Consistently abnormal development is generally caused by contamination somewhere in the system.

Natural seawater must be taken from clean sites. Seawater of unknown purity should be filtered through charcoal.

2A.9 Choosing Eggs

Sea-urchin eggs are not all alike, and some are better suited to certain experiments than others. Sand-dollar eggs take up chemicals much more readily than eggs of ordinary sea urchins and are best for inhibitor studies. Where most sea urchin eggs are apparently impermeable and insensitive to actidione (cyclohexamide), those of the sand dollars are sensitive at very low concentrations (KARNOFSKY and SIMMEL, 1963). *Arbacia* is less permeable to nucleotide analogues such as bromodeoxyuridine or to actinomycin D than *L. pictus*. Usually *Arbacia* is the least permeable to a variety of compounds, *Lytechinus*, *Strongylocentrotus*, and their relatives are moderately permeable, and the sand dollars are the most permeable.

Egg and embryo sensitivity to handling tends to go in reverse order. *Arbacia* eggs and embryo cells are very rugged, while those of the sand dollars are not.

Pigment and yolk granules affect egg usefulness. *Arbacia* eggs are very pigmented. Their deep red color obscures many details of development and may interfere with some isolation procedures. In contrast, *L. pictus* has large, transparent eggs better suited to observing developmental phenomena. Pigment can also be useful. *Arbacia* micromeres have almost no pigment while the rest of the blastomeres do. This aids in micromere identification during large-scale micromere isolation procedures.

If quantities of eggs are desired, *S. purpuratus*, *Paracentrotus lividus*, *Lytechinus variegatus*, and several species of *Tripneustes* give many small eggs. It is not uncommon to have one *S. purpuratus* give 25 ml of eggs, and some give up to 50 ml.

2A.10 Fertilization

Eggs and sperm are brought together by stirring a very small volume of concentrated sperm into an egg suspension or, better, by preparing a dilute sperm suspension and immediately adding part of it to the eggs. Only the least amount of sperm necessary to give complete fertilization should be used. Excess sperm may lead to polyspermy (page 140) and to abnormal development, the first evidence of which is the appearance of more than 2 asters and the subsequent division of the egg into three or four cells rather than two. A ratio of about 50 sperm to one egg is the maximum of many species. After fertilization is complete, excess sperm are removed by allowing the eggs to settle, drawing off the sperm containing water, and adding fresh seawater.

Following fertilization, the fertilization membrane is raised and the hyaline layer formed shortly after; male and female pro-nuclei fuse 10-20 minutes after fertilization, and mitosis and first cleavage occur after one or two hours. The time depends on temperature. Early sea urchin development has a Q_{10} of about 2 for many species. That is, lowering the temperature by $10^{o}C$ doubles development time. The times for the various stages and the cell numbers in early *S. purpuratus* development are given elsewhere (HINEGARDNER, 1967).

The fertilization membrane first appears as a blister at the point of sperm entry and rapidly extends over the egg surface forming a transparent, acellular membrane surrounding the embryo. The synchrony of later developmental stages can often be predicted from the synchrony of membrane formation and the first cleavage.

Some batches of eggs or sperm fail to perform, sometimes because of immotile sperm. Unless the spermatozoa are vigorously active, fertilization will be poor or unsuccessful. More often the eggs are at fault. HARVEY (1956) gives several methods that may improve fertilization.

However, unless the eggs are particularly precious, it is better to spawn another female. Most workers reject a batch of eggs having less than 90-95% fertilization. If large numbers of eggs are not required, single fertilized eggs may be selected by the mouth pipet method of HÖRSTADIUS (page 367) from batches with a low fertilization rate.

2A.11 Removing Extracellular Structures

The unfertilized sea-urchin egg is surrounded by two extracellular layers: the jelly coat, and the vitelline membrane which gives rise to the fertilization membrane. The jelly coat contains the sperm agglutinin and fertilizin (TYLER and TYLER, 1966). Depending on the species, the jelly coat may wash off during handling or remain until the embryo hatches. It usually has the same refraction index as water and is not visable under the microscope. It can sometimes be seen in immature eggs (Fig. 2A.3). Its presence may be indicated by how closely eggs pack or how far particles or sperm are kept from the egg surface. It can usually be stained by a weak solution of Janus green in seawater. The jelly coats of *Echinarachnius* and *Dendraster* contain visible pigment granules.

Although the jelly coat seldom hinders molecules entering the egg or embryo, jelly-free eggs are sometimes desirable. In some species the jelly is removed by passing the eggs through fine mesh, but if it is

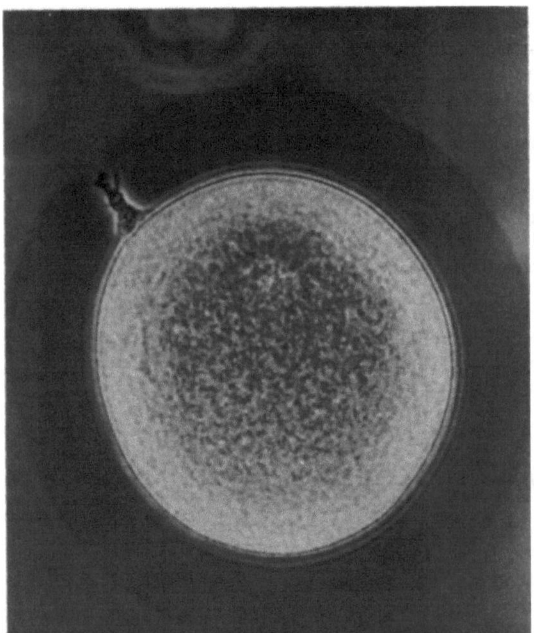

Fig. 2A.3. Jelly coat surrounding immature *Lytechinus pictus* egg is usually invisable in mature egg. Projection into the coat may have been attachment to ovary

more tenacious, it may easily be removed in acid seawater. The pH of a quantity of seawater is adjusted so that an added equal volume of egg suspension gives a final pH between 5.5 and 5.8 (HAGSTRÖM, 1959). The eggs are added and allowed to settle in a container shallow enough to permit settling in 1 to 3 minutes. The acid seawater is then drawn off and enough fresh seawater added to bring the pH back to normal. This brief acid treatment removes the jelly coat without harming the eggs.

At fertilization the cortical granules empty their contents into the space between the cell membrane, and the vitelline membrane which then rises some distance from the egg surface (page 136). Material originating in the cortical granules is deposited on the inner surface of the vitelline membrane now called the fertilization membrane. In the laboratory this membrane is not necessary for normal development of most sea urchin species. In the ocean it probably protects the early embryo from physical damage until cilia develop and the embryo can swim to the surface.

The fertilization membrane is permeable to many compounds, but it can interfere in experimental situations. Ways to remove it, the jelly coat, and the hyaline layer are given in Table 2A.3 and described below.

Before fertilization the vitelline membrane may be removed by pretreatment with urea, glycerol, or proteolytic enzymes, or the fertilization membrane may be removed soon after it is formed. In removing the vitelline membrane, urea (MOORE, 1930) or glycerol (KANE, 1970) creates an inert, isotonic solution free of ions. When a few drops of a thick suspension of unfertilized eggs are added to 50 ml of 1 *M* urea for 2 minutes, fertilization membranes and hyaline layers fail to develop after subsequent fertilization in normal seawater. The eggs cleave properly but instead of remaining as a compact embryo the cells form irregular strings, probably because the hyaline layer did not develop. This procedure has been modified to treat larger volumes of eggs (HULTIN, 1948).

Pretreatment with trypsin or other proteolytic enzymes removes the vitelline membrane without affecting the hyaline layer. Eggs are suspended in seawater containing twice-crystallized trypsin at a final concentration of 0.25 mg/100 ml. A small aliquot of suspension is set aside and every few minutes a sample is washed and fertilized. The rest is allowed to settle by gravity. When microscopic examination of the fertilized samples shows that the fertilization membrane is not being formed, the trypsin solution is drawn off the main egg suspension, and the eggs are washed 3 times with fresh seawater and fertilized. This method tends to prevent clumping (EPEL, 1970).

The fertilization membrane may also be removed from *S. purpuratus* eggs with 5-10 m*M* dithiothreitol (DTT) at pH 9.2 (EPEL et al., 1970). The basic procedure is the same as above. The eggs are left in DTT, and aliquots are removed at intervals, washed three or four times to remove any DTT, and fertilized. When the fertilization membrane no longer forms, the rest of the eggs are washed and fertilized. Eggs tend not to clump after this treatment, and neither trypsin nor DTT seems to affect development.

Proteolytic enzymes may also be used shortly after the fertilization membrane forms (VACQUIER and MAZIA, 1968). Pronase is made up in seawater at a concentration of 2 mg per milliliter. Eggs are fertilized in a known volume of seawater. After about 90 seconds an equal

Table 2A.3. Methods of removing fertilization membrane, jelly coat, and hyaline layer. (From BERG, 1967)

Proteolytic enzymes	Trypsin (0.1%) digests transitional fertilization membrane and hyaline layer of fertilized sand-dollar eggs (*Dendraster*). Trypsin (also papain, ficin, and chymotrypsin) treatment of unfertilized sea-urchin eggs (*Strongylocentrotus*) prevents formation of fertilization membranes.
	Trypsin (0.1%) and chymotrypsin digest jelly layer, fertilization membrane, and hyaline layer on sea-urchin eggs (*Paracentrotus*, *Psammechinus*, and *Echinocardium*).
	Papain (0.1 mg per milliliter) removes hyaline layer of sea-urchin egg (*Arbacia*).
	Protease (0.1 - 1.0%) prevents formation of fertilization membranes and digests transitional fertilization membranes and hyaline layer of sea-urchin eggs (*Lytechinus* and *Strongylocentrotus*).
	Ficin (0.2 - 0.4%, pH 8.2) digests vitelline membrane of sea-urchin egg (*Psammechinus*).
	Hatching enzyme (supernatant from concentrated hatching blastulae) digests fertilization membranes of sea-urchin eggs (*Strongylocentrotus* and *Arbacia*).
Salts	Ca-free seawater, isosmotic NaCl and KCl disperse hyaline layer of sea-urchin egg.
	NaCl removes jelly coat of sea-urchin egg (*Arbacia*).
	Hypertonic seawater treatment causes subsequent dispersal of hyaline layer of sea-urchin egg (*Arbacia*).
Alkaline solutions	Alkaline seawater or alkaline NaCl solution (pH 10.5 by Na_2CO_3) dissolves transitional fertilization membrane of sea-urchin egg (*Lytechinus*).
Acid solutions	Acidified seawater (pH 3.5 - 4.5) dissolves jelly coats of various eggs (echinoderms, molluscs, and annelids).

Sulfhydryl compounds	Cysteine (0.1%, pH 7.8 - 8.0) and thioglycolic acid (0.5%, pH 7.8 - 8.0) dissolve transitional fertilization membranes and prevent formation of hyaline layer on sea-urchin eggs (*Psammechinus* and *Echinocardium*).
	Cysteine (0.1 *M*) treatment of unfertilized sea-urchin eggs (*Strongylocentrotus*) causes subsequent dissolution of fertilization membranes.
	Thioglycolic acid (0.01 *N*) softens fertilization membranes of sea-urchin egg (*Psammechinus*) and renders them digestible by proteolytic enzymes.
	Mercaptoethylgluconamide (0.1%, pH 7.5 - 8.5) prevents transitional fertilization membrane of sea-urchin egg (*Strongylocentrotus*) from hardening.
Nonelectrolytes	Urea, sucrose, and glycerol (1 *M*, pH 7.0) prevent formation of fertilization membrane and hyaline layer in echinoderm eggs (*Dendraster*, *Strongylocentrotus*, *Paracentrotus*, and *Arbacia*).

volume of pronase solution is added, giving a final concentration of 1 mg per milliliter. The pronase must be added just after all the eggs have been fertilized but before the membranes have hardened. If the eggs are washed within a few minutes after the addition of pronase, they will still have a hyaline layer and develop normally. Left in pronase, they will cleave, but the hyaline layer - usually not affected by short pronase treatment - will gradually be digested, allowing the blastomeres to fall apart.

Just after formation, the fertilization membrane of *Paracentrotus* is weak enough to be removed by passing the fertilized eggs through mesh bolting cloth (LINDAHL and LUNDIN, 1948) or by the mouth pipet method of HÖRSTADIUS (page 367). It may be removed from *S. purpuratus* eggs by the bolting cloth method if it is first treated with chemicals (MAZIA et al., 1961), from *E. parma* eggs by forcing them through a hypodermic needle just after fertilization, and from *Lytechinus* eggs with gentle use of a Dounce or Potter Elvehjem homogenizer (BURCHILL and BLOMQUIST, 1969).

After the fertilization membrane has been removed, eggs are usually sticky. To keep them from sticking together, agitation should be minimal. To keep them from sticking to a plastic or glass container, coat the container bottom with 1.5% agar.

2A.12 Normal Development

All echinoids have similar developmental stages. A few species with large eggs develop from egg directly into sea urchin, but most species used in embryological studies go through a feeding pluteus stage. Normal development will be dealt with in Chapter 7.

2A.12.1 Using Artificial Seawater

Until the pluteus stage, it is not necessary to raise embryos in seawater. Many simple salt mixtures such as the following (MAZIA) work just as well: NaCl, 28.32 gm per liter; KCl, 0.77 gm per liter; $MgCl_2 \cdot 6H_2O$, 5.41 gm per liter; $MgSO_4 \cdot 7H_2O$, 7.13 gm per liter; $CaCl_2$, 1.18 gm per liter. After these have dissolved, add $NaHCO_3$, 0.2 gm per liter. If necessary, adjust the pH to 8.2. For calcium-free seawater omit CaCl.

There are also formulae for artificial seawater lacking other ions (CAVANAUGH, 1956). A number of commercial complex salt mixtures come close to duplicating natural seawater, but none equals it. Adult sea urchins and feeding larvae cultured in these products tend to be more susceptible to disease.

2A.12.2 Raising Embryos

Eggs may be raised to plutei with little trouble. The simplest method is to allow the fertilized eggs to settle on the bottom of a flat container such as a finger bowl, beaker, or petri dish. There should be less than a monolayer of eggs on the bottom; if they are too crowded, development will be abnormal. This type of culture is simple, but it limits the number of eggs that may be handled conveniently. Stirring with a 30-rpm clock motor attached to a plastic paddle, allows large

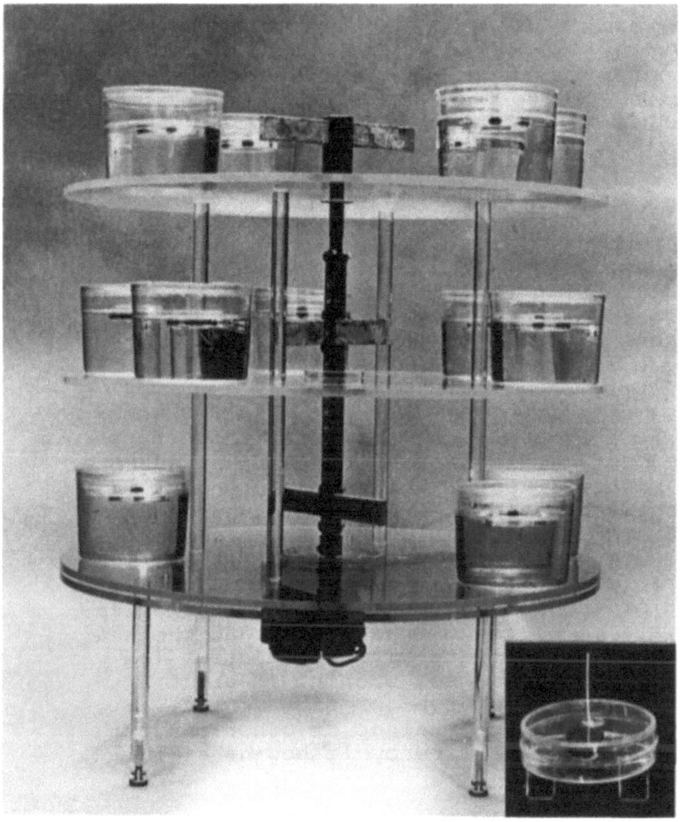

Fig. 2A.4. Magnetic stirring apparatus for raising larvae is 50 cm high and 46 cm in diameter. Bottom shelf rests on ball bearings to allow shelf rotation. Three Alnico V magnets are mounted on center shaft driven by 25-rpm motor. Larvae are grown in polystyrene dishes. Two holes (diameter: about 3/32 inch) are drilled in dish lids - one near edge, other in center - to ventilate culture. Center hole holds stirrer axle. Assembled stirrer (insert) floats in culture dish and consists of small magnet cemented in lids of two 35 x 10-mm polystyrene petri dishes. Piece of monofilament nylon fish line cemented in lid serves as axle. Two paddles are attached to bottom. Paddle design shown minimizes contact between larvae and paddle

numbers of eggs to be raised at concentrations up to 4×10^4 eggs per milliliter.

At pluteus, the embryo's food reserves become exhausted and it degenerates unless fed. If feeding is delayed more than a few days past fertilization, the larvae do not survive. They continue to swim for many days until they become small, ciliated spheres. With proper feeding, most species may be cultured in the laboratory (HINEGARDNER, 1969).

Diatoms were once thought to be good food organisms (HARVEY, 1956), but have proved far inferior to flagellated algae. The green algae *Dunaliella* can be used as food for such species as *Arbacia* and *S. purpura-*

tus. An unidentified *Rhodomonas* - designated 3C by GUILLARD - and algae of the genus *Pyranimonas* can be used for *L. pictus*; and *Cryptomonas*, *Dunaliella* and *Oxyrrhis* for *Psammechinus miliaris* (CZIHAK, 1960). Algae are grown in pure culture, centrifuged out of the culture medium, and washed in seawater before being fed to the larvae. Algae grow well in enriched seawater media such as GUILLARD and RYTHER's (1962) diluted to half strength.

Larvae of *L. pictus*, *A. punctulata*, and - with a slightly different method, many other species - are cultured at a concentration of about 1 specimen per 1.5 ml of seawater. Each day larvae are given about the number of algae they will eat that day. Fig. 2A.4 illustrates one type of apparatus for providing the agitation that most larvae require for healthy growth: a 25-rpm magnet stirrer in which the slave magnets are lateral to the master magnet instead of above it, permit a large number of culture dishes to be stirred (HINEGARDNER, 1969).

In clean, well-fed cultures, metamorphosis takes place in 3-4 weeks for *L. pictus* and *Arbacia* and in 4-6 weeks for *P. miliaris*. In most species metamorphosis is not spontaneous; mature larvae must be exposed to a substrate containing a thin bacterial scum. While metamorphosis is not fully understood, the observable events are rapid: in about an hour, a swimming larva containing the echinoid rudiment becomes a small sea urchin with 5 tube feet (CAMERON and HINEGARDNER, 1974). *P. miliaris* takes several hours (CZIHAK, 1960). Most of the larval tissue becomes a lump on top of the animal, major tissue reorganization occurs, and in 8-10 days a complete gut has formed and the sea urchin begins to feed. A number of substrate-dividing diatoms (as yet unidentified) serve as food for the young sea urchin. In about 6 months it reaches a diameter of 1 cm and develops ripe gonads, at which time it feeds on large, multicellular algae or other large food. The urchin can be spawned, and the life cycle begins again.

References

ANDERSON, E., 1968. Oocyte differentiation in the sea urchin Arbacia punctulata, with particular reference to the origin of cortical granules and their participation in the cortical reaction. J. Cell. Biol., 37, 514-539.

BERG, W.E., 1967. Some experimental techniques for eggs and embryos of marine invertebrates. In: Methods in Developmental Biology (F.H. Wilt and N.K. Wessells, eds.), p. 716-767. New York: Thomas Y. Crowell Co.

BOOLOOTIAN, R.A., 1966. Reproductive Physiology. In: Physiology of Echinodermata (R.A. Boolootian, ed.), p. 561-613. New York: Interscience Publishers.

BURCHILL, B.R., BLOMQUIST, C.H., 1969. Removal of fertilization membranes from sea urchin (Lytechinus pictus) eggs. Experientia, 25, 540-541.

CAMERON, R.A., HINEGARDNER, R.T., 1974. Initiation of metamorphosis in laboratory cultured sea urchins. Biol. Bull. 146, 335-342.

CAVANAUGH, G.M., 1956. Formulae and methods V. of the Marine Biological Laboratory Chemical Room. Woods Hole, Mass.: Marine Biological Laboratory.

CITKOWITZ, E., 1971. The hyaline layer: its isolation and role in echinoderm development. Dev. Biol. 24.

COSTELLO, D.P., DAVIDSON, M.E., EGGERS, A., FOX, M.H., HENLEY, C., 1957. Methods for obtaining and handling marine eggs and embryos. Woods Hole, Mass.: Marine Biological Laboratory.

CZIHAK, G., 1960. Untersuchungen über die Coelomanlagen und die Metamorphose des Pluteus von Psammechinus miliaris. Zool. Jahrb. Abt. Anat. Ontog. Tiere 78, 235-256.
ENDO, Y., 1961. Changes in the cortical layer of sea urchin eggs at fertilization as studied with the electron microscope. I Clypeaster japanicus. Exp. Cell Res., 25, 383-397.
EPEL, D., 1970. Methods for removal of the vitelline membrane of sea urchin eggs. II. Controlled exposure to trypsin to eliminate post-fertilization clumping of embryos. Exp. Cell Res., 61, 69-70.
EPEL, D., WEAVER, A.M., MAZIA, D., 1970. Methods for removal of the vitelline membrane of sea urchin eggs. I. Use of Dithiothreitol (Cleland Reagent). Exp. Cell Res., 61, 64-68.
GUILLARD, R.R.L, RYTHER, J.H., 1963. Studies on marine planktonic diatomes. I. Cyclotella nana Hustedt and Detonula confervacea (Cleve) Gran. Can. J. Microbiol., 8, 229-239.
HAGSTRÖM, B.E., 1959. Further experiments on jelly-free sea urchin eggs. Exp. Cell Res., 17, 256.
HARVEY, E.B., 1956. The American Arbacia and other sea urchins. Princeton, N.J.: Princeton University Press.
HINEGARDNER, R.T., 1967. Echinoderms. In: Methods in Developmental Biology (F.H. Wilt and N.K. Wessels, eds.), p. 139-155. New York: Thomas Y. Crowell Co.
HINEGARDNER, R.T., 1969. Growth and development of the laboratory cultured sea urchin. Biol. Bull., 137, 465-475.
HÖRSTADIUS, S., 1973. Experimental embryology of echinoderms. London: Oxford Univ. Press.
HULTIN, T., 1948. Species specificity in fertilization reaction. I. The role of the vitelline membrane of sea urchin eggs in species specificity. Arkiu für Zoologi, 40A (12), 1-9.
HYMEN, L.H., 1955. The Invertebrates: Echinodermata the coelomate bilateria, Vol. IV. New York: McGraw-Hill Book Co., Inc.
KANE, R.E., 1970. Direct isolation of the hyaline layer protein released from the cortical granules of the sea urchin egg at fertilization. J. Cell Biol. 45, 615-622.
KARNOFSKY, D.A., SIMMEL, E.B., 1963. Effects of growth inhibiting chemicals on the sand dollar embryos Echinorachnius parma. Prog. Exp. Tumor Res. 3, 254-295.
LINDAHL, P.E., LUNDIN, J., 1948. Removal of the fertilization membranes from large quantities of sea urchin eggs. Science, 108, 481-482.
MAZIA, D., MITCHISON, J.M., MEDINA, H., HARRIS, P., 1961. The direct isolation of the mitotic apparatus. J. Biophys. Biochem. Cytology, 10, 467.
McBRIDE, E.W., 1903. The development of Echinus esculentus, together with some points in the development of E. miliaris and E. acutus. Phil. Trans. Roy. Soc. London, Series B, 195, 285-327.
MOORE, A.R., 1930. Fertilization and development without membrane formation in the egg of the sea urchin, Strongylocentrotus purpuratus. Protoplasma, 9, 9-17.
SHAPIRO, H., 1935. The validity of the centrifuge method for estimating aggregate cell volume in suspensions of the egg of the sea urchin, Arbacia punctulata. Biol. Bull., 68, 363-377.
SPOTTE, S.H., 1970. Fish and invertebrate culture, water management in closed systems. New York: Wiley-Interscience.
TYLER, A., 1949. A simple, non-injurious method for inducing spawning of sea urchins and sand dollars. Collect. Net., 19, 19-20.
TYLER, A., TYLER, B.S., 1966. The gametes; some procedure and properties. In: Physiology of Echinodermata (R.A. Boolootian, ed.), p. 639-682. New York: Interscience Publishers.
VACQUIER, V.D., MAZIA, D., 1968. Twinning of sea urchin embryos by treatment with dithiothreitol. Exp. Cell Res., 52, 459-468.

2B. Handling Japanese Sea Urchins and Their Embryos

K. Osanai

2B.1 General Handling

2B.1.1 Physiologic Salt Solution

A common physiologically balanced salt solution for marine animals, natural seawater must be used within several days after being drawn from the sea because its organic substances begin to be decomposed by bacteria. Kept in containers for more than one month, its organic substances are completely decomposed. Natural seawater can be used for culture after being filtered through active carbon particles and adjusted to pH 8.2 with sodium bicarbonate.

Artificial seawater - first designed by HERBST (1904) and modified by MOTOMURA (1938) - may be used instead of natural seawater. Bivalent cation-free seawater is prepared by substituting NaCl for $CaCl_2$ and Na_2SO_4 for $MgSO_4$ (Table 2B.1).

Table 2B.1. Formulas for artificial seawater

Chemicals	HERBST	Ca^{++}-free (Grams per liter)	Ca^{++}- and Mg^{++}-free (Grams per liter)
NaCl	26.3	26.9	26.9
KCl	0.7	0.7	0.7
$MgSO_4 \cdot 7H_2O$	11.9	11.9	
$CaCl_2 \cdot 2H_2O$	1.5		
$Na_2SO_4 \cdot 10H_2O$			10.4
$NaHCO_3$	0.5	0.5	0.5

2B.1.2 pH Indicator

If no strict pH value is required, color indicators may be convenient. Phenol red (saturated aqueous solution) is used to indicate alkalinity and methyl red (0.1% solution in ethyl alcohol) to indicate acidity. A few drops in 1000 ml of seawater are not harmful to sea urchin gametes and embryos. Color indicators may also be used to detect pH change during experiments. When the pH of an isotonic solution of potassium chloride or urea must be adjusted to that of normal seawater, the required pH is obtained by adding to the solution sodium bicarbonate and a small amount of phenol red until the coloration is the same as normal seawater containing the same amount of dye.

2B.1.3 Culture Vessel for Embryos

For culturing small numbers of embryos, Petri dishes - 6 cm in diameter and 20 ml in volume - may be used. An inverted microscope is convenient to observe living eggs and embryos without taking off the cover of the dish, the bottom of which must be flat.

2B.1.4 Fixatives for Gametes, Embryos, and Larvae

BAKER's formalin-calcium chloride solution is used to fix gametes and developing embryos. The shapes of the embryos and blastomeres are well preserved, and the pigments or vitally stained granules do not fade. The solution formula is: formalin, 10 ml; calcium chloride (10% anhydrous), 10 ml; and distilled water, 80 ml. Precipitated calcium carbonate is kept bottled (BAKER, 1956).

It has been found that the 10% formalin solution usually used to preserve specimens sometimes deforms early, cleaving sea-urchin embryos, while BAKER's solution deforms little. In the colorless BAKER's solution, the color of fixed specimens is nearly natural and pigment substances are not removed.

A 10% seawater solution of neutralized formalin is used for plutei. In paraffin sectioning, embryos fixed with BAKER's solution are often deformed by alcohol dehydration. The cortical granules of unfertilized eggs are well preserved with MOTOMURA's fixative No. 1 (MOTOMURA, 1957; OSANAI, 1964b). MOTOMURA (1958, 1960a, 1960b, 1963) designed a series of fixatives for mucopolysaccharides and adapted some of them to jelly substances, cortical granules, and mucosubstances of the embryonic cells and blastocoeles of sea urchins (Table 2B.2). The most widely used fixative for studying nucleic acids cytochemically is a 3:1 mixture of alcohol and acetic acid.

Table 2B.2. Fixative formulas for sea-urchin eggs and embryos. (From MOTOMURA, 1958, 1960a, 1960b, 1963)

Chemicals[a]	Fixatives				
	I	IV	V	VI	VII
$Cd(NO_3)_2 \cdot 4H_2O$	5				
CdI_2			5	5	5
HgI_2			3	3	
Iodine					1
Picric acid	1				
$Na_2H_3IO_6$		2			
Acetic acid		4			
Formalin	10	15	15	15	15
H_2O	90	85	85		85
Buffer solution[b]				85	

[a]Units of crystalized chemicals in grams; of liquids, milliliters.
[b]Made by dissolving 0.5 gm of veronal and 0.6 gm of sodium diethylbarbiturate in 100 ml of distilled water.

2B.2 Echinoids as Experimental Materials

Many species of echinoids - sea urchins, heart urchins, and cake urchins, or sand dollars - live in the seas around the islands of Japan. Their habitat, breeding season, and gamete characteristics vary. The species commonly used in Japan for embryologic experiments are discussed here, including where they occur, their habitat, how to collect them, their breeding season, and their egg characteristics.

2B.2.1 Regular Echinoids

Anthocidaris crassispina (A. AGASSIZ) occurs along the coast of southern Japan from Kyushu to Tokyo Bay among reefs in shallow seas. It is gathered at low tide by scooping or diving. It breeds at Misaki from the middle of June to August, at Sugashima from the end of June to August, at Naruto from July to early August, and at Nagato in Yamaguchi prefecture and Seto from May to August. Eggs are opaque, pink, about 90μ in diameter, and very numerous. Fertilization membrane elevation is low.

Glyptocidaris crenularis (A. AGASSIZ) occurs only in northern Japan from Hakodate to Inubô-zaki on the east coast and to Sado Island on the west. It is found on hard sea bottom 10-150 m down and on small stones or rocky bottom 10-20 m down at Asamushi. It is obtained by trawling or dredging. It breeds at Asamushi from late February to April. Eggs are transparent, slightly salmon-red in color, about 125μ in diameter, very numerous, and highly sensitive to rising temperature. They develop abnormally above 15°C. Surrounding the egg is a very thick coat of jelly.

Hemicentrotus (Strongylocentrotus) pulcherrimus (A. AGASSIZ) occurs only in Japan from Hakodate to Kagoshima under stones in water up to 45 m deep. It is scooped up with a net attached to a long pole and with a glass-bottom box. Breeding at Asamushi is from late December to early May, at Misaki and Naruto from January to April, at Sugashima from January to April, at Seto from January to March, and on the Japan Sea in Yamaguchi prefecture from October to March. Eggs are semi-transparent, about 95μ in diameter, and orange. The cortex contains pigment granules. Eggs are resistant to chemicals and rising temperature. Adults are small and numerous.

Mespilia globulus (LINNE) occurs in southern Japan as far north as the Bôso Peninsula on conglomerate or gravel bottom in water up to 60 m deep. It is gathered by diving. It breeds at Sugashima from July to August, and at Seto from June to September. Transparent eggs make it possible to observe nuclei.

Parasalenia gratiosa (A. AGASSIZ) occurs in southern Japan in rocky areas in water up to 70 m deep. It is obtained at low tide or by diving. It breeds at Misaki from late October to December, at Sugashima from October to early December, and on Tachibana Bay in Tokushima prefecture in November. Eggs are opaque, about 94μ in diameter, and very numerous.

Strongylocentrotus intermedius (A. AGASSIZ) is very common in northern Japan as far south as Onagawa on the Pacific coast and Syônai, Yamagata prefecture, on the Japan Sea. It is found on boulders, in muddy sand with eel grasses, and in water up to 35 m deep. It is scooped up with a long pole-net. It breeds at Shinori, Hokkaido, from September to October, and at Asamushi from October to November. Eggs are transparent, orange, 84-94μ in diameter, and numerous.

Strongylocentrotus nudus (A. AGASSIZ) is common in northern Japan from Hokkaido to Sagami Bay on the Pacific coast. It is found on rocky bottom up to 180 m deep, on boulders with eel grasses, and at Asamushi up to 30 m deep. It is gathered by scooping, trawling, or fishing with a special basket. It breeds at Shinori, Hokkaido, from September to October and at Asamushi from September to early November. Eggs are very numerous. Adults are large and plentiful.

Temnopleurus hardwiki (GRAY) occurs in northern Japan from Hakodate to Toba in areas of eel grass with red algae on muddy sand in shallow bays. It is obtained by dredging, trawling, or diving. It breeds at Asamushi from July to early September. Eggs are transparent, about 100μ in diameter, and contain brown pigment granules. They are not abundant but develop quickly. They are highly sensitive to reagents such as chemicals that cause parthenogenesis.

Temnopleurus toreumaticus (LESKE) occurs from central to southern Japan on sandy bottom up to 45 m deep. It is gathered by dredging, trawling, or diving. Breeding is at Misaki in August, at Sugashima from June to early September, and at Naruto from August to September. Eggs are similar to those of *T. hardwiki*.

Texopneustes pileolus (LAMARCK) occurs on the Pacific coast of southern Japan as far north as Sagami Sea on rocks up to 90 m deep. It is obtained by diving. It breeds at Sugashima from May to the middle of July, at Seto from April to August, and at Tokushima from the middle of October to November. Eggs are transparent and numerous.

2B.2.2 Irregular Echinoids (Heart Urchins and Sand Dollars)

Echinocardium cordatum (PENNANT) occurs from Hokkaido to Kyushu usually buried in muddy or sandy bottom up to 230 m deep and at Asamushi in sandy bottom 1-2 m deep and muddy bottom 20-30 m deep. It is reached by dredging or by a scoop attached to a long pole. It breeds at Asamushi from the middle of April to July. Eggs are transparent.

Astericlypeus manni (VERRILL) occurs in southern Japan on sandy bottom in shallow water. It is obtained by dredging. It breeds at Misaki from the middle of August to early September, and at Sugashima in June. Eggs are semitransparent and about 150μ in diameter.

Clypeaster japonicus (DÖDERLEIN) is very common in southern Japan on gravel bottom up to 75 m deep. It is reached by diving or is picked up at low tide. It breeds at Misaki from late June to July, at Sugashima from July to August, and at Seto from June to August. Eggs are transparent and 110μ in diameter.

Peronella japonica (MORTENSEN) occurs in southern Japan on sandy or gravel bottom up to 50 m deep. It is obtained by dredging or diving or is picked up at low tide. It breeds at Misaki in April, and at Sugashima from July to April. Eggs are opaque, about 240μ in diameter, and few. They develop rapidly and metamorphose 70-80 hours after fertilization.

Sea-urchin gonads are used for food in Japan. The five species of commercial value are *Anthocidaris crassispina, Hemicentrotus pulcherrimus, Pseudocentrotus depressus, Strongylocentrotus intermedius*, and *Strongylocentrotus nudus*. The annual yield of Japanese sea urchins is about 20,000 tons. Sea urchins are caught by scooping at low tide, diving, trawling, and dredging. At Ohma, Aomori prefecture, they are caught in flat, round baskets baited with brown algae and lowered to the sea bottom. In this manner the sea-urchin habitat is not disturbed or the sea bottom destroyed.

2B.3 Maintenance

In wet paper or fresh algae and packed in a wooden box, sea urchins can survive for one day without special cooling except in summer. For longer periods they must be kept in cool seawater. Water temperature too high causes adults to shed gametes. In the marine laboratory sea urchins are kept in aquaria with running seawater. If apparatus is not available, they may be put in the bottom of a shallow dish about 30 cm in diameter. The dish is filled with seawater and refrigerated at about 5°C. The water is changed every day. Sea urchins survive at least a week without feeding, and their eggs develop normally after fertilization.

Running seawater aquaria (FUJI, 1967) for *S. intermedius* adults are shown in Figs. 2B.1 and 2B.2. Young sea urchins - less than 10 mm in test diameter - eat mainly diatoms and later algae. An adult specimen of *S. intermedius* - about 50 mm in diameter - eats 1%-6% of its body weight daily (FUJI, 1967). Brown algae such as *Sargassum thunbergii*, *Sargassum seratifolium*, and *Ecklomina cava* are used for sea-urchin food.

2B.4 Sexual Dimorphism

Generally it is very hard to determine the sex of a sea urchin by external characteristics. TAHARA, OKADA, and KOBAYASHI (1958, 1960) studied sexual dimorphisms in the 11 Japanese species and classified them as *Mespilia globulus* or *Tripneustes gratilla*, depending on the shape of the genital papillae (Fig. 2B.3 and Table 2B.3). In *H. pulcherrimus*, MOTOMURA (1941) found the tube feet on the oral side to be yellow in the female and white or faded purple in the male.

Table 2B.3. Sexual dimorphism in Japanese sea urchins. (From TAHARA, OKADA, and KOBAYASHI, 1958, 1960)

Type	Shape of Genital Papillae	Species
Mespilia	Male: Short and conical Female: Flat; below body surface	*Anthocidaris crassispina* *Hemicentrotus pulcherrimus* *Mespilia globulus* *Pseudocentrotus depressus* *Temnopleurus toreumaticus* *Texopneustes pileolus*
Tripneustes	Male: Remarkably long Female: Short and stumpy	*Diadema setosum* *Echinocardium cordatum* *Echinometra mathaei* *Echinostrephus aciculatus* *Tripneustes gratilla*

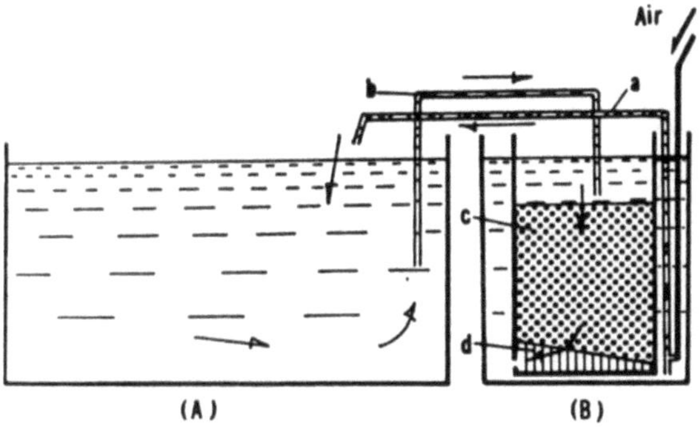

Fig. 2B.1. Aquarium (A) with circulating seawater system: (B), filter chamber: a, air lift; b, siphon; c, sand; d, glass wool. Arrows indicate water circulation. (From FUJI, 1967)

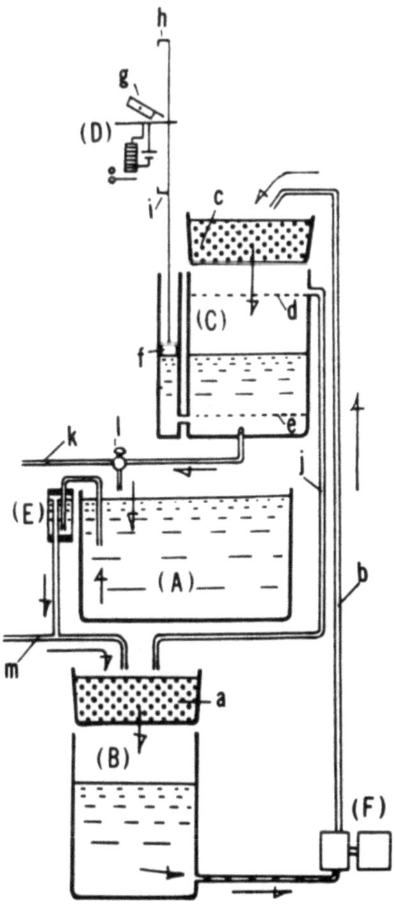

Fig. 2B.2. Aquarium (A) with upper and lower tanks. Overflow water from constant-level siphon (E) is led into lower tank(B) through lower filter chamber (a) and sifted to upper tank (C) through feedpipe (b) and upper filter chamber (c) by feedpump (F). Upper tank has float switch (D) to open and close electric circuit in response to rise or fall of water level in upper tank. Feedpump is operated by float motion (f), keeping water in upper tank between upper (d) and lower (e) levels and providing continuous water supply for aquarium; g, mercury switch; h, bar to close electric circuit; i, bar to open electric circuit; j, overflow pipe; k, feedpipe; l, stopclock; m, water shoot. (From FUJI, 1967)

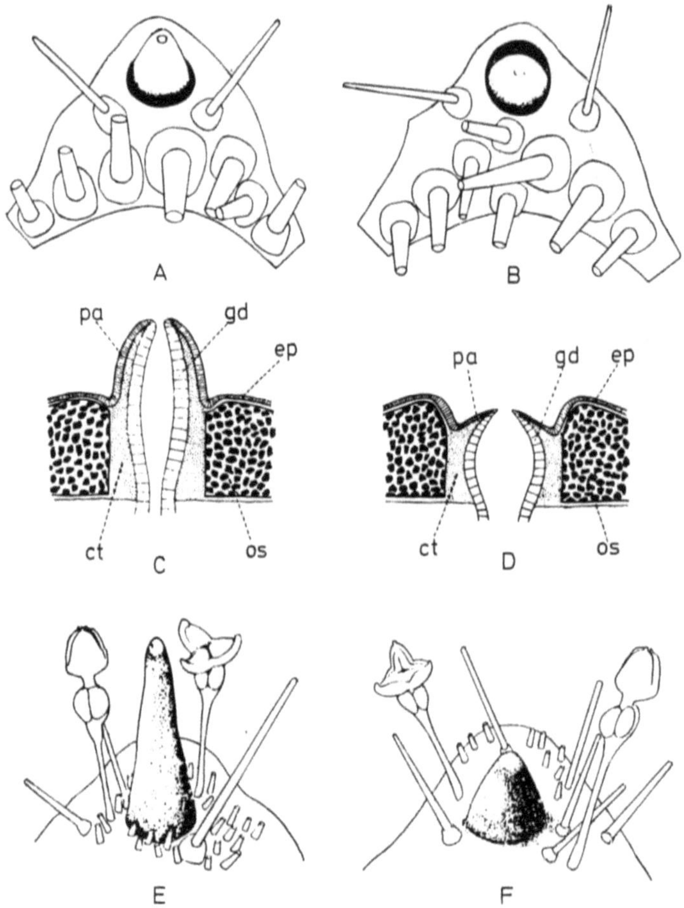

Fig. 2B.3. Genital papillae of *Mespilia globulus* (male, A, C; female, B,D) and *Tripneustes gratilla* (male, E; female, F). Sagittal section of genital papillae (C, D) is shown; ct, connective tissue; ep, epidermis; gd, genital duct wall; os, genital plate ossicle; pa, wall of genital papilla. (From TAHARA, OKADA, and KOBAYASHI, 1958)

2B.5 Artificial Spawning

Potassium chloride is widely used in Japan to induce shedding of gametes. The ARISTOTLE lantern is removed by cutting around the peristomial membrane with scissors, the sea urchin is placed upside down on a 20-30 ml beaker filled with seawater. Body fluid is removed and 0.5 *M* potassium chloride solution (pH 8.2) poured into the body cavity. The gametes are shed into seawater and deposited on the bottom of the beaker. When the shedding stops, the supernatant fluid is taken away and the eggs washed several times in pure seawater before using (IIDA, 1942).

This method yields a large number of gametes, but the sea urchin is killed. If its survival is required, spawning must be induced by

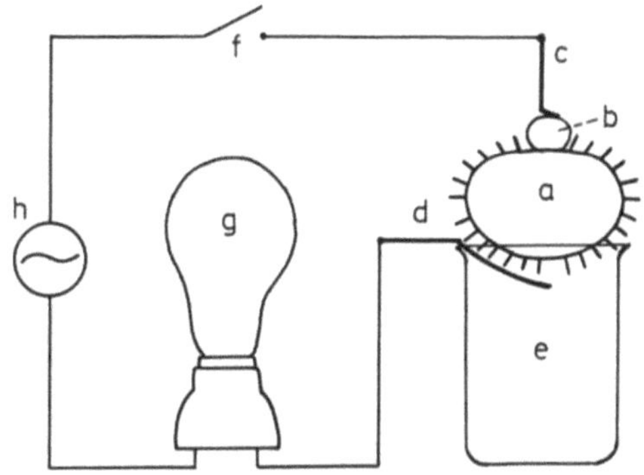

Fig. 2B.4. Simplified apparatus to stimulate sea-urchin spawning electrically; a, sea urchin; b, cotton wool wet with seawater; c and d, stimulating lead or silver electrodes; e, beaker of seawater; f, key; g, 40- to 100-W bulb; h, 100-V electrical source, 50 or 60 cps. (Modified from IWATA, 1962)

gentler methods such as chemical injection or electrical stimulation. The latter was first used by IWATA (1950) in Japanese sea urchins and later by HARVEY (1954, 1956) in American species. Shedding begins by placing the electrodes at any two points on the shell of *Arbacia punctulata* but at the oral and aboral sites in Japanese sea urchins (IWATA, 1962). In the electrical apparatus simplified by IWATA (1962)(Figs. 2B.4 and 2B.5), current strength is adjusted by changing the capacity of the electrical lamp inserted in the stimulating circuit.

Fig. 2B.5. *Hemicentrotus pulcherrimus* eggs are discharged by electrical stimulation

The gametic discharge induced by electrical stimulation or potassium chloride lasts as long as the stimulation continues. That induced by acetylcholine falls off abruptly a few minutes after injection but may be made long lasting by pretreatment with eserine. Gamete discharge induced by acetylcholine is inhibited by tubercurarine, hexamethinium, and magnesium sulphate, while electrically or potassium-induced discharge is not. Inducement by acetylcholine requires excitation in the neuromuscular junction, while electrically or potassium-induced discharge directly induces gonadal muscle contraction (IWATA and FUKASE, 1964a, 1964b).

2B.6 Gametes

2B.6.1 Storing Gametes

Stored at low temperature (5-10°C), eggs keep their ability to be fertilized quite a while. Even without cooling, diluted sperm keep their power to fertilize if placed in seawater adjusted to pH 6 with diluted hydrochloric acid.

2B.6.2 Jelly Coat

The Chinese (or Indian) ink method is commonly used to detect the jelly coat surrounding the unfertilized egg. One drop of egg suspension is mixed on a slide glass with one drop of Chinese ink suspension prepared with seawater. A cover glass is then placed over it. The colloidal carbon particles of Chinese ink cannot penetrate the jelly layer which remains a clear zone around the egg.

In methylene blue solution, the jelly coat abruptly precipitates into a thin membraneous structure adhering closely to the vitelline membrane. When the eggs are exposed to 0.0005% seawater solution of methylene blue for 30-60 seconds and then inseminated in normal seawater, part of the cytoplasm is discharged through the entrance point of the spermatozoon. The discharged cytoplasm draws back again, leaving a small membraneous projection indicating where the sperm entered (YAMAMOTO, 1965).

2B.6.3 Removing the Vitelline Membrane

In unfertilized *S. nudus* eggs, the jelly coat, the vitelline membrane are removed by 0.5 *M* sodium thiocyanate solution (pH 8.2) (MOTOMURA, 1953). Isotonic urea solution dissolves the jelly coat, and the vitelline membrane, and the cortical granules. The membrane can also be removed by proteolytic enzymes such as trypsin and pronase. When unfertilized eggs are treated with urea or proteolytic enzyme for a few minutes, their ability to be fertilized is greatly reduced, but it returns if treatment is prolonged (Fig. 2B.6). The reduced ability to be fertilized - which lasts as long as the parthenogenetic activation takes - seems to be caused by a sperm-blocking mechanism that occurs with egg activation (OSANAI and TAMURA, 1970). A high percentage of eggs treated with 0.1% seawater solution of pronase P for 60-120 min are able to be fertilized. The pronase treatment sometimes causes exogastrulated plutei.

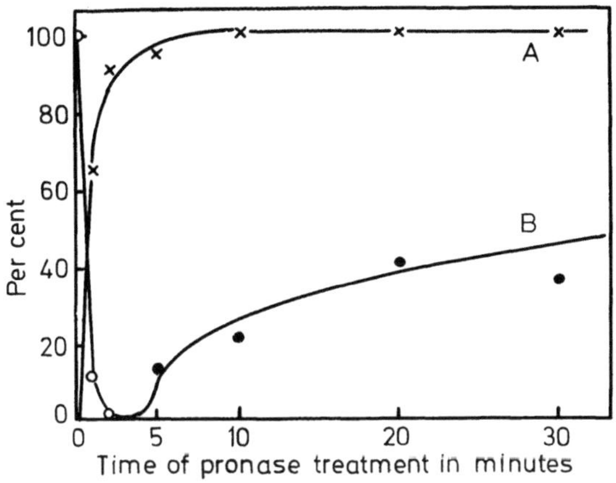

Fig. 2B.6. Parthenogenetic activation (A) and fertilization (B) of sea-urchin eggs pretreated with 0.1% seawater solution of pronase P (*Strongylocentrotus nudus*). Circles indicate eggs with fertilization membrane; black dots, eggs without. Fertilization membrane is not elevated in eggs inseminated after pronase treatment longer than 5 minutes. (From OSANAI and TAMURA, 1970)

2B.6.4 Removing the Fertilization Membrane

To remove the fertilization membrane, place the fertilized eggs in a calcium-free salt or isotonic urea solution before the membrane is toughened and shake them in a test tube half filled with the egg suspension. Or an inhibitor such as sodium sulfite, which does not block fertilization, may be used to remove the membrane. The eggs are fertilized in a mixture of 100 ml of seawater and 1 ml of M/3 sodium sulfite, causing a thin membrane to develop with wide perivitelline space. Thirty minutes after insemination, the eggs are transferred to normal seawater and shaken in a test tube (MOTOMURA, 1954). Its membrane removed, the egg forms a hyaline layer and develops normally (OSANAI, 1969).

2B.6.5 Egg Pigments

Unfertilized *H. pulcherrimus* eggs contain two kinds of orange pigment: pigment granules embedded in the cortical cytoplasm and pigment distributed homogeneously throughout the endoplasm. The pigment granules seem to cling to the inner surface of the egg. If the unfertilized eggs are placed in a mixture of 7 parts 1 *M* sucrose solution and one part seawater, a layer of colorless cortical granules - lying between the layer of pigment granules and the egg surface - becomes visible. Even after centrifuging, the pigment granules maintain a rigid, immobile cortex (MOTOMURA, 1941).

In *T. hardwiki*, brown pigment granules - dispersed in the ooplasm of unfertilized eggs - are sedimented toward the centrifugal pole while the cortical cytoplasm is immobile at a centrifugal force of 30,000 *G* for 1 minute, indicating that the cortex does not contain pigment granules. After fertilization, some of the granules become fixed in the cortex and the rest migrate inward and gather around the aster (OSANAI, 1964a, 1969).

Pigment granules in the cortex of fertilized eggs help to distinguish the old surface from the new. They are embedded only in the old cortex of the fertilized egg, not in the cortex formed during cleavage furrow formation (MOTOMURA, 1941; ENDO, 1966). Pigment substance or granule migration in the inner cytoplasm indicates endoplasmic activation (OSANAI, 1964a, 1969).

2B.7 Artificial Fertilization

Sea-urchin eggs may be fertilized at the beginning or the end of the breeding season. Membrane formation and development of underripe eggs are strongly influenced by pretreatment with periodate. Underripe *Psammechinus miliaris* eggs inseminated in a mixture of 2 ml seawater and 0.1 ml 0.001 *N* sodium periodate showed a 68%-94% increase in membrane development (RUNNSTRÖM and KRISZAT, 1950). Overripe eggs may be fertilized in calcium-free seawater (OKADA, 1950).

2B.8 Development and Growth

Without special feeding, sea urchins develop into swimming plutei a few days after fertilization. Their rate of development depends on species and temperature (Table 2B.4).

Table 2B.4. Timetable of early development

Stage	*Hemicentrotus pulcherrimus* 10-15°C[a]	*Temnopleurus toreumaticus* 23-28°C[b]	*Glyptocidaris crenularis* 10°C[c]
First cleavage	1.5-2.0 hr	40 min	3 hr, 10 min
Two-cell stage	2.0-2.5 hr	45 min	
Second cleavage			4 hr, 10 min
Four-cell stage	3-4 hr	70 min	
Fourth cleavage			7 hr, 40 min
Sixteen-cell stage	5-6 hr	2 hr	
Hatching	23 hr	4.5-5 hr	27 hr
Mesenchymal blastula		8 hr	36 hr
Early gastrula	32 hr	10 hr	53 hr
Midgastrula	42 hr	12 hr	
Late gastrula	52 hr	15 hr	67 hr
Pyramid larva	76 hr	18 hr	
Maximum pluteus without feeding	120 hr	30 hr	150 hr

[a]From OKADA and MIYAUCHI, 1954.

[b]From OKADA and MIYAUCHI, 1958.

[c]From FUKUSHI, 1960a.

Table 2B.5. Sea urchins from plutei to imagoes

Species	Temperature	Length of Time Before Metamorphosed Imagoes Appear (Days)	Foods
Anthocidaris crassispina[a]	26°-27°C	23	*Chaetoceros simplex*
Glyptocidaris crenularis[b]	10°C	36	*Peridinium* sp.
Hemicentrotus pulcherrimus[a]	15°-20°C	45	*Chaetoceros calcitrans*
Mespilia globulus[c]		35	Diatoms adhering to *Zostera marina*
Pseudocentrotus depressus[d]	19°-23°C	40	*Chaetoceros simplex* *Skeletonema costatum* *Chlamydomonas* sp.
Strongylocentrotus intermedius[e]	5°-19°C		*Phadactylum tricornutum*
Strongylocentrotus nudus[e]	8°-18°C		*Phadactylum tricornutum*
Temnopleurus hardwiki[f]		36	*Peridinium* sp. *Licmophora* sp.
Temnopleurus toreumaticus[g]	22°-30°C	32	

[a]From NAKAMURA and INOUE, 1965; cited by MATSUI, 1966.
[b]From FUKUSHI, 1960a.
[c]From ONODA, 1942.
[d]From YAMABE, 1962; cited by MATSUI, 1966.
[e]From KAWAMURA, 1970.
[f]From FUKUSHI, 1960b.
[g]From MITSUYA and GAMO, 1966; personal communication from ISHIKAWA.

The plutei must be fed for further development and growth. The larvae can complete their metamorphosis without special foods if their water is replaced every day with fresh seawater containing such natural foods as diatoms (FUKUSHI, 1960a, 1960b).

For raising *H. pulcherrimus* and *A. crenularis*, NAKAMURA and INOUE (1965) used as a culture vessel a 2-liter, flat-bottom flask aerated by a pump. They used *Chaetoceros calcitrans* for food (food density: 10^4 cells per milliliter). Until the sea urchins reached a test diameter of approximately 8 mm, they ate only adhering diatoms. Then they ate mainly seaweed. In *Strongylocentrotus intermedius* , FUJI (1967) noted changes of eating habits as changes in growth occurred. Larvae density is usually one to ten per milliliter of seawater. Data on larvae and imagoes are given in Tables 2B.5 and 2B.6.

Table 2B.6. Sea urchins from plutei to adults. (From NAKAMURA and INOUE, 1965; cited by MATSUI, 1966)

	Length of Time After Fertilization
Hemicentrotus pulcherrimus	
Pluteus; six-arm period	9 days
Eight-arm period	22 days
Initial tube feet appear	39 days
Sinking and adhering	45 days
Young sea urchin	67 days
Test diameter: 1 mm	74 days
Test diameter: 4-5 mm	4 mo
Test diameter: 8-10 mm	6 mo
Test diameter: 14.5-19.5 mm	10 mo
Test diameter: 18.5-26.5 mm	13 mo
Anthocidaris crassispina	
Pluteus; eight-arm period	7 days
Tube feet and spines appear	20 days
Sinking and adhering	23 days
Test diameter: 1 mm	33 days
Test diameter: 2 mm	56 days
Test diameter: 6 mm	5 mo
Test diameter: 8 mm	6 mo
Test diameter: 9.5 mm	8 mo

Acknowledgement

The section "Echinoids as Experimental Materials" is summarized from data reported by FUKUSHI (1960a), HIRAI (1963), KAWAMURA (1970), KOBAYASHI (1969), KOBAYASHI and NAKAMURA (1967), MATSUI (1966), and UCHINOMI (1954, 1960, 1965), and communicated personally by HIRAMOTO, Tokyo Institute of Technology; ISHIKAWA, Sugashima Marine Biological Station, Nagoya University; KATSURA, Marine Station for Experimental Biology, Tokushima University; NAKAMURA, Yamaguchi-ken Gaikai Suisan Shikenjô; and NUMAKUNAI, Marine Biological Station of Asamushi, Tôhoku University.

References

BAKER, J.R., 1956. Cytological Technique. 2nd Ed. London: Methuen Co., Ltd.

FUJI, A., 1967. Ecological studies on the growth and food consumption of Japanese common littoral sea urchin, Strongylocentrotus intermedius (A. AGASSIZ). Mem. Fac. Fish. Hokkaido Univ., 15, 83-160.

FUKUSHI, T., 1960a. The external features of the development of the sea urchin, Glyptocidaris crenularis (A. AGASSIZ). Bull. Mar. Biol. St. Asamushi, Tôhoku Univ., 10, 57-63.

FUKUSHI, T., 1960b. The formation of the echinus rudiment and the development of the larval form in the sea urchin, Temnopleurus hardwiki. Bull. Mar. Biol. St. Asamushi, Tôhoku Univ., 10, 65-72.

HARVEY, E.B., 1954. Electrical method of determining the sex of sea urchins. Nature, 173, 86.

HARVEY, E.B., 1956. The American Arbacia and Other Sea Urchins. Princeton: Princeton Univ. Press.

HIRAI, E., 1963. On the breeding seasons of invertebrates in the neighborhood of the Marine Biological Station of Asamushi. Sci. Rep. Tôhoku Univ., Ser. IV (Biol.), 29, 369-375.

IIDA, T., 1942. Uni no ran o ransô o kiridasazuni toru hôhô (A method obtaining sea urchin eggs without dissecting ovaries). Zool. Mag., 54, 280.

IWATA, K.S., 1950. A method of determining the sex of sea urchins and of obtaining by electric stimulation. Annot. Zool. Jap., 23, 39-42.

IWATA, K.S., 1962. A simplified electric method of determining the sex of sea urchins. Zool. Mag., 71, 301-302.

IWATA, K.S., FUKASE, H., 1964a. Artificial spawning in sea urchins by acetylcholine. Biol. J. Okayama Univ., 10, 51-56.

IWATA, K.S., FUKASE, H., 1964b. Comparison of discharge of the gametes by three artificial means in sea urchins. Biol. J. Okayama Univ., 10, 57-64.

KAWAMURA, K., 1970. On the development of the planktonic larvae of Japanese sea urchins, Strongylocentrotus intermedius and S. nudus. Sci. Rep. Hokkaido Fish. Exp. St., 12, 25-32.

KOBAYASHI, N., 1969. Spawning periodicity of sea urchins at Seto. III. Tripneustes gratilla, Echinometra mathaei, Anthocidaris crassispina and Echinostrephus aciculatus. Sci. Eng. Rev. Doshisha Univ., 9, 254-269.

KOBAYASHI, N., NAKAMURA, K., 1967. Spawning periodicity of sea urchins at Seto. II. Diadema setosum. Pub. Seto Mar. Biol. Lab., 15, 173-184.

MATSUI, I., 1966. Uni no Zôshoku (The Propagation of the Sea Urchins). Tokyo: Nippon Shuisanshigen Hogo Kyokai.

MOTOMURA, I., 1938. Effect of some salt solutions on the parthenogenetic membrane formation of sea urchin eggs. Sci. Rep. Tôhoku Univ. Ser. IV (Biol.), 13, 85-88.

MOTOMURA, I., 1941a. Materials of the fertilization membrane in the eggs of echinoderms. Sci. Rep. Tôhoku Univ. Ser. IV (Biol.), 16, 345-363.

MOTOMURA, I., 1941b. The sexual character of the sea urchin, Strongylocentrotus pulcherrimus (A. AGASSIZ). Sci. Rep. Tôhoku Univ. Ser. IV (Biol.), 16, 431.

MOTOMURA, I., 1953. Secretion of the sperm aggulutinin in the fertilized and denuded eggs of the sea urchin, Strongylocentrotus nudus. Sci. Rep. Tôhoku Univ. Ser. IV (Biol.), 20, 93-97.

MOTOMURA, I., 1954. Inhibition and acceleration of the toughening of the fertilization membrane in the sea urchin's eggs. Sci. Rep. Tôhoku Univ. Ser. IV (Biol.), 20, 158-162.

MOTOMURA, I., 1957. On the nature and localization of the third factor for the toughening of the fertilization membrane of the sea urchin egg. Sci. Rep. Tôhoku Univ. Ser. IV (Biol.), 23, 168-181.
MOTOMURA, I., 1958. Secretion of mucosubstance in the cleaving eggs of the sea urchin. Bull. Mar. Biol. St. Asamushi, 9, 79.
MOTOMURA, I., 1960a. Secretion of mucosubstance in the gastrula of the sea urchin embryo. Bull. Mar. Biol. St. Asamushi, 10, 165-169.
MOTOMURA, I., 1960b. On the structure of the cortical granule in the sea urchin egg. Sci. Rep. Tôhoku Univ. Ser. IV (Biol.), 26, 367-374.
MOTOMURA, I., 1963. Fixatives for electron microscopy. Bull. Mar. Biol. St. Asamushi, 11, 249-252.
OKADA, K., 1950. An improved method of artificial insemination on some marine eggs. Sci. Rep. Tôhoku Univ. Ser. IV (Biol.), 18, 262-269.
OKADA, K., MIYAUCHI, H., 1954. Normal table of the early developmental stages on the sea urchin, Hemicentrotus pulcherrimus. J. Gakugei, Tokushima Univ., 5, 57-68.
OKADA, K., MIYAUCHI, H., 1958. Sanshôuni no shokihassei kijunzuhyô (Normal table of the early development of the sea urchin, Temnopleurus toreumaticus). J. Gakugei, Tokushima Univ., 8, 59-71.
ONODA, K., 1932. Koshitaka-uni no yôsei to sono hentai (Larvae and their metamorphosis of Mespilia globulus). Zool. Mag., 44, 20-21.
OSANAI, K., 1964a. Ecto- and endoplasmic relation in the partially activated egg of the sea urchin. Sci. Rep. Tôhoku Univ. Ser. IV (Biol.), 30, 105-117.
OSANAI, K., 1964b. Cortical granule as a body containing cyto-fertilizin in the sea urchin egg. Sci. Rep. Tôhoku Univ. Ser. IV (Biol.), 30, 119-131.
OSANAI, K., 1969. Behavior of pigment granules during early development in the eggs of sea urchin, Temnopleurus hardwiki. Ann. Rep. Fac. Ed. Univ. Iwate, 29, Pt. 3, 35-39.
OSANAI, K., TAMURA, N., 1970. Uni-seishi chûshutsu-butsu no seirisayô (Physiological natures of the sperm extract of the sea urchin). Jap. J. Dev. Biol., 24, 80-81.
RUNNSTRÖM, J., KRISZAT, G., 1950. Action of periodate on the surface reactions attending the activation of the egg of the sea urchin, Psammechinus miliaris. Exp. Cell Res., 1, 355-370.
TAHARA, Y., OKADA, M., KOBAYASHI, N., 1958. Secondary sexual characters in Japanese sea urchins. Pub. Seto Mar. Biol. Lab., 7, 165-172.
TAHARA, Y., OKADA, M., KOBAYASHI, N., 1960. Further notes on the sexual dimorphisms in Japanese sea urchins. Pub. Seto Mar. Biol. Lab., 8, 183-189.
UTINOMI, H., 1954. A check list of echinoids found in the Kii regions. Pub. Seto Mar. Biol. Lab., 3, 339-358.
UTINOMI, H., 1960. Echinoides from Hokkaido and the neighbouring subarctic waters. Pub. Seto Mar. Biol. Lab., 8, 337-350.
UTINOMI, H., 1965. Uni-kô (Echinoidea). New Illustrated Encyclopedia of the Fauna of Japan 3. Tokyo: Hokuryu-kan Pub. Co.
YAMAMOTO, M., 1965. An experimental study on the cleavage of sea urchin eggs by means of the formation of the egg protuberance. Sci. Rep. Tôhoku Univ. Ser. IV (Biol.), 31, 55-62.

2C. Maturation Periods of European Sea Urchin Species*

Species	Location	Time of Year
Arbacia lixula	Naples	April-November
	Villefranche	September-November
Brissopsis lyrifera	Kristineberg	July-August
Echinocardium cordatum	Kristineberg	June-September
	Roscoff	April-August
Echinocardium flavescens	Bergen	February-September
	Kristineberg	June-August
Echinocardium pennatifidum	Kristineberg	June-August
Echinocyamus pusillus	Bergen	June-August
	Kristineberg	June-July
Echinus acutus	Bergen	February-May
Echinus esculentus	Bergen	February-May
	Kristineberg	April-May
	Plymouth	March-May
	Roscoff	March-June
	Tromsö	May-June
Genocidaris maculata	Naples	March-June
Paracentrotus lividus	Banyuls	March-July
	Naples	All Year
	Roscoff	June-August
Psammechinus microtuberculatus	Naples	All Year
Psammechinus miliaris	Bergen	June-August
Z-form	Kristineberg	June-August
S-form	Kristineberg	July-November
	Plymouth	May-September
	Roscoff	April-August
Spatangus purpureus	Bergen	May-July
	Kristineberg	June-August
Sphaerechinus granularis	Naples	April-December
	Roscoff	January-October
Strongylocentrotus droebachiensis	Bergen	January-April
	Kristineberg	March
	Tromsö	February-May
	Trondheim	February-April
Strongylocentrotus pallidus	Bergen	February-April
	Tromsö	February-May
	Trondheim	February-April

*G. Czihak, B. Hagström, S. Hörstadius, and S. Lönning

3. Gametogenesis

J. Piatigorsky

The studies on gametogenesis in sea urchins have included a variety of approaches, such as gross morphologic and biochemical investigations of the gonads at different times of the annual reproductive cycle and detailed analyses of structural, physiologic, and molecular changes that occur during growth of the germ cells. Investigations on oogenesis are especially germane to the understanding of the events that take place at fertilization, which has been studied so extensively in sea urchins. The interest of gametogenesis thus extends beyond simply the series of changes that occur during differentiation of the germ cells. Problems related to gametogenesis in sea urchins include the environmental control of cellular processes, the establishment of dormancy in preparation for embryonic life, and the process of early development itself.

The purpose of this chapter is to present a panorama of the numerous and varied studies conducted on gametogenesis in sea urchins. I have chosen to include differing opinions and reports where such exist, rather than leaving the impression that most of the important aspects of gametogenesis are understood. There is much that needs to be learned about every aspect of gametogenesis, and it is hoped that this review will point out some of the areas that need further clarification. Most of the remarks are confined to sea urchins. Reference can be made to a number of relatively recent reviews and monographs for general discussions of gametogenesis in other animals (TYLER, 1955; BRACHET, 1960; BISHOP and WALTON, 1960; RAVEN, 1961; AUSTIN, 1961, 1965; ROOSEN-RUNGE, 1962; ANDRE, 1963; MONROY, 1965; WILLIAMS, 1965; GRANT, 1965; DAVIDSON, 1968; BEATTY, 1970; NELSON, 1971).

3.1 Gonad Biology

3.1.1 Structure

Regular sea urchins have 5 gonads, and most irregular sea urchins have 3 or 4 (HYMAN, 1955). Each gonad is attached to the interambulacral plates by perivisceral epithelium and is connected to the exterior by a gonoduct that passes through the genital plate. Mature gametes are shed through the gonoducts and gonopores. Gonads have many acini opening into the gonoduct.

Fig. 3.1. Electron micrograph of boundary of acinus of testes; PE, perivisceral epithelium; BL_1 and BL_2, basement laminae; SMF, smooth muscle fiber; CC, fibers of connective tissue; Sg, portions of spermatogonia. *Arbacia punctulata*. x 50,600. Inset: Light micrograph of portion of acinus of testes; PE, perivisceral epithelium; SMF, smooth muscle fiber; Sd, spermatids; IC, interstitial cells; Sz, spermatozoa. *Strongylocentrotus purpuratus*. Epon-embedded, toluidine blue-stained. x 400. (From LONGO and ANDERSON, 1969) ▶

The histologic appearance of the gonad wall has been studied in detail in *Arbacia punctulata* (WILSON, 1940). It consists of 5 distinct layers: a flat, ciliated epithelium; a collagenous and reticular connective tissue layer; parallel bands of smooth muscle; another layer of

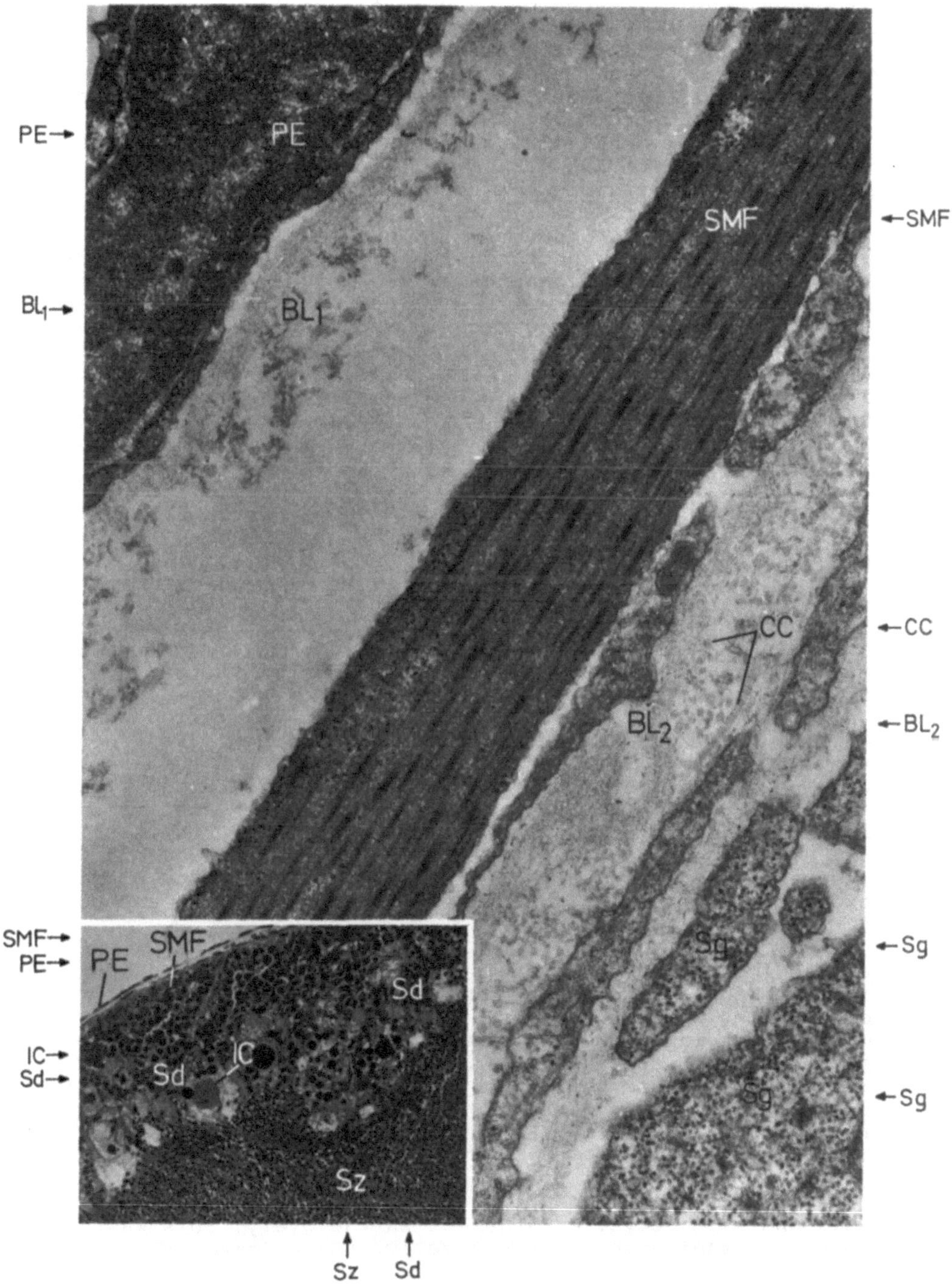

connective tissue containing scattered nervous elements; and an inner germinal epithelium. The germinal epithelium varies in prominence during the annual reproductive cycle. Nongerminal cells - or nutritive phagocytes - are present among the germinal cells (CAULLERY, 1925; LIEBMAN, 1950; HOLLAND and GIESE, 1965).

The spermatocytes develop asynchronously in the testis, and the mature spermatozoa become concentrated in the center of the lumen (Fig. 3.1). The ultrastructural appearance of the testis wall - similar to that of the ovary - is also shown. The outer coelomic epithelium is a single layer of squamous cells with desmosome-like connections. The cells have microvilli situated around a flagellum, an elongated nucleus, mitochondria, and a prominent Golgi apparatus. On the testicular side, the perivisceral epithelium adheres closely to a basement lamina. Smooth muscle fibers are visible between the layers of collagen.

A section of a ripe ovary of *A. punctulata* is shown in Fig. 3.2. The oocytes also do not develop synchronously in the ovary, and all stages of maturation are present simultaneously. The larger, mature oocytes are situated closer to the lumen of the acinus. The ripe eggs - easily distinguished by their small nucleus in contrast to the large germinal vesicle of the primary oocyte - are found in the center of the lumen. This gradient of maturity of germ cells within the acinus is typical of most sea urchins, although there may be some exceptions. It has been reported that in *Mespilia globulus* oogonia migrate from the outer wall of the ovary through the muscle layer into the lumen where they divide, develop, and then return to the periphery and mature (TENNENT and ITO, 1941).

3.1.2 Reproductive Cycle

Each year most sea urchins have one reproductive cycle (HYMAN, 1955; HARVEY, 1956) consisting of the initiation of gonadal activities, growth and maturation of the germ cells, spawning, and recession and quiescence of the gonads (GIESE, 1959). Environmental factors - such as fluctuations in temperature, length of day, food supply, and phase of moon - act as regulatory controls during the annual reproductive cycle (BOOLOTIAN, 1966). When the external conditions are kept constant in the laboratory - temperature, illumination, and food supply - ripe sea urchins remain gravid throughout the year (TYLER and TYLER, 1966a; PIATIGORSKY and TYLER, 1967). No single environmental factor can be consistently correlated with the reproductive cycles of all sea urchins, however. Probably multiple factors are involved, or different sea urchins respond to different stimuli. The mechanisms by which exogenous factors influence the reproductive cycle are not understood.

During the breeding season, the gonads become larger. This is conventionally measured as a relative increase in gonadal volume or weight and is called the gonad index (BOOLOTIAN, 1966). The gonad index of males and females behaves similarly during the annual reproductive season for most species (Fig. 3.3). Histologic studies have shown that the higher gonad indices are accompanied by a thickening of the germinal layer of the gonad (YOSHIDA, 1952; FUJI, 1960a, 1960b, 1960c; HOLLAND and GIESE, 1965; PEARSE and GIESE, 1966; HOLLAND, 1967; CHATLYNNE, 1969).

Rising gonad index is associated with spermatogonial multiplication in the testis and with oocyte enlargement in the ovary. This was shown directly in an autoradiographic investigation using *Strongylocentrotus*

(a)

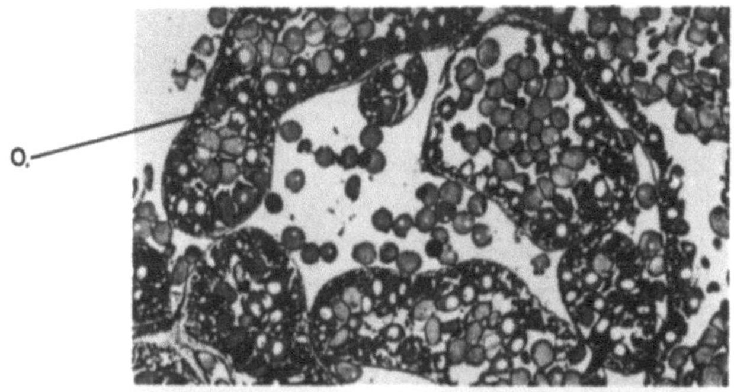

(b)

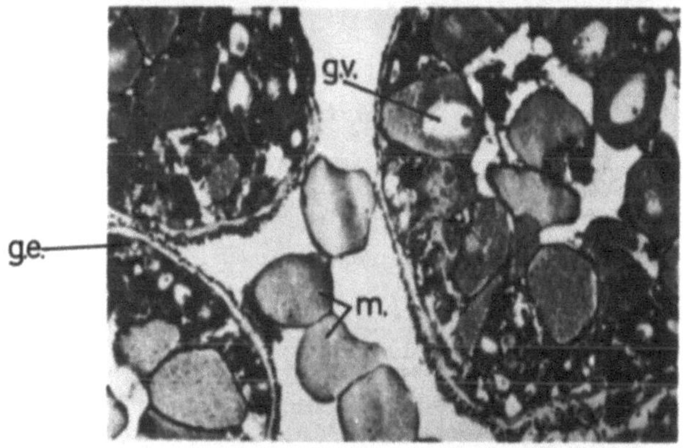

(c)

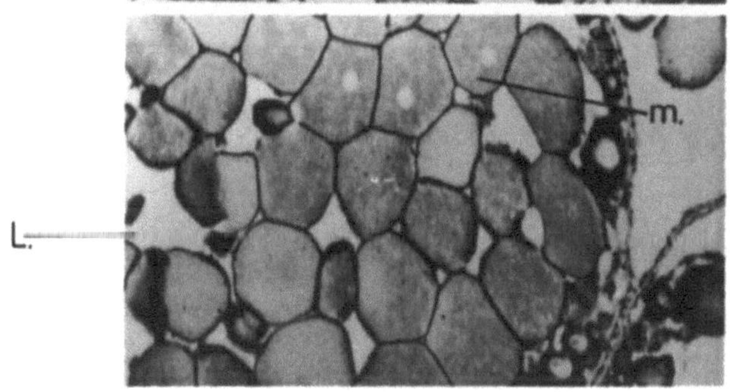

Fig. 3.2a-c. Histologic sections of *Arbacia punctulata* ovaries; photographed at increasing magnifications. (a) Survey field, including several acini. Eggs outside ovarian wall (o) escaped during dissection and fixation. Darkly stained layers of oocytes line inner wall. Larger, less basophilic ootids lie in central lumen. (x 43). (b) Wall region at higher magnification. Layer of small oocytes and oogonia - germinal epithelium (g.e.) - lines inner wall. Wall itself appears 3-layered. Connective tissue and smooth muscle separate 2 epithelia. Occasional large oocyte in this region is distinguished by its very large germinal vesicle nucleus (g.v.) with prominent nucleolus. Mature eggs (m.) lie in extraacinar space. (x 200). (c) Densely packed mature eggs (m.) distinguished by light basophilia - compared with that of small oocytes - and small pronucleus filling lumen (L.). (x 257). (From GROSS, MALKIN, and HUBBARD, 1965)

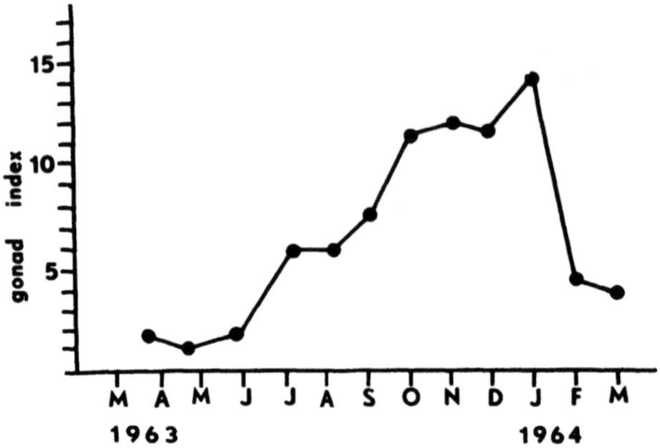

Fig. 3.3. Average gonad indices of *Strongylocentrotus purpuratus* sampled monthly during 1963-1964 annual reproductive cycle. Males and females were considered together for point calculation. Gonad index represents gonad weight and was calculated by the following relation: gonad weight/urchin weight x 100. (From HOLLAND and GIESE, 1965)

purpuratus (HOLLAND and GIESE, 1965). An intracoelomic injection of H^3-thymidine during the reproductive months resulted in the effective labeling of spermatozoa but very little labeling of oocyte nuclei. When animals were injected with H^3-thymidine well in advance of the reproductive season, there were some labeled nuclei in primitive spermatogonia and in small oogonia and preleptotene primary oocytes. Thus, there is a slow continuation of the mitotic phase of gametogenesis during the quiescent period of the reproductive cycle.

3.1.3 Seasonal and Biochemical Gonad Changes

There is no evidence that a single substance or group of substances has a causative or regulatory role in gonadal changes during the reproductive cycle. In general, all the organic constituents increase to different extents during the reproductive months (BOOLOTIAN, 1966; GIESE, 1966). Table 3.1 gives one set of data obtained for *S. purpuratus*, but similar results have not been obtained consistently for all species. For example, STOTT (1931) found that glycogen decreases during maturation in *Echinus esculentus*, GREENFIELD et al. (1958) indicated that lipid may be the greatest variant in *Strongylocentrotus franciscanus*, and PEARSE and GIESE (1966) showed that there was no seasonal change in the percentages of lipid, carbohydrate, protein, and total nitrogen in *Sterechinus neumayeri*. Another variation occurs in the lipid-to-protein ratio which rises in the testis and falls in the ovary of *S. purpuratus* but decreases in both testis and ovary of *S. franciscanus* as the gonad index increases (GIESE, 1966).

In contrast to the gonads, the constituents of the perivisceral fluid do not vary in a consistent fashion during the reproductive cycle (LASKER and GIESE, 1954; BENNETT and GIESE, 1955). It is probable, then, that the biochemical changes of the gonads reflect the accumulation of mature gametes rather than changes in the intercellular material.

There have been a number of studies on the variation of different chemical substances of the gonads during the reproductive season

Table 3.1. Increase in organic constituents of gonads of purple sea urchin (*Strongylocentrotus purpuratus*) during growth from shrunken to maximal size. (From GIESE et al., 1959)

	Male			Female		
	Spent	Gravid	Relative Increase	Spent	Gravid	Relative Increase
Gonad index	1.42	21.6	15.2X	1.18	21.8	18.5X
Total weight (arbitrary; gm)	1	15.2	15.2X	1	18.5	18.5X
Total nitrogen (mg)	84	1170	14.0X	65.5	1369	21.0X
Nonprotein nitrogen (mg)	29.9	265	8.9X	26.6	328	12.3X
Protein (mg)	336.5	5639	16.7X	243	6512	26.7X
Lipid (mg)	174	1854	10.4X	146	3663	25.1X
Glycogen (mg)	35	989	28.2X	21.8	917.6	42.1X
Reducing sugar (mg)	0.036	5.16	143.0X	0.07	6.84	98.0X
RNA (mg)	23	152	6.6X	31	832.5	26.8X
DNA (mg)	44	1368	31.1X	0.014	0.120	8.6X

(RUSSO, 1926, *Paracentrotus lividus*; STOTT, 1931, *E. esculentus*; GREENFIELD et al., 1958, *S. franciscanus*; GIESE et al., 1959, *S. purpuratus*; GIESE, 1961, *Allocentrotus fragilis*; GIESE et al., 1964, *Stomopneustes variolaris*; PEARSE and GIESE, 1966, *S. neumayeri*).

3.2 Germ Cells

3.2.1 Origin

Primordial germ cells have been reported in outgrowths of the circular genital cord (HAMANN, 1887; TENNENT and ITO, 1941), in the left posterior coelomic lining at the time of metamorphosis (MACBRIDE, 1903), and in association with the axial gland in sea urchins only several mm in diameter (RUSSO, 1894; DELAVAULT, 1966). These structures are all of mesodermal origin. Primordial germ cells first appear as migratory groups of mesenchyme in the vicinity of the hydroporic canal in another echinoderm, the hermaphroditic sea star *Asterina gibbosa* (DELAVAULT, 1966). Apparently the germ cells of sea urchins arise from mesenchymal derivatives relatively late in development in comparison with other organisms (TYLER, 1955).

It has been suggested that the outer epithelium of the ovary of *Echinometra lucunter* contributes to the supply of germ cells in the adult (TENNENT, GARDINER, and SMITH, 1931). The outer epithelial cells were seen frequently in mitosis, the smallest germinal cells were of similar size and appearance as the outer epithelial cells, and nuclei could be found in positions suggesting their migration through the muscle fibers to the germinal epithelium. These observations were substantiated in *M. globulus* (TENNENT and ITO, 1941). On the other hand, WILSON (1940) reported that the early germ cells resembled connective tissue cells in *A. punctulata*. More recent studies do not support the view that germ cells come from the outer epithelium of the gonad (HOLLAND and GIESE, 1965; LONGO and ANDERSON, 1969).

3.2.2 Sex Determination

Sex in sea urchins is genetically determined; the sex chromosomes of males are either XY or XO (MAKINO, 1951; HARVEY, 1956; DELAVAULT, 1966). Sex hormones (estrogens and progesterone) have been isolated from the ovaries of sea urchins (DONAHUE, 1939, 1940; BOTTICELLI, HISAW, and WOTIZ, 1961). Possibly these are involved in the differentiation of germ cells. External factors also may influence the sexual differentiation of the germ cells. For example, temperature - especially "cold shocks" followed by warmer waters - may produce hermaphroditic sea urchins (BOOLOTIAN, 1960; MOORE et al., 1963a, 1963b).

It has been suggested that hermaphroditism indicates a seasonal dimorphism in echinoids; an interconversion between male and female (NEEFS, 1937, 1952; REVERBERI, 1940, 1947). BOOLOTIAN and MOORE (1956) have reviewed this possibility and rejected it, as did CAULLERY (1925) after the same idea was originally expressed by GIARD (1900) who examined hermaphroditic examples of *Echinocardium cordatum*.

3.3 Immature Gonads

Immature sea urchin gonads have been examined by FUJI (1960a, 1960b) in *Strongylocentrotus intermedius* and *S. nudus*. The former, with test diameters under 15 mm, and the latter, with test diameters under 30 mm, were found to possess neuter gonads. The gonads were small, elongated, and semitransparent with a very thin wall. However, HOLLAND and GIESE (1965) could distinguish males from females in immature specimens of *S. purpuratus* that weighed less than 3 g and had a test diameter of 16 mm. The immature ovaries possessed oogonia and preleptotene as well as spirene-stage and dictyotene-stage primary oocytes. The largest histologic sections of a dictyotene primary oocyte were 15μ x 12μ, the largest germinal vesicle was 10μ, and the largest nucleolus was 3μ. The large primary oocytes in a mature, gravid ovary were 65μ x 45μ, their largest germinal vesicle was 28μ x 22μ, and their largest nucleolus was 8μ. The immature testes contained only primitive spermatogonia. After exposing the immature gonads to H^3-thymidine for one hour, the nuclei of many oogonia and preleptotene primary oocytes and of some primitive spermatogonia were labeled.

Early observations by MOORE (1935) indicated that *E. cordatum* becomes sexually mature the second year after metamorphosis. HINEGARDNER (1969) showed that it takes 6 months or less to raise *Lytechinus pictus* to sexual maturity under laboratory conditions. At this young stage, however, only a few hundred ripe eggs can be obtained.

3.4 Spermatogenesis

Once a primitive spermatogonium stops dividing mitotically, it enters meiosis and becomes a primary spermatocyte. The first meiotic division produces 2 secondary spermatocytes, and the second division results in 4 haploid spermatids. These differentiate to form mature spermatozoa. Spermatogonial formation of spermatids is called spermatocytogenesis, and subsequent development to mature spermatozoa is termed spermiogenesis. A detailed discussion of meiosis and spermatogenesis in general can be found in reviews by others (WILSON, 1925; BISHOP and WALTON, 1960; BRACHET, 1960; ROOSEN-RUNGE, 1962; ANDRE, 1963; NELSON, 1971).

3.4.1 Spermatogonia

Scattered primitive spermatogonia are present in groups - or cell nests - at the base of the germinal layer during the nonreproductive months in *S. purpuratus* (HOLLAND and GIESE, 1965). They are slightly basophilic, and their nuclei are about 5μ in diameter and contain 1 or 2 nucleoli. As the breeding season approaches, basophilic germ cells with a nucleus 3-4μ in diameter appear. HOLLAND and GIESE (1965) speculate that these may be analogous to intermediate and type B spermatogonia of mammals (BISHOP and WALTON, 1960; ROOSEN-RUNGE, 1962). Intermediate spermatogonia divide to give rise to type B spermatogonia which in turn divide to form 2 primary spermatocytes. It is not known why a spermatogonium stops dividing mitotically and enters meiosis. The number of spermatogonial divisions that take place in sea urchins is undetermined. In mammals in which this has been studied extensively, it ranges from 3 to 8 (BISHOP and WALTON, 1960).

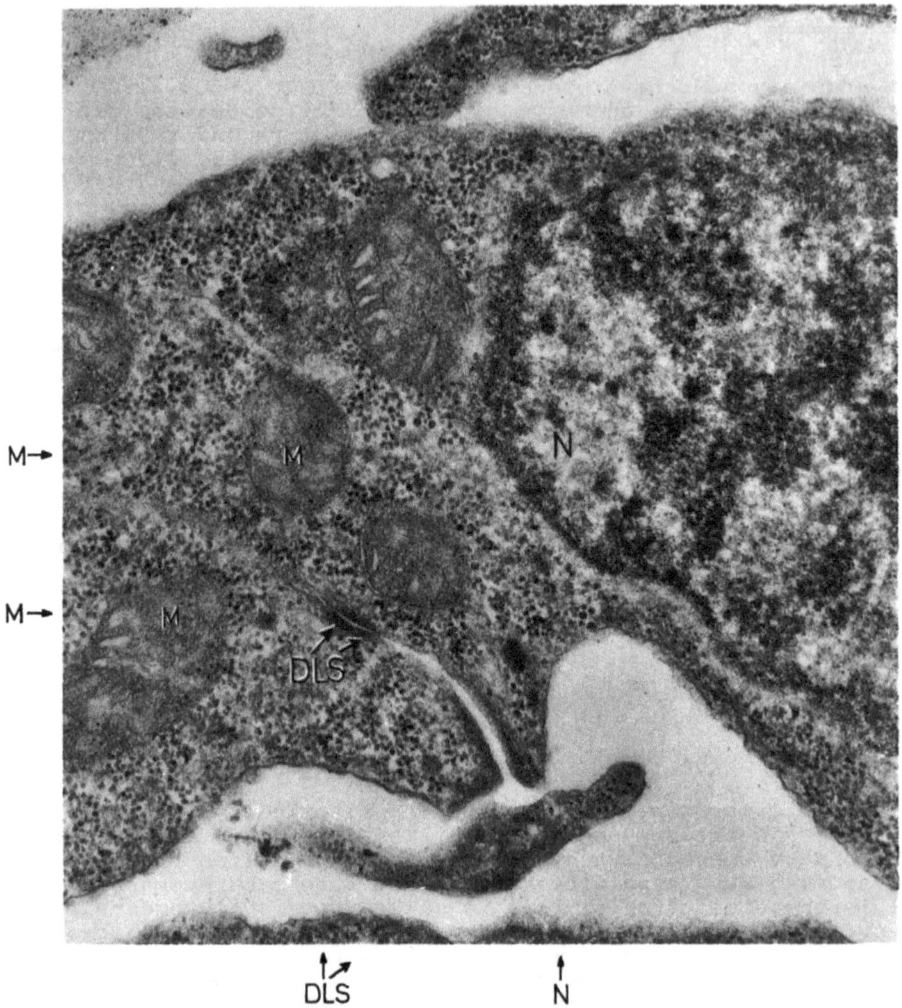

Fig. 3.4. Electron micrograph of 2 *Arbacia punctulata* spermatogonia associated by desmosome-like structure (DLS); N, nucleus; M, mitochondria. x 51,000. (From LONGO and ANDERSON, 1969)

The large nucleus of a spermatogonium has a perforated envelope and often contains one, and perhaps more, nucleoli (LONGO and ANDERSON, 1969). Chromatin can be seen as electron opaque masses at the periphery of the nucleus, in aggregates throughout the interior of the nucleus, and around the nucleolus. The cytoplasm contains ribosomes, rough endoplasmic reticulum, mitochondria, vesicular structures, lipid droplets, a Golgi apparatus, and patches of granular material, especially adjacent to the nucleus. Two perpendicularly oriented centrioles are located close to the plasma membrane. The future proximal centriole of the mature spermatozoan is associated with a flagellum. A striated fibrous rootlet extends from this centriole. Around the spokes, each centriole has 9 satellites that terminate at a dense spherical body. Spermatogonia are often associated with desmosome-like connections (Fig. 3.4).

3.4.2 Spermatocytes

In the sea urchin, spermatocytes are found adjacent to dividing spermatogonia. The spermatocytes are closer to the lumen of the testis than are the spermatogonia and can be distinguished from spermatogonia in *S. intermedius* and *S. nudus* by their smaller size and decreased affinity for haematoxylin (FUJI, 1960a). In *S. purpuratus*, spermatocytes are reported to have thread-like chromatin or stubby chromosomes instead of spermatogonia (HOLLAND and GIESE, 1965). The primary spermatocytes of *Hemicentrotus pulcherrimus* and *Anthocidaris crassispina* have thicker chromosomes than the secondary spermatocytes (NISHIKAWA, 1961). This would be expected since the former is tetraploid and the latter diploid.

There is an interesting difference between spermatocytes of sea urchins and those of mammals. In mammals, both nuclear and cytoplasmic volume increases (ROOSEN-RUNGE, 1962), while in sea urchins the nuclear diameter of 3-4μ is retained (HOLLAND and GIESE, 1965) and the cytoplasmic volume is decreased (FUJI, 1960a; HOLLAND, personal cummunication). A detailed morphologic study of primary and secondary spermatocytes has not been published for sea urchins as it has for mammals (e.g. NICANDER and PLOEN, 1969, for the rabbit).

3.4.3 Spermatids

Transformation of spermatids into mature spermatozoa has been studied with the electron microscope by LONGO and ANDERSON (1969) in *S. purpuratus* and *A. punctulata*. The nuclear matrix coalesces to form dense chromatin aggregates that grow by continued deposition of material. As the chromatin condenses, the nucleus elongates and becomes conical. The basic constituents of the mature sperm nucleus of Swedish sea urchins are granules of about 100 Å frequently arranged linearly (AFZELIUS, 1955a). As the chromatin becomes compact, a fibrous complex associated with the distal centriole extends into the nucleus, forming a conical depression called the centriolar fossa. The fibrous complex then degenerates.

Microtubules have been implicated as a causative factor for the elongation of the nucleus during spermatid maturation in a number of different species (FAWCETT, ANDERSON, and PHILLIPS, 1971), but after investigating spermatogenesis in mammals, birds, insects, and annelids, these investigations conclude "that the shape of the sperm head is probably not a consequence of external modeling by pressures applied to the condensing spermatid nucleus by microtubules in the perikaryal cytoplasm, or by filaments in the ectoplasm of the supporting cells. Instead, it is concluded that the shape of the nucleus must be largely determined from within, by a specific genetically controlled pattern of aggregation of the molecular subunits of DNA and protein during condensation of the chromatin."

Saccules from the Golgi apparatus presumably fuse to form a granular acrosomal vesicle that migrates to the apical region of the spermatid. The vesicle is 0.25μ in diameter in *S. purpuratus* and 0.20μ in *A. punctulata*. The nuclear membrane underlying the vesicle indents, fills with a reticular material, and becomes the subacrosomal region (DAN, 1956; AFZELIUS and MURRAY, 1957).

The acrosome contains acid and alkaline phosphatases (ANDERSON, 1968b), sulfatases (HATHAWAY and TYLER, 1958; HATHAWAY and METZ, 1961), enzymes capable of splitting the jelly coat of the egg (HATHAWAY and WARREN, 1961; ISAKA, KANATANI, and SUZUKI, 1966) and actin (TILNEY et al., 1973). It is not known when these are synthesized during spermatogenesis.

Mitochondria become concentrated at the base of the nucleus and fuse together to form one or several large mitochondria that surround the centrioles and anterior region of the flagellum. Lipid bodies are present in the midpiece of *A. punctulata* but not in *S. purpuratus*.

Developing spermatids of sea urchins are joined and presumably have cytoplasmic exchange. This also occurs in other organisms (BISHOP and WALTON, 1960). Large portions of the cytoplasm containing ribosomes, glycogen, and other materials are sloughed off. These remnants are probably digested by the wandering nutritive phagocytes in the testis.

After the fibrous complex disappears, the proximal centriole moves into the centriolar fossa and the distal centriole becomes fixed between the nucleus and the mitochondria in the midpiece. The long axis of the proximal centriole becomes aligned with the axis of the spermatozoan, continuous with the flagellum which has the typical fibril 9 + 2 pattern (BRADFIELD, 1953; AFZELIUS, 1959; LONGO and ANDERSON, 1969).

3.4.4 DNA Synthesis: Duration and Control of Spermatogenesis

The duration of spermatogenesis has been investigated by HOLLAND and GIESE (1965) in *S. purpuratus* by an autoradiographic study of the testis after injecting H^3-thymidine. Since both spermatogonia and primary spermatocytes synthesize DNA, this method lends itself only to determining the interval between the primary spermatocyte stage and the mature spermatozoan. The minimum time for a preleptotene primary spermatocyte to become a spermatid is 3 to 6 days. According to the authors, the primary spermatocyte probably has an extended first meiotic prophase, while the secondary spermatocyte quickly divides into 2 early spermatids. Following meiosis, it takes about 4 more days for spermiogenesis to become complete. In other tests with *A. punctulata*, males labeled with H^3-thymidine for 28 days had large numbers of radioactive spermatozoa as expected (ZEITZ, FERGUSON, and GARFINKEL, 1969). This is similar to another invertebrate, *Drosophila melanogaster*, in which spermatocyte maturation takes 4 days and spermiogenesis 5 days (CHANDLEY and BATEMAN, 1962). In the grasshopper the total time for spermatogenesis is approximately 28 days (MUCKENTHALER, 1964), and in mammals it takes 30 to 40 days (BISHOP and WALTON, 1960).

There are apparently several mechanisms that control the rate of sperm production (HOLLAND and GIESE, 1965). One occurs before DNA synthesis has terminated in primary spermatocytes, as inferred from the observation that the time between DNA synthesis in spermatocytes and their subsequent development to mature spermatozoans is constant when measured in 5 months of the reproductive cycle. Another possible control of the rate of accumulation of mature spermatozoa involves the nutritive phagocytes which ingest spermatozoa during the nonreproductive months, similar to the phagocytic role of Sertoli cells in the testes of mammals (ROOSEN-RUNGE, 1962).

An interesting variation is seen in the cidaroid sea urchin *Stylocidaris affinis* in which a primitive form of echinoderm spermatogenesis may occur (HOLLAND, 1967). Primary spermatocytes accumulate throughout the year but do not develop into spermatozoa until just before shedding. This differs from other sea urchins in which the primary spermatocytes do not accumulate in the testis before maturation into spermatozoa.

A small band of satellite DNA has been found in the sperm of *Lytechinus variegatus* (STAFFORD and GUILD, 1969; PATTERSON and STAFFORD, 1970) and *Arbacia lixula* (SPONAR, SORMOVA, and GLISIN, 1970), representing less than 0.5% of the total DNA. The density of satellite DNAs is 1.722 in *Lytechinus* and 1.718 in *Arbacia*, heavier than the nuclear DNAs of 1.695 and 1.698, respectively. The satellite DNA density does not correspond to mitochondrial DNA which has a density of 1.703 (PIKO, TYLER, and VINOGRAD, 1967). Its high guanosine and cytosine content (63%) and its affinity for hybridization with ribosomal RNA indicate that it is ribosomal DNA.

3.4.5 RNA Synthesis

Surprisingly, RNA synthesis during spermatogenesis has not been studied in sea urchins. In other organisms - notably insects (HENDERSON, 1963, 1964; MUCKENTHALER, 1964; DAS, SIEGEL, and ALFERT, 1965; CLAYBROOK and BLOCK, 1967; ZALOKAR, 1965) and mice (MONESI, 1965a, 1965b) - RNA synthesis decreases when the chromosomes condense and are coiled during spermiogenesis. The heteropycnotic X and Y chromosomes are inactive and remain condensed throughout meiosis.

Isolated chromatin of *Arbacia punctulata* (PAOLETTI and HUANG, 1969) and *Strongylocentrotus purpuratus* (OZAKI, 1971) has very little capacity to support RNA synthesis in vitro. This is attributed to the basic proteins present in the sperm nucleus (see below). After removal of these basic proteins by treatment with salt, the in vitro template activity of the chromatin approaches that of free DNA. Sperm chromatin has a biphasic melting profile (in contrast to that of embryos) (PAOLETTI and HUANG, 1969; OZAKI, 1971). The Tm of each phase is higher than that of free DNA. According to OZAKI (1971), this suggests that "DNA bound in 2 different ways rather than a mixture of free and bound DNA is present in the chromatin." The low template activity of chromatin could be attributed to such an absence of free DNA.

3.4.6 Protein Synthesis

There is a noticeable absence of studies on protein synthesis during spermatogenesis in sea urchins. One study indicates that sorbitol dehydrogenase activity decreases during spermatogenesis (BISHOP, 1968), but it is not known if this represents changes in synthesis or just in enzyme activity.

The mouse investigations of MONESI (1965b) have shown that protein synthesis continues when RNA synthesis ceases during meiosis and spermiogenesis. Developing spermatogonia and spermatocytes apparently synthesize stable messenger RNAs which may persist during spermatogenesis.

Many proteins have been identified in mature sperm, including the antifertilizin of the plasma membrane, the contractile and enzymatic proteins of the tail, the nuclear proteins, the hydrolytic enzymes that function during fertilization, and the respiratory enzymes of the midpiece (TYLER and TYLER, 1966a, 1966b).

3.4.7 Nuclear Proteins

In many of the organisms investigated, proteins in the spermatid nucleus become more basic and enriched with arginine during spermiogenesis (BLOCH, 1969). This has not been specifically studied in sea urchins. The nuclear proteins of sea-urchin sperm are generally more like histones than protamines (KOSSEL and STAUDT, 1926; KOSSEL, 1928; HULTIN and HERNE, 1949; HAMER, 1955; VENDRELY, 1957; PALAU, RUIZ-CARRILLO, and SUBIRANA, 1969; PAOLETTI and HUANG, 1969; OZAKI, 1971). Electrophoretic analysis reveals only 4 or 5 sperm histones (PAOLETTI and HUANG, 1969; OZAKI, 1971; EASTON and CHALKLEY, 1972). Differences between histones from sperm and from sea-urchin embryos have been reported: Sperm histone contains more arginine than embryo histone (HNILICA, 1967; PALAU, RUIZ, and SUBIRANA, 1969; PAOLETTI and HUANG, 1969; OZAKI, 1971), and the basic protein of sperm chromatin has a very fast electrophoretic mobility that resembles salmon sperm protamine (OZAKI, 1971). The electrophoretic mobility of histone fractions F1, F2b and F2-a2 in sperm differ radically from these fractions in embryos and, sperm histones have less microheterogeneity on polyacrylamide gels than do embryo histones (EASTON and CHALKLEY, 1972).

Not all proteins in the sea-urchin sperm nucleus are basic. BERNSTEIN and MAZIA (1952) found up to 25% of nonhistone protein in the sperm nucleus of *S. purpuratus*. More recent estimates (OZAKI, 1971) indicate a mass ratio of nonhistone protein to DNA of 0.13 in the isolated chromatin, in contrast to a mass ratio of histone protein to DNA of 1.02.

The function of sperm nuclear proteins is unknown. Such possibilities as chromatin condensation, streamlining the cell, inhibiting genetic transcription, erasing cell developmental history, and chromosome protection have been considered (BLOCH, 1969).

3.5 Oogenesis

The major morphologic, physiologic, and biochemical aspects of oogenesis in sea urchins will be considered here. Some of the descriptive studies of sea-urchin oogenesis are given in Table 3.2.

3.5.1 Oogonia

Oogonia are about 5-7μ in diameter (VERHEY and MOYER, 1967a; MILLONIG, BOSCO, and GIAMBERTONE, 1968) and have a large, oval nucleus filled with low-density fibrillar material. The nucleus has a nonporous envelope of 2-unit membranes separated by a space with an average width of 425 Å. There is one large nucleolus, and the chromatin is granular and basophilic and tends to clump (TENNENT and ITO, 1941). The cytoplasm makes up about half the cell volume and contains mainly ribosomes, a few mitochondria, some lipid inclusions, ergastoplasmic lamellae, and Golgi elements.

3.5.2 Primary Oocytes

The cells produced by the final oogonial division are primary oocytes that enter meiotic prophase. The primary oocyte evidently does not grow significantly until the diplotene stage when synapsis and crossing over are finished. The chromosomal changes during oogenesis have been studied in detail in *M. globulus* (TENNENT and ITO, 1941). The growth stage begins when the chromatin is in the diffusion stage. This involves thinning, elongation, and separation of the tetraploid chromo-

Table 3.2. Morphologic studies on oogenesis in sea urchins

Species	Histologic	Ultrastructural
Echinus esculentus	BRYCE (1902)	AFZELIUS (1955b, 1956a, 1956b, 1957a)
Echinocardium cordatum	CAULLERY (1925)	AFZELIUS (1956a, 1957a)
Echinocardium favescens		AFZELIUS (1957a)
Echinometra lucunter	MILLER and SMITH (1931) TENNENT, GARDINER, and SMITH (1931)	
Mespilia globulus	TENNENT and ITO (1941)	
Diadema setosum	YOSHIDA (1952)	
Strongylocentrotus nudus *Strongylocentrotus intermedius*	FUJI (1960a)	
Lytechinus variegatus	COWDEN (1963)	VERHEY and MOYER (1967a)
Arbacia punctulata	ESPER (1965)	VERHEY and MOYER (1967a) ANDERSON (1968) BAL et al. (1968)
Strongylocentrotus purpuratus	HOLLAND and GIESE (1965) CHATLYNNE (1969)	
Stylocidaris affinis	HOLLAND (1967)	
Psammechinus miliaris *Strongylocentrotus droebachiensis* *Spatangus purpureus*		AFZELIUS (1955b, 1956a, 1956b, 1957a) LÖNNING (1964)
Lytechinus pictus		VERHEY and MOYER (1967a)
Arbacia lixula	MONNE and HARDE (1951)	MILLONIG (1967) MILLONIG, BOSCO, and GIAMBERTONE (1969)
Paracentrotus lividus	MONNE and HARDE (1951)	MILLONIG, BOSCO, and GIAMBERTONE (1969)
Hemicentrotus pulcherrimus *Pseudocentrotus depressus*		TAKASHIMA (1960) TAKASHIMA and TAKASHIMA (1966) SAKAI (1956a, 1956b, 1956c)
Brissopsis lyrifera		AFZELIUS (1955b) LÖNNING (1964)
Temnopleurus toreumaticus *Heliocidaris crassispina*		TAKASHIMA (1960)

Fig. 3.5. Section through ooplasm of young *Arbacia punctulata* oocyte; GC, Golgi complex; CG_0, presumptive cortical granules; CG_1, miniature cortical granule; CG, cortical granule; ER, cisternae of endoplasmic reticulum; YB, yolk bodies. x 12,000. (From ANDERSON, 1968a)

somes which lose their basophilia. As the chromatin takes on this diffuse appearance, the nucleus acquires a lightly basophilic network from the formation of irregular connections between adjacent chromosomes. Lampbrush chromosomes exist in the sea-urchin primary oocyte and in oocytes of other organisms during the growth stage (DAVIDSON, 1968).

Pores or annuli averaging 1000-1200 Å in diameter appear at regular intervals on the primary oocyte nuclear membrane (AFZELIUS, 1955b; VERHEY and MOYER, 1967a; MILLONIG, BOSCO, and GIAMBERTONE, 1968). The larger oocytes have 40 to 80 densely packed annuli per square micron. The nuclear pores have a complex structure. A tangential section of pore rim has alternating dense granules and vesicles. Pores appear to be traversed by a cylinder that extends about 600 Å into the nucleus and about 200 Å into the cytoplasm. These may possibly be 2 cylinders joined within the pore. The inner diameter of the cylinder is about 500 Å, the outer diameter 1000 Å. Nuclear pores form throughout the primary oocyte stage. Their function may be related to transporting materials across the nuclear membrane.

Cortical granules are so named because they are present at the cell surface of mature, unfertilized eggs (HARVEY, 1956). They liberate their contents to the exterior by fusing with the plasma membrane after insemination and participate in the formation of the fertilization membrane (MONROY, 1965). They are lamellar bodies slightly less than 1μ in diameter (AFZELIUS, 1956b; RUNNSTRÖM, 1966). Their fine structural appearance differs from species to species. Cytochemically cortical granules have been shown to contain sulfated mucopolysaccharides and protein (MONNE and SLAUTTERBACK, 1950; MONNE and HARDE, 1951; IMMERS, 1960; ANDERSON, 1968a). It has been estimated that there are about 30,000 cortical granules in a mature egg of *Psammechinus miliaris* (RUNNSTRÖM, MONNE, and WICKLUNG, 1946).

Initial cytochemical tests showed that the cortical granules originate in the endoplasm and move to the periphery on maturation of the oocyte (MOTOMURA, 1936, 1941; RUNNSTRÖM and MONNE, 1945; MONNE and HARDE, 1951). This has been confirmed in different species of sea urchins by electron microscopy (McCULLOCH, 1952; AFZELIUS, 1956a, KUROZUMI, 1957; TAKASHIMA, 1960; LÖNNING, 1963; RUNNSTRÖM, 1966; VERHEY and MOYER, 1967a; ANDERSON, 1968b). The results of autoradiographic experiments with S^{35}-sulfate are consistent with the migration of cortical granules at maturation (IMMERS, 1961). Labeled sulfate was incorporated at random in small oocytes, but in later stages the labeled area was limited to the cortex. These experiments do not demonstrate directly that sulfate is incorporated into the cortical granules but may reflect a more rapid degradation of sulfated macromolecules in the endoplasm of older oocytes.

The formation of cortical granules involves a fusion of vesicles. In *Brissopsis lyrifera*, vacuoles 0.12μ in diameter appear to coalesce to form lacunae in which the granules develop (RUNNSTRÖM, 1966). The vesicular component of the Golgi apparatus is pinched from the tips of the saccules and transformed into presumptive cortical granules in *A. punctulata* (ANDERSON, 1968a). These coalesce and develop into mature granules bounded by a unit membrane (Fig. 3.5).

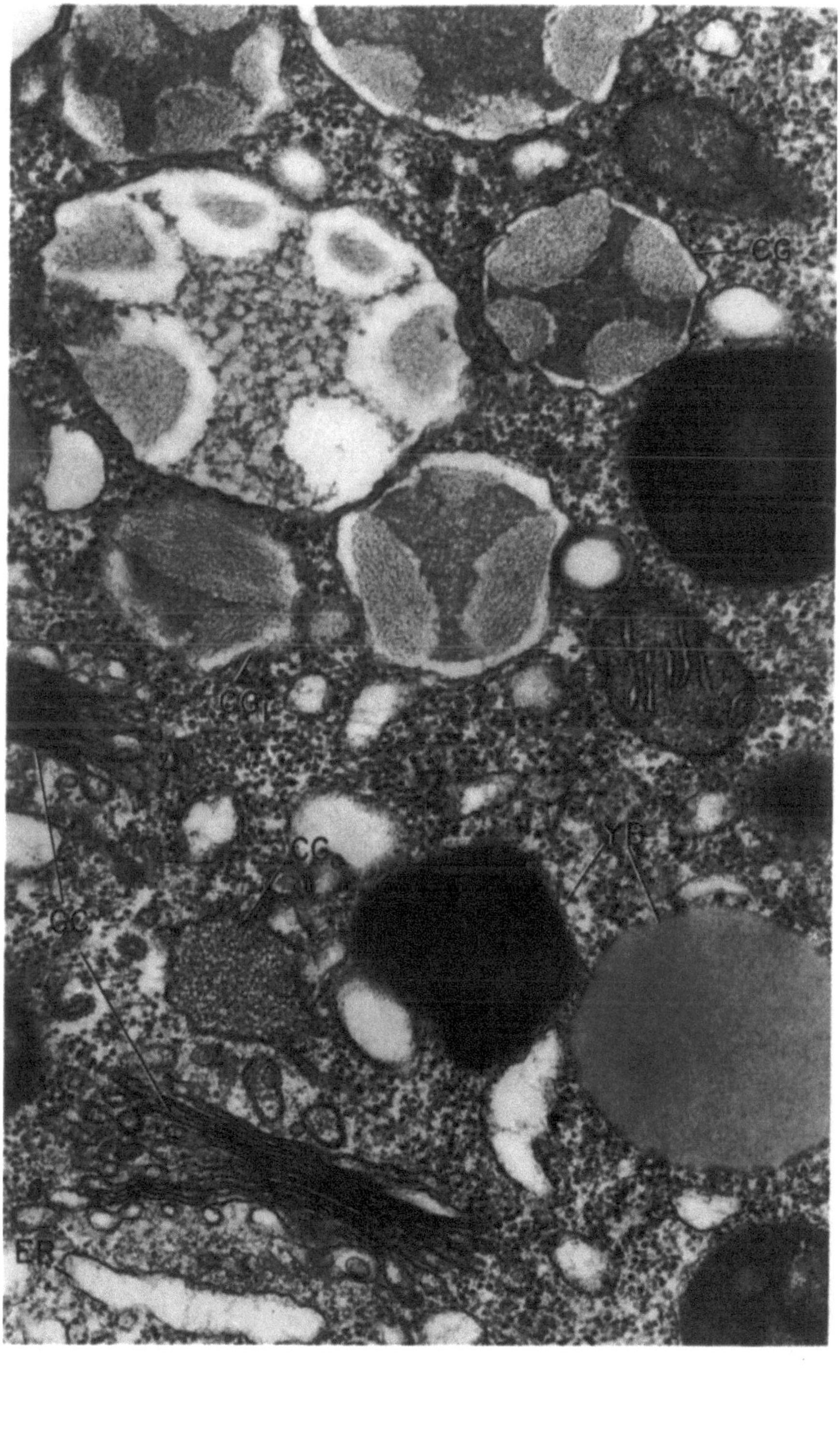
CG
CG
CG
YB
GC
ER

How cortical granules migrate to the oocyte surface is not known. MONNE and HARDE (1951) have speculated that "the movement is due to centrifugal currents produced upon condensation of the cytoplasm", while RUNNSTRÖM (1963) believes that changes in the partition of substances between phases cause the granules to move to the surface. Such changes could occur when the germinal vesicle breaks down. To oocytes of *B. lyrifera*, RUNNSTRÖM (1966) added periodate which caused the endoplasm to separate from the surface, suggesting that this is an experimental induction of phase separation that results in an association of the cortical granules with the cell surface. By whatever mechanism, the cortical granules attach to the plasma membrane so firmly that they cannot be displaced by centrifugation as other cell components are (HARVEY, 1956; MERCER and WOLPERT, 1962).

Two types of transient, cytoplasmic structures have been described in *A. lixula* and *P. lividus* (MILLONIG, BOSCO and GIAMBERTONE, 1968): One is a "dense spherical body of about 1μ which is frequently surrounded by a mesh of strands of similar material." These do not appear granular. The other "is composed of granules of about 300 Å and of a homogenous intergranular substance." These structures are observed mostly in the smaller, previtellogenic oocytes and occasionally in oogenia. They are not present near the nucleus. Their origin and function are unknown.

Annulate lamellae are parallel arrays of double membranes that have the same ultrastructural appearance as the nuclear envelope (AFZELIUS, 1955b, 1956b). They appear during the terminal stages of oogenesis. They were first observed in sea urchins as parallel fibrils about 1000 Å wide and 5-10μ long in the clear zone of stratified eggs of *A. punctulata* (McCULLOCH, 1952). They are frequently associated with granulated structures called heavy bodies (Fig. 3.6). Annulate lamellae are found in different organisms and cells such as snail ootestes, clam oocytes, and developing amphibian pancreas (SWIFT, 1956).

The formation of annulate lamellae involves either a delamination of the entire nuclear envelope (SWIFT, 1956; REBHUN, 1956; MERRIAM, 1959) or a fusion of vesicles derived from the outer nuclear membrane (KESSEL, 1963, 1964, 1965; HSU, 1967). The similarity in the structures of the nuclear envelope and the annulate lamellae suggests that the nuclear envelope delaminates to form the annulate lamellae (VERHEY and MOYER, 1967a; BAL et al., 1968)(Fig. 3.7). It is probable that the disruption of the nuclear membrane at maturation contributes to the formation of annulate lamellae.

The function of the annulate lamellae is not known. It is likely that their association with RNA is important (SWIFT, 1956; REBHUN, 1956; AFZELIUS, 1957a, 1957b). BAL et al. (1968) speculate that annulate lamellae may contain ribosomal precursors which are assembled after fertilization, but neither the types of RNA attached to annulate lamellae nor their fate during development is known. The annulate lamellae disperse at the first cleavage of the zygote (HARRIS, 1967).

Heavy bodies contain granules of 150 Å and are surrounded - but generally not completely enclosed - by annulate lamellae. AFZELIUS (1957a) first described heavy bodies in late oocytes and mature eggs of 6 species of European sea urchins. He called them that because they were in the centrifugal end of stratified eggs. They do not always accumulate at the heavy pole of centrifuged eggs. They are found in the light, centripetal half in eggs of *A. punctulata* (GROSS, PHILPOTT, and NASS, 1960) and *S. purpuratus* (HARRIS, 1967). Heavy bodies range from 0.5μ to 3μ in diameter and number about 1000 to 1500 in mature eggs of *E. esculentus* and *P. miliaris* (AFZELIUS, 1957a) and *S. purpuratus*

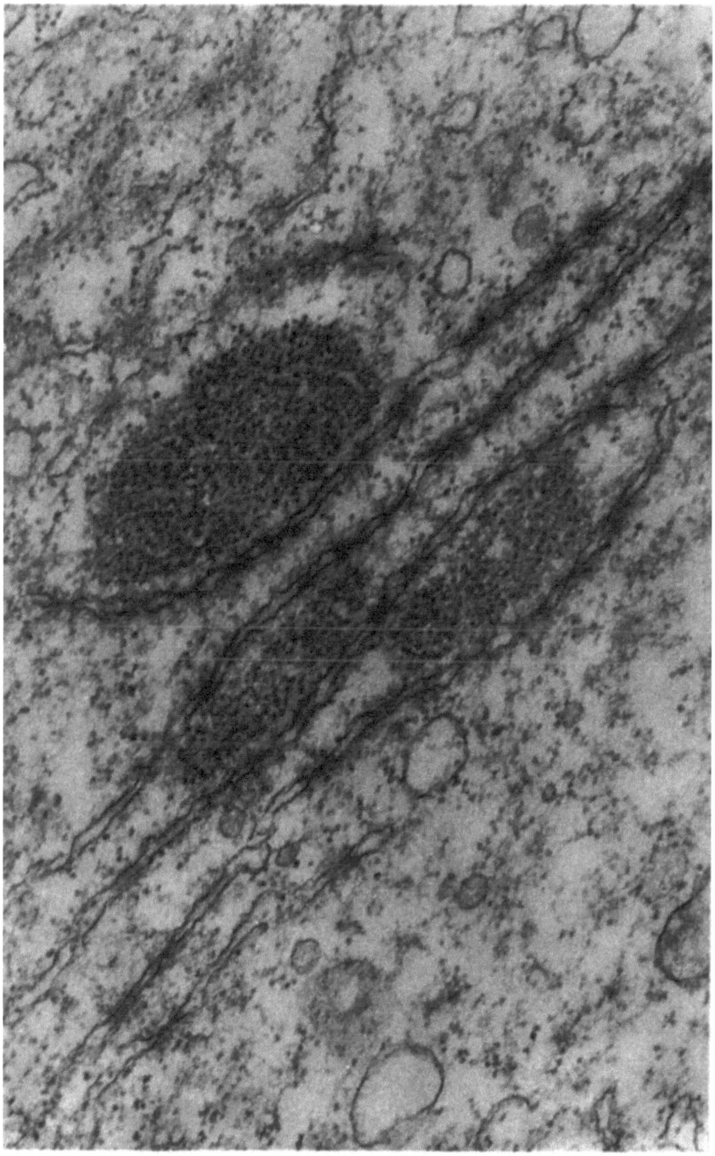

Fig. 3.6. Section through heavy-body layer of centrifuged and stratified unfertilized *Strongylocentrotus purpuratus* egg. Parallel membranes are annulate lamellae, and granular masses are heavy bodies. x 35,000. (Courtesy of P. HARRIS)

(HARRIS, 1967). Histochemical and enzymatic tests have shown that they contain RNA (AFZELIUS, 1957a; PASTEELS, CASTIAUX, and VANDERMEERSSCHE, 1958; SANCHEZ, 1966, 1968a; CONWAY and METZ, 1970) and basic protein (SANCHEZ and LANET, 1966; SANCHEZ, 1968) but not DNA (AFZELIUS, 1957a; CONWAY and METZ, 1970)(Fig. 3.8).

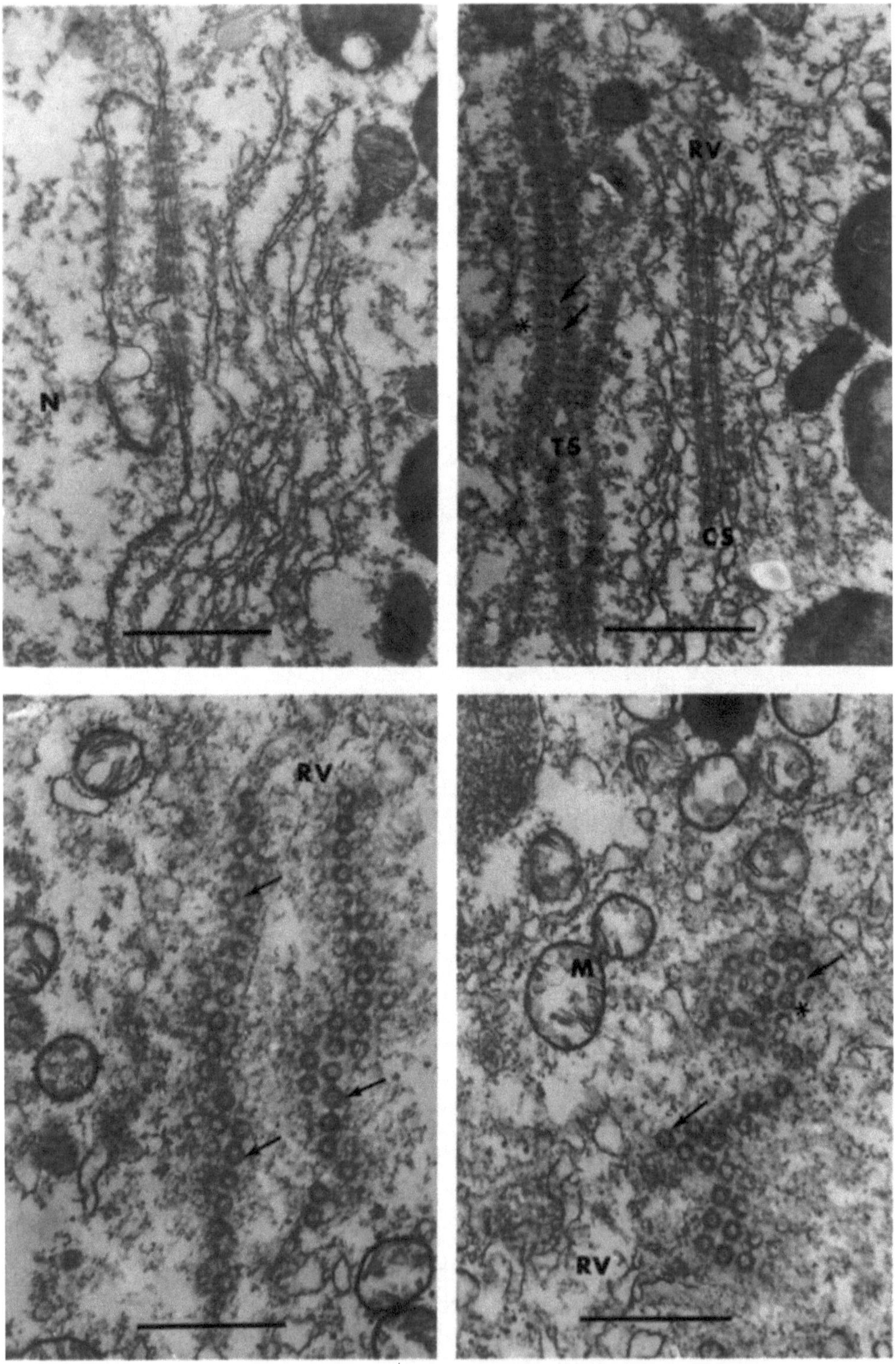
N
RV
*
TS
CS
RV
M
*
RV

◀ Fig. 3.7. Upper left: Intranuclear vesicle with annuli in register with those in nuclear envelope. Such intranuclear vesicles suggest delamination of annulate lamellae from nuclear envelope. Upper right: Annulate lamellae in cytoplasm. Asterisk indicates tangential section of annulus showing that central core consists of two vesicles. Arrows indicate fibers joining two adjacent rows of annuli. Note similarity between tangential section (TS) of annulate lamellae and that of nuclear envelope. Lower left: Tangential section of annulate lamellae. Arrows indicate annuli in which details of rim and central core may be seen. Lower right: Tangential section of annulate lamellae. Arrows indicate annuli in which details of rim and central core may be seen. In annulus at left of asterisk, four fibers connecting rim and central core are faintly discernible. Constriction in mitochondrion suggests division. x 23,000. N, nucleus; RV, rough-surfaced vesicles; CS, cross section; TS, tangential section; M, mitochondrion. *Arbacia punctulata*. (From VERHEY and MOYER, 1967a)

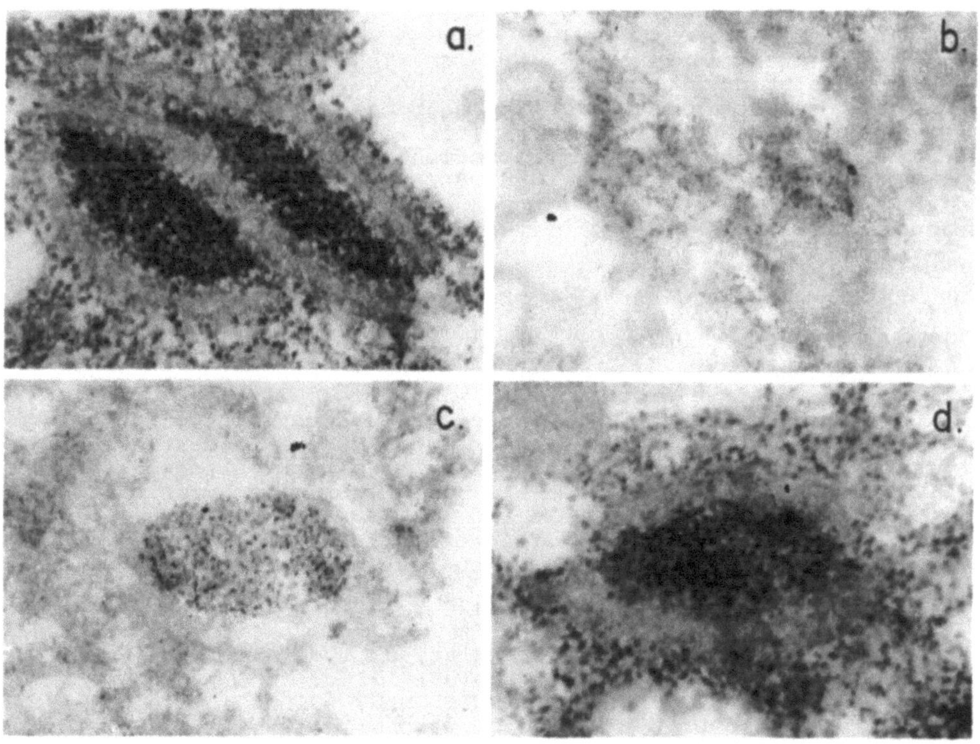

Fig. 3.8a-d. Indium trichloride staining of nucleic acids in thin sections of heavy bodies in *Lytechinus variegatus*. x 74,000. (a) Stained control. (b) Unstained control. (c) Ribonuclease treatment prior to staining. (d) Deoxyribonuclease treatment prior to staining. (From CONWAY, 1971)

The manner in which heavy bodies form has been reviewed by HARRIS (1969). They begin to appear late in oogenesis, just prior to and during the breakdown of the germinal vesicle, as do the annulate lamellae. Structures resembling nascent forms of heavy bodies can be found as outpocketings of the nuclear membrane (AFZELIUS, 1957a), suggesting that they are delaminated from the nuclear membrane. This is consistent with their close association with annulate lamellae. They are often found close to the nucleus. A recent study, however, indicates that heavy body formation in *A. punctulata* may follow nuclear maturation (CONWAY and METZ, 1974). It has been suggested that the 150-Å particles may be synthesized by the annulate lamellae (MERRIAM, 1959) or cytoplasmic particles may become trapped in the annulate lamellae after their formation (VERHEY and MOYER, 1967a).

Although heavy bodies are probably derived from the nucleus, they are not extruded nucleoli. Heavy bodies have a greater affinity for toluidine blue and dissolve more readily in seawater than do nucleoli (AFZELIUS, 1957a) and are unaffected by fixation with neutral osmium tetroxide in seawater that extracts most of the nucleolar material (HARRIS, 1969). The nucleolar-like bodies or minor nucleoli - present in the nucleus of mature eggs - have smaller grains than the heavy bodies (HARRIS, 1969).

Since heavy bodies contain RNA, their function is of special interest. One possibility is that they contain messenger RNA - presumably attached to ribosomes - that functions during early development (HARRIS, 1967, 1969). This is consistent with the fact that heavy bodies disintegrate by metaphase of the first cleavage. Heavy aggregates of ribosomes have been isolated from homogenates of unfertilized eggs relatively inactive in protein synthesis until after fertilization (CANDELAS and IVERSON, 1966; PIATIGORSKY, 1968; PIATIGORSKY and TYLER, 1970). It has been suggested that these may represent suppressed polyribosomes. Perhaps their intracellular source is the heavy bodies.

Another hypothesis suggests that heavy bodies contain messenger RNA that is important for later development (SANCHEZ, 1966, 1968). This idea was proposed on the basis of experiments with dactinomycin (actinomycin D) injected into *P. lividus* females. Interruption of RNA synthesis for 1 to 3 days results in eggs with abnormally shaped heavy bodies, some of which appear hollow. These eggs can develop to the blastula stage but do not gastrulate. RNA synthesis during development was not examined. Possibly the dactinomycin pretreatment affected the nuclear (SLATER and SPIEGELMAN, 1970; WILT, 1970) or mitochondrial (CRAIG, 1970; CHAMBERLAIN, 1970; CRAIG and PIATIGORSKY, 1971) RNA synthesis during early development. It is essential to purify and analyze the RNA and protein of the heavy bodies to clarify their function.

Yolk platelets are membrane-limited, granular, ellipsoidal or spherical bodies approximately 1μ in diameter (AFZELIUS, 1956b; GROSS, PHILPOTT, and NASS, 1960; VERHEY and MOYER, 1967a; MILLONIG, BOSCO, and GIAMBERTONE, 1968). They contain protein, mucopolysaccharide, and probably sialic acid (MONNE and HARDE, 1951; SAKAI, 1956a, 1956b, 1956c; PERLMAN, BOSTROM, and VESTERMAK, 1959; COWDEN, 1962; ESPER, 1965). Most investigators agree that yolk accumulation occurs during the later stages of oogenesis, although yolk granules evidently appear in small primary oocytes of several species of Japanese sea urchins (SAKAI, 1956a, 1956b, 1956c; TAKASHIMA, 1960).

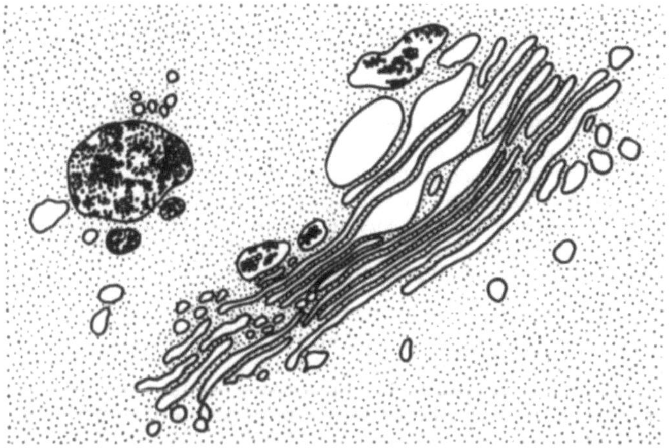

Fig. 3.9. Interpretation of formation of yolk granules taken from electron micrographs of *Strongylocentrotus droebachiensis* oocytes. Cytoplasm outside Golgi double membrane is dotted to stress that no anastomoses exist among double membrane interspaces. Reticulum is easily misconceived when single membranes of two pairs are interpreted as double membrane. Expanded vesicle - filled with granular material and associated with cluster of smaller vesicles - is thought to be stage in yolk granule formation. (From AFZELIUS, 1956b)

Initial studies on yolk formation in sea urchins were by PARAT (1928) who thought the "vacuome", which he equated with the Golgi apparatus, synthesizes yolk. Electron microscopic studies confirmed that the Golgi apparatus participates in yolk formation (AFZELIUS, 1956b, 1957a) (Fig. 3.9). The formative stages of yolk globules precipitate silver, as does the Golgi apparatus (CELI, D'ESTE, and TETI, 1967).

Intense ribosomal activity is implicated in yolk synthesis. During the vitellogenesis stage, many rough-surfaced membranes are present in the cytoplasm, and ribosomes are found attached to the unit membrane of yolk granules (VERHEY and MOYER, 1967a; BAL et al., 1968). It has even been suggested that the globules arise by an aggregation of ribonucleoprotein particles (CELI, D'ESTE, and TETI, 1967). Basophilic yolk nuclei, or yolk bodies, can be seen to consist of rough-surfaced, circular lamellae with mitochondria, yolk, and cortical granules in the center. This has raised the possibility that these structures may be transformed into yolk (MILLONIG, 1958; AFZELIUS, 1957a).

There is evidence that yolk precursor material is ingested by the oocyte in small pinocytotic vesicles (TAKASHIMA and TAKASHIMA, 1966). This occurs in many other organisms (TYLER, 1955; RAVEN, 1961). The vesicles appear to coalesce and form large vacuoles whose substance is transferred into smaller vesicles and rough-surfaced cysternae of the endoplasmic reticulum. The Golgi complex multiplies in the region of these vesicles and cysternae and presumably is involved in their transformation into yolk. This possible mode of yolk formation combines the accumulation of yolk from the environment and its synthesis by the oocyte. If this is the mechanism of yolk formation, isolated oocytes should not be able to initiate or to sustain yolk synthesis. This has not been explored.

Each yolk particle may contain as much DNA as a mitochondrion (PIKO, TYLER, and VINOGRAD, 1967). The yolk DNA has the same properties as mitochondrial DNA and may be due to mitochondrial contamination of the yolk fraction during the isolation procedures (BERGER, 1968; PIKO, personal communication). A recent report indicates that yolk has DNA that replicates after fertilization (ANDERSON, 1969). DNA could become associated with yolk either by the transformation of mitochondria into yolk or by the incorporation of DNA from surrounding cells.

The gelatinous coat, jelly layer, or fertilizin consists mainly of a homogeneous, sulfated mucopolysaccharide with a molecular weight of at least 300,000 and an axial ratio of 20:1 or 30:1 (TYLER, 1949b; TYLER, 1956; TYLER and TYLER, 1966a). Electron diffraction study of the isolated gelatinous coat of *A. punctulata* suggests "the presence of crystallites oriented either parallel or at random" (INOUE, COUSINEAU, and KRUPA, 1971). It is present on the surface of the egg and is progressively accumulated during the later stages of oogenesis. This is evident from measurements made on shed oocytes of *A. punctulata* (HARVEY, 1956). Oocytes with a diameter of 60μ had a jelly layer of only 3.2μ; larger oocytes and mature eggs (74μ in diameter) had a layer about 30μ thick. Fertilizin of mature eggs has been labeled with S^{35}-sulfate during the last 6 days of oogenesis in *Arbacia* (TYLER and HATHAWAY, 1958). Labeled sulfate has also been incorporated into the jelly coat of *P. lividus* as soon as 4 hours after injecting the female (IMMERS, 1961).

The mechanism by which the gelatinous coat is made is not known. It is generally presumed that the oocyte makes its own jelly layer (TYLER, 1948), but nongerminal cells may participate (VASSEUR, 1952). Electron microscopic investigation has provided some putative evidence that the cortical granules may be involved in the secretion of the gelatinous coat (TAKASHIMA, 1960). Although this is consistent with other reports indicating that cortical granules contain fertilizin (MOTOMURA, 1960; OSANAI, 1964), it appears unlikely because the jelly appears before the cortical granules have migrated to the periphery.

The properties of the jelly layer change as oogenesis progresses. The jelly coat of oocytes is only rarely precipitated by anti-jelly serum prepared in rabbits using ripe eggs as antigens, while that of mature eggs has a strong precipitin reaction (PERLMAN, 1956). PERLMAN speculated that the antigenic groups may be unmasked during maturation. The jelly layer of resting oocytes is not depolymerized by spermatozoa or precipitated by basic dyes (toluidine blue and brilliant cresyl blue) and protamin sulfate as it is in mature eggs (RUNNSTRÖM, 1964). The jelly of oocytes has a greater ability to reduce the dyes than the jelly of the mature eggs and is less easily washed off or disrupted by passage through gauze. RUNNSTRÖM suggested that basic proteins mask specific sites on the jelly coat and are removed by proteolysis at maturation.

3.5.3 Nucleolus

Nucleoli grow larger as oocytes grow larger (SUBRAMANIAM, 1934; AFZELIUS, 1957a; ESPER, 1965; HOLLAND and GIESE, 1965; VERHEY and MOYER, 1967a; MILLONIG, BOSCO, and GIAMBERTONE, 1969; CHATLYNNE, 1969). There are 3 periods of intensive growth of the nucleolus during oogenesis

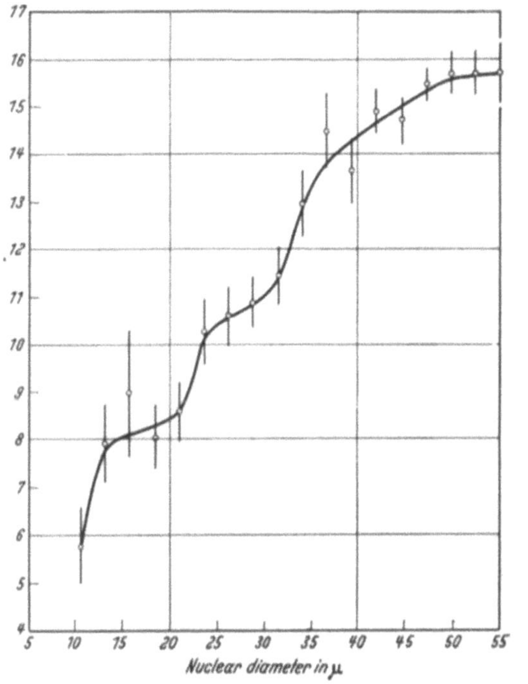

Fig. 3.10. Relation between size of nucleus and size of nucleolus is shown in *Psammechinus miliaris* oocytes. Measurements were performed in vivo with estimated error for single measurement of 0.5μ. Approximately 8 measurements were performed each size class of nucleus. Nucleolus has 3 periods of size constancy during growth of nucleus. Same curve is obtained when size of nucleolus is plotted against size of oocyte. (From AFZELIUS, 1957a)

(Fig. 3.10). The nucleolus may extrude material into the cytoplasm (SUBRAMANIAM, 1934; TENNENT and ITO, 1941). Electron microscopic studies however have not supported the view that large particles leave the oocyte nucleolus (AFZELIUS, 1957a; MILLONIG, BOSCO, and GIAMBERTONE, 1968).

AFZELIUS (1957a) was the first to describe the ultrastructure of the nucleolus of sea-urchin oocytes. It is diffusely granulated with a few cavities of 0.1μ to 1.0μ. The granules are at least 200 Å in diameter. The nucleolus consists of a core with a dense meshwork of fibrous material and a granular cortex (VERHEY and MOYER, 1967a).

Changes occur in the nucleolus during oogenesis (MILLONIG, 1967; MILLONIG, BOSCO, and GIAMBERTONE, 1968). In the oogonium the nucleolus is in contact with the inner nuclear membrane, while in the previtellogenic oocyte it is found in the center of the nucleus. More ribosome-like particles appear in the cortex, and the central core develops dense strands of fibrillar material. Granules appear in the inner core at the end of the previtellogenic stage. During vitellogenesis the core and cortex look granulated. One or more vacuoles appear in the center with a granular-fibrillar content. Histochemical studies show that the vacuoles become larger and more numerous (CHATLYNNE, 1969) and grow at the same rate as the nucleolar diameter (ESPER, 1965).

The nucleolus of a sea-urchin oocyte contains protein, RNA, mucopolysaccharide, and lipid (CASPERSSON and SCHULTZ, 1940; AFZELIUS, 1957a; ESPER, 1965; CHATLYNNE, 1969; VINCENT, 1955). The cytochemical studies of ESPER (1965) have demonstrated a ring of DNA around the nucleolus and that the nucleolar vacuole contains bound lipid at its first appearance but very little RNA or protein. The content of RNA and protein decreases in the nucleolus while that of mucopolysaccharide and lipids increases as the oocyte enlarges and the nucleolar vacuole grows. The meaning of this is not known. At maturation the nucleolar cortex loses all of its staining characteristics and disappears. It may dissolve in the nucleoplasm or may be destroyed by the growing vacuole (MILLONIG, BOSCO, and GIAMBERTONE, 1968). How much of the nucleolus enters the cytoplasm is unknown.

The major function of the nucleolus in animal cells is to synthesize ribosomal RNA (PERRY, 1965, 1967; DARNELL, 1968; MADEN, 1968). Nuclei of amphibian oocytes contain DNA that has ribosomal RNA cistrons (BROWN and DAWID, 1968; GALL, 1968; DAVIDSON, 1968). This has not yet been shown directly in sea-urchin oocytes, however. Recent investigations on the oocytes of the gephyrean worm *Urechis caupo* have shown that the fibrillar core of the nucleolus synthesizes 38 S precursor ribosomal RNA and cleaves it to the 30 S precursor ribosomal RNA which is transported and stored with its associated proteins in the granular cortex of the nucleolus (DAS et al., 1970). It is likely that this occurs in sea urchins, too.

Spherical granules or minor nucleoli 1-2μ in diameter appear in the nucleus during vitellogenesis (MONNE, 1945; AFZELIUS, 1957a; MILLONIG, 1967; MILLONIG, BOSCO, and GIAMBERTONE, 1968; HARRIS, 1969). They contain densely packed fibrils less than 80 Å wide. There can be up to 30 minor nucleoli per nucleus. Their properties are somewhat different from those of the nucleoli. Minor nucleoli are less soluble in isotonic sodium chloride than the typical nucleoli (AFZELIUS, 1957a) and, according to histochemical investigations, are rich in basic proteins but perhaps less so in RNA (SANCHEZ and LANET, 1966; SANCHEZ, 1968). After maturation, minor nucleoli increase in number and can be seen adhering to the nuclear membrane. The function of the minor nucleoli is not known.

3.5.4 Oocyte Growth

The greatest increase in volume of the growing oocyte occurs before vitellogenesis (COWDEN, 1962; ESPER, 1965; VERHEY and MOYER, 1967a; GRANT, 1965). Previtellogenic oocytes of *A. lixula* and of *P. lividus* grow from 5μ to 50μ in diameter (MILLONIG, BOSCO, and GIAMBERTONE, 1968). Vitellogenesis and maturation result in an ootid 70μ in diameter.

Changes in the diameter of oocytes have been measured at different times of the annual reproductive cycle (Table 3.3). The terminology regarding the ovarian stage is from FUJI (1960a). The data taken from the other investigators have been adapted to fit these stages. The measurements on *Strongylocentrotus* oocytes were all made on histologic sections while those on *Arbacia* were made on living material. The histologic measurements are probably somewhat low due to shrinkage during fixation. After maturation of the oocyte, the nucleus is reduced from 28μ x 22μ to 9μ in diameter in *Strongylocentrotus* (HOLLAND and GIESE, 1965) and from 38.4μ to 11.5μ in diameter in *Arbacia* (HARVEY, 1956).

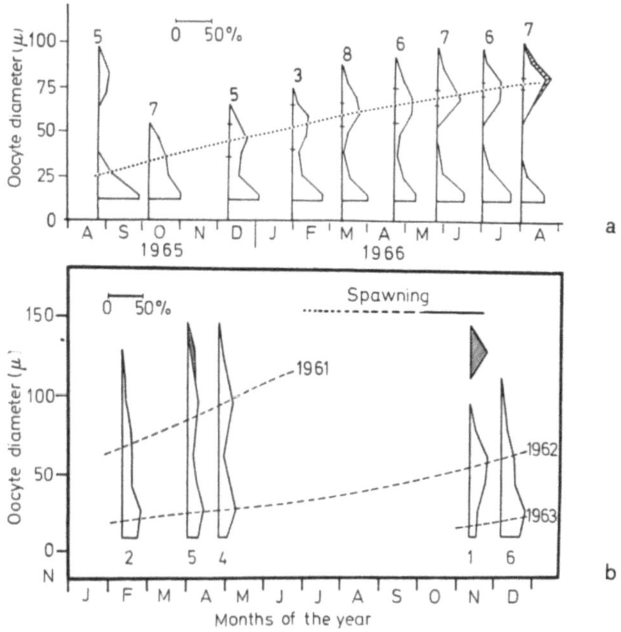

Fig. 3.11a and b. (a) Polygons show frequencies (see scale) of primary oocyte diameters in ovaries of *Stylocidaris affinis* averaged from samples collected August 1965 - August 1966. Nine average-frequency polygons were constructed from cytometric data from number of animals indicated above polygons. Mean diameter (plus or minus one SD) of group of growing primary oocytes is included for most polygons. Hatched area of polygon at far right indicates ripe eggs. Dotted line through means describes progressive increase in diameter of group of growing primary oocytes. Note continuous presence of population of very small oocytes. (From HOLLAND, 1967). (b) Polygons show oocyte frequencies (see scale) of different-size oocytes in ovaries of *Sterechinus neumayeri* averaged from samples collected from McMurdo Sound at different periods during 1961. Number of females averaged into each polygon is given as N. Hatched areas of polygons represent ova; unhatched areas, primary oocytes. Dotted line indicates rate of oocyte growth; year is when ova of each oocyte generation presumably spawned. (From PEARSE and GIESE, 1966)

Similar growth changes during the reproductive cycle occur in the antarctic sea urchin *Sterechinus neumayeri* (PEARSE and GIESE, 1966) and in the cidaroid sea urchin *Stylocidaris affinis* (HOLLAND, 1967)(Fig. 3.11). Small primary oocytes remain in the ovary all year. It is not known why all the small oocytes do not develop in the same reproductive cycle. The growing oocytes gradually increase in diameter until they are shed as mature ova. HOLLAND (1967) has calculated that the increase in the average volume of the oocytes of *Stylocidaris* between December and August (time of spawning) was approximately linear at a rate of 200μ^3 per day. It is interesting that in *Stylocidaris* - in contrast to other species - mature eggs to not accumulate in the ovary. The primary oocytes apparently mature immediately before being spawned. HOLLAND has suggested that this is a retention of a primitive echinoderm characteristic.

Table 3.3. Average diameters (μ) of oocytes, nuclei, and nucleoli during stages of annual reproductive cycles of *Strongylocentrotus* and *Arbacia*

		Growth Stage of Ovary[a]				
Species	Structure	Recovering-Spent	Growing	Premature	Mature	Reference
S. intermedius[b]	Oocyte	5 - 15	40 - 60	80 - 140 x 40 - 80 (oval) 80 x 100	80 x 100	FUJI (1960a)
S. nudus[b]	Nucleus		20 - 30	40	40	
S. purpuratus[b]	Oocyte	15 - 12	30 x 25	45 x 30	65 x 45	HOLLAND and GIESE (1965)
	Nucleus	9	15	22 x 18	28 x 22	
	Nucleolus	3	5	6	8	
	Oocyte	10 - 20	20 - 30	50 - 60	50 - 60	CHATLYNNE (1969)
A. punctulata[c]	Oocyte	14.3	33	60	74	HARVEY (1956)
	Nucleus	12.8	22	30	38.4	
	Nucleolus	4.8	9.6	10	11.2	

[a] From FUJI, 1960a.

[b] Histologic preparation.

[c] Living preparation.

Oocytes become increasingly basophilic during growth indicating significant RNA synthesis (COWDEN, 1962; ESPER, 1965). VERHEY and MOYER (1967a) have quantitated the increase in ribosomes (Table 3.4). An oocyte in the vitellogenesis stage has 100 times more ribosomes than an oogonium. Others have estimated that there are as many as 4,000 to 6,000 times more ribosomes in full-grown oocytes than in oogonia (MILLONIG, BOSCO, and GIAMBERTONE, 1968). This difference may be due in part to the choice of the average size of an oogonium and that of the average size of a terminal oocyte. There is also an increase in cytoplasmic basic proteins during the growth of the oocyte (TALEPOROS, 1959; DAVENPORT and DAVENPORT, 1966; SANCHEZ, 1968). This may also reflect an increasing number of ribosomes. The increase in mitochondria, lipids, yolk granules, cortical granules, and pigment vesicles during oogenesis is given in Table 3.5.

It has been suggested that nongerminal ovarian accessory cells - predominantly the nutritive phagocytes - transfer materials into the growing oocytes (MILLER and SMITH, 1931; MOORE, 1935; LIEBMAN, 1950). Histologic observations indicate a very close association between oocytes and accessory cells. The idea that ovarian cells transfer materials into the oocyte is supported indirectly by the fact that the nutritive phagocytes grow and become globulated during the reproductive months (CAULLERY, 1925; HOLLAND and GIESE, 1965; CHATLYNNE, 1969). The contents of the accessory cells apparently may differ with species. It has been reported that the accessory cells of *A. punctulata*, *L. variegatus*, and *L. pictus* do not contain glycogen (VERHEY and MOYER, 1967b), while those of *S. purpuratus* do (CHATLYNNE, 1969).

Table 3.4. Changes in ribosome numbers and concentration at two stages of sea-urchin oogenesis. (From VERHEY and MOYER, 1967a)

Stage	Ribosomes per μm^3 Cytoplasm	Ribosomes per Cell
Oogonium	3,600	6,542,900
Oocyte[a]	7,800	675,838,200
Ratio of increase between stages	2.2	103

[a]Oocytes were in vitellogenesis stage.

There is some doubt whether or not the accessory cells contribute significantly to the growth of the oocyte. Histologic and ultrastructural studies indicate that material is not transferred into the oocyte (COWDEN, 1962; HOLLAND and GIESE, 1965; VERHEY and MOYER, 1967a; CHATLYNNE, 1969). Transfer of materials certainly does not account for all of the growth because oocytes isolated from the ovary are very active in RNA and protein synthesis (PIATIGORSKY, OZAKI, and TYLER, 1967). The role of accessory cells in oocyte growth remains undetermined.

3.5.5 Maturation

Sea-urchin oocytes undergo both maturation divisions of the chromosomes while still in the ovary, differing from many other species in which oogenesis is arrested in the ovary either before the first (star-

Table 3.5. Size, volume, and number of cytoplasmic inclusions in sea-urchin eggs at two stages in oogenesis. (From VERHEY and MOYER, 1967a)

Type of Inclusion	Average Size		Average Volume		Number per Cell		Total Volume Occupied per Cell		Ratio of Increase in Number per Cell
	Oogonium	Oocyte[a]	Oogonium	Oocyte[a]	Oogonium	Oocyte[a]	Oogonium	Oocyte[a]	
Mitochondria	0.29μ x 0.4μ	0.29μ x 0.4μ	$0.15\mu^3$	$0.15\mu^3$	393	19,827	$59.4\mu^3$	$2,994\mu^3$	50
Lipid	0.381μ (rad)	0.381μ (rad)	$0.27\mu^3$	$0.27\mu^3$	22	3,512	$5.9\mu^3$	$941\mu^3$	160
Yolk	not present	Spherical: 0.32μ (rad) Prolate spheroid: 0.32μ x 0.58μ	not present	$0.19\mu^3$	not present	20,218	not present	$3,902\mu^3$	
Cortical granules	not present	0.61μ (rad)	not present	$1.01\mu^3$	not present	4,277	not present	$4,303\mu^3$	
Pigment vesicles	not present	0.44μ (rad)	not present	$0.36\mu^3$	not present	2,479	not present	$892\mu^3$	

[a]Oocytes were in vitellogenesis stage.

fish, some molluscs, polychaetes), during the first (ascidians), or during the second (most vertebrates) meiotic division (MONROY, 1965). A detailed description of chromosome behavior during oogenesis in *M. globulus* is given by TENNENT and ITO (1941).

Maturing oocytes expel a diploid polar body during the first meiotic division and a haploid polar body during the second division. The fate of the centrioles after meiosis is not known. None has been identified by electron microscopy in the mature egg. The course of events during nuclear maturation in oocytes of *A. punctulata* at 23°C is as follows (HARVEY, 1956): "Oocyte reaches full size (74µ in diameter); germinal vesicle approaches cell wall; nucleolus disappears; 1 1/4 hours later the germinal vesicle breaks down; red pigment granules numerous near breaking germinal vesicle; mitochondria accumulate opposite side of disruption of germinal vesicle; 2-2 1/2 hours later the first polar body given off; 1-1 1/2 hours later the second polar body given off; during process of polar body emission, the egg flattens in the axis of the polar bodies for about 7 minutes; 2 hours later the mature nucleus is formed."

The polar bodies are given off in the region of the micropyle, a funnel-shaped opening in the gelatinous coat of the egg (BOVERI, 1901; JENKINSON, 1911; TENNENT, TAYLOR, and WHITAKER, 1929; TENNENT and ITO, 1941; HARVEY, 1956). The position of the micropyle is not fixed with respect to the nucleus (HOADLEY, 1934). Since the polar bodies may come off in any relation to the stratifications of centrifuged eggs, the position of the micropyle is probably established by the original polarity of the egg (HARVEY, 1956). This position - the animal pole - is the free end of the egg (JENKINSON, 1911; LINDAHL, 1932; HÖRSTADIUS, 1939; TENNENT and ITO, 1941) and not the point of attachment to the ovarian wall - the vetal pole - as originally believed (BOVERI, 1901).

The classic treatise by WILSON (1895) on the egg of *Toxopneustes (Lytechinus) variegatus* was probably the first demonstration that oocytes with an intact germinal vesicle develop surface papillae and become polyspermic when fertilized (Fig. 3.12). The vitelline membrane does not elevate and further development is arrested. WILSON pointed out that "these peculiarities are of great interest as showing how profoundly the egg-constitution is affected by the changes taking place in the nucleus at the time the polar bodies are formed."

Fertilized oocytes become polyspermic and form papillae when the germinal vesicle is intact, when in metaphase of the first or second meiotic division, or even when in anaphase of the second division (BRACHET, 1922; RUNNSTRÖM and MONNE, 1945). The degree of polyspermy is proportional to the extent of maturation. It is likely that the basic cause of the polyspermy is the failure of the inseminated oocytes to raise fertilization membranes. The papillae may reach a length of 100µ and are amoeboid (RUNNSTRÖM, 1963). Their formation is not affected by 2,4-dinitrophenol and does not require ATP synthesis (RUNNSTRÖM, 1963). The papillae develop 1 to 5 minutes after insemination and retract 5 to 30 minutes later, depending on the species (HAGSTRÖM and LÖNNING, 1961). Other studies on fertilized oocytes of sea urchins (SUGIYAMA, 1938; ROTHSCHILD and SWANN, 1949) and of sand dollars (SEIFRIZ, 1926) have shown similar results.

The ability of fertilized oocytes to form asters develops during nuclear maturation. Microcinematographic observations have shown that the sperm head rotates continuously but does not swell in the oocyte as in the mature egg (HAGSTRÖM and LÖNNING, 1961). Asters do not form in primary oocytes but can develop after the germinal vesicle has disrupted (HAGSTRÖM and LÖNNING, 1961; FRANKLIN, 1965).

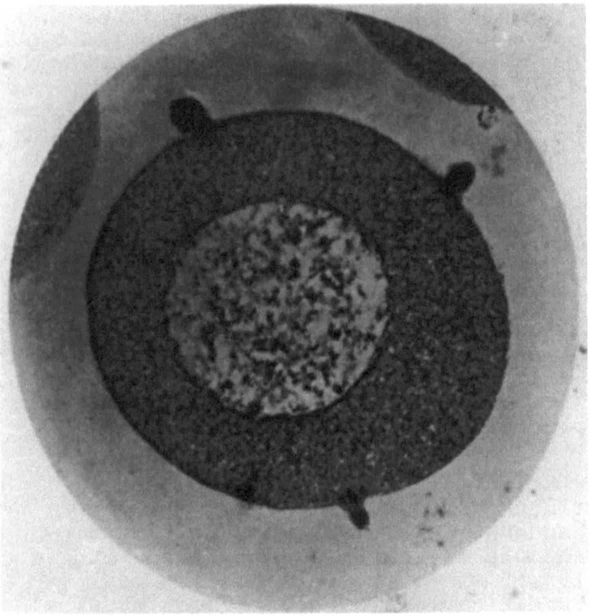

Fig. 3.12. Histologic section of inseminated primary oocyte of *Toxopneustes (Lytechinus) variegatus*. Surface papillae indicate sperm entrance cones. x 950-1000. (From WILSON, 1895)

Sperm penetration into oocytes involves fusion of the sperm and egg plasma membrane, as in the fertilization of mature eggs (HAGSTRÖM and LÖNNING, 1961; FRANKLIN and METZ, 1962; RUNNSTRÖM, 1963; FRANKLIN, 1965). These ultrastructural studies have also shown that the cortical granules - scattered in the cytoplasm rather than adjacent to the membrane - do not break down. A region develops - appearing under phase-contrast as a "corona" around the germinal vesicle - in which yolk granules disappear. The regular structure of the annulate lamellae becomes disorganized, and the nuclear membrane folds. When the papillae retract, the organization of the uninseminated oocyte is regained.

The restrictions for the specificity of fertilization are present in the oocyte even though the requirements for monospermy are not met before maturation is complete. This would be expected since the gelatinous coat is present in the larger oocytes and the specificity toward homologous sperm resides in the fertilizin in the jelly and possibly in the plasma membrane (TYLER and TYLER, 1966b). Tests showed that mature eggs had the same specificity towards sperm as did oocytes when 5 species of ova were inseminated with 10 different species of spermatozoa (HAGSTRÖM and LÖNNING, 1961). Homologous sperm caused more protrusions to develop than did heterologous sperm. One study however suggests that *A. punctulata* oocytes are penetrated more effectively by sperm from the sand dollar *Mellita quinquiesperforata* than are the mature eggs (FRANKLIN, 1965).

Even though inseminated oocytes are polyspermic, they are less receptive to sperm than are mature eggs. It takes a higher sperm concentration to penetrate an oocyte than a mature egg (HAGSTRÖM and LÖNNING, 1961).

There are changes in a number of other surface properties that occur on cytoplasmic maturation (RUNNSTRÖM, 1949, 1966). These include changes in light refraction and birefringence (RUNNSTRÖM, MONNE, and WICKLUND, 1946; MONROY, 1948), sensitivity to treatment with proteolytic and lipolytic enzymes (MINGANTI, 1954) and morphologic appearance (RUNNSTRÖM and MONNE, 1945; McCULLOCH, 1952; LÖNNING, 1964). The oocyte has a relatively wrinkled surface with irregular cytoplasmic filaments and protrusions, while the mature egg is smoother and has regular papillae (WOLPERT and MERCER, 1961). Tests on the extent of detachment of the vitelline membrane from the cell surface in hypertonic seawater suggest that the membrane adheres more tightly to the cortical layer after the formation of the first poly body and show that treatment with hypertonic water causes oocytes to shrink with a smooth surface; mature eggs shrink with a wrinkled surface (RUNNSTRÖM and MONNE, 1945).

Even after both polar bodies have been formed, it has been reported that the unfertilized egg exists for some hours in an underripe condition (RUNNSTRÖM, 1949). Changes in the properties of aging eggs have been studied by PASPALEFF (1927) and by GOLDFARB (1935a, 1935b, 1937). Underripe eggs form a smooth surface more slowly than ripe eggs after treatment with hypertonic seawater (RUNNSTRÖM and MONNE, 1945). Serum, amino acids, and chelating agents such as versene or periodate overcome this response (RUNNSTRÖM and KRIZAT, 1950; WICKLUND, KRIZAT, and RUNNSTRÖM, 1953; WICKLUND, 1954). Underripe haploid eggs are not artificially activated as readily as completely mature eggs (WICKLUND, 1947) and exhibit unusually strong contractions, poor membrane elevation, and abnormal cleavage after fertilization (RUNNSTRÖM, 1948). The bases for these phenomena are not clear.

3.5.6 Respiration

There have been very few studies on the respiration of oocytes in sea urchins due to the difficulty of obtaining isolated oocytes. The first study indicated that the respiratory rate and the ability to oxidize *p*-phenylenediamine of oocytes of *A. punctulata* is less than that of mature eggs (BOELL et al., 1940). LINDAHL and HOLTER (1941) used a Cartesian diver to show that primary oocytes of *P. lividus* consume oxygen about 3 times faster than mature eggs and possibly even faster than fertilized eggs. Respiration rate drops markedly when the germinal vesicle breaks down. Experiments with *P. miliaris* later confirmed that respiration decreases at maturation (BOREI, 1948, 1949). In these tests it was shown that mature eggs of this species decrease their rate of respiration 2.5-fold 4 to 6 hours after being shed into seawater, at which time they reach a basal level. Oocytes also showed a decreasing respiration rate similar to that of the ripe eggs. Experiments were conducted 4 hours after obtaining the gametes. Measurement of the oxygen consumption by oocytes (germinal vesicle intact), underripe eggs (polar bodies expelled but not yet normally fertilizable), mature eggs (fertilizable), and overripe eggs (no longer fertilizable) gave the following results in terms of microliters of oxygen per cell per hour (BOREI, 1948): oocyte, 0.66×10^{-4}; underripe egg, 0.56×10^{-4}; ripe egg, 0.51×10^{-4}; overripe egg, 0.45×10^{-4}. BOREI attempted to measure the oxygen consumption of oocytes immersed in ovarian fluid by obtaining the cells with a micropipet introduced through the ovary wall. In spite of the facts that the oocytes were contaminated with other cell types and were densely packed in the Cartesian diver, the results were similar to those obtained in seawater.

The respiration rate of shed, unfertilized eggs does not decrease in all species. The respiratory level of unfertilized eggs of *A. punctulata*

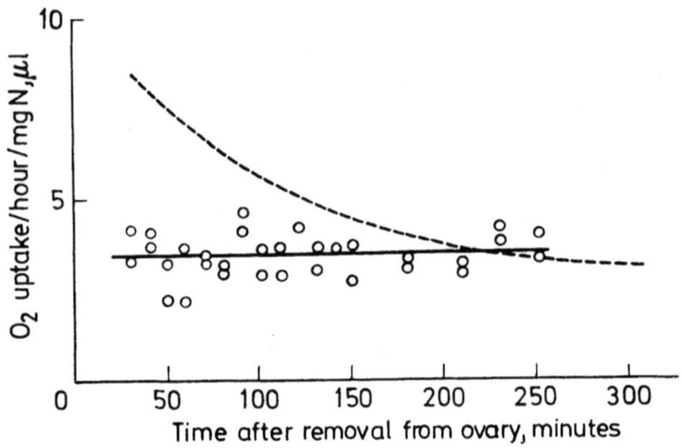

Fig. 3.13. Oxygen uptake of unfertilized *Pseudocentrotus depressus* eggs at 20°C. Dotted line represents *Psammechinus miliaris* and is redrawn from BOREI (1948a) for comparison. (From YASUMASU and NAKANO, 1963)

remains constant for several days if kept uncontaminated (TYLER, RICCI, and HOROWITZ, 1938). The oxygen consumption by unfertilized eggs of *H. pulcherrimus*, *Pseudocentrotus depressus*, and *A. crassispina* does not decrease after spawning (YASAMASU and NAKANO, 1963; OHNISHI and SUGIYAMA, 1963)(Fig. 3.13). The reason for the difference in behavior of the eggs of *Psammechinus* is not known. MONROY (1965) speculated that it may be due to the extreme salinity in which these Swedish sea urchins live.

Information about the possible mechanism of suppression of respiration in unfertilized eggs comes from studies on the increase in oxygen consumption that occurs at fertilization which was first described by WARBURG (1908). Among the ideas advanced to explain the low level of respiration in unfertilized eggs are a physical separation of cytochrome oxidase from cytochrome c, the presence of an inhibitor of cytochrome oxidase, a shortage of phosphate acceptors, inhibition of the pentose and hexose phosphate pathways, a paucity of TPN and TPNH, and the physical state of mitochondria (MONROY, 1965).

3.5.7 Egg Restitution after Shedding

Under natural conditions sea urchins occasionally shed ripe gametes several times during the reproductive months, but spawning probably occurs only once each breeding season (BOOLOTIAN, 1966). In the laboratory, however, ripe sea urchins may be induced to spawn.

Gravid female sea urchins force-shed with KCl produce more ripe eggs (TYLER, 1949b; TYLER and TYLER, 1966a). Two weeks of restitution may produce as many eggs as were shed initially. In experiments with *A. punctulata*, *L. pictus*, and *S. purpuratus*, TYLER obtained 4 successive batches of eggs from individual sea urchins. The original quantity of eggs was recovered each time. Animals opened after the first shedding had practically all their ripe eggs spawned, indicating that the new eggs must have matured during the experiment.

The restitution of eggs is a useful method to label materials synthesized during oogenesis. This was first used to label the gelatinous coat (fertilizin) of *A. punctulata* with S^{35}-sulfate (TYLER and HATHAWAY, 1958). RNA (PIATIGORSKY, 1965; GROSS, MALKIN, and HUBBARD, 1965;

PIATIGORSKY and TYLER, 1967), protein (OZAKI, PIATIGORSKY, and TYLER, 1966), and DNA (HOLLAND and GIESE, 1965; PIKO, TYLER, and VINOGRAD, 1967) of unfertilized eggs have since been labeled during oogenesis by long-term incubations in vivo. When sea urchins are injected with labeled substances, almost all the radioactivity is retained by the organism. ERB and MAURER (1962) injected H^3-lysine into *P. miliaris* and found that only 0.1% of the label was expelled in 4 hours. In longer experiments, PIATIGORSKY and TYLER (1967) showed that only about 5% of injected H^3-uridine escaped from *L. pictus* in 24 hours; the rest remained in the animal as tested for 3 weeks.

3.5.8 DNA Synthesis

DNA is present in the nucleus and in the cytoplasm of sea-urchin eggs (TYLER and TYLER, 1966a; PIKO, TYLER, and VINOGRAD, 1967; HARRIS, 1969). The cytoplasm contains 5 to 10 times more DNA than the nucleus (Fig. 3.14).

Only the nuclei of oogonia and small preleptotene primary oocytes synthesize nuclear DNA (NIGON and GILLOT, 1965; HOLLAND and GIESE, 1965). The long-term experiments of HOLLAND and GIESE have shown that

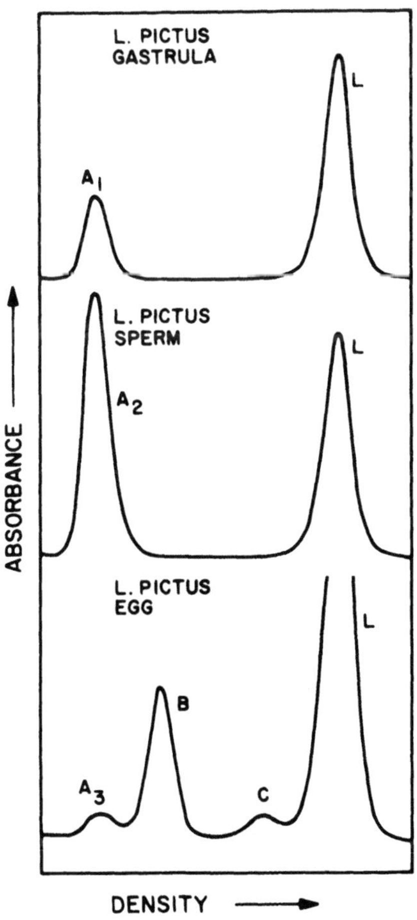

Fig. 3.14. Tracings of direct scans at 265μm of buoyant density bands (in CsCl) of 3 preparations of DNA of *Lytechinus pictus* eggs and sperm after 25 hours of centrifugation in same rotor at 44,770 rpm in Beckman model E centrifuge. L, density marker DNA (1.731 gm/cm^3) of *Micrococcus lysodeikticus*. A_1, A_2, and A_3, DNAs identified as nuclear and with similar buoyant densities of 1.693. B, DNA identified as derived from mitochondria and yolk and with buoyant density of 1.703. C, Unidentified nucleic acid band of buoyant density 1.719. Sperm DNA does not show satellite band due to small amount analyzed. (From PIKO, TYLER, and VINOGRAD, 1967)

nuclear DNA synthesis takes place even during the nonreproductive months. Factors other than DNA synthesis apparently regulate the growth of the oocyte. Growing primary oocytes, secondary oocytes, and mature ova do not have labeled nuclei even many months after a female sea urchin (*S. purpuratus*) is injected with H^3-thymidine. A primary oocyte evidently undergoes a long incubation period - perhaps one reproductive cycle - between nuclear DNA synthesis and growth.

Labeled DNA has been obtained from regenerated eggs of *L. pictus* (PIKO, TYLER, and VINOGRAD, 1967). Females were injected with 2 successive doses of H^3-thymidine (200μCi, 6 Ci/mmole) at one-month intervals. Thirty days later, 100,000 eggs were obtained which contained 0.1% of the injected radioactivity in DNA with a density of 1.70 gm/cm^3 in cesium chloride. This contrasts with a buoyant density of 1.693 for nuclear DNA. Experiments with eggs of *A. punctulata* labeled with H^3-thymidine during oogenesis have shown that the labeled DNA is mitochondrial DNA (MATSUMOTO and PIKO, 1971). This explains the density difference between labeled and nuclear DNA. Mitochondrial DNA consists of duplex molecules about 5μ in perimeter (Fig. 3.15). About 90% of the mitochondrial DNA occur as single circles and the remainder are interlinked (catenated) dimers or higher oligomers (PIKO et al., 1968; MATSUMOTO et al., 1974).

A third species of DNA with a density of 1.719 has been extracted from ripe eggs of *L. pictus* (PIKO, TYLER, and VINOGRAD, 1967). This was not reported labeled during oogenesis and was not found in later experiments (PIKO et al., 1968). Its high density suggests that it may correspond to the satellite of ribosomal DNA in sperm (STAFFORD and GUILD, 1969; PATTERSON and STAFFORD, 1970; SPONAR, SORMOVA, and GLISIN, 1970). It is not known if ribosomal DNA is amplified in sea-urchin oocytes. Another possibility is that the DNA with a density of 1.719 was the genetic material of an endosymbiont reported in eggs of *P. lividus* (ANDERSON, 1969).

3.5.9 RNA Synthesis

RNA synthesis takes place in oocytes of all sizes. This has been observed autoradiographically when the ovaries were excised and labeled in vitro (FICQ, 1964), when isolated oocytes were labeled in vitro (PIATIGORSKY, OZAKI, and TYLER, 1967), and when intact sea urchins were injected with radioactive RNA precursors and the ovaries labeled in vivo (GROSS, MALKIN, and HUBBARD, 1965; SANCHEZ, 1968).

In short-term experiments, labeling is most intense in the nucleolus and nucleoplasm. Longer incubations result in the appearance of cytoplasmic radioactivity. Mature eggs in the central region of the lumen are rarely labeled in short-term tests. They become radioactive when labeling continues over several days in the intact animal. This presumably represents the maturation of radioactive oocytes.

Inhibition of RNA synthesis during oogenesis produces morphologic abnormalities (SANCHEZ, 1966, 1968). Three days after injection of dactinomycin into *P. lividus*, young oocytes have fibrillar formation in the cytoplasm while older ones have twisted lamellae adjacent to the outer surface of the nuclear membrane. The heavy bodies are few and abnormal. RNA and basic proteins accumulate in the nucleus, the membrane of which becomes excessively folded. Leptotene chromosomes become more condensed. Evidently the synthesis of RNA has a major role in the structural organization of the oocyte. Extensive treatment with dactinomycin could have any number of side effects, however.

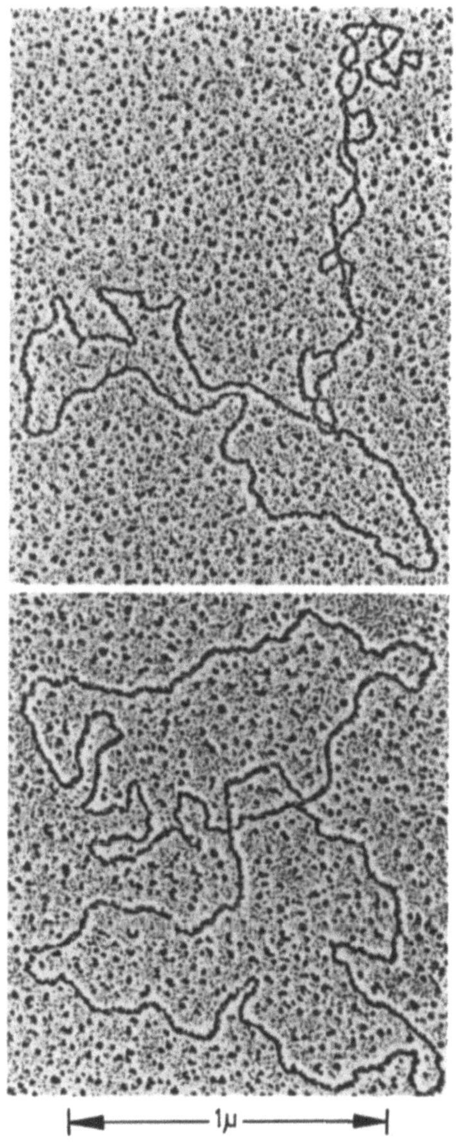

Fig. 3.15. Electron micrographs of selected catenated molecules from Fig. 3.14 (band B) of unfertilized egg (*Lytechinus pictus*) DNA. Upper: Dimer with one relaxed (open) and one twisted circle. Lower: Dimer with 2 relaxed circles. Relaxed configuration presumably indicates that molecule has suffered at least one nick or lesion. Catenated forms of mitochondrial DNA represent only about 10% of the population; others are single circles. (From PIKO, TYLER, and VINOGRAD, 1967)

The synthesis and conservation of ribosomal RNA occur throughout oogenesis in sea urchins (GROSS, MALKIN, and HUBBARD, 1965; PIATIGORSKY and TYLER, 1967) as they do in amphibians (BROWN and LITTNA, 1964). The incorporation of labeled RNA precursors is mostly into the mature ribosomal RNAs after incubation of one week or longer. There is also some incorporation into low molecular weight RNAs around 4 S. The distribution of labeled RNAs isolated from ripe eggs obtained 33 days after injecting a spawned female *L. pictus* with H^3-5-uridine is shown in Fig. 3.16; 70%-80% of the label sediments with the 2 species of ribosomal RNAs, and 5%-10% with the low molecular weight RNAs in the 4 S region. The latter probably consists of several types of low molecular weight RNAs (DARNELL, 1968). Similar results are found

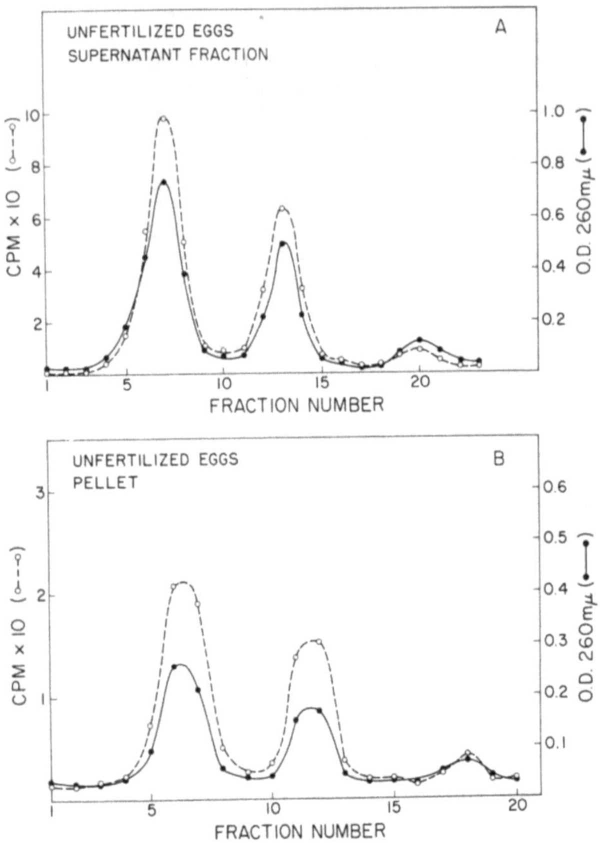

Fig. 3.16. Sedimentation pattern of RNA from supernatant fraction (A) and low-speed pellet (B) of homogenate of RNA-labeled unfertilized eggs of *Lytechinus pictus*. Spawned female was injected with 150μCi of H^3-5-uridine (specific activity 25,000 Ci/*M*) and shed 33 days later. Labeled, unfertilized eggs were washed several times by centrifugation in 0.55 *M* KCl, homogenized in 3 volumes of 0.01 *M* Na acetate buffer at pH 5.0, centrifuged at 10,000 *g* for 10 minutes, and supernatant fraction set aside at 0°C. Pellet was washed twice by centrifugation with 20-30 volumes of homogenization buffer, all supernatant fractions combined, and pellet resuspended in about 40 volumes of buffer. Preparations were phenol-extracted at 4°C and treated with DNase (RNase-free, Worthington, 15 μg/ml, 4°C, 30 minutes). Samples (0.3 ml) of labeled extracts were centrifuged through linear 5%-20% sucrose density-gradient in 0.01 *M* Na acetate and 0.1 *M* NaCl at pH 5.0 at 37,000 rpm for 5 hours at 5°-10°C. Three-drop (about 0.20-ml) fractions were collected from each tube and diluted with equal volume of distilled water. The 260-μm absorption was determined in each fraction. Radioactivity measurements were made by scintillation counting. (From PIATIGORSKY and TYLER, 1967)

after 4 months of labeling (PIATIGORSKY and TYLER, 1970). Since the labeled nucleotide pool is about 95% depleted 25 to 30 days after the injection of isotope (Fig. 3.17), ribosomal RNA synthesized by small oocytes is probably conserved during oogenesis.

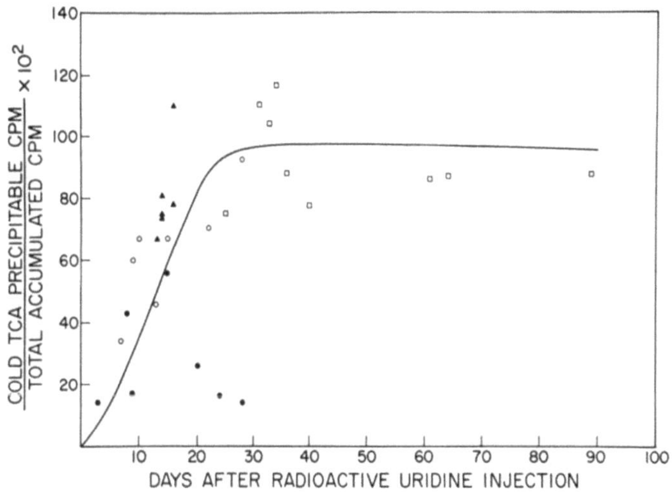

Fig. 3.17. Percent recovery of injected isotope found in mature eggs of *Strongylocentrotus purpuratus* and *Lytechinus pictus* labeled during oogenesis. Sea urchins were spawned with KCl, injected with labeled uridine, and incubated at 20°C. At designated time after injection of radioactive uridine, mature eggs were shed with potassium chloride and assayed for total radioactivity by scintillation counting. Each point specified percentage of radioactivity - relative to that administered to animal - recovered in shed eggs. ● *L. pictus*, C^{14}-2-uridine; ○ *L. pictus*, H^3-6-uridine; □ *L. pictus*, H^3-5-uridine; ▲ *S. purpuratus*, H^3-uniformly-labeled-uridine. (From PIATIGORSKY and TYLER, 1967)

The maturation of 26S and 18S ribosomal RNAs occurs by a non-conservative degradation of a 32S precursor ribosomal RNA which is synthesized in the oocyte nucleolus (SCONZO et al., 1972; GIUDICE et al., 1973). The 32S RNA is specifically cleaved in the nucleolus to a 27S and a 21S RNA, which in turn is further degraded to give the 26S and 18S ribosomal RNAs, respectively. Experiments with purified nucleoli from isolated vitellogenic oocytes of *P. lividus* (GIUDICE et al., 1972) suggested that nucleolar proteins protect the 32S precursor ribosomal RNA from random degradation by ribonuclease, leaving it exposed at specific sites for its proper maturational cleavages (GIUDICE et al., 1973). Ribosomal RNA maturation appears to proceed more slowly in sea urchin oocytes than in developing gastrulae (SCONZO et al., 1972).

A great deal of ribosomal RNA synthesis during oogenesis appears to be universal among animals (PIATIGORSKY, OZAKI, and TYLER, 1967; DAVIDSON, 1968), and it is consistent with the intense labeling of the nucleolus (PERRY, 1965; DARNELL, 1968). The multiplication of active nucleoli does not occur in sea urchins as it does in amphibians (BROWN and DAWID, 1968; GALL, 1968) and fish (VINCENT et al., 1968). The proliferation of nucleoli indicates an amplification of ribosomal DNA. Oocytes of several invertebrate species with a single large nucleolus have also amplified ribosomal DNA cistrons (LIMA-de-FARIA, BIRNSTIEL, and JAWORSKA, 1969; GALL, MACGREGOR, and KIDSTON, 1969; DAWID and BROWN, 1970). Sea-urchin oocytes may therefore synthesize additional copies of ribosomal DNA.

Experiments suggest that appreciable amounts of nonribosomal RNA are made at the terminal stages of oogenesis. Even though 10%-20% of the

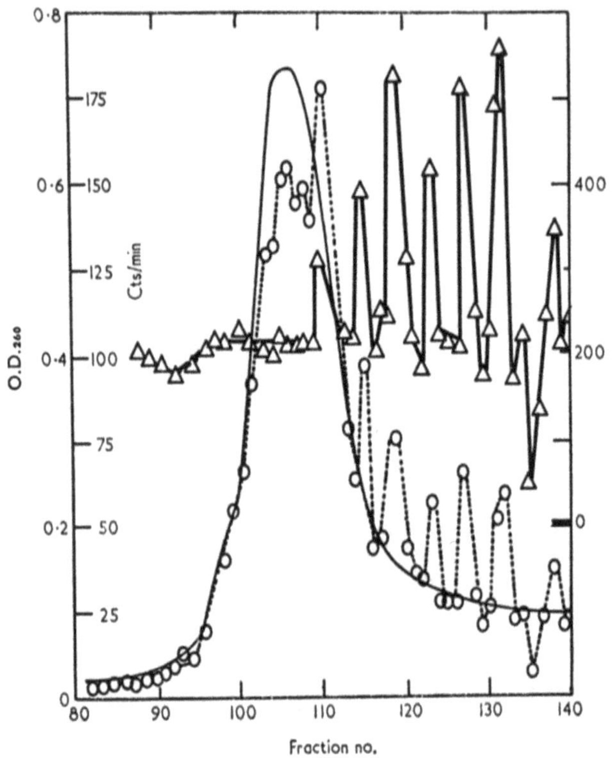

Fig. 3.18. Methylated albumin kieselguhr chromatographic profile of RNA from unfertilized *Arbacia punctulata* eggs labeled with H^3-uridine during last week of oogenesis. In region plotted, eluting salt gradient corresponds to linear increase from 0.7 to 1.2 *M* NaCl. Optical densities of effluent were determined from continuous record - accurately registered (solid line) - and from measurements on individual fractions used for computation of specific activities. Peak optical density represents ribosomal RNA. Radioactivities plotted were determined by directly counting effluent fractions in Bray's solution. Open circles and dotted line: counts; triangles and thicker line: calculated specific activities. Specific activities and optical densities fell to very low values on both sides of region shown (more than 90% of radioactive material eluted). Fraction volume was 2.2 ml. Note relative constancy of specific activity on ascending side of optical density peak. (From GROSS, MALKIN, and HUBBARD, 1965)

labeled RNA is polydisperse after short-term (GROSS, MALKIN, and HUBBARD, 1965) and long-term (PIATIGORSKY and TYLER, 1967) experiments, this RNA has a higher specific activity than the ribosomal RNA only after relatively short incubations. The RNA with high specific activity may be separated from the ribosomal RNA by methylated albumin kieselguhr chromatography (Fig. 3.18). At least 1.5% of the total radioactivity hybridizes with DNA from homologous sperm in the presence of a 350% excess of unlabeled ribosomal RNA.

Three additional species of RNA have been detected in unfertilized eggs of *S. purpuratus* by polyacrylamide gel electrophoresis (SLATER and SPIEGELMAN, 1970; DAIGNEAULT, BELLEMARE, and COUSINEAU, 1970; DUBOIS et al., 1971). These are approximately 20 S, 12 S and 9 S. When these RNAs are examined during development, the 20-S component progressively diminishes and is gone by gastrulation, while the 12-S and 9-S species remain in the same relative proportions. The 12-S and 9-S RNAs are no longer present in late prism or pluteus stages of development. The evidence indicates that the 12-S and 9-S RNAs are not degradation products of higher molecular weight RNA species.

The function of these 3 species is not known. Apparently the 12-S and 9-S RNAs are found exclusively with the yolk granules of fractionated homogenates (DUBOIS et al., 1971). These investigators speculate that "perhaps the yolk 12 S and 9 S RNA are in some way related to the structural organization of these granules, and disappear gradually as is the fate of their lipoprotein content." On the other hand, there is some evidence that the 20-S and 12-S RNAs may be synthesized on mitochondrial templates (VESCO and PENMAN, 1969; SWANSON and DAWID, 1970). There are prominent peaks of RNA of approximately 20 S and 12-15 S synthesized by mitochondria in activated nonnucleate egg fragments of sea urchins (CRAIG, 1970; CHAMBERLAIN, 1970). In starfish oocytes, mitochondria synthesize 15 to 19-S RNA during maturation (LaMARCA, SHIPPEE, and SCHUETZ, 1971). Still another possibility is that the 20-S, 12-S, and 9-S RNAs are messenger RNA, although messenger RNAs found in unfertilized eggs are associated with small ribonucleoprotein particles and not with yolk (see below).

The rate of RNA synthesis appears to decrease appreciably at maturation, as indicated in the autoradiographic experiments above. The failure to detect RNA synthesis in unfertilized eggs is due - at least in part - to the low permeability of RNA precursors into mature eggs (WHITELEY, 1949; LITCHFIELD and WHITELEY, 1959; PIATIGORSKY and WHITELEY, 1965; MITCHISON and CUMMINS, 1966). When this large permeability difference is taken into account, calculations indicate that the unfertilized egg may be synthesizing more RNA than previously realized (SIEKEVITZ, MAGGIO, and CATALANO, 1966). The degree of inhibition of RNA synthesis that occurs at maturation has not been definitively established.

The compartmentalization of RNAs synthesized during oogenesis has been investigated by fractionation of homogenates of labeled unfertilized eggs (PIATIGORSKY, 1968; PIATIGORSKY and TYLER, 1970). The labeled ribosomal RNA is mainly in single ribosomes, although some is found in ribosomal aggregates, about two-thirds of which are associated with cellular membranes. At present it is not certain whether the aggregates in homogenates represent adventitious clumps of ribosomes or arrested polyribosomes. The possibility that these are heavy bodies has already been discussed. Polydisperse RNA is found on ribonucleoprotein particles smaller than ribosomes (Fig. 3.19). These may be comparable to the informosomes synthesized during early development (SPIRIN and NEMER, 1965; SPIRIN, 1966). Recent experiments indicate that these ribonucleoprotein particles in unfertilized eggs contain 9S histone messenger RNA (SKOULTCHI and GROSS, 1973). It is also probable that they contain tubulin messenger RNA (RAFF et al., 1971, 1972).

3.5.10 Protein Synthesis

Autoradiographic experiments have demonstrated that ovarian oocytes of all sizes synthesize protein (IMMERS, 1961; ERB and MAURER, 1962; FICQ, 1964). This is also true of isolated oocytes (PIATIGORSKY, OZAKI, and TYLER, 1967). Amino acids are incorporated into the nucleolus, nucleoplasm, and cytoplasm. After 30 minutes to 4 hours of label-

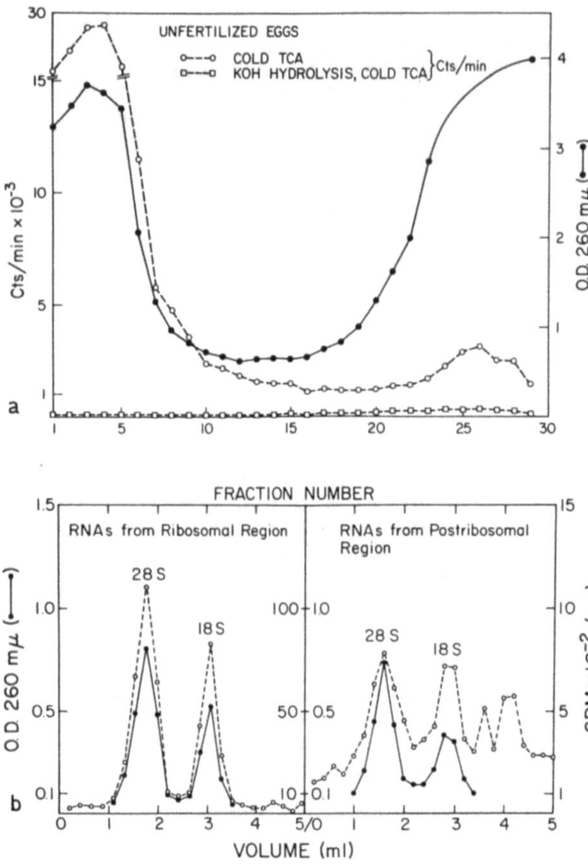

Fig. 3.19a and b. (a) Sedimentation pattern of postribosomal region of unfertilized *Lytechinus pictus* eggs labeled with H^3-5-uridine (150-200μCi injected, 25 Ci/mmole) during oogenesis for 119 days. Shed eggs were washed twice in 0.55 *M* KCl, homogenized in 2 volumes of buffer, centrifuged at 10,000 rpm for 10 minutes in No. 40 rotor of Spinco ultracentrifuge, and 0.3 ml of supernatant fraction layered on 4.83 ml of linear 15%-30% (w/v) sucrose gradient. Centrifugation was for 6 hours at 30,000 rpm in SW 39 rotor, and 0.18-ml fractions were collected with peristaltic pump, diluted with equal volumes of distilled water, and 260-nm absorption determined. Half the material was assayed directly for cold 5% TCA-precipitable radioactivity and other half subjected to alkaline hydrolysis (0.3 *N* KOH, 18 hours, 37°C) and assayed for cold 5% TCA-precipitable label. (b) Sedimentation pattern of labeled RNAs phenol-extracted from ribosomal region and from postribosomal region of sucrose gradients of homogenates of unfertilized *L. pictus* eggs labeled with H^3-5-uridine (150μCi injected, 25 Ci/mmole) during oogenesis for 122 days. Ribosomal region consisted of 4 fractions comprising peak of single ribosomes. Postribosomal region consisted of fractions 11-22 of sucrose gradient centrifugation similar to that shown above. Nonradioactive ribosomes were added to pooled fractions; regions were phenol-extracted and examined by sucrose gradient centrifugation. Fractions of 0.2 ml were collected with peristaltic pump, 260-μm absorption was determined, and cold 5% TCA-precipitable radioactivity assayed. (From PIATIGORSKY and TYLER, 1970)

ing with H^3-lysine or H^3-leucine, the density of grains is about the same in the cytoplasm and in the nucleus, but it is 2 to 4 times greater in the nucleolus than in the nucleoplasm (ERB and MAURER, 1962). It is not known if this represents nucleolar protein synthesis or accumulation of labeled protein in the nucleolus.

Puromycin inhibits protein synthesis in sea-urchin oocytes (FICQ, 1964; MACKINTOSH and BELL, 1970), but protein synthesis continues in the presence of dactinomycin (FICQ, 1964; PIATIGORSKY, OZAKI, and TYLER, 1967). This has been tested for as long as 25 hours. Thus, oocyte messenger RNA does not turn over rapidly. There have not been any studies on the effect of dactinomycin on the synthesis of specific proteins in oocytes. It is not known if messenger RNAs for all the proteins are equally stable.

One study has recently been performed on the synthesis of specific proteins during oogenesis. Vitellogenic oocytes of *P. lividus* were shown to synthesize histones (COGNETTI, SPINELLI, and VIVOLI, 1974). It was calculated that one isolated, vitellogenic oocyte synthesizes approximately 0.187 pg of an arginine-rich histone in 4 hours. These authors point out that histone synthesis during oogenesis differs from the usual coupling of histone synthesis with DNA synthesis during the cell cycle. It is not known if the histones synthesized during oogenesis utilize the same messenger RNAs that are activated after fertilization or if they are degraded at maturation.

At maturation the rate of protein synthesis decreases but is not eliminated (MACKINTOSH and BELL, 1967; TYLER, PIATIGORSKY, and OZAKI, 1966; EPEL, 1967; TYLER, TYLER, and PIATIGORSKY, 1969). The difference in C^{14}-valine incorporation into protein by an isolated oocyte and by a mature egg is shown in Fig. 3.20. The greater incorporation by oocytes cannot be attributed entirely to differences in the uptake of valine

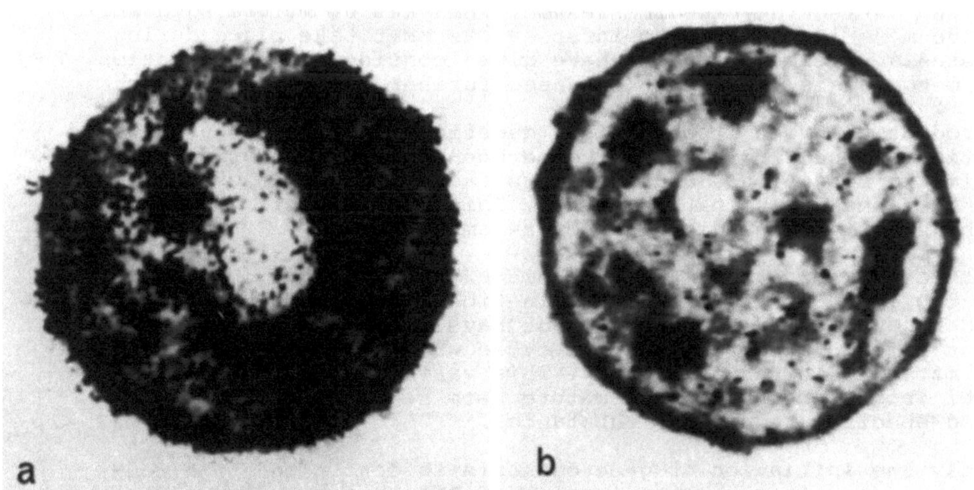

Fig. 3.20a and b. Autoradiographs of 5-μ histologic sections of oocyte (a) and mature egg (b) of *Lytechinus pictus* labeled together in suspension in artificial seawater with C^{14}-valine for 1 hour and developed with Kodak Nuclear Track Emulsion NTB3 for 4 days. (From PIATIGORSKY, OZAKI, and TYLER, 1967)

because unfertilized eggs accumulate amino acids quite well (TYLER, PIATIGORSKY, and OZAKI, 1966) even though uptake is restricted (MITCHISON and CUMMINS, 1966).

The mechanism by which the oocyte inhibits protein synthesis at maturation is not known. All the components for protein synthesis appear to be present in the unfertilized egg (MONROY, 1965; TYLER, 1967; GROSS, 1968), and protein synthesis is greatly stimulated after fertilization, even in the absence of new RNA synthesis (GROSS and COUSINEAU, 1963, 1964; GROSS, MALKIN, and MOYER, 1964). The unfertilized egg thus contains considerable messenger RNA (SLATER and SPIEGELMAN, 1966; GROSS et al., 1973). The possibility that ribosomes are activated after fertilization has received considerable attention (HULTIN, 1961, 1964; MONROY, MAGGIO, and RINALDI, 1965; CANDELAS and IVERSON, 1966; BACKSTRÖM, 1966; PIATIGORSKY and TYLER, 1968, 1970; HILLE, 1974). Other mechanisms for the regulation of protein synthesis in sea-urchin eggs have been proposed, such as inhibition of peptide chain initiation (MACKINTOSH and BELL, 1969), soluble inhibitors of messenger translation (CANDELAS and IVERSON, 1966; TIMOURIAN, 1967), suppressed transfer enzymes (CASTANEDA and TYLER, 1968), inactive messenger RNA containing ribonucleoprotein particles smaller than ribosomes (SPIRIN and NEMER, 1965; SPIRIN, 1966), and inhibited messenger RNA-protein complexes larger than ribosomes (MANO and NAGANO, 1966; STAVY and GROSS, 1967). Further discussion of this topic is beyond the scope of the present chapter.

3.6 Conclusions

The maturation of sea-urchin germ cells has been studied by examining the gonads at different times of the reproductive cycle and by investigating individual gametes in the gonads at various stages of development during the breeding season. Analyses of mature eggs and sperm have made it possible to infer events that take place during gametogenesis. Although studies have given considerable information about gametogenesis, many problems need further examination.

The reproductive cycle itself raises questions that remain unanswered. The initiation and termination of gametogenesis can be correlated with climatic conditions, and exogenous factors may even exert some control over the determination of sex. This has a genetic basis in sea urchins as in other dioecious organisms.

Very little is known about the mechanisms that regulate the reproductive cycle and control sex determination. Sex hormones have been detected in ovaries of sea urchins but have not been examined at different times of the reproductive cycle when relative amounts of organic material in the gonad vary. This variation is probably the result of increasing numbers of mature germ cells, and it has not suggested specific regulatory substances.

Apparently the initiation of gametogenesis is controlled to a considerable extent after the termination of DNA synthesis in oocytes and spermatocytes. This is deducted from the observations that there is some DNA synthesis during the quiescent period of the reproductive cycle without the accumulation of mature gametes and that preleptotene primary oocytes do not grow immediately after DNA synthesis. Their growth probably occurs one year after the final round of DNA synthesis. This is in marked contrast to the production of spermatozoa. A primary spermatocyte becomes a spermātid 3 to 6 days after DNA synthesis and takes about 4 more days to become a mature spermatozoan.

The ultrastructural changes during spermatogenesis - especially during spermiogenesis - have been studied much more extensively than the biochemical changes. The development of the acrosome, mature nucleus, midpiece, and tail have been investigated with the electron microscope. The fine structure of spermatogonia, primary spermatocytes, and secondary spermatocytes have been given only cursory treatment. While some specific proteins have been localized in different organelles - such as phosphatases in the acrosome, basic proteins in the nucleus, and lipids in the midpiece - the progressive sequestering of these materials has not been studied.

There is a conspicuous absence of investigations on RNA and protein synthesis during sea-urchin spermatogenesis. Such fundamental questions as the duration of RNA and protein synthesis, the stability of messenger RNAs, or the appearance of specific proteins during the maturation of spermatocytes and spermatids have not been explored. More attention has been given to characterizing the components of the mature spermatozoan. These studies have increased our knowledge about the basic proteins in the nucleus, have identified the presence of enzymes and structural proteins that interact with homologous eggs, and have revealed a high-density species of satellite DNA that seems to be ribosomal DNA.

There have been more studies on oogenesis than on spermatogenesis. The origin, chemical composition, ultrastructural appearance, and fate of many egg structures - such as the heavy bodies, annulate lamellae, cortical granules, and yolk particles - have been investigated, but their mode of formation and function in oogenesis or early development are still controversial. Nothing is known about such structures as spherical bodies except their morphologic description. The extent to which primary oocytes receive materials from accessory cells or from the interstitial environment, or the nature of such accumulated substances, has not yet been established.

The growing primary oocyte synthesizes mitochondrial DNA, ribosomal, polydisperse, and low-molecular-weight RNAs, mucopolysaccharides, and proteins. The synthesis of large amounts of ribosomal RNA is consistent with the presence of a prominent and continually enlarging nucleolus in the primary oocyte. Protein synthesis continues in the absence of RNA synthesis for at least one day, indicating that messenger RNAs in the oocyte are relatively stable. Ribosomal, polydisperse, and low-molecular-weight RNAs are conserved in the mature egg. The polydisperse RNAs contain messenger RNA. That for histones have been isolated from the mature egg. Protein synthesis experiments after fertilization indicate that tubulin messenger RNA also exists in the ripe egg.

There is very little information about the synthesis of specific proteins and other macromolecules during oogenesis. The most studied macromolecule is that of the jelly coat, a sulfated acidic mucopolysaccharide made near the end of oogenesis. It is presumed that the jelly coat is synthesized by the oocyte itself rather than by accessory ovarian cells. Preliminary experiments have been launched to study histone synthesis during oogenesis. Many important questions concerning protein synthesis during oogenesis remain unanswered. For example, it is not known if there are transient oocyte specific proteins, if different proteins are synthesized at different stages of oogenesis, if the messenger RNAs for all proteins are made immediately before their translation during oogenesis, or if messenger RNAs utilized after fertilization are also translated during oogenesis. The advent of recent techniques for obtaining isolated sea urchin oocytes at different stages of growth give us reason to expect that these and other key problems of oogenesis will soon be approached experimentally.

References

AFZELIUS, B.A., 1955a. The fine structure of the sea urchin spermatozoa as revealed by the electron microscope. Z. Zellforsch. 42, 134-148.

AFZELIUS, B.A., 1955b. The ultrastructure of the nuclear membrane of the sea urchin oocyte as studied with the electron microscope. Exp. Cell Res., 8, 147-158.

AFZELIUS, B.A., 1956a. The ulstrastructure of the cortical granules and their products in the sea urchin egg as studied with the electron microscope. Exp. Cell Res., 10, 257-285.

AFZELIUS, B.A., 1956b. Electron microscopy of Golgi elements in sea urchin eggs. Exp. Cell Res., 11, 67-85.

AFZELIUS, B.A., 1957a. Electron microscopy of the basophilic structures of the sea urchin egg. Z. Zellforsch., 45, 660-675.

AFZELIUS, B.A., 1957b. Electron microscopy on sea urchin gametes. Uppsala: Almquist and Wiksells Baktryckeri AB.

AFZELIUS, B.A., 1959. Electron microscopy of the sperm tail; results obtained with a new fixative. J. Biophys. Biochem. Cytol., 5, 269-278.

AFZELIUS, B.A., MURRAY, A., 1957. The acrosomal reaction of spermatozoa during fertilization or treatment with egg water. Exp. Cell Res., 12, 325-337.

ANDERSON, E., 1968a. Oocyte differentiation in the sea urchin Arbacia punctulata, with particular reference to the origin of cortical granules and their participation in the cortical reaction. J. Cell Biol., 37, 514-539.

ANDERSON, W., 1968b. Cytochemistry of sea urchin gametes. III. Acid and alkaline phosphatase activity of spermatozoa and fertilization. J. Ultrastr. Res., 25, 1-14.

ANDERSON, W., 1969. Nuclear and cytoplasmic DNA synthesis during early embryogenesis of Paracentrotus lividus. J. Ultrastr. Res., 26, 95-110.

ANDRE, J., 1963. Some aspects of specialization in sperm, 91-115. In: The General Physiology of Cell Specialization (D. Mazia and A. Tyler, eds.). New York, San Francisco, Toronto, London: McGraw Hill Book Co., Inc.

AUSTIN, C.R., 1961. The Mammalian Egg. Springfield, Ill.: Charles C. Thomas.

AUSTIN, C.R., 1965. Fertilization. Englewood Cliffs, N.J.: Prentice-Hall, Inc.

BÄCKSTRÖM, S., 1966. A complex between basic proteins and acid polysaccharides in sea urchin oocytes and eggs. Acta Embryol. Morph. Exp., 9, 37-43.

BAL, A.K., JUBINVILLE, F., COUSINEAU, G.H., INOUE, S., 1968. Origin and fate of annulate lamellae in Arbacia punctulata eggs. J. Ultrastr. Res., 25, 15-28.

BEATTY, R.A., 1970. The genetics of the mammalian gamete. Biol. Rev., 45, 73-119.

BENNETT, J., GIESE, A.C., 1955. The annual reproductive and nutritional cycles in two western sea urchins. Biol. Bull., 109, 226-237.

BERGER, E.R., 1968. A quantitative study on sea urchin egg mitochondria in relation to their DNA content. J. Cell Biol., 39, 12a-13a.
BERNSTEIN, M.H., MAZIA, D., 1952. The desoxyribonucleoprotein of sea urchin sperm. 1. Isolation and analysis. Biochim. Biophys. Acta 10, 600-606.
BISHOP, D.W., 1968. Testicular enzymes as fingerprints in the study of spermatogenesis, 261-286. In: Perspectives in Reproduction and Sexual Behavior (M. Diamond, ed.). Bloomington, London: Indiana University Press.
BISHOP, M.W.H., WALTON, A., 1960. Spermatogenesis and the structure of mammalian spermatozoa, 1-129. In: Marshall's Physiology of Reproduction, vol. I, part 2 (A.S. Parkes, ed.). London: Longmans, Green and Co., Ltd.
BLOCH, D.P., 1969. A catalog of sperm histones. Genet. Suppl., 61, 93-111.
BOELL, E., CHAMBERS, J.R., GLANCY, E.A., STERN, K.G., 1940. Oxidase activity and respiration of cells and cell fragments. Biol. Bull., 79, 352.
BOOLOTIAN, R.A., 1960. The effect of temperature on gonadal growth of Strongylocentrotus purpuratus. Anat. Rec., 137, 342-343.
BOOLOTIAN, R.A., 1966. Reproductive physiology, 561-613. In: Physiology of Echinodermata (R.A. Boolotian, ed.). New York, London, Sydney: John Wiley and Sons.
BOOLOTIAN, R.A., MOORE, A.R., 1956. Hermaphroditism in Echinoids. Biol. Bull., 111, 328-335.
BOREI, H., 1948. Respiration of oocytes, unfertilized eggs and fertilized eggs from Psammechinus and Asterias. Biol. Bull., 95, 124-150.
BOREI, H., 1949. Independence of post-fertilization respiration in the sea-urchin egg from the level of respiration before fertilization. Biol. Bull., 96, 117-122.
BOTTICELLI, C.R., HISAW, F.L., Jr., WOTIZ, H.H., 1961. Estrogens and progesterone in the sea urchin (Strongylocentrotus franciscanus) and pecten (Pecten hericius). Proc. Soc. exp. Biol. and Med., 106, 887-889.
BOVERI, T., 1901. Die Polarität von Ovocyte, Ei und Larve des Strongylocentrotus lividus. Zool. Jahrb. Abt. f. Anat., 14, 630-653.
BRACHET, A., 1922. Recherches sur la fécondation prématurée de l'oeuf d'oursin (Paracentrotus lividus). Arch. Biol., 32, 205-248.
BRACHET, J., 1960. Gametogenesis, 1-44. In: Biochemistry of Development. New York, Los Angeles, London, Paris: Pergamon Press.
BRADFIELD, J.R.G., 1953. New features of protoplasmic structure observed in recent electron microscope studies. Quart. J. Microscop. Sci., 94, 351-367.
BROWN, D.D., DAWID, I., 1968. Specific gene amplification in oocytes. Science, 160, 272-280.
BROWN, D.D., LITTNA, E., 1964. Variations in the synthesis of stable RNA's during oogenesis and development of Xenopus laevis. J. Mol. Biol., 8, 688-695.
BRYCE, T.H., 1902. Maturation of the ovum in Echinus esculentus. Quart. J. Micro. Sci., 46, 177-224.
CANDELAS, G.C., IVERSON, R.M., 1966. Evidence for translational level control of protein synthesis in the development of sea-urchin eggs. Biochem. Biophys. Res. Comm., 24, 867-871.
CASPERSSON, T., SCHULTZ, J., 1940. Ribonucleic acids in both nucleus and cytoplasm, and the function of the nucleolus. Proc. Nat. Acad. Sci. U.S., 26, 507-515.
CASTANEDA, M., TYLER, A., 1968. The lability of in vitro amino acid incorporating systems of sea urchin eggs in relation to nuclease release and other factors. Biochim. Biophys. Acta, 166, 741-744.

CAULLERY, M., 1925. Sur la structure et le functionnement des gonades chez le echinides. Trav. Sta. Zool. Wimereux, 9, 21-35.

CELI, A.D., D'ESTE, L., TETI, D., 1967. La vitellogensi in Arbacia lixula e Sphaerechinus granularis (Echin. Echin.). Experientia 23, 433-434.

CHAMBERLAIN, J.P., 1970. RNA synthesis in anucleate fragments and normal embryos of the sea urchin, Arbacia punctulata. Biochim. Biophys. Acta, 213, 183-193.

CHANDLEY, A.C., BATEMAN, A.J., 1962. Timing of spermatogenesis in Drosophila melanogaster using tritiated thymidine. Nature, 193, 299-300.

CHATLYNNE, L.G., 1969. A histochemical study of oogenesis in the sea urchin, Strongylocentrotus purpuratus. Biol. Bull., 136, 167-184.

CLAYBROOK, C.J., BLOCH, D.P., 1967. Synthesis of ribonucleic acid and histone change during spermatogenesis in the grasshopper Chortophaga viridifasciata. Nature, 215, 966-969.

COGNETTI, G., SPINELLI, G., VIVOLI, A., 1974. Synthesis of histones during sea urchin oogenesis. Biochim. Biophys. Acta 349, 447-455.

CONWAY, C.M., 1971. Evidence for RNA in the heavy bodies of sea-urchin eggs. J. Cell Biol., 51, 889-893.

CONWAY, C.M., METZ, C.B., 1970. Cytochemical demonstration of RNA in heavy bodies of sea-urchin eggs. J. Cell Biol., 47, 40a.

CONWAY, C.M., METZ, C.B., 1974. In vitro maturation of Arbacia punctulata oocytes and initiation of heavy body formation. An ultrastructure study. Cell Tiss. Res. 150, 271-279.

COWDEN, R.R., 1962. RNA and yolk synthesis in growing oocytes of the sea urchin, Lytechinus variegatus. Exp. Cell Res., 28, 600-604.

CRAIG, S.P., 1970. Synthesis of RNA in non-nucleate fragments of sea-urchin eggs. J. Mol. Biol., 47, 615-618.

CRAIG, S.P., PIATIGORSKY, J., 1971. Protein synthesis and development in the absence of cytoplasmic RNA synthesis in non-nucleate egg fragments and embryos of sea urchins: Effect of ethidium bromide. Develop. Biol. 24, 214-232.

DAIGNEAULT, R., BELLEMARE, G., COUSINEAU, G.H., 1970. A 12-S and 9-S RNA species in Strongylocentrotus purpuratus eggs. Biochim. Biophys. Acta, 224, 256-259.

DAN, J.C., 1956. The acrosome reaction. Internat. Rev. Cytol., 5, 365-393.

DARNELL, J.E., Jr., 1968. Ribonucleic acids from animal cells. Bacteriol. Rev., 32, 262-290.

DAS, N.K., MICOU-EASTWOOD, J., RAMAMURTHY, G., ALFERT, M., 1970. Sites of synthesis and processing of ribosomal RNA precursors within the nucleolus of Urechis caupo eggs. Proc. Nat. Acad. Sci. U.S., 67, 968-975.

DAS, N.K., SIEGEL, E.P., ALFERT, M., 1965. Synthetic activities during spermatogenesis in the locust. J. Cell Biol., 25, 387-395.

DAVENPORT, R., DAVENPORT, J.C., 1966. A cytochemical study of cytoplasmic basic proteins in echinoderm oogenesis. Exp. Cell Res., 42, 429-437.

DAVIDSON, E.H., 1968. Gene Activity in Early Development. New York, London: Academic Press.

DAWID, I.B., BROWN, D.D., 1970. The mitochondrial and ribosomal DNA components of oocytes of Urechis caupo. Develop. Biol., 22, 1-14.

DELAVAULT, R., 1966. Determinism of sex, 615-638. In: Physiology of Echinodermata (R.A. Boolotian, ed.). New York, London, Sydney: John Wiley and Sons.

DONAHUE, J.K., 1939. Estrogenic properties of extracts of ovaries of certain marine invertebrates. Proc. Soc. exp. Biol. and Med., 42, 220-222.

DONAHUE, J.K., 1940. Occurrence of estrogens in the ovaries of certain marine invertebrates. Endocrinology, 27, 149-152.

DUBOIS, R., DUPRE, M., DENKER, I., INOUE, S., COUSINEAU, G.H., 1971. 12 S and 9 S RNA species. Localization and fate during development of Strongylocentrotus purpuratus eggs. Exp. Cell Res., 68, 197-204.
EASTON, D., CHALKLEY, R., 1972. A high-resolution electrophoretic analysis of the histones from embryos and sperm of Arbacia punctulata. Exp. Cell Res. 72, 502-508.
EPEL, D., 1967. Protein synthesis in sea-urchin eggs: A "late" response to fertilization. Proc. Nat. Acad. Sci. U.S., 57, 899-906.
ERB, V.W., MAURER, W., 1962. Autoradiographische Untersuchungen über den Eiweißstoffwechsel von Oocyten und Eizellen. Z. Naturforsch., 17B, 268-273.
ESPER, H., 1965. Studies on the nucleolar vacuole in the oogenesis of Arbacia punctulata. Exp. Cell Res., 38, 85-96.
FAWCETT, D.W., ANDERSON, W.A., PHILLIPS, D.M., 1971. Morphogenetic factors influencing the shape of the sperm head. Develop. Biol., 26, 220-251.
FICQ, A., 1964. Effets de l'actinomycine D et de la puromycine sur le métabolisme de l'oocyte en croissance. Exp. Cell Res., 34, 581-594.
FRANKLIN, L.E., 1965. Morphology of gamete membrane fusion and of sperm entry into oocytes of the sea urchin. J. Cell Biol., 25, 81-100.
FRANKLIN, L., METZ, C.B., 1962. Electron microscope study of sperm entry into sea urchin oocytes. Biol. Bull., 123, 473.
FUJI, A., 1960a. Studies on the biology of the sea urchin. I. Superficial and histological gonadal changes in gametogenic process of two sea urchins, Strongylocentrotus nudus and S. intermedius. Bull. Fac. Fish. Hokkaido Univ., 11 (1), 1-14.
FUJI, A., 1960b. Studies on the biology of the sea urchin. II. Size at first maturity and sexuality of two sea urchins, Strongylocentrotus nudus and S. intermedius. Bull. Fac. Fish. Hokkaido Univ., 11 (2), 43-48.
FUJI, A., 1960c. Studies on the biology of the sea urchin. III. Reproductive cycle of two sea urchins, Strongylocentrotus nudus and S. intermedius. Bull. Fac. Fish. Hokkaido Univ., 11 (2), 49-57.
GALL, J.C., 1968. Differential synthesis of the genes for ribosomal RNA during amphibian oogenesis. Proc. Nat. Acad. Sci. U.S., 60, 553-560.
GALL, J.G., MACGREGOR, H.C., KIDSTON, M.E., 1969. Gene amplification in the oocytes of dytiscid water beetles. Chromosoma, 26, 169-187.
GIARD, A., 1900. A propos de la parthenogenese artificielle des oeufs d'echinodermes. C.R. Soc. Biol. Paris, 52, 761-764.
GIESE, A.C., 1959. Comparative physiology: annual reproductive cycles of marine invertebrates. A. Rev. Physiol., 21, 547-576.
GIESE, A.C., 1961. Further studies on Allocentrotus fragilis, a deep-sea echinoid. Biol. Bull., 121, 141-150.
GIESE, A.C., 1966. On the biochemical constitution of some echinoderms, 757-796. In: Physiology of Echinodermata (R.A. Boolotian, ed.). New York, London, Sydney: John Wiley and Sons.
GIESE, A.C., GREENFIELD, L., HUANG, H., FARMANFARMAIAN, A., BOOLOTIAN, R., LASKER, R., 1959. Organic productivity in the reproductive cycle of the purple sea urchin. Biol. Bull., 116, 49-58.
GIESE, A.C., KRISHNASWAMY, S., VASU, B.S., LAWRENCE, J., 1964. Reproductive and biochemical studies on a sea urchin, Stomopneustes variolaris, from Madras Harbor. Comp. Biochem. Physiol., 13, 367-380.
GIUDICE, G., PIRONNE, A.M., ROCCHERI, M., TRAPANI, M., 1973. Maturational cleavage of nucleolar ribosomal RNA precursor can be catalyzed by non-specific endonuclease. Biochim. Biophys. Acta 319, 72-80.
GIUDICE, G., SCONZO, G., BONO, A., ALBANESE, I., 1972. Studies on sea urchin oocytes. I. Purification and cell fractionation. Exp. Cell Res. 72, 90-94.

GOLDFARB, A.J., 1935a. Change in size and shape of aging eggs (Arbacia punctulata). Biol. Bull., 68, 180-190.
GOLDFARB, A.J., 1935b. Vicosity changes in aging unfertilized eggs of Arbacia punctulata. Biol. Bull., 68, 191-206.
GOLDFARB, A.J., 1937. Effects of age on Arbacia eggs. VI. Increased stretching and bursting of egg membrane in hypotonic sea water. Physiol. Zool., 10, 59-70.
GRANT, P., 1965. Informational molecules and embryonic development, 483-593. In: The Biochemistry of Animal Development, vol. I (R. Weber, ed.). New York, London: Academic Press.
GREENFIELD, L., GIESE, A.C., FARMANFARMAIAN, A., BOOLOTIAN, R.A., 1958. Cyclic biochemical changes in several echinoderms. J. Exp. Zool., 139, 507-524.
GROSS, P.R., 1968. Biochemistry of differentiation. Ann. Rev. Biochem., 37, 631-660.
GROSS, P.R., COUSINEAU, G.H., 1963. Effects of actinomycin D on macromolecular synthesis and early development in sea-urchin eggs. Biochem. Biophys. Res. Comm., 4, 321-326.
GROSS, P.R., COUSINEAU, G.H., 1964. Macromolecular synthesis and the influence of actinomycin on early development. Exp. Cell Res., 33, 368-395.
GROSS, K.W., JACOBS-LORENA, M., BAGLIONI, C., GROSS, P.R., 1973. Cell-free translation of maternal messenger RNA from sea urchin eggs. Proc. Nat. Acad. Sci. U.S. 70, 2614-2618.
GROSS, P.R., MALKIN, L.I., HUBBARD, M., 1965. Synthesis of RNA during oogenesis in the sea urchin. J. Mol. Biol., 13, 463-481.
GROSS, P.R., MALKIN, L.I., MOYER, W.A., 1964. Templates for the first proteins of embryonic development. Proc. Nat. Acad. Sci. U.S., 51, 407-414.
GROSS, P.R., PHILPOTT, D.E., NASS, S., 1960. Electron microscopy of the centrifuged sea urchin egg, with a note on the structure of the ground cytoplasm. J. Biophys. Biochem. Cytol., 7, 135-142.
HAGSTRÖM, B.E., LÖNNING, S., 1961. Studies on the species specificity of Echinoderms. Sarsia, 4, 5-19.
HAMANN, O., 1887. Beiträge zur Histologie der Echinodermen. Jen. Zeits. N.F., 14, 87-266.
HAMER, D., 1955. The composition of the basic proteins of echinoderm sperm. Biol. Bull., 108, 35-39.
HARRIS, P., 1967. Structural changes following fertilization in the sea-urchin egg. Exp. Cell Res., 48, 569-581.
HARRIS, P., 1969. Relation of fine structure to biochemical changes in developing sea-urchin eggs and zygotes, 315-340. In: The Cell Cycle. Gene-Enzyme Interactions (G.M. Padilla, G.L. Whitson, I.L. Cameron, eds.). New York, London: Academic Press.
HARVEY, E.B., 1956. The American Arbacia and other sea urchins. Princeton, N.J.: Princeton University Press.
HATHAWAY, R.R., METZ, C.B., 1961. Interactions between Arbacia sperm and S^{35}-labeled fertilizin. Biol. Bull., 120, 360-369.
HATHAWAY, R.R., TYLER, A., 1958. Evidence for the splitting-off of S^{35}-labeled sulfate from the fertilizin of Arbacia eggs upon spontaneous reversal of sperm-agglutination. Biol. Bull., 115, 337.
HATHAWAY, R.R., WARREN, L., 1961. Further investigation of egg jelly dispersal by Arbacia sperm extract. Biol. Bull., 121, 416-417.
HENDERSON, S.A., 1963. Differential ribonucleic acid synthesis of X and autosomes during meiosis. Nature, 200, 1235.
HENDERSON, S.A., 1964. RNA synthesis during male meiosis and spermiogenesis. Chromosoma, 15, 345-366.
HILLE, M.B., 1974. Inhibitor of protein synthesis isolated from ribosomes of unfertilized eggs and embryos of sea urchins. Nature 249, 556-558.

HINEGARDNER, R.T., 1969. Growth and development of the laboratory cultured sea urchin. Biol. Bull., 137, 465-475.
HNILICA, L.S., 1967. Proteins of the cell nucleus. Prog. Nuc. Acid. Res. Mol. Biol., 7, 25-106.
HOADLEY, L., 1934. The relation between the position of the female pronucleus and the polar bodies in the unfertilized egg of Arbacia punctulata. Biol. Bull., 67, 220-222.
HOLLAND, N.D., 1967. Gametogenesis during the annual reproductive cycle in a cidaroid sea urchin (Stylocidaris affinis). Biol. Bull., 133, 578-590.
HOLLAND, N.D., GIESE, A.C., 1965. An autoradiographic investigation of the gonads of the purple sea urchin (Strongylocentrotus purpuratus). Biol. Bull., 128, 241-258.
HÖRSTADIUS, S., 1939. The mechanics of sea urchin development, studied by operative methods. Biol. Rev., 14, 132-179.
HSU, W.S., 1967. The origin of annulate lamellae in the oocyte of the ascidian, Boltenia villosa Stimpson. Z. Zellforsch. Microskop. Anat., 82, 376-390.
HULTIN, T., 1961. Activation of ribosomes in sea-urchin eggs in response to fertilization. Exp. Cell Res., 25, 405-417.
HULTIN, T., 1964. On the mechanism of ribosomal activation in newly fertilized sea-urchin eggs. Develop. Biol., 10, 305-328.
HULTIN, T., HERNE, R., 1959. Amino acid analysis of a basic protein fraction from sperm nuclei of some different invertebrates. Ark. Kemi, 26A (2), 1-8.
HYMAN, L.H., 1955. The Invertebrates: Echinodermata, vol. IV. New York, Toronto, London: McGraw Hill Book Co., Inc.
IMMERS, J., 1960. Studies on cytoplasmic components of sea-urchin eggs stratified by centrifugation. Exp. Cell Res., 19, 499-514.
IMMERS, J., 1961. Comparative study of the localization of incorporated ^{14}C-labeled amino acids and $^{35}SO_4$ in the sea urchin ovary, egg and embryo. Exp. Cell Res., 24, 356-378.
INOUE, S., COUSINEAU, G.H., KRUPA, P.L., 1971. Jelly coat material of Arbacia eggs: an electron diffraction study. Biol. Bull., 141, 391.
ISAKA, S., KANATANI, H., SUZUKI, N., 1966. Jelly dispersing enzyme obtained from spermatozoa of sea urchin, Anthocidaris crassispina. Exp. Cell Res., 44, 66-72.
IWATA, K., 1950. A method of determining the sex of sea urchin and of obtaining eggs by electric stimulation. Annot. Zool. Jap., 23, 39-42.
JENKINSON, J.W., 1911. On the origin of the polar and bilateral structure of the egg of the sea urchin. Arch. f. Entw. Mech., 32, 699-716.
KESSEL, R.G., 1963. Electron microscope studies on the origin of annulate lamellae in oocytes of Necturus. J. Cell Biol., 19, 391-414.
KESSEL, R.G., 1964. Electron microscope studies on oocytes of an echinoderm, Thyone briareus, with species reference to the origin and structure of annulate lamellae. J. Ultrastr. Res., 10, 498-514.
KESSEL, R.G., 1965. Intranuclear and cytoplasmic annulate lamellae in tunicate oocytes. J. Cell Biol., 24, 471-487.
KOSSEL, A., 1928. The Protamines and Histones. London, New York: Longmans Green and Co.
KOSSEL, R., STAUDT, W., 1926. Zur Kenntnis der basischen Proteine. Z. Physiol. Chem., 159, 172-173.
KUROZUMI, K., 1957. Electron microscopic investigations on the fine structure of egg cells. I. On resting cell structure; II. On the mitotic figures. Acta Anat. Nipponica, 32, 175-226.
LaMARCA,. M.J., SHIPPEE, E.L., SCHUETZ, A.W., 1971. RNA synthesis during oocyte maturation in Arterias forbesi. Biol. Bull., 141, 394.
LASKER, R., GIESE, A.C., 1954. Nutrition of the sea urchin, Strongylocentrotus purpuratus. Biol. Bull., 106, 328-340.

LIEBMAN, E., 1950. The leucocytes of Arbacia punctulata. Biol. Bull., 98, 46-59.
LIMA-de-FARIA, A., BIRNSTIEL, M., JAWORSKA, H., 1969. Amplification of ribosomal cistrons in the heterochromatins of Acheta. Genetics, 61, Suppl., 145-159.
LINDAHL, P.E., 1932. Zur Kenntnis des Ovarialeies bei dem Seeigel. Arch. f. Entw., 126, 373-390.
LINDAHL, P.E., HOLTER, H., 1941. Über die Atmung der Oxozyten erster Ordnung von Paracentrotus lividus und ihre Veränderung während der Reifung. C. R. Tr. Lab. Carlsberg Ser. Chim., 24, 49-57.
LITCHFIELD, J.B., WHITELEY, A.H., 1959. Studies on the mechanism of phosphate accumulation by sea urchin embryos. Biol. Bull., 117, 133-149.
LONGO, F.J., ANDERSON, E., 1969. Sperm differentiation in the sea urchins Arbacia punctulata and Strongylocentrotus purpuratus. J. Ultrastr. Res., 27, 486-509.
LÖNNING, S., 1963. Electron microscopical studies of oocytes and eggs from the sea urchin Brissopsis lyrifera. Sarsia, 11, 11-16.
LÖNNING, S., 1964. Studies on the ultrastructure of sea urchin oocytes and the changes induced by insemination. Sarsia, 15, 9-15.
MACBRIDE, E.W., 1903. The development of Echinus esculentus together with some points in the development of E. miliaris and E. acutus. Roy. Soc. London, Phil. Trans. B 195, 285-327.
MACKINTOSH, F.R., BELL, E., 1967. Stimulation of protein synthesis in unfertilized sea urchin eggs by prior metabolic inhibition. Biochem. Biophys. Res. Comm., 27, 425-430.
MACKINTOSH, F.R., BELL, E., 1970. Reversible response to puromycin and some characteristics of the uptake and use of amino acids by unfertilized sea-urchin eggs. Biol. Bull., 139, 296-303.
MACKINTOSH, F.R., BELL, E., 1969. Regulation of protein synthesis in sea-urchin eggs. J. Mol. Biol., 41, 365-380.
MADEN, B.E.H., 1968. Ribosome formation in animal cells. Nature, 219, 685-689.
MAKINO, S., 1951. An atlas of the chromosome numbers in animals. Ames, Iowa: Iowa State College Press.
MANO, Y., NAGANO, H., 1966. Release of maternal RNA from some particles as a mechanism of activation of protein synthesis by fertilization in sea-urchin eggs. Biochim. Biophys. Res. Comm., 25, 210-215.
MATSUMOTO, L., KASAMATUSU, H., PIKO, L., VINOGRAD, J., 1974. Mitochondrial DNA replication in sea urchin oocytes. J. Cell Biol. 63, 146-159.
MATSUMOTO, L., PIKO, L., 1971. In vivo radioactive labeling of mitochondrial DNA in Arbacia punctulata oocytes. Biol. Bull., 141, 397.
McCULLOCH, D., 1952. Note on the origin of the cortical granules in Arbacia punctulata eggs. Exp. Cell Res., 3, 605-607.
MERCER, E.H., WOLPERT, L., 1962. An electron microscope study in the cortex of the sea urchin (Psammechinus miliaris) egg. Exp. Cell Res., 27, 1-13.
MERRIAM, R.W., 1959. The origin and fate of annulate lamellae in maturing sand dollar eggs. J. Biophys. Biochem. Cytol., 5, 117-121.
MILLER, R.A., SMITH, H.B., 1931. Observations on the formation of the egg of Echinometra lucunter. Paps. Tortugas Lab. Carnegie Inst. Washington, 27 (413), 47-52.
MILLONIG, G., 1958. Die submikroskopische Struktur des Dotterkernes in den Eiern von Arbacia lixula. Mikroskopie, 13, 239-246.
MILLONIG, G., 1967. The morphological changes of the nucleoli during oogenesis and embryogenesis. Atti Acc. Naz. Lincei, Classe Scien. Fis. Mat. Nat., 104, 113-115.
MILLONIG, G., BOSCO, M., GIAMBERTONE, L., 1968. Fine structure analysis of oogenesis in sea urchins. J. Exp. Zool., 169, 293-314.

MINGANTI, A., 1954. Studies on the surface layers of oocytes and mature eggs of Psammechinus miliaris. Exp. Cell Res., 7, 1-14.
MITCHISON, J.M., CUMMINS, J.E., 1966. The uptake of valine and cytidine by sea urchin embryos and its relation to the cell surface. J. Cell Sci., 1, 35-47.
MONESI, V., 1965a. Differential rate of ribonucleic acid synthesis in the autosomes and sex chromosomes during male meiosis in the mouse. Chromosoma, 17, 11-21.
MONESI, V., 1965b. Synthetic activities during spermatogenesis in the mouse. Exp. Cell Res., 39, 197-224.
MONNE, L., 1945. Investigations into the structure of the cytoplasm. Arkiv. f. Zool., 36A (23), 1-28.
MONNE, L., HÅRDE, S., 1951. On the cortical granules of the sea-urchin egg. Arkiv. f. Zool. (2) 1, 487-498.
MONNE, L., SLAUTTERBACK, D.B., 1950. Differential staining of various polysaccharides in sea-urchin eggs. Exp. Cell Res., 1, 477-491.
MONROY, A., 1948. Cortical changes accompanying maturation in sea-urchin egg. Experientia, 4, 353-355.
MONROY, A., 1965. Chemistry and Physiology of Fertilization. New York, Chicago, San Francisco, Toronto, London: Holt, Rinehart and Winston.
MONROY, A., MAGGIO, R., RINALDI, A.M., 1965. Experimentally induced activation of the unfertilized sea-urchin egg. Proc. Nat. Acad. Sci. U.S., 54, 107-111.
MOORE, H.B., 1935. The biology of Echinocardium cordatum. J. Mar. Biol. Assoc., 20, 655-671.
MOORE, H.B., JUTARE, T., BAUER, J.C., JONES, J.A., 1963a. The biology of Lytechinus variegatus. Bull. Mar. Sci. Gulf Caribb., 13, 23-53.
MOORE, H.B., JUTARE, T., JONES, J.A., McPHERSON, B.F., ROPER, C.F.E., 1963b. A contribution to the biology of Tripneustes esculentus. Bull. Mar. Sci. Gulf Caribb., 13, 267-281.
MOTOMURA, I., 1936. Notes on the cytoplasmic structure of the egg of a sea urchin Strongylocentrotus pulcherrimus. Zool. Mag., 48, 753-758.
MOTOMURA, I., 1941. Materials on the fertilization membrane in the eggs of Echinoderms. Sci. Rep. Tohoku Imp. Univ. IV B 16, 345-363.
MOTOMURA, I., 1960. On the structure of the cortical granule in the sea-urchin egg. Sci. Rep. Tohoku Univ. 4th Ser. (Biol.), 26, 367-374.
MUCKENTHALER, F.A., 1964. Autoradiographic study of nucleic acid synthesis during spermatogenesis in the grasshopper, Melanoplus differentialis. Exp. Cell Res., 35, 531-547.
NEEFS, Y., 1937. Sur divers cas d'hermaphroditisme chez Arbacia lixula. C. R. Acad. Sci., 204, 900-902.
NEEFS, Y., 1952. Sur le cycle sexuel de Sphaerechinus granularis. C. R. Acad. Sci., 234, 2233-2235.
NELSON, L., 1971. Differentiation of the male germ cell, 243-278. In: Developmental Aspects of the Cell Cycle (I.L. Cameron, G.M. Padilla, E.F. Zimmerman, eds.). New York, London: Academic Press.
NICANDER, L., PLOËN, L., 1969. Fine structure of spermatogonia and primary spermatocytes in rabbits. Z. Zellforsch., 99, 221-234.
NIGON, V., GILLOT, S., 1963. Etude radiographique du metabolisme des acides nucleiques et des proteines au cours du developpement de l'oeuf Arbacia lixula. Cahiers Biol. Mar., 4, 277-298.
NISHIKAWA, S., 1961. Notes on the chromosomes of two species of echinoderms Hemicentrotus pulcherrimus and Anthocidaris crassispina (A. Agassiz). Zool. Mag. Tokyo, 70, 425-428.
OHNISHI, T., SUGIYAMA, M., 1963. Polarographic studies of oxygen uptake of sea-urchin eggs. Embryologia, 8, 79-88.
OSANAI, K., 1964. Cortical granule as a body containing cyto-fertilizin

in the sea-urchin egg. Sci. Rep. Tohoku Univ. 4th Ser. (Biol.), 30, 119-131.
OZAKI, H., 1971. Developmental studies of sea urchin chromatin. Chromatin isolated from spermatozoa of the sea urchin Strongylocentrotus purpuratus. Develop. Biol., 26, 209-219.
OZAKI, H., PIATIGORSKY, J., TYLER, A., 1966. Electrophoretic patterns of soluble proteins synthesized during oogenesis and early development of sea urchin. Ann. Rep., Calif. Inst. of Tech., 44-45.
PALAU, J., RUIZ-CARRILLO, A., SUBIRANA, J.A., 1969. Histones from the sperm of the sea urchin Arbacia lixula. Europ. J. Biochem., 7, 209-213.
PALMER, L., 1937. The shedding reaction in Arbacia punctulata. Physiol. Zool., 10, 352-367.
PAOLETTI, R.A., HUANG, R.C., 1969. Characterization of sea urchin sperm chromatin and its basic proteins. Biochem., 8, 1615-1625.
PARAT, M., 1928. Contribution à l'etude morphologique et physiologique du cytoplasme. Chondriome, vacuome (appareil de Golgi) enclaves, etc., pH, oxydases, rH de la cellule animale. Arch. Anat. Microsc., 24, 73-357.
PASPALEFF, G., 1927. Über Protoplasmareifung bei Seeigeleiern. Pubb. Staz. Zool. Napoli, 8, 1-70.
PASTEELS, J.J., CASTIAUX, P., VANDERMEERSSCHE, G., 1958. Ultrastructure du cytoplasme et distribution de l'acide ribonucleique dans l'oeuf fecondé tant normal que centrifugé de Paracentrotus lividus. Arch. Biol., 69, 627-643.
PATTERSON, J.B., STAFFORD, D.W., 1970. Sea urchin satellite deoxyribonucleic acid. Its large scale isolation and hybridization with homologous ribosomal ribonucleic acid. Biochem., 9, 1278-1283.
PEARSE, J.S., GIESE, A.C., 1966. Food, reproduction and organic constitution on the common Antarctic echinoid Sterechinus neumayeri (Meissner). Biol. Bull., 130, 387-401.
PERLMAN, P., 1956. Responses of unfertilized sea-urchin eggs to antiserum. Exp. Cell Res., 10, 324-353.
PERLMAN, P., BOSTROM, H., VESTERMAK, A., 1959. Sialic acids in the gametes of the sea urchin. Exp. Cell Res., 17, 439-446.
PERRY, R.P., 1965. The nucleolus and the synthesis of ribosomes. Nat. Cancer Inst. Monograph, 18, 325-340.
PERRY, R.P., 1967. The nucleolus and the synthesis of ribosomes. Prog. Nuc. Acid Res., 6, 219-257.
PIATIGORSKY, J., 1965. Labeled uridine uptake by maturing oocytes in sea urchins. Ann. Rep., Calif. Inst. of Tech., p. 65.
PIATIGORSKY, J., 1968. Ribonuclease and trypsin treatment of ribosomes and polyribosomes from sea-urchin eggs. Biochim. Biophys. Acta, 166, 142-155.
PIATIGORSKY, J., OZAKI, H., TYLER, A., 1967. RNA- and protein-synthesizing capacity of isolated oocytes of the sea urchin Lytechinus pictus. Develop. Biol., 15, 1-22.
PIATIGORSKY, J., TYLER, A., 1967. Radioactive labeling of RNAs of sea-urchin eggs during oogenesis. Biol. Bull., 133, 229-244.
PIATIGORSKY, J., TYLER, A., 1970. Changes upon fertilization in the distribution of RNA-containing particles in sea-urchin eggs. Develop. Biol., 21, 13-28.
PIATIGORSKY, J., WHITELEY, A.H., 1965. A change in permeability and uptake of ^{14}C uridine in response to fertilization in Strongylocentrotus purpuratus eggs. Biochim. Biophys. Acta, 108, 404-418.
PIKO, L., BLAIR, D.G., TYLER, A., VINOGRAD, J., 1968. Cytoplasmic DNA in the unfertilized sea-urchin egg: physical properties of circular mitochondrial DNA and the occurrence of catenated forms. Proc. Nat. Acad. Sci. U.S., 59, 838-845.
PIKO, L., TYLER, A., VINOGRAD, J., 1967. Amount, location, priming

capacity, circularity and other properties of cytoplasmic DNA in sea-urchin eggs. Biol. Bull., 132, 68-90.
RAFF, R.A., COLOT, H.V., SELVIG, S.E., GROSS, P.R., 1972. Oogenetic origin of messenger RNA for embryonic synthesis of microtubule proteins. Nature, 235, 211-214.
RAFF, R.A., GREENHOUSE, G., GROSS, K.W., GROSS, P.R., 1971. Synthesis and storage of microtubule proteins by sea urchin embryos. J. Cell Biol., 50, 516-527.
RAVEN, C.P., 1961. Oogenesis: The storage of developmental information. New York, Los Angeles, London, Paris: Pergamon Press.
REBHUN, L.I., 1956. Electron microscopy of basophilic structures of some invertebrate oocytes. I. Periodic lamellae and the nuclear envelope. J. Biophys. Biochem. Cytol., 2, 93-104.
REVERBERI, G., 1940. Su due case ermafroditismo. Boll. Zool., 11, 11-15.
REVERBERI, G., 1947. Ancora null' ermafroditismo nei ricci di mare. Boll. Zool., 14, 65-73.
ROOSEN-RUNGE, E.C., 1962. The process of spermatogenesis in mammals. Biol. Rev., 37, 343-377.
ROTHSCHILD, L., SWANN, M.M., 1949. The fertilization reaction in the sea-urchin egg. A propagated response to sperm attachment. J. Exp. Biol., 26, 164-176.
RUNNSTRÖM, J., 1948. Membrane formation in different stages of cytoplasmic maturation of the sea-urchin egg. Arkiv. f. Zool., 40A (19), 1-6.
RUNNSTRÖM, J., 1949. The mechanics of fertilization in metazoa. Advan. Enzymology, 9, 241-327.
RUNNSTRÖM, J., 1963. Sperm-induced protrusions in sea-urchin oocytes: A study of phase separation and mixing in living cytoplasm. Develop. Biol., 7, 38-50.
RUNNSTRÖM, J., 1964. On some properties of the jelly coat in oocytes and mature eggs of sea urchins. A study of phase-dependent changes of metaplastic layers in the cell surface. Biol. Bull., 127, 132-153.
RUNNSTRÖM, J., 1966. The vitelline membrane and cortical particles in sea-urchin eggs and their function in maturation and fertilization. Advan. Morphogenesis, 5, 221-325.
RUNNSTRÖM, J., KRISZAT, G., 1950. Action of periodate on the surface reactions attending the activation of the egg of the sea urchin, Psammechinus miliaris. Exp. Cell Res., 1, 355-370.
RUNNSTRÖM, J., MONNE, L., 1945. On some properties of the surface layers of immature and mature sea-urchin eggs, especially the changes accompanying nuclear and cytoplasmic maturation. Arkiv. f. Zool., 36A (18), 1-27.
RUNNSTRÖM, J., MONNE, L., WICKLUND, E., 1946. Studies on the surface layers and the formation of the fertilization membrane in sea-urchin eggs. J. Colloid Sci., 1, 421-452.
RUSSO, A., 1894. Sul sistema genitale e madreporico degli Echinidi regolari. Boll. Soc. Nat. Napoli, 8, 90-109.
RUSSO, G., 1926. Ricerche sulla costituzione chimica delle ghiandole genitali. I. Gli aminoacidi del testiculo di Strongylocentrotus lividus, in diversi periodi del circlo funzionale dell' organo. Archo. Sci. Biol., 8, 161-181.
SAKAI, T., 1956a. Cytological studies on the oogenesis of sea urchin: Report I. On the granular changes by using nuclear fixatives. Bull. Exp. Biol., 6, 87-96.
SAKAI, T., 1956b. Cytological studies on the oogenesis of sea urchin: Report II. On the granular changes fixed with plasma fixatives. Bull. Exp. Biol., 6, 157-164.
SAKAI, T., 1956c. Cytological studies on the oogenesis of sea urchin: Report III. On the cytochemical observations. Bull. Exp. Biol., 6, 165-174.

SANCHEZ, S., 1966. Inhibition du developpement de l'oeuf d'oursin apres action de l'actinomycine D sur l'ovocyte. Compt. Rend. Soc. Biol., 160, 143-145.
SANCHEZ, S., 1968a. Effets de l'actinomycine D sur les constituents cellulaires et la metabolisme de l'ARN de l'ovocyte d'oursin (Paracentrotus lividus). Exp. Cell Res., 50, 19-31.
SANCHEZ, S., LANET, A., 1966. Distribution des proteines basiques nucleaires et cytoplasmiques dans l'ovocyte en croissance de Paracentrotus lividus. Compt. Rend. Soc. Biol., 160, 145-147.
SCONZO, G., BONO, A., ALBANESE, I., GIUDICE, G., 1972. Studies on sea urchin oocytes. II. Synthesis of RNA during oogenesis. Exp. Cell Res. 72, 95-100.
SEIFRIZ, W., 1926. Protoplasmic papillae of Echinarachimus oocytes. Protoplasma 1, 1-14.
SIEKEVITZ, P., MAGGIO, R., CATALANO, C., 1966. Some properties of a rapidly labeled ribonucleic acid species in Sphaerechinus granularis. Biochim. Biophys. Acta, 129, 145-156.
SKOULTCHI, A., GROSS, P.R., 1973. Maternal histone messenger RNA: Detection by molecular hybridization. Proc. Nat. Acad. Sci. U.S. 70, 2840-2844.
SLATER, D.W., SPIEGELMAN, S., 1966. An estimation of genetic messages in the unfertilized echinoid egg. Proc. Nat. Acad. Sci. U.S. 56, 164-170.
SLATER, D.W., SPIEGELMAN, S., 1970. Transcriptive expression during sea urchin embryogenesis. Biochim. Biophys. Acta, 213, 194-207.
SPIRIN, A.S., 1966. On "masked" forms of messenger RNA in early embryogenesis and in other differentiating systems, 1-38. In: Current Topics of Developmental Biology, vol. 1 (A.A. Moscona, A. Monroy, eds.). New York, London: Academic Press.
SPIRIN, A.S., NEMER, M., 1965. Messenger RNA in early sea urchin embryos: Cytoplasmic particles. Science, 150, 214-217.
SPONAR, J., SORMOVA, Z., GLISIN, V., 1970. Sea urchin satellite DNA. Its isolation, some properties and relative amount at different stages of sea urchin development. FEBS Letters, 11, 254-256.
STAFFORD, D.W., GUILD, W.R., 1969. Satellite DNA from sea urchin sperm. Exp. Cell Res., 55, 347-350.
STAVY, L., GROSS, P.R., 1967. The protein-synthetic lesion in unfertilized eggs. Proc. Nat. Acad. Sci. U.S., 57, 735-742.
STOTT, F.C., 1931. The spawning of Echinus esculentus and some changes in gonad composition. J. Exp. Biol., 8, 133-150.
SUBRAMANIAM, M.K., 1934. The oogenesis of Salmacis bicolor (Agassiz) with a suggestion as to the function of Golgi bodies. Proc. Indian Acad. Sci., 1, 291-317.
SUGIYAMA, M., 1938. Experiments on the formation of protoplasmic protrusions from sea urchin oocytes. J. Fac. Sci. Imp. Univ. Tokyo Sect. IV 5, 127-131.
SWANSON, R.F., DAWID, I.B., 1970. The mitochondrial ribosome of Xenopus laevis. Proc. Nat. Acad. Sci. U.S., 66, 117-124.
SWIFT, H., 1956. The fine structure of annulate lamellae. J. Biophys. Biochem. Cytol., 2, 415-418.
TAKASHIMA, Y., 1960. Studies on the ultrastructure of the cortical granules in sea-urchin eggs. Tokushima J. exp. Med., 6, 341-349.
TAKASHIMA, Y., TAKASHIMA, R., 1966. Electron microscope investigations on the modes of yolk and pigment formation in sea urchin oocytes. Okajimas Fol. anat. jap., 42, 249-264.
TALEPOROS, P.P., 1959. Cytoplasmic 'histones' and 'protomines' in the egg of the sea urchin Strongylocentrotus purpuratus. J. Histochem. Cytochem., 7, 322.
TENNENT, D.H., GARDINER, M.S., SMITH, D.E., 1931. A cytological and biochemical study of the ovaries of the sea urchin Echinometra

lucunter. Paps. Tortugas Lab. Carnegie Inst., Washington, 27 (413), 1-46.
TENNENT, D.H., ITO, I., 1941. A study of the oogenesis of Mespilia globulus (Linné). J. Morph., 69, 347-404.
TENNENT, D.H., TAYLOR, C.V., WHITAKER, D.M., 1929. An investigation on organization in a sea-urchin egg. Paps. Tortugas Lab. Carnegie Inst., Washington, 26 (391), 1-104.
TILNEY, L.G., HATANO, S., ISHIKAWA, H., MOOSEKER, M.S., 1973. The polymerization of actin: It's role in the generation of the acrosomal process of certain echinoderm sperm. J. Cell Biol. 59, 109-126.
TIMOURIAN, H., 1967. Protein synthesis in sea-urchin eggs. I. Fertilization induced changes in subcellular fractions. Develop. Biol., 16, 594-611.
TYLER, A., 1948. Fertilization and immunity. Physiol. Rev., 28, 180-219.
TYLER, A., 1949a. Properties of fertilizin and related substances of eggs and sperm of marine animals. Am. Nat., 83, 195-219.
TYLER, A., 1949b. A simple, non-injurious method for inducing repeated spawning of sea urchins and sand dollars. Collect. Net, 19, 19-20.
TYLER, A., 1955. Gametogenesis, fertilization and parthenogenesis, 170-212. In: Analysis of Development (B.A. Willier, P.A. Weiss, V. Hamburger, eds.). Philadelphia: W.B. Saunders Co.
TYLER, A., 1956. Physico-chemical properties of the fertilizins of the sea urchin Arbacia punctulata and the sand dollar Echinarachinus parma. Exp. Cell Res., 10, 377-386.
TYLER, A., 1967. Masked messenger RNA and cytoplasmic DNA in relation to protein synthesis and processes of fertilization and determination in embryonic development. Develop. Biol. Suppl., 1, 170-226.
TYLER, A., HATHAWAY, R.R., 1958. Production of S^{35}-labeled fertilizin in eggs of Arbacia punctulata. Biol. Bull., 115, 369.
TYLER, A., PIATIGORSKY, J., OZAKI, H., 1966. Influence of individual amino acids on uptake and incorporation of valine, glutamic acid and arginine by unfertilized and fertilized sea-urchin eggs. Biol. Bull., 131, 204-217.
TYLER, A., RICCI, N., HOROWITZ, N.H., 1938. The respiration and fertilizable life of Arbacia eggs under sterile and non-sterile conditions. J. Exp. Zool., 79, 129-143.
TYLER, A., TYLER, B.S., 1966a. The gametes; some procedures and properties, 639-682. In: Physiology of Echinodermata (R.A. Boolotian, ed.). New York, London, Sydney: John Wiley and Sons.
TYLER, A., TYLER, B.S., 1966b. Physiology of fertilization and early development, 683-741. In: Physiology of Echinodermata (R.A. Boolotian, ed.). New York, London, Sydney: John Wiley and Sons.
TYLER, A., TYLER, B.S., PIATIGORSKY, J., 1969. Protein synthesis by unfertilized eggs of sea urchins. Biol. Bull., 134, 209-219.
VASSEUR, E., 1952. The chemistry and physiology of the jelly coat of the sea-urchin egg. Stockholm, Kihlströms-Tryck AB 1-32.
VENDRELY, R., 1957. Données récentes sur la chimie de l'ADN et des desoxyribonucleoproteines. Archiv des Julius Klaus-Stift. Vererbungsforsch. Sozialanthropol. u. Rassenhyz., 32, 538-553.
VERHEY, C.A., MOYER, F.H., 1967a. Fine structural changes during sea urchin oogenesis. J. Exp. Zool., 164, 195-226.
VERHEY, C.A., MOYER, F.H., 1967b. The role of accessory cells in sea urchin oogenesis. Am. Zool., 7, 754.
VESCO, C., PENMAN, S., 1969. The cytoplasmic RNA of Hela cells: New discrete species associated with mitochondria. Proc. Nat. Acad. Sci. U.S., 62, 218-225.
VINCENT, W.S., 1955. Structure and chemistry of nucleoli. Internat. Rev. Cytol., 4, 269-298.

VINCENT, W.S., HALVORSON, H.O., CHEN, H.R., SHIN, D., 1968. Ribosomal RNA cistrons in single and multinucleolate oocytes. Biol. Bull., 135, 441.
WARBURG, O., 1908. Beobachtungen über die Oxydationsprozesse im Seeigelei. Hoppe-Seylers Z. f. Physiol. Chem., 57, 1-16.
WHITELEY, A.H., 1949. The phosphorus compounds of sea urchin eggs and the uptake of radio-phosphate upon fertilization. Am. Nat., 83, 249-267.
WICKLUND, E., 1947. The action of clupein on the unfertilized sea-urchin eggs and its influence on the fertilization of these eggs. Arkiv f. Zool., 40A (5), 1-18.
WICKLUND, E., 1954. The influence of some substances on the rate of smoothing on non-fertilized sea-urchin eggs in hypertonic medium. Arkiv. f. Zool., (2) 7, 97-108.
WICKLUND, E., KRISZAT, G., RUNNSTRÖM, J., 1953. Influence of certain substances on the fluidity of the cytoplasm and the fertilizability of the egg of the sea urchin, Psammechinus miliaris. J. Embryol. exp. Morph., 1, 319-325.
WILLIAMS, J., 1965. Chemical constitution and metabolic activities of animal eggs, 13-71. In: The Biochemistry of Animal Development, vol. I (R. Weber, ed.). New York, London: Academic Press.
WILSON, E.B., 1895. An Atlas on the Fertilization and Karyokinesis of the Ovum. New York, London: MacMilland and Co.
WILSON, E.B., 1928. The Cell in Development and Heredity. New York: MacMilland and Co.
WILSON, L.P., 1940. Histology of the gonad wall of Arbacia punctulata. J. Morph., 66, 463-479.
WILT, F.H., 1970. The acceleration of ribonucleic acid synthesis in cleaving sea-urchin embryos. Develop. Biol., 23, 444-455.
WOLPERT, L., MERCER, E.H., 1961. An electron microscope study of fertilization of the sea-urchin egg Psammechinus miliaris. Exp. Cell Res., 22, 45-55.
YASUMASU, I., NAKANO, I., 1963. Respiratory level of sea-urchin eggs before and after fertilization. Biol. Bull., 125, 182-187.
YOSHIDA, M., 1952. Some observations on the maturation of the sea urchin, Diadema setosum. Annotnes Zool. Japan, 25, 265-271.
ZALOKAR, M., 1965. Etude de la formation de l'acide ribonucleique et des proteines chez les insectes. Rev. Suisse Zool., 72, 241-262.
ZEITZ, L., FERGUSON, R., GARFINKEL, E., 1969. Incorporation of ^{3}H-TdR and ^{3}H-BUdR into sperm DNA of Arbacia punctulata after coelomic injection. Exp. Cell Res., 56, 159-163.

4. Fertilization

M. Ishikawa

Fertilization takes place when the spermatozoon activates the egg and when the maternal and paternal hereditary factors combine. O. HERTWIG (1876) ascribed the egg activation in fertilization to the fusion of gamete nuclei (syngamy). BOVERI (1887) discussed this fusion from a cytologic point of view and proved it to be inconsistent with the phenomena of parthenogenesis, partial fertilization, merogony, and gynogenesis, in none of which does activation involve the process of syngamy. He found that what is essential to the fertilization phenomenon is the introduction and subsequent behavior of the sperm centriole.

WILSON (1925) proposed that "physiological and cytological aspects of fertilization involve a series of problems of which the most important are: (1) the general conditions for fertilization; (2) the union of the gametes and the fertilization-reaction; (3) the behavior of the gamete-nuclei, and the origin of the chromosomes; (4) the origin and further development of the sperm-center, cleavage-centers and the associated cytoplasmic structures; and (5) the formation of mitochondria."

Today one may define fertilization as the series of processes by which the spermatozoon initiates and participates in the development of the egg, including all the steps from the approach of the spermatozoon to the fusion of the pronuclei within the egg. These are the events that we shall consider here.

4.1. Activity of Spermatozoa

Sea-urchin spermatozoa move in vigorous, counter clockwise circles until they make contact with the egg (LILLIE, 1919), and one of them normally enters (YANAGIMACHI, 1953). The fertilization membrane elevated, the supernumerary spermatozoa gradually diminish their activity on the egg surface (Fig. 4.1).

In a calcium-free medium in which no fertilization can occur, no boring motion takes place. When a little calcium is added, spermatozoa start boring and fertilization begins (Fig. 4.2). YANAGIMACHI (1953) concluded that the active boring movement of the spermatozoa is prerequisite to fertilization of the sea-urchin egg and that calcium ions influence boring. Today it is generally accepted that the spermatozoon does not bore its way into the egg by the lash-like movement of the flagellum.

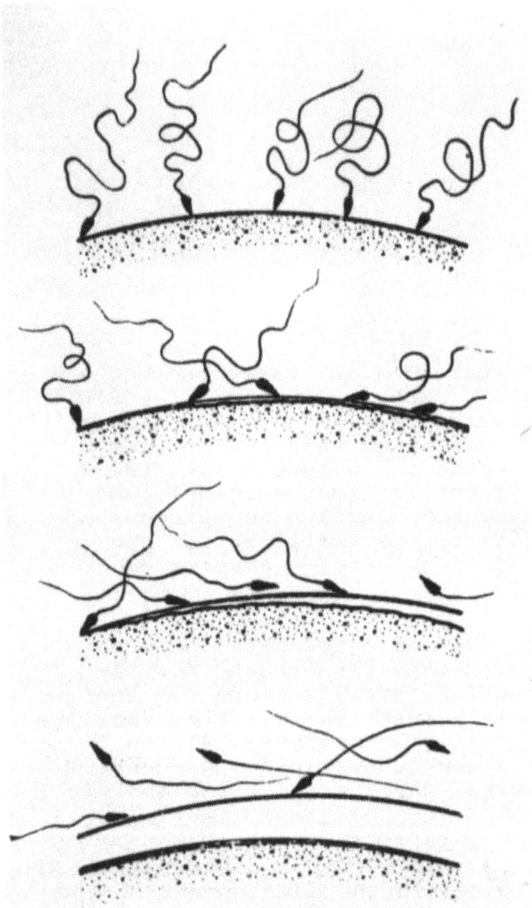

Fig. 4.1. Boring motion of supernumerary spermatozoa decreases when fertilization membrane is lifted. (From YANAGIMACHI, 1953)

4.2 Acrosome Reaction

In the tip of the sea-urchin spermatozoon, RETZIUS (1910) noticed a small particle which BOWEN (1924) suggested might be secreted by the acrosome. The acrosome - or apical body of the spermatozoon - is sometimes called the perforatorium because of the former belief that it was used for boring into the egg. POPA (1927) was the first to report that "in fresh preparations of spermatozoa of *Arbacia* in egg-water, elimination of a granular substance through the pointed apex of the spermatozoon can be observed quite clearly. ... The adhesion of the spermatozoa to one another or other objects is brought about by this granular substance. ... Very probably, under the influence of fertilizin or some substance dissolved in 'egg-water', spermatozoa eliminate a sticky substance through the micropyle in the apex, which as long as it still adheres to the apex of the spermatozoa keeps them agglutinated. ... After it is lost from the tip, the spermatozoa again spread out in the fluid."

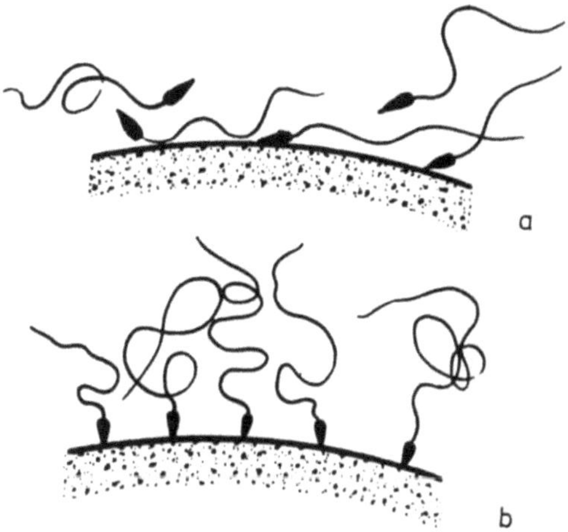

Fig. 4.2a and b. Activating effect of calcium on spermatozoa. (a) Calcium-deficient condition. (b) Just after adding calcium. (From YANAGIMACHI, 1953)

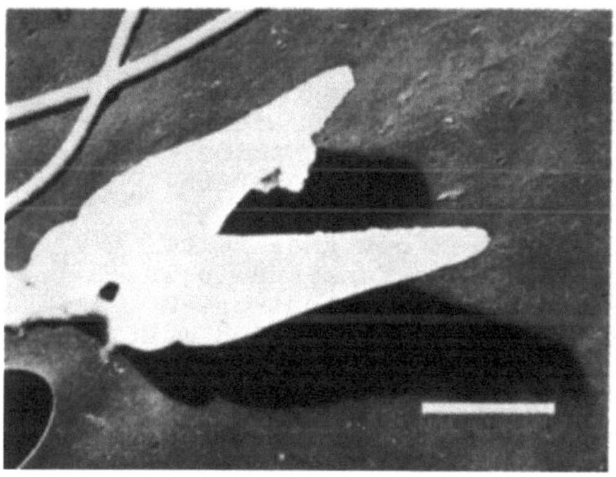

Fig. 4.3

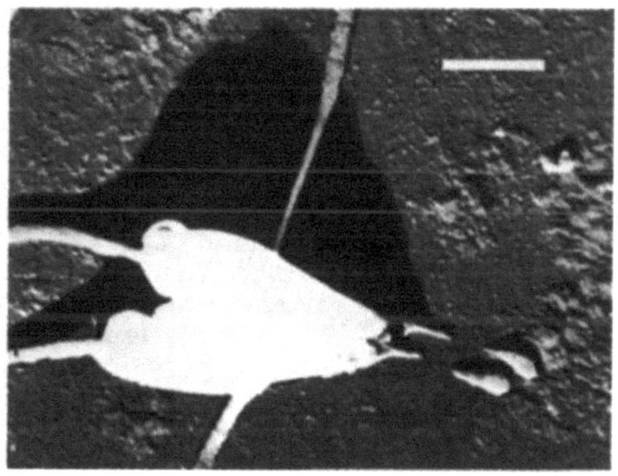

Fig. 4.4

Fig. 4.3. *Hemicentrotus pulcherrimus* spermatozoa in seawater fixed with neutralized formalin. Chrome-shadowed electron micrograph. (From DAN, 1954. By permission of Biol. Bull.)

Fig. 4.4. *Hemicentrotus pulcherrimus* spermatozoa fixed 2 seconds after adding homologous egg water. Acrosome filament is commonly broken in preparations fixed immediately after adding egg water. (From DAN, 1954. By permission of Biol. Bull.)

DAN (1952) studied the acrosomes of sea-urchin spermatozoa by phase contrast microscopy: "...in freshly suspended spermatozoa treated with egg-water, a particle projecting some microns from the sperm's tip which appeared as a flabby gelatinous projection was seen." By electronmicroscopy, it appeared that "part of the acrosome substance disperses to a considerable extent, while a central core or fibre of some sort (which may be a separate structure that appears secondarily), possessing greater cohesiveness and dispersing less readily in seawater, remains attached to the former acrosomal region" (Figs. 4.3-4.5).

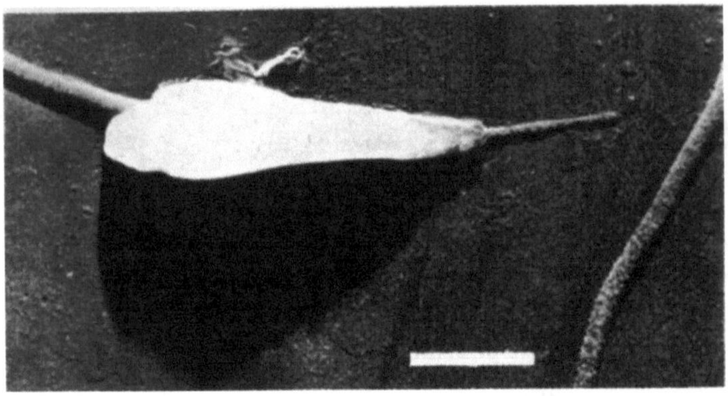

Fig. 4.5. *Hemicentrotus pulcherrimus* spermatozoa fixed 3 minutes after adding egg water and 30 seconds after reversing agglutination. (From DAN, 1954. By permission of Biol. Bull.)

DAN (1954) showed that calcium deficiency prevents acrosomal reaction of sea-urchin spermatozoa in response to stimuli, tends to increase the agglutinating effect, causing sperm to form clusters on contact with egg-water, and prevents loss of fertilizing capacity after agglutination is reversed. DAN concluded that "(a) the agglutination reaction and the acrosome reaction must be considered to be separate phenomena, both occurring in response to the stimulus of egg-water, but not causally interrelated; (b) in view of the fact that fertilizing capacity is retained to full extent by sperm suspensions in which the acrosomes remain intact, and is at least partially lost after artificially induced acrosome reaction, this acrosome reaction must play an important role in the process of fertilization."

In the sea-urchin acrosomal reaction, a rapid change in the more hyaline region underlying the apical granule produces a filament that pushes the granular material on its apical pole. The filament usually breaks when dried in preparations fixed 2 to 5 seconds after egg-water is added but less often if fixed 20 to 30 seconds later, indicating that tensile strength increases in seawater or perhaps with time (DAN, 1956).

The ultra-fine structure of the acrosomal region in sea-urchin spermatozoa has been studied by AFZELIUS (1955), ROTHSCHILD (1956a), BERNSTEIN (1962), DAN, OHORI, and KUSHIDA (1964), FRANKLIN (1965), and LONGO and ANDERSON (1968, 1970). The most conspicuous elements of the acrosome are the mass of homogeneously particulate material forming the apically located acrosomal granule (Figs. 4.6 A-C, g and 4.7 A-E, g) and a loose clump of unoriented, fibrous material in an osmiophobic matrix (Fig. 4.6 A-f, m) filling a pouch-shaped depression in the apical end of the nucleus (DAN, OHORI, and KUSHIDA, 1964).

Another substance still unknown (Fig. 4.6 A-d) forms a wide band around the anterior part of the sperm head immediately beneath the plasma membrane, encircling the distal part of the nucleus and the adjacent flanks of the acrosomal granule. The acrosomal granule of the sea-urchin spermatozoon is capped by a thin trigger layer, around the inner side of which are several layers of membrane precursor substances (Fig. 4.6 A-a, a', b, and c). The structural changes during acrosomal reaction (Fig. 4.7) may be due to the increased permeability of the sperm plasma membrane (DAN, OHORI, and KUSHIDA, 1964).

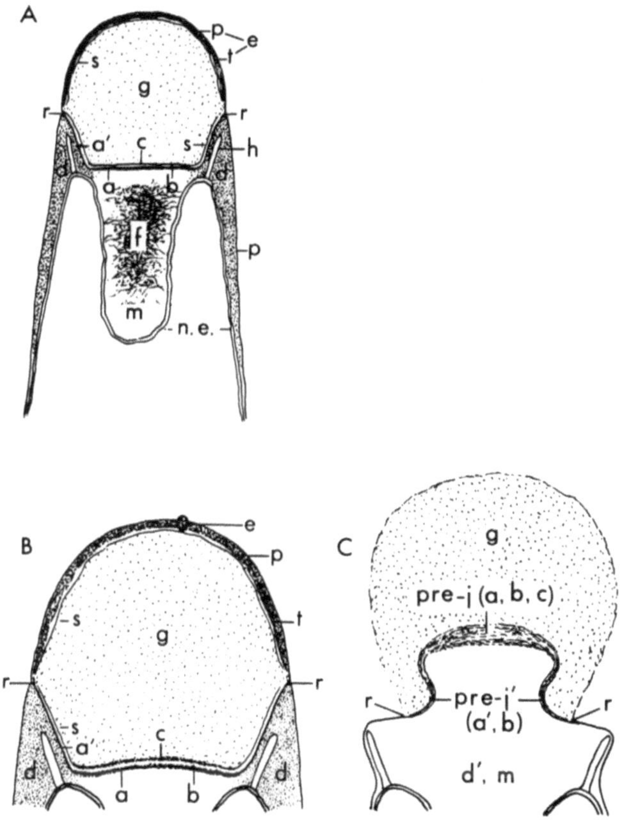

Fig. 4.6A-C. (A) Median longitudinal section through intact sea urchin acrosome. (B, C) Apical part of acrosome showing early changes in precursor layers leading to formation of acrosomal process membrane. (B) Intact acrosome. (C) Stage in mid reaction immediately after breakdown of e. Precursor layers a, b, and c have interacted to form premembrane of process shaft. Layers a' and b form premembrane of process base. Degraded d substance (d') in central lumen of acrosome. (From DAN, OHORI, and KUSHIDA, 1964. By permission of Academic Press.) (Abbreviations as in Fig. 4.7)

Acrosomal filaments in spermatozoa of *Echinocardium cordatum* and *Strongylocentrotus purpuratus* have been studied by ROTHSCHILD and TYLER (1955). *Echinocardium* has 2 types of spermatozoa: a knob on a short stalk, and a filament more than 1.5 times as long as the rest of the head. In *Strongylocentrotus* the longest process observed is about half the length of the head.

In the sea urchin *Diadema setosum*, spawning occurs only at full moon (FOX, 1924; YOSHIDA, 1952). Although spermatozoa obtained 3 or 4 days before the spawning date can be made to agglutinate by using egg-water, most of them do not undergo acrosomal reaction. All the spermatozoa react when they are tested on the day of spawning (DAN, 1956). DAN (1956) observed that "*Diadema setosum* has spermatozoa in which the acrosome region consists of a large, spherical, refringent knob; the sphere disappears on addition of egg-water, and only a stubby clump of apparently fibrous material projecting from the tip of the spermatozoon remains." The relationship between spawning habit and success of the acrosomal reaction suggests that for the acrosomal reaction to take place, spermatozoa must be fully mature in a physiologic sense (DAN, 1956).

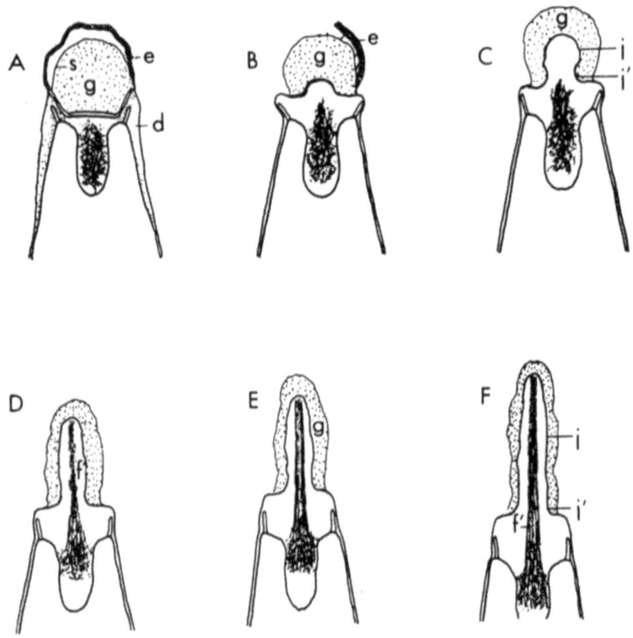

Fig. 4.7A-F. Course of acrosomal process formation. (A-D) Stages found in 1-second fixations. (A) Apical membrane (e) expands away from granule surface (s) before disintegration; d is partially dissolved. (B) Apical membrane disintegrates; d' moves into acrosomal lumen; contents of lumen expand; precursor layers at inner side of granule interact to form process membrane. (C) Fibrous material being rearranged in long axis of acrosomal lumen; process extends through granule center. (D) Process surface consists of uniform membrane continuous with plasma membrane covering nucleus; fibrous core roughly organized in long axis of process lumen. (E) 2 seconds; process about 0.2 μm long; granule substance spread uniformly over process surface. (F) 8 seconds; process about 0.3 - 0.35 μm long. Abbreviations used: a, a', b, c, precursors of acrosomal process membrane; d, particulate material; d', degraded form of d; e, compound membrane (p + t); f, fibrous precursor of axial core; f', axial core; g, acrosomal granule; h, horn; j, surface membrane of acrosomal process shaft; j', surface membrane of acrosomal process base; m, osmiophobic material in nuclear depression; n.e., nuclear envelope; p, plasma membrane; pre-j, process premembrane; pre-j', base process membrane; r-r, circular locus of junction of a' and plasma membrane; s, condensed surface of acrosomal granule; t, trigger layer lining plasma membrane apical to r-r. (From DAN, OHORI, and KUSHIDA, 1964. By permission of Academic Press)

4.3 Spermatozoon Penetration

4.3.1 Sperm and Egg Meet

HULTIN (1948a, 1948b) first showed that the vitelline membrane of sea-urchin eggs is important to the species specificity of fertilization. Immunologic analysis of fertilization in sea-urchin eggs has been extensive (PERLMANN, 1954, 1956, 1957; PERLMANN and PERLMANN, 1957a, 1957b; BAXANDALL, PERLMANN, and AFZELIUS, 1964a, 1964b; BAXANDALL, 1966). Unfertilized eggs - treated with anti-egg serum prepared by injecting egg homogenates into the rabbit - have shown strongly depressed fertilizability, due mainly to one of the egg-surface antigens called the F-antigen which may be a specific, sperm-receptor carbohydrate of the egg cortex (RUNNSTRÖM, HAGSTRÖM, and PERLMANN, 1959). The other antigens - the A- and C-antigens - cause parthenogenetic activation and cortical damage, respectively, and also contribute to fertilizability depression (PERLMANN and PERLMANN, 1957). Using ferritin-labeled antibodies and electron microscopy, BAXANDALL, PERLMANN, and AFZELIUS (1964b) found that A- and F-antigens appear closely associated with the vitelline membrane on which antigen-antibody interactions may block the sperm receptor sites. A- and F-antigens are thought to be vital to the initial interaction between egg and sperm.

After immunologic studies on fertilization, METZ and THOMPSON (1967) concluded that "the fertilization-inhibiting action of normal, multivalent (7S) anti-egg homogenate antibody depends upon formation of cross-linked lattices between neighboring antigens on the egg surface, not on the blocking of specific antigenic sites by complementary antibody, because the papain-digested univalent (3.5S) antibody fails to affect the fertilizability of the egg, and also the eggs which have been rendered unfertilizable by treatment with multivalent anti-egg serum recover considerable fertilizability following treatment with protease. ... Failure of the univalent antibody to inhibit fertilization of eggs indicates that at least the vitelline membrane surface-antigens are not essential in fertilization and implies that such antigenic sites do not interact with sperm substance in fertilization."

Table 4.1. Inhibitory effect of pancreatine on sperm-egg bonding without impairment of receptivity for egg activation. (From AKETA, 1967)

Treated with 0.4% pancreatine seawater for 20 minutes	Untreated (control)
4 spermatozoa adhered to 26 eggs	Many supernumerary spermatozoa adhered to eggs
No membrane formation	100% membrane formation
20% cleavage	100% cleavage
Effect of urea activation 100%; monaster formed	Effect of urea activation 100%; monaster formed

Using a proteolytic enzyme such as pancreatine to remove or partially destroy the vitelline membrane of *Hemicentrotus* eggs causes considerable loss of sperm-binding capacity and a corresponding loss of egg fertilizability (AKETA, 1967)(Table 4.1). Attempting to isolate the sperm-

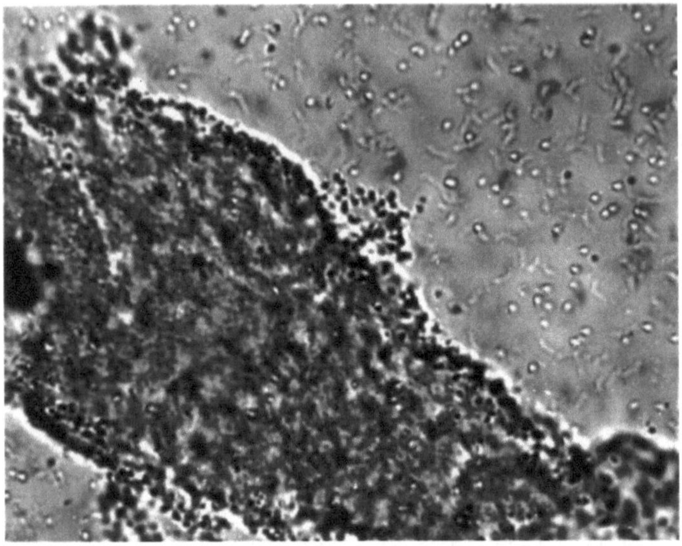

Fig. 4.8. Sperm adhering to precipitate prepared from vitelline membrane of *Hemicentrotus* eggs. (From AKETA, 1967)

egg binding substance from the vitelline membrane, AKETA (1967) wrote: "When the fertilization membrane of the egg began to elevate after insemination, the egg suspension was mixed with 10 times volume of a molar solution of urea. It was then filtered through a fine bolting silk-cloth. The filtrate was centrifuged and the supernatant containing broken pieces of the vitelline membrane was collected. To this supernatant, NaOH was added to the final concentration of 0.1 *N*. The vitelline membrane became dissolved by this treatment. The solution was dialysed against tap water. When the final solution was brought to pH 3.8 with 0.2 *N* HCl, white precipitates appeared. They were collected by centrifugation, and the precipitates were washed several times with distilled water."

AKETA also suggested a simpler isolation method: "Unfertilized eggs were treated with a 1-molar urea solution for 5 to 10 minutes, and the vitelline membrane was dissolved (MOTOMURA, 1941). ... After dialysis over-night, precipitate was collected at pH 3.8." AKETA tested the sperm-binding capacity of the preparation thus obtained. When fresh spermatozoa were added to the seawater containing the precipitate, spermatozoa adhered to the surface of the precipitate (Fig. 4.8).

In another experiment, AKETA used a fine pipet to blow air through seawater containing binding substance. When fresh spermatozoa were added to the bubbled medium, they gathered on the surface of the bubbles (Fig. 4.9). Under calcium-free conditions or with pure seawater, this phenomenon did not occur.

AKETA and ONITAKE (1969) studied the vitelline membrane at the time of fertilization and found that "when jellyless eggs of *Hemicentrotus pulcherrimus* were pretreated with antiserum against sperm-binding protein which was made by the isolated vitelline membrane of homologous species, they could not bind sperm at all and became unfertilizable

Fig. 4.9. Sperm adhering to air bubbles in seawater containing binding substance. (From AKETA, 1967)

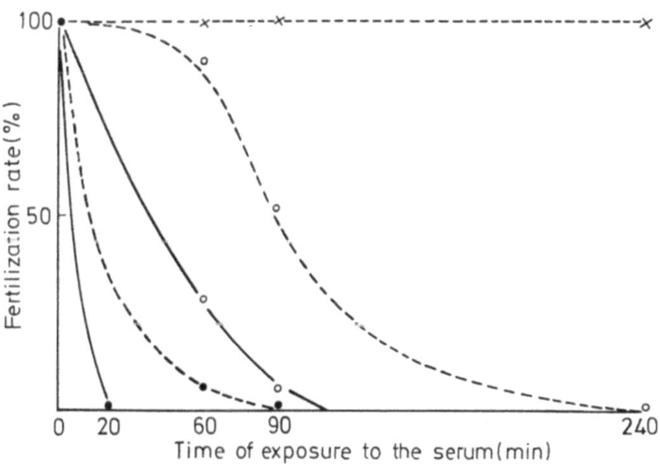

Fig. 4.10. Fertilization inhibited by antiserum against sperm-binding protein of homologous sea urchin eggs (*Hemicentrotus pulcherrimus*). x--x, normal serum; ●—●, 1/2 diluted antiserum; ●--●, 1/4 diluted antiserum; o—o, 1/8 diluted antiserum; o--o, 1/16 diluted antiserum. (From AKETA and ONITAKE, 1969. By permission of Academic Press)

(Fig. 4.10). ... Antisera against sperm-binding protein from *Heliocidaris crassispina* or *Temnopleurus toreumaticus* blocked sperm penetration into the egg of the heterologous species, *Hemicentrotus*, but did not inhibit its sperm-binding capacity" (Fig. 4.11).

In the sea urchin *Arbacia punctulata*, FRANKLIN (1965) used the electron microscope to investigate morphologic processes related to the initial contact of spermatozoa with eggs. He concluded that "contact of sperm

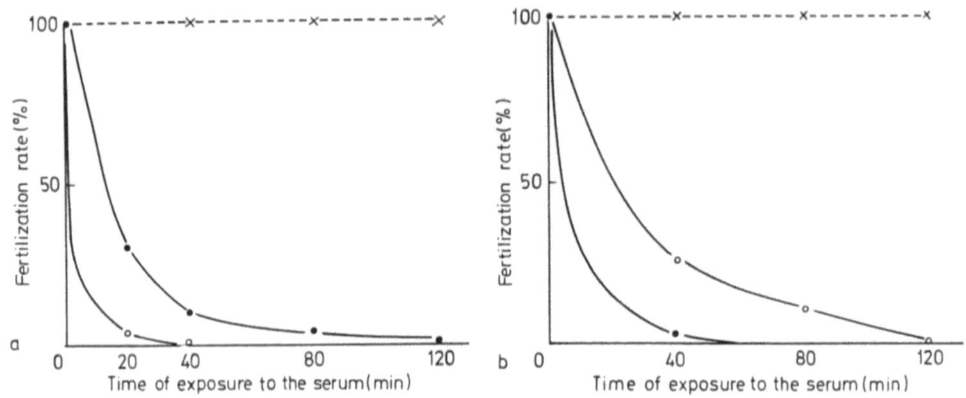

Fig. 4.11 a and b. Fertilization inhibited by antiserum against sperm-binding protein of heterologous sea urchin eggs. (a) *Heliocidaris crassispina*. (b) *Temnopleurus toreumaticus*. x--x, normal serum; o—o, 1/4 diluted anti-*Heliocidaris* serum; •—•, 1/4 diluted anti-*Temnopleurus* serum. (From AKETA and ONITAKE, 1969. By permission of Academic Press)

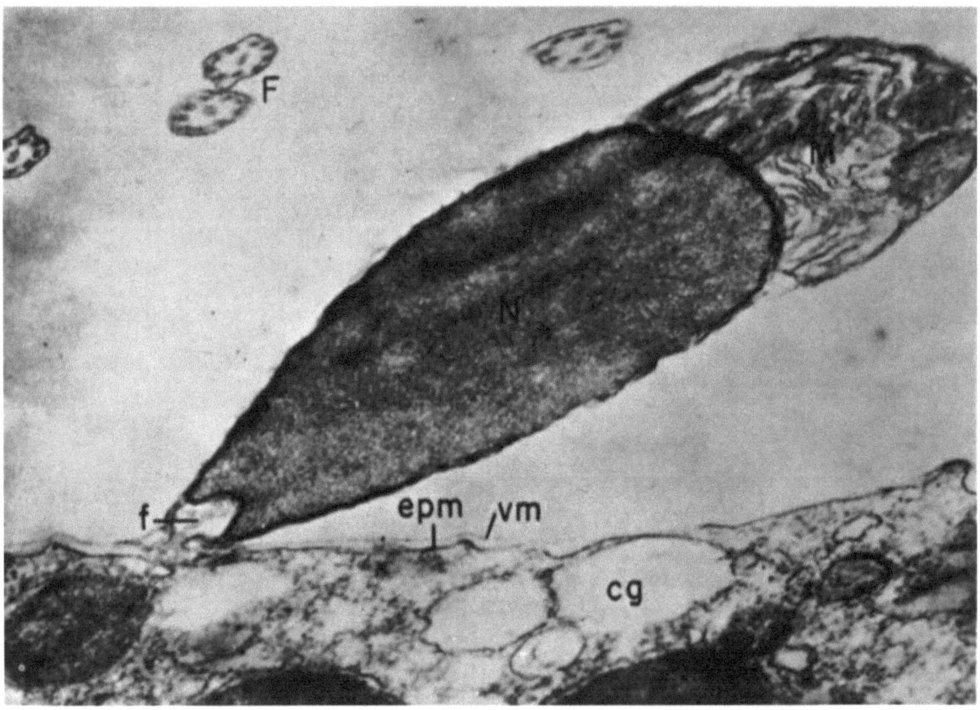

Fig. 4.12. Acrosomal tubule of a spermatozoon in contact with mature egg surface. Note fibrils (f) in acrosomal tubule and thin vitelline membrane (vm) of the egg, *Arbacia punctulata*. cg, cortical granule, epm, egg plasma membrane, F, sperm flagellum. (From FRANKLIN, 1965. By permission of Rockefeller University Press)

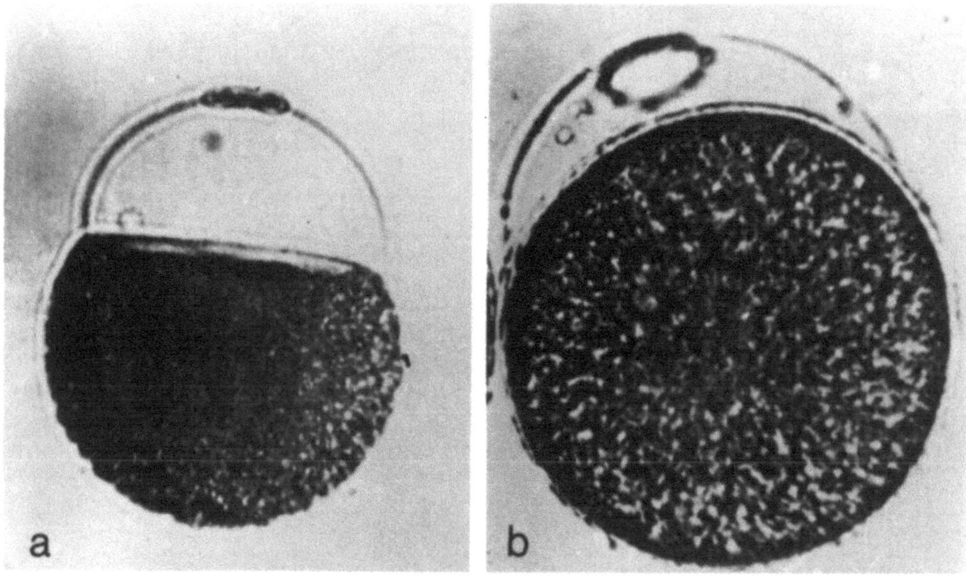

Fig. 4.13 a and b. (a) Side view of sperm-entry mark. (b) Sperm-entry mark with ring-like structure. (From ISHIKAWA, 1960)

with mature eggs or oocytes was found even in specimens fixed within 10 to 20 seconds after insemination. ... The apex of the acrosomal tubule first seems to be pressed into the egg surface so that the thick-walled region of the acrosomal tubule, still surrounded by acrosomal granule material, is on a level with the egg surface" (Fig. 4.12). LONGO and ANDERSON (1970) examined contact between sperm and nicotine-treated *Arbacia* eggs and noted that "the acrosome is associated with some flocculent material at the egg surface".

4.3.2 Sperm Passage through Egg Vitelline Membrane

In fertilized sea-urchin eggs transferred immediately after insemination to an isotonic salt solution containing a large number of calcium ions, a ring-like structure called the sperm-entry mark (Fig. 4.13) appeared on the fertilization membrane (ISHIKAWA, 1958). The calcium-rich solution was prepared by adding 10 ml of 10/27 *M* $CaCl_2$ to 90 ml of isotonic salt solution (100 ml of 5/9 *M* NaCl plus 2.1 ml of 5/9 *M* KCl). The sperm-entry mark on the fertilization membrane was related to the membrane-hardening mechanism after fertilization.

The temporal relationship of membrane hardening, third-factor secretion - one of the factors related to membrane hardening (MOTOMURA, 1950) - and sperm-entry mark formation is shown (Fig. 4.14). When fertilized eggs in stage III (Fig. 4.14, 32 to 45 seconds after insemination) are transferred to the calcium-rich solution, sperm-entry mark formation proceeds (Fig. 4.14 A-G) while the membrane is only

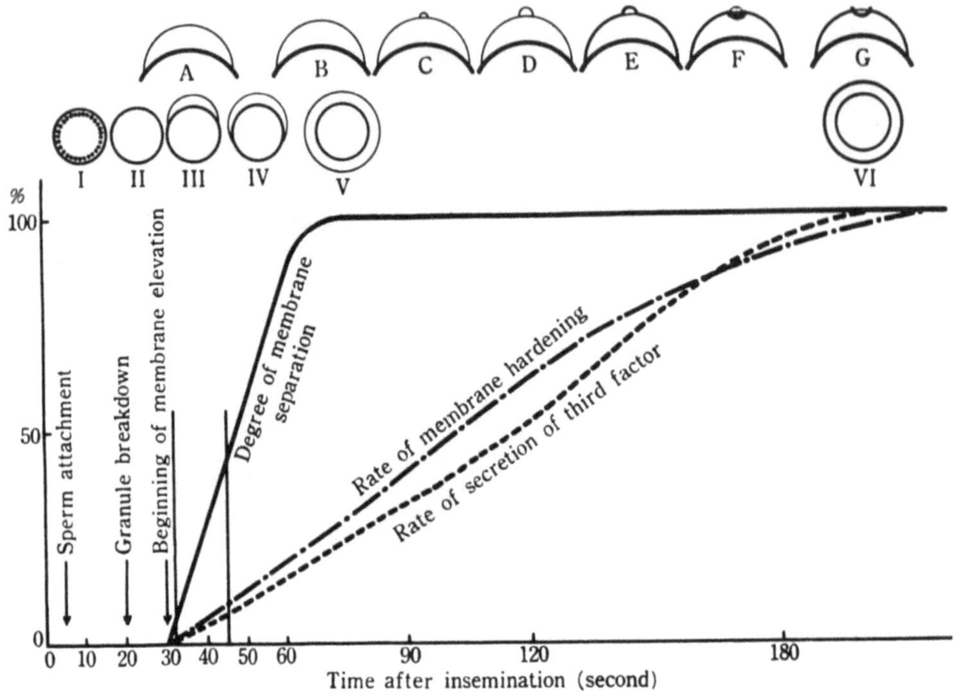

Fig. 4.14. Time relations among sperm-entry mark formation, membrane hardening, and third-factor secretion. (A-G) Formation stages of sperm-entry mark. I-VI: Membrane formation stages in ordinary seawater. Data of membrane hardening and third-factor secretion were adopted from SUGIYAMA (1938a) and MOTOMURA (1950), respectively. (From ISHIKAWA, 1960)

partly elevated (covering less than one-third of the egg surface) and barely hardened. When eggs are transferred to the calcium-rich solution before or after stages II or IV, no sperm-entry mark formation occurs.

About 1 minute after proper transfer to the calcium-rich solution, a tiny cone-like protrusion appeared on the fertilization membrane (Fig. 4.14 C). The size and height of the protrusion gradually increased (Fig. 4.14 D and E), after which calcium precipitated on the inside and around the base of the protrusion (Fig. 4.14 E and F). The protrusion sank, and the ring-like structure of the sperm-entry mark was complete (Fig. 4.14 G). The inside of the sperm-entry mark looked granular, while the fertilization membrane beside it was smooth (Fig. 4.15). In a dispermic egg, 2 sperm-entry marks appeared on the fertilization membrane.

Membrane hardening (SUGIYAMA, 1938a, 1938b) and third-factor secretion (MOTOMURA, 1950) reach a peak at 3 to 5 minutes at which time sperm-entry mark formation is also complete. "...at the time of fertiliza-

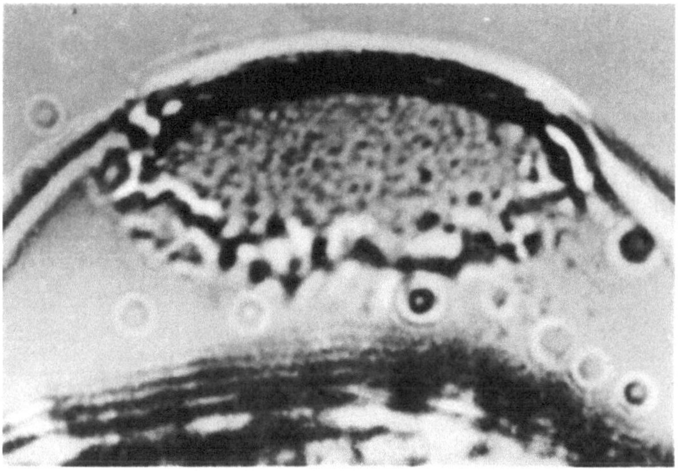

Fig. 4.15. Magnified sperm-entry mark showing granular membrane. (From ISHIKAWA, 1960)

tion, the vitelline membrane may be affected by a lytic substance which is ejected from the acrosome of the spermatozoon, and the injured part of the vitelline membrane may be soon recovered in the ordinary seawater accompanying the formation of the fertilization membrane, but it persists in the Ca-rich solution, forming the sperm-entry mark on the fertilization membrane" (ISHIKAWA, 1960).

4.3.3 Gamete Membrane Fusion

As a rule, the sperm acrosomal tubule and the plasma membrane of the mature egg fuse 20 to 30 seconds after insemination (FRANKLIN, 1965), everting the acrosomal tubule and, by exposing a new surface of it, bringing fibrils and other of its structures into contact with the cortical cytoplasm of the mature egg or oocyte (Figs. 4.16 and 4.17).

4.3.4 Spermatozoon Incorporation

Under the electron microscope, FRANKLIN (1965) observed sperm entry in the oocytes of *A. punctulata* and *Lytechinus variegatus*. Examination of thin sections of oocytes fixed at intervals of 30 to 40 seconds after insemination revealed several ultrastructural changes in the gametes. At the early stages of sperm incorporation, small vesicles were evident along the regions of contact of sperm nucleus and oocyte cytoplasm (Fig. 4.18). Later, the sperm nucleus expanded and the lateral nuclear membrane seemed to disappear (Fig. 4.19). As shown by ROTHSCHILD (1956b), basal and centriolar parts of the sperm nuclear membrane and the apical part of the nucleus did not change during or after sperm entry (Fig. 4.20). The sperm nucleus and middle piece were later incorporated into the ground substance of the fertilization cone, and the sperm-plasma membrane was absent from the organelles (Figs. 4.19 and 4.20).

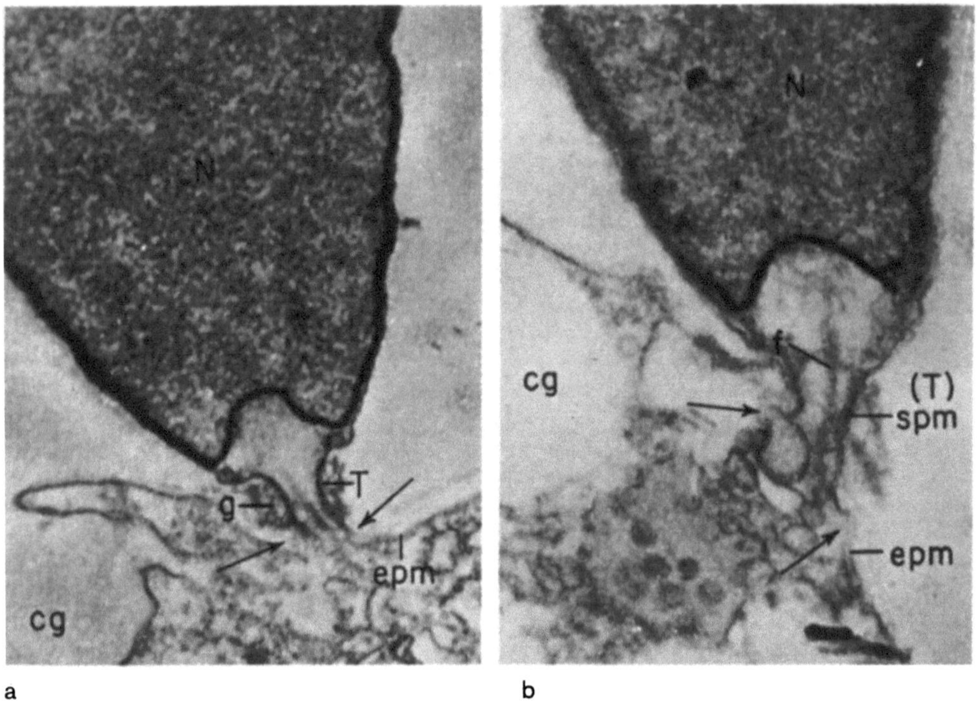

Fig. 4.16 a and b. Initial fusion of sperm acrosomal tubule (T) and plasma membrane (epm) of mature egg. Acrosomal tubule membrane is continuous with sperm plasma membrane (spm). Arrows indicate fusion sites. cg, cortical granule; f, fibrils within acrosomal tubule; g, acrosomal granule; N, nucleus. *Arbacia punctulata*. (From FRANKLIN, 1965. By permission of Rockefeller University Press)

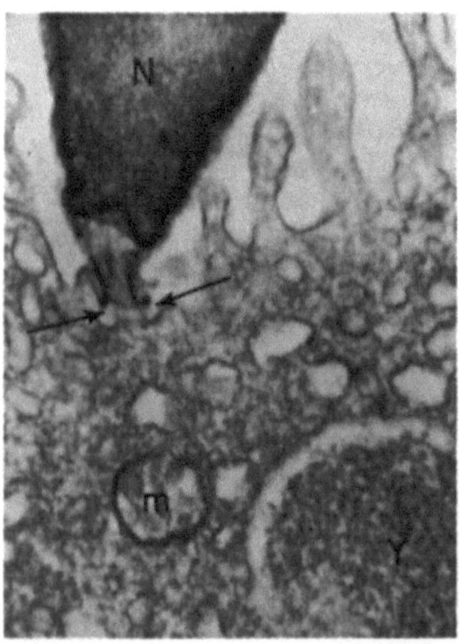

Fig. 4.17. Initial fusion of sperm acrosomal tubule and plasma membrane of oocyte. Arrows indicate fusion sites. N, nucleus; m, mitochondrion; Y, yolk granule. *Arbacia punctulata*. (From FRANKLIN, 1965. By permission of Rockefeller University Press)

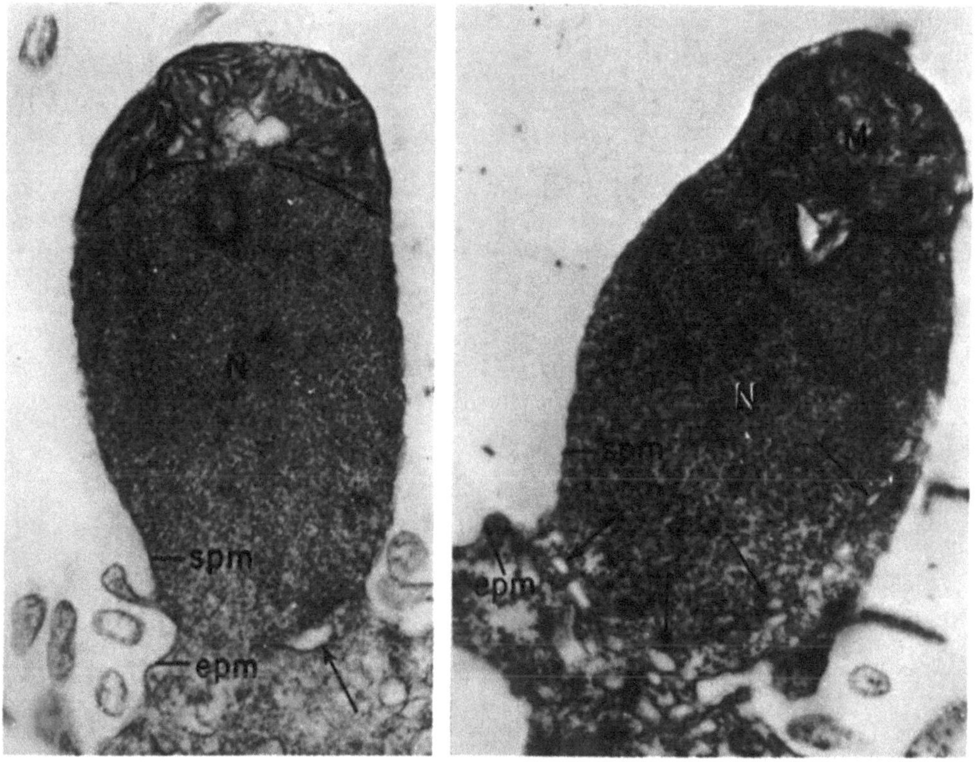

a b

Fig. 4.18 a and b. Early stages of sperm nucleus entry. Ooocyte and sperm membranes are continuous. Small vesicles are evident where sperm nucleus and oocyte cytoplasm are in contact (arrows). epm, egg plasma membrane; M, middle piece; N, nucleus; spm, sperm plasma membrane. *Arbacia punctulata*. (From FRANKLIN, 1965. By permission of Rockefeller University Press)

4.3.5 Sperm Tail Engulfment

Although in animals the entire sperm generally enters the egg, at least part of the tail in the sea urchin and starfish is left outside (WILSON, 1925). In the sea urchin the complete tail is usually taken into the egg cytoplasm 2 to 4 minutes after penetration starts (DAN, 1950). Using the electron microscope, TAKASHIMA and TAKASHIMA (1960) observed that on a sea-urchin sperm that had partly entered the fertilization cone of a nicotine-treated mature egg, the tail membrane was still intact. Speaking of an inseminated oocyte, FRANKLIN (1965) said, "portions of sperm tails deeper in the oocyte were surrounded by small membranous vesicles. ... in a hybrid cross between *Millita quinquiesperforata* (sperm) and *Arbacia punctulata* (oocytes), sperm flagella that had just been incorporated were also surrounded by such vesicles."

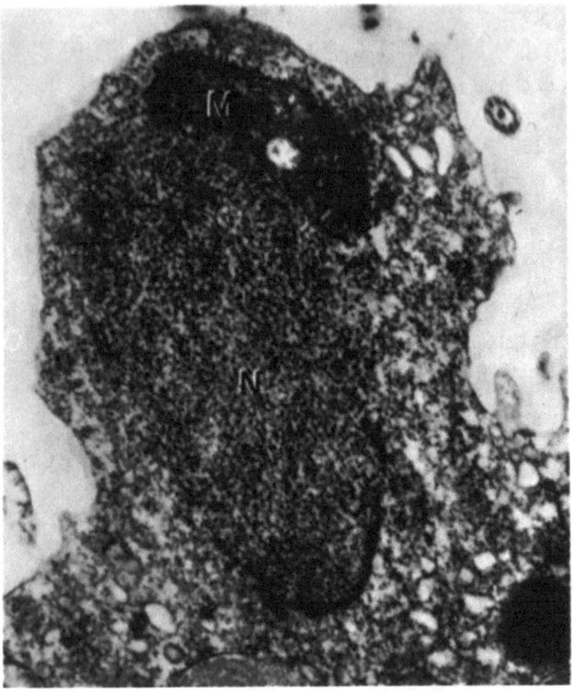

Fig. 4.19. Incorporated sperm nucleus and middle piece; sperm plasma membrane is absent. N, nucleus; M, middle piece. *Arbacia punctulata*. (From FRANKLIN, 1965. By permission of Rockefeller University Press)

4.3.6 Fertilization Cone Formation

Shortly after the sperm attaches to the egg surface, a cytoplasmic protrusion appears and subsequently develops into the fertilization cone which lasts 5 to 15 minutes after sperm penetration begins. In eggs subjected to such conditions as heat, cold, and nicotine solution, abnormal fertilization cone production and endurance may be induced (DAN, 1950a). RUNNSTRÖM (1963), LÖNNING (1964), and FRANKLIN (1965) have described oocyte fertilization cones. "Soon after sperm migration the oocyte cytoplasm exhibits 3 phases (phase I: greater part of the cone consists of ground cytoplasm; phase II: rich in dispersed endoplasmic vesicles; phase III: rich in granular components)" (RUNNSTRÖM, 1963).

In sea-urchin oocytes, the fertilization cone is generally resorbed in 10 to 20 minutes but may last more than 30 minutes (HAGSTRÖM and

---

Fig. 4.20. Sperm migration to cytoplasmic organelle zone. Centrioles (C) and middle piece (M) maintain original positions in relation to sperm nucleus. FC, fertilization cone; N, nucleus; p, pigment granule; Y, yolk granule. *Arbacia punctulata*. (From FRANKLIN, 1965. By permission of Rockefeller University Press) ▶

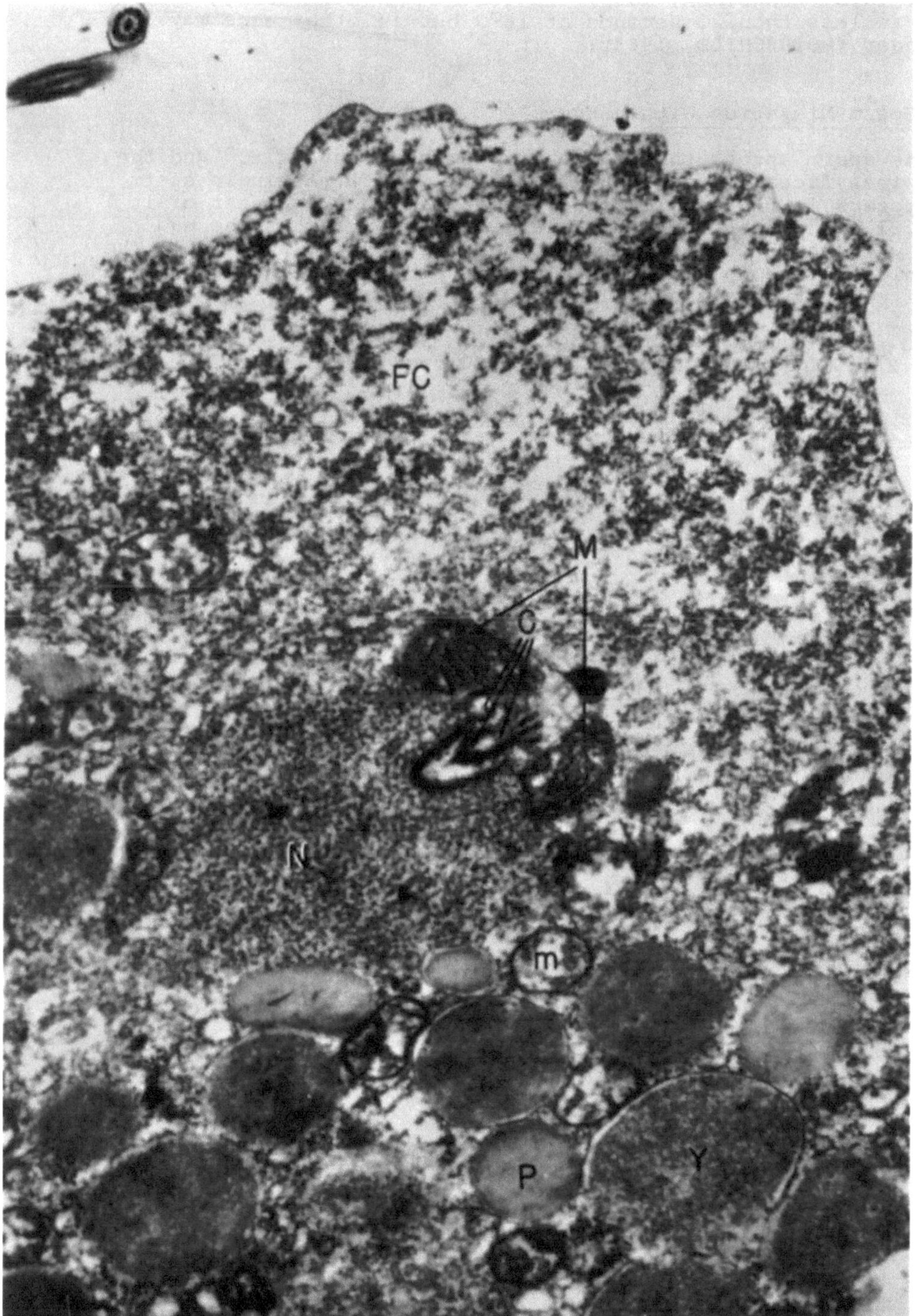
FC
M
C
N
m
P
Y

LÖNNING, 1963; RUNNSTRÖM, 1963). In *Psammechinus miliaris* eggs, it disappears in less than 20 seconds at 18°C but in other eggs may last much longer (ROTHSCHILD, 1965c).

4.3.7 Sperm Migration Within Egg

After the sperm enters the egg, the sperm head turns 180° and the middle piece faces the egg center (WILSON, 1925). A single sperm aster appears in the middle piece, sometimes in the center of it (Fig. 4.21). The sperm nucleus enlarges and approaches the egg nucleus with which it eventually unites (Figs. 4.22 and 4.23). The 2 pronuclei fuse completely, forming the synkarion. The maternal and paternal constituents are no longer discernible.

When the sperm and egg pronuclei migrate through the egg cytoplasm (Fig. 4.24), the sperm pronucleus usually moves in a curve: its pene-

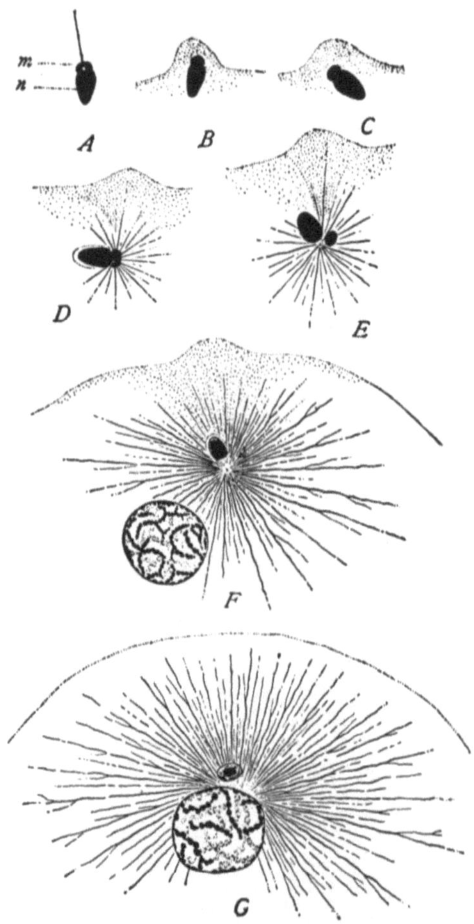

Fig. 4.21. Sperm head entrance and rotation and sperm aster formation in sea urchin *Toxopneustes*. (From WILSON, 1925)

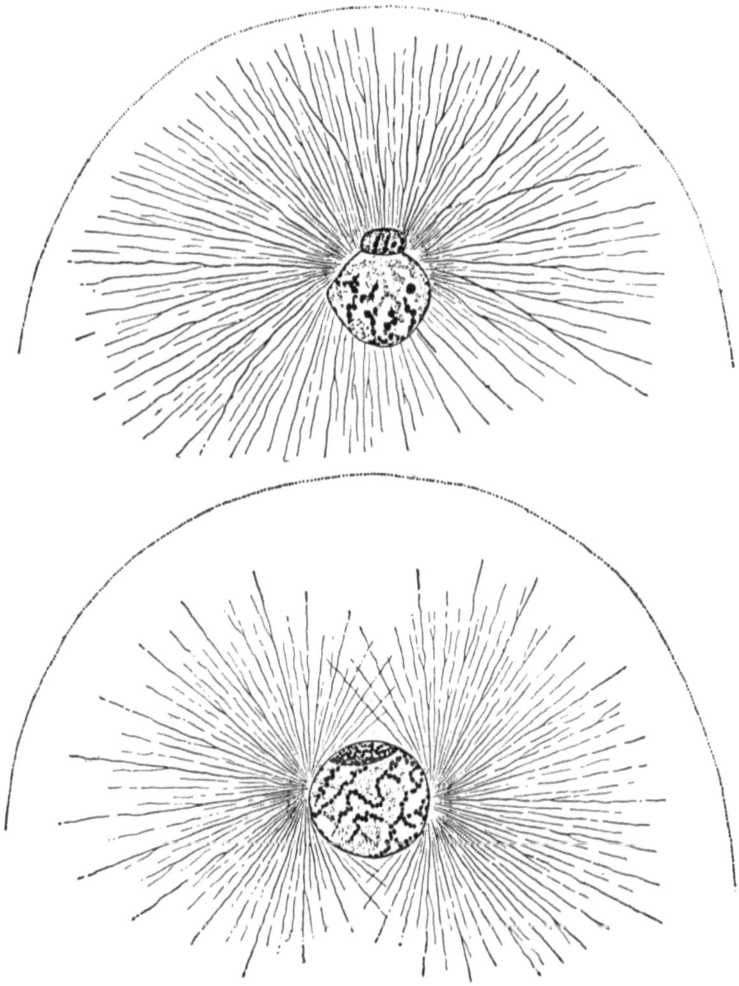

Fig. 4.22. Gamete-nuclei conjugation and sperm aster division in sea urchin *Toxopneustes*. (From WILSON, 1925)

tration path is nearly vertical to the surface, and its copulation path takes it to the point of union with the egg pronucleus (WILSON, 1925). The united pronuclei follow the cleavage path to the first cleavage nucleus.

"The sperm nucleus entered into the egg first moves inward following a penetration path that has no constant relation to the position of the egg pronucleus and is approximately but never exactly radial. ... After penetration of a certain distance its direction changes slightly to that of the copulation path, which again is directed not precisely towards the egg pronucleus but towards a meeting point where it comes in contact with the egg pronucleus" (WILSON, 1925).

Unaffected by the position of the egg pronucleus the penetration path is determined by the relationship between the sperm pronucleus and the egg cytoplasm. The copulation path is influenced by interactions

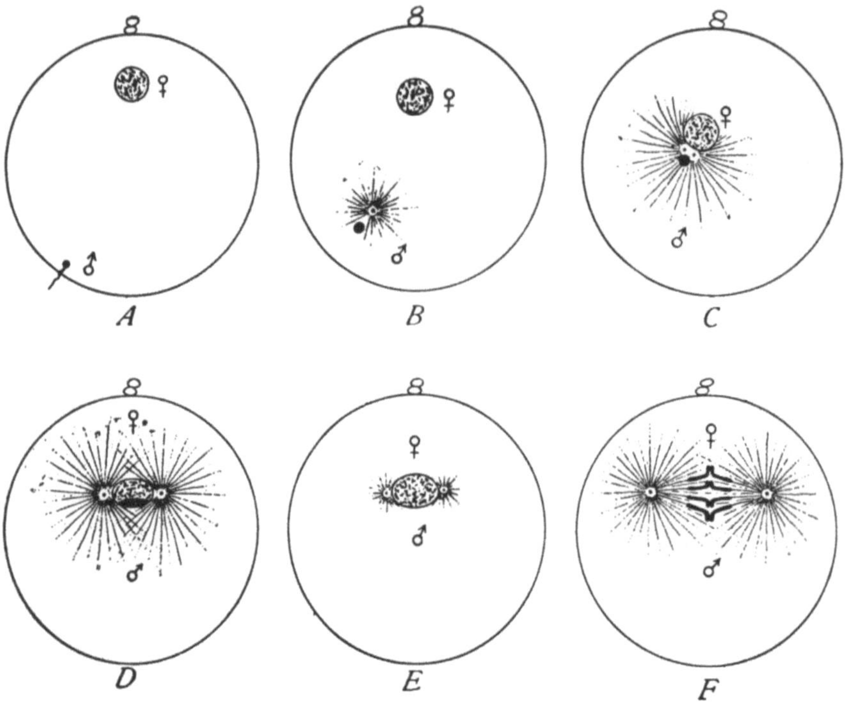

Fig. 4.23A-F. Sea urchin fertilization. A, mature egg, entrance of sperm; B and C, approach of pronuclei, division of sperm center; D, sperm aster divided, fusion of pronuclei; E, fusion-nucleus in pause, reduction of asters; F, first cleavage figure. (From WILSON, 1925)

between the pronuclei (Fig. 4.24 A-D), probably not actual attraction. As the synkarion moves to its final position, it seems to establish an equilibrium in the cell system. Pronuclei motion before fusion was studied by KUHL and KUHL (1949) using time-lapse cinematography.

The sperm nuclei of oocytes sometimes rotate 180° within 2 1/2 minutes after insemination, while the centriole and middle piece of the incorporated sperm retain their original positions in relation to the sperm pronucleus (FRANKLIN, 1965). Pigment granules, yolk platelets, and mitochondria of the egg cytoplasm become more compact after sperm entry.

Pressure of 5000 pounds per square inch applied for 30 to 60 minutes and Colcemid (deacetyl-n-methylcolchicine) in concentrations of at least 2.7×10^{-5} *M* block pronuclear fusion in the egg of *A. punctulata* (ZIMMERMAN and SILBERMAN, 1964; ZIMMERMAN and ZIMMERMAN, 1967): "High pressure and Colcemid may affect the movement of pronuclei by disrupting the microtubule structure of the astral rays."

4.3.8 Sperm Aster Formation

BOVERI first studied aster formation in 1887. HARVEY (1956) reported that the sperm aster is formed 8 minutes after insemination of mature *A. punctulata* eggs at 23°C. Sperm asters do not form in inseminated oocytes of the sea urchin (LÖNNING, 1964; FRANKLIN, 1965).

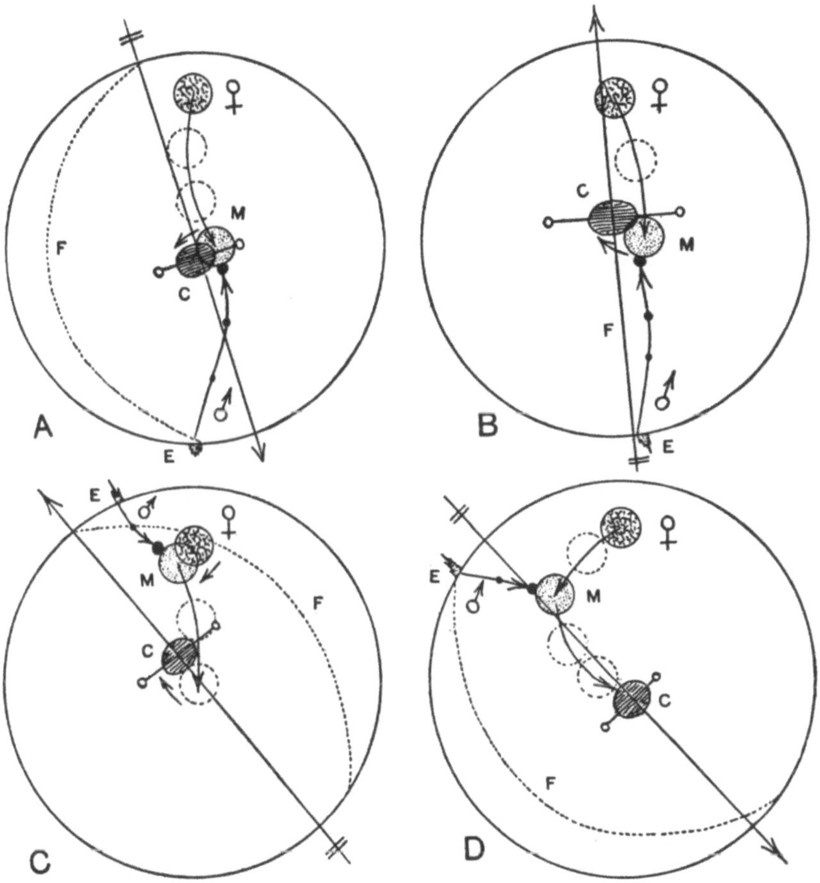

Fig. 4.24A-D. Pronuclei paths in 4 sea urchin *Toxopneustes* eggs. ♀, egg nucleus; E, sperm entrance point. Arrows indicate nuclei paths; M, meeting point. Cleavage-nucleus in final position (C); through it is drawn the axis of the resulting cleavage-figure. Arrow indicates egg axis; arrow point is turned toward upper pole. Curved, dotted lines indicate first cleavage plane which passes near entrance point. (From WILSON, 1925)

When the sperm aster begins to develop, the proximal centriole dissociates from the sperm flagellum, and the 9 triplet tubules that comprise the centrioles give off thin conical satellites (Fig. 4.25) (LONGO and ANDERSON, 1968). Near the centrioles, microtubule segments are short and randomly oriented, and elements of endoplasmic reticulum may occur. Farther away from the centrioles, longer microtubules are separated from each other by cytoplasmic egg components. Initially parallel bundles of microtubules form the conical structure of the sperm aster.

As it continues to develop, the distal centriole dissociates from the sperm pronucleus while the pericentriolar region around the 2 centrioles becomes packed with segments of microtubules and smooth

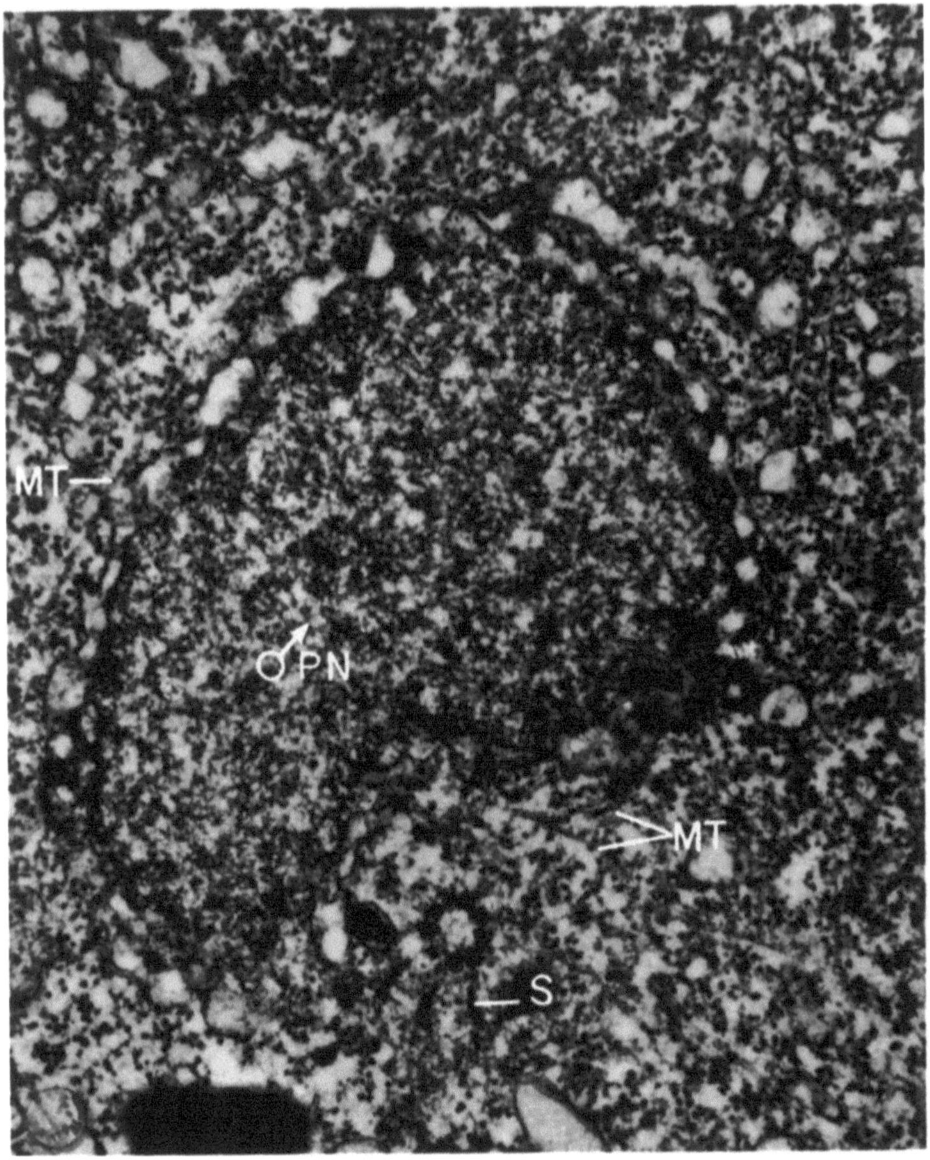

Fig. 4.25. Pericentriolar satellites (S) and microtubules (MT) are adjacent to male pronucleus (♂PN). (From LONGO and ANDERSON, 1968. By permission of Rockefeller University Press)

endoplasmic reticulum (Fig. 4.26) and a centrosphere is formed. Far from the centrosphere the aster is made up of numerous profiles of smooth endoplasmic reticulum and annulate lamellae that extend into the peripheral zone of the egg. In its later stage of development, the sperm aster increases in size, apparently because quantities of microtubules, endoplasmic reticulum, and annulate lamellae accumulate.

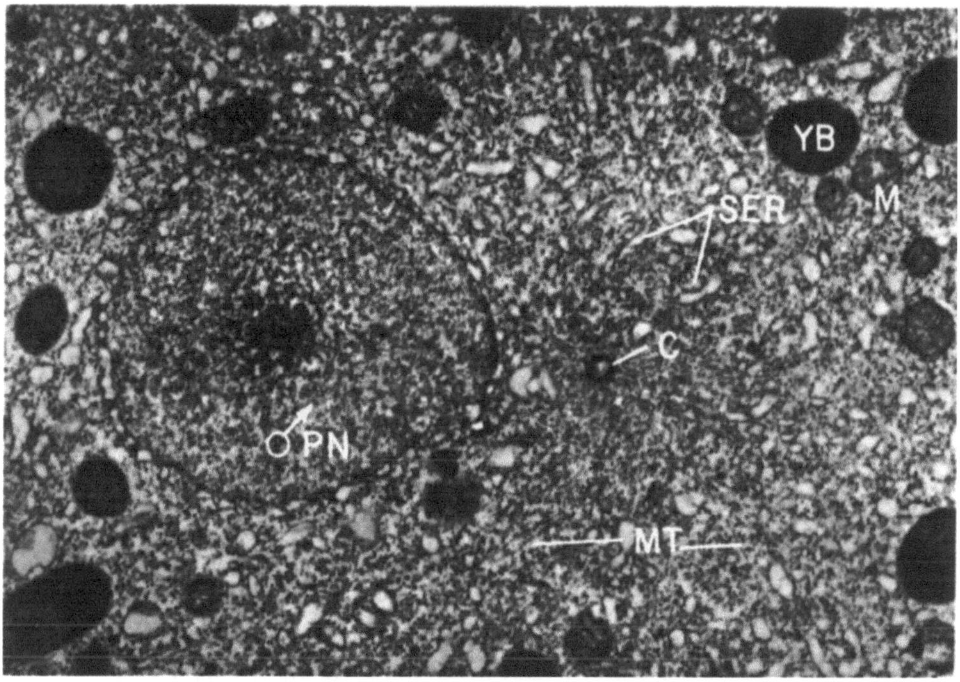

Fig. 4.26. Centriole (C) and sperm aster during migration of male pronucleus (♂PN). Microtubules (MT) and smooth endoplasmic reticulum (SER) surround centriole, forming a centrospheric region and excluding larger ooplasmic constituents such as yolk bodies (YB) and mitochondria (M) from this area. (From LONGO and ANDERSON, 1968. By permission of Rockefeller University Press)

4.3.9 Sperm Pronucleus Development

The endoplasmic reticulum causes the formation of the sperm-pronuclear envelope. It is responsible for the formation of a perforated, lamellar structure (ITO, 1960; HARRIS, 1961; ROBBIN and GONATAS, 1964; PASTEELS, 1965; REBHUN and BERNSTEIN, 1967). Pronuclear development begins after sperm rotation and consists of "(a) the reorganization of the sperm chromatin in the sense of dispersion of the tightly packed chromatin of the sperm head; and (b) the formation of the pronuclear envelope" (LONGO and ANDERSON, 1968). The sperm-pronuclear envelope vesiculates along its lateral margin where the chromatin begins to disperse (Fig. 4.27). The envelope at the apex and base of the sperm pronucleus causes a different degree of chromatin dispersion. Chromatin yields a heart-shaped sperm pronucleus consisting of 3 zones: a perimeter of randomly oriented filaments (Figs. 4.28 and 4.29, FDC); a middle zone of aggregates of filaments about 55 nm wide (Figs. 4.28 and 4.29, CDC); and an inner zone of aggregates of electron-opaque chromatin (Fig. 4.29, CC).

Before the sperm-pronuclear envelope appears, smooth-surfaced vesicles accumulate along the periphery of the dispersing chromatin. "During migration the sperm-pronucleus develops by addition of smooth vesicular elements to the surface of the continuous dispersion of chromatin. ...

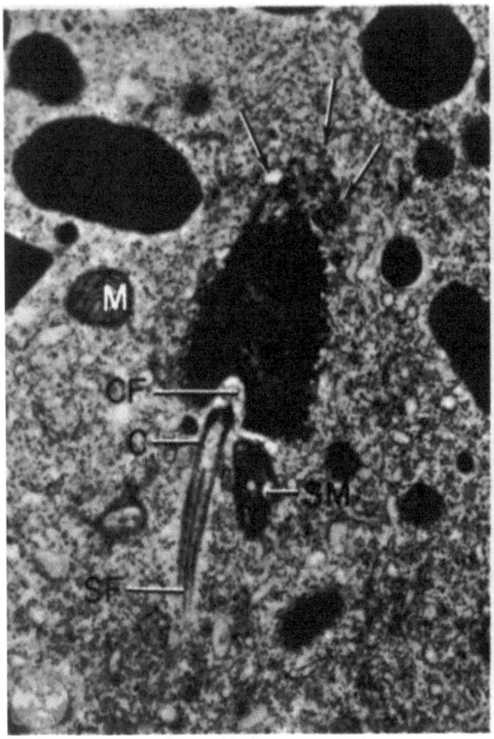

Fig. 4.27

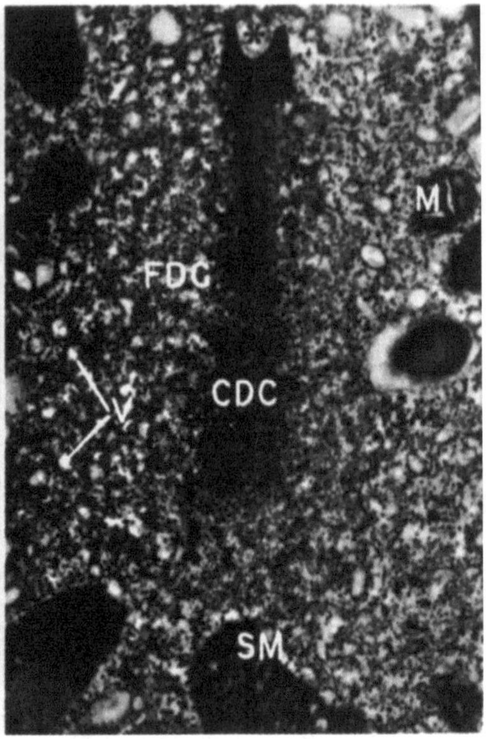

Fig. 4.28

Fig. 4.27. Advanced stage of sperm chromatin dispersal (arrows). Nuclear envelope at centriolar fossa (CF) is intact. SM, sperm mitochondrion; SF, sperm flagellum; C, centriole; M, egg mitochondria. (From LONGO and ANDERSON, 1968. By permission of Rockefeller University Press)

Fig. 4.28. Advanced stage of sperm chromatin dispersal. Vesicles (V) are evident at junction of finely dispersed chromatin (FDC) and ooplasm. Nuclear envelope surrounding anterior region of sperm nucleus (asterisk) is intact. More medial part of sperm nucleus is made up of coarse aggregates of chromatin (CDC). SM, sperm mitochondrion; M, egg mitochondria. (From LONGO and ANDERSON, 1968. By permission of Rockefeller University Press)

While the structural changes of the sperm aster are taking place, the heart-shaped sperm-pronucleus transforms into a spheroid. ... Throughout this period the sperm-pronucleus continues to increase in size" (LONGO and ANDERSON, 1968). Diffusion of sperm-pronucleus chromatin continues until it appears as a delicate mesh of entangled filaments (Fig. 4.30). During chromatin dispersion, the sperm-pronuclear envelope forms as a double membrane.

4.3.10 Egg Pronucleus Development

The nucleus of the unfertilized egg of *S. purpuratus* is usually irregular in shape and becomes even more irregular after fertilization

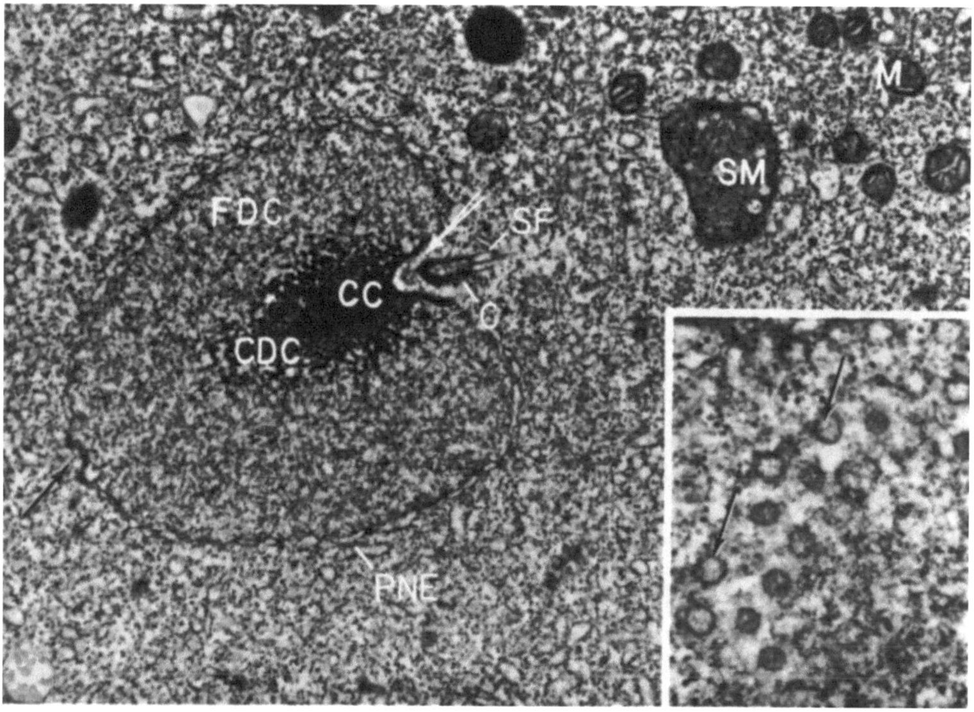

Fig. 4.29. Heart-shaped male pronucleus delimited by pronuclear envelope (PNE) perforated by pores (arrow, insert). Double- and single-stem arrows indicate parts of male pronuclear envelope at centriolar fossa and apex of sperm nucleus, respectively. Layer of electron opaque material is associated with these regions. FDC, finely dispersed chromatin; CDC, coarsely dispersed chromatin; CC, condensed chromatin; SM, sperm mitochondrion; C, centriole; SF, sperm flagellum; M, egg mitochondria. (From LONGO and ANDERSON, 1968. By permission of Rockefeller University Press)

(HARRIS, 1967). The egg pronucleus is surrounded by ribosomes, smooth-surfaced endoplasmic reticulum, and annulate lamellae (Fig. 4.31) (LONGO and ANDERSON, 1968). The egg pronucleoplasm contains fibrillar chromatin and - along the periphery of the pronucleus - a variable number of nucleolus-like bodies (Fig. 4.31a, NL) made up of compact, fine-textured material and granules (Fig. 4.31c, arrows). The nucleolus-like bodies contain RNA because they stain red with methylgreen-pyronin (LONGO and ANDERSON, 1968). They have been found in fertilized and unfertilized eggs of several species of sea urchins (MILLONIG, 1967; HARRIS, 1967a, 1967b).

The pores and the perinuclear space (Fig. 4.31b, PNP, PNS) of the egg pronuclear envelope overlying the nucleolus-like bodies are much larger than those in other parts of the pronuclear envelope (LONGO and ANDERSON, 1968).

When the distance between sperm pronuclei and egg pronuclei is as little as 2 μm, the part of the egg pronucleus nearest the advancing sperm pronucleus flattens and extends finger-like nucleoplasmic projections toward it (Fig. 4.32, NP) (LONGO and ANDERSON, 1968).

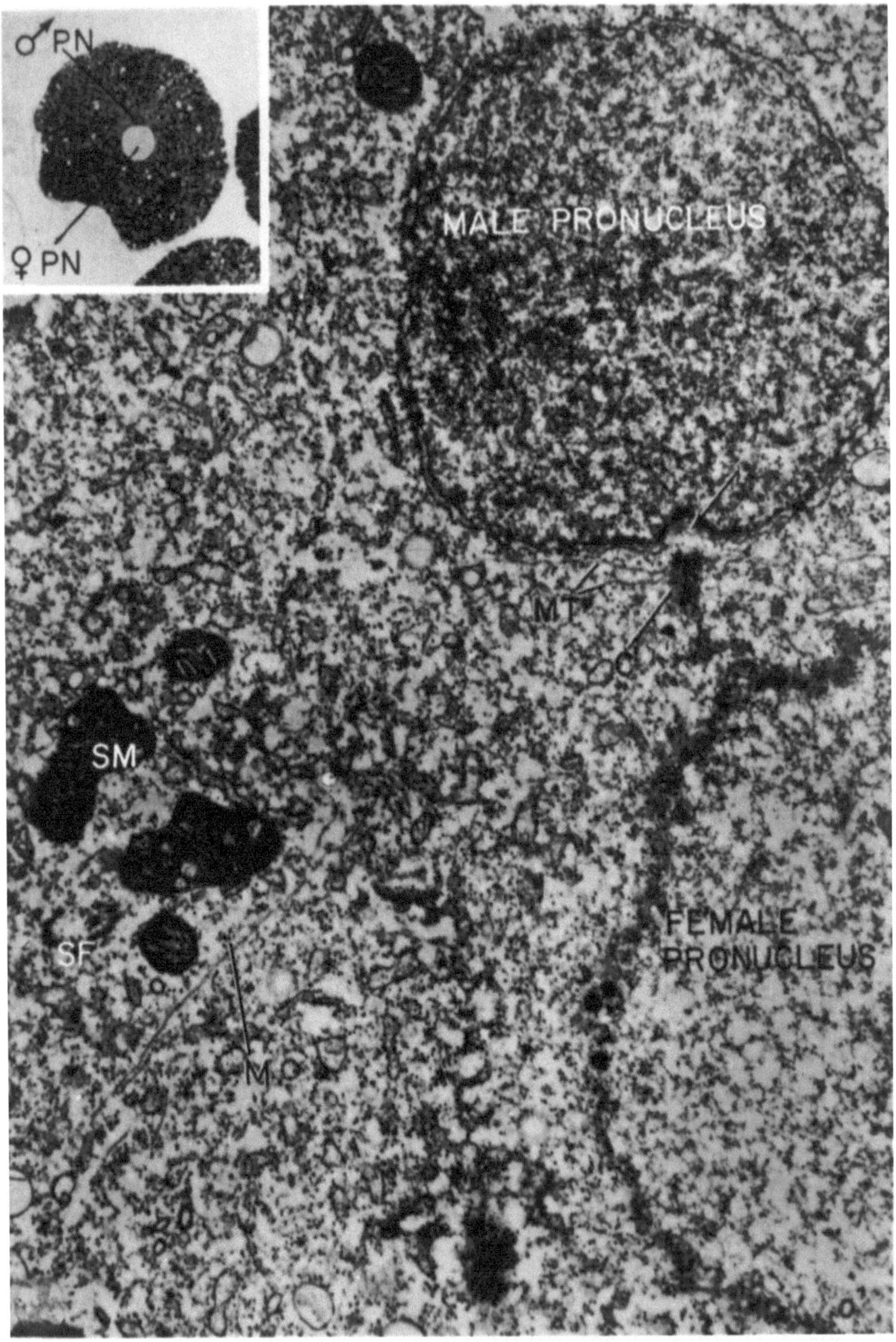
♂ PN
♀ PN
MALE PRONUCLEUS
MT
C
SM
SF
M
FEMALE PRONUCLEUS

◀ Fig. 4.30. Male and female pronuclei before fusing (♂PN and ♀PN, insert). Centriole (C) and associated microtubules (MT) are between pronuclei, while sperm mitochondrion (SM) and part of sperm flagellum (SF) are at one side. Arrow indicates centriolar fossa. (From LONGO and ANDERSON, 1968. By permission of Rockefeller University Press)

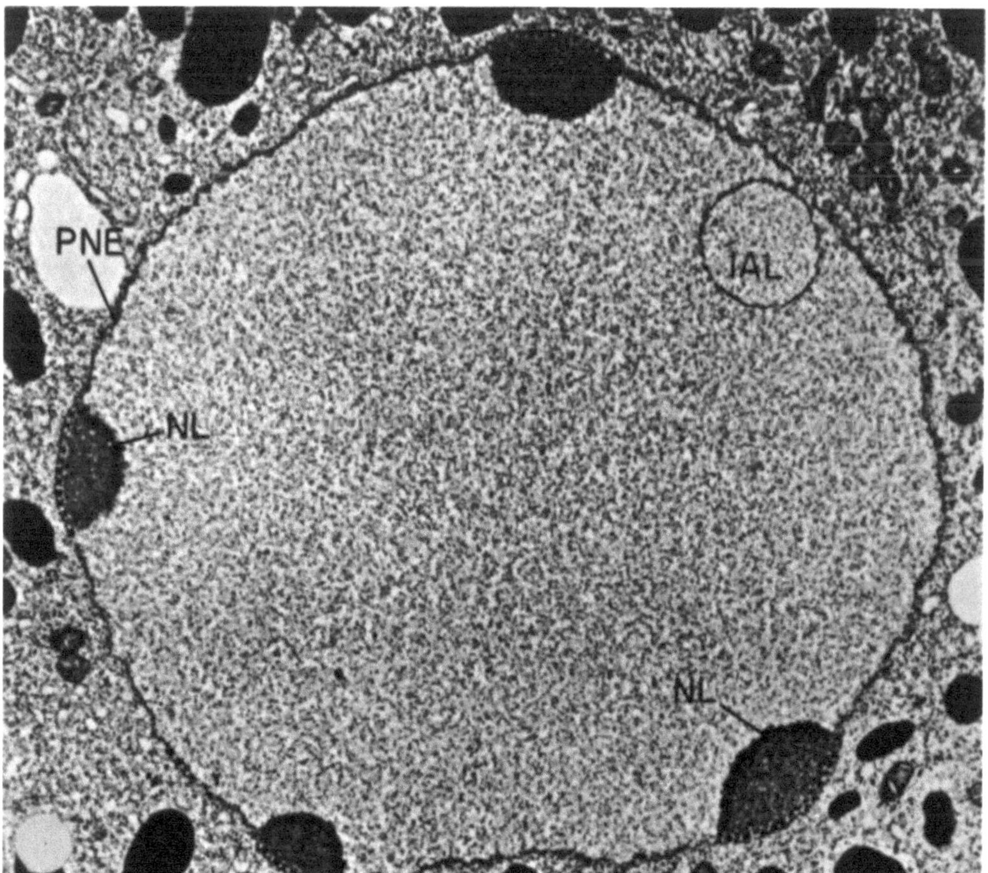

a

Fig. 4.31a. Legend see p. 126

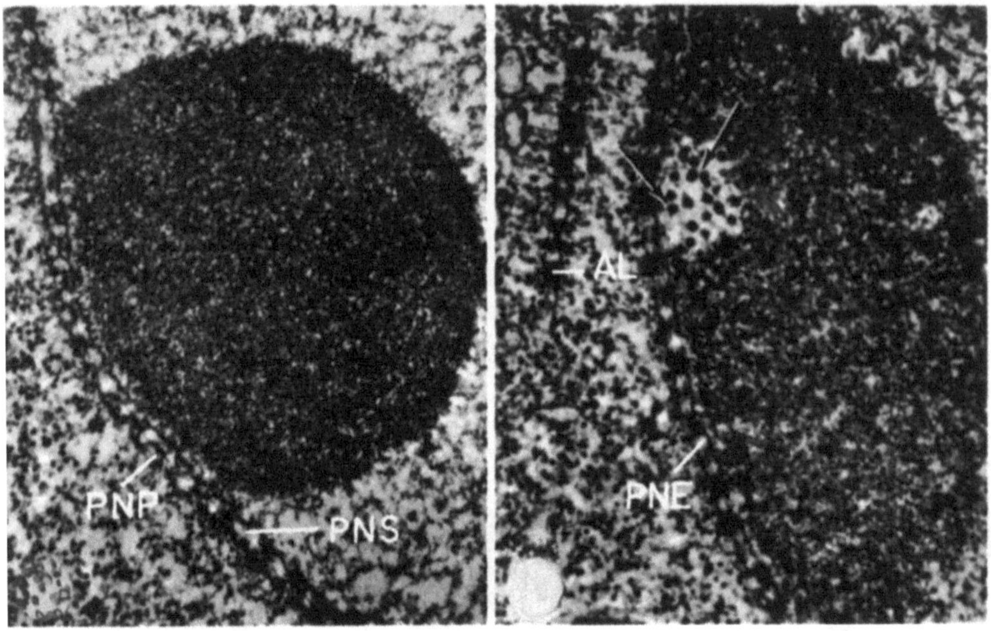

b c

Fig. 4.31a-c. Female pronucleus of inseminated egg containing 4 nucleoli-like structures (NL) and intranuclear annulate lamella (IAL). Perforated pronuclear envelope (PNE) on surface of these aggregations differs from PNE of other regions of pronucleus. Some of nucleoli-like structures contain dense granules (arrows). PNS, perinuclear space; AL, annulate lamellae. (From LONGO and ANDERSON, 1968. By permission of Rockefeller University Press)

Fig. 4.32. Male and female pronuclei meet. Female pronucleus shows flattened and developed nucleoplasmic projections (NP) near advancing male pronucleus. Endoplasmic reticulum (ER) surrounding male pronucleus is continuous with ooplasmic region surrounding female pronucleus. AL, annulate lamellae; NL, nucleolus-like body. (From LONGO and ANDERSON, 1968. By permission of Rockefeller University Press) ▶

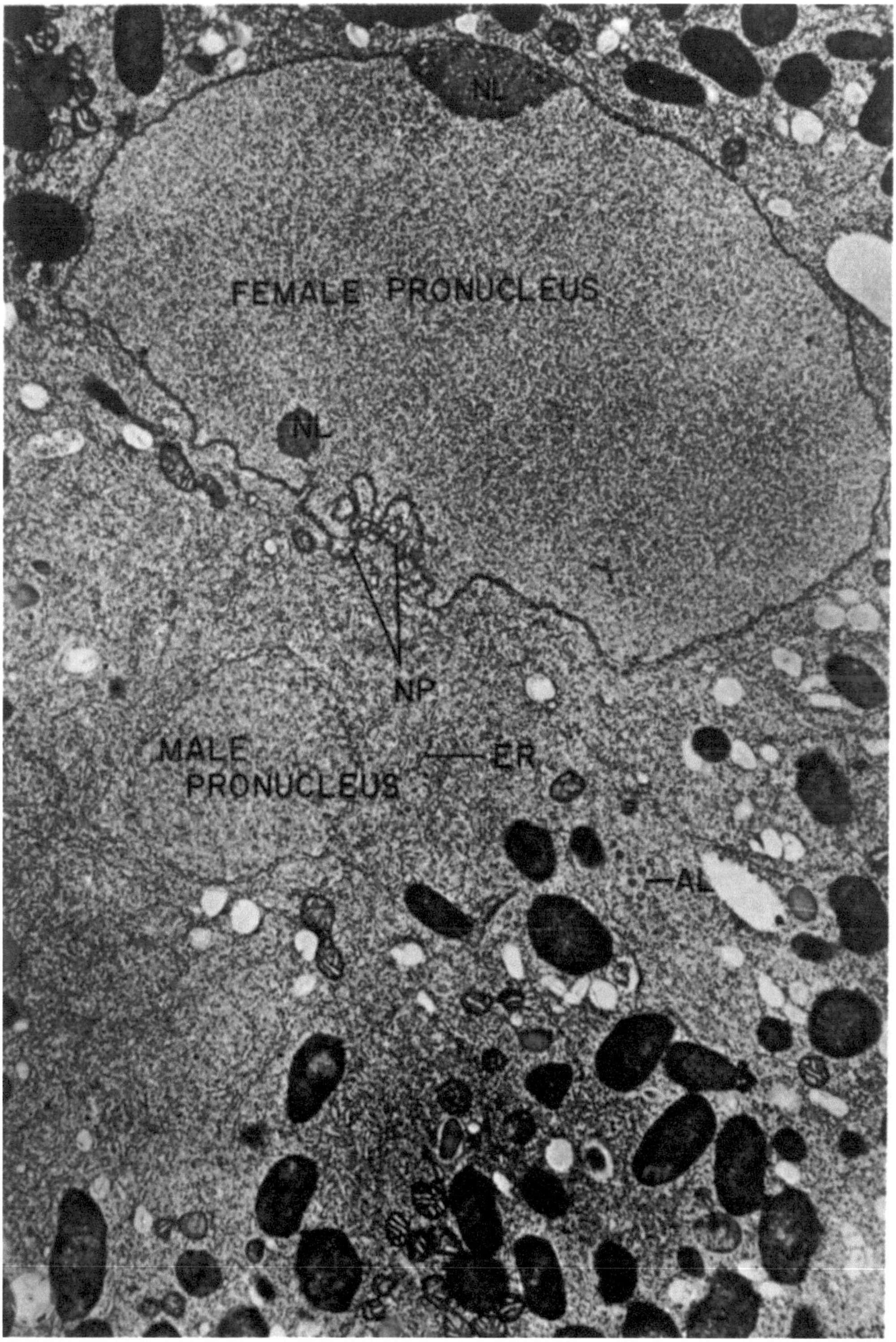
NL
FEMALE PRONUCLEUS
NL
NP
MALE PRONUCLEUS
ER
AL

4.3.11 Pronuclei Fusion

"In cases of dispermy the movement of the second sperm-nucleus is retarded if the first sperm-nucleus comes close to the egg-nucleus" (WILSON, 1925). If the spermatozoon enters the egg in the immediate vicinity of the egg pronucleus, syngamy occurs before a sperm aster forms and "syngamy can take place without intervention of astral rays...some kind of mechanism of attraction other than astral rays is responsible for the union of the pronuclei" (DAN, 1950a, 1950b).

Polyspermic eggs of the sea urchin *T. toreumaticus* were produced by pretreatment with a 1-*M* urea solution, and refertilized eggs were obtained by the method of SUGIYAMA (1951). In both, WILSON's finding on dispermy was confirmed, and it was believed that the pronuclear attraction mechanism might be blocked after 1 sperm pronucleus had conjugated with the egg pronucleus (ISHIKAWA, 1954, 1962).

In another experiment on pronuclear fusion, "unfertilized eggs of *Temnopleurus toreumaticus* were treated with hypertonic seawater (6 ml of 2.5 *N* NaCl plus 50 ml of seawater) for 20 minutes and then returned to ordinary seawater. ... After this treatment, the eggs began to cleave parthenogenetically with intact cortical granules and without a fertilization membrane. ... When the blastomeres were in-

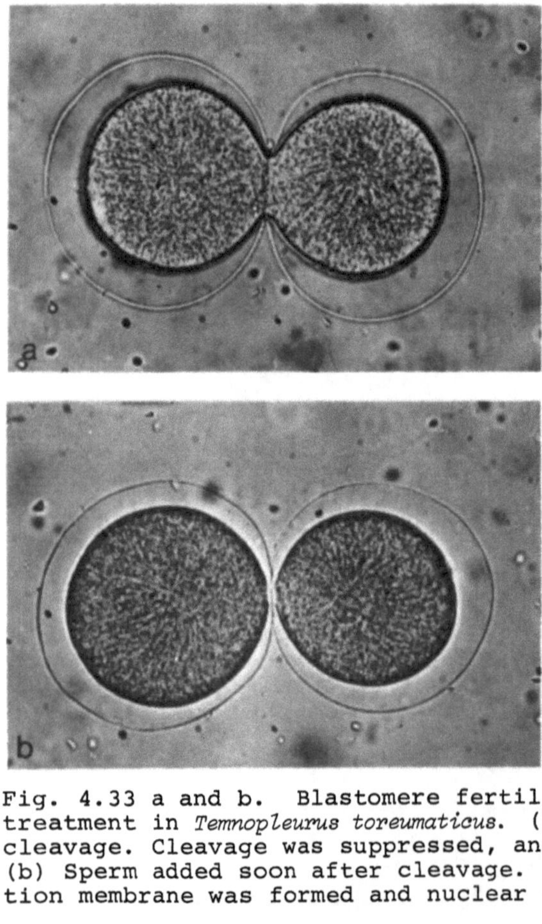

Fig. 4.33 a and b. Blastomere fertilization after parthenogenetic treatment in *Temnopleurus toreumaticus*. (a) Sperm added just before cleavage. Cleavage was suppressed, and nuclear conjugation occurred. (b) Sperm added soon after cleavage. In both blastomeres, fertilization membrane was formed and nuclear conjugation occurred

seminated soon after the first cleavage, a fertilization membrane was formed in both blastomeres (Fig. 4.33), which might still be haploid, and subsequently conjugation of the nuclei of the blastomere and sperm took place" (ISHIKAWA, 1954, 1962).

In electron microscopic studies on the fusion of pronuclei, the area of contact of the pronuclei is first restricted to a small locus at which the outer membranes of the sperm pronuclei and egg pronuclei come in contact. The inner lamina later meet and fuse, forming an internuclear bridge between the 2 pronuclei (Fig. 4.34). The bridge gradually increases in diameter until it becomes an ellipsoid body containing a protuberance that was the sperm pronucleus. This zygote nucleus later becomes a sphere (LONGO and ANDERSON, 1968).

4.4 Partial Fertilization

In fertilized eggs of the sea urchin, *Echinus*, BOVERI (1888) showed that the sperm nucleus may be left behind and the sperm aster approach the egg pronucleus alone, in which case the sperm aster and the egg pronucleus divide, and the first cleavage of the egg takes place without the sperm nucleus. Only 1 of the blastomeres contains a sperm nucleus with which the blastomere nucleus ordinarily conjugates in the 2-cell stage. The nucleus of 1 blastomere is completely maternal and the other biparental. BOVERI (1888) called this partial fertilization.

TEICHMANN (1903) found that partial fertilization takes place when spermatozoa are exposed to weak alkaline seawater (Fig. 4.35), and ISHIKAWA (1954, 1962) confirmed this: "Fresh dry sperm was diluted to 10^{-3} with KOH-seawater (1 ml of N/10 KOH plus 20 ml of seawater, pH 9.8). ... In this suspension, the spermatozoa retained their fertilizing capacity at least for 2 minutes at 28°C. The partial fertilization took place in a high percentage by insemination with the spermatozoa which had been treated with the alkaline medium for 2 minutes" (Table 4.2).

Table 4.2. Partial fertilization with spermatozoa exposed to KOH seawater (1 ml N/10 KOH + 20 ml seawater) at temperature of 28°C. Temporary 3-cell stage shows partial fertilization because cleavage of haploid blastomere in second cleavage generally precedes cleavage of diploid blastomere. (From ISHIKAWA, 1962)

			Second cleavage		
Time of exposure to KOH seawater (min)	Formation of fertilization membrane (%)	First cleavage (%)	Normal division (%)	Division characteristic of partially fertilized eggs (%)	No division (%)
0.5	100	100	100	0	0
1.0	100	100	100	0	0
1.5	100	100	69	31	0
2.0	100	100	60	40	0
3.0	56	54	40	14	46
4.0	14	4	0	0	100
5.0	0	0	0	0	100

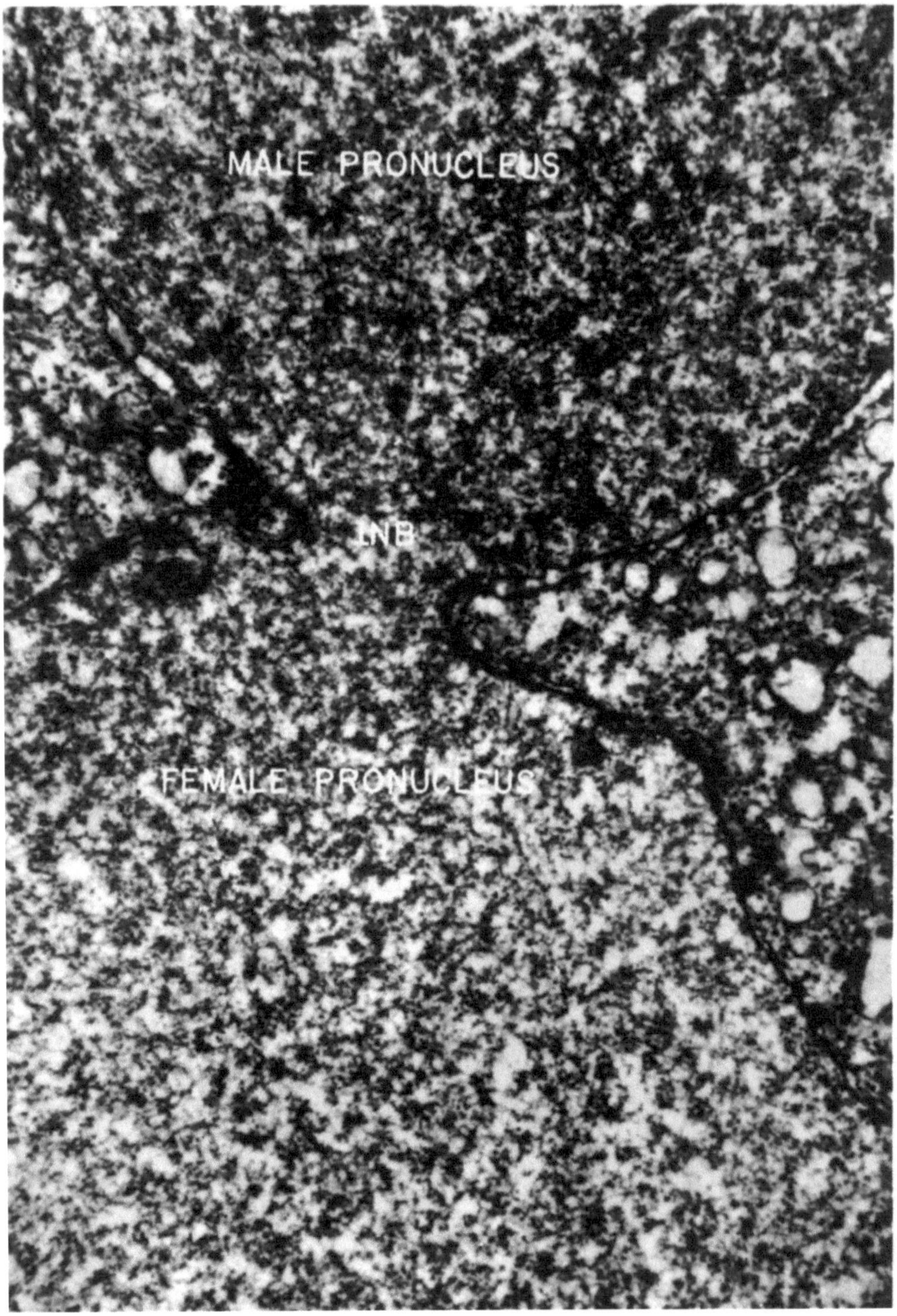
MALE PRONUCLEUS
INB
FEMALE PRONUCLEUS

◀ Fig. 4.34. Completion of coalescence of inner and outer lamina of male and female pronuclear envelopes. Internuclear bridge (INB) connects the 2 pronuclei. (From LONGO and ANDERSON, 1968. By permission of Rockefeller University Press)

Using blastomeres that developed after partial fertilization, ISHIKAWA (1954, 1962) studied the fusion of nuclei of eggs inseminated twice: once with haploid and once with diploid nuclei of blastomeres: "Immediately after the first insemination with alkaline-treated sperms, the fertilization membrane was removed in 1 *M* urea. ... The eggs were washed again with 1 *M* urea for 3 minutes just before the first cleavage, and one of the blastomeres divided into 2 small blastomeres, which were assumed to have the haploid nucleus. ... 10 minutes later, cleavage of the other blastomere, which was considered to be diploid,

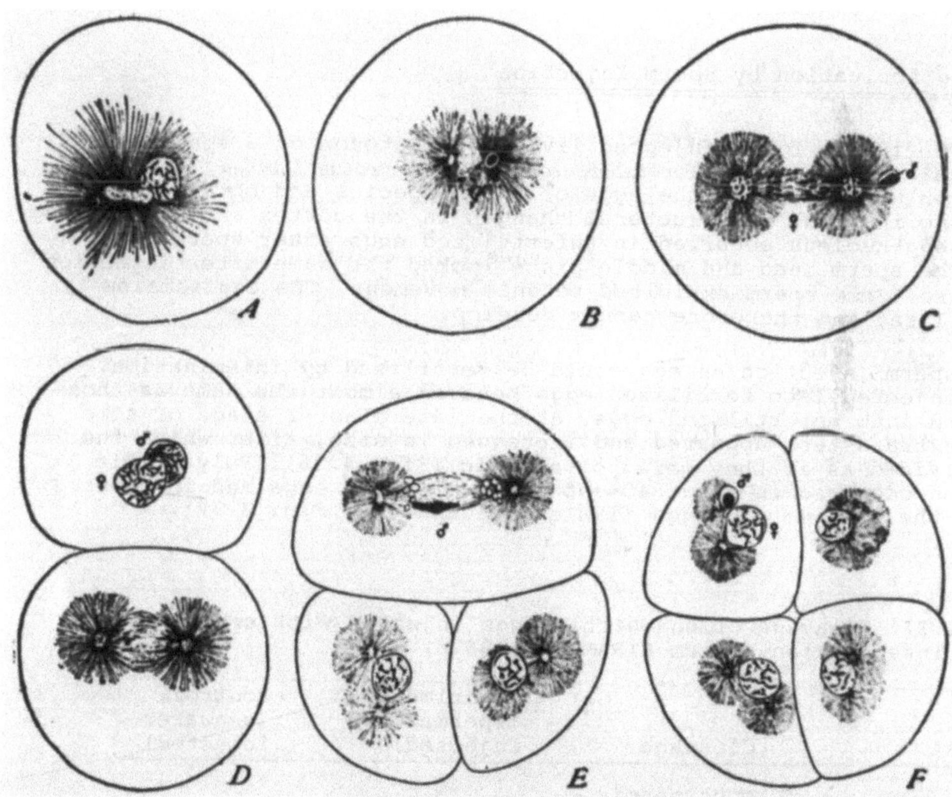

Fig. 4.35A-F. Partial fertilization in sea urchin *Echinus* after sperm were treated with weak KOH solution. (A) Approach of pronuclei, division of sperm aster. (B) Developing amphiaster. (C) Anaphase, sperm nucleus near one pole. (D) 2-cell stage, sperm nucleus conjugated and cleavage nucleus. (E) Late second cleavage, sperm nucleus lagging top. (F) 4-cell stage, sperm nucleus still separate, 3 cells with maternal nuclei only. (From TEICHMANN, 1903)

followed. ... The second insemination with normal sperms was performed before the cleavage of the diploid blastomere, namely at the stage of 3 blastomeres." The sperm nucleus was found to conjugate with the blastomere nucleus in the haploid blastomere but not in the diploid blastomere.

Results of experiments with polyspermic fertilization, refertilization, blastomere fertilization induced by parthenogenetic activation, and partial fertilization, suggest that the phenomenon of nuclear conjugation involves some kind of monospermy-insuring mechanism as well as the polyspermy-preventing mechanism of the egg at the time of sperm entry.

In hybrids between *Sphaerechinus* and *Paracentrotus*, HERBST (1912) observed something resembling partial fertilization; the nuclei were of 2 sizes and in some cases showed a small (maternal) nucleus in 1 blastomere and a large (hybrid) nucleus in the other.

4.5 Egg Activation by Sperm Injection

HIRAMOTO (1962a) first injected living spermatozoa of 3 species of sea urchins (*Psammechinus microtuberculatus*, *Paracentrotus lividus*, and *Clypeaster japonicus*) into the eggs of those species and obtained the following results: No structural changes in the cortex and no swelling of the egg-nucleus occurred in unfertilized eggs after sperm injection. The sperm head and middle piece looked the same after injection as before. Some sperm exhibited tetanic movement. The egg remains unfertilized and therefore cannot develop.

After sperm injection an egg could be fertilized by insemination. Sperm injected into fertilized eggs behaved almost the same as those injected into unfertilized eggs. At the late diaster stage of some eggs, extra asters appeared and increased in size, after which the eggs divided as if they were polyspermic (Fig. 4.36). Polyspermic division occurred in about 40% of the *Psammechinus* eggs and in about 50% of the *Paracentrotus* eggs (Tables 4.3 and 4.4; Fig. 4.37).

Table 4.3. Cleavage of sea-urchin eggs injected with spermatozoa before insemination. (From HIRAMOTO, 1962a)

Species	Cleavage	Experimental (spermatozoa injected)	Control (seawater injected)
Psammechinus microtuberculatus	Monospermic	8	
	Polyspermic	5	
	Total	13	
Paracentrotus lividus	Monospermic	11	9
	Polyspermic	11	1
	Total	22	10

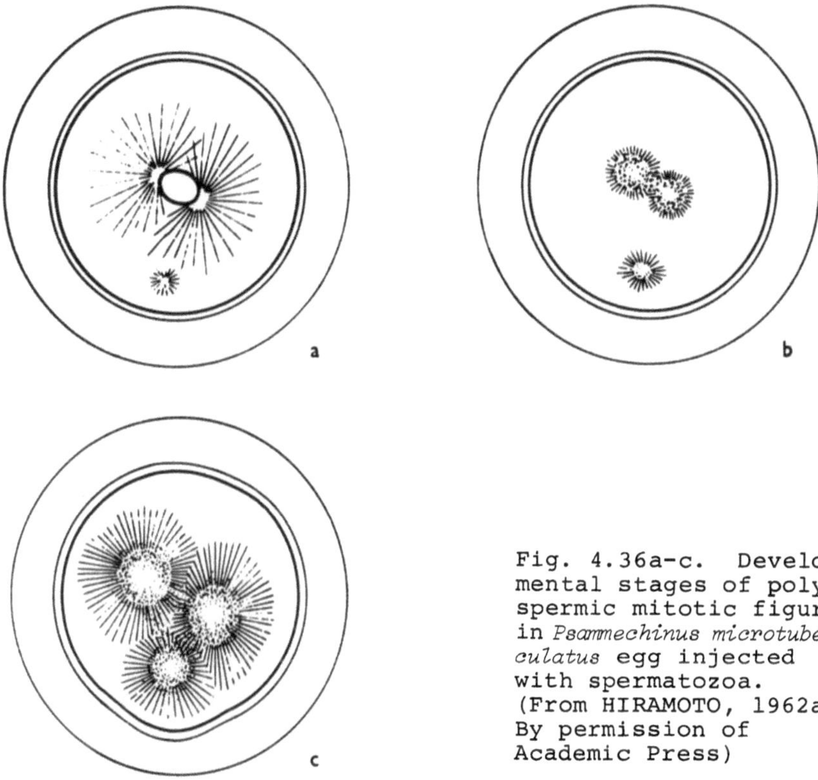

Fig. 4.36a-c. Developmental stages of polyspermic mitotic figure in *Psammechinus microtuberculatus* egg injected with spermatozoa. (From HIRAMOTO, 1962a. By permission of Academic Press)

Table 4.4. Cleavage of sea-urchin eggs injected with spermatozoa after insemination. (From HIRAMOTO, 1962a)

Species	Cleavage	Experimental (spermatozoa injected)	Control: Seawater injected	Control: Injected spermatozoa into perivitelline space	Control: Applied spermatozoa to egg deprived of fertilization membrane
Psammechinus microtuberculatus	Monospermic	19		15	
	Polyspermic	6		0	
	Total	25		15	
Paracentrotus lividus	Monospermic	5	18	16	36
	Polyspermic	7	0	1	1
	Total	12	18	17	37

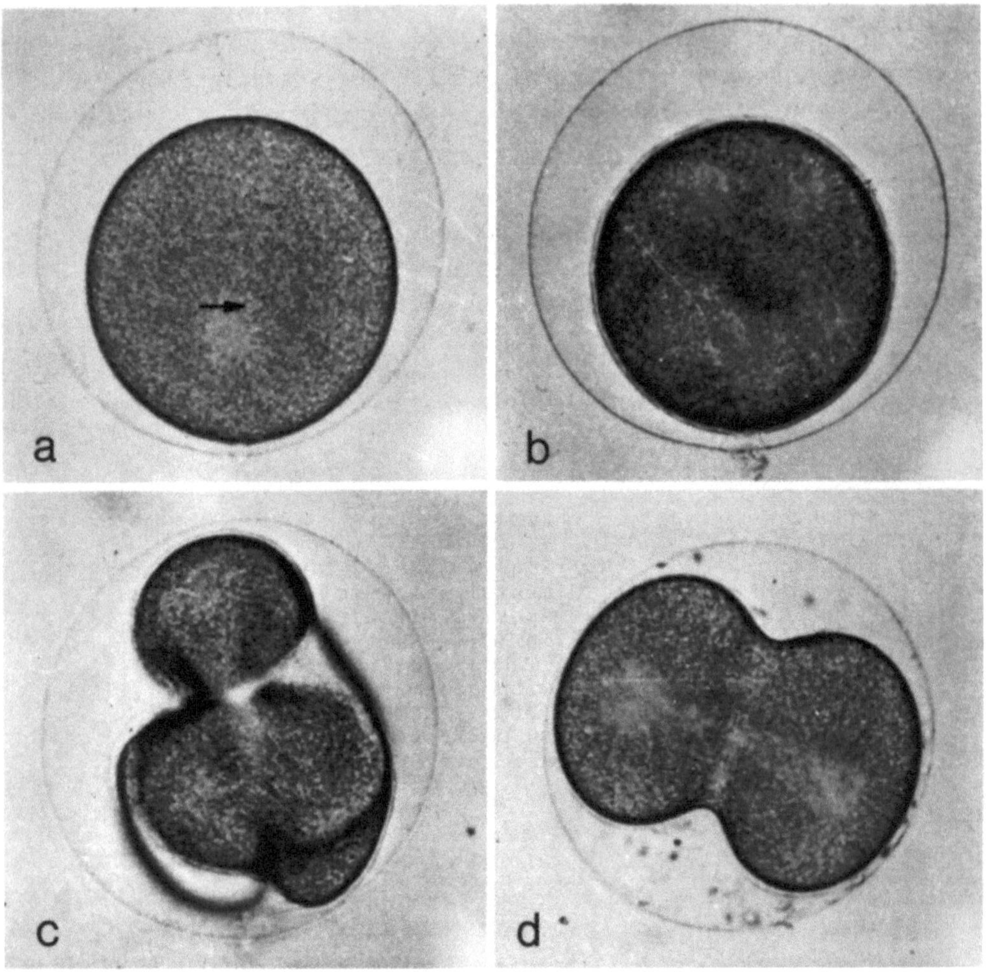

Fig. 4.37a-d. Successive stages of egg injected with spermatozoa after insemination. Injected spermatozoon is barely visible in egg protoplasm (arrow). (d) Normal egg cleavage after spermatozoa were injected into perivitelline space. *Paracentrotus lividus*. (From HIRAMOTO, 1962a. By permission of Academic Press)

4.6 Sperm Role in Egg Activation

Paracentrotus eggs were allowed to interact with spermatozoa for a short time. The eggs were then immersed in hypertonic seawater to kill the sperm. Some eggs - presumably those in which sperm-egg collision was successful - were activated but not fertilized. The egg nucleus swelled, but cleavage did not occur (ROTHSCHILD, 1953).

In gametes of *P. microtuberculatus*, HIRAMOTO (1962b) analyzed the mode of sperm stimulating action at the time of fertilization. With a micropipet he removed sperm at various stages from sperm attachment

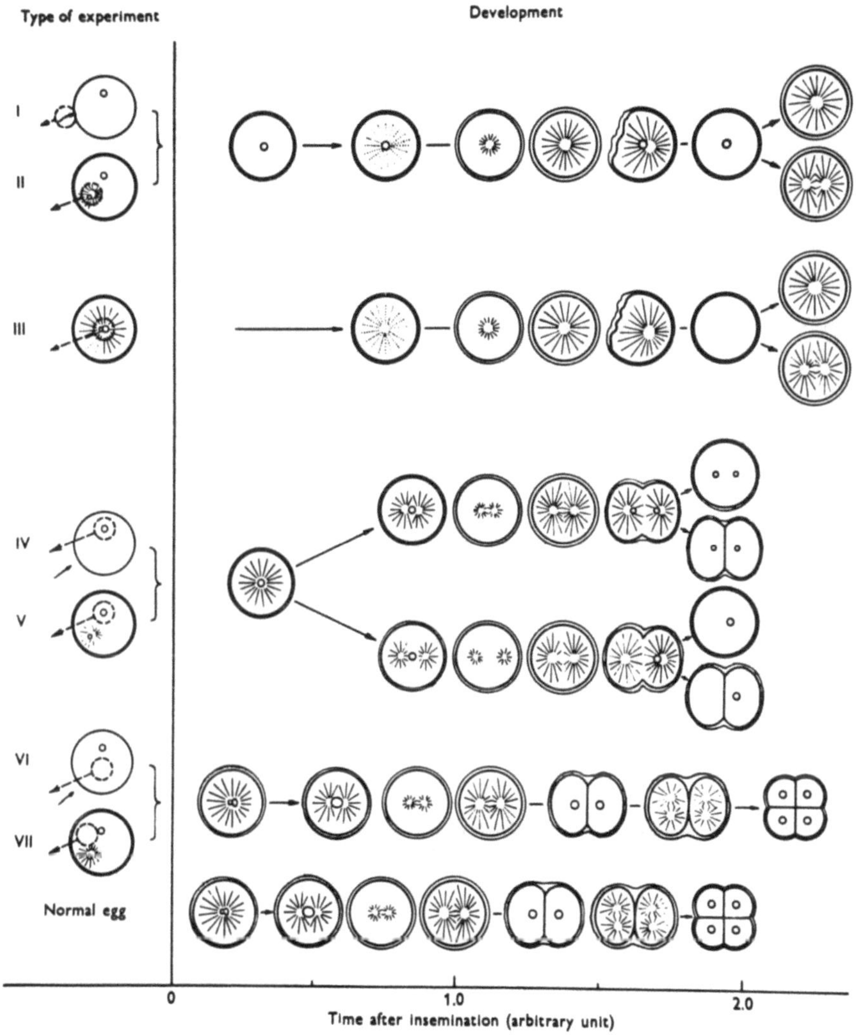

Fig. 4.38. Summary of experimental results. Fertilization membrane omitted. (From HIRAMOTO, 1962b. By permission of Academic Press)

to meeting of the pronuclei. He also removed the egg pronucleus to study its role in fertilization. His results are summarized in Fig. 4.38.

Either the sperm attached to the egg surface or the sperm pronucleus was removed from fertilized eggs (Fig. 4.38, I,II), and sperm pronuclei and egg pronuclei were removed from fertilized eggs (Fig. 4.38, III). In all the eggs a monaster formed and the egg shape and thickness of hyaline layer changed. In some eggs from which the egg pronucleus was removed before or after insemination (Fig. 4.38, IV,V), normal diaster with spindle formed and nuclear division took place; in others there was no spindle between the 2 asters, and nuclear division did not occur. In all eggs an incomplete cleavage furrow appeared between the 2 asters but later receded. The removal of a small amount of egg cytoplasm (Fig. 4.38, VI,VII) scarcely affected the mitotic process. HIRAMOTO (1962b) concluded that "the spermatozoon plays 2 different roles in fertilization: the first is the activation of the egg at the egg-surface, which induces a rhythmic activity in the egg cytoplasm; the second is a contribution to the formation of the dicentric mitotic figure by providing a pair of astral centers in the egg cytoplasm."

4.7 Egg Cortical Response

In fertilization or artificial activation, the cortical response of sea-urchin eggs has long been considered important. LOEB (1913a) stated that "membrane formation and the activation of development of the sea urchin egg is due to cytolysis of the surface or the cortical layer of the egg." In fertilization the breakdown of the cortical granules of the egg begins at the point of sperm entry and spreads over the egg surface like a wave.

Using dark-field illumination, RUNNSTRÖM (1928) noticed that the interference color of the egg surface changes from yellow-red to silver-white 4 to 6 seconds after fertilization. ROTHSCHILD and SWANN (1949) showed that the time of surface change as recorded by dark-field cinematography is about 20 seconds at 18°C in *P. miliaris* eggs fertilized at the equator: "Such a cortical change following fertilization could be due to the diffusion of a substance derived directly or indirectly from the sperm head, and the diffusion might not be intracortical but intracellular, i.e., passing through the interior cytoplasm." RUNNSTRÖM and KRISZAT (1952) felt that the impulse had cortical propagation and spread by chain reaction rather than by pure diffusion.

In the cortical response SUGIYAMA (1953a, 1953b) pointed out 2 mechanisms responsible for the wave-like breakdown of granules at fertilization: "The one is that the breakdown of a cortical granule may be the cause of the breakdown of its neighboring granules. ... Thus their wave-like breakdown could be of a self-propagating nature. ... The other mechanism is that there may occur an antecedent propagating change in the cortex which may cause the granular breakdown, which appears to be wave-like." SUGIYAMA examined these possibilities and found that "sodium choleinate and wasp-venom were effective in inducing the formation of the fertilization membrane, and the mode of response of the eggs was somewhat different from that obtained in experiments with butyric acid or urea solution." Results from sodium choleinate experiments are given in Table 4.5. In the eggs treated with choleinate, local membrane elevation occurred after transfer to seawater: the cortical granules broke down in the part of the cortex with separated membrane. It is not very likely therefore, that local breakdown of the cortical granules causes breakdown of the neighboring granules. Some invisible cortical change of a propagating nature may cause the breakdown of the cortical granules at fertilization. SUGIYAMA (1953a) called the invisible cortical change in the sea-urchin egg the "fertilization wave" after a similar phenomenon that YAMAMOTO (1944) observed in *Oryzias* egg and identified by that name.

SUGIYAMA (1953b) further analyzed the cortical response of the egg: "The jellyless eggs are partially exposed at the equatorial region to various activating reagents, such as wasp-venom, sodium choleinate, urea, glycerine, sucrose, distilled water, detergents (Monogen and

Table 4.5. Fertilization membrane formation by means of sodium choleinate at temperature of 11°C. (From SUGIYAMA, 1953a)

Drops of Na-choleinate solution added to 1cc seawater	Mode of membrane formation
20	Membrane violently separated in 10 sec.
15	Membrane violently separated in 40 sec.
10	"Normal" membrane formed in 50% of eggs in 2 min. Some eggs formed only partially. After 3 min, 92% of eggs had completed membrane elevation.
5	No response in most eggs after 5 min. Partial membrane formed in a few eggs.
4	No response in any eggs after 5 min.

Lipon, which consist chiefly of a mixture of myristyl and lauryl-sulfate, and of alkylbenzene sulfonic acid, respectively) and fatty acids (butyric, propionic and acetic acid), and then are washed with seawater." The unexposed surface of the eggs was then examined for cortical granule breakdown. "When sodium choleinate, wasp-venom and detergents were used, the cortical granules in the exposed part of the egg soon began to break down, but in the compressed sides of the egg (the unexposed part) the granules remained completely intact. ... On the other hand, when the egg surface was partially exposed to fatty acids, distilled water or an isotonic solution of a non-electrolyte, the cortical granules in the exposed part began to break down and the breakdown proceeded in the unexposed part." SUGIYAMA (1953b) concluded that the cortical change caused by treatment with the former group of reagents induces granular breakdown without intervention of the fertilization wave and that the cortical change provoked by the reagents of the latter group is of a propagating nature and is essentially the same as that which follows the entrance of spermatozoa. The cortical change at fertilization and artificial activation is illustrated in Fig. 4.39.

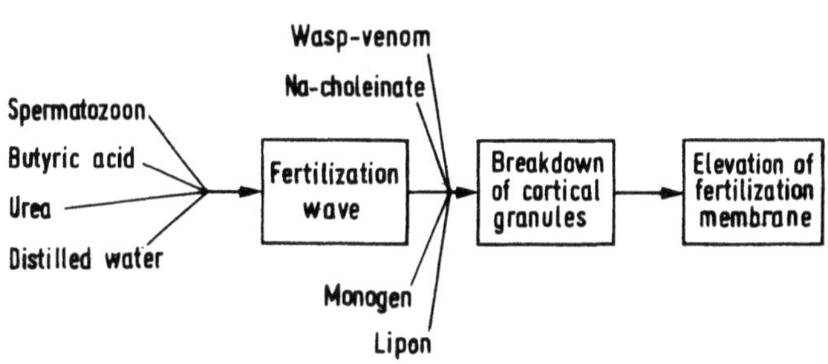

Fig. 4.39. Diagrammatic illustration of cortical change. (From SUGIYAMA, 1953b. By permission of Biol. Bull.)

4.8 Narcotized Egg Responses

SUGIYAMA (1956) found that no cortical change occurs at fertilization if the unfertilized egg of *H. pulcherrimus* is completely narcotized with 6% urethane in seawater (Table 4.6) and found that this response inhibition is reversible. Eggs treated with urethane for 30 seconds, transferred to normal seawater for 12 minutes, and inseminated, nearly all showed a cortical reaction (Table 4.7).

Table 4.6. Cortical change in eggs narcotized with 6% urethane seawater at temperature of 9°C. (From SUGIYAMA, 1956)

Immersion in urethane seawater (sec)	Eggs showing cortical change (%)		
	After insemination	After urea treatment (7 sec)	After butyric acid treatment (1 min)
0	99	98	99
10	98	99	98
20	81	80	76
30	0	0	0

SUGIYAMA (1956) obtained similar results from an experiment on narcotized egg response to artificial activating reagents: Butyric acid and urea solution, for example, do not cause cortical change in completely narcotized eggs; egg response is readily recoverable. The other group of activating reagents - such as sodium choleinate, wasp venom, and Monogen - induce cortical granule breakdown and membrane formation in completely narcotized eggs.

"The cortex of the unfertilized egg is an irritable system and when it is stimulated by a spermatozoon or an activating reagent of the first group, an impulse is provoked and conducted over the entire cortex. ... This is the spreading of the 'fertilization-wave'. ... The narcotic may decrease the irritability of the egg-cortex so that the impulse is inhibited and the fertilization-wave does not take place" (SUGIYAMA, 1956).

But can the fertilization wave spread over the fertilized or activated egg as well as the unfertilized egg? UEHARA and SUGIYAMA (1969) sucked an egg from both sides into 2 capillaries, exposing to Monogen only a ring-shaped area of egg surface (Fig. 4.40); the membrane formed only on the exposed surface. When 1 capillary was removed and sperm was added to the exposed side of the egg, the fertilization membrane spread to the other side, crossing the equatorial region

Table 4.7. Narcosis reversibility in eggs narcotized with 6% urethane seawater at temperature of 9°C. (From SUGIYAMA, 1956)

Treatment	Immersion in urethane seawater 30 sec	Immersion in urethane seawater 30 sec and kept in normal seawater 12 min
Showing cortical change after insemination	0%	98%

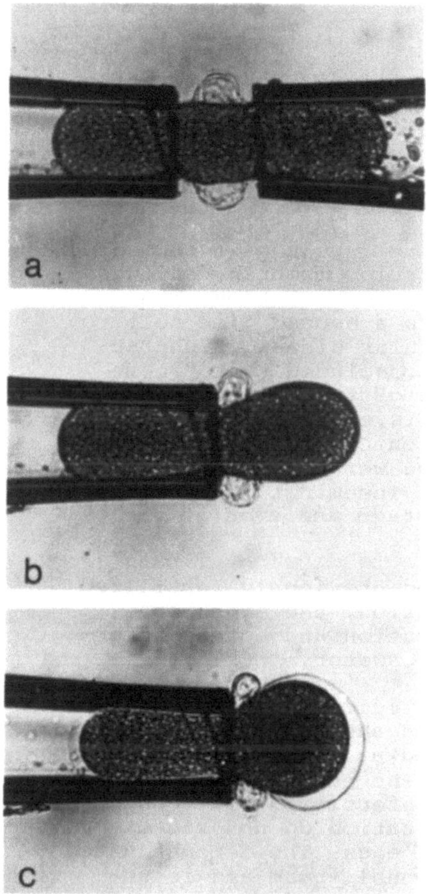

Fig. 4.40a-c

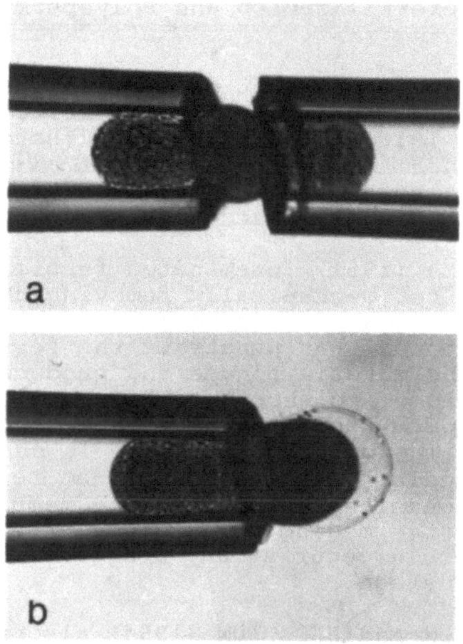

Fig. 4.41 a and b

Fig. 4.40a-c. Fertilization wave spreads across equatorial region in which cortical change was provoked by detergent, Monogen. (a) Egg was sucked into 2 capillaries and treated with 0.075% Monogen-seawater. (b) 1 capillary was removed and sperm added. (c) Fertilization membrane formed on both sides of egg. (From UEHARA and SUGIYAMA, 1969)

Fig. 4.41 a and b. In egg narcotized with urethane solution, fertilization wave is blocked at equatorial region. (a) Egg was sucked into 2 capillaries and ring-shaped area of egg narcotized with 6% urethane seawater. (b) 1 capillary was removed. After insemination, no membrane formed on unexposed region of capillary. (From UEHARA and SUGIYAMA, 1969)

where the cortical change had taken place. When the exposed egg region between 2 capillaries was narcotized with urethane and sperm was added to 1 side of the egg after 1 capillary was removed, fertilization membrane formed only on the egg surface of the sperm-entry side (Fig. 4.41). UEHARA and SUGIYAMA concluded that "the conduction system of the egg is considered to remain workable after the breakdown of the cortical granules, but not after narcosis."

4.9 Refertilization and Polyspermy

It was believed that once an egg had been fertilized, it could not be fertilized again even if the fertilization membrane were removed (LOEB, 1916; LILLIE, 1919), although BURY (1913) had reported that if *Strongylocentrotus lividus* and *Echinus microtuberculatus* eggs that had been fertilized and formed fertilization membranes were kept at 0°C, they could be refertilized.

SUGIYAMA (1951) inseminated fertilized *H. pulcherrimus* eggs a second time after mechanically removing the fertilization membrane and washing the eggs with Ca-Mg-free seawater or 1 *M* urea solution. The spermatozoa that penetrate the fertilized eggs a second time help to form the mitotic figure and produce the irregular cleavages characteristic of polyspermy (Table 4.8). SUGIYAMA concluded that "there is a sperm-excluding substance on the surface of fertilized eggs, and this substance is stable in seawater but is easily lost in Ca-Mg-free media." He also found that refertilization could take place without mechanically removing the membrane if eggs were treated with a Ca-Mg-free medium immediately after the first insemination. Fertilization could occur at any stage of the 1-cell stage and even at the 2-cell stage.

HAGSTRÖM and HAGSTRÖM (1954) also refertilized previously fertilized sea-urchin eggs. Before the initial fertilization, the eggs were treated with trypsin to inhibit membrane elevation and before refertilization were treated with Ca- or Mg-free seawater.

TYLER, MONROY, and METZ (1956) found that fertilized, mechanically demembranated eggs of sea urchins *Lytechinus pictus* and *L. variegatus* can be refertilized if they are heavily reinseminated at least an half hour after the initial fertilization (Table 4.9). The earlier the second insemination is performed, the more readily refertilization occurs. Sperm concentrations that give 100% refertilization do not cause any significant polyspermy in normal unfertilized eggs. The demembranated fertilized egg is capable of accepting additional sperm and is unable to block quickly all but one sperm. "Treatment of the unfertil-

Table 4.8. Fertilized eggs deprived of fertilization membranes and refertilized in Ca-Mg-free medium. (From SUGIYAMA, 1951)

		3 hours after superimposed insemination			
Medium membrane removed in	Sperm concentration in second insemination	Undivided eggs, not fertilized (%)	Normally divided eggs (%)	Polyspermic eggs (%)	Polyspermic eggs in control, not inseminated again (%)
Ca-Mg-free seawater	10^{-2}	1	1	98	1
	10^{-3}	0	5	95	
	10^{-4}	0	80	20	
	10^{-5}	1	98	1	
Urea solution (1 *M*)	10^{-2}	0	0	100	0
	10^{-3}	0	1	99	
	10^{-4}	0	49	51	
	10^{-5}	0	94	6	

Table 4.9. Refertilization of mechanically demembranated *Lytechinus pictus* eggs after fertilization.[a] (From TYLER, MONROY and METZ, 1956)

Experiment	Minutes between first and second inseminations	Sperm (x 10^6) per milliliter of egg suspension		Polyspermy determined at first cleavage at about 2 hours (%)
		First insemination	Second insemination	
1	Not demembranated	1.2	0	1
	Demembranated	1.2	0	3.5
	2 1/2	1.2	1.2	53
	4 1/2	1.2	1.2	51
	10	1.2	1.2	67
	20	1.2	1.2	20
	30	1.2	1.2	13
	40	1.2	1.2	35
2	Not demembranated	28	0	1
	Demembranated	28	0	5.6
	3	28	28	100
	5	28	28	100
	7	28	28	98
	10	28	28	96
	30	28	28	40
3	Not demembranated	2	0	2.5
	Demembranated	2	0	13
	5	2	6	100
	10	2	6	100
	20	2	6	81
	40	2	6	27
4	Not demembranated	1	0	3
	Demembranated	1	0	22
	10	1	0.4	37
	10	1	1	51
	10	1	4	100
	10	1	8	100
	40	1	0.4	32
	40	1	1	35
	40	1	4	76
	40	1	8	65
5	Not demembranated	49	0	2
	Demembranated	49	0	3
	8	49	49	100
	8	49	245	100
	15	49	49	68
	15	49	245	100
	30	49	49	11
	30	49	245	78
	40	49	49	17
	40	49	245	61

[a]Semen in experiments 1 to 4 was diluted initially to 1% in 10^{-3} *M* Versene in seawater and stored for later inseminations at that concentration. Semen in experiment 5 was diluted to 10% in ordinary seawater. Sperm counts were made in experiments 2, 4, and 5. In experiments 1 and 3, sperm was estimated from dilution based on 2 x 10^{10} per milliliter of semen.

ized eggs with trypsin, in order to suppress membrane formation, or with papain at the time of fertilization, so as to cause reversal of membrane elevation, also permits refertilization upon reinsemination without further treatment with Ca-Mg-free seawater or urea solution. ... There is the possible presence of fertilizin-like sperm receptors on the surface of the denuded fertilized egg" (TYLER, MONROY, and METZ, 1956).

A phenomenon similar to refertilization was first reported by LOEB (1915): "Eggs of *Strongylocentrotus purpuratus* that have been artificially activated can be entered by a number of sperms, and also blastomeres can be fertilized if the fertilization membrane is removed." Reversal of parthenogenetic development and subsequent fertilization was observed in *Arbacia* eggs (LOEB, 1913b; WASTENEYS, 1916), but many other investigators (MOORE, 1916, 1917; LILLIE, 1919, 1921; JUST, 1922; LILLIE and JUST, 1924) concluded that it was impossible to superimpose fertilization on parthenogenesis after optimal treatment with the activating agent. In later years, ISHIDA and NAKANO (1950) and NAKANO (1954) reported that artificially activated eggs of *H. pulcherrimus* could be fertilized if the membranes were removed mechanically and the eggs pretreated with Ca-Mg-free seawater. Cleavage of them resembled that of polyspermy.

4.10 Conclusions

Fertilization is the union of an egg cell with a spermatozoon and the fusion of sperm pronuclei and egg pronuclei. The spermatozoon activates the egg, inducing it to develop, while maternal and paternal hereditary characteristics are combined.

One of the ways the spermatozoon reacts to egg water or an egg substance, fertilizin, is by agglutination. First studied by LILLIE (1913, 1919) and TYLER (1941), agglutination was attributed to the interaction between an egg substance called fertilizin and a complementary sperm-surface substance called antifertilizin. This fertilizin-antifertilizin reaction led TYLER to advance his fertilizin theory to explain the sea-urchin fertilization mechanism.

The other way the spermatozoon generally reacts to egg water or fertilizin involves the acrosome. In sea urchins an acrosomal filament forms on the tip of the spermatozoon in reaction to dissolved egg jelly (DAN, 1952). While agglutination and acrosomal reaction are both responses to the stimulus of egg water, they are not causally interrelated. Sperm having intact acrosomes retain their ability to fertilize and lose it when acrosomal reaction is artificially induced. In the sea-urchin spermatozoon the acrosomal reaction is believed to be important to fertilization (DAN, 1954).

To understand the fertilizing behavior of spermatozoa, the electron microscope has been used for ultrastructural research on oocytes and eggs treated with polyspermy-inducing reagents.

The work of FRANKLIN (1965) and LONGO and ANDERSON (1968) on the morphologic changes in the fertilization of sea-urchin gametes is summarized as follows: Gamete contact and fusion and early events in the egg cytoplasm in the oocyte are similar in the mature egg, although in the oocyte the spermatozoa that enter do not develop and complete sperm pronuclei and sperm asters do not form. After a spermatozoon enters a mature egg, the lateral sperm-nuclear region disappears, the apical nuclear envelope remains intact, and the sperm nucleus is incorporated into the ground substance of the fertilization cone.

Within 2 to 2 1/2 minutes after insemination, sperm-nucleus rotation in the cytoplasm sometimes occurs. The sperm aster begins to develop from the centriolar region of the sperm, growing by means of the accumulation of large numbers of microtubules, vesicles of endoplasmic reticulum, and annulated lamellae. When the sperm pronuclei and egg pronuclei are close together, the part of the egg pronucleus nearest the sperm pronucleus flattens and extends finger-like nucleoplasmic projections toward it. The pronuclei touch and fuse, forming an internuclear bridge that grows into the spherical zygote nucleus.

In the partial fertilization described by BOVERI (1888) the sperm aster and egg pronucleus divide without the male pronucleus. The first cleavage of the egg separates 2 blastomeres, one of which contains a female nucleus from a blastomere, and another one contains both a sperm nucleus and a female nucleus. Partial fertilization may be induced by insemination with spermatozoa treated with weak alkaline seawater (TEICHMANN, 1903; ISHIKAWA, 1962).

Living spermatozoa injected into sea-urchin eggs do not activate unfertilized eggs (HIRAMOTO, 1962a). If the injected eggs are fertilized by insemination, they divide like polyspermic ones. Fertilized eggs in which spermatozoa are injected show a high percentage of polyspermic division.

The spermatozoon role in fertilization was examined by removing the spermatozoon - or the sperm pronucleus and egg pronucleus - with a micropipet at stages from sperm attachment to pronuclei meeting (HIRAMOTO, 1962b). It was found that the spermatozoon is important to egg activation at the egg surface. The rhythmic changes induced by the spermatozoon change the egg shape, the hyaline coat, and the monaster.

The first visible and dramatic egg change at fertilization occurs when cortical granule breakdown induces fertilization membrane formation. SUGIYAMA (1953a, 1953b) concluded that at the time the sperm attach to the egg surface and at the time reagents activate the egg artificially, an invisible cortical change of propagating nature - the fertilization wave - passes over the egg and leads to cortical granule breakdown.

After the fertilization membrane is mechanically removed in Ca-Mg-free seawater, sea-urchin eggs may be refertilized at the 2-cell stage or later (SUGIYAMA, 1951). Artificially activated eggs may be refertilized in the same manner (ISHIDA and NAKANO, 1950; NAKANO, 1954). Such refertilization experiments may prove to be very important to the study of sperm attachment and entry and the block to polyspermy.

Acknowledgement

The author is grateful to Dr. J.C. DAN, Dr. M. SUGIYAMA, and Dr. T. UEHARA for their criticism and suggestions.

References

AFZELIUS, B.A., 1955. The fine structure of the sea urchin spermatozoa as revealed by the electron microscope. Z. Zellforsch. 42, 134-148.
AKETA, K., 1967. On the sperm-egg bonding as the initial step of fertilization in the sea urchin. Embryologia 9, 238-245.
AKETA, K., ONITAKE, K., 1969. Effect on fertilization of antiserum against sperm-binding protein from homo- and heterologous sea urchin egg surfaces. Exp. Cell Res. 56, 84-86.
BAXANDALL, J., 1966. The surface reactions associated with fertilization of the sea urchin egg as studied by immunoelectron microscopy. J. Ultrastr. Res. 16, 158-180.
BAXANDALL, J., PERLMANN, P., AFZELIUS, B.A., 1964a. Immuno-electron microscope analysis of the surface layers of the unfertilized sea urchin egg. I. Effects of the antisera on the cell ultrastructure. J. Cell Biol. 23, 609-628.
BAXANDALL, J., PERLMANN, P., AFZELIUS, B.A., 1964b. Immuno-electron microscope analysis of the surface layers of the unfertilized sea urchin egg. II. Localization of surface antigens. J. Cell Biol. 23, 629-650.
BERNSTEIN, M.H., 1962. Normal and reactive morphology of sea urchin spermatozoa. Exp. Cell Res. 27, 197-209.
BOVERI, Th., 1887. Über den Anteil der Spermatozoen an der Teilung des Eies. Sitz. Ges. Morph. u. Physiol. 3, 151-164.
BOVERI, Th., 1888. Über partielle Befruchtung. Sitz. Ges. Morph. u. Physiol. 4, 64-72.
BOWEN, R.H., 1924. On the acrosome of the animal sperm. Anat. Res. 28, 1-13.
BURY, J., 1913. Experimentelle Untersuchungen über die Einwirkung der Temperatur 0°C auf die Entwicklung der Echinideneier. Arch. f. Entw. Mech. 36, 537-594.
DAN, J.C., 1950a. Sperm entrance in echinoderms observed with the phase contrast microscope. Biol. Bull. 99, 399-411.
DAN, J.C., 1950b. Fertilization in the medusan, Spirocodon saltatrix. Biol. Bull. 99, 412-415.
DAN, J.C., 1952. Studies on the acrosome. I. Reaction to egg-water and other stimuli. Biol. Bull. 103, 54-66.
DAN, J.C., 1954. Studies on the acrosome. III. Effect of calcium deficiency. Biol. Bull. 107, 335-349.
DAN, J.C., 1956. The acrosome reaction. Inter. Rev. Cytol. 5, 368-385.
DAN, J.C., OHORI, Y., KUSHIDA, H., 1964. Studies on the acrosome. VII. Formation of the acrosomal process in sea urchin spermatozoa. J. Ultrastr. Res. 11, 508-524.
FOX, H.M., 1924. Lunar periodicity in reproduction. Proc. Roy. Soc. London 95, 523-550.
FRANKLIN, L.E., 1965. Morphology of gamete membrane fusion and of sperm entry into oocytes of the sea urchin. J. Cell Biol. 25, 81-100.
HAGSTRÖM, B., HAGSTRÖM, B., 1954. Re-fertilization of the sea urchin egg. Exp. Cell Res. 6, 491-496.
HAGSTRÖM, B.E., LÖNNING, S., 1961. Studies of the species specificity of echinoderms. Sarsia 4, 5-19.
HARRIS, P.J., 1961. Electron microscope study of mitosis in sea urchin blastomeres. J. Biophys. Biochem. Cytol. 11, 419-431.
HARRIS, P.J., 1967a. Nucleolus-like bodies in sea urchin eggs. Amer. Zool. 7, 753-754.
HARRIS, P.J., 1967b. Structural changes following fertilization in the sea urchin egg: Formation and dissolution of heavy bodies. Exp. Cell Res. 48, 569-581.
HARVEY, E.B., 1956. The American Arbacia and other sea urchins. Princeton Univ. Press, Princeton, N.J.

HERBST, C., 1912. Vererbungsstudien. VII. Die cytologischen Grundlagen der Verschiebung der Vererbungsrichtung nach der mütterlichen Seite. 2. Mitteilung. Arch. f. Entw. Mech. 34, 1-89.
HERTWIG, O., 1876. Beiträge zur Kenntnis der Bildung, Befruchtung und Teilung des tierischen Eies. Morphol. Jahrb., Leipzig 1, 347-434.
HIRAMOTO, Y., 1962a. Microinjection of the live spermatozoa into sea urchin eggs. Exp. Cell Res. 27, 416-426.
HIRAMOTO, Y., 1962b. An analysis of the mechanism of fertilization by means of enucleation of sea urchin eggs. Exp. Cell Res. 28, 323-334.
HULTIN, T., 1948a. Species specificity in fertilization reaction. The role of the vitelline membrane of sea urchin eggs in species specificity. Arkiv Zool. 40A, no. 12, 1-9.
HULTIN, T., 1948b. Species specificity in fertilization reaction. II. Influence of certain factors on the cross-fertilization capacity of Arbacia lixula. Arkiv Zool. 40A, no. 20, 1-8.
ISHIDA, J., NAKANO, E., 1950. Fertilization of activated sea urchin eggs deprived of fertilization membrane by washing with Ca-Mg-free media. Annot. Zool. Japon. 23, 43-48.
ISHIKAWA, M., 1954. Experiments on the conjugation of the nuclei in the re-fertilized egg and the activated egg following superimposed fertilization (in Japanese). Zool. Mag. (Tokyo) 63, 424.
ISHIKAWA, M., 1958. Experimental formation of the mark of sperm-entry and its relation to the toughening of the fertilization membrane in the sea urchin egg. Bull. Mar. Biol. Stat., Asamushi 9, 49-50.
ISHIKAWA, M., 1960. Experimental formation of the mark of sperm-entry on the fertilization membrane in the sea urchin egg. Embryologia 5, 186-193.
ISHIKAWA, M., 1962. Studies on the conjugation of the nuclei in polyspermic eggs and re-fertilized blastomeres of the sea urchin, Temnopleurus toreumaticus. Embryologia 7, 259-265.
ITO, S., 1960. The lamellar systems of cytoplasmic membranes in dividing spermatogenic cells of Drosophila virilis. J. Biophys. Biochem. Cytol. 7, 433-440.
JUST, E.E., 1922. Initiation of development in the egg of Arbacia. I-III. Biol. Bull. 43, 384-422.
KUHL, W., KUHL, G., 1949. Neue Ergebnisse zur Cytodynamik der Befruchtung und Furchung des Eies von Psammechinus miliaris GMEL. Zool. Jb., Abt. Anat. Ontog. 70, 1-59.
LILLIE, F.R., 1913. The mechanism of fertilization. Science 38, 524-528.
LILLIE, F.R., 1919. Problem of fertilization. Chicago: Univ. Chicago Press.
LILLIE, F.R., 1921. Studies of fertilization. IX. On the question of superposition of fertilization on parthenogenesis in Strongylocentrotus purpuratus. Biol. Bull. 40, 23-31.
LILLIE, F.R., JUST, E.E., 1924. Fertilization, pp. 451-506. Cowdry's General Cytology.
LOEB, J., 1913a. Artificial parthenogenesis and fertilization. Chicago: Univ. Chicago Press.
LOEB, J., 1913b. Reversibility in artificial parthenogenesis. Science 38, 749-751.
LOEB, J., 1915. Reversible activation and incomplete membrane formation of the unfertilized eggs of the sea urchin. Biol. Bull. 29, 103-110.
LOEB, J., 1916. The organism as a whole. New York, London: Putnam's and Sons.
LONGO, F.J., ANDERSON, E., 1968. The fine structure of pronuclear development and fusion in the sea urchin, Arbacia punctulata. J. Cell Biol. 39, 339-368.

LONGO, F.J., ANDERSON, E., 1970. The effects of nicotine on fertilization in the sea urchin, Arbacia punctulata. J. Cell Biol. 46, 308-325.
LÖNNING, S., 1964. Studies of the ultrastructure of sea urchin oocyte and the changes induced by insemination. Sarsia 15, 9-15.
METZ, C.B., THOMPSON, P.H., 1967. Effect of papain digested, univalent antibody on the morphology, cleavage and fertilizing capacity of sea urchin eggs. Exp. Cell Res. 45, 433-449.
MILLONIG, G., 1967. The structural changes of the nucleolus during oogenesis and embryogenesis of Arbacia lixula. J. Cell Biol. 35, 177A.
MOORE, C.R., 1916. On the superposition of fertilization on parthenogenesis. Biol. Bull. 31, 137-179.
MOORE, C.R., 1917. On the capacity for fertilization after initiation of development. Biol. Bull. 33, 258-295.
MOTOMURA, I., 1941. Materials of the fertilization membrane in the eggs of Echinoderms. Sci. Rep. Tohoku Imp. Univ., 4th Ser. 16, 345-363.
MOTOMURA, I., 1950. On a new factor for the toughening of the fertilization membrane of the sea urchins. Sci. Rep. Tohoku Univ., 4th Ser. 18, 561-570.
NAKANO, E., 1954. Further studies on the fertilization of activated sea urchin eggs. Jap. J. Zool. 11, 245-251.
PASTEELS, J., 1965. Aspects structuraux de la fécondation vus au microscope électronique. Arch. Biol. Liège 76, 463-509.
PERLMANN, P., 1954. Study on the effect of antisera on unfertilized sea urchin eggs. Exp. Cell Res. 6, 485-490.
PERLMANN, P., 1956. Response of unfertilized sea urchin eggs to antiserum. Exp. Cell Res. 10, 324-353.
PERLMANN, P., 1957. Analysis of the surface structures of the sea urchin egg by means of antibodies. I. Comparative study of the effects of various antisera. Exp. Cell Res. 13, 365-390.
PERLMANN, P., PERLMANN, H., 1957a. Analysis of the surface structures of the sea urchin egg by means of antibodies. II. The J- and A-antigens. Exp. Cell Res. 13, 454-474.
PERLMANN, P., PERLMANN, H., 1957b. Analysis of the surface structures of the sea urchin egg by means of antibodies. III. The C- and F-antigens. Exp. Cell Res. 13, 475-487.
POPA, G.T., 1927. The distribution of substances in the spermatozoon (Arbacia and Nereis). Biol. Bull. 52, 238-257.
REBHUN, L., BERNSTEIN, L., 1967. In vivo sperm aster augmentation by hexylene glycol: Independence of protein synthesis. J. Cell Biol. 35, 111A.
RETZIUS, G., 1910. Zur Kenntnis der Spermien der Echinodermen. Biol. Untersuch. 15, 55-62.
ROBBIN, E., GONATAS, N.K., 1964. The ultrastructure of mammalian cell during the mitotic cycle. J. Cell Biol. 21, 429-463.
ROTHSCHILD, LORD, 1953. The fertilization reaction in the sea urchin. The induction of polyspermy by nicotine. J. Exp. Biol. 30, 57-67.
ROTHSCHILD, LORD, 1956a. Sea-urchin spermatozoa. Endeavour 15, 79-86.
ROTHSCHILD, LORD, 1956b. The fertilizing spermatozoon. Discovery 18, 64-65.
ROTHSCHILD, LORD, 1956c. Fertilization. London: Methuen and Co., Ltd.
ROTHSCHILD, LORD, SWANN, M.M., 1949. The fertilization reaction in the sea urchin egg. A propagated response to sperm attachment. J. Exp. Biol. 26, 164-176.
ROTHSCHILD, LORD, TYLER, A., 1955. Acrosomal filaments in spermatozoa. Exp. Cell Res. Suppl. 3, 304-311.
RUNNSTRÖM, J., 1928. Die Veränderungen der Plasmakolloide bei der Entwicklungserregung des Seeigeleies. Protoplasma 4, 388-514.
RUNNSTRÖM, J., 1963. Sperm-induced protrusions in sea urchin oocytes: A study of phase separation and mixing in living cytoplasm. Develop. Biol. 7, 38-50.

RUNNSTRÖM, J., HAGSTRÖM, B.E., PERLMANN, P., 1959. Fertilization. In: The cell - biochemistry, physiology, morphology (J. Brachet, A.E. Mirsky, eds.), 1, 327-397. New York: Academic Press.
RUNNSTRÖM, J., KRISZAT, G., 1952. The cortical propagation of the activation impulse in the sea urchin egg. Exp. Cell Res. 3, 419-426.
SUGIYAMA, M., 1938a. Effect of some divalent ions upon the membrane development of sea urchin eggs. J. Fac. Sci. Imp. Univ. Tokyo, Sec. 4, 4, 501-508.
SUGIYAMA, M., 1938b. Further studies on the development of the fertilization membrane in sea urchin eggs. Annot. Zool. Japon. 17, 360-364.
SUGIYAMA, M., 1951. Re-fertilization of the fertilized eggs of the sea urchin. Biol. Bull. 101, 335-344.
SUGIYAMA, M., 1953a. Physiological analysis of the cortical response of the sea urchin egg to stimulating reagents. I. Response to sodium choleinate and wasp venom. Biol. Bull. 104, 210-215.
SUGIYAMA, M., 1953b. Physiological analysis of the cortical response of the sea urchin egg to stimulating reagents. II. The propagating or non-propagating nature of the cortical changes induced by various reagents. Biol. Bull. 104, 216-223.
SUGIYAMA, M., 1956. Physiological analysis of the cortical response of the sea urchin egg. Exp. Cell Res. 10, 364-376.
TAKASHIMA, R., TAKASHIMA, Y., 1960. Electron microscope observations on the fertilization phenomenon of sea urchin with special reference to the acrosome filament. Tokushima J. Exp. Med. 6, 334-339.
TEICHMANN, E., 1903. Über die Beziehung zwischen Astrosphären und Furchen. Arch. f. Entw. Mech. 16, 243-327.
TYLER, A., 1941. The role of fertilizin in the fertilization of eggs of the sea urchin and other animals. Biol. Bull. 81, 190-204.
TYLER, A., MONROY, A., METZ, C.B., 1956. Fertilization of fertilized sea urchin eggs. Biol. Bull. 110, 184-195.
UEHARA, T., SUGIYAMA, M., 1969. Propagation of the fertilization-wave on the once-activated surface of the sea urchin egg. Embryologia 10, 356-362.
WASTENEYS, H., 1916. The rate of oxidations in reversed artificial parthenogenesis. J. Biol. Chem. 24, 281-298.
WILSON, E.B., 1925. The cell in development and heredity. New York: Macmillan Co., 3rd ed.
YAMAMOTO, T., 1944. Physiological studies on fertilization and activation of fish eggs. I. Response of the cortical layer of the egg of Oryzias latipes to insemination and to artificial stimulation. Annot. Zool. Japon. 22, 109-125.
YANAGIMACHI, R., 1953. A note on the so-called "boring movement" of sea urchin spermatozoa (in Japanese). Zool. Mag. (Tokyo) 62, 22-26.
YOSHIDA, M., 1952. Some observations on the maturation of the sea-urchin, Diadema setosum. Annot. Zool. Japon. 25, 265-271.
ZIMMERMAN, A.M., SILBERMAN, L., 1964. Further studies on incorporation of H^3-thymidine in Arbacia eggs under hydrostatic pressure. Biol. Bull. 127, 355.
ZIMMERMAN, A.M., ZIMMERMAN, S., 1967. Action of Colcemid in sea urchin eggs. J. Cell Biol. 34, 483-488.

5. Parthenogenetic Activation and Development

M. Ishikawa

In 1887, O. and R. HERTWIG reported that fertilization membrane formation in *Paracentrotus lividus* could be artificially induced by shaking with chloroform, and R. HERTWIG (1895, 1896) found that artificial cleavage resulted from treating the eggs with strychnine. By exposing the eggs to sea water containing NaCl, KCl, or $MgCl_2$, MORGAN (1899) obtained embryos that cleaved to about 64 cells. In 1899, LOEB obtained parthenogenetically developed *Arbacia punctulata* plutei using $MgCl_2$-sea water and, in 1904 improved the method of artificial parthenogenesis by combining the membrane-forming and hypertonic treatments. PASPALEFF (1925) published a study on the cytology of activated eggs.

JUST (1922) and BATAILLON (1929) showed that a single application of a sufficiently strong hypertonic solution induced membrane formation and development to top swimming larvae. HARVEY (1936, 1940) reported treatment with hypertonic sea water for about 20 minutes to be the best method of artificial parthenogenesis for *A. punctulata*. She prepared the hypertonic medium by boiling down sea water to half its volume or by adding 30 gm of NaCl to 1 liter of sea water. Activating agents reported up to 1951 are listed in Table 5.1.

Parthenogenetically activated eggs develop a little more slowly than fertilized eggs. It is usually possible to raise embryos to the pluteus stage, but after that it is very difficult. In 1909 DELAGE raised two parthenogenetic embryos of *Paracentrotus (Strongylocentrotus) lividus* from metamorphosis to young adulthood. They lived for 18 months and grew to 3.1 and 3.6 cm in diameter. SHEARER and LLOYD (1913) reported 15 larvae of *Echinus esculentus* parthenogenetically metamorphosed in about 8 weeks.

5.1 Ways of Inducing Parthenogenetic Development

5.1.1 Double Treatment with Butyric Acid and Hypertonic Sea Water

LOEB (1913) described artificial parthenogenesis as follows: "The unfertilized eggs of *Strongylocentrotus purpuratus* are placed in 50 ml of sea water plus 2.8 ml of N/10 butyric acid at 15°C and left in this solution for 1.5 to 2.5 minutes. All the eggs form membranes when replaced in normal sea water. If they are transferred earlier from the acid to normal sea water, they form no membrane, nor do the eggs form a membrane after remaining too long in the acid, because the acid injures the eggs. After the eggs are transferred from the acid to normal sea water for 15 or 20 minutes, they are placed in hypertonic sea water (a mixture of 50 ml of sea water plus 8 ml of 2.5 *M* NaCl). From this the activated eggs are transferred to normal sea water after 30 to 60 minutes at 15°C, at intervals of 2.5 to 6 minutes. After transference to normal sea water, those eggs which have been just long enough in the hypertonic solution begin to segment and develop. When the length of exposure is correct, the first division occurs as in normal fertilization."

Discussing the preparation of hypertonic sea water, LOEB said: "A mixture of 2.5 *M* NaCl plus KCl + $CaCl_2$ in the proportions in which these salts exist in sea water is still better than 2.5 *M* NaCl, since it is less injurious. There are slight differences in the quantitative aspects of the method according to the species of sea urchin. The eggs of *Arbacia punctulata* which were exposed longer than 25 minutes to the hypertonic solution at 23°C did not develop."

Table 5.1. Agents inducing parthenogenetic activation of sea-urchin eggs[a]

Physical agents
Mechanical agitation
Puncture
Electricity (induction shocks, direct current)
Heat (about 32°C)
Cold (0°-10°C)
Photodynamic action (visible light plus rose bengal or eosin)
Ultraviolet light
X-rays
Chemical agents
Inadequate or diminished oxygen
Hypertonic sea water by evaporation
Hypotonic sea water or distilled water
Salts: NaCl, NaBr, $NaNO_3$, NaI, NaCNS, Na_2SO_4, KCl, KI, KCNS, $CaCl_2$, $MgCl_2$, $BaCl_2$, $SrCl_2$, $HgCl_2$
Acids: HCl, HNO_3, phosphoric acid
CO_2
Acetic acid, lactic acid, butyric acid
Butyric acid + hypertonic sea water (LOEB's double method)
Many other fatty acids
Alkalis and amines
NaOH, KOH, NH_3
Tetra-ethyl ammonium hydroxide
Butylamine, benzylamine, protamine
Fat solvents, esters, and narcotics
Toluol, chloroform, ether, alcohol, propyl alcohol
Benzene, amylene, salicylaldehyde, acetone
Chloretone, urethane, chloral hydrate
Methylacetate, ethylacetate, ethylbutyrate, methylsalicylate
Detergents: Bile salts, soap, detergent
Glucosides: Saponine, solanine, digitalin
Alkaloids: Strychnine, quinine, pilocarpine
Nonelectrolytes: Sucrose, urea, thiourea, glycerine, acetamide
Proteins, enzymes, organ extracts
Egg albumin, blood serum, thrombin from ox blood, ovarian extract, sperm extract, extract of injured tissues, lipolysin, hirudin, papain
Miscellaneous: KCN, picric acid (trinitrophenol), chlorine, iodine, tannine, ammonia, hexaresorcinol, vitamin K (2-methyl-1,4,-naphtoquinone)

[a]From HARVEY, 1956.

Table 5.2. Some organic compounds that induce membrane formation in unfertilized eggs of *Hemicentrotus pulcherrimus*[a]

Chemicals	Optimum Concentration
Benzene[b]	Saturated in sea water
Toluene[b]	Saturated in sea water
Xylene (*o*-, *m*-, *p*-)[b]	Saturated in sea water
Toluidine (*o*-, *p*-)	5 ml saturated toluidine plus 1 ml sea water
Phenol	0.3%
Resorcin	2.5%
Hydroquinone	Saturated in sea water (about 5.9%)
Pyrogallol	12%
Cresol (*o*-, *m*-, *p*-)	2 ml saturated cresol + 8 ml sea water
Thymol	3 ml saturated thymol + 7 ml sea water
Vanillin	0.2%
Benzaldehyde	Saturated in sea water
Salicylaldehyde[b]	Saturated in sea water
Naphthol	2 ml saturated naphthol + 8 ml sea water
Indole	Saturated in sea water
Scatole	Saturated in sea water
Isatine	Saturated in sea water

[a]From ISHIKAWA, 1954a

[b]HERBST (1893) and LOEB (1913) reported that benzene, toluene, xylene, and salicylaldehyde effectively cause membrane formation.

5.1.2 Thymol

In eggs of the sea urchin *Hemicentrotus pulcherrimus*, ISHIKAWA (1954) studied the effects of a number of organic compounds: aromatic hydrocarbons and the derivatives of aniline, phenol, naphthalene, purine, and pyrrole. Those found to induce membrane formation are listed in Table 5.2.

Used with LOEB's double treatment method, thymol induces both membrane formation and development. Unfertilized *Pseudocentrotus depressus* eggs are treated with a mixture of 3 ml of saturated thymol in sea water plus 7 ml of sea water for 50 seconds and then removed to sea water. Membrane formation is thus induced. After 20 minutes, the eggs are immersed in hypertonic sea water (8 ml of 2.5 *M* NaCl plus 50 ml of sea water) for 50 minutes and again returned to sea water. Egg development soon starts, and in a few days top swimming plutei appear.

5.1.3 Chloretone

In sea-urchin eggs YAMAMOTO (1956) reported artificial parthenogenesis induced by double treatment with chloretone (Chlorobutanol) and hypertonic sea water. The "milder" chloretone affects the eggs much as other reagents do. The best way to obtain pluteus larvae is as follows: "When unfertilized eggs of the sea urchins *Hemicentrotus pulcherrimus* or *Pseudocentrotus depressus* are immersed in M/50 chloretone sea water for 3 to 5 minutes, a membrane is formed as in normal fertilization. The activated eggs are rinsed with sea water for about 15 minutes and then treated with weak hypertonic sea water (50 ml of sea water + 8 ml of

Table 5.3. Artificial parthenogenesis in sea-urchin eggs induced by double treatment with chloretone and hypertonic sea water[a]

Time Exposed to Hypertonic Sea Water (min)	*Hemicentrotus pulcherrimus* (11°C)		*Pseudocentrotus depressus* (16°C)	
	Cleavage (%)	Larvae (%)	Cleavage (%)	Larvae (%)
0	19	0	17	0
20	40	0	43	0
30	45	2	49	0
40	51	20	54	19
50	54	36	56	38
60	47	21	49	25

[a]From YAMAMOTO, 1956.

2.5 *M* NaCl) for 40 to 50 minutes and finally transferred to sea water." They undergo relatively normal cleavage and further development. The rates of cleavage and development into swimming blastula larvae are given in Table 5.3. A 14-day-old pluteus obtained by this method has a well-developed mouth, tripartite alimentary tract, anus, and spicules (Fig. 5.1).

5.2 How Parthenogenetic Activation Works

5.2.1 Hypotheses of Artificial Activation of the Egg

It is well known from LOEB's classic work (1913) that although artificial formation of the fertilization membrane induces nuclear and cytoplasmic changes in the sea-urchin egg, cells usually do not divide and disintegration begins in a few hours. Treating eggs with a hypertonic medium after membrane formation and transferring them to normal sea water assures their cleavage and further development. It is possible that cytaster induction is important here (WILSON, 1925). Because the fate of the egg centriole is unknown, there has been considerable speculation as to whether the centers of artificially produced asters come from the last centriole or whether they arise de novo. HARVEY (1936) experimented with artificial activation of enucleated sea-urchin eggs and suggested that asters are produced by centrioles of de novo origin. Electron microscopy has shown a centriole in asters of parthenogenetically activated eggs of *S. purpuratus* (DIRKSEN, 1966). It is also possible that the hypertonic medium acts as an inhibitor - or corrective factor - on the oxidative disintegration initiated by the membrane-forming agent (LOEB, 1913).

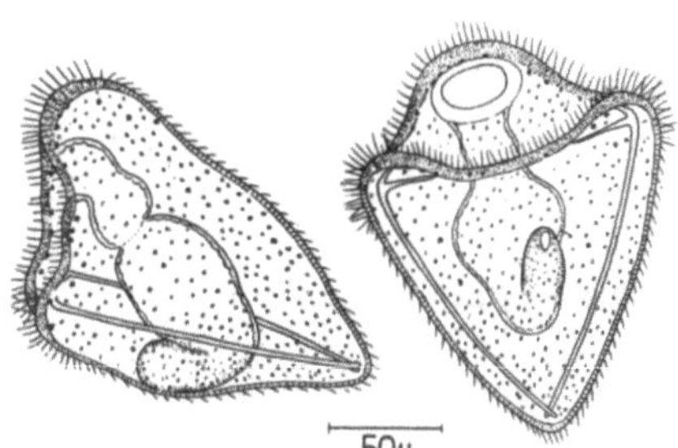

Fig. 5.1. Lateral and ventral views of parthenogenetically developed, 14-day-old pluteus of *Hemicentrotus pulcherrimus*. (From YAMAMOTO, 1956)

RUNNSTRÖM, HAGSTRÖM, and PERLMANN (1959) compared LOEB's dual conception of artificial parthenogenesis with the action of the spermatozoon, the acrosomal region of which they believe may induce cortical changes that, in turn, induce the synthesis of nuclear material and monaster formation. The action of the acrosome thus seems comparable to the butyric acid treatment of LOEB's "improved" method. After the spermatozoon is engulfed, the centrosome plays the active role and organizes the division apparatus. The egg divides, and the centrosome retrogresses. In artificial parthenogenesis it seems to be reactivated in what is evidently a very exacting process requiring a fine balance between breakdown and synthesis.

Table 5.4. Effect of DNP on efficiency of hypertonic treatment in parthenogenetic activation of egg[a]

	Percentage of Developed Eggs					
	(1)[b]		(2)[c]		(1)[b] After 50 min Treatment With (2)[c]	
Time Immersed (min)	2-Cell Stage	Blastula	2-Cell Stage	Blastula	2-Cell Stage	Blastula
20	0	0	0	0	0	0
30	44	4	0	0	38	3
40	58	4	0	0	48	5
50	64	15	0	0	59	13
60	72	35	0	0	63	28

[a]From ISHIKAWA, 1957.

[b]After membrane formation, eggs were treated with hypertonic medium: 8 ml 2.5 *N* NaCl + 50 ml sea water.

[c]After membrane formation, eggs were treated with hypertonic medium: 8 ml 2.5 *N* NaCl + 50 ml 10^{-4} *M* DNP sea water.

Table 5.5. Effect of sodium azide on efficiency of hypertonic treatment in parthenogenetic activation of egg[a]

	Percentage of Cleaved Eggs		
Time Immersed (min)	(1)[b]	(2)[c]	(1)[b] After 50 min Treatment With (2)[c]
20	28	0	10
30	41	0	38
40	60	0	56
50	74	0	58
60	38	0	21

[a]From ISHIKAWA, 1957.

[b]After membrane formation, eggs were treated with hypertonic medium: 8 ml 2.5 *N* NaCl + 50 ml sea water.

[c]After membrane formation, eggs were treated with hypertonic medium: 8 ml 2.5 *N* NaCl + 50 ml 5 x 10^{-3} *M* NaN_3 sea water (pH 7.3).

Table 5.6. Effect of potassium cyanide on efficiency of hypertonic treatment in parthenogenetic activation of egg[a]

Time Immersed (min)	Percentage of Cleaved Eggs		
	(1)[b]	(2)[c]	(1)[b] After 50 min Treatment With (2)[c]
20	2	0	0
30	20	0	22
40	52	0	50
50	64	0	58
60	56	0	48

[a]From ISHIKAWA, 1957.

[b]After membrane formation, eggs were treated with hypertonic medium: 8 ml 2.5 *N* NaCl + 50 ml sea water.

[c]After membrane formation, eggs were treated with hypertonic medium: 8 ml 2.5 *N* NaCl + 50 ml 10^{-4} *M* KCN sea water.

5.2.2 Phosphorus Metabolism and Oxygen Consumption

ISHIKAWA (1954b, 1957) studied the ability of dinitrophenol, sodium azide, and potassium cyanide to induce parthenogenetic development of sea-urchin eggs and found that even small amounts of them rendered ineffective the hypertonic sea-water treatment of *H. pulcherrimus* eggs after artificial membrane formation (Tables 5.4, 5.5, 5.6). Eggs treated with butyric acid-sea water and hypertonic sea water containing one of these reagents and then with hypertonic solution without the inhibiting reagent cleaved in fairly high percentage. LOEB (1906) found the effect of hypertonic solution completely suppressed when potassium cyanide inhibited respiration. It is clear that energy-yielding phosphorus metabolism is essential to effective hypertonic treatment.

In determining the ^{32}P uptake of parthenogenetically activated *S. purpuratus* eggs, LITCHFIELD and WHITELEY (1959) said, "Artificially activated eggs, which were treated with butyric acid alone or butyric acid-hypertonic sea water, exhibit a ^{32}P-accumulation comparable to that of fertilized eggs, and the onset of phosphate accumulation is preceded by a lag period. The increase in uptake does not begin until well after the visible cortical events of membrane elevation have occurred."

ISHIKAWA (1962) studied the changes in labile phosphate content, radioactive phosphate uptake, and oxygen consumption of parthenogenetically activated eggs of *H. pulcherrimus*, *P. depressus*, and *Temnopleurus toreumaticus*. Unfertilized eggs and those activated by butyric acid alone showed no measurable difference in labile phosphate content, but activated eggs in hypertonic sea water showed a marked increase (Table 5.7) which was inhibited by 2,4,-dinitrophenol, sodium azide, potassium cyanide, and monoiodoacetamide (Table 5.8).

Unfertilized eggs showed no ^{32}P uptake, but activated eggs showed rapid ^{32}P penetration. Hypertonic sea water made the ^{32}P penetrate the activated eggs even more rapidly. Experiments on ^{32}P incorporation into labile phosphate compounds gave similar results (Fig. 5.2).

Table 5.7. Labile phosphate content in *Hemicentrotus* eggs during LOEB's double treatment for artificial parthenogenesis at 11.5°C[a]

	Experiment	Inorganic P (μg P/mg N)	Labile P (μg P/mg N)	Inorganic P plus Labile P (μg P/mg N)
Unfertilized eggs	1	19.0	15.3	34.3
	2	19.2	15.4	34.6
	3	19.4	14.6	34.0
Activated eggs	1	19.2	15.9	35.1
	2	19.3	15.7	35.0
	3	19.4	15.0	34.4
Activated eggs immersed in hypertonic sea water	1	12.7	22.3	35.0
	2	14.3	20.9	35.2
	3	14.6	20.2	34.8

[a]From ISHIKAWA, 1962. Eggs activated by butyric acid were kept in sea water for 15 minutes and divided into two groups: one kept in sea water for 50 minutes, the other immersed in hypertonic sea water for 50 minutes.

Table 5.8. Labile phosphate content in *Hemicentrotus* eggs treated with hypertonic sea water containing 2,4-dinitrophenol, sodium azide, potassium cyanide, or monoiodoacetamide at 11°-12°C[a]

	Content of Labile P (μg P/mg N)	
Inhibitor and Its Final Concentration	Eggs Treated With Hypertonic Sea Water With Inhibitor	Eggs Treated With Hypertonic Sea Water Without Inhibitor
Dinitrophenol (10^{-4} *M*)	10.9	20.8
Sodium azide (5×10^{-3} *M*; pH 7.3)	11.4	20.5
Potassium cyanide (10^{-4} *M*)	15.1	21.1
Iodoacetamide (10^{-3} *M*)	15.3	20.8

[a]From ISHIKAWA, 1962. Eggs activated by butyric acid were kept in sea water for 15 minutes and treated with hypertonic sea water with or without metabolic inhibitor for 50 minutes.

In unfertilized eggs treated with hypertonic sea water alone, parthenogenetic cleavage was reported to start without membrane formation (LOEB, 1913), and eggs of *T. toreumaticus* were later found to develop parthenogenetically without cortical granule breakdown (ISHIKAWA, 1954b).

In the latter experiment, unfertilized eggs were treated with hypertonic ^{32}P-sea water (16 ml of 2.5 *M* NaCl plus 100 ml of sea water with 6.6μCi $H_3{}^{32}PO_4$) for 10 to 50 minutes at 27°C. The ^{32}P incorporation which reached a maximum after immersion for 20 minutes, was 40% greater than in eggs immersed in isotonic ^{32}P-sea water (Fig. 5.3).

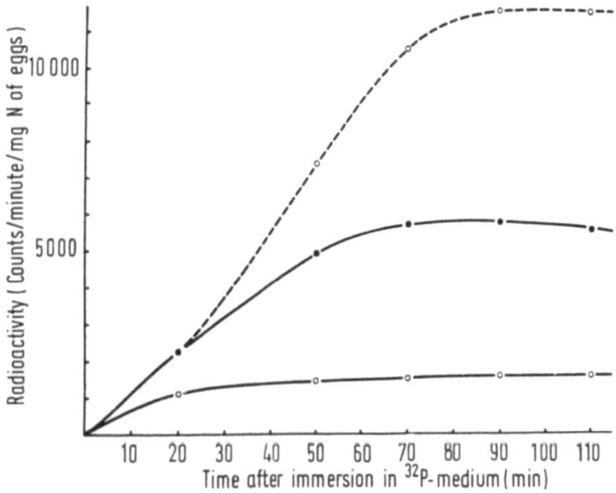

Fig. 5.2. ^{32}P incorporated into labile phosphate in acid-soluble fraction of *Pseudocentrotus* eggs at 18°C. o—o, unfertilized eggs in ^{32}P-sea water; ●—●, eggs in ^{32}P-sea water after thymol activation; o---o, eggs activated and immersed in hypertonic ^{32}P-sea water. (From ISHIKAWA, 1962.)

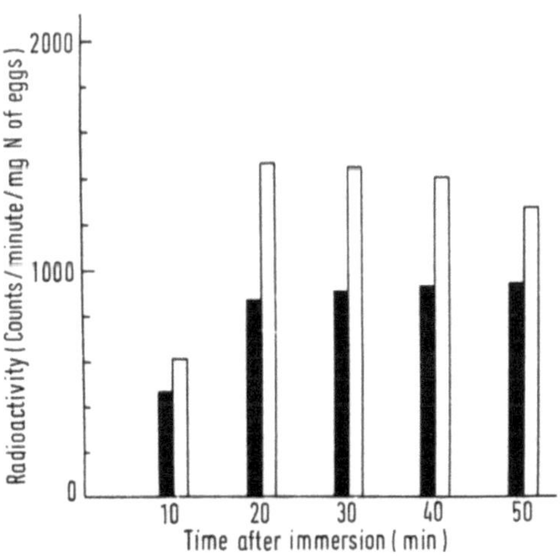

Fig. 5.3. ^{32}P uptake by *Temnopleurus* eggs immersed in isotonic and hypertonic ^{32}P-sea water at 27°C. ▬, unfertilized eggs immersed in ^{32}P-sea water; ▭, unfertilized eggs immersed in hypertonic ^{32}P-sea water. (From ISHIKAWA, 1962.)

In ISHIKAWA's study of the relationship between hypertonicity and ^{32}P uptake, eggs were treated for 20 minutes with ^{32}P-sea water containing various amounts of 2.5 *M* NaCl (2 to 16 ml) in 100 ml of ^{32}P-sea water. Radioactivity was adjusted to 0.055μCi ^{32}P per milliliter of sea water (Fig. 5.4). ^{32}P uptake tended to increase with hypertonicity and was about 40% higher than in untreated eggs. Hypertonicity was optimal for inducing cleavage of unfertilized eggs.

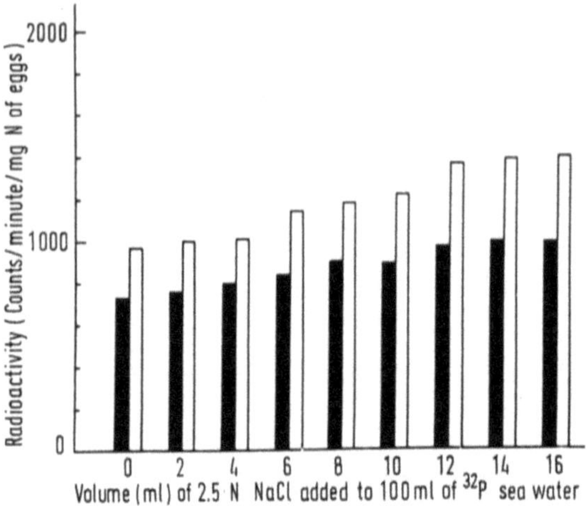

Fig. 5.4. Hypertonic effects on ^{32}P uptake of unfertilized *Temnopleurus* eggs at 27°C. ▬, ^{32}P incorporated into labile phosphate; ▭, ^{32}P incorporated into whole eggs. (From ISHIKAWA, 1962.)

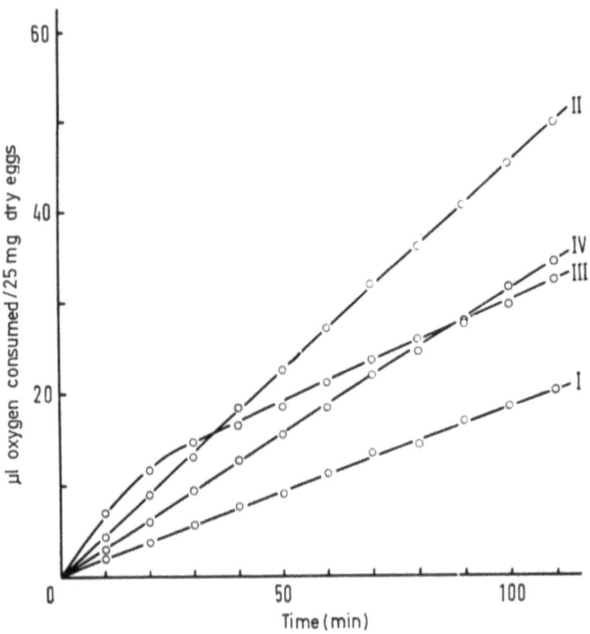

Fig. 5.5. Oxygen consumption of *Temnopleurus* eggs during LOEB's double treatment for artificial parthenogenesis at 28°C. I, unfertilized eggs; II, eggs activated by thymol; III, eggs of II immersed in hypertonic sea water; IV, eggs of III transferred to ordinary sea water. (From ISHIKAWA, 1962.)

The oxygen consumption of eggs activated by butyric acid or thymol - normally increased to that of fertilized eggs - was suppressed immediately or shortly after the eggs were put into hypertonic sea water. When they were transferred to ordinary sea water, oxygen consumption again increased and remained almost constant. From the results of thymol-hypertonic treatment of *Temnopleurus* eggs (Fig. 5.5), it appears that increased activity leading to labile phosphate accumulation is needed to induce parthenogenetic cleavage and that increased phosphate esterification alone is not enough.

5.2.3 Effects of Sulfhydryl Reagents

Studying in unfertilized eggs of *H. pulcherrimus* the effects of sulfhydryl reagents on cleavage induction by means of LOEB's double treatment, ISHIKAWA (1960) followed the 7 courses shown in Fig. 5.6. As sulfhydryl reagents he used monoiodoacetamide (IAA), *p*-chloromercuribenzoate (PCMB), and mercuric chloride.

In Course 1, Series I, cleavage induced by hypertonic treatment was somewhat suppressed in activated eggs exposed to IAA-sea water (10^{-4} *M*) for 20 minutes. In Courses 3-7, Series II, 20 minutes after membrane formation eggs activated by butyric acid treatment were put into hypertonic sea water with and without IAA. Some of the eggs treated with hypertonic IAA-sea water were transferred to 10^{-2} *M* cysteine-sea water (Courses 5 and 6) and the others were placed in ordinary sea water (Courses 3 and 4). Eggs treated with hypertonic sea water without IAA showed a high percentage of cleavage (Table 5.9) while those treated with hypertonic sea water with IAA showed no cleavage even after treatment in cysteine-sea water for 20 minutes (Fig. 5.7A). Cleavage was induced when they were again treated with hypertonic sea water without IAA (Fig. 5.7B).

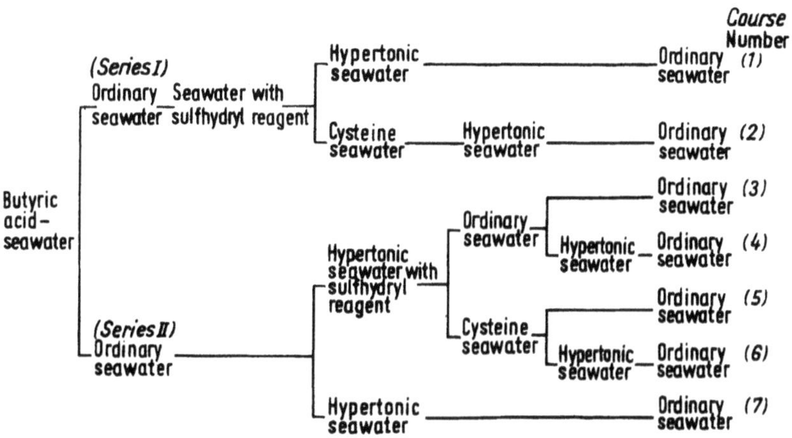

Fig. 5.6. General experimental procedure to test effects of sulfhydryl reagents on cleavage induction by double treatment of LOEB. (From ISHIKAWA, 1960.)

Table 5.9. Effect of IAA percentage of parthenogenetic cleavage in hypertonic sea water at $18^{o}C$[a]

		Course Number	Time of First Hypertonic Treatment (min) 30	40	50
Eggs activated by butyric acid treatment and treated with hypertonic IAA sea water after 20 minutes	Placed in ordinary sea water	(3)	0%	0%	0%
	Followed by second hypertonic treatment for 30 minutes	(4)	0	0	0
	Exposed to cysteine-sea water	(5)	0	0	0
	Exposed to cysteine-sea water, followed by second hypertonic treatment for 30 minutes	(6)	22	36	38
Eggs activated by butyric acid treatment and treated with ordinary hypertonic sea water after 20 minutes		(7)	45	62	52

[a]From ISHIKAWA, 1960. Activation by butyric acid with subsequent hypertonic treatment containing IAA (10^{-4} *M*) is compared with that followed by cysteine (10^{-2} *M*) and second hypertonic treatment.

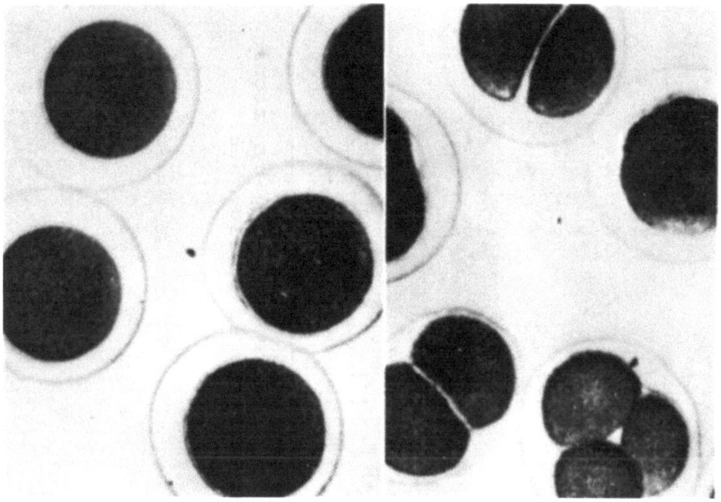

A B

Fig. 5.7. (A) Eggs treated with hypertonic IAA-sea water (Course 3); cleavage did not occur. (B) Eggs treated with hypertonic IAA-sea water, cysteine treatment, and ordinary hypertonic treatment (Course 6); cleavage occurred. (From ISHIKAWA, 1960.)

In Series I, activated eggs exposed to the PCMB solution (7×10^{-4} *M*) immediately after membrane formation showed no cleavage treated only with hypertonic sea water but cleaved if they were washed with cysteine-sea water (10^{-2} *M*) and treated with hypertonic sea water. In Series II, activated eggs treated with hypertonic PCMB-sea water showed no cleavage after treatment in cysteine-sea water (Course 5) but cleaved after the second hypertonic treatment without PCMB (Course 6) (Table 5.10).

In Series I, no cleavage was induced in Course 1 if activated eggs were treated with $HgCl_2$. The hyaline layer became swollen in $HgCl_2$-sea water, and the egg surface roughened after the eggs were returned to ordinary sea water (Table 5.11). Eggs washed with cysteine-sea water did not roughen, and a few of them cleaved when treated with hypertonic sea water (Course 2). Most of the eggs in Courses 3 and 4 failed to cleave, although some formed incomplete furrows (Fig. 5.8A). In Courses 5 and 6, the eggs cleaved after treatment with cysteine-sea water (Fig. 5.8B). It seems apparent that cleavage induced by hypertonic treatment depends to a large extent on the sulfhydryl groups of cytoplasmic proteins.

5.2.4 Uptake of Labeled Amino Acids

NAKANO, GIUDICE, and MONROY (1958) showed methionine-^{35}S incorporation into proteins of eggs activated with butyric acid sea water, and BRACHET, FICQ, and TENCER (1963) reported that artificial activation significantly increased leucine-^{3}H incorporation. DENNY and TYLER (1964) found that artificial activation stimulated protein synthesis in anucleate halves of sea-urchin eggs.

Table 5.10. Effect of PCMB on percentage of parthenogenetic cleavage at 19°C[a]

		Course Number	Time of First Hypertonic Treatment (min) 20	30	40
Eggs treated with PCMB immediately after membrane formation and exposed to hypertonic sea water		(1)	0%	0%	0%
Eggs treated with PCMB immediately after membrane formation and exposed to cysteine-sea water and hypertonic sea water		(2)	35	53	64
Eggs activated by butyric acid treatment and treated with hypertonic PCMB sea water after 20 minutes	Placed in ordinary sea water	(3)	0	0	0
	Followed by second hypertonic treatment for 30 minutes	(4)	0	0	0
	Exposed to cysteine-sea water	(5)	0	0	0
	Exposed to cysteine-sea water and second hypertonic treatment for 30 minutes	(6)	10	25	43
Eggs activated by butyric acid treatment and treated with ordinary hypertonic sea water after 20 minutes		(7)	8	54	78

[a]From ISHIKAWA, 1960. Activated eggs were exposed to PCMB (7×10^{-4} *M*) before or during hypertonic treatment. Cysteine (10^{-2} *M*) and second hypertonic treatments were performed as in Fig. 5.6.

Table 5.11. Effect of $HgCl_2$ on percentage of parthenogenetic cleavage at 19°C[a]

		Course Number	Time of First Hypertonic Treatment (min) 20	30	40
Eggs treated with $HgCl_2$ immediately after membrane formation and exposed to hypertonic sea water		(1)	0%	0%	0%
Eggs treated with $HgCl_2$ immediately after membrane formation and exposed to cysteine-sea water and hypertonic sea water		(2)	0	22	38
Eggs activated by butyric acid treatment and treated with hypertonic $HgCl_2$ sea water after 20 minutes	Placed in ordinary sea water	(3)	0	0	0
	Followed by second hypertonic treatment for 30 minutes	(4)	0	0	0
	Exposed to cysteine-sea water	(5)	0	62	42
	Exposed to cysteine-sea water and second hypertonic treatment for 30 minutes	(6)	68	65	72
Eggs activated by butyric acid treatment and treated with ordinary hypertonic sea water after 20 minutes		(7)	25	48	70

[a] From ISHIKAWA, 1960. Activated eggs were exposed to $HgCl_2$ (10^{-5} *M*) before or during hypertonic treatment. Cysteine (10^{-2} *M*) and second hypertonic treatments were performed as in Fig. 5.6.

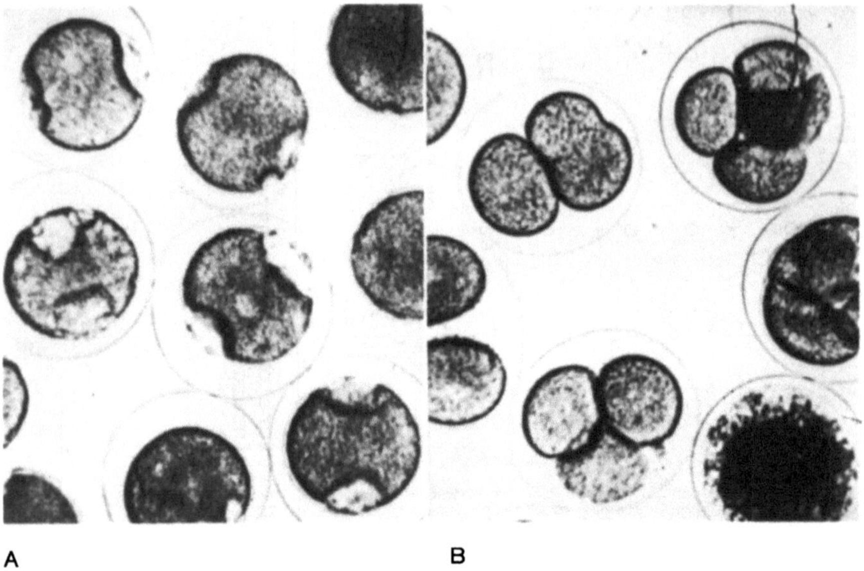

Fig. 5.8. (A) Eggs treated with hypertonic $HgCl_2$-sea water (Course 3); cleavage did not occur. (B) Eggs treated with hypertonic $HgCl_2$-sea water and washed with cysteine-sea water (Course 5); cleavage occurred. (From ISHIKAWA, 1960.)

In autoradiographic experiments on valine-^{14}C and leucine-^{3}H incorporation into sea-urchin eggs activated sufficiently or insufficiently with butyric acid-sea water, NAKASHIMA and SUGIYAMA (1969) showed that insufficient stimulation enhanced amino acid incorporation into eggs but induced no visible cortical change. The incorporation was more than 1.6 times greater than in unfertilized eggs. The results of the valine-^{14}C incorporation experiments are shown in Fig. 5.9 and Table 5.12.

Table 5.12. Valine-^{14}C incorporation into eggs of *Hemicentrotus pulcherrimus* at 18°C[a]

	Average of Grain Counts per $100 \mu m^2$	Ratio to Grain Count in Unfertilized Egg
Unfertilized egg	30.7	–
Insufficiently stimulated egg[b]	51.2	1.66
Fully activated egg[c]	150.2	4.89
Fertilized egg	219.4	7.14

[a]From NAKASHIMA and SUGIYAMA, 1969.

[b]Treated with 4% butyric acid (4 ml N/10 butyric acid plus 96 ml sea water) for 20 seconds.

[c]Treated with 4% butyric acid for 60 seconds.

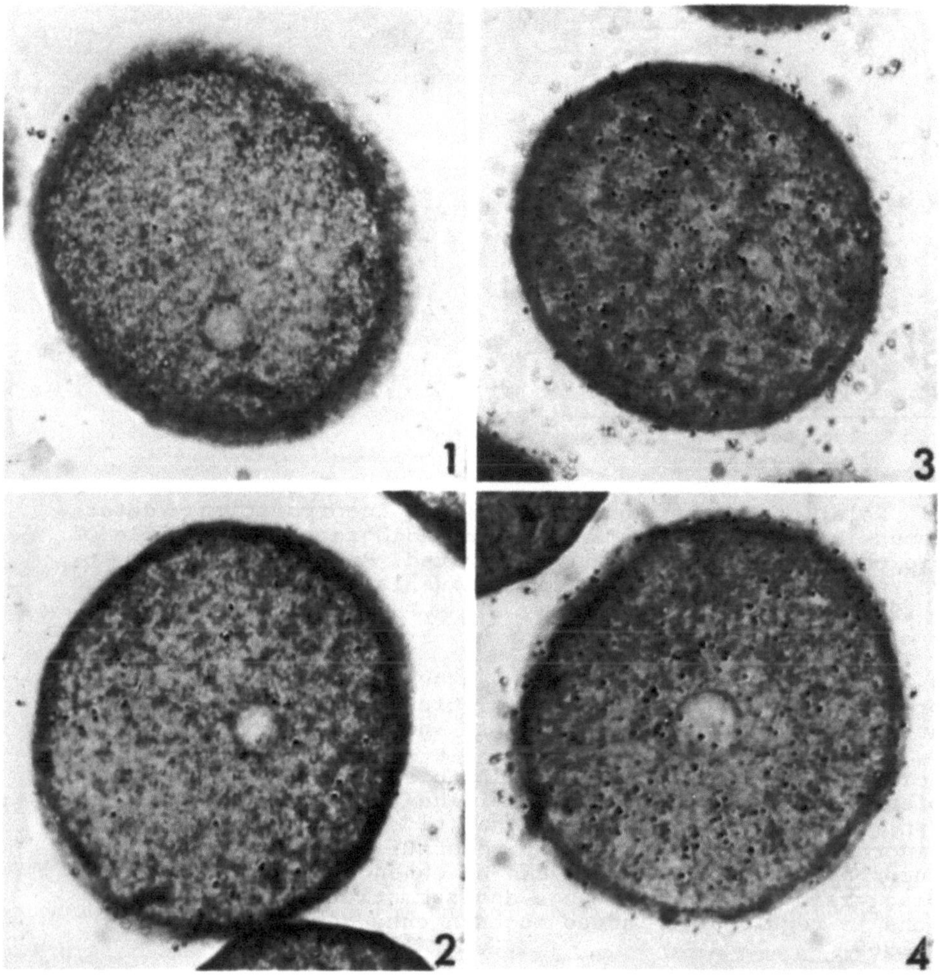

Fig. 5.9. Autoradiographs of eggs of *Hemicentrotus pulcherrimus* showing valine-^{14}C incorporation. (1) Unfertilized eggs; (2) insufficiently stimulated egg; (3) fully activated egg; (4) fertilized egg. (From NAKASHIMA and SUGIYAMA, 1969.)

NAKASHIMA and SUGIYAMA have suggested that "the fact that the incorporation of amino acids is increased without accompanying breakdown of the cortical granules indicates that the increase may be linked to an invisible change, probably the fertilization-wave, which is caused by the insufficient stimulation." It has recently been postulated that m-RNA is present in the unfertilized egg in a masked state and is unmasked by activation.

5.2.5 Uptake of Labeled Thymidine

UTO and SUGIYAMA (1969) studied thymidine-^{3}H incorporation into artificially activated *P. depressus* eggs stimulated with butyric acid-sea

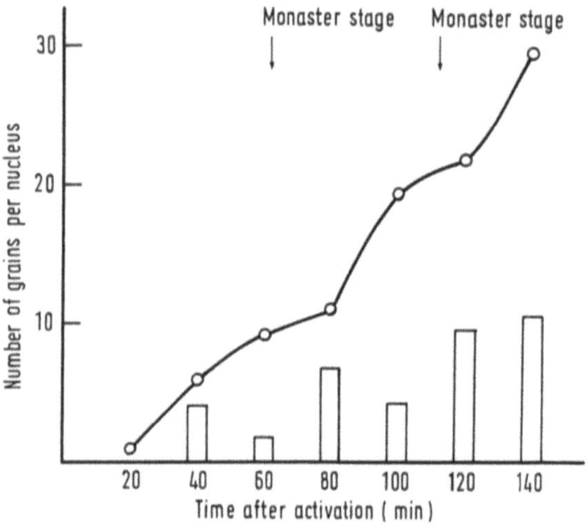

Fig. 5.10. Relationship between thymidine-^{3}H incorporation - determined by counting the grain number in autoradiographs - and cycle of morphologic changes in artificially activated *Pseudocentrotus* eggs at 20°C. Bars, pulse labeling; line, continuous labeling. (From UTO and SUGIYAMA, 1969.)

water and found that when stimulation was sufficient to cause cortical granule breakdown, thymidine-^{3}H incorporation into the nucleus increased at monaster formation (Figs. 5.10 and 5.11).

Butyric acid-sea water stimulation insufficient to cause cortical granule breakdown did not increase thymidine-^{3}H incorporation. When the weak stimulation was repeated several times, thymidine-^{3}H incorporation increased with the number of repetitions (Table 5.13, Fig. 5.12). "Incorporation of thymidine can be enhanced without any accompanying visible cortical change and summation of insufficient stimulations is required to induce such an enhancement" (UTO and SUGIYAMA, 1969).

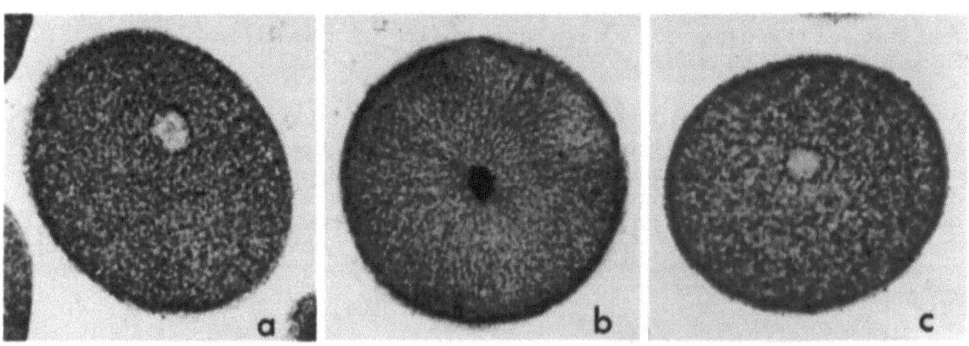

Fig. 5.11. Thymidine-^{3}H incorporation in *Pseudocentrotus* eggs. (a) Insufficiently stimulated egg; (b) artificially activated egg; (c) unfertilized egg. (From UTO and SUGIYAMA, 1969.)

Table 5.13. Thymidine-^{3}H incorporation into *Anthocidaris* eggs incubated five hours at 25°C (5μCi per milliliter)[a]

Unfertilized Eggs	Insufficiently Stimulated Eggs			Fully Activated Eggs
	Counts per Minute per 10^4 Eggs			
	Once	Twice	Three Times	
408	487	738	1780	2850

[a]From UTO and SUGIYAMA, 1969.

Comparing thymidine-^{3}H incorporation in *P. lividus* eggs fertilized or parthenogenetically activated with thymol and KCl and continuously incubated with thymidine-^{3}H, VON LEDEBUR-VILLIGER (1972) fixed equal egg samples at regular intervals, half of each sample prepared for scintillation counting and the other half processed for autoradiography. Fertilized eggs showed a first synthetic phase 20 to 40 minutes after fertilization, corresponding to the time between fusion of the pronuclei and prophase of the first cleavage (Fig. 5.13), followed by a plateau 40 to 60 minutes after fertilization, coinciding with the prophase and metaphase of the first mitotic cycle. During the first anaphase and telophase, the second S-phase took place. The third synthetic phase coincided with the anaphase and telophase of the second mitotic cycle.

Parthenogenetically activated eggs had a low rate of thymidine-^{3}H incorporation until 140 minutes after activation when it increased strikingly. Autoradiographs of each sample determined the percentage of actually replicating nuclei in fertilized and artificially activated eggs. The values from scintillation counting were referred to the percentage of labeled nuclei. Each fertilized egg had incorporated thymidine-^{3}H, while the curve for parthenogenetically activated eggs was distinctly altered, showing a marked plateau from 90 to 170 minutes after activation. The values of that plateau were half those of the first plateau in fertilized eggs. VON LEDEBUR-VILLIGER showed the thymidine-^{3}H-pool to be equal in artificially activated and fertilized eggs: "If thymidine-^{3}H incorporation is a reliable measure for DNA synthesis, this means that the artificially activated eggs

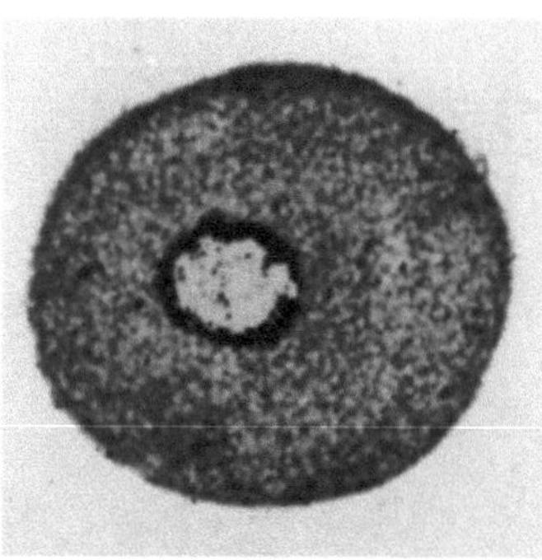

Fig. 5.12. Thymidine-^{3}H incorporation in *Anthocidaris* egg insufficiently stimulated three times. (From UTO and SUGIYAMA, 1969.)

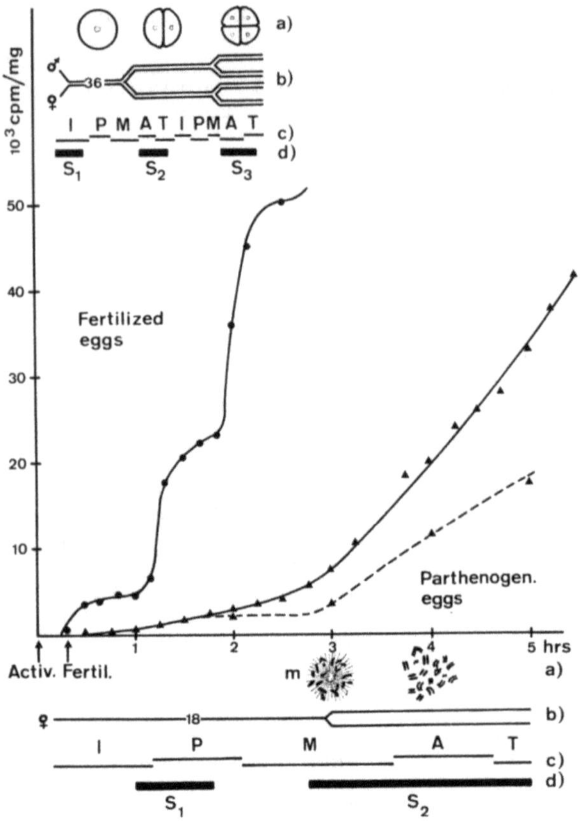

Fig. 5.13. Thymidine-^{3}H incorporation in fertilized and artificially activated *Paracentrotus lividus* eggs continuously incubated with 5-thymidine-^{3}H (3,4μCi/ml, 17,7 Ci/mmole). Dried eggs (counts per minute per milligram) were determined by liquid scintillation (solid lines). These values were corrected for percentage of labeled nuclei as determined in autoradiographs (dotted line). Cytological studies of fertilized eggs (upper left) and artificially activated eggs (bottom) are summarized. (a) Egg stage; m, monaster; (b) number of chromosomes; (c) time and duration of mitotic phases; I, interphase; P, prophase; M, metaphase; A, anaphase; T, telophase; (d) time and duration of DNA synthesis; S_1, first phase; S_2, second phase; S_3, third phase. (From VON LEDEBUR-VILLIGER, 1972.)

containing half the number of chromosomes synthesize half the amount of DNA." The first synthetic phase of parthenogenetically activated eggs begins later and lasts distinctly longer than that of fertilized controls. The increased DNA synthesis begins about three hours after activation and coincides with the formation of the first monaster in parthenogenetically activated eggs.

5.3 Conclusions

A wide variety of chemical and physical agents has been used to activate sea-urchin eggs artificially and to induce formation of the activation membrane which probably has the same properties as the fertilization membrane. Treatment with the membrane-forming agent alone causes a cortical reaction and induces nuclear change and monaster formation, usually leading to cytolysis and occasionally to egg cleavage.

LOEB's double treatment method of artificial parthenogenesis (treatment with a membrane-forming agent and hypertonic sea water) easily induces cleavage and further development. Many of the eggs so treated develop to the pluteus stage, but rearing sea-urchins beyond that is generally very difficult. Several methods of artificial parthenogenesis - using butyric acid, thymol and chloretone as the membrane-forming reagents - have been described. Treatment with hypertonic sea water is then necessary to induce significant cleavage and further development.

Cytasters are believed to be produced by centrioles of de novo origin (HARVEY, 1936), and DIRKSEN (1966) has shown centrioles in the cytasters of artificially activated eggs. It is also believed (LOEB, 1913) that the hypertonic medium acts as an inhibitor - or corrective factor - of the cytolytic disintegration caused by membrane-forming agents. In artificial parthenogenesis, the centriole that remained in the egg cytoplasm after maturation division may be activated by hypertonic treatment and later produce bipolar division centers.

Dinitrophenol, sodium azide, and potassium cyanide suppress the action of hypertonic sea water (ISHIKAWA, 1954, 1957). Inhibition is probably caused by blocking the energy-yielding phosphorus metabolism, increased labile phosphate content is shown in activated eggs in hypertonic solution, and increased ^{32}P incorporation is seen in labile phosphate compounds. The increase of labile phosphate is inhibited by adding dinitrophenol, sodium azide, potassium cyanide, and monoiodoacetamide to the hypertonic solution. ^{32}P uptake tends to increase with hypertonicity. The optimal hypertonicity for inducing cleavage increases the ^{32}P uptake of untreated eggs about 40%. The increased oxygen consumption of activated eggs treated with butyric acid or thymol is strongly suppressed in hypertonic solution and relieved by transferring the eggs to ordinary sea water. Increased activity leading to labile phosphate accumulation may be needed to induce parthenogenetic cleavage.

The inhibitory effect of sulfhydryl reagents on egg response to hypertonic treatment is reversed by treatment with cysteine. It seems apparent that cleavage induced by hypertonic treatment depends largely on the sulfhydryl groups of egg proteins.

After artificial activation, incorporation into the egg proteins of various labeled amino acids - methionine-^{35}S, leucine-^{3}H (-^{14}C), and valine-^{14}C - greatly increases (NAKANO, GIUDICE, and MONROY, 1958; BRACHET, FICQ, and TENCER, 1963; DENNY and TYLER, 1964; NAKASHIMA and SUGIYAMA, 1969). Anucleate halves respond to parthenogenetic activation, and increased amino acid incorporation into proteins may occur without the nucleus (BRACHET et al., 1963). Artificially induced cortical change considerably increases thymidine-^{3}H incorporation into the nucleus.

It seems likely that these results are related to the mechanism that triggers development in artificial parthenogenesis, but the available data are too few for further conclusions to be drawn.

Acknowledgement

The author is grateful to Dr. J.C. DAN, Dr. M. SUGIYAMA, and Dr. T. UEHARA for their criticism and suggestions.

References

BATAILLON, E., 1929. Analyse de la fécondation par la parthénogénèse expérimentale. Arch. f. Entw. 115, 707-778.
BRACHET, J., FICQ, A., TENCER, R., 1963. Amino acid incorporation into proteins of nucleate and anucleate fragments of sea urchin eggs: Effect of parthenogenetic activation. Exptl. Cell Res. 32, 168-170.
DELAGE, Y., 1909. Le sexe chez les oursins issus de parthénegénèse expérimentale. C.R. Acad. Sci. Paris 148, 453-455.
DENNY, P.C., TYLER, A., 1964. Activation of protein biosynthesis in non-nucleate fragments of sea urchin eggs. Biochem. Biophys. Res. Comm. 14, 245-249.
DIRKSEN, E.R., 1966. The presence of centrioles in artificially activated sea urchin eggs. Jour. Biophys. Biochem. Cytol. 11, 244-247.
HARVEY, E.B., 1936. Parthenogenetic merogony or cleavage without nuclei in Arbacia punctulata. Biol. Bull. 71, 101-121.
HARVEY, E.B., 1940. A comparison of the development of nucleate and non-nucleate eggs of Arbacia punctulata. Biol. Bull. 79, 166-187.
HARVEY, E.B., 1956. The American Arbacia and other Sea Urchins. Princeton: Princeton Univ. Press.
HERBST, C., 1893. Über die künstliche Hervorrufung von Dottermembranen an unbefruchteten Seeigeleiern nebst einigen Bemerkungen über die Dotterhautbildung überhaupt. Biol. Zentr. 13, 14.
HERTWIG, O., HERTWIG, R., 1887. Über den Befruchtungs- und Teilungsvorgang des tierischen Eies unter dem Einfluß äußerer Agentien. Jena. Zeitschr. 20, 120-242.
HERTWIG, R., 1895. Über Centrosoma und Centralspindel. Sitzungsb. Ges. Morph. und Physiol., München 11, 41-59.
HERTWIG, R., 1896. Über die Entwicklung des unbefruchteten Seeigeleies. Festschr. f. Gegenbaur, Leipzig 2, 21-86.
ISHIKAWA, M., 1954a. The activation of sea urchin eggs by means of phenols. Embryologia 2, 51-56.
ISHIKAWA, M., 1954b. Effects of hypertonic sea water on the sea urchin egg in the process of artificial parthenogenesis (in Japanese). Zool. Mag. (Tokyo) 63, 161.
ISHIKAWA, M., 1957. Antagonistic effects of some chemicals on the induction of parthenogenetic development of the sea urchin egg by hypertonic treatment. Embryologia 3, 261-266.
ISHIKAWA, M., 1960. The effects of sulfhydryl reagents on artificial parthenogenesis in the sea urchin egg. Embryologia 5, 85-94.
ISHIKAWA, M., 1962. The relation between the incorporation of phosphorus and the induction of cleavage by hypertonic treatment in the sea urchin egg. Embryologia 7, 109-126.
JUST, E.E., 1922. Initiation of development in the egg of Arbacia. I-III. Biol. Bull. 43, 384-422.

LITCHFIELD, J.B., WHITELEY, A.H., 1959. Studies on the mechanism of phosphate accumulation by sea urchin embryos. Biol. Bull. 117, 133-149.
LOEB, J., 1899. On the nature of the process of fertilization and the artificial production of normal larvae (Plutei) from the unfertilized eggs of the sea urchin. Am. Jour. Physiol. 3, 135-138.
LOEB, J., 1904. Über Befruchtung, künstliche Parthenogenese und Cytolyse des Seeigeleies. Pflüger's Arch. f. ges. Physiol. 103, 257-265. Trans. in Univ. Cal. Pub. 2, 73-81 (1905).
LOEB, J., 1906. Versuche über den chemischen Charakter des Befruchtungsvorgangs. Biochem. Zeitschr. 1, 183-206.
LOEB, J., 1913. Artificial Parthenogenesis and Fertilization. Univ. Chicago Press, Chicago.
MORGAN, T.H., 1899. The action of salt solutions on the unfertilized and fertilized eggs of Arbacia and of other animals. Arch. f. Entw. 8, 448-539.
NAKANO, E., GIUDICE, G., MONROY, A., 1958. On the incorporation of S^{35}-methionine in artificially activated sea urchin eggs. Experientia 14, 11-13.
NAKASHIMA, S., SUGIYAMA, M., 1969. Incorporation of amino acids into sea urchin eggs stimulated insufficiently with an activating reagent. Develop. Growth Differ. 11, 115-122.
PASPALEFF, G., 1925. Cytologische Untersuchungen an stimulierten Seeigeleiern. Zellstimulationsforsch. 1, 413-451.
RUNNSTRÖM, J., HAGSTRÖM, B.E., PERLMANN, P., 1959. Fertilization, in The Cell-Biochemistry, Physiology, Morphology. Edited by J. Brachet and A.E. Mirsky, Vol. 1, New York: Academic Press Inc.
SHEARER, C., LLOYD, D.J., 1913. On methods of producing artificial parthenogenesis in Echinus esculentus and the rearing of the parthenogenetic plutei through metamorphosis. Quart. Jour. Microscop. Sci. 58, 523-551.
UTO, N., SUGIYAMA, M., 1969. Incorporation of thymidine-^{3}H into the activated sea urchin egg with special reference to subthreshold stimulation with an activating reagent. Develop. Growth Differ. 11, 123-129.
VON LEDEBUR-VILLIGER, M., 1972. Cytology and nucleic acid synthesis of parthenogenetically activated sea urchin eggs. Exp. Cell Res. 72, 285-308.
WILSON, E.B., 1925. The Cell in Development and Heredity. 3rd edition, New York: Macmillan Co.
YAMAMOTO, T., 1956. Chloretone as a parthenogenetic agent in sea-urchin eggs. Embryologia 3, 81-87.

6. Cytology of Parthenogenetically Activated Sea Urchin Eggs

M. von Ledebur-Villiger

O. and R. HERTWIGS's early discovery of artificial parthenogenesis in sea urchins raised many questions: What is the origin of the centriole, asters and spindle? What cytological changes follow parthenogenetic activation? Do parthenogenetic embryos develop with the haploid or diploid set of chromosomes? R. HERTWIG (1896) found that after treatment with a strychnine solution the nucleus of *Echinus* and *Sphaerechinus* eggs may give rise to a typical mitotic spindle in which the chromosomes divide. MORGAN (1899, 1900) observed artificial astrospheres in the cytoplasm of *Arbacia* eggs treated with hypertonic solutions of NaCl or $MgCl_2$ or with strychnine. The centrioles were thought to arise de novo and possibly to develop from the achromatic part of the nucleus.

In a more detailed study on the cytological changes following artificial activation of *Toxopneustes variegatus* eggs, WILSON (1901) found that eggs give various responses to activating agents. After noting cytaster (the artificial astrospheres of MORGAN) and monaster formation he concluded: "All the facts indicate that the cytasters are of the same nature as the cleavage asters and that their central bodies are of the same nature as centrosomes. The centrosomes are primarily formed de novo. The number of chromosomes is one half that occurring in fertilized eggs, namely 18 instead of 36."

DELAGE (1901) studied the cytology of artificial parthenogenesis in *Paracentrotus lividus* and found the full number of chromosomes restored in parthenogenetic embryos. He observed 9 chromosomes in the unfertilized egg and 18 in the cells of artificially activated embryos. As, however, the haploid number of chromosomes of *P. lividus* is 18, DELAGE's observations supported those who disagreed with him ("les partisans de la non-régulation" as he called them). HINDLE (1910) noted that "The reduced number of chromosomes persists in the cells of parthenogenetic embryos of *Strongylocentrotus purpuratus*" but NEBEL et al. (1937) found the diploid state restored in *Arbacia punctulata* eggs stimulated by monochromatic ultraviolet radiation. DRIESCH (1905) measured the nuclear diameter in cells of parthenogenetic gastrulae and suggested the occurrence of haploid, diploid and tetraploid nuclei.

PASPALEFF (1925) noted 4 ways that artificially activated eggs may develop: haploid mitosis, monaster formation, multipolar mitosis and karyogene spindle formation. In recent studies (VON LEDEBUR-VILLIGER, 1972) eggs of *P. lividus* treated with thymol and hypertonic KCl solution show no visible change immediately after thymol activation but immersion in KCl solution causes the nuclei to swell and migrate toward the center of the egg. About 20 minutes later, small granules appear next to the nuclear membrane and about 1 1/2 hours after activation 18 chromosomes, the haploid set for *P. lividus*, develop as long, thin threads with the nuclear membrane still intact (prophase). From this point on development may follow 4 paths.

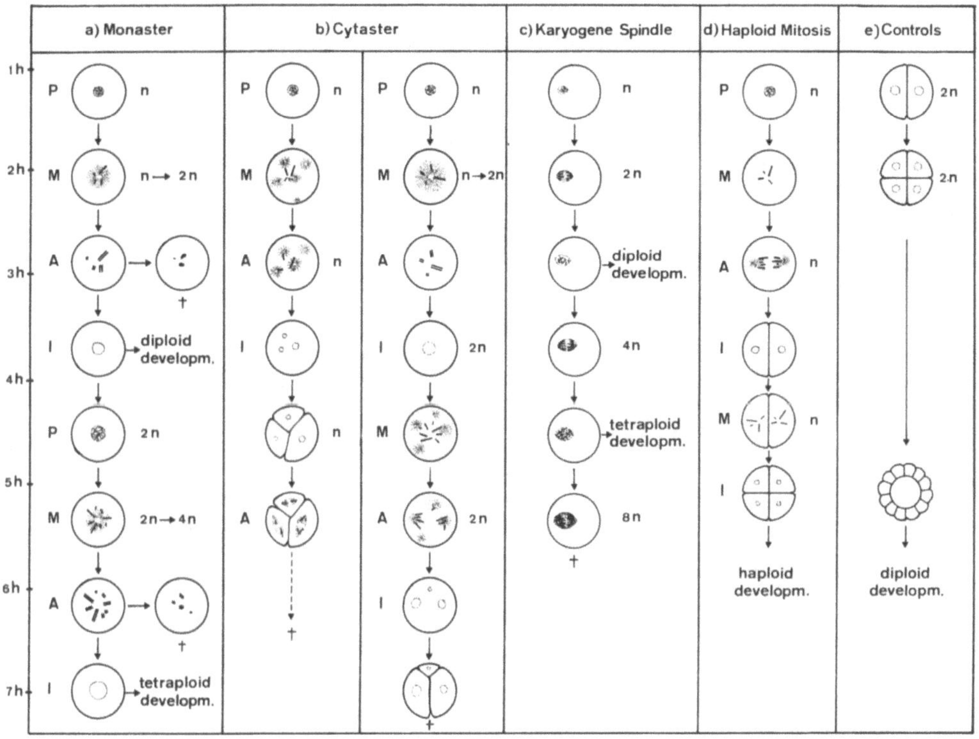

Fig. 6.1. Cytology of parthenogenetically activated *Paracentrotus lividus* eggs. Chronological comparison of the 4 types of parthenogenetic development (a-d) and development of fertilized controls (e). Haploid number of chromosomes is assumed to be 4. Activation membranes and fertilization membranes are omitted. P, prophase; M, metaphase; A, anaphase; I, interphase

6.1 Monaster Formation

After the nuclear membrane disintegrates, a monaster forms at or near the center of the egg (Fig. 6.1a). The centrifugal rays draw the chromosomes toward the periphery of the egg, arranging them in an irregular circle. The chromosomes are now strongly spiralized. About 2 hours after activation chromosomal condensation reverses slightly and the chromosomes begin to divide (metaphase). Still side by side, the daughter chromosomes separate about 1 hour later (anaphase) at which stage the monaster disappears.

The 36 purely maternal chromosomes enter a telophase-like stage and become less visible. Each pair of chromosomes is surrounded by a nuclear vesicle. If the vesicles are far apart, the karyomeres that form do not fuse or only adjacent ones do, leading to a varying number of small nuclei, each containing a different number of chromosomes. These nuclei may undergo several mitotic divisions, sometimes followed by irregular cell divisions, but embryos soon stop developing. If, however, the vesicles are closely packed, they fuse to form a single diploid nucleus, from which a diploid embryo may develop by normal mitosis.

After about 1 1/2 hours, a second monaster often forms and the chromosomes again divide. Far apart, the karyomeres fuse together into small nuclei, and close together into a tetraploid nucleus that undergoes normal mitosis to become a tetraploid blastula.

6.2 Cytaster Formation

Instead of a monaster, several small cytasters may appear in the activated cytoplasm (Fig. 6.1b). Those near the nucleus actively distribute the chromosomes. The mitotic figures exhibit 3, 4, or more poles that cause irregular chromosomal distribution. Competition for chromosomes may take place after the first prophase (haploid set of chromosomes) or, oftener, after the chromosomes of a monaster have divided. The resulting nuclei contain varying numbers of chromosomes and may divide even though the cytoplasm does not, creating a cell having as many as 30 nuclei. The nuclear divisions may be followed by cell divisions by which the larger nuclei are embedded in more of the cytoplasm than the small nuclei. These cells of different sizes divide a few times and then stop developing. It seems that a complete haploid set of chromosomes is essential for normal development.

6.3 Karyogene Spindle Formation

In a few eggs artificial activation induces formation of a karyogene spindle (Fig. 6.1c). Still excentric, the nucleus swells, becomes elliptical in shape and the nuclear membrane finally disappears. While this is going on, the chromosomes condense. A karyogene spindle, the poles of which show no astral rays is built up. The chromosomes divide and are pulled somewhat toward the poles of the spindle. No daughter nuclei are formed, and the spindle apparatus disappears leaving a diploid nucleus, which is able to double its chromosomes again in the same manner. A tetraploid nucleus results. If the process is again repeated, an octoploid nucleus forms. It was not possible to follow separately the further development of these nuclei, as they are indistinguishable from 2n and 4n nuclei arising from a monaster. Octoploid nuclei are not found in viable blastulae.

6.4 Haploid Mitosis

Egg activation sometimes results in normal mitosis involving only the 18 maternal chromosomes (Fig. 6.1d). A normal-looking mitotic apparatus forms, and the chromosomes divide and are distributed to the 2 poles to form 2 haploid daughter nuclei. Cell division follows. From this 2-cell stage a normal haploid blastula may arise.

6.5 Combinations

Artificially activated eggs may also develop in combinations of these ways. For example, 1 blastomere may contain a monaster and several cytasters or, in later stages, monaster chromosomes may divide in only some of the blastomeres of an embryo. Mosaic embryos result.

Activated eggs of the early spawning season (October or November) contain examples of all 4 types of development, while late batches (April or May) develop mostly by monaster or by intranuclear chromosomal division without forming any mitotic apparatus. The chromosomes are not distributed to daughter nuclei, which explains why very few activated eggs develop into swimming blastulae at the end of the reproduction season.

Chromosomal reproduction takes about the same time in monasters, cytasters, and haploid mitosis and slightly less time in karyogene spindles. All 4 types of parthenogenetic development take longer than fertilized controls.

In studies on the cytology and DNA content of nuclei from hatched parthenogenetic blastulae (VON LEDEBUR-VILLIGER, 1972), embryos were allowed to develop to the hatched blastula stage. Healthy blastulae, which, in contrast to abnormal blastulae, swim actively near the surface were fixed, stained by the Feulgen method, and their nucleic DNA content was measured spectrophotometrically. A single blastula showed either a few large cells having large nuclei, or many small cells having small nuclei or cells and nuclei of various sizes (mosaic embryos). The balance between cell number and size thus keeps embryo size remarkably constant.

The DNA measurements include prophase and metaphase nuclei having replicated amounts of DNA, anaphase nuclei having nonreplicated DNA values, and interphase nuclei having DNA values ranging from nonreplicated to replicated amounts of DNA. Values from interphase nuclei from parthenogenetic blastulae range over the whole spectrum (Fig. 6.2a), while those from the metaphase and prophase nuclei could be divided into 3 groups (Fig. 6.2b), the mean values of which were the terms of a geometric progression forming ratios of approximately 1:2:4. Where the metaphase plate chromosomes of parthenogenetic blastulae could be counted, the corresponding DNA content was measured, confirming that Group I represented the DNA content of haploid nuclei, Group II the DNA value of diploid nuclei, and Group III the DNA amount of tetraploid nuclei. The nuclei were chosen at random making it possible to conclude that in the parthenogenetic blastulae examined most of the nuclei had the tetraploid set of chromosomes and that nuclei with the haploid set were relatively rare (VON LEDEBUR-VILLIGER, 1972).

The DNA values for interphase, metaphase and anaphase nuclei in normal control blastulae (Fig. 6.2c and d) showed a relation between the means of the metaphase values (replicated amount of DNA) and the anaphase values (nonreplicated amount) very near the theoretical value of 2. The 2n values of the control nuclei were lower than the 2n values of the parthenogenetic blastulae, probably because of different optimal hydrolysis times. The rather wide variation in both parthenogenetic and control nuclei might have been partly due to different amounts of cytoplasm adhering to the nuclei and interfering with absorption during spectrophotometric measurements.

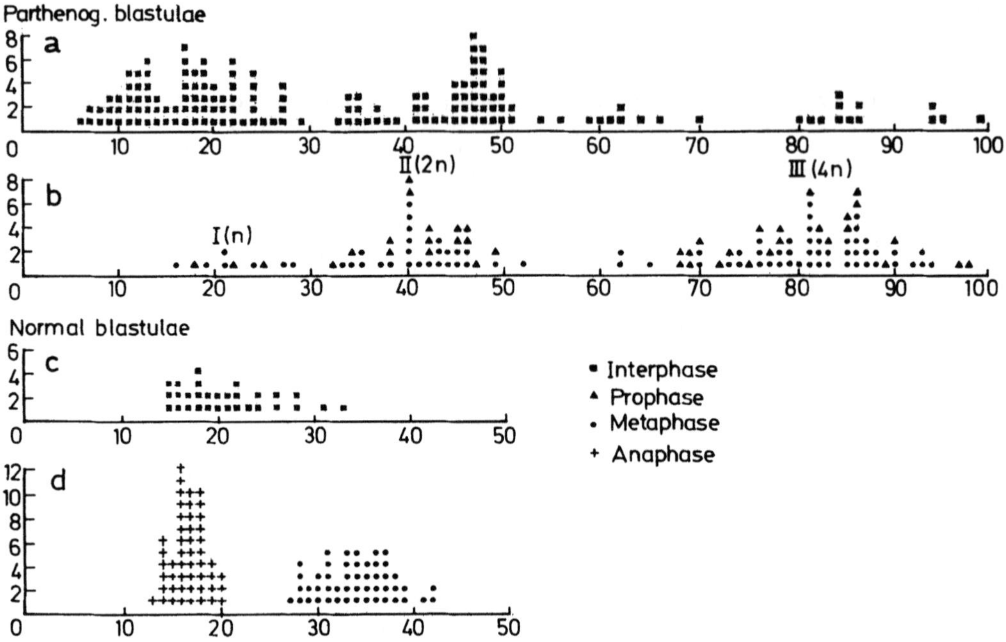

Fig. 6.2. Distribution of Feulgen values of nuclei from parthenogenetic blastulae (24 hours after activation) and control blastulae (15 hours after fertilization). (a) Interphase nuclei from parthenogenetic blastulae; (b) prophase and metaphase nuclei from parthenogenetic blastulae; (c) interphase nuclei from controls; (d) metaphase and anaphase nuclei from controls. Hydrolysis was performed in 4N HCl at 25°C for 90 minutes. Abscissa, arbitrary units of extinction integrals; ordinate, frequency of measured values

Several types of parthenogenetic blastulae have been observed (VON LEDEBUR-VILLIGER, 1972). The non-mosaic type contains nuclei of 1n, 2n or 4n values while the mosaic type contains nuclei of 1n and 2n, or 2n and 4n values. Blastulae having 1n, 2n and 4n nuclei are exceptional. In a mosaic blastula the ratio between the 2 types of nuclei is not constant, varying from 1:1 to about 1:9, and indicating that mosaic types may arise at various stages of early development. Of the 6 types of blastulae found in 1 batch of activated eggs (Table 6.1), most were non-mosaic tetraploid.

Table 6.1. Frequency of 6 types of parthenogenetic blastulae found in 1 batch of activated eggs

	Non-mosaic types			Mosaic types		
Ploidy	1n	2n	4n	1n/2n	2n/4n	1n/2n/4n
Frequency	4	18	42	7	10	1
Percentage	5	22	51	9	12	1

Electron microscopic studies have shown that centrioles in artificially activated sea-urchin eggs have a structure and electron density similar to centrioles in fertilized eggs. DIRKSEN (1961) found centrioles in the center of cytasters, and SACHS and ANDERSON (1970) observed 1, 2, or 3 centrioles in artificially activated eggs before the first nuclear division, but not in ripe sea-urchin eggs.

Considering whether the centers of artificially induced asters derive from the egg centriole or are formed de novo, SACHS and ANDERSON suggested that the egg centriole persists in the form of "submicroscopic" precursors. After activating nonnucleated fragments of sea-urchin eggs, HARVEY (1936) assumed that the centrioles formed under those conditions arose de novo. HERTWIG (1896) and MORGAN (1900) also believed that centrioles in artificially activated sea-urchin eggs are formed de novo.

Artificially activated *P. lividus* eggs collected at the beginning of the spawning season showed asters and spindles (VON LEDEBUR-VILLIGER, 1972). Of these eggs, 30% - 40% reached the swimming blastula stage. Activated eggs from the end of the spawning season rarely cleaved and only 1% - 3% developed into blastulae. How an egg develops seems to depend on its age. As sea-urchin eggs complete meiosis before fertilization or activation, cleavage success after activation may depend on the age of the egg centriole. Normally suppressed by the presence of a sperm centriole, the activation of the egg centriole may occur through parthenogenetic treatment, especially if the stimulation is not too long after the second meiotic division. This applies to centrioles of eggs that cleave normally in haploid mitosis and to those of eggs that cleave after chromosomal replication in a monaster. The centrioles that DIRKSEN (1961) found in the center of cytasters may be of a different origin.

References

DELAGE, Y., 1901. Etudes expérimentales sur la maturation cytoplasmique et sur la parthénogenèse artificielle chez les Echinodermes. Arch. Zool. exp. 9, 285-326.

DIRKSEN, E.R., 1961. The presence of centrioles in artificially activated sea-urchin eggs. J. Biophys. Biochem. Cytol. 11, 244-247.

DRIESCH, H., 1905. Zur Cytologie parthenogenetischer Larven von Strongylocentrotus. Wilhelm Roux' Arch. Entwicklungsmech. Organismen 19, 648.

HARVEY, E.B., 1936. Parthenogenetic merogony or cleavage without nuclei in Arbacia punctulata. Biol. Bull. 71, 101-121.

HERTWIG, O., HERTWIG, R., 1887. Über den Befruchtungs- und Teilungsvorgang des tierischen Eies unter dem Einfluß äußerer Agentien. Jena. Z. Naturwiss. 13, 120-241.

HERTWIG, R., 1896. Über die Entwicklung des unbefruchteten Seeigeleies. Festschr. Gegenbaur II, 21-86.

HINDLE, E., 1910. A cytological study of artificial parthenogenesis in Strongylocentrotus purpuratus. Wilhelm Roux' Arch. Entwicklungsmech. Organismen 31, 145-163.

VON LEDEBUR-VILLIGER, M., 1972. Cytology and nucleic acid synthesis of parthenogenetically activated sea-urchin eggs. Exp. Cell Res. 72, 285-308.

MORGAN, T.H., 1899. The action of salt-solutions on the unfertilized and fertilized eggs of Arbacia and other mammals. Wilhelm Roux' Arch. Entwicklungsmech. Organismen 8, 448-539.

MORGAN, T.H., 1900. Further studies on the action of salt-solutions and of other agents on the eggs of Arbacia. Wilhelm Roux' Arch. Entwicklungsmech. Organismen 10, 489-524.
NEBEL, B.R., HARVEY, E.B., HOLLAENDER, A., 1937. The cytology of Arbacia punctulata activated by monochromatic ultraviolet radiation. Biol. Bull. 73, 365-366.
PASPALEFF, G., 1925. Cytologische Untersuchungen an stimulierten Seeigeleiern. Zellstimulationsforsch. 1, 413-448.
SACHS, M.I., ANDERSON, E., 1970. A cytological study of artificial parthenogenesis in the sea urchin Arbacia punctulata. J. Cell Biol. 47, 140-158.
WILSON, E.B., 1901. Experimental studies in cytology. I. A cytological study of artificial parthenogenesis in sea-urchin eggs. Wilhelm Roux' Arch. Entwicklungsmech. Organismen 12, 529-589.

7. Normal Development to Metamorphosis

K. Okazaki

The sea-urchin egg completes reduction division within the ovary (Fig. 7.1 [1-2]). The unfertilized egg is a spherical cell about 90μm in diameter, surrounded by a transparent jelly coat about 30μm thick. The refractive index of the jelly coat is the same as that of seawater, making it invisible by ordinary microscope. In a suspension of India ink, however, it remains transparent because the ink particles do not penetrate it. It is easily stained with vital stains such as Janus green. Spermatozoa pass through the jelly coat at any point and enter the egg surface wherever they happen to reach it. The jelly is composed of polysaccharide and has a strong agglutinating effect on sperm, but its role in fertilization has not been fully explained. Even when the jelly coat is removed by seawater acidified at pH 5, fertilization and egg development proceed normally.

After spawning, the polarity of the mature egg (the animal-vegetal axis) cannot be recognized except in eggs of *Paracentrotus (Strongylocentrotus) lividus* in certain habitats. In these eggs the pigment band encircles the vegetal half except the equitorial and polar regions (SELENKA, 1883; BOVERI, 1901a, 1901b). The polarity is already laid down in the ovary. The animal pole - where the polar bodies are formed when the eggs are still in the ovary - is antipolar to the point of attachment to the ovary wall.

The egg pronucleus lies eccentrically in the cytoplasm (Fig. 7.1 [2]) which is crowded with small particles, about half of which are yolk granules. The rest are mitochondria, pigment granules, and various minute bodies. Directly under the surface membrane is a compact layer of special granules about 1μm in diameter called cortical granules, cortical particles, or cortical bodies. Electron microscopy shows they are surrounded by a membrane 50 Å thick containing two kinds of electron-dense materials embedded in a less dense matrix. The egg surface is covered by two membranes, both 150-200 Å thick: the outer or vitelline membrane and the inner or plasma membrane which is in a very close contact with the membranes of the cortical granules.

7.0.1 Fertilization

When the spermatozoon makes contact with the egg surface, an impulse is transmitted around the egg from the point of contact, reaching the opposite side of the egg in 20 seconds. The cortical granules swell, and their membranes and the plasma membrane at the point of close contact burst, separating the vitelline membrane from the egg surface. As the cortical granules open, the first electron-dense components are released and deposited on the separated vitelline membrane, now called the fertilization membrane (Fig. 7.1 [3]). The second electron-dense components of the cortical granules left behind in the open granules gradually come out and form close to the egg surface the hyaline layer which, several minutes later, light microscopy shows to be translucent and several microns thick (Fig. 7.1 [4]).

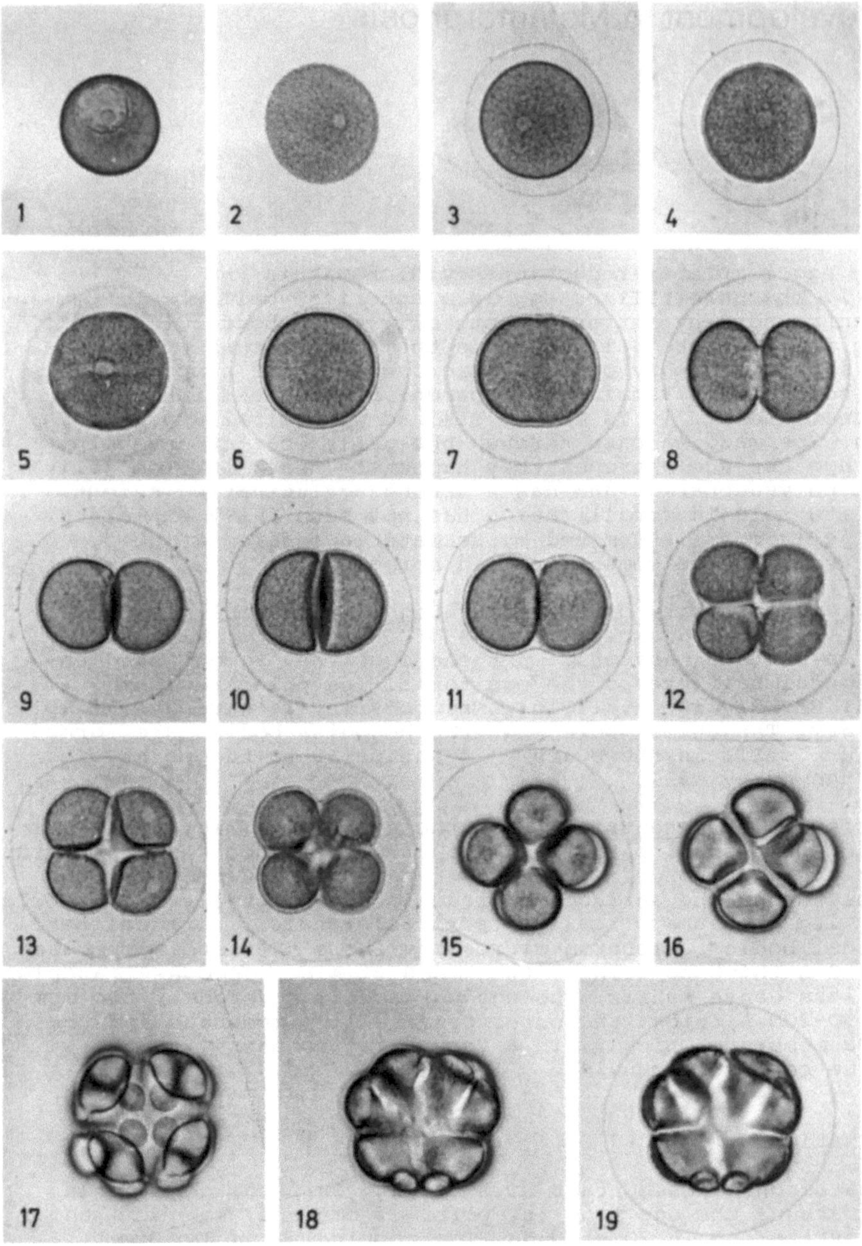
1
2
3
4
5
6
7
8
9
10
11
12
13
14
15
16
17
18
19

◀ Fig. 7.1. Developmental process of *Mespilia globulus*. Larvae are raised at room temperature: 1-62, 27°-28°C; 63-71, 25°-27°C. 1, immature egg; 2, unfertilized egg; 3, 3 min after fertilization; 4, 15 min, hyaline layer visible; 5, clear streak stage, 20 min; 6, 30 min, diaster visible and hyaline layer thickens; 7, 32 min; 8, 34 min; 9, immediately after cleavage, 40 min; 10, resting 2-cell stage, 50 min; 11, blastomeres become round before next cleavage, 54 min; 12, immediately after second cleavage, 58 min; 13, resting 4-cell stage, 63 min; 14, just before third cleavage, 1 hr 10 min, (14-17 seen from animal or vegetal pole); 15, immediately after third cleavage, 1 hr 15 min; 16, resting 8-cell stage, 1 hr 20 min; 17-19, 16-cell stage, 1 hr 30-40 min; 17, vegetal view, immediately after fourth cleavage. 18-19, resting stage, lateral view of same embryo, different loci

---

As the vitelline membrane rises to become the fertilization membrane, the fertilized egg surface - covered by the hyaline layer - is formed by fusion of the plasma membrane and the emptied membranes of the cortical granules. Intergranular portions are transformed into cytoplasmic projections penetrating the hyaline layer (see page 279) and corresponding to the attachment fibers described by DAN and ONO (1952). Meanwhile a spermatozoon is "drawn" into the egg, but before the sperm head has disappeared inside, a small bulge of hyaline cytoplasm - the fertilization cone - forms on the egg surface at the point of sperm entry, usually vanishing in a few minutes, although in *Pseudocentrotus* it remains for hours.

A few minutes after sperm entry, the sperm head stops, begins to swell, and changes into the sperm pronucleus. At the same time the middle piece breaks down and releases the centriole, around which egg cytoplasm forms the sperm monaster. The centriole stays close to the sperm pronucleus and lies at the center of the monaster which becomes larger as its rays elongate. When the ray tips reach the egg pronucleus, it suddenly moves toward the monaster center and the sperm pronucleus. The two pronuclei remain in contact and unite to form a single nucleus, the synkarion which has the diploid number of chromosomes.

7.0.2 Nuclear Division and Cleavage

After the two pronuclei unite (syngamy), the monaster flattens in the direction of the egg axis and the monaster center changes to a disc that appears as a bright streak when seen from the side. This is called the clear-streak stage, with the synkarion always at the center of the streak (Fig. 7.1 [5]). Although the biologic meaning of this stage is unknown, the direction of the streak - perpendicular to the egg axis - is the earliest indication of the main axis in eggs without a pigment ring.

The streak soon vanishes and the synkarion enlarges. While the nucleus grows, the nuclear membrane disappears and nuclear division begins (Fig. 7.1 [6]). Although the metaphase diaster attains considerable size as shown by section (WILSON, 1895) or by isolation of the mitotic apparatus (DAN and NAKAJIMA, 1956), the astral rays are often hard to see in living material. The astral center is a bright, dumbbell-shaped

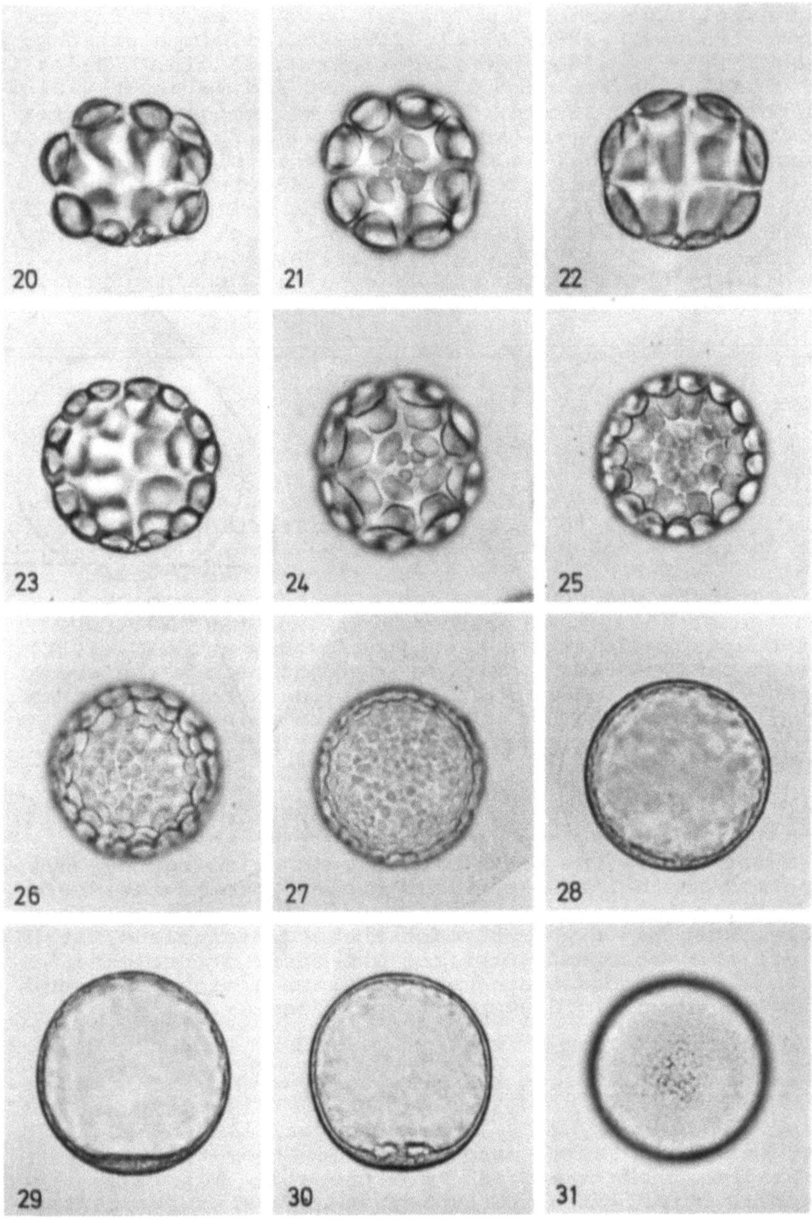

Fig. 7.1 (continued). 20-22, 32 cell stage, 1 hr 50 min - 2 hr; 21, vegetal view; 23-24, after sixth division, 2 hr 15 min; 25, after seventh division, 2 hr 40 min; 26, after eighth division, 3 hr 15 min; 27, after ninth division, 4 hr; 28, after tenth division, 5 hr, embryo rotating in fertilization membrane. 29, blastula after hatching, 6 hr; 30-31, very early mesenchymal blastula, primary mesenchymal cells beginning to form, 7 hr; 30, lateral view; 31, vegetal view

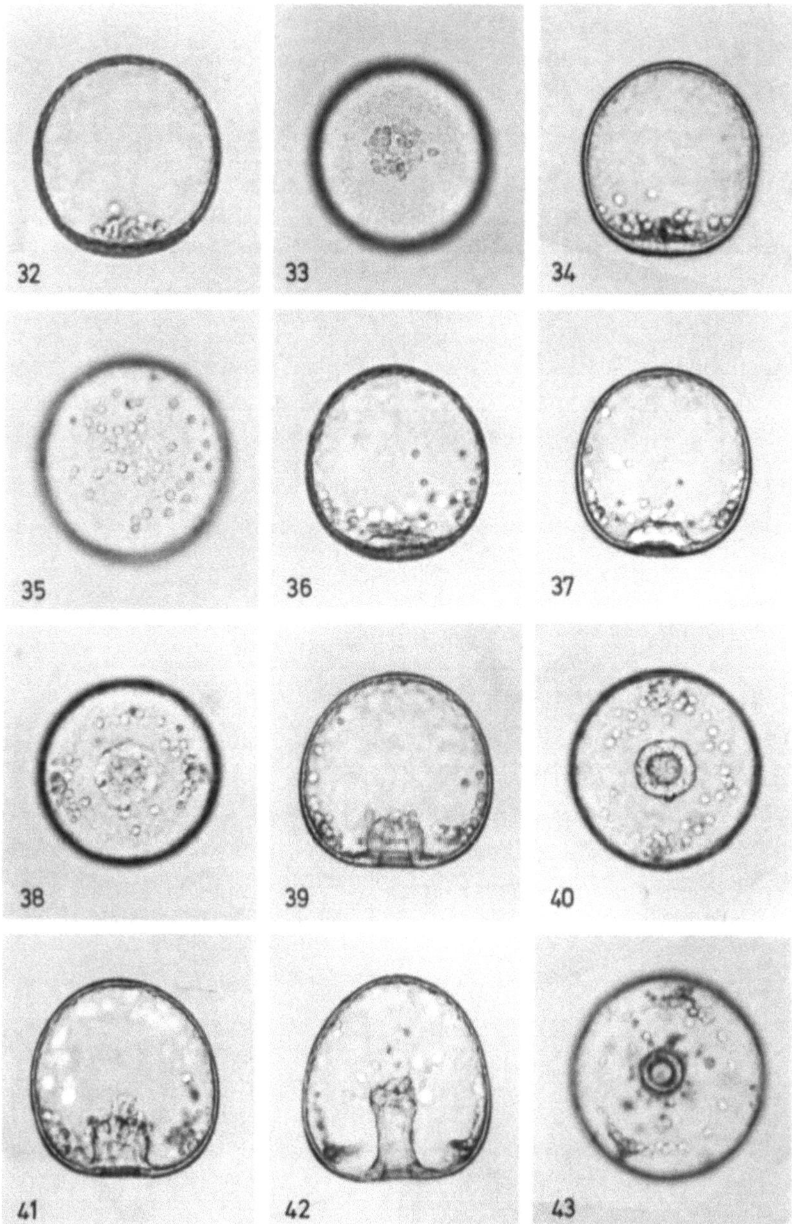

Fig. 7.1 (continued). 32-33, early mesenchymal blastula, 8 hr; 34-35, midmesenchymal blastula, 9 hr; 36, late mesenchymal blastula, 10 hr; 37-40, early gastrula, 11 hr; 41-43, midgastrula, 12 hr

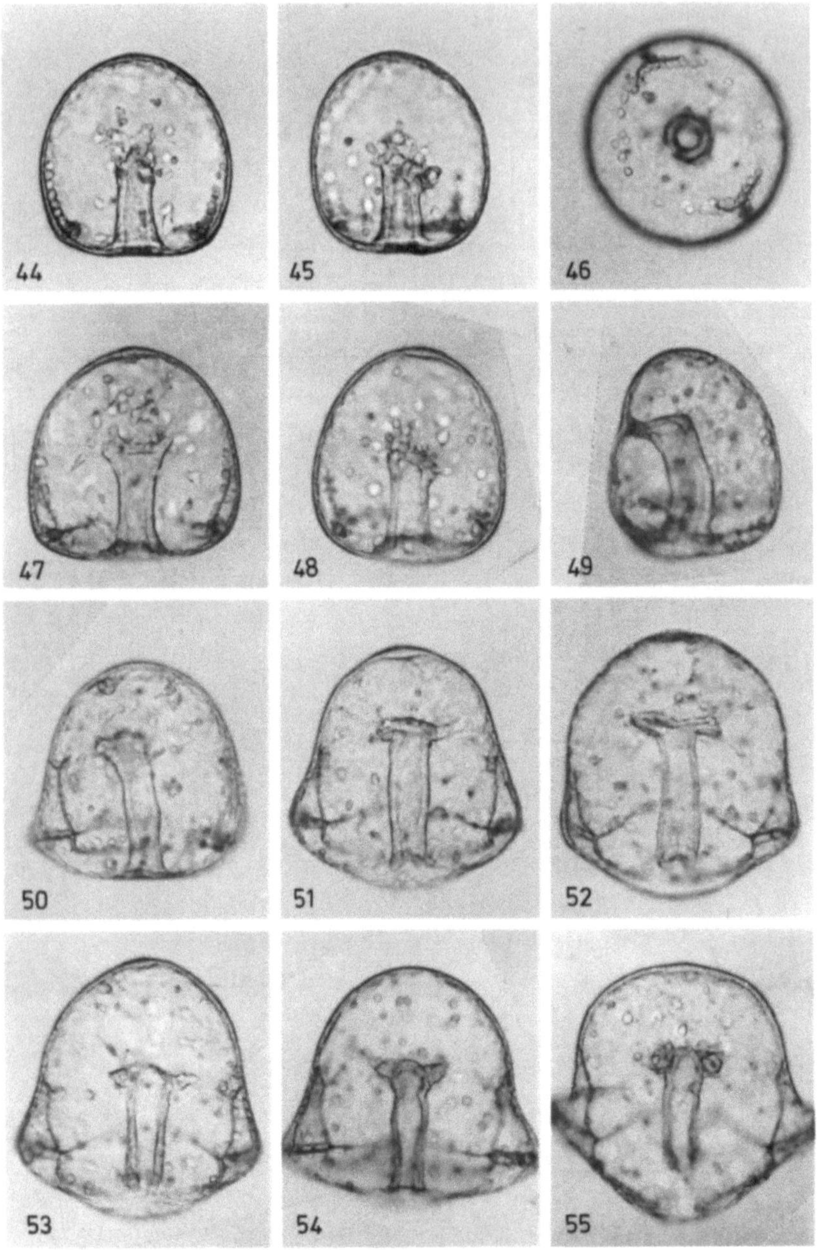

Fig. 7.1 (continued). 44-46, late gastrula, gastrulation completed, 13 hr; 47-49, early prism stage, blastocoelic volume increased after gastrulation, 14 1/2 hr; 49, archenteron tip in contact with stomodeum invagination; 50-52, midprism stage, archenteron tip forms bilobed space, 16 hr; 53-55, late prism, pair of coelomic vesicles are separated from archenteron which begins to be constricted, 18 hr

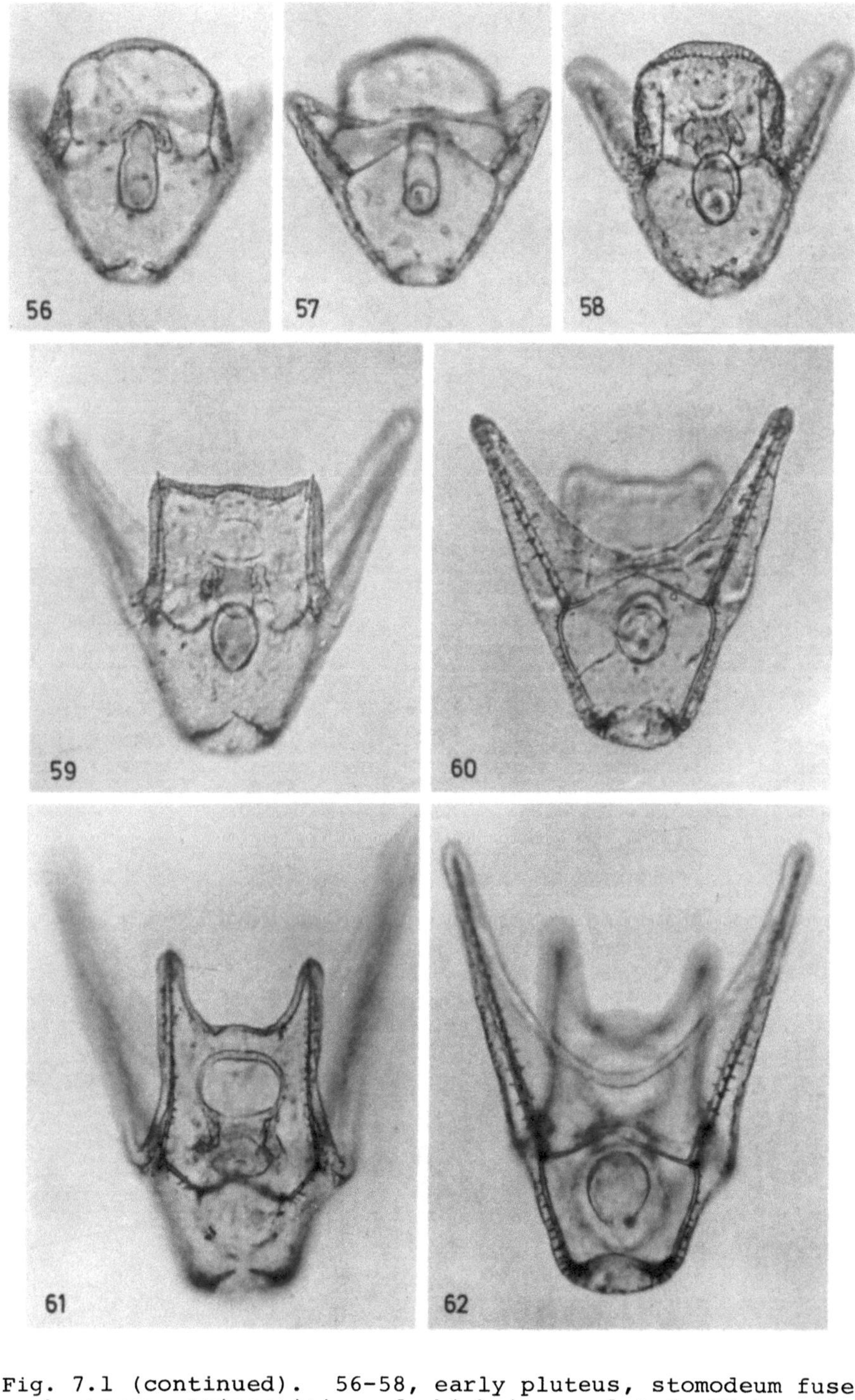

Fig. 7.1 (continued). 56-58, early pluteus, stomodeum fuses with archenteron, tripartition of which is completed, 20 hr; 59-60, pluteus with two arms, 24 hr; 61-62, pluteus with four arms, 32 hr

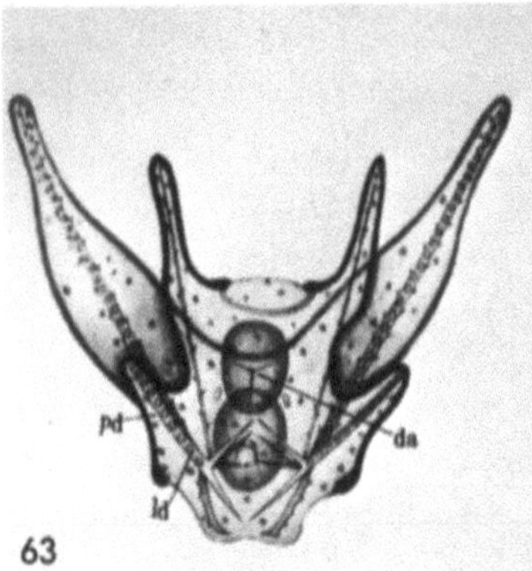
pd
da
ld

63

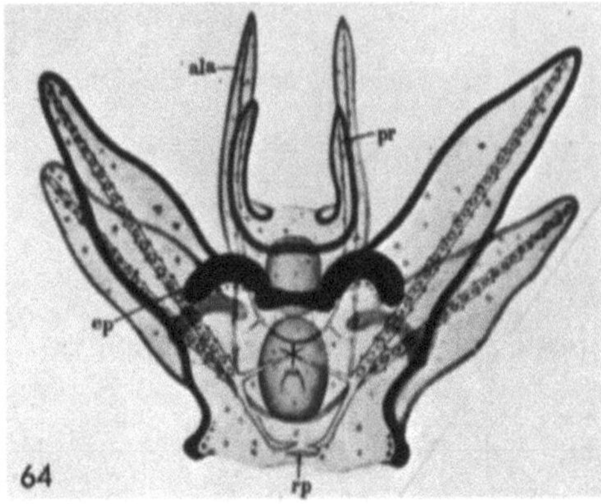
ala
pr
ep
rp

64

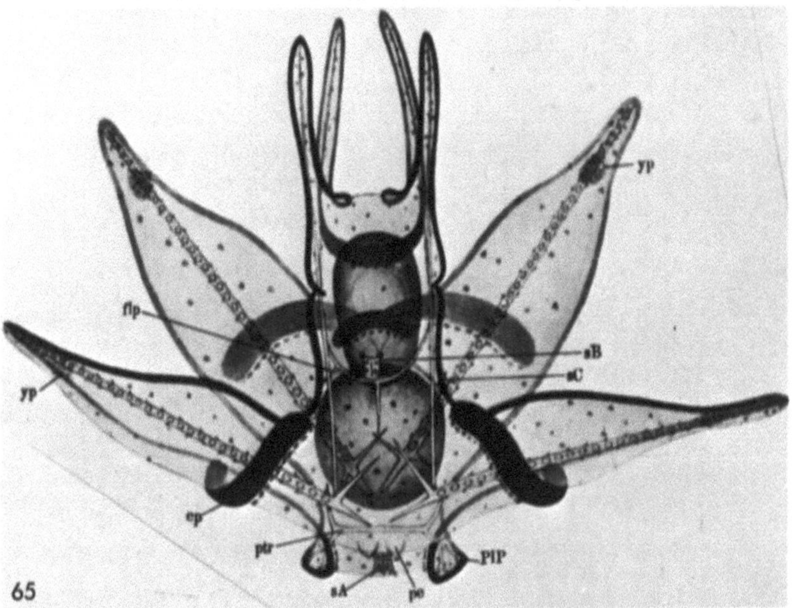
yp
flp
sB
sC
yp
ep
ptr
PlP
sA
po

65

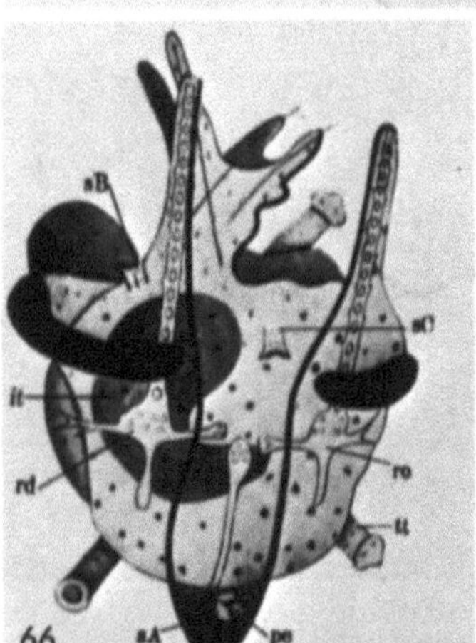
sB
sC
it
rd
ro
ti
sA
po

66

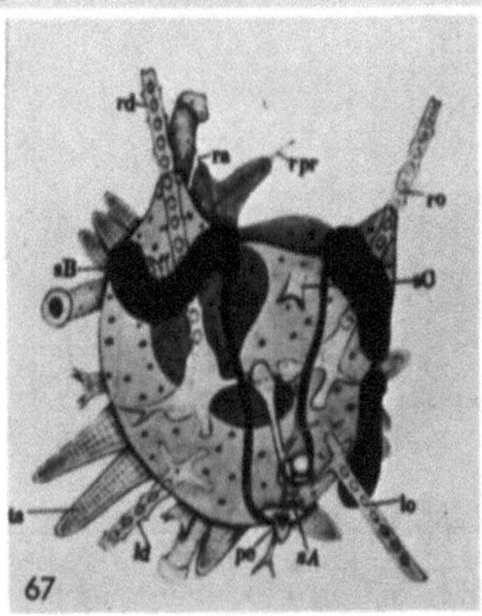
rd
ra
rpr
ro
sB
sC
ta
lo
ld
po
sA

67

Fig. 7.1 (continued). 63, pluteus 8 days old, dorsal view; 64, pluteus 15 days old, ventral view; 65, pluteus 30 days old, dorsal view, entering stage of metamorphosis; 66, pluteus beginning metamorphosis, 35 days old, three of five azygous tube feet are projecting; 67, same as 66, right-side view, absorption of pluteus arms almost complete, many adult spines have appeared

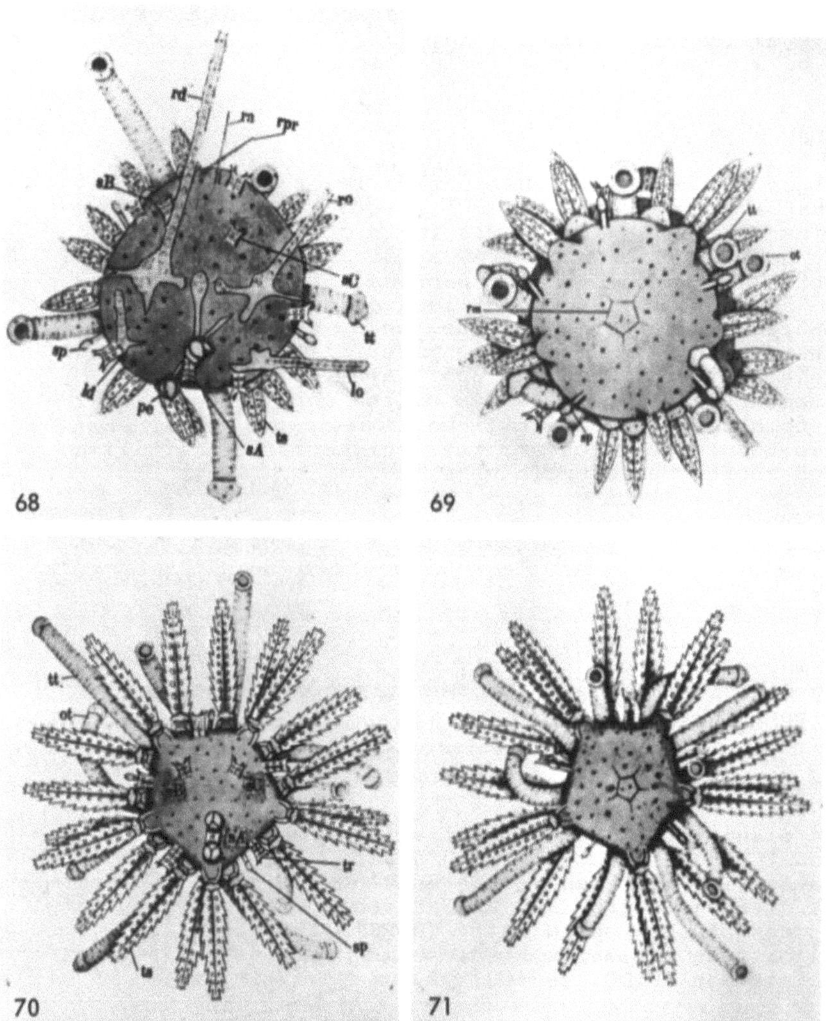

Fig. 7.1 (continued). 68, young sea urchin just after metamorphosis, 35 days old, aboral view, 69, same as 68, oral view; 70, imago 3 days after metamorphosis, aboral view; 71, same as 70, oral view. ala, anterolateral arm; da, dorsal arch; ep, epaulet; flp, first lateral process; it, intestine; ld, left posterodorsal rod; lo, left postoral rod; ot, primary ordinary tube foot; pd, posterodorsal arm; pe, pedicellaria; plp, posterior lateral process; pr, preoral arm; ptr, posterior transverse rod; ra, right anterolateral rod; rd, right posterodorsal rod; rm, rudiment of the mouth; ro, right postoral rod; rp, rudiment of the posterior transverse rod; rpr, right preoral rod; sA,sB,sC, spine A,B,C; sp, sphaeridium; tr, tetraradiate spine; ts, typical spine; tt, terminal tube foot; yp, yellowish green patch. (63-71 from ONODA, 1936)

region. The dumbbell elongates and the two asters separate. When the diaster fills the whole cell, nuclear division is complete and daughter nuclei are formed.

The cell then constricts in the equatorial plane of the spindle and cleavage begins as the egg becomes elongated in the direction of the spindle axis and the egg outline becomes ovoid (Fig. 7.1 [7]). The cleavage furrow appears in the middle of the long dimension bisecting the spindle, and the egg cell becomes cocoon-shaped and divides into two daughter cells called blastomeres (Fig. 7.1 [7-10]). The process of cleavage will be discussed in details in chapter 10.

7.0.3 Cortical Layer (Cortex)

In unfertilized eggs the cortex is the surface layer with cortical granules that burst and disappear at fertilization. Although electron microscopic studies have failed to show a layer of ground plasm at the periphery of the fertilized egg, many studies have indicated the presence of a cortex having physical properties different from the cell interior. In fertilized and unfertilized eggs the cortex is indicated when granules in the peripheral region of the cell are not affected by ordinary centrifugation while materials of the central plasma are easily stratified (see page 335). By micromanipulation HIRAMOTO (1957) demonstrated the presence of the cortical gel layer and measured its thickness. In unfertilized *Hemicentrotus pulcherrimus* eggs the cortex is about 3μm thick and gets thicker after fertilization, reaching 4μm shortly before cleavage.

7.1 Cleavage Stage

A fertilized egg may divide synchronously ten times at about hourly intervals. During division the cells become arranged in the form of a hollow sphere, the blastula.

7.1.1 Cleavage Pattern

The first clavage plane includes the polar axis which divides into two blastomeres of equal size. It is generally agreed that the position of the first cleavage is unrelated to the entrance point of the spermatozoon. The first cleavage furrow appears at random, as shown by vital staining on the side opposite sperm entry (HÖRSTADIUS, 1928, 1939) or by the fertilization cone in *Pseudocentrotus* which remains for several hours after fertilization (ENDO, 1954).

It is not yet known whether or not the first cleavage is related to the larval plane of symmetry. Some workers think the two coincide, others that the first cleavage plane is frontal, and still others that the two are unrelated (HÖRSTADIUS, 1939).

The cleavage furrow may come in unilaterally, animal-pole side first, and the dividing cell pass through a heart-shaped stage. Such a cleavage pattern occurs when the spindle is formed closer to the animal pole instead of in the center. Heart-shaped cleavage is normally found in *Astriclypeus manni* (DAN and DAN, 1947) and abnormally in many species.

The second cleavage is also equal and also includes the egg axis (Fig. 7.2B), around which the resulting four blastomeres are grouped. The central space between the blastomeres is the beginning of the blastocoel (Fig. 7.1 [13]).

Although the third cleavage is again equal, the plane of cleavage is at a right angle to the egg axis and divides the four blastomeres into upper and lower tiers (Figs. 7.1 [15-16] and 7.2C). If the furrows of the third cleavage are strictly equatorial, the resulting eight blastomeres are all the same size, making distinction between animal and vegetal poles impossible. In some species the furrows cut farther toward the vegetal pole so that the four animal blastomeres are larger than the vegetal ones.

The fourth cleavage is characteristically unequal. The four upper cells of the 8-cell stage divide equally and meridionally, and eight blastomeres of the same size ring the egg axis. The four lower cells cleave unequally, and four large upper cells and four small ones converge toward the vegetal pole (Fig. 7.1 [17-19]). In the 16-cell stage, the blastomeres are arranged in three layers: an upper layer containing eight medium-sized mesomeres, a middle layer containing four large macromeres, and a lower layer containing four small micromeres (Fig. 7.2D).

The fourth cleavage is often asynchronous. The vegetal quartet may divide several minutes ealier than the animal quartet causing the embryo to pass through a 12-cell stage. Such cleavage time lag varies in different batches of eggs but occurs in most (HARVEY, 1956; OKAZAKI, unpublished; CZIHAK, unpublished).

In the fifth cleavage, the mesomeres, macromeres, and micromeres divide asynchronously. A short 20- and 28-cell stage precedes the 32-cell stage. The four macromeres divide first, then the 8 mesomeres, and then the 4 micromeres.

If mitosis is triggered by a cytoplasmic factor, the blastomeres having more cytoplasm are to be expected to start cleavage first (CZIHAK, 1971). This is supported by the fact that in a hybrid, cleavage times corresponded to the cleavage time of the female (MOORE, 1933). HÅGSTROM and LÖNNING (1965) published a timetable for the cleavage of *Echinocyamus pusillus* from the 16-cell stage on.

The ring of 8 mesomeres of the previous stage divides into upper and lower layers of equal size (an_1 and an_2, Fig. 7.2E). The macromeres divide meridionally, forming a single ring of 8 cells, and the micromeres bud off smaller blastomeres toward the vegetal pole. The embryo as a whole consists of 32 cells arranged in five tiers (Fig. 7.1 [20-22]). For micromere derivatives, the following abbreviations will be used: lm (large micromere) for the large daughter cell and sm (small micromere) for the small daughter cell.

BOVERI (1901a) showed that the descendants of the micromeres of the 16-cell stage develop into the primary mesenchyme cells that form spicules. (In Fig. 7.2, micromeres and spicules are indicated in black). Based on time-lapse cinematographic studies of *Mespilia* embryos, ENDO (1966) stated that only lm descendants of the 32-cell stage develop into primary mesenchymal cells while sm descendants become pharynx.

At the sixth cleavage, an_1 and an_2 divide horizontally into 4 layers of 8 cells. The macromeres also divide horizontally into an upper

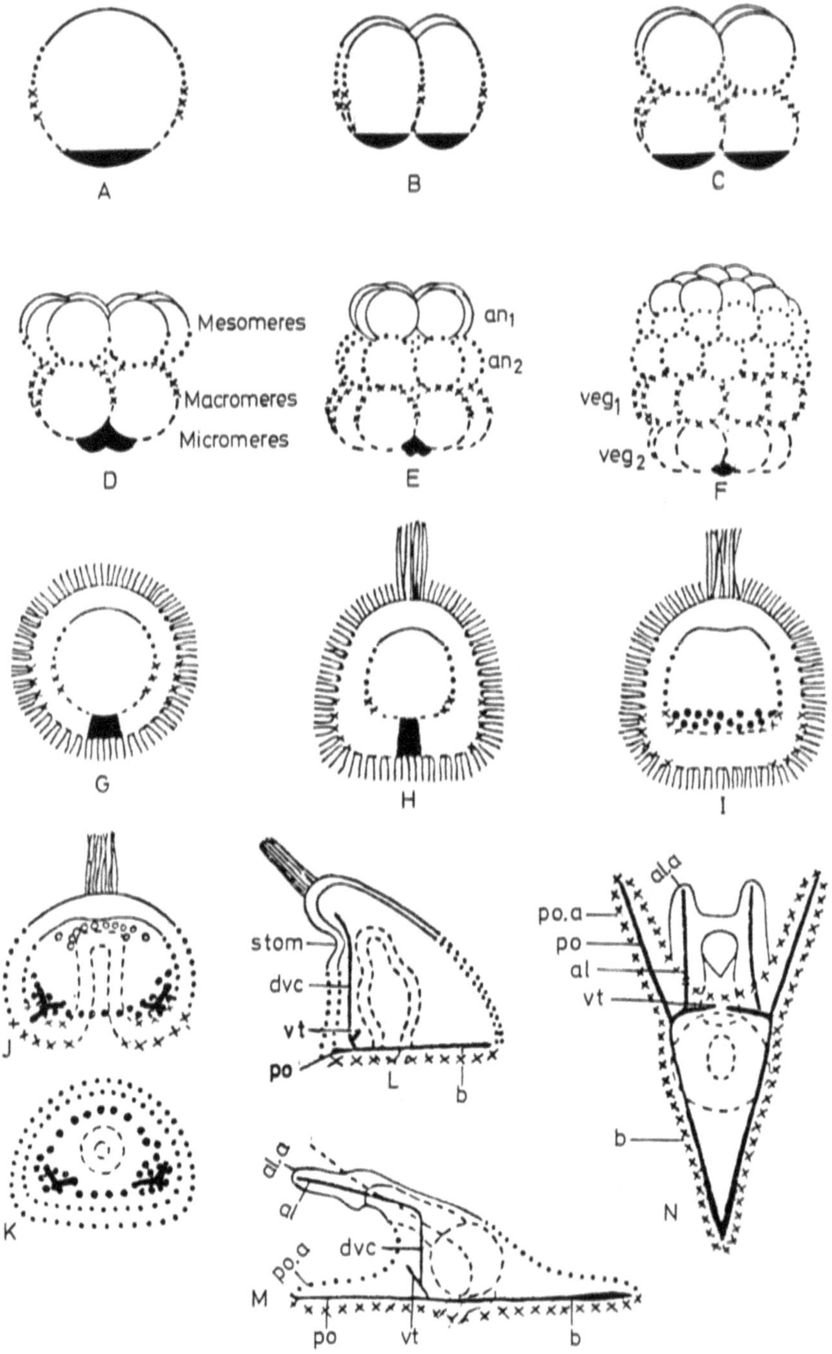
A
B
C
Mesomeres
Macromeres
Micromeres
D
an1
an2
E
veg1
veg2
F
G
H
I
J
K
stom
dvc
vt
po
L
b
al.a
po.a
po
al
vt
b
N
al.a
al
dvc
po.a
M
po
vt
b

◀ Fig. 7.2. Normal development of *Paracentrotus lividus*. an_1 (solid lines); an_2 (dotted lines); veg_1 (crosses); veg_2 (broken lines); micromeres (black). (A) Uncleaved egg; (B) 4-cell stage; (C) 8-cell stage; (D) 16-cell stage; (E) 32-cell stage; (F) 64-cell stage; (G) young blastula; (H) blastula after hatching with apical organ before formation of primary mesenchyme; (I) mesenchymal blastula; (J) gastrula, secondary mesenchyme and two triradiate spicules are formed; (K) transverse optical section of same gastrula, bilateral symmetry established; (L) prism stage, stomodeum invaginating; (M) pluteus with four arms seen from left side, broken line indicates egg axis; (N) pluteus from anal side. al, anterolateral rod; al.a, anterolateral arm; b, body rod; dvc, dorsoventral connecting rod; po, postoral rod; po.a, post-oral arm; stom, stomodeum; vt, ventral transverse rod. (From HÖRSTADIUS, 1939)

---

layer (veg_1) and a lower layer (veg_2) (Fig. 7.2F; Table 7.1). In cell lineage studies with vitally stained blastomere groups (Fig. 7.2), an_1, an_2, and veg_1 become ectoderm while veg_2 (stippled) becomes endoderm and secondary mesenchymal cells (HÖRSTADIUS, 1935). In this cleavage lm divide later than upper layers, and sm fail to divide. After the sixth cleavage embryo cells number 56-60 (Fig. 7.1[24]). The 64-cell stage is only conceptual and not actually present. At the time sm divide, descendants of mesomeres and macromeres have undergone the seventh or eighth cleavage. As the cell sets cleave asynchronously, the number of the cleavage cycle refers to the cleavage of macromere descendants.

Between the 20- and 60-cell stages an embryo is called a morula (meaning mulberry) because it is composed of a larger number of still discrete blastomeres and has a bumpy appearance resembling a mulberry (Fig. 7.1[20-24]). From the fourth cleavage on, the term horizontal cleavage is inaccurate because cleavage is no longer parallel to the equatorial plane of the embryo but radial to its surface.

After the sixth cleavage, the time and pattern of cleavage become hard to follow because the number of cells increases, cleavage becomes asynchronous, and cleaved cell arrangement is disorderly. ENDO (1966) followed the cleavage pattern of the vegetal half of the *Mespilia* embryo until the tenth cleavage. In *Mespilia*, the blastocoel is extremely large, the cytoplasm is very transparent, and the blastomeres are in regular array. According to ENDO, veg_1 and veg_2 divide 4 times after the sixth cleavage, alternating horizontal and meridional cleavage, forming 8 32-cell tiers. Four lm divide meridionally after the sixth cleavage (Fig. 7.1 [24-25]), horizontally after the seventh (Fig. 7.1 [25-26]), and meridionally at about the ninth, finally becoming 2 16-cell layers. The sm divide once meridionally, forming a ring of 8 cells (Table 7.1). After the tenth cleavage, the vegetal half of the embryo totals 296 cells (32 x 8 + 16 x 2 + 8) while the animal half totals 512 cells: 808 cells altogether. MACBRIDE (1914) also counted 808 cells in the *Echinus esculentus* embryo at the same stage. After the tenth cleavage, the cleavage rate declines and mitotic stages are rare. The number of cells at the pluteus stage is only slightly higher, and cell division is random.

As a rule, blastomeres become round just before and just after cleavage and flat in the resting intervals (Fig. 7.1[9-16]). Even in the morula stage, although the embryo surface is especially rough during cleavage, it becomes smoother between cleavages. After the seventh cleavage, as the blastomeres become smaller and the embryo surface smoother - especially during interphase - the embryo is called a blastula and the central cavity a blastocoel (Fig. 7.1 [25-28]).

Table 7.1. Cleavage pattern and cell tiers

Developmental Fate		Ectodermal Wall	Archenteron, Coelomic Vesicle, Secondary Mesenchyma		Primary Mesenchyma	Pharynx
Cleavage Number	Cells In Embryo	Mesomeres	Macromeres		Micromeres	
4	16	8	4		4	
5	32	8 (an_1) 8 (an_2)	8		4 (lm)	4 (sm)
6	56- 60	32	8 (veg_1)	8 (veg_2)	8	
7	108-116	64	16	16	8 8	
8	212-216	128	16 16	16 16		8
9	424	256	32 32	32 32	16 16	
10	808	512	32 32 32 32	32 32 32 32		

7.1.2 Micromere at 16-Cell Stage

In *P. lividus* eggs, the vegetal half of which has a pigment band, pigmented macromeres and unpigmented micromeres are segregated at the fourth cleavage. The eggs of *Arbacia punctulata* have red pigment granules and *H. pulcherrimus* orange - evenly distributed in the cortex around the periphery of the egg cytoplasm. Initial uniform pigment distribution changes at the fourth cleavage: 8 blastomeres on the animal side have equal amounts of pigment, while the macromeres on the vegetal side receive most of the pigment and the micromeres very little (HARVEY, 1956; UEMURA, personal communication). The pigment granules are embedded in the cortex so that unequal pigment distribution may indicate that macromeres and micromeres have different types of cortex (DAN, 1954b).

The difference between the two asters is striking. If the metaphase mitotic apparatus is isolated from *Clypeaster* eggs cleaving into macromeres and micromeres for the 16-cell stage, the shape of the macromere aster is a radiate sphere, while the micromere aster is flat on the side of the vegetal pole and looks cut off (DAN and NAKAJIMA, 1956). This suggests that an effector at the cortex of the vegetal pole may influence normal astral ray development before the fourth cleavage. MORGAN (1893) observed that the red pigment in the *Arbacia* egg receded from the vegetal pole at the 8-cell stage.

7.1.3 Hyaline Layer

At fertilization the hyaline layer is formed by breakdown products of the cortical granules. In a few minutes its thickness increases to several microns and is clearly visible even by ordinary microscope. HERBST (1900) discovered that when calcium ions were removed from the medium, the hyaline layer dissolved and the blastomeres fell apart. HERBST also believed that the many fine processes on the blastomere surface were transformed from the hyaline layer when the calcium ions were removed, as he saw no processes when the hyaline layer was intact.

In *Mespilia globulus* DAN and ONO (1952) saw many fine processes between the hyaline layer and the blastomere surface when the layer was lifted off a few minutes before cleavage. From the staining reaction by neutral red they considered these processes to be cytoplasmic and identifiable with the processes seen in a calcium-free medium. As evidence of strong adhesion between blastomeres and the hyaline layer, they noted that if *Mespilia* embryos from which the fertilization membranes had been removed were put into hypotonic seawater at early cleavage stages (from the 4- to the 16-cell stage), the hyaline layer became inflated and the blastomeres were carried with it away from each other. DAN and ONO suggested that individual blastomeres are fixed to the inner surface of the hyaline layer by fine processes called attachment fibers. Electron micrographs later showed that on the egg surface between the points where cortical granules burst at fertilization, numerous cytoplasmic projections form and penetrate into the hyaline layer (ENDO, 1961; WOLPERT and MERCER, 1963; KANE and STEPHENS, 1969).

The hyaline layer is a kind of calcium gel composed mainly of polysaccharide and protein (VASSEUR, 1948; NAKANO and OHASHI, 1954; NAKANO, 1956; VACQUIER, 1969). It is relatively inelastic and freely permeable to molecules of subcolloidal dimensions such as sugar or salt ions but impermeable to colloidal substances such as gum arabic or gelatin (DAN and ONO, 1952). Its principal protein component, insoluble in calcium-containing seawater, was named hyalin by STEPHENS and KANE (1970). At relatively low concentrations of divalent ions, it forms a clear gel made soluble by removing the divalent cations. But when a little calcium is added, the gel forms again.

This unique protein was first isolated from whole-egg homogenate (KANE and HERSH, 1959) and identified by immunologic analysis as a major component of the hyaline layer (YAZAKI, 1968). Hyalin may be extracted from unfertilized eggs or their isolated cortex but not from fertilized eggs from which the hyaline layer has been removed (KANE and STEPHENS, 1969). It may be extracted from eggs treated immediately after fertilization before the hyaline layer forms (KANE, 1970). Most hyalin seems to be in the cortical granules in solated form and changes into the hyaline gel layer on contact with divalent cations of seawater after cortical granule breakdown.

7.2 The Blastula before Hatching

Although no gross form changes are seen in the blastula before hatching, its wall undergoes subtle morphologic and physiologic changes. Individually fastened to the hyaline layer - a common envelope - the blastomeres establish connections among themselves and change into tissue. Before cilia develop, septate desmosomes form near the outer edge of the blastula wall, and basement membrane develops along the inner surface. The blastula wall becomes impermeable to sugar molecules, and hatching enzyme is secreted.

7.2.1 Septate Desmosomes (Septate Terminal Bars)

The blastomere surface is covered by a plasma membrane about 70 Å thick. Until the seventh division cycle, blastomeres have random contact with each other, and the apposed surfaces are smooth without any particular structure (Fig. 7.1[25]). After the eighth division in *M. globulus* (Fig. 7.1[26]) and 4 hours before hatching in *P. miliaris*, a specialized area about 0.5μm thick appears on the confronting plasma membranes at the outer edge of the blastula wall. Plasma membrane cross sections in this region become denser, parallel, and regularly separated by 150-200 Å of electron-opaque material. Small bars - septate desmosomes - develop in *Mespilia* after the ninth division (Fig. 7.1[27]). After the tenth division in *Mespilia* (Fig. 7.1[28]) and 1 hour before hatching in *Psammechinus*, the septate desmosomes are fully formed and clearly defined as a series of about a dozen parallel plates - the cross section of which resembles a ladder - about as dense as the plasma membrane and sometimes continuous with it. The ladder-like structures appear in all transverse sections indicating that the plate probably encircles the outer edge of the cell.

BALINSKY (1959) first described septate desmosomes in 3 African sea urchins: *Tripneustes gratilla* (L.), *Toxopneustes pileolus* (Lam.), and *Salmacis bicolor* Agassiz. Data have been accumulated on several other species (WOLPERT and MERCER, 1963 on *P. miliaris*; ENDO, 1966 on *M. globulus*; and GIBBINS et al., 1969 on *A. punctulata*).

7.2.2 Basement Membrane

While the septate desmosomes develop, a thin, continuous layer covers the inner surface of the blastula wall. Staining reaction to Alcian blue or Hale's reagent suggests that the layer probably contains acid polysaccharide. WOLPERT and MERCER (1963) described it as fuzzy, but other electron micrographs have shown it to be more clearly defined (OKAZAKI and NIIJIMA, 1964; ENDO, 1966; GIBBINS et al., 1969). By a simple technique using distilled water, it may be isolated from the embryo as a complete balloon. The balloon is dissolved by proteolytic enzymes, indicating that the layer may be a protein and polysaccharide complex. It is considered a basement membrane because the balloon - isolated after the gastrula stage - carries mesenchymal cells and spicules in regular position and retains the indentation corresponding to the archenteron, clearly showing that the layer lines the ectodermal and entodermal walls.

7.2.3 Loss of Permeability to Sucrose

MOORE et al. (1939, 1940) discovered that the *Dendraster excentricus* blastula wall that had been freely permeable to sucrose molecules became impermeable at the tenth cleavage. In embryos raised in pure seawater and transferred to a mixture of seawater and isotonic sucrose solution before the tenth cleavage, the blastocoel volume is not changed by sugar passing freely into the blastocoel, but after the tenth cleavage the blastulae are deflated and the blastula wall crumples. In embryos kept in sugar-containing seawater since early cleavage stage and returned to pure seawater before the tenth division, blastocoel volume is not changed by sugar leakage, while in embryos returned to pure seawater after the tenth cleavage, blastocoel volume is increased.

In *Mespilia* DAN (1952) repeated MOORE's experiment and confirmed that permeability change of the blastula wall occurred as early as the ninth cleavage. ENDO (1966) examined DAN's experiment by continuously observing individual embryos and noted that change in permeability to sucrose was detectable after the eighth cleavage and became complete toward the tenth cleavage.

As the cells are continuously impermeable to sugar molecules, it is possible that sugar is hindered in the hyaline layer, in the intercellular space including septate desmosomes, or in the basement membrane. Although the hyaline layer is permeable to sucrose at early cleavage stages, WOLPERT and MERCER (1963) felt that it might act as a barrier to sugar molecules because of the structural and chemical changes at time of hatching (IMMERS, 1961; WOLPERT and MERCER, 1963). But the hatching enzyme - the molecular weight of which exceeds 10,000 (YASUMASU, personal communication) - must pass through the hyaline layer to attack the fertilization membrane and hatch.

DAN (1960) and ENDO (1966) suggested that septate desmosomes might prevent the movement of sugar molecules because the ladder steps are at a right angle to the direction a diffusing substance would pass and because the structure is just starting to form when permeability to sugar is blocked. OKAZAKI and NIIJIMA (1964) observed that the basement membrane might play an important part in obstructing sugar passage as the entire blastula wall remains impermeable to sugar even when the septate desmosomes dissolve locally or break when the primary mesenchymal cells escape from the blastula wall. The basement membrane seems to be elastic and sticky, however, so that a break would immediately be repaired.

7.2.4 Spherical Shape and Increased Volume of Blastula

During the divisions of the fertilized egg, the blastomeres form a single layer enclosing the blastocoel (Fig. 7.1 [3-28]). At each cleavage the embryo enlarges, not by cell growth, however, but by blastocoel expansion. But why do the successive divisions give rise to a hollow sphere instead of a compact, random cell cluster, and what causes the blastocoel expansion?

The hyaline layer of *Mespilia* expands just before the first cleavage when seawater accumulates between the hyaline layer and the egg surface (Fig. 7.1 [6]), probably as the result of a colloidal secretion that exerts osmotic pressure. After each cleavage the accumulated fluid flows into a lens-shaped gap or segmentation cavity that forms between the blastomeres (Fig. 7.1[6, 10, 11, 14]). The inside volume of the hyaline envelope increases by steps, and as the envelope is inflated, the blastomeres attached to it are carried away from each other, leaving a space at the center (Fig. 7.1 [13, 19, 22])(DAN, 1952; DAN and ONO, 1952; DAN, 1960).

GUSTAFSON and WOLPERT did not consider osmosis as being responsible for the increased blastocoel volume. Rather, cell'membrane tension increases at each cleavage, and as the plane of cleavage is radial to the blastula wall, when the cells elongate and constrict they exert on each other a tangential force that expands the blastula wall area and stretches the hyaline layer. After many divisions, the daughter cells must lie on a plane because of attachment to the hyaline layer, which makes the blastula wall more expanded in area and thinner (WOLPERT and GUSTAFSON, 1961b; GUSTAFSON and WOLPERT, 1963a, 1967). These investigators do not explain how the blastula wall remains a hollow sphere without wrinkling (for further discussion, see page 236).

While osmosis may or may not aid in blastula formation, in the blastocoel are colloidal substances composed mainly of mucopolysaccharides and secreted by the blastomeres (MONNE and SLAUTTERBACK, 1950, 1952; MONNE and HÅRDE, 1951; IMMERS, 1956, 1961; MOTOMURA, 1967). And the plane of cleavage radial to the blastula wall is essential to blastula formation in the sea-urchin embryo. A tangential cleavage plane would cause a layer more than one cell thick. What determines the plane of cleavage is not clear, but it is known that the surface properties and cytoplasmic contents of cells are radially polarized. The outer surface adhering to the hyaline layer is highly convoluted with many processes, while the inner surface is smooth and covered by the basement membrane. Cilia are formed on the outer surfaces, the edges of which are joined by the septate desmosomes. The Golgi region lies near the outer surface, and the rough endoplasmic reticula are adjacent to the inner surface (WOLPERT and MERCER, 1963; GIBBINS et al., 1969).

But what causes the radial polarity of the cells? At the periphery of the egg cytoplasm of the *H. pulcherrimus* egg is a row of orange-colored granules difficult to see in white light but more easily visible under a Wratten No. 49 filter that changes the orange color to dark red. At the 2-cell stage, these granules appear on the free surfaces of the blastomeres but not on the apposed surfaces between them (Fig. 7.3A). Pigment localization remains throughout the cleavage stages (Fig. 7.3B, C). Even after hatching, the granules lie beneath the ciliated surface at the periphery of the larva (MOTOMURA, 1935; DAN and DAN, 1940; DAN, 1960), indicating that the original cortical material remains in that superficial position despite cytoplasmic movement and increased surface in each cleavage. The cell surface mechanically bound to the overlying hyaline layer keeps the original cortex from coming to the

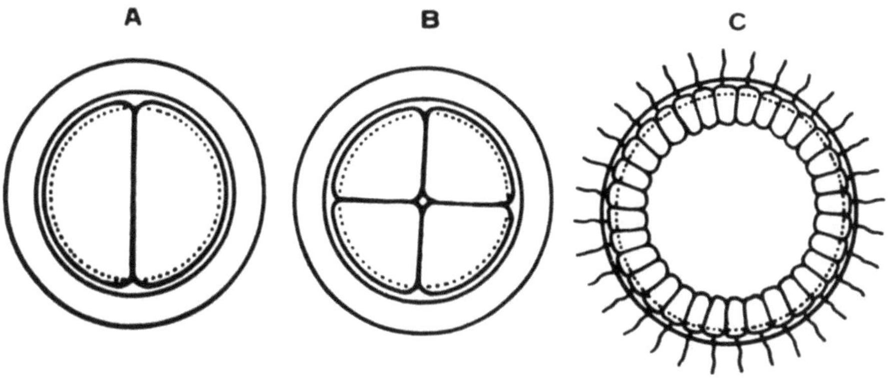

Fig. 7.3. Superficial position of orange-colored granules (dots) during development of *Hemicentrotus pulcherrimus*. (A) Interphase of 2-cell stage; outer circle, fertilization membrane; inner circle, hyaline layer. (B) Resting 4-cell stage. (C) Ciliated blastula after hatching. (From DAN, 1960)

inside. When the cells divide, the cortex becomes stretched at the furrow region (DAN, 1954a, 1954b, 1960).

7.2.5 Cilia

At the time of the tenth division cycle, cilia develop on the blastomeres - one cilium on each blastomere - and the metachronal beating of the cilia causes the embryo to rotate within the fertilization membrane. The cilia penetrate the hyaline layer and into the perivitelline space. Some kind of conducting system - possibly the septate desmosomes, judging from when they are formed and how they are shaped - causes the metachronism of the ciliary beat.

To detach the cilia from the embryo, the embryo is placed in 1 *M* sodium acetate or hypertonic seawater, osmolarity of which is increased to twice that of natural seawater by addition of NaCl. In natural seawater the cilia that regenerate on the deciliated embryo in 1 hour have normal length and activity (AUCLAIR and SIEGEL, 1966; IWAIKAWA, 1967).

7.2.6 Hatching

Embryos escape from the fertilization membrane after the tenth cleavage, but hatching enzyme secretion begins as early as the eighth cleavage and peaks at the ninth. Hatching enzyme in the sea-urchin embryo was discovered in *H. pulcherrimus* (ISHIDA, 1936) and confirmed in *A. punctulata* (KOPAC, 1941) and *P. lividus* and *Sphaerechinus granularis* (ISHIDA, 1967). In *Hemicentrotus* and *Anthocidaris* embryos YASUMASU (1958, 1961) purified and crystallized it. Measured by the light-scattering method, the molecular weight of the crystalline enzyme is about 13,000 (YASUMASU, personal communication).

In the *Hemicentrotus* embryo the hatching enzyme becomes detectable by homogenizing it about 4 1/2 hours before hatching. Culture conditions were the same as those in which hatching takes place 10 1/2 hours after fertilization. The enzyme is bound to the intracellular granules as mitochondria but shifts to the soluble fraction of the cytoplasm at 3 hours before hatching, increases rather rapidly, and emerges from the embryo (YASUMASU, 1960).

When the cilia rotate the embryo in the fertilization membrane before hatching, as they do in many sea-urchin species, it may seem that the force of the ciliary beat tears the membrane. Actually, however, the membrane is destroyed by a purely chemical process. The fertilization membrane of the *Pseudocentrotus* embryo normally disappears before the embryo starts to rotate, and that of the *Hemicentrotus* embryo dissolves on schedule even when calcium ion deficiency keeps the cilia from developing. In such species as *Strongylocentrotus droebachiensis*, *Peronella japonica* Mortensen, and *Echinarachnius mirabilis*, even the primary mesenchymal cells may form when hatching is delayed by other embryonic changes.

7.3 The Blastula after Hatching

Immediately after hatching, the blastula is practically spherical in shape, and in some species the position of the primary egg axis is impossible to determine. The blastula soon lengthens in the egg-axis direction, and the vegetal region of the blastula wall thickens. The posterior wall is flattened where the primary mesenchymal cells separate and migrate into the blastocoel, moving toward the animal pole along the blastula wall and finally becoming arranged in a characteristic ring-like pattern.

7.3.1 Change in Form of Blastula Wall

The thickening of the blastula wall in the vegetal region generally gives the blastula a pear-shaped contour (Fig. 7.4A[b], B[d]). Larvae with spacious blastocoels (such as *Mespilia*) retain their roundness (Fig. 7.1[29]). The posterior wall of the blastula flattens, and the cells around the vegetal pole - the presumptive primary mesenchymal cells - separate, leaving clefts between them (Fig. 7.4B[d, e]). At the same time, the cell layer at the animal pole thickens and changes into a sensory element - or apical organ - having a tuft of long, immotile cilia - or apical tuft - overlying a cell layer known as the apical plate (Fig. 7.2H). These changes in the blastula wall may be brought about by altered local adhesion and contact between cells (GUSTAFSON and WOLPERT, 1962; OKAZAKI et al., 1962; GUSTAFSON and WOLPERT, 1963a, 1967). For further discussion cf. GUSTAFSON's article, page 237.

Fig. 7.4. Main changes in general shape of larvae from hatching to early prism stage. (A) *Pseudocentrotus depressus*. (From OKAZAKI, 1956a). (B) *Psammechinus miliaris*. (From GUSTAFSON and WOLPERT, 1962) ▶

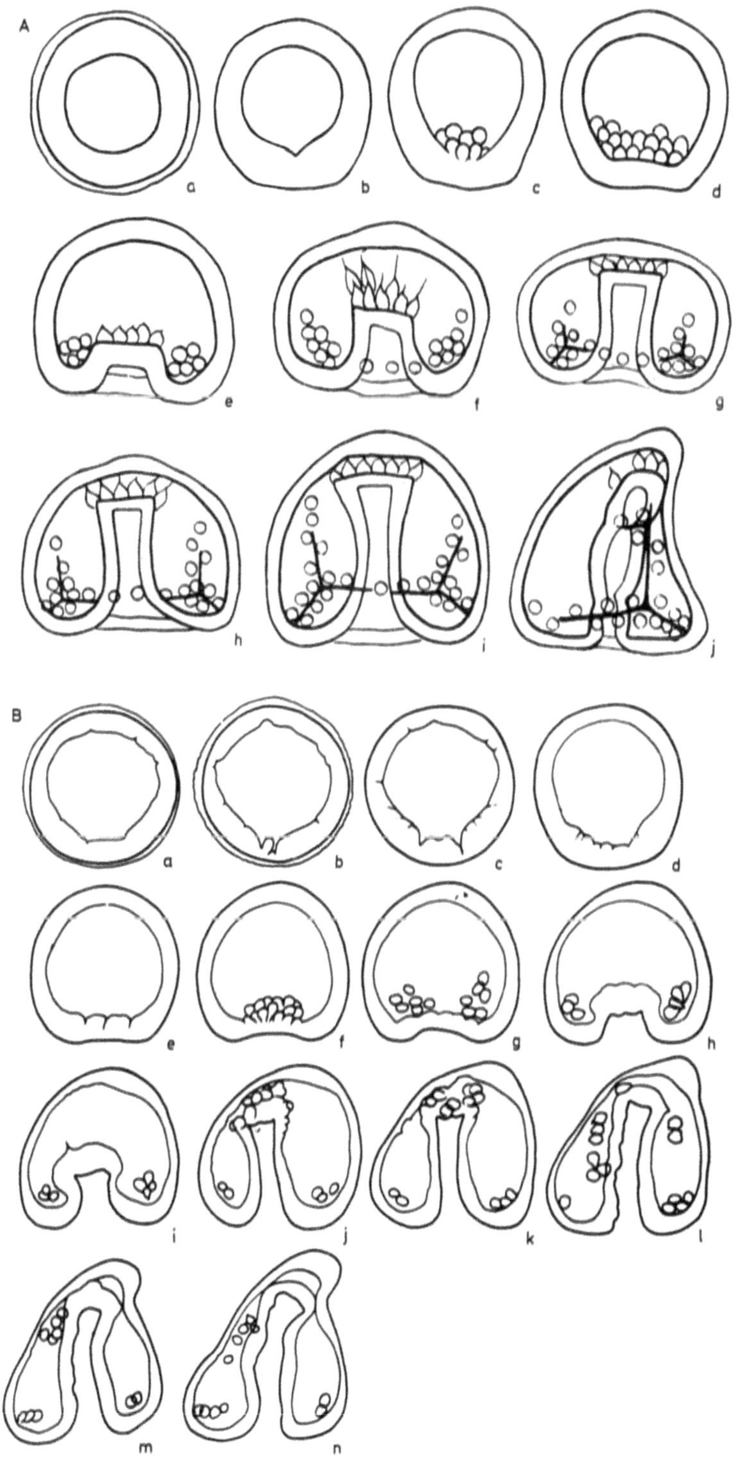
A
a
b
c
d
e
f
g
h
i
j
B
a
b
c
d
e
f
g
h
i
j
k
l
m
n

7.3.2 Release of the Primary Mesenchymal Cells from Blastula Wall

A few hours after hatching, cells around the vegetal pole show lively pulsatory activity toward the blastocoel, and from the plane in which they have been arranged bulge out like mushrooms and push out their pilei (Fig. 7.1 [30, 31]). As the pilei grow larger, the cells still in the plane become thinner, allowing the cells in the wall to slip out into the blastocoel (Figs. 7.1 [32, 33]; 7.4A[c], B[f]). The cells freed are primary mesenchymal cells and the larvae with them are mesenchymal blastulae.

To be liberated from the blastula wall, the primary mesenchymal cells must overcome the following: the attachment fibers that bind them to the hyaline layer; the cell junctions of the septate desmosomes; and the migration through the basement membrane that delimits the blastocoel. Electron micrographs show that cells escape from the hyaline layer by pulling off the attachment fibers; that septate desmosomes and cilia disappear when the pulsing motion begins; and that pseudopodial activity penetrates the basement membrane (GIBBINS et al., 1969).

Primary mesenchymal cell liberation takes place at the center of the posterior wall, and the surrounding cells draw together to close the gaps. As more and more cells are liberated, those in the posterior wall shift toward the center. The liberated cells become spherical and cluster around the liberation point. When about 30 primary mesenchymal cells have entered the blastocoel, the earlier cells are pushed toward the periphery (Fig. 7.1 [33]), and the posterior blastula wall is occupied by the descendants of the veg_2-cells (Fig. 7.2 [I]). The loosened excess hyaline layer is lifted up in blisters or pouches at the center of the posterior wall and is sometimes torn to pieces (RUNNSTRÖM, 1935; DAN, 1952).

7.3.3 Movement of Primary Mesenchymal Cells

The primary mesénchymal cells clustered around the vegetal pole send out fine, straight pseudopodia by which they separate into groups, beginning with the cells at the periphery of the cluster (Figs. 7.1 [34-36]; 7.4A[d], B[g]). The first cells often attach their pseudopodial tips to the ectodermal wall, while the cells in groups are chained together by their pseudopodia. The cells move in groups toward the animal pole, and the general direction of pseudopodial activity is along the longitudinal meridians of the larvae. The movement of individual cells is intermittent and rather rapid, which may mean that cell movement is due to the contraction of the attached pseudopodia.

When the cells have reached the level characteristic of their species, they sever pseudopodial connections, form a typical pattern, reestablish connections around the larval axis instead of along it, and form the organic matrix for the spicules.

7.3.4 Number of Primary Mesenchymal Cells

BOVERI (1901a) believed that all micromere descendants at the 16-cell stage migrate into the blastocoel as primary mesenchymal cells, but ENDO (1966) found in *Mespilia* that only lm descendants migrate while sm descendants remain at the center of the posterior wall. The final number of descendants of lm is 32 and of sm, 8.

Shortly after their release from the vegetal wall, the primary mesenchymal cells in *Clypeaster* and *Mespilia* number about 30 (OKAZAKI, 1960, 1965). About 50-60 have been reported in other species, but in what stage is not known. *Mespilia* and *Clypeaster* have 50-60 cells when spicules form.

Whether or not primary mesenchymal cells divide after leaving the blastula wall is not definite. Some workers have never found mitotic figures among primary mesenchymal cells (GUSTAFSON and WOLPERT, 1961; GIBBINS et al., 1969), but others have (AGRELL, 1954; OKAZAKI, 1960; ENDO, 1966).

7.4 Gastrula Stage

In summer sea urchins of Japan, gastrulation begins at about 28°C approximately 10 hours after fertilization, and in winter sea urchins at about 20°C approximately 20 hours after fertilization. At the same time, primary mesenchymal cells take on a well-defined pattern and form spicules. *Paracentrotus* and *Psammechinus* start gastrulation at about 18 hours at 20°C.

7.4.1 Distribution Pattern of the Primary Mesenchymal Cells

In species having spacious blastocoels, primary mesenchymal cells migrate clear away from the posterior wall into the blastocoel - sometimes more than halfway across the cavity toward the animal pole (Fig. 7.1 [37]) - and then back to the vegetal side to form an annular pattern along the inner surface of the ectodermal wall (Figs. 7.1[38], 7.4A[f]). The mesenchymal ring forms parallel to the posterior wall of the larva: close to it in larvae having small blastocoels and farther away from it in larvae having large blastocoels. In larvae vegetalized by LiCl or some other means, the mesenchymal ring shifts toward the animal pole according to the degree of vegetalization (HERBST, 1896; RUNNSTRÖM, 1928; HÖRSTADIUS, 1936a; GUSTAFSON and WOLPERT, 1961; OKAZAKI et al., 1962; CZIHAK, 1962a).

The cells of the mesenchymal ring tend to group at the right and left corners of the ventral side and to draw in nearby cells to form 2 triangles - mesenchymal aggregates - and give rise to the spicules (Fig. 7.5A, B). If a larva is seen along the animal-vegetal axis, radii from the larval center to the ventral wall including the 2 mesenchymal aggregates form an angle of about 140° (Fig. 7.5B), while those on the dorsal wall form a 220° angle. Such larvae acquire both bilaterality and dorsoventrality (DAN, 1960).

Three rows of primary mesenchymal cells extend from the 3 apices of the mesenchymal aggregate: 1 ventrolaterally from the apex facing the animal pole, and 2 transversely, connecting the triangles by their remaining apices, 1 around the dorsal side and the other around the ventral side (Fig. 7.5A). These rows of cells become arranged into a more orderly fashion into chains, completing the mesenchymal ring.

This basic pattern of primary mesenchymal cells is common to all sea urchins. For several hours the cell arrangement remains unchanged; then the spicular rudiment grows into a triradiate spicule and later lengthens its 3 arms toward the apices of the cell aggregate.

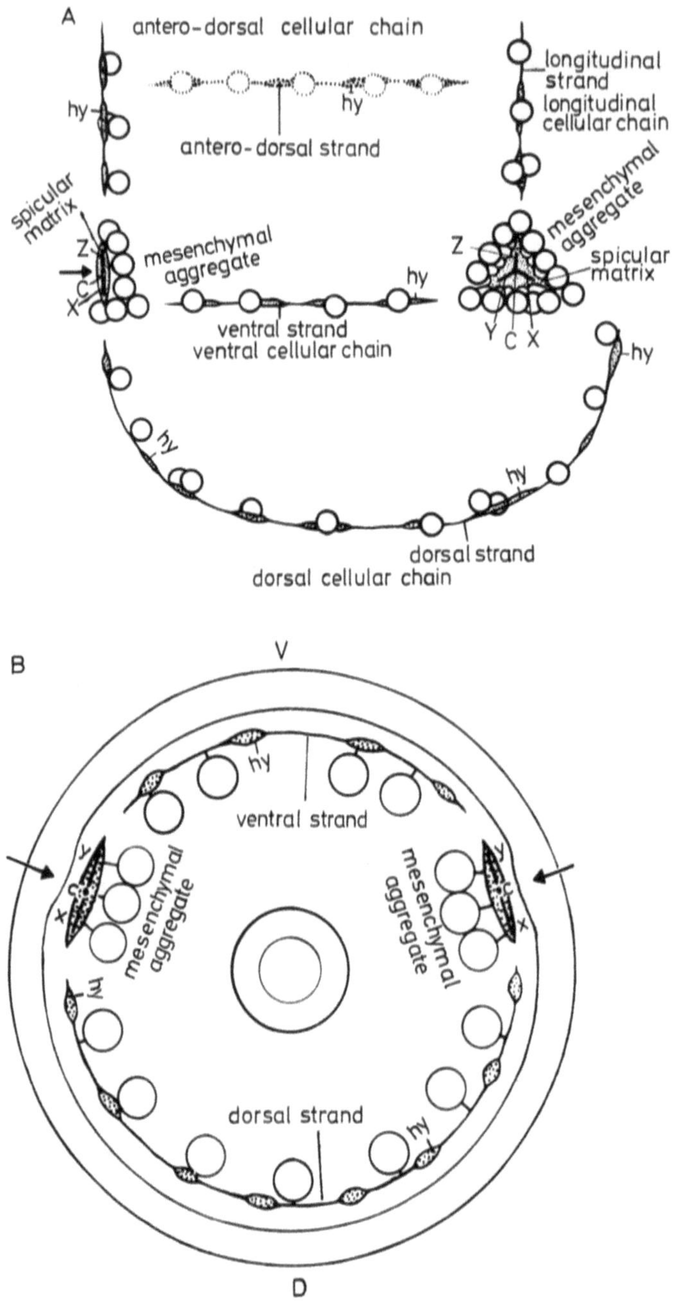

The number of primary mesenchymal cells number then 50 or more, 10-15 cells make up each mesenchymal aggregate from which cell chain extends toward the animal pole. About 10 cells on the dorsal side and several cells on the ventral side span each pair of aggregates. In some species several more cells are scattered anterodorsally (Fig. 7.5A) in parallel to the mesenchymal ring to form a new chain.

C

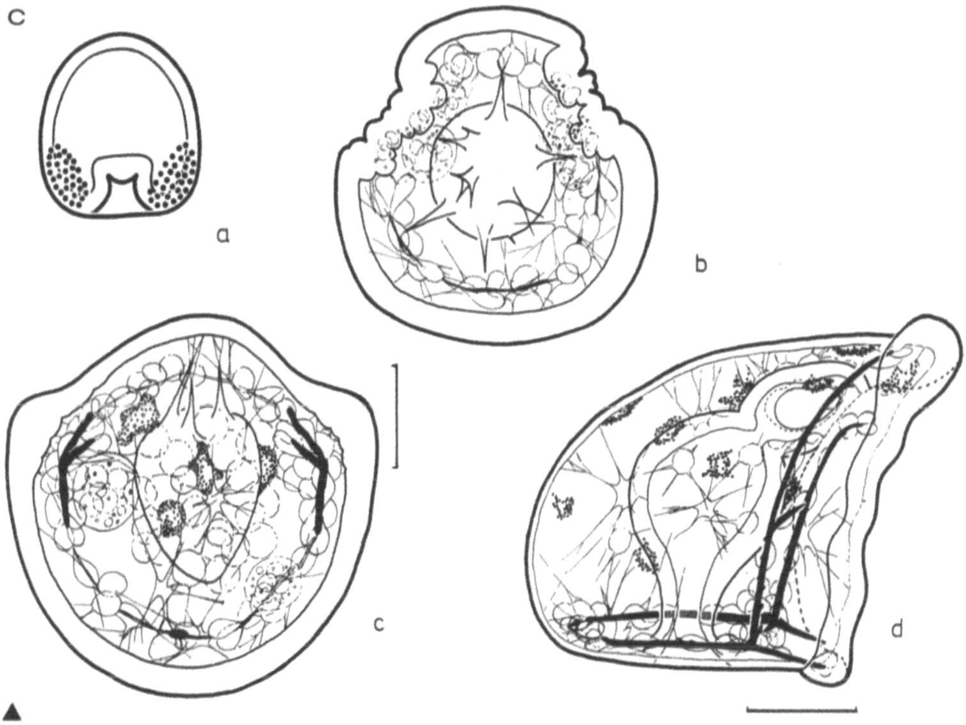

◀ Fig. 7.5A-C. Topographic relationship between arrangement of primary mesenchymal cells and organic matrices of spicule. (A) Full portrait. (From OKAZAKI, 1960). (B) Ventral view (original). 3 arms of triradiate spicule (X, Y, Z) converge at C. Same symbols are used in Figs. 7.9 and 7.10. Arrows indicate optical axis of spicule. D, dorsal side of larva; V, ventral side; hy, hyaloplasmic mass. (C) UV-irradiation of the attachment zones for the primary mesenchymal cells. (From CZIHAK, 1962b). a, big dark points showing the irradiation field; b, some hours after irradiation the absorption of destructed cells is clearly visible; c, later a new attachment zone is differentiated and the point of spicule formation is a good measure for demonstration of the dorsal shift of the attachment zone. The bars show the difference between the normal position of the attachment zone and the new one; d, lateral view of the same larva

How primary mesenchymal cells become arranged in a well-defined pattern has long been of interest. Generally agreed that it is governed by the ectoderm, it is, however, not known how it is achieved (DRIESCH, 1896; HERBST, 1896; SCHMIDT, 1904; RUNNSTRÖM, 1928; HÖRSTADIUS, 1928; VON UBISCH, 1937). GUSTAFSON and WOLPERT (1961) and OKAZAKI et al. (1962) noted the following correlation between the mode of cell contact in the ectodermal wall and the pattern of mesenchymal distribution.

In the ectodermal wall of swimming blastulae, from a circular zone of tall cells taller peaks project into the blastocoel at the right and left corners of the ventral side. On the projecting peaks the migrating mesenchymal cells form mesenchymal aggregates, while some of the cells remain along the ridges connecting the two peaks (Fig. 7.5B).

The more adhesive the cells are, the taller they must be to achieve maximum contact (SCHMITT, 1941). It is possible that the ectoderm acts as the template of the mesenchymal pattern, trapping the mesenchymal cells by increased adhesion, but neither local variations of adhesiveness in the ectodermal wall (and their cause) nor the means by which the mesenchymal cells move toward greater adhesiveness has been demonstrated. If the attachment points are destroyed, the mesenchymal cells shift dorsally (Fig. 7.5C).

7.4.2 Spicular Organic Matrix

Although the presence of clear cytoplasm enveloping the spicule was reported by several workers, the origin and role of the cytoplasm remained obscure for a long time (THEEL, 1892; WOODLAND, 1906; PRENANT, 1926a, 1926b; VON UBISCH, 1937). Later investigations showed it to be an organic matrix for the spicule and formed by fusion of pseudopodal material of the primary mesenchymal cells (OKAZAKI, 1960, 1965; WOLPERT and GUSTAFSON, 1961a).

When the pair of mesenchymal aggregates forms, most of its cells send out fine, straight pseudopods in all directions (Fig. 7.6A), and within 30 minutes a large, hyaloplasmic mass forms on each pseudopod of some of the aggregate cells. Fine processes radiating from the masses join them together and they fuse (Fig. 7.6B-D). As the mother cells form new hyaloplasmic masses and the number of cells forming hyaloplasmic masses increases, the fusion product of the masses becomes larger at the mesenchymal aggregate center in which a small, calcareous granule forms as a spicular rudiment (Fig. 7.6E, F). The triangular shape of the encasing matrix at this point roughly resembles the shape of the mesenchymal aggregate. Since the matrix is very flat and close to the mesenchymal aggregate in the direction of the ectoderm, it looks like a sheath between the mesenchyme and the ectoderm from the side (Fig. 7.6F). Fine matrix processes connect it to the mesenchymal cells and to the inner surface of the ectoderm.

As the spicular rudiment becomes triradiate, the triangular matrix extends its three apices by fusing newly formed hyaloplasmic masses and itself becomes triradiate from its closeness to the growing spicule. When the triradiate spicular arms are about 20μm long, the matrix is seen only as a thin layer with fine processes along the spicular surface (Fig. 7.6G).

Fig. 7.6. Spicular matrix formation in *Clypeaster japonicus*. (A) Cells in mesenchymal aggregate form fine straight pseudopods. (B) Cells form hyaloplasmic masses on their pseudopods. (C) Hyaloplasmic masses are connected by fine processes originating in hyaloplasmic masses. (D) Hyaloplasmic masses coalesce by contracting the fine processes. (E) Small spicular rudiment appears at center of matrix. (F) Profile of matrix when spicular rudiment develops. (G) Spicular matrix when spicular rudiment develops into triradiate spicule, matrix is apposed to surface of spicule and extends beyond its tips. Arrows indicate fine processes connecting matrix and hyaloplasmic masses in strands. a, animal pole side; d, dorsal side; v, ventral side. (From OKAZAKI, 1960) ▶

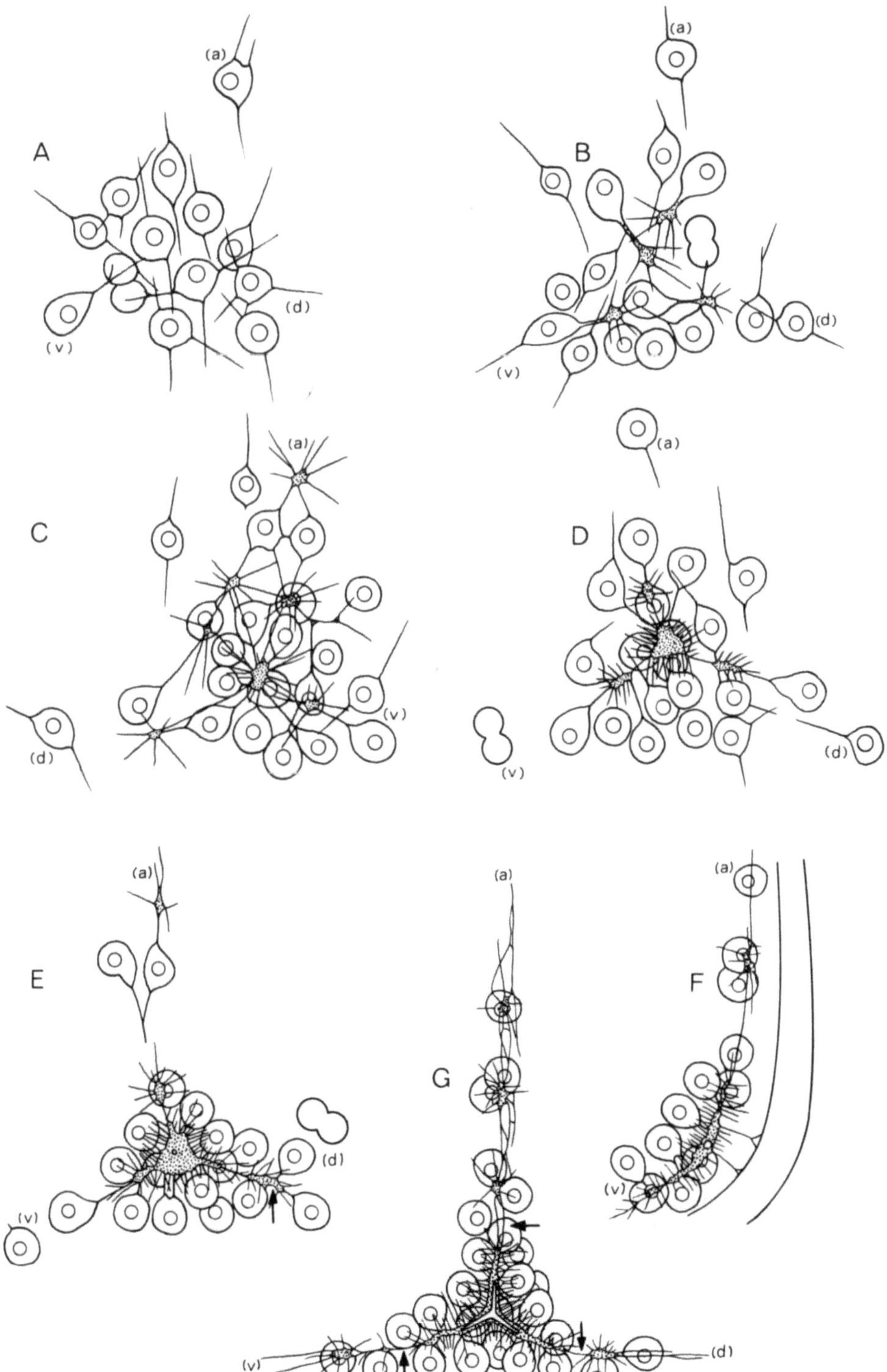
A
(a)
(d)
(v)
B
(a)
(d)
(v)
C
(a)
(v)
(d)
D
(a)
(v)
(d)
E
(a)
(d)
(v)
G
(a)
(v)
(d)
F
(a)
(v)

While the spicular matrix forms in the mesenchymal aggregate, the hyaloplasmic masses of the pseudopodia of component cells are distributed linearly in the cell chains (Fig. 7.5A,B). These hyaloplasmic masses are strung together by their fine processes to form a hyaloplasmic strand.

The proximal ends of the three strands running along the longitudinal, dorsal, and ventral cell chains are joined with the 3 arms of the spicular matrix, and a pair of matrices lengthen their arms by adding hyaloplasmic masses to the strands until the matrices are united in a single system.

Judging from its staining reaction, the spicular matrix contains acid mucopolysaccharide (OKAZAKI and NIIJIMA, 1964) and alkaline phosphatase (HSIAO and FUJII, 1963). If the skeletons are isolated from plutei and mineral components are dissolved by acid, the sheath-like structures remain in the same shape as the initial skeletons and show metachromasia by toluidine blue. The sheath-like structure obtained from *Pseudocentrotus* plutei weighs roughly 1/100 of the initial skeleton and contains 36% protein, 2.3% hexose, 2.4% amino sugar, and 3.6% reducing sugar (UEMURA, unpublished).

7.4.3 Gastrulation

After the primary mesenchymal cells are released, the blastula wall of the vegetal side (the posterior wall) is made up of cells derived from veg_2-cells of the 64-cell stage (Fig. 7.2[I]). Vital staining shows that veg_2-cell descendants are the presumptive endoderm preparing for gastrulation while the primary mesenchymal cells form the mesenchymal ring (HÖRSTADIUS, 1935, 1936b, 1939). The first sign of gastrulation is the closer aggregation of presumptive endodermal cells. The circular aggregate - called the endodermal plate - is clearly distinguishable from the relatively flat wall of veg_1-descendants (Fig. 7.1[36]). Gastrulation occurs in 2 phases (DAN and OKAZAKI, 1956; GUSTAFSON and KINNANDER, 1956; OKAZAKI, 1956a; KINNANDER and GUSTAFSON, 1960).

The first phase of gastrulation is invagination of the endodermal plate in toto into the blastocoel, up to a depth of about one-third of the blastocoel height in *P. miliaris* and *Pseudocentrotus depressus* and about one-fifth of the blastocoel height in *Clypeaster japonicus* and *M. globulus* which have larger blastocoels. As a result, a longitudinal section of the invaginated plate - or archenteron - resembles a low, flat-topped table (Fig. 7.1[37-39]; 7.4A[e], B[h]). When invagination starts, the cells in the endodermal plate have a great many plasma lobes on the surface facing the blastocoel. During invagination they show a pulsing motion that becomes confined to the archenteron tip. When the archenteron invaginates to a depth of one-fifth to one-third of the blastocoel height, the pulsing motion is reduced and fine pseudopodia appear at the archenteron tip (GUSTAFSON and KINNANDER, 1956; KINNANDER and GUSTAFSON, 1960)(Fig. 8.3), indicating that the second phase of gastrulation has begun (Fig. 7.1[39-41]). Invagination seems to originate in the endodermal plate, for when the animal half of a young gastrula is cut away, the endodermal plate continues to invaginate at the bottom of the open cup of vegetal cells (MOORE and BURT, 1939).

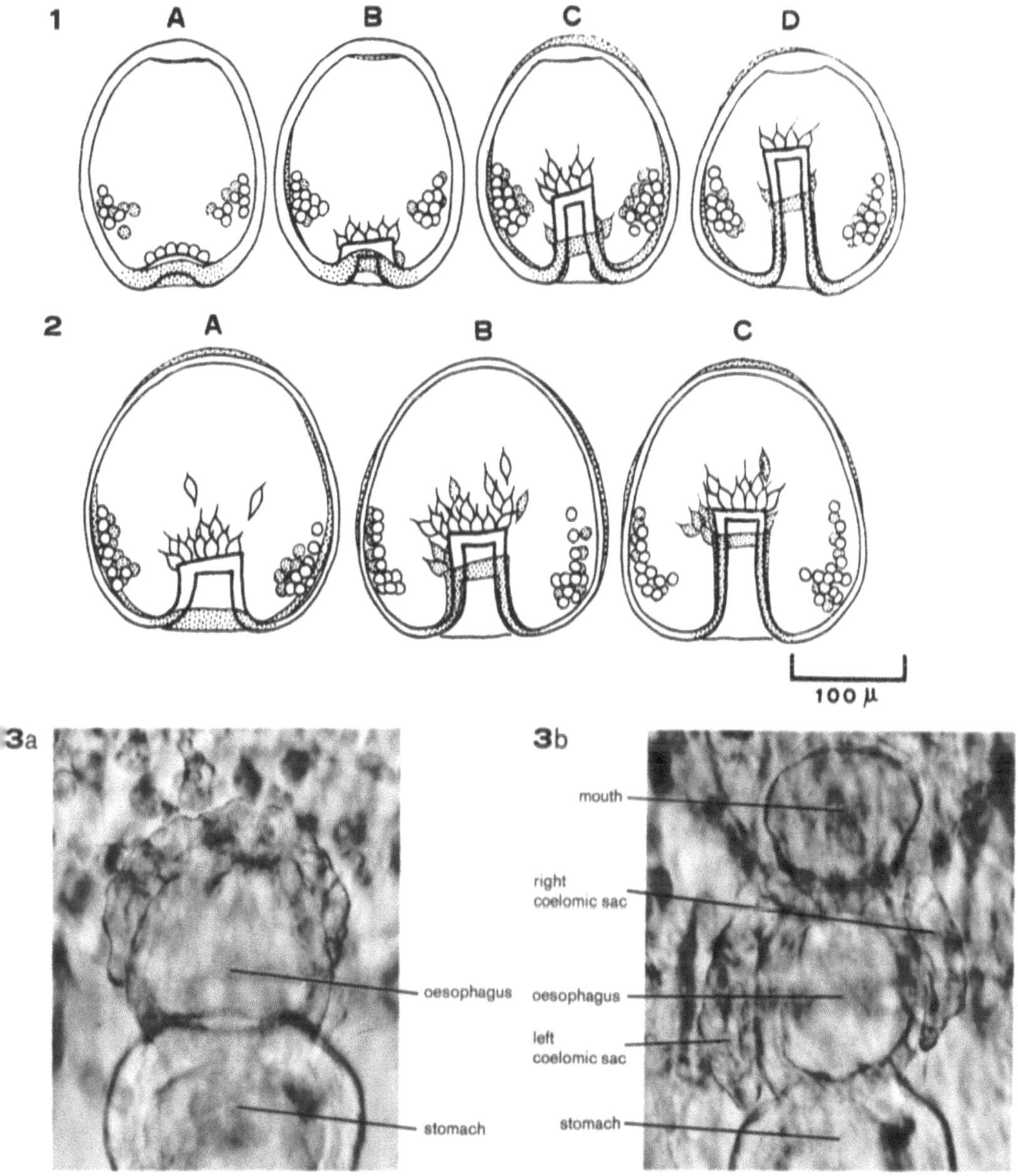

Fig. 7.7. Single larva during gastrulation. (1) *Clypeaster japonicus*. (2) *Mespilia globulus*. Dots represent preceding stage. From endodermal plate formation to completion of gastrulation, contours of larvae and positions of primary mesenchymal cells show little change, while archenterons lengthen noticeably and wall becomes thinner. (From DAN and OKAZAKI, 1956). (3) Two stages of development of coelomic sacs in *Psammechinus miliaris*. (a) Uniform cell mass detaches archenteron tip which bears constriction separating esophagus and stomach. (b) Contact between stomodeum and esophagus divides mass of coelomic cells into larger coelomic sac (left) and smaller one (right). (From CZIHAK, 1962b)

In the second phase of gastrulation, fine pseudopods sent out from the secondary mesenchymal cells at the archenteron tip traverse the blastocoel until they reach the ectodermal wall close to the animal pole. According to GUSTAFSON and WOLPERT (1962), the pseudopodal tips first reach the ectodermal wall on the dorsal side but shift to the animal pole and reach the place on the ventral wall where stomodeal invagination later occurs (Fig. 7.4B[j-n]). If the blastocoel is too large for the pseudopod of a single cell to cover it, some of the secondary mesenchymal cells leave the archenteron tip and bridge the space between it and the ectodermal wall (Fig. 7.1[42-44]).

When the pseudopodal tips reach the ectodermal wall on the side of the animal pole, they anchor there and pull the archenteron which becomes taller and slenderer as its wall thins (Figs. 7.1[41-45]; 7.4A[f-g], B [i-k]) (Second phase of gastrulation). The contour of the larva and the position of the primary mesenchymal cells are little affected (DAN and OKAZAKI, 1956)(Fig. 7.7[1, 2]), indicating that the second phase of gastrulation is caused not by increased volume of endodermal mass or by veg_1-descendants added by emboly but by stretching of the invaginated veg_2-descendants. As a result of pseudopodal pull, the archenteron tip becomes squarish while the ectodermal wall where the pseudopodal tips are concentrated is often indented.

In *Pseudocentrotus*, the gastrula is often flattened along the animal-vegetal axis as the result of pseudopodal contraction (Fig. 7.4A [e, g]). If the pseudopodal connection is experimentally suppressed, the archenteron - normal in the first phase of gastrulation - does not get any longer and turns outward, causing the larva to become an exogastrula. Developed in this manner, an exogastrula is different from one due to vegetalization caused by LiCl or other means; in the former the larva develops into a normal pluteus except that the gut is everted (DAN and OKAZAKI, 1956; OKAZAKI, 1956a).

While the archenteron becomes longer, the blastopore becomes narrower and the posterior wall of the gastrula is occupied by the descendants of veg_1-cells (Figs. 7.1[40, 43, 46], 7.2[J]).

After the second phase of gastrulation, the larval volume which has remained constant suddenly increases (Figs. 7.1[44, 47]; 7.4A[g, h]). Some of the secondary mesenchymal cells move to new positions around the archenteron where their long pseudopods span the distance between archenteron and ectoderm or primary mesenchymal cells. The other secondary mesenchymal cells remain on the archenteron tip and form coelomic vesicles (Fig. 7.1[47, 48, 50, 51]). Some of the dispersed cells contain pigment granules.

7.5 Prism and Pluteus Stages

In the prism stage the ventral side of the gastrula - which becomes the oral surface - flattens, and the dorsal side curves toward it. The animal pole of the gastrula inclines toward the ventral side and after inclining and rounding off becomes the oral lobe. The posterior wall, the center of which is the blastopore, spreads out while the spicule lengthens. The bend in the animal-vegetal axis gives the larva the shape of a right triangle, from which comes the term prism stage. From prism to pluteus, the contour changes, the stomodeum is invaginated, the digestive tract is completed, the coelom separates, and the spicules grow.

7.5.1 Stomodeum Invagination and Digestive Tract Development

After the pseudopodal contact is made between the ventral ectoderm and the archenteron tip, the archenteron tilts toward the flattened ventral wall where, just under the oral lobe, a shallow depression appears - the stomodeum - composed of descendants of an_1 cells (Figs. 7.1[49]; 7.2[L]; 7.4A[j], B[m,n]). The stomodeum deepens with the pulsing motion of its inner surface until it reaches the blind archenteron tip (GUSTAFSON and WOLPERT, 1963b). An opening through the stomodeal invagination and the archenteron tip becomes the larval mouth (Fig. 7.1[49]), around which is a band of strong cilia.

The stomodeum was thought to be a response to the approaching archenteron tip, but the stomodeum may be formed in exogastrulae and in the absence of the archenteron by treating the an_1 layer with vegetal cell component or LiCl (HÖRSTADIUS, 1939, 1949).

While the stomodeal invagination is becoming the larval mouth in the ectoderm, the front of the archenteron, which is continuous with the mouth, becomes a narrow tube, or esophagus; the middle expands into the stomach; and the rear becomes the slender intestine (Fig. 7.1 [54-58]). This tripartition seems to be a sort of self-differentiation, as the everted gut of exogastrulae also separates into three parts. It is believed that the junction between the mouth and the esophagus - the pharynx - is composed of sm descendants at the 32-cell stage (ENDO, 1966). The blastopore remains unchanged and serves as the larval anus.

7.5.2 Coelom Formation

When the archenteron tip joins the stomodeum, the secondary mesenchymal cells at the archenteron tip form 2 lateral masses from which develop a pair of small coelomic vesicles at the sides of the esophagus.

Instead of dividing into 2 cell masses, in such species as *E. esculentus*, *P. miliaris*, and *M. globulus* the secondary mesenchymal cells at the archenteron tip become a thin-walled layer forming a flat, laterally bilobed space that later separates from the digestive tract as a pair of sacs, the left usually larger than the right (MACBRIDE, 1914; GUSTAFSON and WOLPERT, 1963b; ENDO, 1966; CZIHAK, 1962b)(Figs. 7.1 [51-53], 7.7[3]).

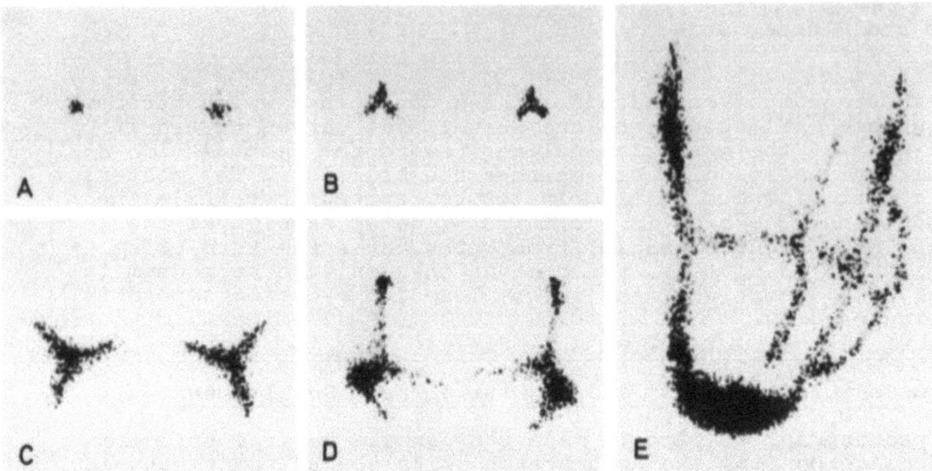

Fig. 7.8. Autoradiographs of *Pseudocentrotus depressus* spicules. Larvae are incubated in ^{45}Ca seawater (1μCi/ml) for 2 hr at early gastrula (A), midgastrula (B), late gastrula (C), prism (D), and pluteus (E). ^{45}Ca is incorporated directly from environmental seawater, through ectodermal wall, into spicule. (From NAKANO et al., 1963)

7.5.3 Spiculae Growth

The shape of the pluteus skeleton shows considerable taxonomic variation, the skeletal rods being simple or spiny or fenestrated or variously branched (MORTENSEN, 1921, 1931, 1937, 1938). Growth of the spicule, however, is governed by a common basic rule irrespective of the final shape of the skeleton. Primary mesenchymal cells form the spicular matrix by fusing their pseudopodal materials. The spicule grows by accumulated mineral deposits in the matrix, its shape being determined by the shape of the matrix, which is determined by the cell distribution pattern. The spicule is composed of magnesian calcite in a Mg/Ca ratio of about 5 to 100 (OKAZAKI, 1970). Spicular calcium is taken up directly from environmental seawater without passing through the digestive tract. Radioactive calcium has shown that calcium uptake is extremely low before the mesenchymal blastula stage but that it increases as the spicule grows. Such calcium accumulates only in the spicule (NAKANO et al., 1963 (Fig. 7.8).

A small, calcareous granule in the central part of the spicular matrix (Figs. 7.6E; 7.9A$_1$, A$_2$), the spicular rudiment develops into a triradiate spicule through hexahedral, spherical, and triangular stages. Three radii form an angle of almost 120° between each other in a plane parallel to the thickest part of the ectodermal wall (RUNNSTRÖM, 1931; OKAZAKI, 1970)(Fig. 7.5B). The initial position of the triradiate spicule is unrelated to the larval axis, but by the time the radii are about 5μm long, one points toward the animal pole and the other 2 toward the dorsal and ventral sides of the larva (Fig. 7.9B$_1$, B$_2$). This is the result of matrix movement when new hyaloplasmic masses join or when cells attach to the matrix. The increasing number of fine processes anchor the matrix to the ectodermal wall.

By the time when the spicular radii lengthen beyond the original matrix in the mesenchymal aggregate, more matrix material is supplied by amalgamation of hyaloplasmic masses in the strands formed by the cell

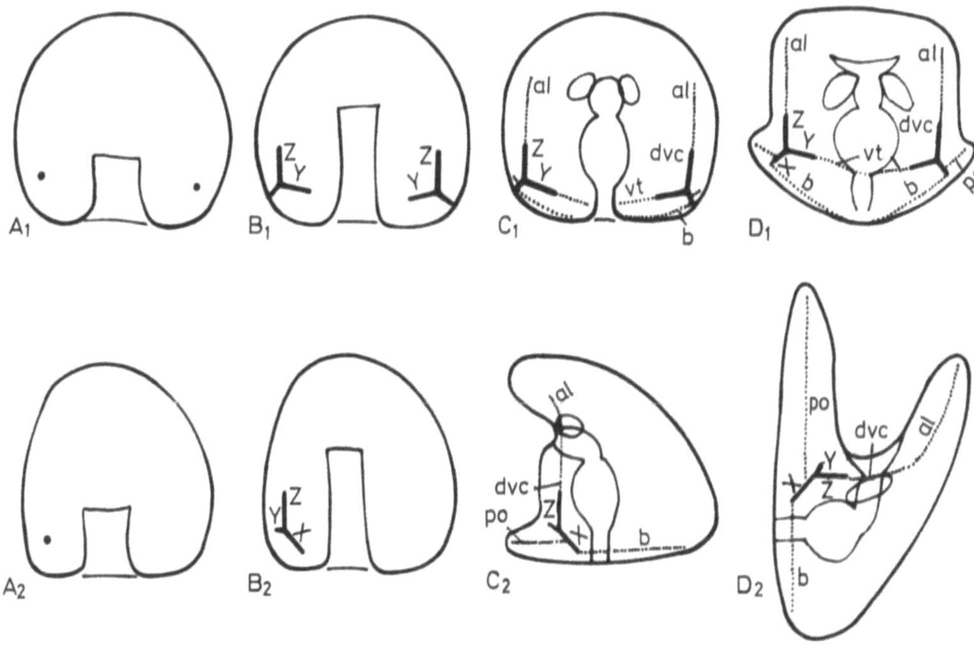

Fig. 7.9. Development of spicules. A_1, B_1, ventral view; C_1, D_1, oral view; A_2, B_2, C_2, D_2, lateral view. Rudimental granules (A_1, A_2) develop into triradiate spicules (B_1, B_2) with three arms (X, Y, Z) in the same plane. Body rod (b) and anterolateral rod (al) are formed by curved extensions of X and Z, respectively. Y lengthens and straightens into ventral-transverse rod (vt). Postoral rod (po) is branch of X; dvc, dorsoventral connecting rod. (From OKAZAKI, 1970)

chains (Figs. 7.5, 7.6G). The spicular radius (Z in Fig. 7.9[B_1]) to the dorsal side of the larva bend in right angle to the plane of the initial triradiate spicule (Figs. 7.1[46], 7.5B, 7.9) and then stretches out along the posterior larval wall to form the body rod (Figs. 7.1[50-52], 7.9, 7.10). Near the point of the bending, a new rod - the postoral rod - grows in a direction opposite to the body rod (Figs. 7.9, 7.10). In species in which the postoral rods are fenestrated, near the junction of the 3 radii 3 parallel branches form, one on each radius. The branches lengthen and form crossbars uniting the branches into a fenestrated rod (Fig. 7.10B).

The spicular radius to the ventral side of the larva (Figs. 7.5B, 7.9, 7.10) lengthens as the ventral transverse rod (Figs. 7.2[L-N], Y in 7.9D_1). The third radius straightens for a while to form the dorsoventral connecting rod (Figs. 7.2[L-M]; 7.9C, D), and then bends in an obtuse angle toward the ventral side (Figs. 7.9, 7.10). After bending straightens out again and becomes the anterolateral rod. In some species the third radius bifurcates and the 2 branches curve away from each other, one lengthening into the anterolateral rod and the other into the recurrent rod (Fig. 7.10B). As the spicule grows, cells in the mesenchymal aggregate disperse along the skeletal rods and approach the tips of the elongating rods.

On the ventral side of the larva, two planes of the triradiate spicule meet (Fig. 7.5B). The body rods come together dorsally and separate ventrally. As the spicular planes turn when the ventral wall flattens, the angles between the body rods become more acute and the contour of the posterior wall changes to an isosceles triangle (Fig. 7.1[57]). This is the beginning of the pluteus stage.

Skeletal pattern of the pluteus stage is basically the same as in prism stage. The body rods, however, lengthen until they meet at the aboral side of the larva where they thicken into a club- or fork-like shape. Between the postoral rods - more than 100μm long - the angle becomes narrower. The 2 anterolateral rods grow parallel to each other, supporting the apical region of the larva - or oral lobe - which flattens dorsoventrally and widens bilaterally to form 2 anterolateral arms (Figs. 7.1[61-62], 7.2N, 7.12B). The ventral transverse rods meet at the center of the ventral side and fuse (Fig. 7.11B). The recurrent rods lengthen almost parallel to the body rods and sometimes form anastomosed branches in a basket pattern.

Lateral spines cover the body rods, postoral rods, and anterolateral rods (Fig. 7.1[59-62]). How the rods thicken and branch and the number and shape of the lateral spines vary from species to species.

7.5.4 Ectoderm and Skeleton

The arm rudiments are formed independent from the skeleton but do not develop far (HÖRSTADIUS, 1939; OKAZAKI, 1956a, 1956b). When the post-

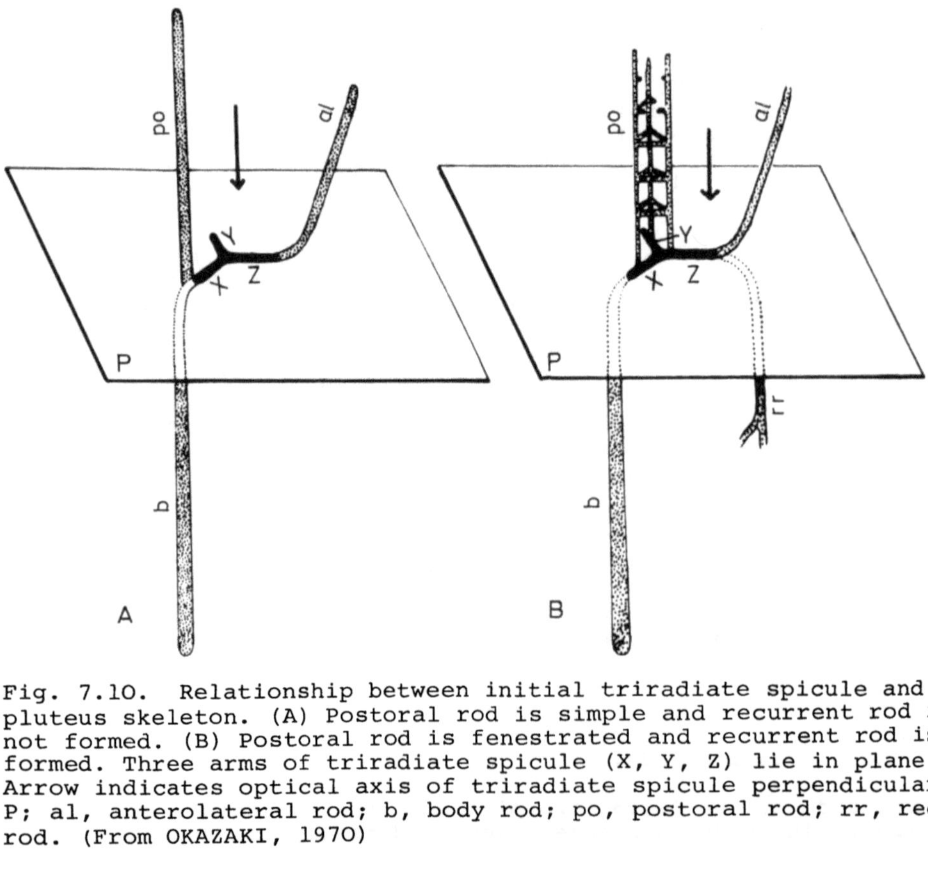

Fig. 7.10. Relationship between initial triradiate spicule and pluteus skeleton. (A) Postoral rod is simple and recurrent rod is not formed. (B) Postoral rod is fenestrated and recurrent rod is formed. Three arms of triradiate spicule (X, Y, Z) lie in plane (P). Arrow indicates optical axis of triradiate spicule perpendicular to P; al, anterolateral rod; b, body rod; po, postoral rod; rr, recurrent rod. (From OKAZAKI, 1970)

oral rod development is suppressed in larvae of *Paracentrotus*, for example, the ectoderm that will become the postoral arm, forms a protrusion that eventually wrinkles and falls off. While the skeletal rod is undoubtedly necessary for normal arm development, an arm does not always grow where a skeletal rod reaches the ectoderm.

The spicular matrix is formed following the cell distribution pattern which correlates with local difference in the thickness of the ectodermal wall. Even so, the ectodermal wall does not determine spicular shape. In germ-layer chimeras in which micromeres were exchanged between the embryos of different species of sea urchins, the micromeres formed skeletons characteristic of their original species (VON UBISCH, 1937, 1939). The development of skeletal rods is described in chapter 15.

In another study micromeres isolated from the other blastomeres at the 16-cell stage and cultured in vitro differentiated into spicule-forming cells and formed skeletons similar to those of the pluteus. Descendants of micromeres isolated from *Hemicentrotus* embryos formed skeletons with simple rods while those from *Anthocidaris* formed fenestrated rods, both characteristic of their species (OKAZAKI, 1971a,

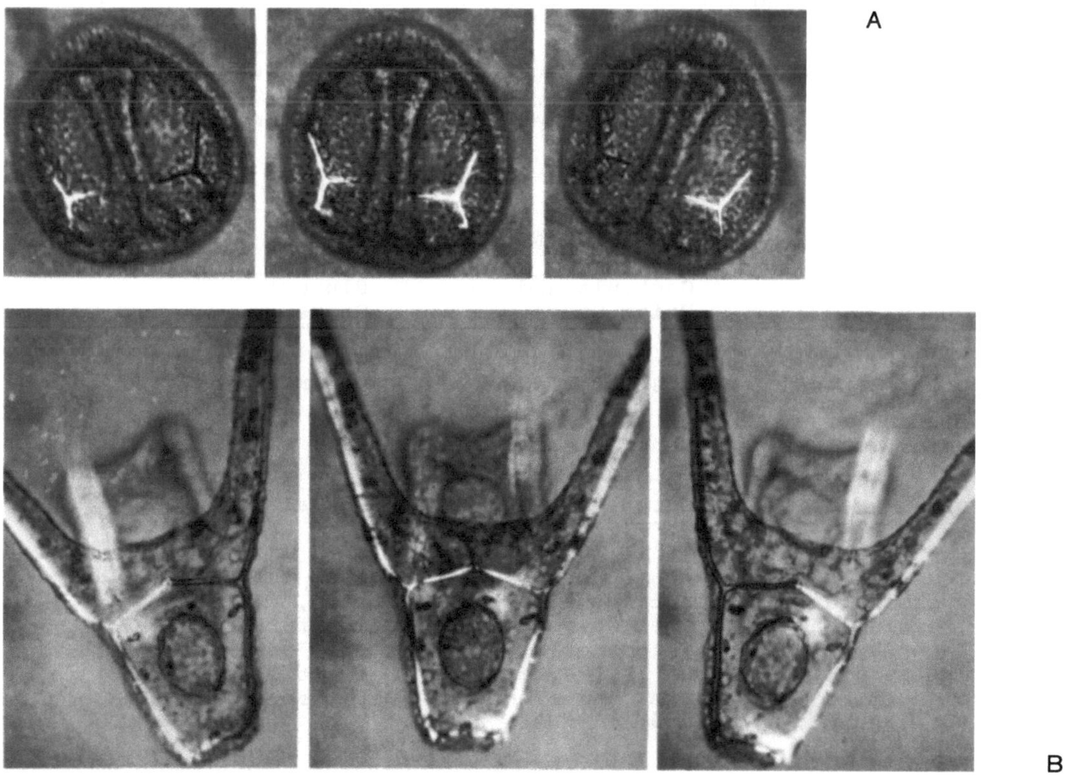

Fig. 7.11. Birefringence of spicule seen by polarization microscope with rotating microscope stage. (A) Late gastrula; spicule alternately brightens and darkens independently. (B) Pluteus with four arms; each skeleton behaves as single crystal in spite of its complicated architecture. (From OKAZAKI, 1970)

1971b). The ectoderm determines the position and orientation of the larval skeleton.

The magnesian calcite make up of the skeleton is strongly birefringent. Polarization microscopy shows that in all stages of development the skeleton behaves optically as a single crystal (Fig. 7.11) and that the 3 radii of the initial triradiate spicule have a common optical axis that remains unchanged during spicular growth. However intricate the skeletal shape, the magnesian calcite molecules are arranged uniaxially throughout the skeleton. The optical axis of the spicule is perpendicular to the plane of the initial triradiate spicule so that if the triradiate spicule is looked down on façade, it does not show birefringence. As a result, the optical axis of the pluteus skeleton is parallel to the morphologic long axis of the body rod (Fig. 7.10).

7.5.5 Four-Armed Pluteus

A few days after fertilization, the larva develops into a four-armed pluteus with a pair of long postoral arms, the tips of which are pigmented, and a pair of much shorter anterolateral arms (Figs. 7.1[61-62], 7.12.1B). The digestive tract begins to function, the mouth opens wide, the esophagus becomes muscular, the stomach develops into a large sac having long cilia, and the slender intestine curves along the stomach to the larval anus on the ventral side. The larval axis forms almost a right angle with the gastrula (Fig. 7.2M), and the larva begins a swimming motion. To describe the pluteus which swims with its pointed, aboral end down and its four arms up, a different set of terms to denote the sides of the organism is used in classical embryology (Table 7.2).

Table 7.2. Terms describing sides of larva

Side of Larva	Gastrula and Prism Stage	Pluteus Stage
Side including blastopore or anus	Posterior side	Ventral side
Side including stomodeum or mouth	Ventral side	Oral side
Remaining curved side	Dorsal side	Dorsal side

The pluteus was so called by JOHANNES MÜLLER because the shape of its larva resembles an easel. Until the four-armed stage, plutei grow without feeding, although the larvae do not develop further unless they obtain such food as diatoms or other planktonic organisms. Raising larvae is difficult, but if they are nourished it is possible to make them form additional arms and eventually metamorphose (ONODA, 1931, 1936; HARVEY, 1949, 1956; CZIHAK, 1960).

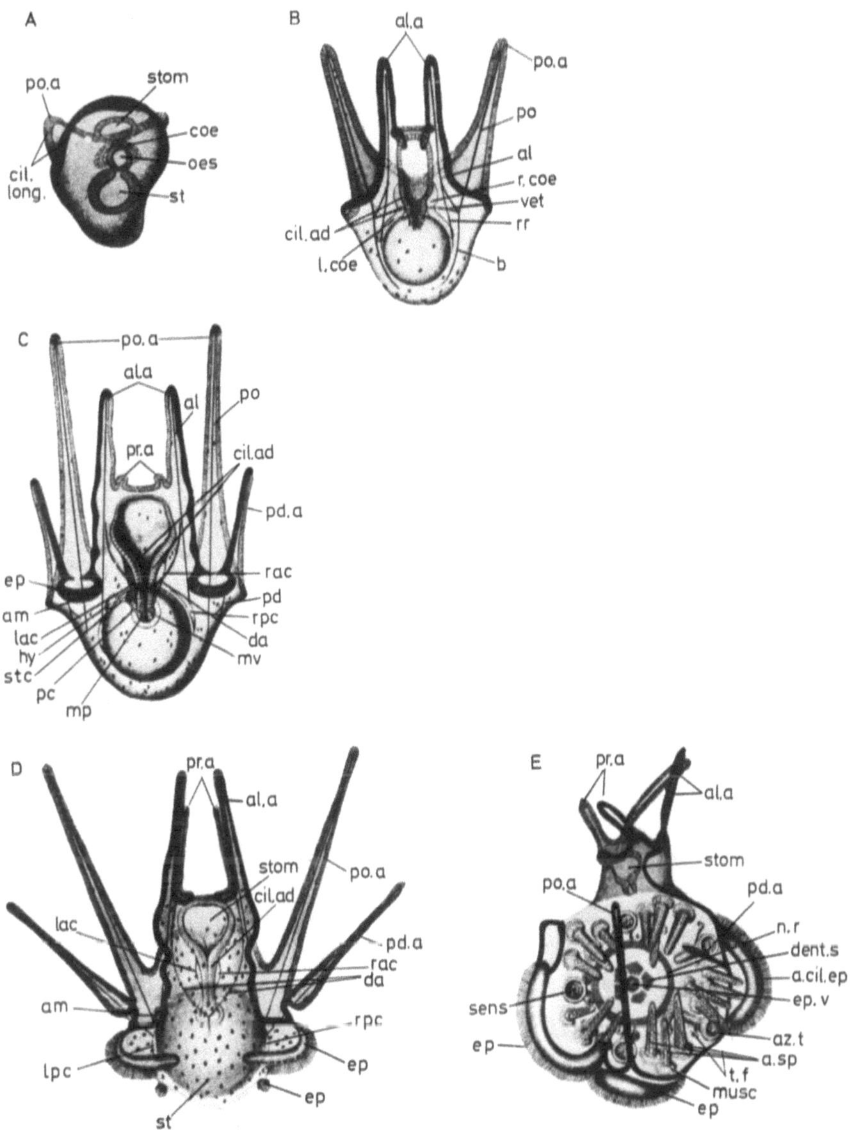

Fig. 7.12.1. *Echinus esculentus* development from prism stage to metamorphosis. (A) Prism stage, 3 days old. (B) Four-armed pluteus, 7 days old; dorsal view. (C) Beginning of eight-armed pluteus, 11 days old; dorsal view. (D) Pluteus with eight fully developed arms, 23 days old; dorsal view. (E) Pluteus beginning to metamorphose, 50 days old; left-side view. al, anterolateral rod; al.a, anterolateral arm; am, amniotic invagination; a.sp, adult spine; az.t, azygous tube foot (primary tube foot); b, body rod; cil.ad, adoral band of cilia; cil. long, longitudinal band of cilia; coe, coelomic vesicle, da, dorsal arch; dent.s, dental sac; ep, epaulette; ep.v, epineural veil; hy, hydrocoel; l.coe, left coelomic sac; lac, left anterior coelom; lpc, left posterior coelom; mp, madreporic pore; musc, muscles of adult spines; mv, madreporic vesicle; nr, nerve ring; oes, esophagus; pc, pore canal; pd, posterodorsal rod; pd.a, posterodorsal arm; po, postoral rod; po.a, postoral arm; pr.a, preoral arm; rac, right anterior coelom; r.coe, right coelomic sac; rpc, right posterior coelom; rr, recurrent rod; st, stomach; stc, stone canal; sens, sense organ; at, apex of azygous tube foot; stom, stomodeum; tf, rudiments of first pair of tube feet; vet, ventral transverse rod. (From MACBRIDE, 1914)

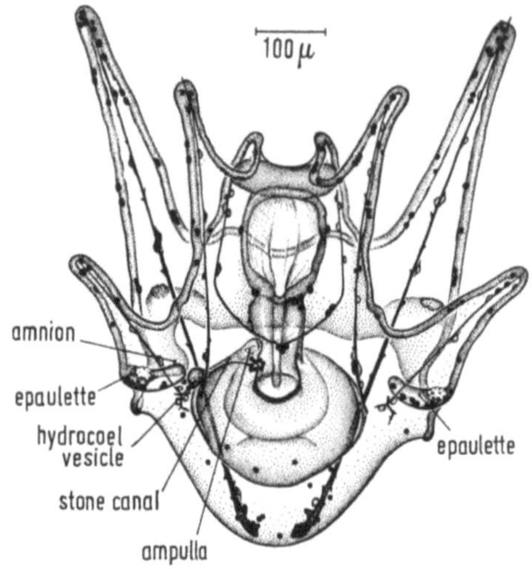

Fig. 7.12. 2A. *Psammechinus*-pluteus, 12 days old. The acute apex has been withdrawn, the apical rods are partially dissolved and near the mouth new arms with supporting skeletal rods (the preoral arms) have been formed. The epaulettes are now separated into band-like groups of ciliated cells. In the left coelom the hydrocoel-vesicle has been formed and opposite to it the ectodermal epithelium begins to sink in to form the amnion. Beyond the hydropore the stone canal is dilated and called the ampulla

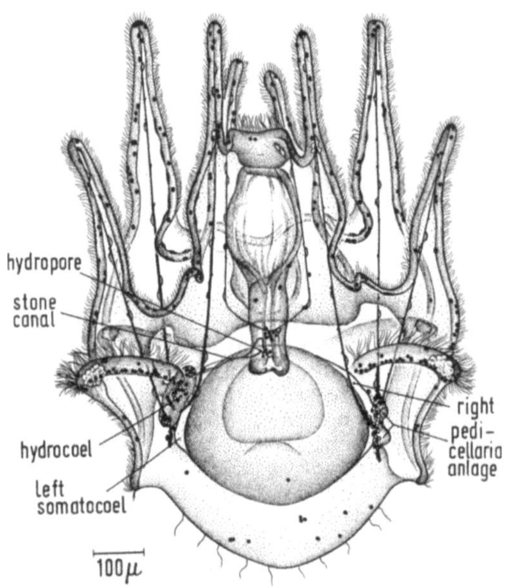

Fig. 7.12. 2B. A pluteus of *Psammechinus*, 14 days old, with 8 arms and apex rods completely withdrawn. The hydrocoel has developed 5 protrusions growing further to form the primary tentacles. Opposite the anlage of the imaginal disk (echinus rudiment) the anlagen of the pedicellaria have been formed

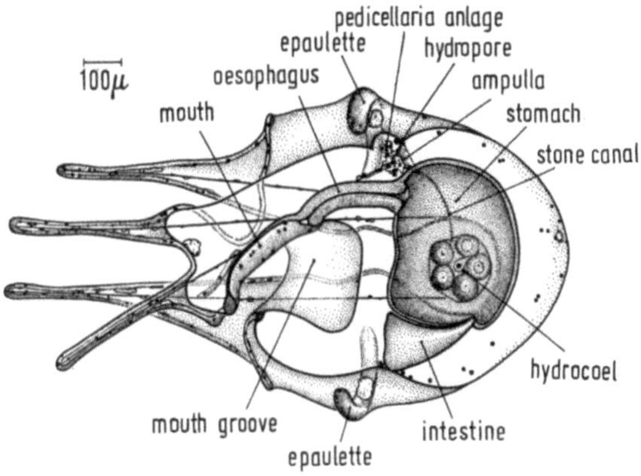

Fig. 7.12. 2C. *Psammechinus*-pluteus of the same stage as the previous one in longitudinal section. The imaginal disk of this individual is located on the right side, and thus the pluteus is of the -+ type (situs inversus, cf. Fig. 7.18). Near the hydropore an anlage of pedicellaria

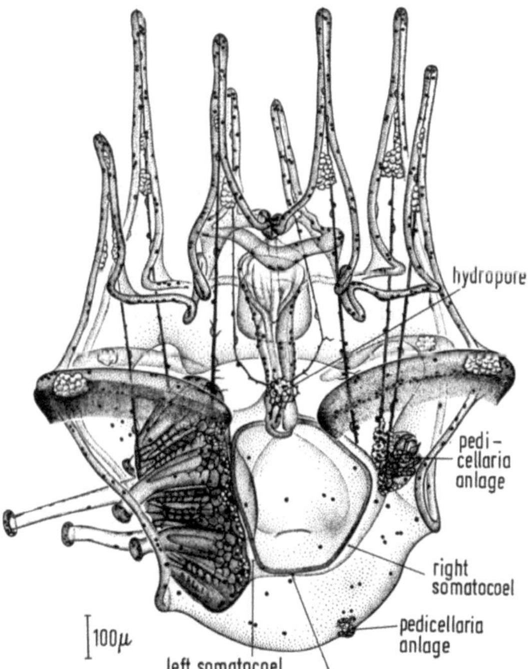

Fig. 7.12. 2D. Late pluteus stage of *Psammechinus*, 28 days old, with fully developed imaginal disk. The long primary tentacles often come out from the amnion in which the fenestrated larval spines with 4 tips are also visible. The first groups of pedicellaria appear together with larval spines opposite the imaginal disk. Another group of pedicellaria has developed near the apex. In the sagittal plane the 2 somatocoels are in contact, forming the mesentery. Using the primary feet the pluteus is able to walk on the substratum

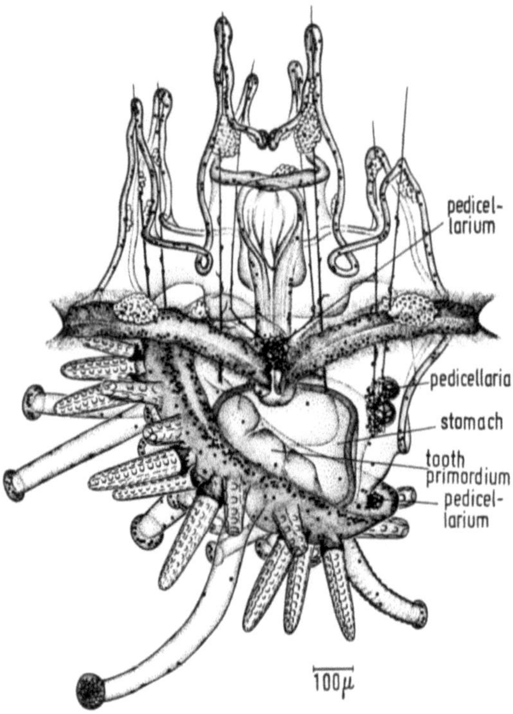

Fig. 7.12. 2E. First stage of metamorphosis in a 33-day-old pluteus of *Psammechinus*. The imaginal disk has begun to unroll and the ectodermal epithelium of the pluteus is beginning to shrink, first recognizable by its withdrawing from the arm rods. Besides the larval spines with 4 tips along the border line of the imaginal disk some of the definitive spines with acute tip have already been formed. The 4 tips of the larval spines are not yet developed to the extent of those in Fig. 7.12.2D: this is one of the frequent individual variations. Note the teeth primordia visible by the impression on the stomach

7.6 Late Larval Stage and Metamorphosis

Temperature, food supply, and species determine how many weeks or months a four-armed pluteus requires to develop into a young sea urchin. The main external changes of the larva during this period are that the number of arms increases and bands of cilia - or epaulettes - form. The principal internal changes are that the hydrocoel develops when the coelomic vesicle of the left side constricts, and the ectodermal wall opposite the hydrocoel forms a depression - the amniotic invagination. Combined, the hydrocoel and the amniotic invagination make up the echinus rudiment - an imaginal disc - in which the adult mouth, the lantern, and the ring nerve are formed. The young sea urchin thus develops encased in the larval body and breaks through the roof of the amniotic cavity at metamorphosis.

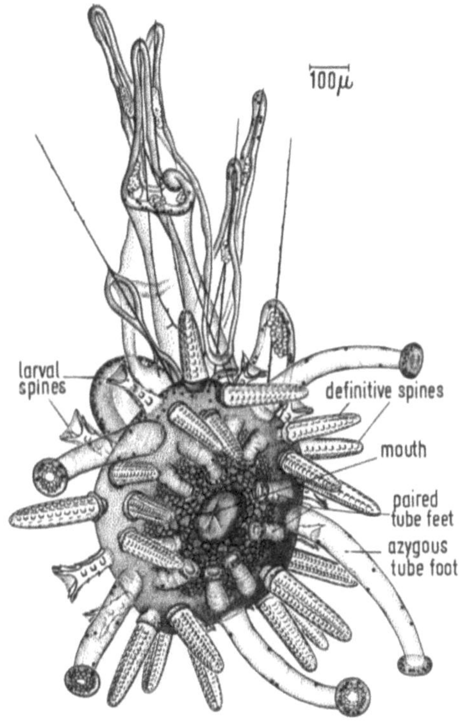

F

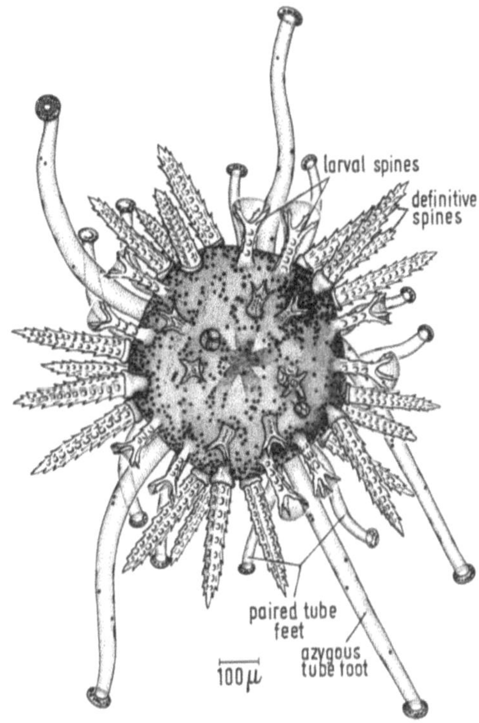

G

Fig. 7.12. 2F. Next step of metamorphosis, one day later. The shrinking of the pluteus ectoderm has proceeded. The former imaginal disk is seen from the new ventral side. Between each primary tentacle and the mouth 2 definitive podia develop and besides the 2 larval spines in each sector 4 definitive spines develop

Fig. 7.12. 2G. Young sea urchin of the same species immediately after metamorphosis, dorsal view. By the shrinkage of the ectodermal epithelium of the pluteus 2 pedicellaria and 7 larval spines developed in the same anlagen have become situated on the dorsal half of the sea urchin. Fig. 7.12.2A - 7.12.2G were drawn by DIERL, FREIBERG and CZIHAK from living specimens. (From CZIHAK, 1960)

7.6.1 External Change

The number of arms of the pluteus depends on the species. As a rule the four-armed pluteus acquires two more pairs of arms; the first, the posterodorsal arms, develop near the postoral arms while the second, the preoral arms, lengthen on the oral lobe between the anterolateral arms (Figs. 7.1[63-65]; 7.12.1C, D). In some species the postoral and posterodorsal rods are fenestrated. The preoral arms are supported by the forks of a V-shaped calcareous center called the dorsal arch, situated in the middorsal line above the esophagus (Figs. 7.12.1C, D; 7.12.2A-D).

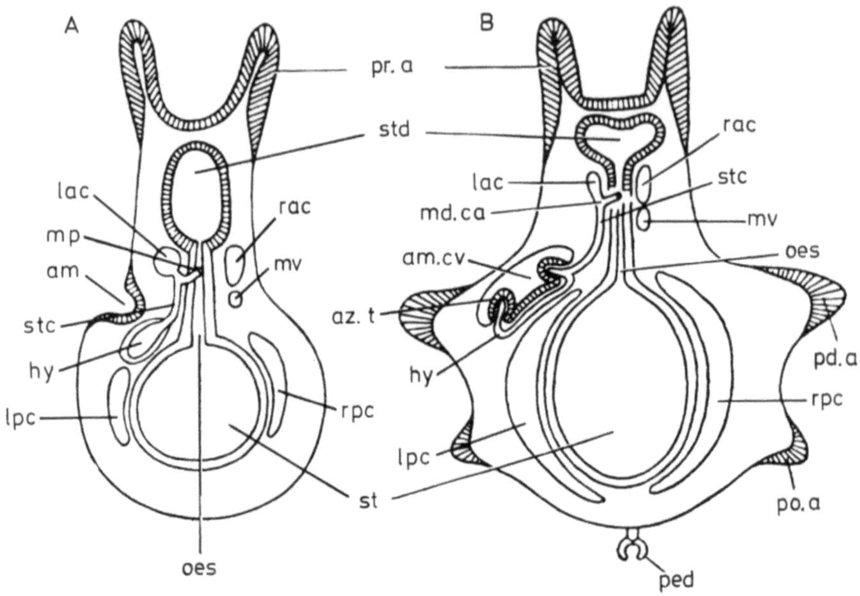

Fig. 7.13. Coelom, hydrocoel, and amniotic cavity. am, amniotic invagination; am.cv, amniotic cavity; az.t, rudiment of azygous tube foot; hy, hydrocoel; lac, left anterior coelom; lpc, left posterior coelom; md.ca, madreporic canal; mp, madreporic pore; mv, madreporic vesicle; oes, esophagus; pd.a, posterodorsal arm; ped, pedicellaria; po.a, postoral arm; pr.a, preoral arm; rac, right anterior coelom; rpc, right posterior coelom; st, stomach; stc, stone canal; stom, stomodeum. (From MACBRIDE, 1914; DAN, 1957)

The third pair, the anterodorsal arms, grow near the anterolateral ones, while the fourth pair, the postlateral arms, grow at the sides of the aboral end of the larva. It has been reported that these additional arms are supported by the skeletal rods that develop independent of the initial triradiate spicule, but reexamination of their optical axes is needed. While the additional arms develop, horizontal bands of cilia - or epaulettes - appear on the ectodermal ridges (Figs. 7.1[65], 7.12.1C, D, 7.12.2A).

7.6.2 Hydrocoel

In the prism stage when the archenteron and the stomodeum connect, a pair of coelomic vesicles is formed (Fig. 7.7[3]). When the larva reaches the pluteus stage with four arms fully developed, the coelomic vesicles develop into 2 slender tubes along the esophagus and the stomach each of which acquires in the middle a constriction that divides it into anterior and posterior parts (Figs. 7.1[61]; 7.7[3b]; 7.12.1B, C; 7.12.2; 7.13A). Meanwhile the anterior part of the left coelom buds off a posterior vesicle - or hydrocoel - which gives rise to the adult water vascular system. The hydrocoel remains connected by a narrow neck - or stone canal (Figs. 7.12.1C, 7.12.2A-C, 7.13A, 7.14A). The anterior part swells where it joins the stone canal, and the swelling - or ampulla - becomes the axial sinus (Fig. 7.15).

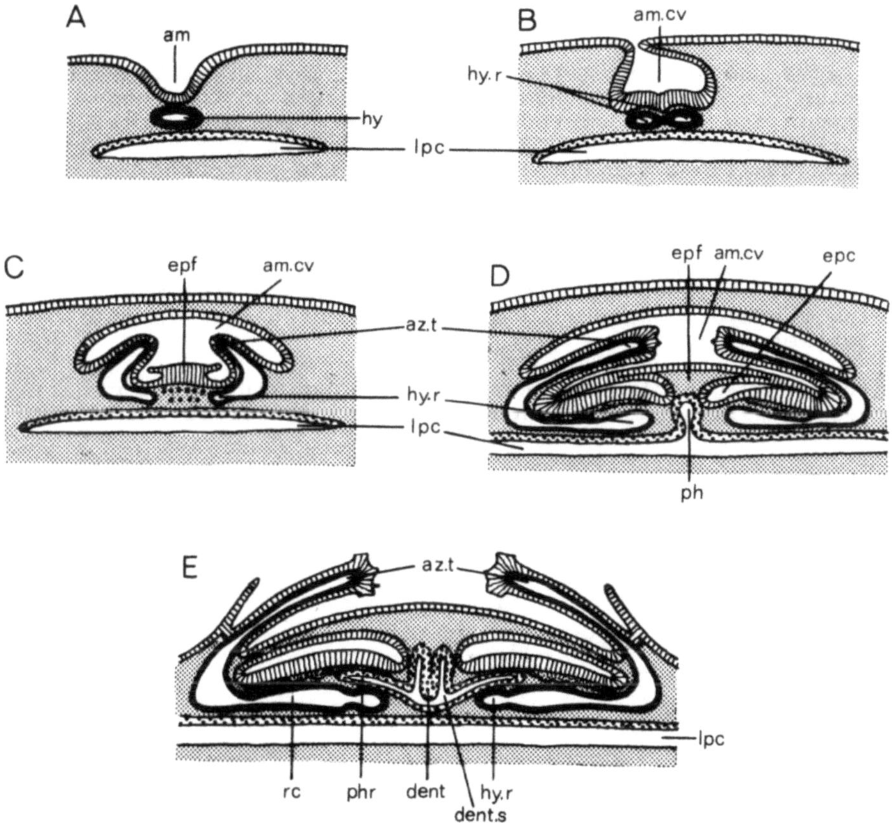

Fig. 7.14. Transverse sections through echinus rudiment of *Echinus esculentus* plutei 21 to 50 days old. (A) Echinus rudiment just formed by apposition of amniotic cavity and hydrocoel. (B) Hydrocoel becomes a ring; amniotic cavity is closing. (C) Amniotic cavity is closed; primary tube feet (azygous tube feet) of hydrocoel protrude into amniotic cavity; epineural folds are forming. (D) Epineural folds are united, covering epineural cavity; perihemal rudiment is formed. (E) Amniotic roof bursts and primary tube feet protrude; perihemal rudiment developed into dental sac and radial perihemal canal. am, amniotic invagination; am.cv, amniotic cavity; az.t, primary tube feet; dent, rudiment of tooth; dent.s, dental sac; epc, epineural cavity; epf, epineural fold; hy, hydrocoel; hy.r, hydrocoel ring; lpc, left posterior coelom; ph, perihemal rudiment; phr, perihemal radial canal; r.c, radial canal of water vascular system. (From MACBRIDE, 1914)

The anterior part of the right coelomic vesicle buds a little later than that of the left, and the connection between the two is severed. The isolated small vesicle becomes the madreporic vesicle (Figs. 7.12.1C, 7.13A). Toward the dorsal side of the larva the anterior chamber of the left coelom sends a tube - the pore or hydropore canal - which opens to the outside - madrepore or hydropore (Figs. 7.12.1C, 7.13A).

7.6.3 Echinus Rudiment

While the last arms and the epaulettes develop on the larval body, a shallow depression - or amniotic invagination - forms in the ectodermal wall between the left posterodorsal and left postoral arms, directly above the hydrocoel (Figs. 7.13A; 7.14A). The amniotic invagination expands like a sac until it touches the hydrocoel.

From above and to the left of the larva, the hydrocoel looks circular; seen in cross section, it is a flat, oval vesicle, the wall of which is a single layer of columnar cells. The posterior border of the hydrocoel first develops two lobes that lengthen until their tips meet and fuse, giving the hydrocoel a ring shape into which the amniotic invagination settles.

After the amniotic invagination becomes fairly large, it is cut off from the ectodermal wall as a flattened sac, the interior of which

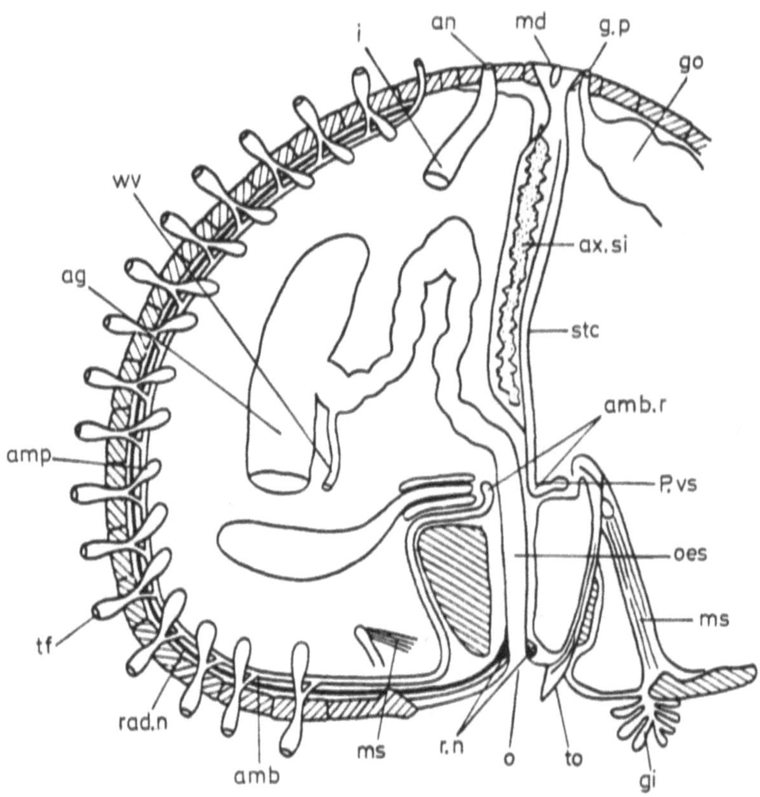

Fig. 7.15. Median cross-section of adult sea urchin. ag, anterior gut; amb, radial ambulacral canal; amb.r, ambulacral ring; amp, ampulla; an, anus; ax.si, axial sinus; gi, gill; go, gonad; g.p, genital pore; i, intestine; md, madrepore; ms, muscle; o, mouth; oes, esophagus; P.vs, Polian vesicle; rad.n, radial nerve; r.n, nerve ring; stc, stone canal; tf, tube foot; to, tooth; wv, water vessel. (From IIJIMA and DAN, 1957)

is called the amniotic cavity. Its outer wall - or amnion - is thin, while its inner wall touching the hydrocoel is thick (Figs. 7.13B, 7.14B). In the amniotic cavity and the hydrocoel - the echinus rudiment - the adult mouth and its surrounding structures are formed.

The hydrocoel ring enlarges, and the 5 outgrowths that protrude as finger-shaped lobes are the rudiments of 5 radial water-vascular canals and their primary terminal azygous tube feet that become the tips of the 5 adult ambulacral zones (Figs. 7.14C, D, E; 7.15). Pushed up by the lobes, the thick floor of the amniotic cavity becomes the ectodermal covering of the 5 tube feet (Fig. 7.14C).

The left posterior coelom extends anteriorly, encircling the hydrocoel and the stomach, and pushes out 5 lobes - or dental sacs - that alternate with the 5 hydrocoel lobes. The dental sacs are the primordia of the lantern. The floor of the amniotic cavity between the azygous tube feet and over the dental sacs thickens into 5 epineural ridges that elevate and spread laterally, giving a T-shape appearance in section (Fig. 7.14C, D). The wings of the ridges unite to form a second floor over the original floor of the amniotic cavity, creating 2 thin, epineural, ectodermal layers (Fig. 7.14D). The top layer covering the base of the tube feet becomes the epineural veil (Fig. 7.12.1E).

Between the lower epineural layer and the original floor of the amniotic cavity is the epineural cavity (Fig. 7.14D) from which radiate 5 canals (Fig. 7.14E) that develop into epineural canals containing the radial nerves, and the epineural ring sinus, containing the ring nerve. The radial nerves arise from the inside of the epineural cavity derived from the original floor of the amniotic cavity (Figs. 7.12E, 7.15).

From the side of the amnion toward the inside of the body, the structures discussed are (1) 5 tube feet protruding into the amniotic cavity, (2) the epineural space circumscribed by the bases of the tube feet, (3) the ring nerve, (4) 5 teeth alternating with the tube feet, and (5) at the lowest level, the hydrocoel ring that forms the water-vascular system. Adult sea-urchin anatomy is quite similar.

In the dental sacs cut off from the left posterior coelom, teeth begin to form - one tooth in each sac - as thickenings of the wall (Fig. 7.14E). The sacs put out tubular perihemal canals, and lobulations from the sacs develop around the hydrocoel and fuse to form a continuous coelomic cavity, the center of which the esophagus breaks through and in which the lantern develops. Ocular plates with tetraradiate spines (juvenile type) appear in the echinus rudiment (Fig. 7.1[70]). When adult calcareous matter develops in *Mespilia*, the accumulation of green pigment makes the echinus rudiment conspicuous in the living animal (ONODA, 1936).

The azygous tube feet extend into the amniotic cavity and acquire terminal discs supported by internal calcareous rings. Paired tube feet develop near the azygous ones (Fig. 7.12F). The oral surface of the young sea urchin develops on the left side of the pluteus, while the aboral surface develops on the right posterior side.

The pluteus digestive tract becomes that of the adult, but the mouth and anus have to be renewed. The right and left posterior coeloms meet to create the mesentary, and on the larval body typical definite spines, genital plates with pedicellariae, pedicellariae alone, and parts of the adult test develop (Figs. 7.1[70], 7.12D-F). Genital and terminal plates come from the skeletal material on the inside ends of the pluteus arms.

A detailed presentation of the differentiation of adult organs and calcareous matter is omitted here, since there is an enormous amount of classical works on various species of sea urchins reviewed by HYMAN (1955) and HARVEY (1956).

7.6.4 Metamorphosis

As metamorphosis approaches, the pluteus becomes sluggish, gets heavier, and sinks to the sea bottom (Fig. 7.12D). The outer wall of the amniotic cavity ruptures, and the tube feet and spines emerge in about an hour (Fig. 7.12D-F).

The newly metamorphosed sea urchin stays on the sea bottom by means of tube feet that rotate the body axis 90°, bringing the left side of the larval body to the substratum. The ectodermal wall of the larval arm is absorbed into the general body surface, and the skeletal rods supporting the arms are cast off (Fig. 7.12D-F). The larval mouth and anus severed, the young sea urchin has only the larval digestive tract. The adult mouth and anus develop later.

To cover the young sea urchin, the epidermis of the inner surface of the amniotic cavity expands, and the 5 branches of the ambulacral ring reaching into the tube feet extend, pushing the azygous tube feet ahead of them. The paired tube feet follow (Fig. 7.12.2F, G), and the radial ambulacral zones of the adult form. By the time the sea urchin reaches the adult stage, the azygous tube feet have been pushed all the way to the dorsal side.

To quote from CZIHAK (1971), "Situs inversus, that is the occurrence of a larger coelomic sac at the right side and development of the hydrocoel at the right side, has frequently been observed. In these cases the vestibulum also forms at the right side and development therefore can proceed in a normal way. Pedicellaria are then formed first at the left side.

In some cases development of a hydrocoel was found on both sides. This can be interpreted as overdevelopment of the coelomic sacs on both sides or fission of a uniform anlage into two equal parts. The two hydrocoels develop normally into two imaginal discs and two ventral sides of sea urchins form Siamese twins after metamorphosis. In this case of course no pedicellaria-anlagen can be found at the place of the imaginal discs. But two of them are formed when the imaginal disc is lacking on both sides; this may be considered the consequence of the presence of two small coelomic sacs on both sides. According to the occurrence of the hydrycoel + - (normal), - + (situs inversus), + + (twins) and - - types of plutei have been distinguished" (Fig. 7.18).

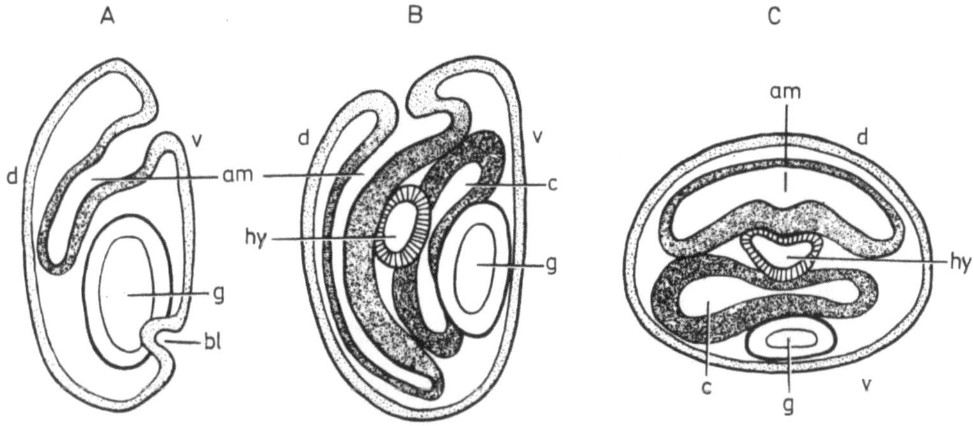

Fig. 7.16. Section of successive stages showing position of amniotic cavity and hydrocoel in larvae of *Peronella japonica* Mortensen. (A) Prism stage, 18 hr after fertilization at 28°C; longitudinal sagittal section. (B) Pluteus, 26 hr; amniotic opening shifted toward dorsal side and blastopore closed; longitudinal sagittal section. (C) Same as B; transverse section. am, amniotic cavity; bl, blastopore; c, coelom; d, dorsal side; g, gut; hy, hydrocoel; v, ventral side. (From OKAZAKI and DAN, 1954)

Fig. 7.17. Development of *Peronella japonica* Mortensen. Larvae raised at room temperature (28°-28.5°C). 1, sperm; 2, unfertilized egg about 300μm in diameter; heavily laden with yolk, opaquely pink; 3, 20 min after fertilization, hyaline layer extremely thin and not visible here; 4, 2-cell stage, 40 min; 5, 4-cell stage, 1 hr; 6, 8-cell stage, 1 hr 20 min; 7, 16-cell stage, 1 hr 40 min; 8, 32-cell stage, 2 hr; 9, morula stage, 3 hr; 10-12, young blastula, folded blastular wall, 4 hr; 13, blastula, expanded blastular wall, 5 hr; 14, primary mesenchymal cells differentiate before hatching, 6 hr; 15, mesenchymal blastula immediately after hatching, 7 hr; 16, late mesenchymal blastula, 9 hr; 17, gastrula, invagination at posterior wall, 10 hr; 18-20, late gastrula, 13 hr, stomodeum-like invagination on ventral side is amniotic; 18, dorsal view; 19, ventral view; 20, lateral view; 21-22, so-called prism stage, 18 hr, amniotic invagination extends along dorsal side of entoderm and forms cavity; 21, dorsal view; 22, ventral view; 23-25, pluteus stage, 24 hr, fully formed pluteus of this species typically has only two postoral arms, number of arms may vary from none to four without affecting later development or metamorphosis, at this stage hydrocoel begins to form from the coelomic sac in median position; 26-28, late pluteus stage, 40 hr, adult tube feet and spines have already differentiated bilaterally and symmetrically in relation to pluteus median plane, tube feet tips seen through wall of larval body; 26, ventral view; 27, dorsal view; 28, lateral view; 29-31, pluteus beginning to metamorphose, 50-60 hr; 29, ventral view; 30, lateral view; 31, ventral view; 32-34, young sea urchin just after metamorphosis, 60 hr; 32, aboral view; 33, oral view; 34, lateral view; 35-37, imago 3 days after metamorphosis; 35, oral view; 36, aboral view; 37, lateral view; 38-40, imago 7 days after metamorphosis; 38, aboral view; 39, lateral view; 40, oral view

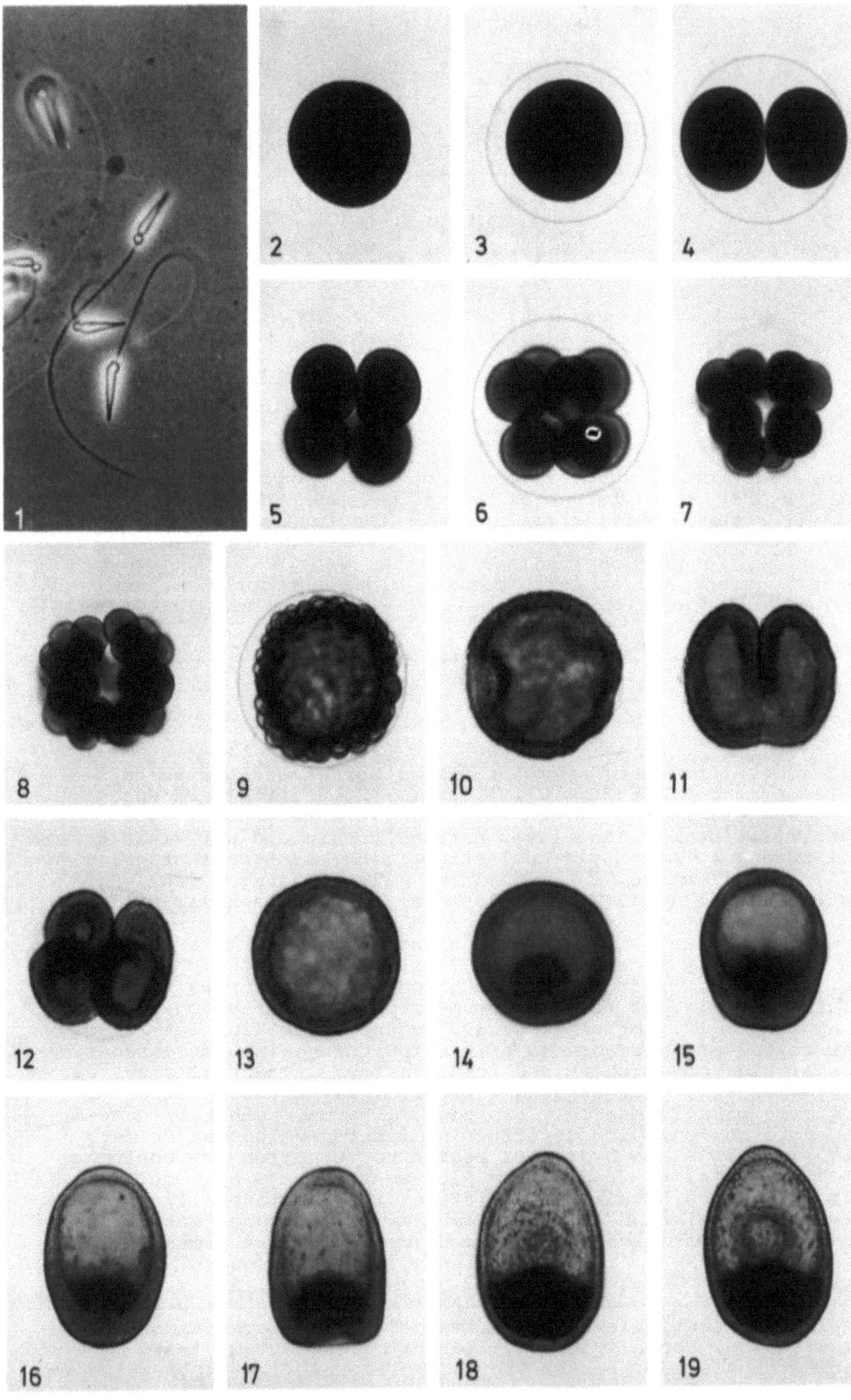

Fig. 7.17. Legend see p. 223

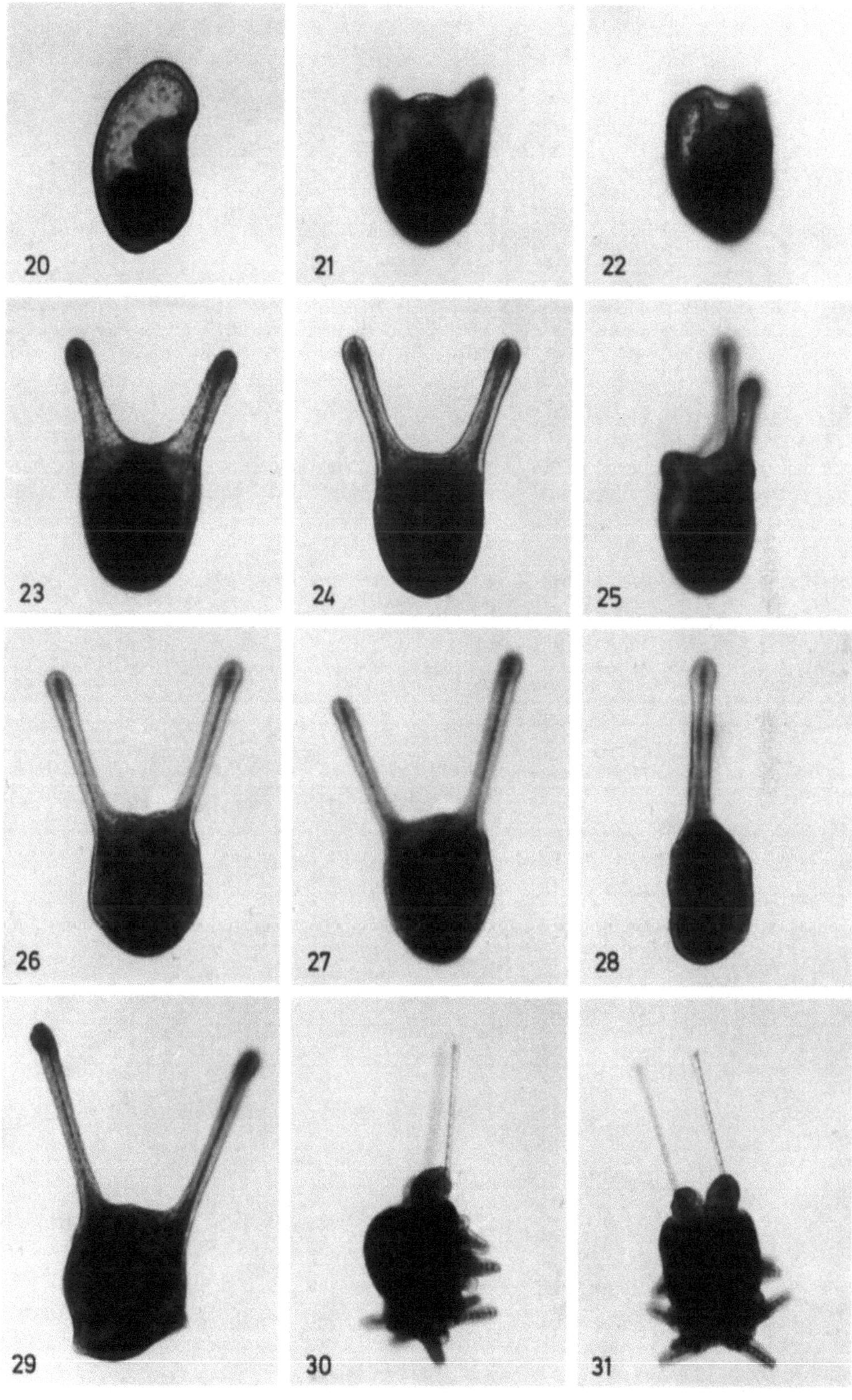

Fig. 7.17. Legend see p. 223

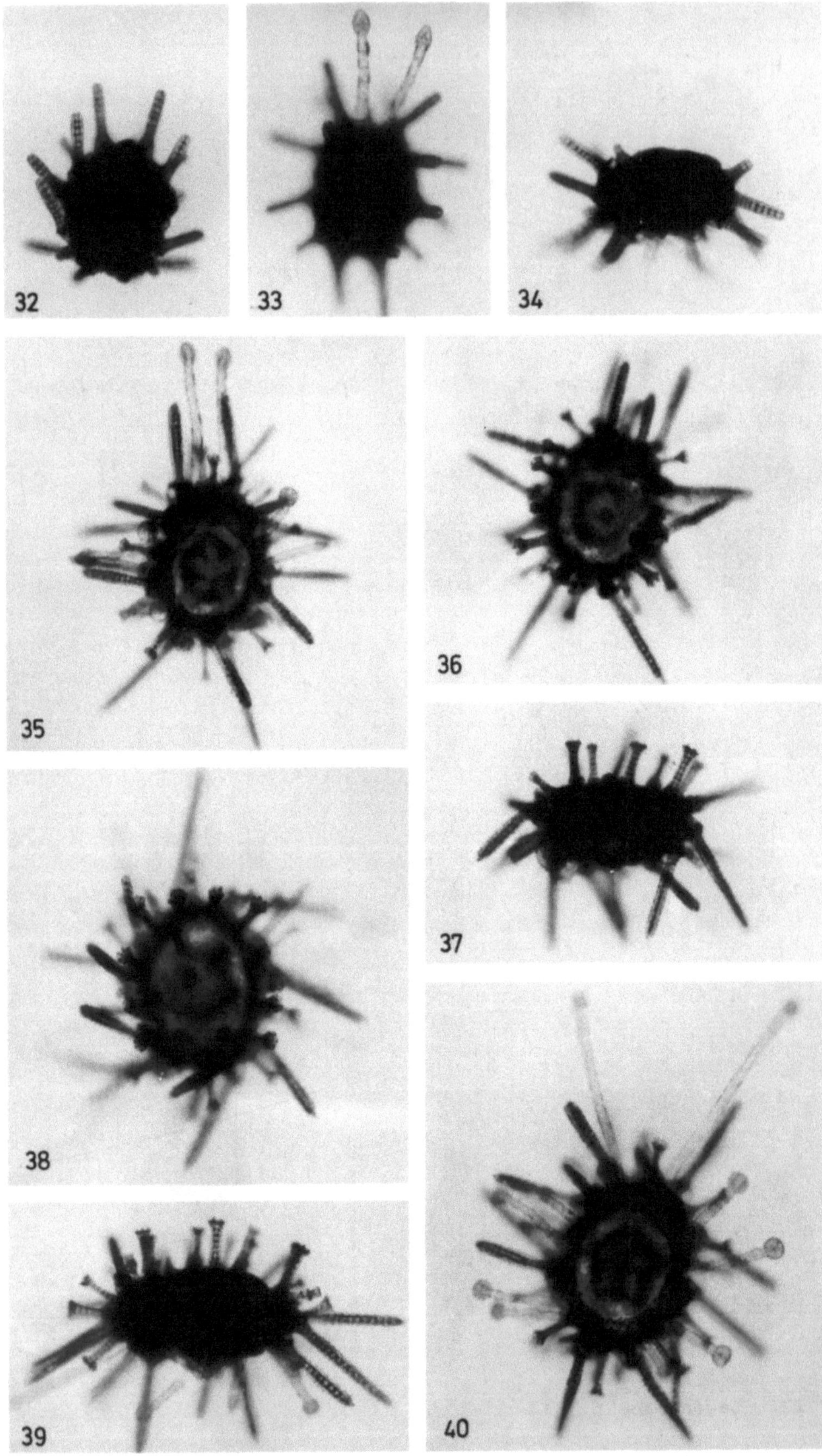
32
33
34
35
36
37
38
39
40

7.6.5 Metamorphosis of *Peronella japonica* Mortensen

In regular sea urchins metamorphosis requires long culturing, which probably explains why there have been so few experimental studies on the subject. Notable exceptions are the works of SHEARER et al. (1914) on the metamorphosis of hybrids, and those of HARVEY (1949, 1956) on the metamorphosis of centrifuged fragments. In the sand dollar *Peronella japonica* Mortensen, however, metamorphosis is completed in three to four days (MORTENSEN, 1921; OKAZAKI and DAN, 1954)(Fig. 7.17). The larva is shaped like a pluteus but has only 2 arms. It has no functional digestive tract so does not need to be fed until it metamorphoses. The amniotic invagination - at the side of the stomodeum of other species - and the hydrocoel form in the middle of the larval body (Fig. 7.16). The amniotic invagination shifts toward the dorsal side of the pluteus, and the anterodorsal surface becomes the oral surface of the adult. After metamorphosis sand dollars survive for about 10 days without being fed. During this time the number of spines and tube feet increases.

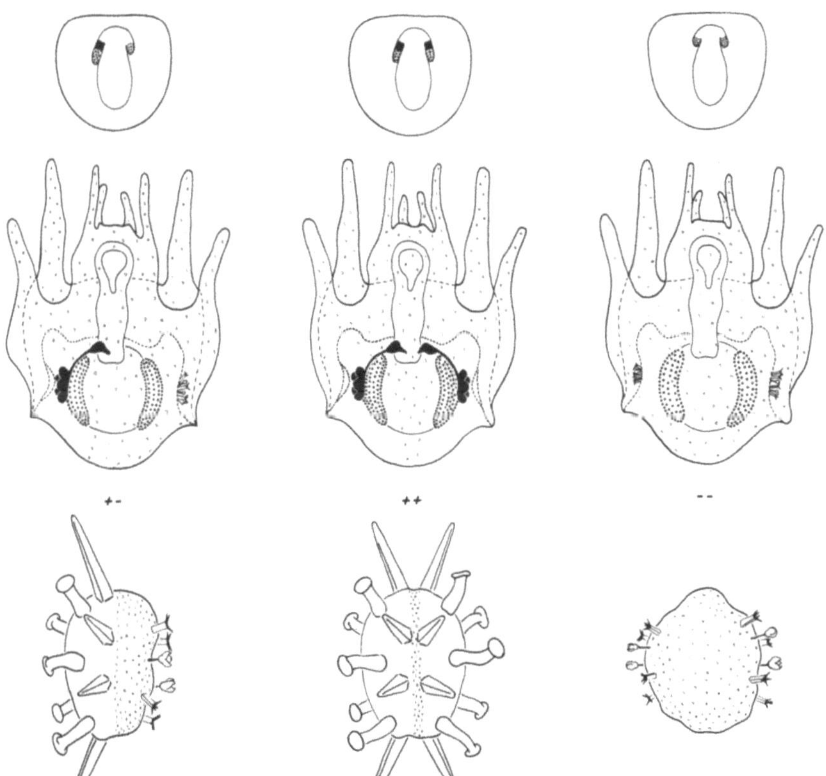

Fig. 7.18. Development of the coelomic sacs in gastrulae and plutei; (a) in normal plutei (+- type) having a hydrocoel only on the left side, (b) in plutei with double echinus rudiment (++ type) having hydrocoels on both sides and (c) in plutei without hydrocoel (-- type). (From CZIHAK, 1970)

References

AGRELL, I., 1954. A mitotic gradient in the sea urchin embryo during gastrulation. Ark. Zool. 6, 213-217.

AUCLAIR, W., SIEGEL, B.W., 1966. Cilia regeneration in the sea urchin embryo: evidence for a pool of ciliary proteins. Science 154, 913-915.

BALINSKY, B.I., 1959. An electron microscopic investigation of the mechanisms of adhesion of the cells in sea urchin blastula and gastrula. Exp. Cell Res. 16, 429-433.

BOVERI, T., 1901a. Die Polarität von Oocyte, Ei und Larve des Strongylocentrotus lividus. Zool. Jahrb. Abt. Anat. Ont. 14, 630-653.

BOVERI, T., 1901b. Über die Polarität des Seeigeleies. Verh. Phys.-med. Ges. Würzburg 34, 145-176.

CZIHAK, G., 1960. Untersuchungen über die Coelomanlagen und die Metamorphose des Pluteus von Psammechinus miliaris (GMELIN). Zool. Jb. Anat. 78 235-279.

CZIHAK, G., 1960. Pseudoradiärsymmetrische Seeigelplutei. Roux' Arch. Entwicklungsmech. 152 593-601.

CZIHAK, G., 1962a. Entwicklungsphysiologie der Echinodermen. Fortschr. Zool. 14 237-267.

CZIHAK, G., 1962b. Entwicklungsphysiologische Untersuchungen an Echiniden. (Topochemie der Blastula und Gastrula, Entwicklung der Bilateral- und der Coelomdivertikel). Roux' Arch. Entwicklungsmech. 154, 29-211.

DAN, K., 1952. Cyto-embryological studies of sea urchins. II. Blastula stage. Biol. Bull. 102, 74-89.

DAN, K., 1954a. Further study on the formation of the "new membrane" in the eggs of the sea urchin, Hemicentrotus (Strongylocentrotus) pulcherrimus. Embryologia 2, 99-114.

DAN, K., 1954b. The cortical movement in Arbacia punctulata eggs through cleavage cycles. Embryologia 2, 115-122.

DAN, K., 1957. Invertebrate Embryology, 199-212. Tokyo: Bai Fu Kan Press.

DAN, K., 1960. Cyto-embryology of echinoderms and amphibia. Inter. Rev. Cytol. 9, 321-367.

DAN, K., DAN, J.C., 1940. Behavior of the cell surface during cleavage III. On the formation of new surface in the eggs of Strongylocentrotus pulcherrimus. Biol. Bull. 78, 486-501.

DAN, K., DAN, J.C., 1947. Behavior of the cell surface during cleavage VII. On the division mechanism of cells with excentric nuclei. Biol. Bull. 93, 139-162.

DAN, K., ONO, T., 1952. Cyto-embryological studies of sea urchins. I. The means of fixation of the mutual positions among the blastomeres of sea urchin larvae. Biol. Bull. 102, 58-73.

DAN, K., NAKAJIMA, T., 1956. On the morphology of the mitotic apparatus isolated from Echinoderm eggs. Embryologia 3, 187-200.

DAN, K., OKAZAKI, K., 1956. Cyto-embryological studies of sea urchins. III. Role of the secondary mesenchyme cells in the formation of the primitive gut in sea urchin larvae. Biol. Bull. 110, 29-42.

DRIESCH, M., 1896. Die taktische Reizbarkeit der Mesenchymzellen von Echinus microtuberculatus. Arch. Entwicklungsmech. 3, 362-380.

ENDO, Y., 1954. Relation of sperm entrance point and the first cleavage plane to polarity in sea urchin eggs. Zool. Mag. 63, 164-165.

ENDO, Y., 1961. Changes in the cortical layer of sea urchin eggs at fertilization as studied with the electron microscope. Exp. Cell Res. 25, 383-397.

ENDO, Y., 1966. Development and Differentiation, 1-61. Tokyo: Iwanami Shoten.

GIBBINS, J.R., TILNEY, L.G., PORTER, K.R., 1969. Microtubules in the formation and development of the primary mesenchyme in Arbacia punctulata. J. Cell Biol. 41, 201-226.
GUSTAFSON, T., KINNANDER, H., 1956. Microaquaria for time-lapse cinematographic studies of morphogenesis in swimming larvae and observations on sea urchin gastrulation. Exp. Cell Res. 11, 36-51.
GUSTAFSON, T., WOLPERT, L., 1961. Studies on the cellular basis of morphogenesis in the sea urchin embryo. Directed movements of primary mesenchyme cells in normal and vegetalized larvae. Exp. Cell Res. 24, 64-79.
GUSTAFSON, T., WOLPERT, L., 1962. Cellular mechanism in the morphogenesis of the sea urchin larva. Change in shape of cell sheets. Exp. Cell Res. 27, 260-279.
GUSTAFSON, T., WOLPERT, L., 1963a. The cellular basis of morphogenesis and sea urchin development. Inter. Rev. Cytol. 15, 139-214.
GUSTAFSON, T., WOLPERT, L., 1963b. Studies on the cellular basis of morphogenesis in the sea urchin embryo. Formation of the coelom, the mouth, and the primary pore-canal. Exp. Cell Res. 29, 561-582.
GUSTAFSON, T., WOLPERT, L., 1967. Cellular movement and contact in sea urchin morphogenesis. Biol. Rev. 42, 442-498.
HAGSTRÖM, B.E., LÖNNING, S., 1965. Studies on cleavage and development of isolated sea urchin blastomeres. Sarsia 18, 1-9.
HARVEY, E.B., 1949. The growth and metamorphosis of the Arbacia punctulata pluteus, and late development of the white halves of centrifuged eggs. Biol. Bull. 97, 287-299.
HARVEY, E.B., 1956. American Arbacia and other sea urchins. Princeton, N.J.: Princeton University Press.
HERBST, C., 1896. Experimentelle Untersuchungen über den Einfluß der veränderten chemischen Zusammensetzung des umgebenden Mediums auf die Entwicklung der Tiere. III. Über das Ineinandergreifen von normaler Gastrulation und Lithiumentwicklung. Arch. Entwicklungsmech. 2, 455-516.
HERBST, C., 1900. Über das Auseinandergehen von Furchungs- und Gewebezellen in Kalk-freiem Medium. Roux' Arch. Entwicklungsmech. 9, 424-463.
HIRAMOTO, Y., 1957. The thickness of the cortex and the refractive index of the protoplasm in sea urchin eggs. Embryologia 3, 361-374.
HÖRSTADIUS, S., 1928. Über die Determination des Keimes bei Echinodermen. Acta Zool. 9, 1-191.
HÖRSTADIUS, S., 1935. Über die Determination im Verlaufe der Eiachse bei Seeigeln. Pubbl. Staz. Zool. Napoli 14, 251-479.
HÖRSTADIUS, S., 1936a. Über die zeitliche Determination im Keim von Paracentrotus lividus. Arch. Entwicklungsmech. 135, 1-39.
HÖRSTADIUS, S., 1936b. Weitere Studien über die Determination im Verlaufe der Eiachse bei Seeigeln. Arch. Entwicklungsmech. 135, 40-68.
HÖRSTADIUS, S., 1939. The mechanics of sea urchin development, studied by operative methods. Biol. Rev. 14, 132-179.
HÖRSTADIUS, S., 1949. Experimental researches on the developmental physiology of the sea urchin. Pubbl. Staz. Zool. Napoli, Suppl. 21, 131-172.
HSIAO, S.C., FUJII, W.K., 1963. Early ontogenic changes in the concentration of alkaline phosphatase in Hawaiian sea urchins. Exp. Cell Res. 32, 217-231.
HYMAN, L.H., 1955. The Invertebrates: Echinodermata, vol. IV, 413-588. New York, Toronto, London: McGraw-Hill Book Co.
IMMERS, J., 1956. Changes in acid mucopolysaccharides attending the fertilization and development of the sea urchin. Ark. Zool. 9, 367-375.

IMMERS, J., 1961. Comparative study of the localization of incorporated ^{14}C-labeled amino acids and $^{35}SO_4$ in the sea urchin ovary, egg, and embryo. Exp. Cell Res. 24, 356-378.
ISHIDA, J., 1936. An enzyme dissolving the fertilization membrane of sea urchin eggs. Annot. Zool. Japon. 15, 453-457.
ISHIDA, J., 1954. Function of egg surface during fertilization in sea urchins. Cyto-chemistry Symposium 2, 65-80.
ISHIDA, J., 1967. Cell Biology, 613-631. Tokyo: Asakura Shoten.
IWAIKAWA, Y., 1967. Regeneration of cilia in the sea urchin embryo. Embryologia 9, 287-294.
KANE, R.E., 1970. Direct isolation of the hyaline layer protein released from the cortical granules on the sea urchin egg at fertilization. J. Cell Biol. 45, 615-622.
KANE, R.E., HERSH, R.T., 1959. The isolation and preliminary characterization of a major soluble protein of the sea urchin egg. Exp. Cell Res., 16, 59-69.
KANE, R.E., STEPHENS, R.E., 1969. A comparative study of the isolation of the cortex and the role of the calcium-insoluble protein in several species of sea urchin eggs. J. Cell Biol. 41, 133-144.
KINNANDER, H., GUSTAFSON, T., 1960. Further studies on the cellular basis of gastrulation in the sea urchin larva. Exp. Cell Res. 19, 278-290.
KOPAC, M.J., 1941. Disintegration of the fertilization membrane of Arbacia by the action of an "enzyme". J. Cell Comp. Physiol. 18, 215-220.
LILLIE, F.R., 1913. The mechanism of fertilization. Science 38, 524-528.
MACBRIDE, E.W., 1914. Text-book of embryology, vol. I. Invertebrata, 504-529. London: Macmillan and Co., Ltd.
MONNE, L., SLAUTTERBACK, D.B., 1950. Differential staining of various polysaccharides in sea urchin. Exp. Cell Res. 1, 477-491.
MONNE, L., HÅRDE, S., 1951. On the formation of the blastocoel and similar embryonic cavities. Ark. Zool. 1, 463-469.
MONNE, L., SLAUTTERBACK, D.B., 1952. On the staining of the cytoplasm with the Schiff reagent during the development of the eggs of Paracentrotus lividus. Ark. Zool. 3, 349-356.
MOORE, A.R., 1933. Is cleavage rate a function of the cytoplasm or of the nucleus? J. Exp. Biol. 10, 230-236.
MOORE, A.R., 1940. Osmotic and structural properties of the blastular wall in Dendraster excentricus. J. Exp. Zool. 84, 73-79.
MOORE, A.R., BURT, A.S., 1939. On the locus and nature of the forces causing gastrulation in the embryos of Dendraster excentricus. J. Exp. Zool. 82, 159-168.
MORGAN, T.H., 1893. Experimental studies on Echinoderm eggs. Anat. Anz. 9, 141-152.
MORTENSEN, Th., 1921. Studies of the development and larval forms of echinoderms, 109-117. Copenhagen.
MORTENSEN, Th., 1931. Contributions to the study of the development and larval forms of echinoderms. I-II. Kgl. Dansk. Vidensk. Selsk. Skr. Naturv. og Math. Ser. 9, 4, (1).
MORTENSEN, Th., 1937. Contributions to the study of the development and larval forms of echinoderms. III. Kgl. Dansk. Vidensk. Selsk. Skr. Naturv. og Math. Ser. 9, 7, (1).
MORTENSEN, Th., 1938. Contributions to the study of the development and larval forms of echinoderms. IV. Kgl. Dansk. Vidensk. Selsk. Skr. Naturv. og Math. Ser. 9, 7, (3).
MOTOMURA, I., 1935. Determination of the embryonic axis in the eggs of amphibia and echinoderms. Sci. Rept. Tohoku Imp. Univ. Ser. IV, 10, 211-254.
MOTOMURA, I., 1967. Formation of diastema in the cleaving egg of the sea urchin. Sci. Rept. Tohoku Univ. Ser. IV, 33, 135-142.

NAKANO, E., 1956. Physiological studies on re-fertilization of the sea urchin egg. Embryologia 3, 139-165.
NAKANO, E., OOHASHI, S., 1954. On the carbohydrate component of the jelly coat and related substances of eggs from Japanese sea urchins. Embryologia 2, 81-85.
NAKANO, E., OKAZAKI, K., IWAMATSU, T., 1963. Accumulation of radioactive calcium in larvae of the sea urchin Pseudocentrotus depressus. Biol. Bull. 125, 125-132.
OKAZAKI, K., 1956a. Exogastrulation induced by calcium deficiency in the sea urchin, Pseudocentrotus depressus. Embryologia 3, 23-36.
OKAZAKI, K., 1956b. Skeleton formation of sea urchin larvae. I. Effect of Ca concentration of the medium. Biol. Bull. 110, 320-333.
OKAZAKI, K., 1960. Skeleton formation of sea urchin larvae. II. Organic matrix of the spicule. Embryologia, 5, 283-320.
OKAZAKI, K., 1965. Skeleton formation of sea urchin larvae. V. Continuous observation of the process of matrix formation. Exp. Cell Res. 40, 585-596.
OKAZAKI, K., 1970. Growth of the spicule in sea urchin larvae. Collagen Symposium VIII., 113-129.
OKAZAKI, K., 1971a. In vitro culture of the micromeres and primary mesenchyme cells isolated from sea urchin embryos and larvae. In: Cells in Early Development (Jap. Soc. Dev. Biologists, ed.), 188-225. Tokyo: Iwanami Shoten Publ.
OKAZAKI, K., 1971b. Spicule formation in sea urchin larvae; observations in vivo and in vitro. Symp. Cell Biol. (Japan) 22, 163-171.
OKAZAKI, K., DAN, K., 1954. The metamorphosis of partial larvae of Peronella japonica Mortensen, a sand dollar. Biol. Bull. 106, 83-99.
OKAZAKI, K., FUKUSHI, T., DAN, K., 1962. Cyto-embryological studies of sea urchins. IV. Correlation between the shape of the ectodermal cells and the arrangement of the primary mesenchyme cells in sea urchin larvae. Acta Embryol. Morph. Exp. 5, 17-31.
OKAZAKI, K., NIIJIMA, L., 1964. Basement membrane in sea urchin larvae. Embryologia 8, 89-100.
ONODA, K., 1931. Notes on the development of Heliocidaris crassispina with special reference to the structure of the larval body. Mem. Coll. Sci. Kyoto Imp. Univ. Ser. B, 7, 103-134.
ONODA, K., 1936. Note on the development of some Japanese Echinoids with special reference to the structure of the larval body. Japan. J. Zool. 6, 637-654.
PRENANT, M., 1926a. Sur le déterminisme de la forme spiculaire chez les larves d'oursins. Compt. Rend. Soc. Biol. 94, 433-435.
PRENANT, M., 1926b. L'etude cytologique du calcaire. III. Observations sur le déterminisme de la forme spiculaire chez les larves pluteus d'oursins. Bull. Biol. France Belgique 60, 522-560.
RUNNSTRÖM, J., 1928. Zur experimentellen Analyse der Wirkung des Lithiums auf den Seeigelkeim. Acta Zool. 9, 365-424.
RUNNSTRÖM, J., 1931. Zur Entwicklungsmechanik des Skelettmusters bei dem Seeigelkeim. Arch. Entwicklungsmech., 124, 273-297.
RUNNSTRÖM, J., 1935. An analysis of the action of lithium on sea urchin development. Biol. Bull. 68, 378-383.
RUNNSTRÖM, J., HAGSTRÖM, B.E., PERLMANN, P., 1959. Fertilization. In: The Cell - Biochemistry, Physiology, Morphology (J. Brachet, A.E. Mirsky, eds.), 327-397. New York: Academic Press, Inc.
SCHMIDT, H., 1904. Zur Kenntnis der Larvenentwicklung von Echinus microtuberculatus. Verh. Phys-Med. Ges. Würzburg 36, 297-336.
SCHMITT, F.O., 1941. Some protein patterns in cells. Growth 5 (Suppl.), 1-20.
SHEARER, C., DE MORGAN, W., FUCHS, H., 1914. On the experimental hybridization of Echinoids. Phil. Trans. Roy. Soc. London B, 204, 255-362.

STEPHENS, R.E., KANE, R.E., 1970. Some properties of hyalin. The calcium-insoluble protein of the hyaline layer of the sea urchin egg. J. Cell Biol. 44, 611-617.
SELENKA, E., 1883. Studien über Entwicklungsgeschichte der Tiere. 2. Die Keimblätter der Echinodermen. Wiesbaden.
THEEL, H., 1892. On the development of Echinocyamus pusillus. Nova Acta Reg. Soc. Sci. Upsaliensis Ser. III, 15, 1-57.
TYLER, A., 1941. The role of fertilizin in the fertilization of eggs of the sea urchin and other animals. Biol. Bull. 81, 190-204.
VON UBISCH, L., 1937. Die normale Skelettbildung bei Echinocyamus pusillus und Psammechinus miliaris und die Bedeutung dieser Vorgänge für die Analyse der Skelette von Keimblatt-Chimären. Z. Wiss. Zool. 149, 402-476.
VON UBISCH, L., 1939. Keimblattchimärenforschung an Seeigellarven. Biol. Rev. 14, 88-103.
VASSEUR, E. 1948. Chemical studies on the jelly coat of the sea urchin egg. Acta Chem. Scand., 2, 900-913.
VACQUIER, V.D., 1969. The isolation and preliminary analysis of the hyaline layer of sea urchin eggs. Exp. Cell Res. 54, 140-142.
WILSON, E.B., 1895. An atlas on the fertilization and karyokinesis of the ovum, 1-32. New York: Columbia University Press, Macmillan and Co.
WOLPERT, L., GUSTAFSON, T., 1961a. Studies on the cellular basis of morphogenesis of the sea urchin embryo. Development of the skeletal pattern. Exp. Cell Res. 25, 311-325.
WOLPERT, L., GUSTAFSON, T., 1961b. Studies of the cellular basis of morphogenesis of the sea urchin embryo. The formation of the blastula. Exp. Cell Res. 25, 374-382.
WOLPERT, L., MERCER, E.H., 1963. An electron microscope study of the development of the blastula of the sea urchin embryo and its radial polarity. Exp. Cell Res. 30, 280-300.
WOODLAND, W., 1906. Studies in spicule formation. III. On the mode of formation of spicular skeleton in the pluteus of Echinus esculentus. Quart. J. Microscop. Sci. 49, 305-325.
YASUMASU, I., 1958. Hatching enzyme of the sea urchins, Hemicentrotus pulcherrimus and Heliocidaris crassispina. Bull. Mar. Biol. Stat. Asamushi 9, 83-84.
YASUMASU, I., 1960. Quantitative determination of hatching enzyme activity of sea urchin blastulae. J. Fac. Sci. Univ. Tokyo Sec. IV, 9, 39-47.
YASUMASU, I., 1961. Crystallization of hatching enzyme of the sea urchin Anthocidaris crassispina. Sci. Papers of College of Gen. Educ. Univ. Tokyo 11, 275-280.
YAZAKI, I., 1968. Immunological analysis of the calcium precipitable protein of the sea urchin eggs. I. Hyaline layer substance. Embryologia 10, 131-141.

8. Cellular Behavior and Cytochemistry in Early Stages of Development

T. Gustafson

In a broad sense, developmental cytology and cytochemistry include all molecular (biochemical) and ultrastructural aspects of the cell in a developing organism. The behavior of the cell indicates its complex biochemical and biophysical nature and bridges the gap between the molecular events caused by the genes and the subsequent development of the embryo.

8.1 The Sea-Urchin Embryo in Studies of Morphogenetic Cell Behavior

For a number of reasons the sea-urchin embryo is very useful for studying cell behavior during morphogenesis (DAN, 1960; GUSTAFSON and WOLPERT, 1963b, 1967). The early larva has relatively few cells, 1.000 - 2.000 during its most dramatic changes in shape, and little growth and cell division during this time. It is transparent and its cell sheets, one cell-layer thick, allow every cell to be seen. Vital staining makes it possible to trace the larval regions back to the blastula wall (HÖRSTADIUS, 1935, 1939), and the fate map adds the details (Fig. 8.1). There is little cell reshuffling; morphogenesis may be described as deformation of the blastula wall without extensive changes in the positions of the cells as they relate to each other (Fig. 8.2). An exception is the release, migration, and attachment of the mesenchymal cells.

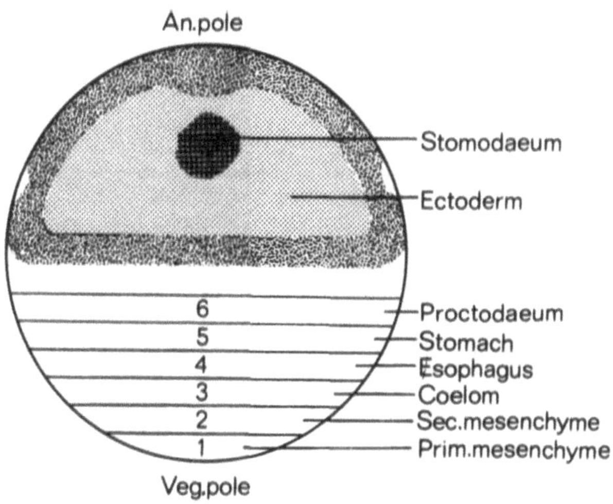

Fig. 8.1. Presumptive fate map of the egg, extrapolating the vital-staining results of HÖRSTADIUS. Zones 1-6 are much overscaled. (From GUSTAFSON, 1964)

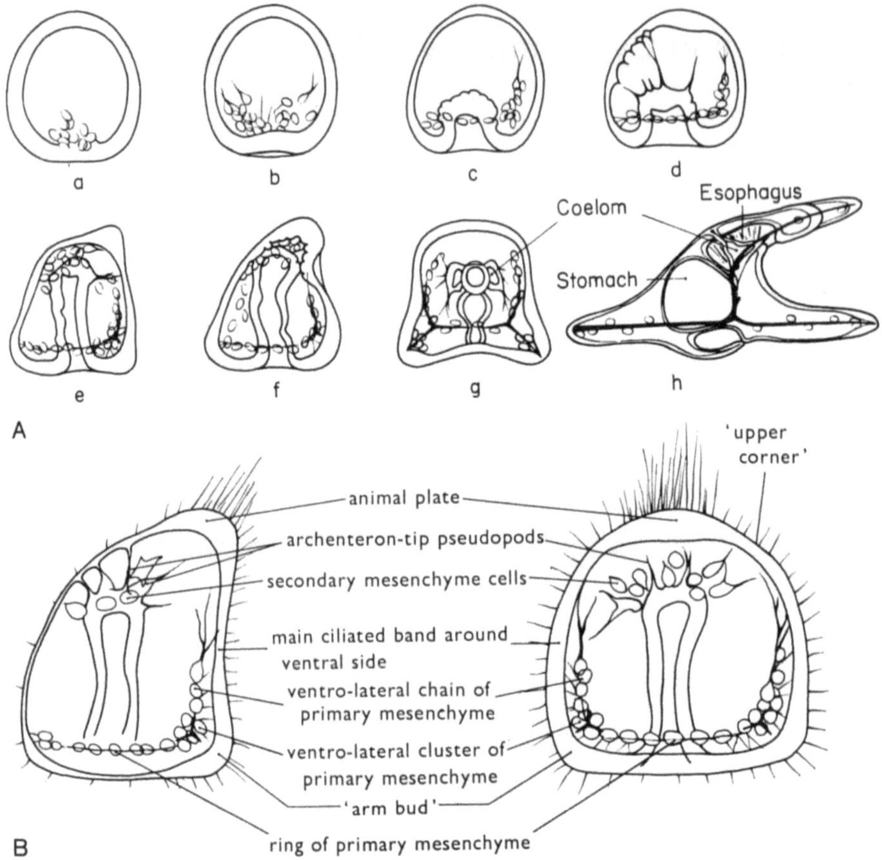

Fig. 8.2 A and B. (A) Early mesenchymal blastula of *Psammechinus miliaris* (a) develops into early pluteus (h). Embryo is seen from ventral side (g). Other views are from lateral side. (B) Advanced gastrula of *P. miliaris* shows primary mesenchymal cell distribution and relationship to main ciliated band. (From GUSTAFSON, 1963, 1964)

Not so complex as to be overwhelming, the sea-urchin larva is complicated enough to pose many interesting morphogenetic questions. While its development seems remote from that of the vertebrates, the kind of interactions that dominate classical sea-urchin embryology also occurs in vertebrate development (NIEUWKOOP, 1969; PATERSON, 1957; BÄCKSTRÖM, 1954). Neural induction, the main interest of many vertebrate embryologists, appears to have its counterpart in the sea-urchin larva. GARSTANG (1894) advanced his auricularia theory in the belief that the echinoderms may be distant relations of the vertebrates, but WILLMER (1970) disagreed.

8.2 Morphogenetic Cell Behavior Analysis

In studying the swimming sea-urchin larva by means of time-lapse cinematography, the larva must be kept in constant position beneath the

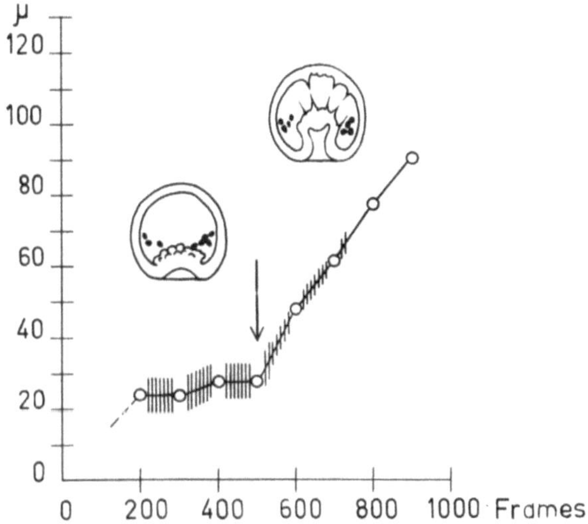

Fig. 8.3. Invagination of archenteron of *Psammechinus miliaris*. Curve indicates archenteron cavity length plotted against time. Extent of pulsatory activity is proportional to length of cross-lines on curve. Arrow indicates appearance of first pseudopods at archenteron tip. Two-step invagination and correlation between onset of second step and appearance of pseudopods are shown. One thousand frames correspond to 5-hr, 35 min. (From GUSTAFSON and WOLPERT, 1961a)

cover slip. The larva may be trapped in a nylon net on which calcium carbonate crystals have been grown (GUSTAFSON and KINNANDER, 1956). At higher magnifications the larva may be placed in paraffin oil or fluorocarbon liquid, with a minimum of seawater, on a cover slip and the cover slip placed in position (GUSTAFSON, 1963; MAZIA and RUBY, 1967; VACQUIER, 1968). The larva may be treated with calcium-free seawater until the hyaline layer dissolves and the cells produce pseudopods that anchor the larva to the slide (OKAZAKI, 1965). Focus between exposures may be altered automatically, after which the film may be divided into two sequences, each representing a single focus (GUSTAFSON and KINNANDER, 1960). Time-lapse films of a larva placed in highly viscous solution of methylated cellulose (metocel) may show very slow rotation and hence reveal overall change in shape.

The films often require detailed analysis. Plotting the length of the archenteron rudiment against time, for instance, shows that invagination occurs in two phases at different speed of elongation (Fig. 8.3) suggesting that two different cellular mechanisms are involved. The first phase is marked by pronounced pulsatory cell activity in the archenteron rudiment, particularly at its tip, and in the second phase pseudopods form (GUSTAFSON and KINNANDER, 1956; KINNANDER and GUSTAFSON, 1960).

With time-lapse studies other structural investigations are conducted. A simple but important method is to compress the larva under a cover slip so that the attachments between cells are revealed. Sucrose solutions are used for the same purpose because the blastula becomes impermeable to sucrose at time of hatching (DAN, 1960). Larvae raised in sucrose-seawater until hatching and then transferred to pure seawater swell osmotically causing partial separation of the cells (GUSTAFSON and WOLPERT, 1962). Enzymatic treatment, calcium-free seawater, and

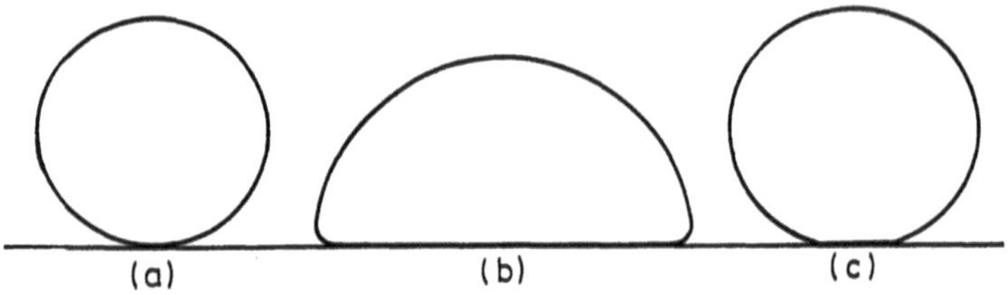

Fig. 8.4 a-c. Change in cell-membrane adhesion and tension affects contact between cell and substrate. (a) No adhesion or tension in membrane. (b) Strong adhesion between cell and substrate causes increased cell-contact with the substrate. (c) Strong tension causes cell-membrane contraction and reduced contact with substrate. (From GUSTAFSON and WOLPERT, 1963b)

dithiothreitol have been used to study cell contacts (DAN, 1960; VACQUIER and MAZIA, 1968a, VACQUIER and MAZIA, 1968b; VACQUIER, 1968).

Ultrastructural investigations of the developing sea-urchin larva, emphasizing regional differentiation and morphogenesis, have so far only begun. BALINSKY (1959), WOLPERT and MERCER (1963), AGRELL (1966), and VACQUIER (1968) have studied mostly cell contacts and cell polarity during blastula formation, while GIBBINS, TILNEY and PORTER (1969) and TILNEY and GIBBINS (1969) have examined microtubular elements in the vegetal region during mesenchymal formation and early gastrulation. BERG, TAYLOR, and HUMPHREYS (1962) and BERG and LONG (1964) have studied early mitochondria.

Classical workers emphasized the role of mechanical forces in morphogenetic events (GUSTAFSON and WOLPERT, 1963b). These forces are based on complicated physiologic cell activities and biosynthetic reactions just as the mechanical force of muscle is based on biochemical and biophysical fact. Morphogenetic models have been suggested for explaining observed cellular phenomena that may account for morphogenetic events, although the cause of cell shape change still remains unexplained (GUSTAFSON and WOLPERT, 1963b, 1967)(Figs. 8.4, 8.5). Because contact between cells is influenced by several factors, not the least of which is cell motility, the concept of cell adhesiveness should be examined with care (ABERCROMBIE, 1964; WEISS, 1967; GUSTAFSON, 1969a).

8.3 Cleavage and Blastula Formation

The astral relaxation theory holds that tension at the cell membrane is first uniform throughout the cell (WOLPERT, 1960, 1966). Cytokinesis is thought to result from polar-surface relaxation and decreased resistance to deformation which allows the equatorial region of the cell, while retaining its contractility, to constrict and divide the cell. Similar to the constriction mechanism proposed by MARSLAND and LANDAU (1954) is the fact that polar relaxation is believed to be caused by asters. It is equally possible that they induce contractility at the equator, as experimental polar amputations have indicated passive surface behavior in the polar regions (RAPPAPORT, 1969). In the polar and equatorial regions, the sol-gel states of the cytoplasm change in opposite ways. Certain cytoplasmic granules periodically release a solvating (relaxing), heparine-like substance (KINOSHITA, 1969) and

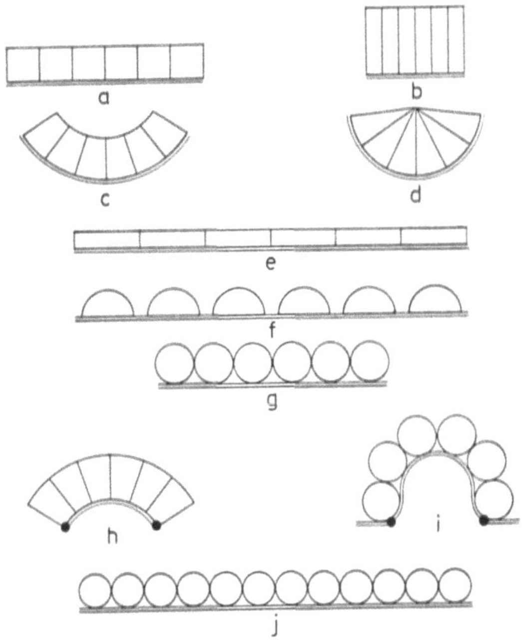

Fig. 8.5 a-j. Models of GUSTAFSON and WOLPERT to illustrate the effect of changes in contact between cells, and between cells and a supporting membrane, on the form of a cell sheet. ..."For convenience only 6 adjacent cells are considered. The cross-sectional area of the cells is fixed. (a) Moderate contact between the cells and between the cells and the supporting membrane. (b) Increased contact between the cells and loss of contact with the supporting membrane. The cells become columnar. (c) The effect of increased contact between the cells if the cells do not alter their contact with the supporting membrane. The sheet becomes curved. (d) The effect of a further increase in contact between cells, compared to that in (c), if the cells do not lose their contact with the supporting membrane. The sheet is maximally curved. (e) The effect of stretching the supporting membrane if the cells are able to increase their contact with it. The cells are flattened. (f) The effect of stretching the supporting membrane if the cells are able to increase their contact with the membrane but not to the same extent as in (e), and lose contact with one another. The cells tend to round up due to the tension in the membrane. (The free edges in the other figures are not drawn rounded since they are still in contact with other cells in the sheet.) (g) The effect of the cells in (a) rounding up. The sheet becomes longer. (h) The effect of the cells in (a) reducing their contact if the ends of the sheet are fixed - it becomes curved. (i) The effect of rounding up of the cells in (a) if the ends of the sheet are fixed. (j) The effect of doubling the number of cells in (g) by division. The sheet becomes longer." ... (From GUSTAFSON and WOLPERT, 1963b)

their pattern of distribution changes with the cell cycle in a way that suggests their participation in the periodic, local cytoplasmic solvation (KINOSHITA and YASAKI, 1967). Partition of these cytoplasmic granules has been described (REHBUN, 1964) which indeed supports the view that the mitotic mechanism may be involved. In some species the attachment fibres in the furrow region seem to be important at the end of cleavage.

A slight cleavage synchrony beginning at the 8-cell stage disappears at about the 32-cell stage. Mitotic waves spread upward from the vegetal pole during blastula formation, and a vegetal-animal gradient in mitotic activity reappears in the late blastula (AGRELL, 1953, 1956a, 1956b, 1958, 1960, 1964; AGRELL and PERSSON, 1956). But it is possible that the absence of synchrony indicates different animal and vegetal mitotic rates that may be related to primary determination along the egg axis (HAGSTRÖM, 1963; HAGSTRÖM and LÖNNING, 1963, 1964, 1965, 1966, 1967).

Early synchrony may be maintained by filopodial connections (VACQUIER, 1968), and cytoplasmic continuity may be involved in the attachments between micromeres and the adjacent cells (HAGSTRÖM and LÖNNING, 1969). The nature of the attachments and the biochemical characteristic of the micromeres have been discussed at length (COWDEN and LEHMAN, 1953; AGRELL and PERSSON, 1956; CZIHAK, 1965b, 1966, 1968; CZIHAK, WITTMANN and HINDENNACH, 1957; HÖRSTADIUS, JOSEFSSON and RUNNSTRÖM, 1967; BÄCKSTRÖM, 1969; CZIHAK and HÖRSTADIUS, 1970).

A great number of microvilli, or attachment fibers, attach the egg surface to the hyaline layer, the purpose of which is similar to the basal membrane of the epithelium. The attachment fibers maintain blastomere arrangement during cleavage and prevent changes in cell position (DAN, 1960)...."Blastula formation may be regarded as a problem in packing of cells and may be accounted for in mainly geometrical and mechanical terms. General considerations of the effect of uniform changes in adhesion and tension in the cell membrane, on the form of a tube or sphere made up of a single sheet of cells, indicate that the changes in cell shape at cleavage, the radial plane of cleavage, and the attachment of the cells to the hyaline layer are sufficient to account for the main features of blastula formation. The increase in blastocoel volume is due to the reduction in thickness of the blastula wall at successive cleavages"....(WOLPERT and GUSTAFSON, 1961b). It has also been indicated that an osmotically active substance (colloid) rhythmically secreted into the blastocoel may account for blastocoel enlargement (DAN, 1960).

A half-blastula may close and form a small blastula. Observations suggest that inter-cell attachments vary from species to species and may compensate for weak hyaline layer attachments. Experimental studies have involved inter-cell attachments weakened by treatment with dithiothreitol; hyaline layer attachments dissolved or weakened by treatment with pronase; and hyaline layer dissolved and inter-cell attachments weakened by treatment with calcium-free seawater (VACQUIER and MAZIA, 1968a, 1968b; VACQUIER, 1968). Eggs from which the hyaline layer was removed by dithiothreitol stick together by means of their attachment fibers in the presence of calcium ions.

As the cleavage furrow forms during cleavage, long inter-cell attachments on opposite sides of it intertwine and pull the blastomeres together. The microtubular elements of these attachments are similar to those of filopods and are inhibited by dithiothreitol (VACQUIER, 1968).

In the sand dollar, cells pull together as they are compressed by the hyaline layer, and break through each other's electrostatic repulsion barrier (cf. WEISS, 1967; PETHICA, 1961). The attachment fibers first make contact by penetrating this barrier.

The tightly compressed cell surfaces in the plane of cleavage become rounded (VACQUIER and MAZIA, 1968a). In the sea-urchin blastula the radial cell surfaces seldom touch except at mitosis. When the cell surfaces are tightly compressed, no cementing substance can be seen.

First observed by ANDREWS (1897), filopods between blastomeres may connect any cell with any other cell. Close contact between the cells makes the filopods hard to see except in calcium-free seawater where they are easily studied by means of phase microscopy and time-lapse cinematography (DAN, 1960; VACQUIER, 1968). Small, bulbous swellings move along the filopodial axis. Some filopods are intertwined; some are branched and when they reach another blastomere fan out over its surface and seem to attach. Filopods may be so strong that they pull out the cell surfaces in blebs. The filopods may hold the blastomeres together or, as electrophysiological experiments suggest, they may be the structural basis for cytoplasmic continuity among cells. Filopod size, number and activity increase when the embryo reaches the 32-cell stage.

About 5 hours after fertilization, a new 0.1μm-thick, extracellular layer, the fuzzy layer, lines the blastocoel and covers the smooth, inner ends of the cells (WOLPERT and MERCER, 1963). It contains a particulate matter that may be the precursor of fibrillar elements in the blastocoel and that histochemical and autoradiographic studies suggest may be composed of secreted acid mucopolysaccharides, the sulphate groups of which are masked by additional proteins (MONNE and HÅRDE, 1951; IMMERS, 1956a, 1961, 1962). The action of the acid polysaccharides, their sulphate groups not masked, changes the hyaline layer to the hyaline membrane at hatching (WOLPERT and MERCER, 1963).

The cells of the blastula wall lie between and are attached to 2 sheets of extracellular material, their radial surfaces seldom touching (DAN, 1960; GUSTAFSON, 1963, 1969a). Lead-stainable material between the cells is shown in electron micrographs (WOLPERT and MERCER, 1963). Between the living blastomeres are gaps, and time-lapse films show pronounced surface motility. Inter-cell attachment seems to occur near the extracellular layers at inner and outer contact points at later development (GUSTAFSON, 1963). Inner contact points may be due to interdigitation of the cell surface (BALINSKY, 1959), while the septate desmosomes that appear at the outer points may be caused by periodic precipitations of some material where the blastocoel meets the hyaline layer (BALINSKY, 1959; DAN, 1960; WOLPERT and MERCER, 1963; AGRELL, 1966). The desmosomes that surround the cell base may cause the blastula to become impermeable to sucrose at hatching (DAN, 1960), but this change may also be related to the structural change of the hyaline layer (GUSTAFSON and WOLPERT, 1962; WOLPERT and MERCER, 1963).

The radial polarity of blastula cells increases gradually, and by the time of hatching is most pronounced (Fig. 8.6). It is possible that these polarizations result from an initial polarization caused by the original external surface complex (the hyaline layer attached to the adjacent cell surface) and new internal surfaces formed at cleavage (WOLPERT and MERCER, 1963). Cell surface increases 100-fold up to a stage with 1.000 cells. Material secreted by the cells may also play a role. Unpolarized itself, this build-up might react with the hyaline layer, changing its structure and forming desmosomes. Further polar distribution of internal organella such as the Golgi's apparatus and the endoplasmic reticulum may possibly reflect internal and external mucoprotein coat differences resulting in an electrical potential across the blastula wall and leading to redistribution of cytoplasmic

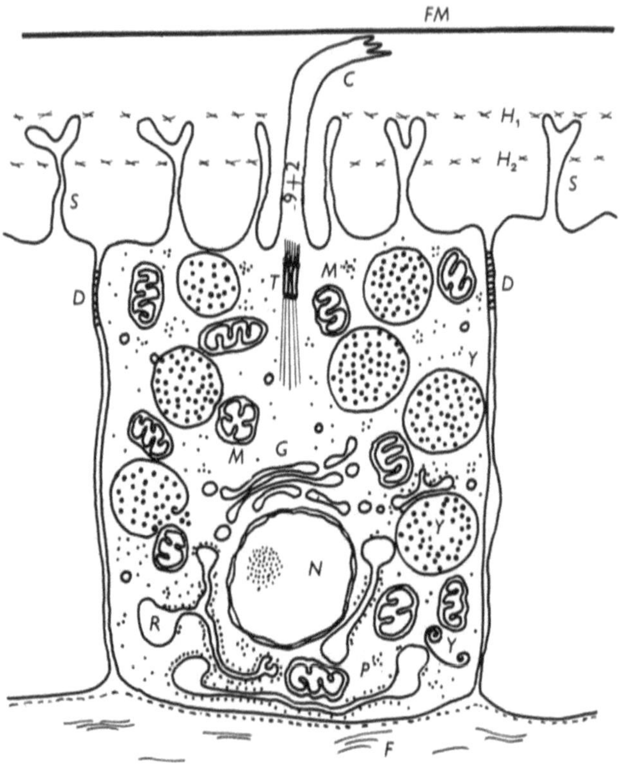

Fig. 8.6. Electron microscopy shows principal structures of blastula wall just before hatching. Sketch shows internal structures of 1 cell and its relationship to adjacent cells and to fertilization membrane (FM). To base of body (T) of cilium (C) is attached customary group of 9 plus 2 filaments extending along cilium and bunch of rootlet fibrils within cells. H1 and H2 indicate hyaline membrane condensations. Cell processes (S) extend beyond H2 to second layer H1. Bands of septate desmosomes (D) encircle each cell. M, mitochondria; Y, yolk granules, some opened as at Y prime showing membrane curling scroll-like; N, nucleus; G, golgi cluster; R, reticulum of particle-covered membranes; P, ribosomes; F, fibrils of inner coat bordering blastocoele. (From WOLPERT and MERCER, 1963)

elements. Graft experiments have shown that these differences may be the basis for the physiologic polarity of the blastula wall (HÖRSTADIUS, 1928). As mitotic cells round up on the hyaline layer and cell pulsatory activity is restricted to the inner surfaces, the relationship between cell-surface polarization and cell behavior in the blastula becomes clear (WOLPERT and MERCER, 1963).

The different cell contacts in the inner and outer ends of the cells, the strong hyaline layer attachments, and the delicacy of the fuzzy layer do not bar the mesenchymal elements and suggest that the inner contact points are more fragile than the outer ones (GUSTAFSON, 1963, 1969). Outer contact disruption, occasional disruption of all cell contact, contact stabilization, and increased contact surface are characteristic morphogenetic features.

8.4 Morphogenetic Cell Behavior in the Ectoderm

Intensified inter-cell contact causes increased localized thickness and curvature of the cell sheet (Fig. 8.5c) if the contacts with the supporting membrane covering the outer, convex surface remain unchanged (GUSTAFSON and WOLPERT, 1962, 1963b, 1967). In addition to light microscopic and preliminary electron microscopic studies further ultrastructural observation and direct measurement of cell adhesion strength are required.

Inter-cell attachments increase in many regions of the ectoderm. According to observations on the animal plate of slightly compressed larvae, the cells appear to zip together from the outside in, thickening and increasing the curvature of that region (Figs. 8.2, 8.7). The extremely rounded animal plate in radialized larvae may become completely detached (Fig. 8.7A). The increased, outward curvature of the animal plate causes the stomodeum region to turn inward, forming a pocket; other activities may also be involved (GUSTAFSON and KINNANDER, 1960;

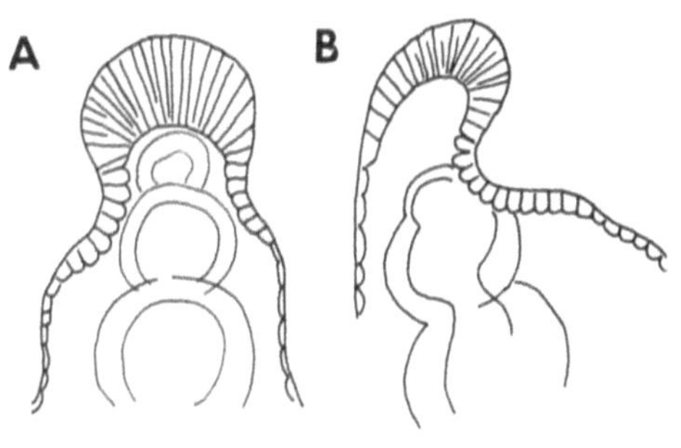

Fig. 8.7 A and B. Relationship between increased cell contact and cell-sheet curvature. (A) Animal plate of radialized larva. (B) Median larval section shows thickened and sharply curving animal plate. Oral invagination is in contact with coelomic rudiment. (From WOLPERT and GUSTAFSON, 1962)

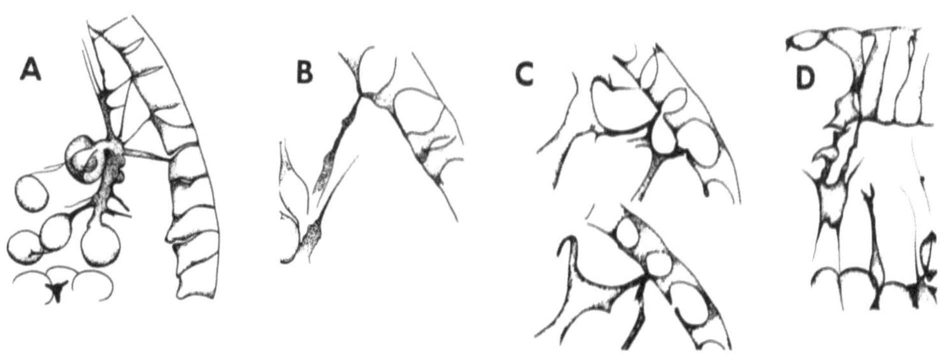

Fig. 8.8 A-D. Mesenchymal pseudopods attached to ectoderm. (A) Those from syncytial cluster of primary mesenchymal cells attached to ectoderm in region of developing ventral ciliated band. (B) Archenteron tip pseudopods. (C) Secondary mesenchymal cells. (D) Pseudopods attached to animal plate where cells were partly separated by larval compression. (From GUSTAFSON, 1963)

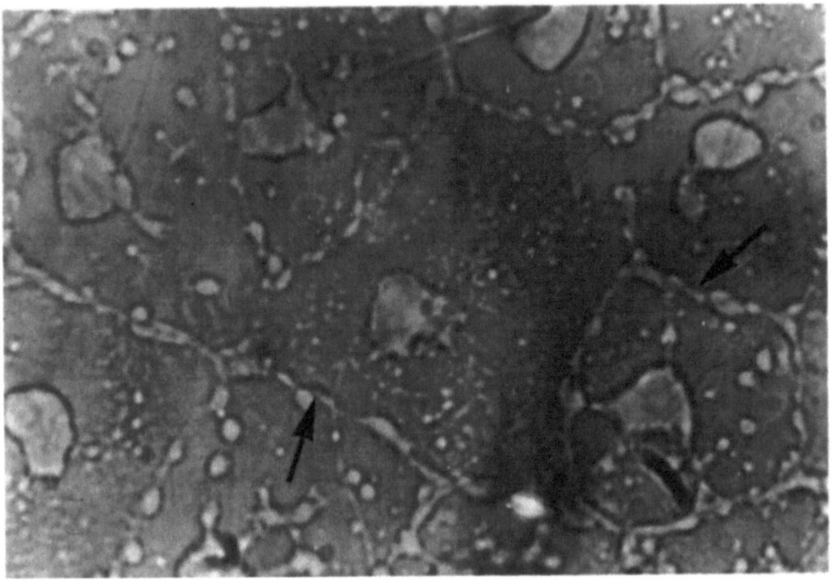

Fig. 8.9. Surface view of thin ectoderm from anal side of pluteus larva of *Psammechinus miliaris*. Arrows indicate incomplete separation of adjacent cells (Zeiss-Nomarski interference contrast)

GUSTAFSON and WOLPERT, 1962, 1963b, 1967; GUSTAFSON, 1969a)(Fig. 8.7B). In the ciliated band region continuous with the animal plate, intercell connections and band curvature increase, flattening the ventral side (Figs. 8.2B, 8.8A).

Changes are often limited to maintaining or stabilizing inner contact points as pigment cells may occupy the space between the ciliated cells. In other parts of the ectoderm, cells may flatten when the inner contact points break, as may happen in more advanced stages of development, frequently as the result of the hydrostatic pressure of water swallowed and then ultrafiltrated from the intestine into the body cavity. Many neuropharmacologic agents affect swallowing, and cause excessive larval swelling or body shrinkage. Where the ectoderm is very thin, cells appear to be in direct contact with the cytoplasm (Fig. 8.9).

The morphogenesis of the ectoderm largely expresses variations in cell contact as the effects of sucrose show (HÖRSTADIUS, 1959, 1961). After hatching occurs sucrose cannot penetrate the body wall. If sucrose is added, it creates an osmotic effect that causes a body shrinkage and tighter cell packing.

Pluteus arm formation has been investigated by means of time-lapse cinematography (GUSTAFSON and WOLPERT, 1961b), and it was found that ..."arm elongation is connected with growth of the skeleton. There is on the other hand no evidence of autonomous elongation. Growth of the skeleton results in tension in the arm ectoderm resulting in its extension. The tension is transmitted to the ectoderm by means of a plug of mesenchyme cells attached to the skeleton tip and constantly following its extension. Consequently, when the plug is absent and the skeleton tip projects strongly, no extension occurs. In addition to extension

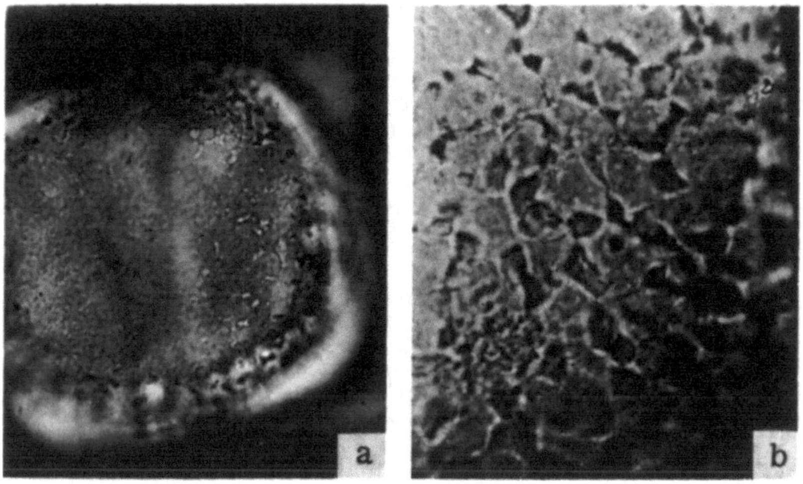

Fig. 8.10 a and b. (a) Ventral view of dark, irregularly delineated spots indicating spontaneously formed spaces between ectoderm cells of gastrula of *Psammechinus miliaris*. Such spaces generally appear near cells packed in ciliated bands. (b) Osmotic swelling causes spaces between cells of blastula wall; higher magnification than (a). Pseudopodal activity tends to close the spaces. (From GUSTAFSON and WOLPERT, 1962)

by the skeleton, the pattern of arm formation depends on an apparent invagination and anchoring of the ectoderm in various regions"....

At times ectodermal cells are extremely active (Fig. 8.10). The pseudopod-like structures between them, very similar to the ones in osmotically swollen blastulae, result from mechanical stress and extension of the cell sheet in connection with ventral flattening of the larva (GUSTAFSON and WOLPERT, 1962). This cell activity is normally checked by close contact between cells in the cell sheet. At present it seems unlikely that the local contraction of the inner and outer ends of individual cells are more important morphogenetically than changes in cellular adhesion.

8.5 Morphogenetic Cell Behavior in the Mesendodermal and Mesenchymal Region

The cell behavior in this region of the larva is partly the reverse of that in the animal region. Cell contacts in the most vegetal region are broken at an early stage by cell activity rather than by external forces such as hydrostatic extension of the body wall. Decreased adhesion may allow tension near the surface, such as tension caused by microtubular elements, to round the cells (Fig. 8.4). Cell-contact breaking may also be aided by irregular movements of the cells (GUSTAFSON and WOLPERT, 1963b, 1967).

If the cells are still attached to the supporting membrane (in Fig. 8.5 it is the hyaline layer) invagination would occur for geometric and mechanical reasons. If the cells are rounded and detach from the supporting membrane, they would tend to leave the sheet individually (GUSTAFSON and WOLPERT, 1963b, 1967). Evidence indicates that morphogenetic model is applicable.

The acid polysaccharides surrounding the inner ends of cells in the vegetal region may possibly make the cells slippery and break the cell contacts (IMMERS and RUNNSTRÖM, 1965). Acid polysaccharides inserted into the cell membrane may cause it to expand, which of course would affect subsequent cell movement. It is possible that they cause redistribution of the cytoplasm as seems to happen in blastula cell polarization (WOLPERT and MERCER, 1963).

The motility of vegetal cells increases markedly. At hatching, it is restricted to the outer pole of the cells and is linked to the activity of the cilia. In zone 1 and in the adjacent region (Fig. 8.1), the center of motility shifts to the opposite end of the cells where pulsatory lobes and filopods appear (GUSTAFSON and KINNANDER, 1956, 1960). Ultrastructural investigations have shown that microtubular elements associated with the roots of the cilia are displaced inward where it is possible that they round the cells and cause pulsatory activity (GIBBINS, TILNEY, and PORTER, 1969; TILNEY and GIBBINS, 1969). The pseudopods that later appear contain similar microtubular elements. But the cause of vegetal motility, whether microtubular elements redistributed by acid polysaccharides; monoamines; or cell-surface movement due to decreased adhesion between the cells, remains obscure.

The pulsatory activity that begins in zone 1 and the extensive cell-attachment breakage there makes it possible for the cells to enter the blastocoel. After that, new mechanisms are apparently necessary for further cell migration. Thin, bristle-like filopods shoot out from the pulsatory cell surfaces as fast as 30μm in 3 minutes. They explore the blastocoel wall, sometimes bend, and may show transient, wave-like motions at their base. If the pseudopodal tip does not attach itself to the blastocoel wall, the filopod may collapse and retract, although blastocoel wall probing may continue. The shortening of filopods attached firmly to the blastocoel wall is the basis for primary mesenchymal cell migration.

After being dispersed, the primary mesenchymal cells form a ring in the lower part of the ectoderm. This ring thickens ventrolaterally and forms branches toward the animal pole. The cell filopods fuse into syncytial, cable-like structures from which filopodal branches continue to explore the blastocoel wall and shorten (OKAZAKI, 1960, 1965; GUSTAFSON and WOLPERT, 1961c; WOLPERT and GUSTAFSON, 1961a; GIBBINS, TILNEY, and PORTER, 1969; HAGSTRÖM and LÖNNING, 1969). The filopodal contacts with the ectoderm are not permanent but slowly continue to break and reform (GUSTAFSON and WOLPERT, 1961c; WOLPERT and GUSTAFSON, 1961a).

The mesenchymal cell arrangement pattern seems to be due to random pseudopodal exploration of substrates that have varying degrees of adhesiveness (GUSTAFSON, 1961, 1963, 1964; GUSTAFSON and WOLPERT, 1961c, 1963b, 1967). The cells line up, presumably along the zones where the attachment possibilities are greatest (Fig. 8.2B), although ectoderm adhesiveness is by no means restricted to sharp zones. Experimental tissue cells were spread on a palladium surface shadowed on cellulose acetate to which the cells were nonadherent (CARTER, 1965). Graded shadowing caused varying surface adhesiveness. All the cells migrated to the most adhesive end of the substratum where masking of the cellulose acetate was strongest.

Formation of the complicated skeletal system of the larva reflects primary mesenchymal cell distribution as the skeleton is laid down within the syncytial structures (OKAZAKI, 1960, 1965; WOLPERT and GUSTAFSON, 1961a; GUSTAFSON and WOLPERT, 1963b, 1967). The degree of skeletal complexity corresponds to the degree of filopod mobility (v. UBISCH, 1932, 1937, 1950, 1957). In transplantation experiments the skeletal structure of the larva was determined by the species of the mesenchyme. When the viscosity of the filopodal cytoplasm was reduced experimentally, its motility and the skeletal branching were increased.

The length of the period of high filopodal motility may also be important (GUSTAFSON, 1969a). If the filopodal contacts break and reform extensively in advanced stages, then filopods may attach themselves at sites that, for one reason or another, were not available to the earlier pseudopods.

The conditions under which the cells in zone 1 are released seem to occur again later in adjacent zones, particularly in zone 2 (Fig. 8.1). The inner ends of the cells are covered by polysaccharides (IMMERS, 1961), pulsatory activity is pronounced, and the inner contact points between the cells tend to break. The cells will not be released, though, because they are still attached to the hyaline layer and joined by the outer contact points (Fig. 8.5). Their outer attachments will make the cells pear-shaped, and invagination will occur.

The roots of early invagination lie in the invaginating material as shown by experiments in which the invaginatory movements still occur in a microsurgically isolated vegetal fragment (MOORE and BURT, 1959). The change in cell shape can bring about only a hemispherical invagination. In animals in which the archenteron region is large enough to fill the blastocoel, invagination still occurs, but the sea-urchin larva requires that the small archenteron rudiment be extended.

After the pronounced pulsatory activity in zone 2, filopods form at the archenteron tip, suggesting a relationship between the filopods and the pulsatory lobes, or lobopods, that they replace. As in zone 1, pulsatory activity is followed by filopod formation. By exploratory movements, attachments to the blastula wall, and filopodal shortening and detachment followed by new cycles of filopodal activity, the archenteron rudiment becomes tubular (GUSTAFSON and KINNANDER, 1956, 1960; GUSTAFSON and WOLPERT, 1963b, 1967)(Fig. 8.3). Treatment with seawater having reduced calcium content or with pancreatin has been shown to cause defective gastrulation in which the filopods are detached from the ectoderm (DAN and OKAZAKI, 1956).

In gastrulation in vegetalized larvae the excess, protruding archenteron has little to do with invagination failure (GUSTAFSON and WOLPERT, 1961a, 1963b). The ectoderm is undersized, but it is delineated from the archenteron rudiment in an ordinary way, and the invagination of the most vegetal part of the mesendoderm rudiment may start as in a normal larva (Fig. 8.2A,G,H). The term exogastrulation should be avoided because it may imply invaginatory movement in the wrong direction.

When the archenteron extends through the blastocoel, the filopod-forming cells detach and disperse, apparently with the aid of the filopodal attachments (GUSTAFSON and KINNANDER, 1956). The cells released from the archenteron tip form the secondary mesenchyme. Most of the cells form pigment granules or vacuoles and are called pigment cells. The filopods bend the archenteron toward the oral region (GUSTAFSON and KINNANDER, 1960).

While the cells that form the coelom show pulsatory and filopodal activity (Fig. 8.11), most of them remain attached to the cell sheet (GUSTAFSON and WOLPERT, 1961a, 1963a, 1963b, 1967; CZIHAK, 1962b, 1965a; GUSTAFSON, 1969a).

A

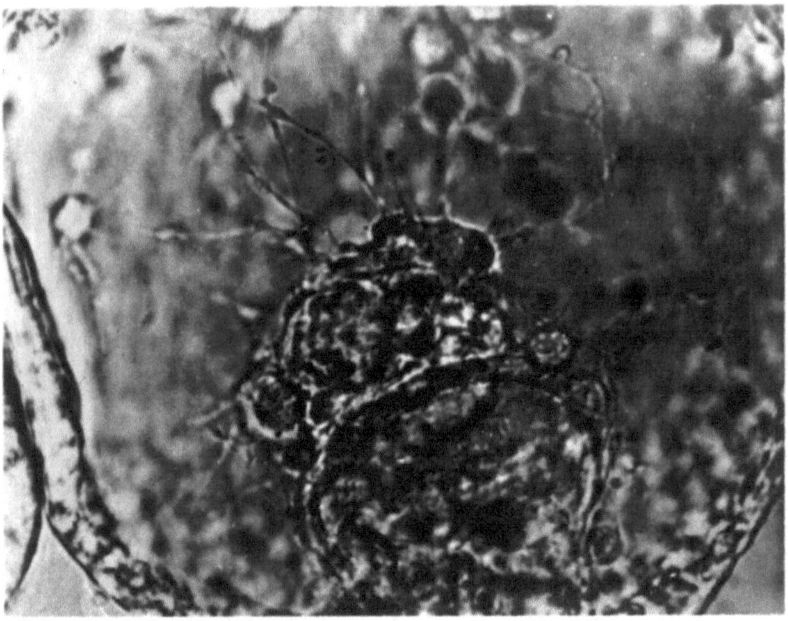

B

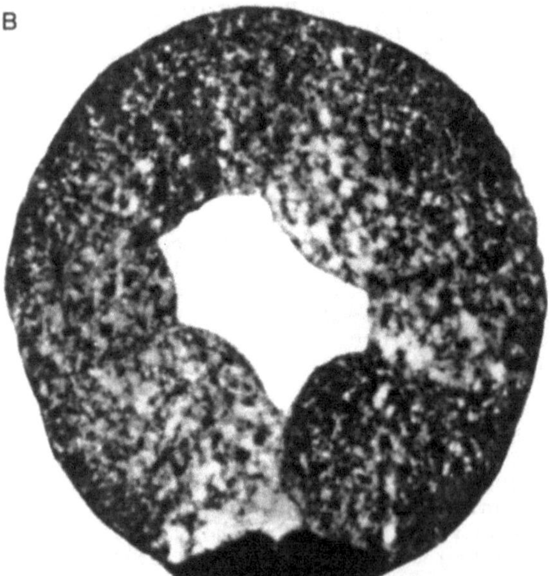

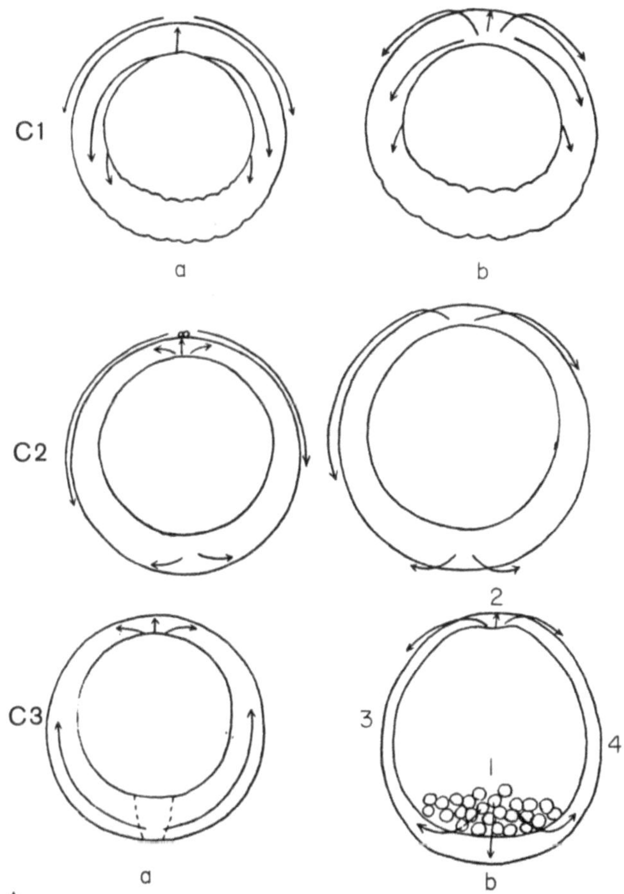

◀ Fig. 8.11 A-C. (A) Unpaired coelom rudiment on top of partially invaginated archenteron in vegetalized *Psammechinus miliaris* larva with radiating pseudopods. (B) Section of 16-cell stage of *P. miliaris* stained with Giemsa stain. Mesomeres and macromeres are in prophase. Dark micromeres with high RNA content are in interphase. (C) Reduction gradients of sea-urchin species in various stages of development. Reduction intensity declines in directions of arrows. (C_1) Early blastula: a, *Strongylocentrotus*; b, *Dendraster*. (C_2) Late blastula: *Patiria*; reduction more rapid on what is probably ventral side. (C_3) a, late blastula, *Strongylocentrotus*; b, mesenchymal blastula, *Strongylocentrotus*. (From GUSTAFSON and WOLPERT, 1963b; AGRELL, 1958; CHILD, 1941)

Between the parts of the invaginated region, constrictions occur (GUSTAFSON and WOLPERT, 1967). It is possible that cellular adhesion and the other factors that determine contact between archenteron tube cells are initially the same from the archenteron tip to the ectoderm. Cell packing would maintain the contact between them, but some reduction of adhesion or other contact-promoting factor between cells at the archenteron tip would increase contact between the cells below it and constriction would occur. This would be similar to a blastula cut in half in which contact between cells at the free edge is increased. The opening constricts and a hollow sphere develops.

8.6 Cytologic Aspects of Filopodal Attachments

The primary mesenchyme is attached at thick-walled zones, the inside of which is covered by polysaccharides believed to aid the pseudopodal attachments (MOTOMURA, 1960; OKAZAKI, FUKUSHI, and DAN, 1962; IMMERS and RUNNSTRÖM, 1971). The filopods attach at the inner contact points between the cells (GUSTAFSON, 1963)(Fig. 8.8) where an adhesive holding the cells together may hold the filopods together too (OKAZAKI, FUKUSHI and DAN, 1962).

Another possibility is that the inner contact points are good substrates for attachment because the contacts tend to disrupt and the overlying fuzzy layer may break under stress in the body wall, exposing clean surfaces; it might be the clean surface that serves as substrate for pseudopods (GUSTAFSON, 1969a). Mesenchymal cells do not attach to the animal plate, and the strong attachments between ectodermal cells leave little chance for spontaneous, partial ruptures. Filopods, however, attach to the animal plate after it ruptures in larvae compressed under a cover slip (Fig. 8.8D).

It is also possible that the inner contact points are composed of cell protrusions that meet and form a meshwork of connections with a barrier of electrostatic repulsion weaker than a continuous layer. Most filopodal attachment is at the tip, possibly because thin processes easily break the electrostatic repulsion barrier (GUSTAFSON, 1969a).

In all these possibilities the inner contact points are positive, which explains why the primary mesenchymal cells attach along the ciliated band where the number of contact points per unit volume is higher than in thinner-walled regions.

The inner cell junctions in the archenteron wall do not serve as a substrate for pseudopodal attachment in early development, possibly because the cell sheet there is not subjected to mechanical stress until the motility of the primary mesenchymal cells decreases. It has also been found that the wall of intestinal compartments extended by water is a suitable substrate for neurons of high motility. Cell-contact specificity may be less important than is often assumed (GUSTAFSON, 1969a).

8.7 Morphogenetic Cell Behavior during Further Development of the Mesendoderm

The later morphogenesis of the archenteron includes coelomic formation and subdivision into 2 compartments; madreporic duct extension from the left coelom; from coelomic pseudopods muscular system formation around the esophagus; mouth formation; and extension of thick-walled intestinal compartments into thin-walled vesicles (GUSTAFSON and WOLPERT, 1963a, 1963b, 1967; CZIHAK, 1965; GUSTAFSON, 1969a). Despite considerable increase in the morphologic complexity of the larva during these events and during formation of the larval primitive nervous system, no new morphogenetic cell activity seems to be involved.

8.8 Biochemical Differentiation of the Blastula and Gastrula

Graded biochemical differences along the animal-vegetal axis of the larva, including respiratory activity, RNA-formation, and protein synthesis will be discussed. Further information about regional and temporal biochemical differentiation is available in BÄCKSTRÖM (1957), CZIHAK (1962a, 1962b), MONROY and MAGGIO (1964), LALLIER (1964), RUNNSTRÖM et al. (1964), and FUDGE-MASTRANGELO (1966).

One of the earliest regional changes is the distribution of cytochrome oxidase activity (CZIHAK, 1963). The amount of RNA in the micromeres increases during early interphase (Fig. 8.11B) and decreases toward prophase, probably as the result of altered cell metabolism (AGRELL, 1958). Similar changes may occur in all embryo cells, but in the macromeres and mesomeres increased RNA, diluted in more cytoplasm, is not demonstrable by cytochemical means. Less cytoplasm in relation to nuclear material may give the micromeres a higher nucleic acid metabolism than the other embryo cells (AGRELL, 1958; COWDEN and LEHMAN, 1963). In the regional differentiation in early stages, the gradient in mitotic activity may reflect biochemical differences between the animal and vegetal regions and aid further biochemical changes of the embryo cells (AGRELL, 1958).

The distribution of dehydrogenase activity has been studied by stain reduction in vitally and supravitally stained material kept under anaerobic conditions (CHILD, 1936, 1941a; HÖRSTADIUS, 1952, 1953; LALLIER, 1958, 1964; BÄCKSTRÖM, 1959; CZIHAK, 1962b, 1963; MONROY and MAGGIO, 1964; GUSTAFSON, 1965, 1969b). Among the findings were an animal-vegetal gradient in reducing capability, a reduction gradient, in early cleavage stages and in the young blastula, and a vegetal-animal gradient in the late blastula (Fig. 8.11C). Marked reducing activity in early stages was found vary in proportion to the potential size of the ectoderm, determined by comparing normal, animalized and vegetalized material, while reducing activity in the late blastula varied in proportion to the potential size of the mesendoderm (HÖRSTADIUS, 1952, 1955).

It is also possible that the reducing capability is related to the rate of reoxidation of NADPH, a factor limiting the activity of the hexose monophosphate shunt (BÄCKSTRÖM, 1959). In this case the gradient may reflect the distribution of transdehydrogenases, diaphorase, or some other link in the respiratory pathway. Early regional variations of the Nadi reaction have been studied (CHILD, 1941b, 1944; CZIHAK, 1963), and the role of biosynthetic reactions involving NADPH should be considered.

Possibly related to the reduction gradients are RNA synthesis variations demonstrated using labeled adenine and autoradiography with or without section pretreatment with ribonuclease (MARKMAN, 1961a, 1961b). Except for extreme incorporation in RNA in the micromeres, the first reduction gradient and RNA labeling distribution are similar (AGRELL, 1958; COWDEN and LEHMAN, 1963; CZIHAK, 1965b). Compatible findings resulted from studies on adenine incorporation into animal and vegetal halves by means of autoradiography and Geiger-counter readings (MARKMAN, 1961c).

Experiments in which the Hale ferric acetate reagent was used to determine the distribution of strongly acidic groups indicated that the interphase nuclei in the animal region of the mesenchymal blastula and gastrula contain stainable, nuclease-sensitive material and that in other regions the staining reaction is very weak (IMMERS, 1956b; MARKMAN, 1957; RUNNSTRÖM, 1961; RUNNSTRÖM et al., 1964; IMMERS, MARKMAN, and RUNNSTRÖM, 1967, 1969; RUNNSTRÖM and IMMERS, 1970; IMMERS and RUNNSTRÖM, 1971). A negative Hale reaction indicated that most of the chromatine was condensed and repressed because of histone binding... "At early stage, the Hale staining of the interphase nuclei is not influenced by exposure to DNase or RNase. If, however, the sections are pretreated with trypsin and thereafter exposed to a mixture of DNase and RNase, the staining capacity of the interphase nuclei is abolished. In stages from the late blastula stage onwards, DNase decreases or abolishes the Hale reaction without any previous pretreatment. RNase may affect but does not alone abolish the staining. The interference from the data is that Fe^{3+} in the Hale reagent competes with nuclear proteins for reactive sites in the nucleic acids. These sites may correspond to phosphate groups".... (MARKMAN and RUNNSTRÖM, 1969).

In all cells stainability may be induced by pretreatment with an alkaline buffer. Section pretreatment with a buffer or buffered trypsin solution before the Hale procedure induces interphase nucleic staining that varies according to buffer pH, stage of development, and position in the embryo. In later developmental stages a less alkaline buffer induces stainability, which suggests that the genetic material becomes able to serve as a template in DNA-dependent RNA synthesis (IMMERS, MARKMAN, and RUNNSTRÖM, 1969).

Darkfield observations indicate that some transcription process begins in the nuclei in the animal region and that those with ultrastructural differences have varying Hale stainability (RUNNSTRÖM, 1967). Beginning in the ectoderm, the basic proteins that suppress protein synthesis at the translational level in the cytoplasm decrease in concentration in the late blastula and gastrula (BÄCKSTRÖM, 1965).

The gradient in RNA synthesis seems to be accompanied by variable protein synthesis intensity. It was shown by autoradiographic studies and Geiger-counter measurements of labeled leucine incorporation into animal and vegetal halves isolated at the 16-cell stage (MARKMAN, 1961a, 1961b, 1961c). This conclusion was later contradicted (BERG, 1965) and reaffirmed (MARKMAN, 1967).

The gradients in the RNA synthesis are also accompanied by enzyme activity and number of mitochondria that increase at gastrulation and are reduced in vegetalized material. Observations in vital-stained material are in line with this (GUSTAFSON, 1965, 1969b; GUSTAFSON and LENICQUE, 1952, 1955)(Fig. 8.12A). The first peak in number of mitochondria stainable with Nile blue sulphate is highest in the ectodermal region in proportion to the size of the potential ectoderm.

Micromere implantation in animal halves isolated at the 16-cell stage has been shown to inhibit mitochondrial development in the vegetal region of the fragment (LENICQUE, HÖRSTADIUS, and GUSTAFSON, 1953), although variable mitochondrial distribution in vital-stained material has been known to be lacking (SHAVER, 1955, 1956), possible because Nile blue sulphate is hard to dissolve in high salinity seawater in which the calcium-salt of the stain readily precipitates, or because the stainability of lipid droplets, less than that of mitochondria, may have interfered with the counts.

Fixed material investigated with the light microscope or the electron microscope failed to show mitochondrial gradients (MORI et al., 1958; BERG, TAYLOR, and HUMPHREYS, 1962; BERG and LONG, 1964). Reported to undergo temporal as well as regional change in appearance, mitochondrial gradients have been confirmed using Janus green B stain (CZIHAK, 1961, 1962b, 1963)(Fig. 8.12). Also substantiated was the finding (CHILD, 1941b) that cytochrome oxidase is most active in the oral region of the ectoderm (CZIHAK, 1963)(Fig. 8.12B). Regional differences in cytochrome oxidase activity probably parallel those in the blastula and are visible during early cleavage, indicating that those in mitochondrial populations in the late blastula may be traced back to very early stages of development (CZIHAK, 1963).

Early cytochrome oxidase activity and mitochondrial development have been shown to be important to morphogenesis. Cytochrome oxidase inhibited 4 to 8 hours after fertilization causes vegetalization similar to that caused by LiCl (CZIHAK, 1963). Chloramphenicol, an agent known to inhibit protein synthesis of mitochondria, also has a vegetalizing effect (HÖRSTADIUS, 1963; LALLIER, 1962).

The heavy vital staining of mitochondria may be due to active stain uptake followed by precipitation as a calcium salt. Such inhibitors of respiration and oxidative phosphorylation as amytal, 2,4-dinitrophenol, and pentachlorophenol have been shown to inhibit stain uptake. The mitochondrial gradients may reflect the distribution of particularly active mitochondria, while nonstainable and weakly stainable mitochondria in the vegetal region may not yet be able to carry out oxidative phosphorylation at a sufficient rate. This is supported by microrespirometric studies comparing the effect of 2,4-dinitrophenol on animal and vegetal fragments (LINDAHL, 1940; IMMERS and RUNNSTRÖM, 1960; CZIHAK, 1963; GUSTAFSON, 1965, 1969b; DE VINCENTIIS, HÖRSTADIUS, and RUNNSTRÖM, 1966; DE VINCENTIIS and RUNNSTRÖM, 1967).

In vegetalized larvae the relationship between density peaks of stainable mitochondria populations and the body wall is constant. Ectodermal and endodermal cells are joined by double contact points while wall cells in the intermediate regions are joined by single contact points. A similar relationship between nucleic stainability by the Hale method and the wall structure has been described (IMMERS and RUNNSTRÖM, 1971).

Probably due to de novo synthesis, abrupt enzymatic differences between adjacent regions occur during later differentiation. For example, the prospective ectoderm shows very little NAD-diaphorase or cytochrome C-reductase activity, while the primary mesenchymal cells show great activity from the time they enter the blastocoel until the skeleton is formed (CZIHAK, 1962b)(Fig. 8.12C). When the coelomic rudiment divides into 2 sacs and the esophagus separates from the stomach by constriction, the esterases show increased unspecific activity that peaks at about 32 hours of age (CZIHAK, 1962b)(Fig. 8.12D).

8.9 Biochemical Activity and Morphogenetic Movement

Morphogenesis in the vegetal region seems to be based largely on initiating cell motility and other microtubular activity, starting in zone 1 (Fig. 8.1) and spreading to the adjacent zones. The muscular elements of the coelom-esophagus complex (Fig. 8.13) are formed by filopods (Fig. 8.11), the rapid contractions of which may have the same source as slow filopodal activity in early development and the pulsatory movements.

Among a number of neuropharmacological agents studied, serotonin (5-hydroxytryptamine) was shown to have a strong, stimulating effect on so-called coelomic movement: contraction of the muscular strands that pull the coelomic sacs and esophagus downward (GUSTAFSON and TONEBY, 1970, 1971). Most other biogenic amines inhibit coelomic movement. Antiserotonins counteract stimulation by serotonin, and one of them (gramine) brings about a rapidly reversible effect. The serotonin precursors tryptophan and 5-hydroxytryptophan are strongly stimulating after some delay. Experiments with lipid-soluble cholinolytics and physostigmine indicate that acetylcholine is involved in muscular

A1 A2

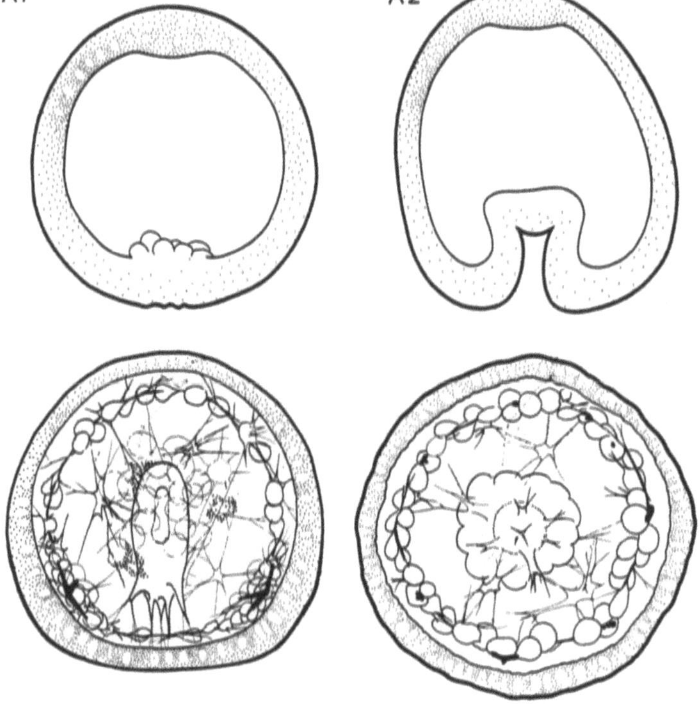

Side view

Seen from the apical pole

B

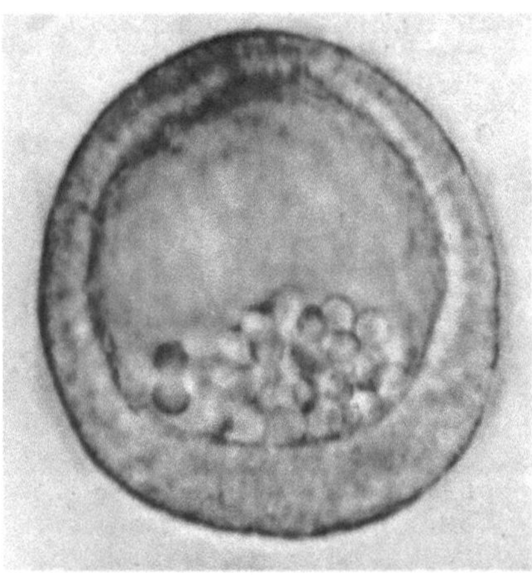

Side view

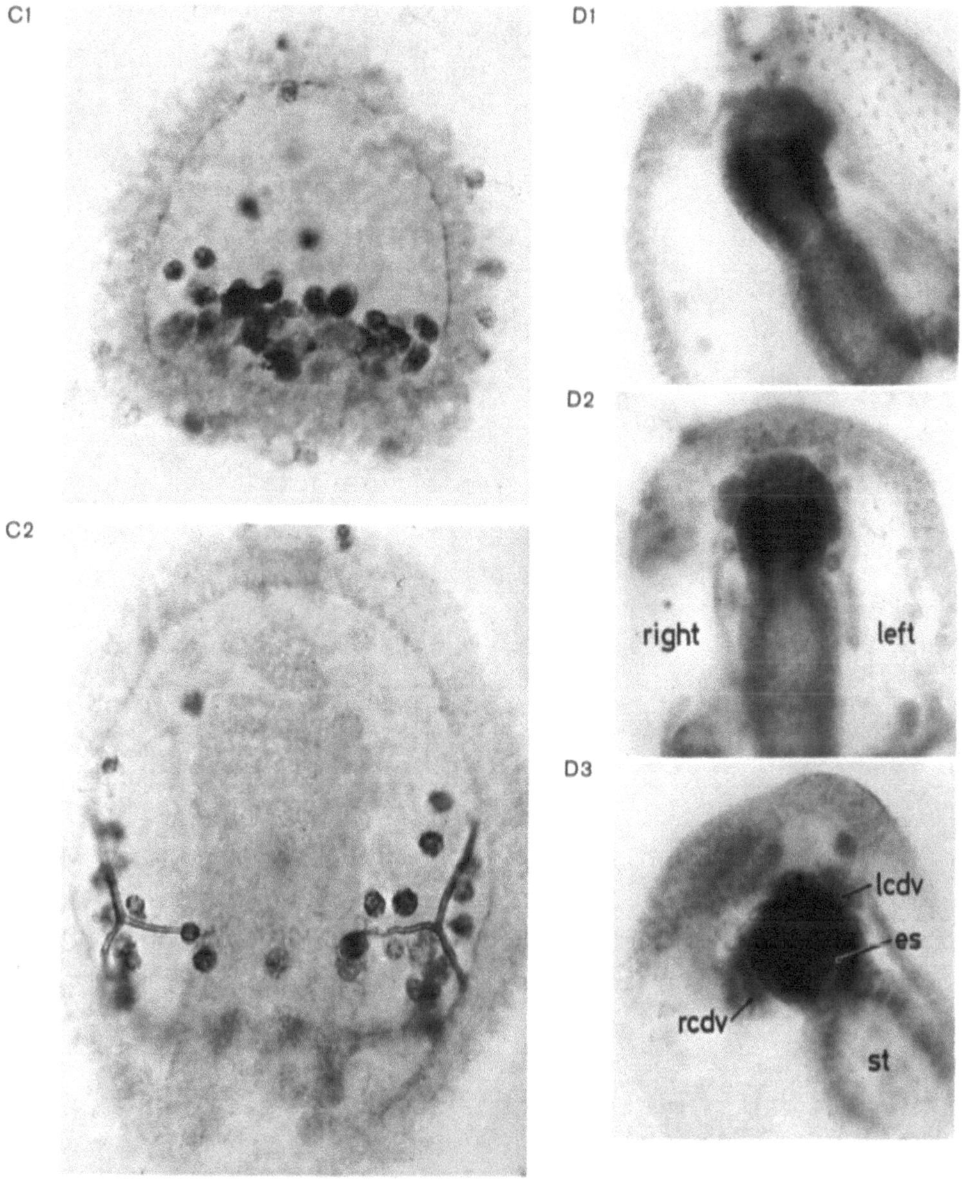

Fig. 8.12 A-D. (A 1 and 2) Mitochondria in early development of *Psammechinus miliaris* stained with Janus green B. (1) Early mesenchymal blastula. (2) Early gastrula; both side and top view. (B) Nadi-reaction gradient in 12-h mesenchymal blastula of *Paracentrotus lividus* is most active in oral part of animal region. (C 1 and 2) Cytochrome C-reductase activity in mesenchymal blastula and gastrula of *P. lividus* is confined, almost entirely to primary mesenchymal cells as they enter blastocoele and form the skeleton. (D 1 - D 3) Unspecific esterase activity when coelom subdivides into 2 sacs and esophagus separates from intestine; (1) lateral view; (2) dorsal view; (3) oral view, rcdv, right coelomic diverticulum; st, stomach; es, esophagus; lcdv, left coelomic diverticulum. (From CZIHAK, 1962b, 1963)

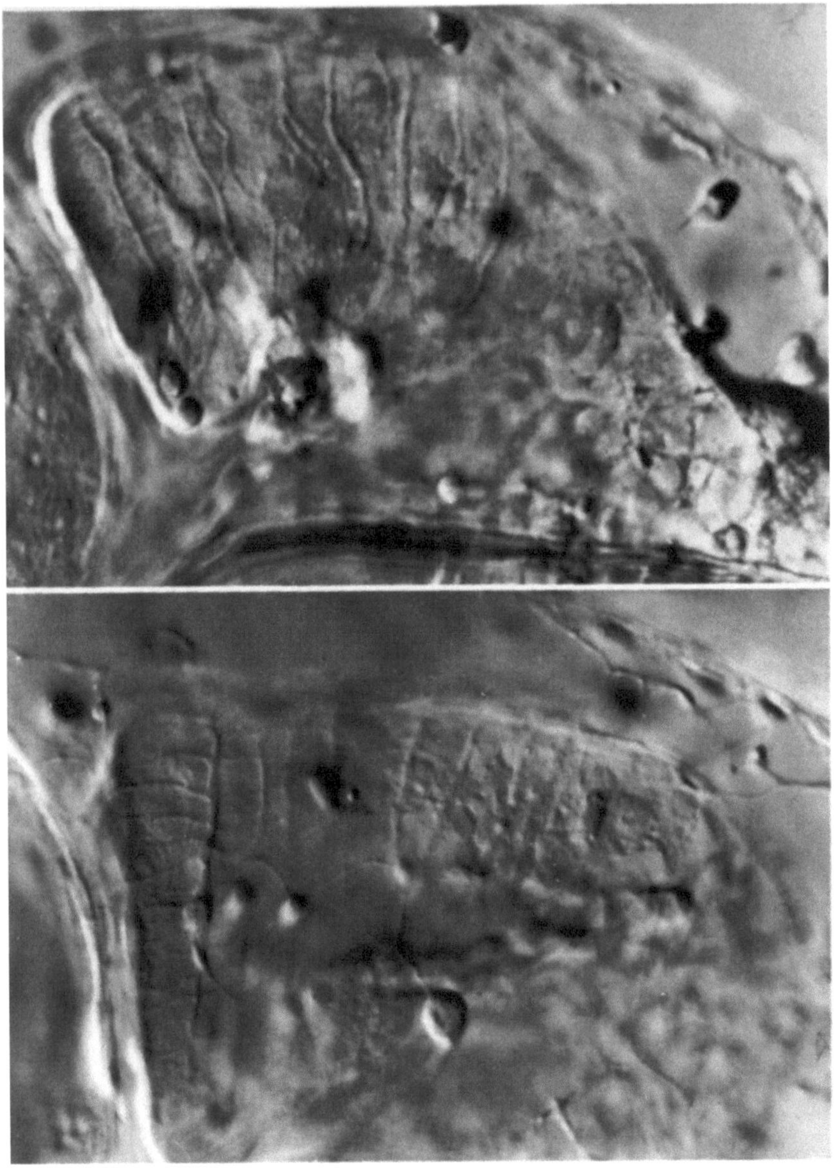

Fig. 8.13. Muscular strands around esophagus of 4-day-old plutei of *Psammechinus miliaris* (Zeiss-Nomarski interference contrast; from GUSTAFSON, 1969a)

movement while quaternary nitrogen cholinolytics and cholinesterase inhibitors are not, probably because they are unable to penetrate.

It is entirely possible that these neuropharmacological agents determine morphogenetic movement, because serotonin, acetylcholine, and acetylcholine esterase occur in early development (AUGUSTINSSON and GUSTAFSON, 1949; BUZNIKOV, 1963, 1966, 1967; BUZNIKOV, CHUDAKOVA, and

ZWEZDINA, 1964; BUZNIKOV and BERDYSHEVA, 1966; BUZNIKOV, ZWEZDINA, and MARKEEVA, 1966; CHUDAKOVA, BERDYSHEVA, and BUZNIKOV, 1966; BERDYSHEVA and MARKOVA, 1967; BUZNIKOV et al., 1968). Studies on the effect of serotonin antagonists and cholinolytics were begun before hatching so that permeability changes would not interfere (GUSTAFSON and TONEBY, 1970, 1971). A number of serotonin antagonists resulted in delay of release of the primary mesenchyme and invagination. The onset of the second phase of gastrulation was sometimes delayed but once started might proceed at a normal rate. In several instances exogenous serotonin weakened the inhibitory effect of the antiserotonins.

If it is true that serotonin formation in the vegetal region initiates cell release, it follows that morphogenesis is hampered by interference with the enzymes that biosynthesize serotonin: tryptophan hydroxylase and 5-hydroxytryptophan decarboxylase. Able to chelate cupric and ferric ions, an inhibitor of a hydroxylation reaction in the synthesis of dopamine inhibits tryptophan hydroxylation as well. Such an inhibitor used against the subsequent decarboxylation of 5-hydroxytryptophan strongly suppresses gastrulation. It was also found that exogenous 5-hydroxytryptophan reduces inhibition caused by intended hydroxylase inhibitor diethyldithiocarbamate (Fig. 8.14).

In addition to activating muscular movement in the pluteus larva, serotonin may also initiate cell motility during morphogenesis. Monoamines have been shown to be important in early cleavage (HOFFMANN and HOFFMANN, 1958; BUZNIKOV et al., 1970).

Morphogenetic results agree with the occurrence and quantitative variations of serotonin in early development (BUZNIKOV, CHUDAKOVA, and ZWEZDINA, 1964). During early gastrulation a series of peaks in the larval serotonin level have been shown by pharmacological bioassay (Fig. 8.15) and by thin-layer chromatography (TONEBY, personal communication). Fluorescence-histochemical studies reveal in the early archenteron a yellow-green fluorescence possibly due to the yellow fluorescence of serotonin against a green fluorescent background. Whether dopamine is involved is undetermined.

While serotonin antagonists are particularly active during early morphogenetic changes, secondary invagination, once it begins, may proceed at a normal pace. It is possible that serotonin plays a minor role in controlling secondary invagination. The pseudopodal elements of the archenteron tip contract as soon they attach, possibly because they receive from the ectoderm an impulse similar to a synaptic impulse. Because the ectoderm contains cholinesterase, that may play a role for the activity of the cilia, and may contain acetylcholine that influences muscular activity of the larva, it is conceivable that a cholinergic mechanism controls pseudopodal activity. Pralidoxime dodecyliodide, a lipid-soluble acetylcholine analogue, strongly inhibits gastrulation as do a number of tertiary cholinolytic substances (GUSTAFSON and TONEBY, 1970). Their inhibitory effect is most evident during the second phase of gastrulation.

Reappearing in the advanced gastrula and remaining at a high level during the early development that follows (BUZNIKOV, CHUDAKOVA, and ZWEZDINA, 1964), serotonin may be formed by the pigmented secondary mesenchyme, the cells of the archenteron tip during early invagination The red pigment granules of these cells show chromaffine reaction, and the granules often show yellow fluorescence if the monoamine-specific fluorescence-histochemical method is used (GUSTAFSON and TONEBY, 1971). Pigmentation increases when tryptophan is added to

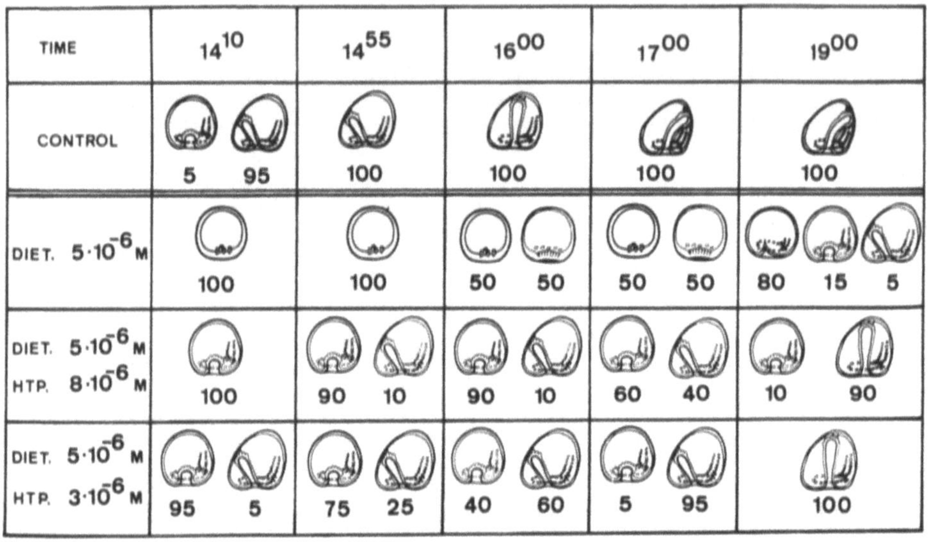

Fig. 8.14. Effects of 5-hydroxytryptophan (HTP) on larvae treated with diethyldithiocarbamate (Diet) and frequency (percent) of larvae observed. (From GUSTAFSON and TONEBY, 1970)

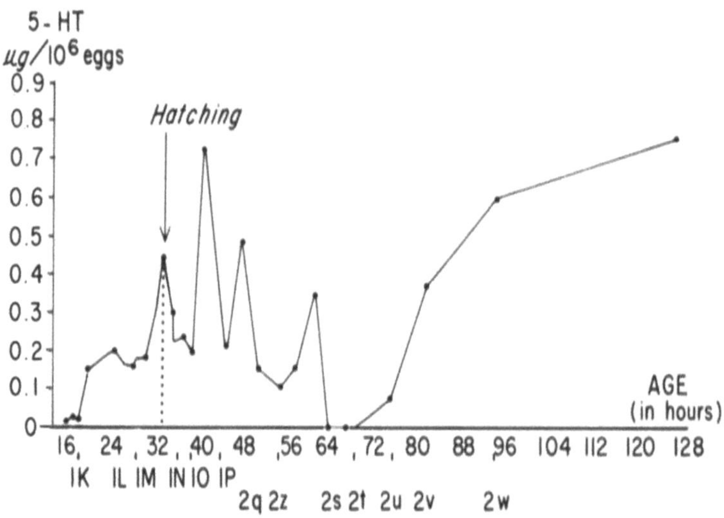

Fig. 8.15. Changes in the content of serotonin (5-hydroxytryptamine [5-HT]) in developing embryos of *Strongylocentrotus droebachiensis*. Early peaks occur when primary mesenchyme forms and steady rise later marks archenteron invagination and subsequent developmental processes. (From BUZNIKOV, 1967)

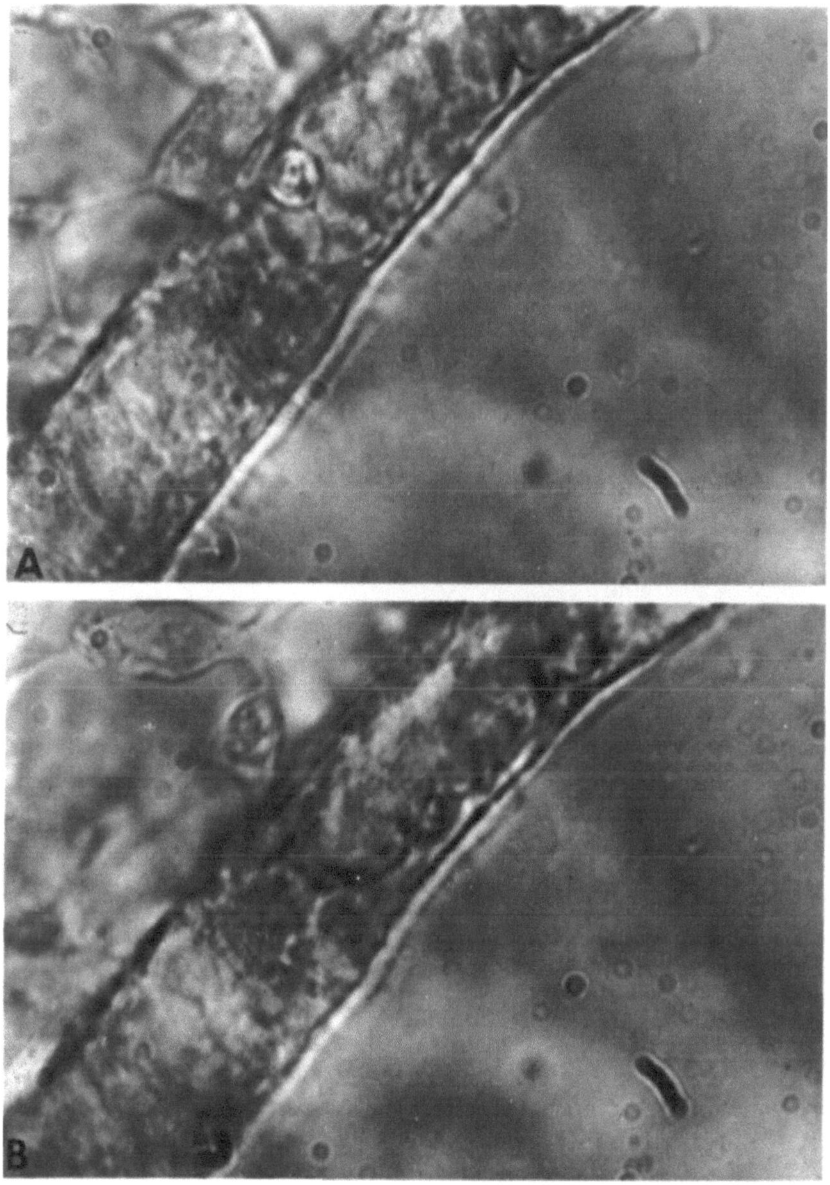

Fig. 8.16 A and B. Neurone-like cells are released from ciliated band of 70-hr *Psammechinus miliaris* pluteus. (A) Rounded cell appears in ciliated band. (B) 2 minutes later, rounded cell appears in body cavity and cell (neuron) distribution has changed. (From GUSTAFSON, 1969a)

seawater, indicating that pigmentation may be closely linked to the metabolic events related to serotonin (RYBERG, personal communication). Some of the pigment cells settle in the ciliated band of the pluteus where one of their functions may be to release serotonin to the ciliated cells. Cilia are stimulated by serotonin and inhibited by anti-

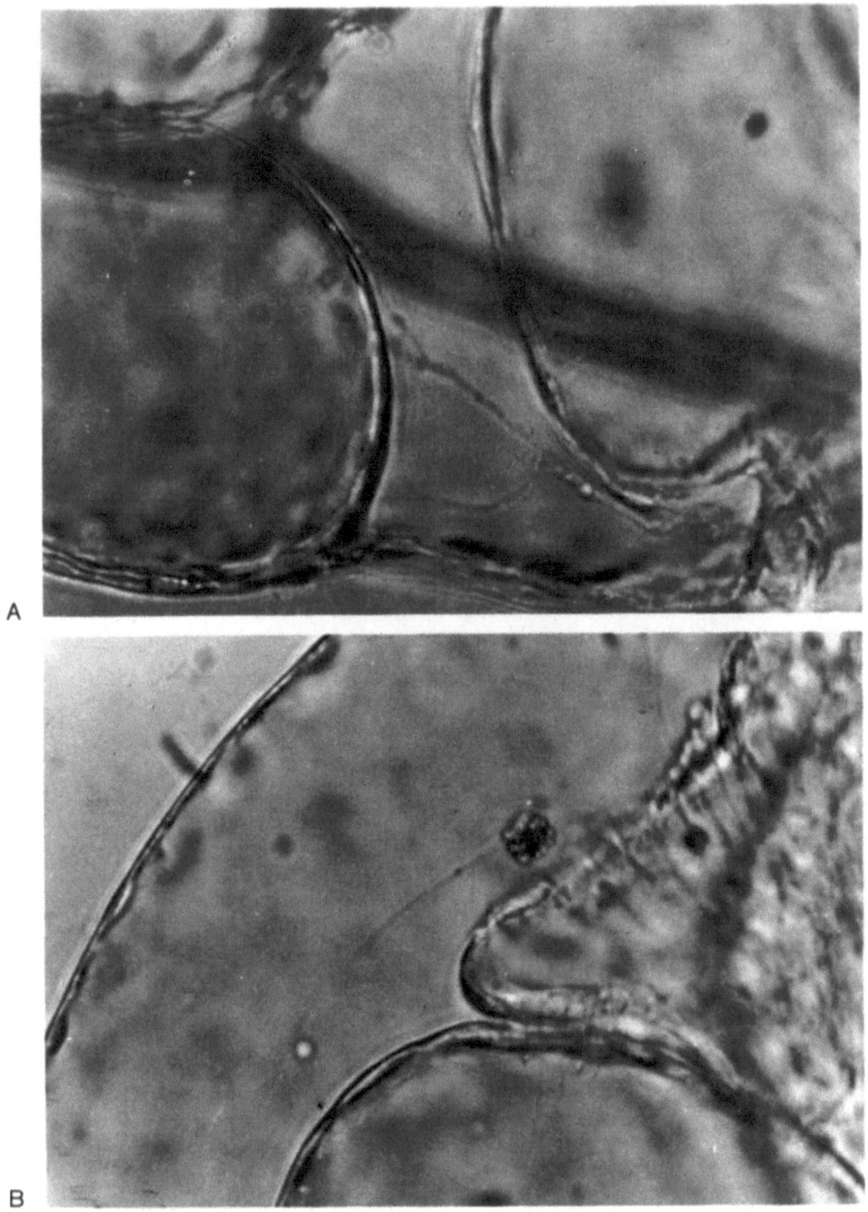

Fig. 8.17 A and B. Lateral views of neuron-like cells of *Psammechinus miliaris* plutei. (A) Neuron extends from ciliated band toward proctodeum and lower part of stomach. (B) Larval pigment cell is attached to neuron on dorsal side of esophagus. Tiny granule in axon was quickly moved backward. (From GUSTAFSON, 1969a)

serotonins. The motility is low in larvae treated with diethyldithiocarbamate without serotonin or 5-hydroxytryptophane added to the seawater. Sometimes called a scavenger, the pigment cell may produce serotonin using substrate obtained from adjacent cells.

Next to the pigment cells in the young pluteus larva, the attachments of ciliated cells break, the ciliated cells become rounded, begin to pulsate, and squeeze into the body cavity like mesenchymal cells (Fig. 8.16). They send out pseudopod-like processes that form a pattern resembling a nervous system (GUSTAFSON and TONEBY, 1971) (Fig. 8.17). The neuron-like cells are often green, from the pigment cells they apparently receive red pigment that quickly turns to green. Green pigment granules in the pigment cells are sometimes intermingled with the red. Below the epithelium in an old pluteus, pigment cells are green, while those within the epithelium are red. Together with pigment, the neuron-like cells presumably receive serotonin from the pigment cells.

The neuron-like cells soon fuse with new pigment cells in motion in the body cavity and take up pigment granules by means of axon-like processes that have a rapid, cytoplasmic flow. In other neurons in cell culture, this flow is stimulated by serotonin (GUSTAFSON and TONEBY, 1971). The granules swell, break, and change color, and under fluorescence-histochemical analysis are shown to contain serotonin.

Inasmuch as serotonin induces muscular contraction, serotonin in the neuron-like cells may serve as a transmitter. These cells are regularly distributed, and in advanced plutei their processes are connected to the coelomic muscles in an elaborate, synaptic manner. Around the endings attached to the stomach wall contractions may be observed.

METCHNIKOFF (1884) and SEMON (1888) described a close relationship between ciliated structures and other elements in echinoderm larvae. Their neural nature is particularly well documented in SEMON's studies on the development of *Synapta* during metamorphosis.

It seems very likely that neurons are formed from cells in the ciliated band and that the pigment cells and their serotonin are involved. This is borne out in the advanced gastrula and in the prism stage in which treatment with tryptophan induces the formation of enormous clumps of neuroblasts that spin out axon-like extensions as in normal larvae (RYBERG, in preparation).

References

ABERCROMBIE, M., 1964. Cell contacts in morphogenesis. Arch. Biol. (Liège) 75, 351-367.

AGRELL, I., 1953. A mitotic gradient in the sea-urchin embryo during gastrulation. Arkiv f. Zool. (2), 6, 213-217.

AGRELL, I., 1956a. A mitotic gradient as the cause of the early differentiation in the sea-urchin embryo. In: Bertil Hanström zoological papers in honour of his sixty-fifth birthday (K.G. Wingstrand, ed.), pp. 27-34. Lund: Zoological Institute, University of Lund.

AGRELL, I., 1956b. The synchronous mitotic rhythm and the effect of lithium during the early development of the sea-urchin embryo. Acta Zool. (Stockholm) 37, 53-60.

AGRELL, I., 1958. A cytoplasmic production of ribonucleic acid during the cell cycle of the micromeres in the sea-urchin embryo. Arkiv f. Zool. (2), 435-440.

AGRELL, I., 1960. Detection of a graded mitotic activity within the sea-urchin embryo through the use of oestradiol. Arkiv f. Zool. (2), 12, 411-414.

AGRELL, I., 1964. Natural division synchrony and mitotic gradients in metazoan tissues. In: Synchrony in cell division and growth (E. Zeuthen, ed.), pp. 39-67. New York: John Wiley & Sons, Inc.
AGRELL, I., 1966. Membrane formation and membrane contacts during the development of the sea-urchin embryo. Z. Zellforsch. 72, 12-21.
AGRELL, I., PERSSON, H., 1956. Changes in the amount of nucleic acids and free nucleotides during early embryonic development of sea urchins. Nature 178, 1398-1399.
ANDREWS, E.A., 1897. Hammars ectoplasmic layer. Amer. Natur. 31, 1027-1032.
AUGUSTINSSON, K.-B., GUSTAFSON, T., 1949. Cholinesterase in developing sea-urchin eggs. J. Cell Comp. Physiol. 34, 311-321.
BÄCKSTRÖM, S., 1954. Morphogenetic effects of lithium on the embryonic development of Xenopus. Arkiv f. Zool. (2) 6, 527-536.
BÄCKSTRÖM, S., 1957. Content and distribution of ascorbic acid in sea urchin embryos of different developmental trends. Exp. Cell Res. 13, 333-340.
BÄCKSTRÖM, S., 1959. Reduction of blue tetrazolium in developing sea urchin eggs after addition of various substrates and phosphopyridine nucleotides. Exp. Cell Res. 18, 357-363.
BÄCKSTRÖM, S., 1965. Basic protein in the sea urchin embryo (Paracentrotus lividus). Acta Embryol. Morphol. Exp. 8, 20-31.
BÄCKSTRÖM, S., 1969. Passage de substances radioactives entre un greffon marqué et son hôte non marqué. Etude autoradiographique de l'embryon d'oursin soumis à la transplantation. C. R. Acad. Sc. (Paris) 269, 1684-1685.
BALINSKY, B.L., 1959. An electron microscopic investigation of the mechanisms of adhesion of the cells in a sea urchin blastula and gastrula. Exp. Cell Res. 16, 429-433.
BERDYSHEVA, L.V., MARKOVA, L.N., 1967. Some cytological observations on the action of antagonists of acetylcholine, catecholamines and serotonin in fertilized sea urchin eggs. Tsitologiya 9, 912-921.
BERG, W.E., 1965. Rates of protein synthesis in whole and half embryos of the sea urchin. Exp. Cell Res. 40, 469-489.
BERG, W.E., LONG, N.D., 1964. Regional differences of mitochondrial size in the sea urchin embryo. Exp. Cell Res. 33, 422-437.
BERG, W.E., TAYLOR, D.A., HUMPHREYS, W.J., 1962. Distribution of mitochondria in echinoderm embryos as determined by electron microscopy. Develop. Biol. 4, 165-176.
BUZNIKOV, G.A., 1963. Tryptamine derivatives applied to the study of the role played by 5-hydroxytryptamine (serotonin) in the embryonic development of Invertebrata. Dokl. Akad. Nauk. SSSR 152, 1270-1272.
BUZNIKOV, G.A., 1966. Participation of acetylcholine, adrenaline and noradrenaline in preneural embryogenesis of Echinodermata. Zh. evoluz. Biokhim. Fiziol. 2, 23-30.
BUZNIKOV, G.A., 1967. Low Molecular Weight Regulators in Early Embryogenesis. Moscow: Nauka.
BUZNIKOV, G.A., BERDYSHEVA, L.V., 1966. Variations in the functional activity of neurohormones in embryos of Paracentrotus lividus. Dokl. Akad. Nauk. SSSR 167, 486-488.
BUZNIKOV, G.A., CHUDAKOVA, I.V., BERDYSHEVA, L.V., VYAZMINA, N.M., 1968. The role of neurohumors in early embryogenesis. II. Acetylcholine and catecholamine content in developing embryos of sea urchin. J. Embryol. Exp. Morphol. 20, 119-128.
BUZNIKOV, G.A., CHUDAKOVA, I.V., ZVEZDINA, N.D., 1964. The role of neurohumours in early embryogenesis. I. Serotonin content of developing embryos of sea urchin and loach. J. Embryol. Exp. Morphol. 12, 563-573.
BUZNIKOV, G.A., KOST, A.N., KUCHEROVA, N.F., MNDZHOYAN, A.L., SUVOROV, N.N., BERDYSHEVA, L.V., 1970. The role of neurohumours in early

embryogenesis. III. Pharmacological analysis of the role of neurohumours in cleavage divisions. J. Embryol. Exp. Morphol. 23, 549-569.
BUZNIKOV, G.A., ZVEZDINA, N.D., MARKEEVA, R.G., 1966. On the possible participation of serotonin and other neurohormones in the regulation of protein biosynthesis (experiments carried out on egg-cells of sea urchins). Dokl. Akad. Nauk. SSSR 166, 1252-1255.
CARTER, S.B., 1965. Principles of cell motility: the direction of cell movement and cancer invasion. Nature 208, 1183-1187.
CHILD, C.M., 1944. Developmental pattern in the starfish Patiria miniata, as indicated by indophenol. Physiol. Zool. 17, 129-151.
CHILD, C.M., 1936. Differential reduction of vital dyes in the early development of echinoderms. Wilhelm Roux Arch. Entwicklungsmech. Organismen 135, 426-456.
CHILD, C.M., 1941a. Patterns and Problems of Development. Chicago: Univ. of Chicago.
CHILD, C.M., 1941b. Formation and reduction of indophenol blue in development of an echinoderm. Proc. Nat. Acad. Sci. U.S. 27, 523.
CHUDAKOVA, I.V., BERDYSHEVA, L.V., BUZNIKOV, G.A., 1966. Changes in acetylcholine concentration during the mitotic cycle of fertilized sea urchin eggs. Tsitologiya 8, 105-107.
COWDEN, R.R., LEHMAN, H.E., 1963. A cytochemical study of differentiation in early echinoid development. Growth 27, 185-197.
CZIHAK, G., 1961. Ein neuer Gradient in der Pluteusentwicklung. Wilhelm Roux Arch. Entwicklungsmech. Organismen 153, 353-356.
CZIHAK, G., 1962a. Entwicklungsphysiologie der Echinodermen. Fortschr. Zool. 14, 238-267.
CZIHAK, G., 1962b. Entwicklungsphysiologische Untersuchungen an Echiniden. (Topochemie der Blastula und Gastrula. Entwicklung der Bilateral- und Radiärsymmetrie und der Coelomdivertikel). Wilhelm Roux Arch. Entwicklungsmech. Organismen 154, 29-55.
CZIHAK, G., 1963. Entwicklungsphysiologische Untersuchungen an Echiniden. (Verteilung und Bedeutung der Cytochromoxydase). Wilhelm Roux Arch. Entwicklungsmech. Organismen 154, 272-295.
CZIHAK, G., 1965a. Entwicklungsphysiologische Untersuchungen an Echiniden. Experimentelle Analyse der Coelomentwicklung. Wilhelm Roux Arch. Entwicklungsmech. Organismen 155, 709-729.
CZIHAK, G., 1965b. Entwicklungsphysiologische Untersuchungen an Echiniden. (Ribonucleinsäure-Synthese in den Mikromeren und Entodermdifferentzierung. Ein Beitrag zum Problem der Induktion). Wilhelm Roux Arch. Entwicklungsmech. Organismen 156, 504-524.
CZIHAK, G., 1965c. Evidences for inductive properties of the micromere-RNA in sea urchin embryos. Naturwissenschaften 6, 141-142.
CZIHAK, G., 1965d. Entoderminduktion bei Echiniden. Verh. Deut. Zool. Ges. Graz, 162-165.
CZIHAK, G., 1966. Entwicklungsphysiologische Untersuchung an Echiniden. (Zerstörung der Mikromeren durch UV-Bestrahlung, ein Beitrag zum Problem der Induktion und Regulierung). Wilhelm Roux Arch. Entwicklungsmech. Organismen 157, 199-211.
CZIHAK, G., 1968. Molekularbiologische Aspekte der frühen Embryonalentwicklung von Seeigeln. Pubbl. Sta. Zool. Napoli 36, 321-345.
CZIHAK, G., HÖRSTADIUS, S., 1970. Transplantation of RNA-labeled micromeres into animal halves of sea urchin embryos. A contribution to the problem of embryonic induction. Develop. Biol. 22, 15-30.
CZIHAK, G., WITTMANN, H.G., HINDENNACH, I., 1967. Uridineinbau in die Nucleinsäuren von Furchungsstadien der Eier des Seeigels Paracentrotus lividus. Z. Naturforsch. 22, 1176-1182.
DAN, K., 1960. Cyto-embryology of echinoderms and amphibia. In: Int. Rev. Cytol. 9 (G.H. Bourne, J.F. Danielli, eds.), pp. 321-367. New York: Academic Press.
DAN, K., OKAZAKI, K., 1956. Cyto-embryological studies of sea urchins. III. Role of the secondary mesenchyme cells in the formation of the

primitive gut in sea urchin larvae. Biol. Bull. (Woods Hole) 110, 29-42.
DE VINCENTIIS, M., HÖRSTADIUS, S., RUNNSTRÖM, J., 1966. Studies on controlled and released respiration in animal and vegetal halves of the embryo of the sea urchin, Paracentrotus lividus. Exp. Cell Res. 41, 535-544.
DE VINCENTIIS, M., RUNNSTRÖM, J., 1967. Exp. Cell Res. 45, 681-689.
FUDGE-MASTRANGELO, M., 1966. Analysis of the vegetalizing action of tyrosine on the sea urchin embryo. J. Exp. Zool. 161, 109-128.
GARSTANG, W., 1894. Preliminary note on a new theory of the phylogeny of the Chordata. Zool. Anz. 17, 122-125.
GIBBINS, J.R., TILNEY, L.G., PORTER, K.R., 1969. Microtubules in the formation and development of the primary mesenchyme in Arbacia punctulata. I. The distribution of microtubules. J. Cell Biol. 41, 201-226.
GUSTAFSON, T., 1961. Studies on the cellular basis of morphogenesis in the sea urchin. In: Biological structure and function, vol. II (J.W. Goodwin, O. Lindberg, eds.), pp. 497-508. New York: Academic Press.
GUSTAFSON, T., 1963. Cellular mechanisms in the morphogenesis of the sea urchin embryo. Cell contacts within the ectoderm and between mesenchyme and ectoderm cells. Exp. Cell Res. 32, 570-589.
GUSTAFSON, T., 1964. The role and activities of pseudopodia during morphogenesis of the sea urchin larva. In: Primitive motile systems (R.D. Allen, N. Kamiya, eds.), pp. 333-348. New York: Academic Press.
GUSTAFSON, T., 1965. Morphogenetic significance of biochemical patterns in sea urchin embryos. In: The biochemistry of development, vol. I (R. Weber, ed.), pp. 139-202. New York: Academic Press.
GUSTAFSON, T., 1969a. Cell recognition and cell contacts during sea urchin development. In: Cellular recognition (R.T. Smith, R.A. Good, eds.), pp. 47-60. New York: Meredith Co.
GUSTAFSON, T., 1969b. Fertilization and development. In: Chemical Zoology (M. Florkin, B.T. Scheer, eds.), pp. 149-206. New York: Academic Press.
GUSTAFSON, T., KINNANDER, H., 1956. Microaquaria for time-lapse cinematographic studies of morphogenesis in swimming larvae and observations on sea urchin gastrulation. Exp. Cell Res. 11, 36-51.
GUSTAFSON, T., KINNANDER, H., 1960. Cellular mechanisms in morphogenesis of the sea urchin gastrula. The oral contact. Exp. Cell Res. 21, 361-373.
GUSTAFSON, T., LENICQUE, P., 1952. Studies on mitochondria in the developing sea urchin egg. Exp. Cell Res. 3, 251-274.
GUSTAFSON, T., LENICQUE, P., 1955. Studies on mitochondria in early cleavage stages of the sea urchin egg. Exp. Cell Res. 8, 114-117.
GUSTAFSON, T., TONEBY, M., 1970. On the role of serotonin and acetylcholine in sea urchin morphogenesis. Exp. Cell Res. 62, 102-117.
GUSTAFSON, T., TONEBY, M., in press. How genes control morphogenesis. Am. Sci.
GUSTAFSON, T., WOLPERT, L., 1961a. Studies on the cellular basis of morphogenesis in the sea urchin embryo. Gastrulation in vegetalized larvae. Exp. Cell Res. 22, 437-449.
GUSTAFSON, T., WOLPERT, L., 1961b. Cellular mechanisms in the morphogenesis of the sea urchin larva. The formation of arms. Exp. Cell Res. 22, 509-520.
GUSTAFSON, T., WOLPERT, L., 1961c. Studies on the cellular basis of morphogenesis in the sea urchin embryo. Directed movements of primary mesenchyme cells in normal and vegetalized larvae. Exp. Cell Res. 24, 64-79.
GUSTAFSON, T., WOLPERT, L., 1962. Cellular mechanisms in the morphogenesis of the sea urchin larva. Change in shape of cell sheets. Exp. Cell Res. 27, 260-279.

GUSTAFSON, T., WOLPERT, L., 1963a. Studies on the cellular basis of morphogenesis in the sea urchin embryo. Formation of the coelom, the mouth and the primary pore-canal. Exp. Cell Res. 29, 561-582.
GUSTAFSON, T., WOLPERT, L., 1963b. The cellular basis of morphogenesis and sea urchin development. In: Int. Rev. Cytol. 15 (G.H. Bourne, J.F. Danielli, eds.), pp. 139-214. New York: Academic Press.
GUSTAFSON, T., WOLPERT, L., 1967. Cellular movement and contact in sea urchin morphogenesis. Biol. Rev. 42, 441-498.
HAGSTRÖM, B.E., 1963. The effect of lithium and o-iodosobenzoic acid on the early development of the sea urchin egg. Biol. Bull. (Woods Hole) 124, 55-64.
HAGSTRÖM, B.E., LÖNNING, S., 1963. The effect of trypsin on the early development of the sea urchin egg. Arkiv f. Zool. (2) 15, 377-380.
HAGSTRÖM, B.E., LÖNNING, S., 1964. The rate of development in isolated halves of sea urchin embryos. Sarsia 15, 17-22.
HAGSTRÖM, B.E., LÖNNING, S., 1965. Studies of cleavage and development of isolated sea urchin blastomeres. Sarsia 18, 1-9.
HAGSTRÖM, B.E., LÖNNING, S., 1966. Analysis of the effect of dinitrophenol on cleavage and development of the sea urchin embryo. Protoplasma 62, 246-254.
HAGSTRÖM, B.E., LÖNNING, S., 1967. Cytological and morphological studies on the action of lithium on the development of the sea urchin embryo. Wilhelm Roux Arch. Entwicklungsmech. Organismen 158, 1-12.
HAGSTRÖM, B.E., LÖNNING, S., 1969. Time-lapse and electron microscopic studies of sea urchin micromeres. Protoplasma 68, 271-288.
HOFMANN, H., HOFMANN, E., 1958. Über die pharmakologische Hemmung der Zellteilungsvorgänge durch Derivate des Phenothiazin und ihre Kombinationen mit Narkotika nach Versuchen am Seeigelei. Pubbl. Sta. Zool. Napoli 30, 347-357.
HOLTER, H., LINDAHL, P.E., 1940. Beiträge zur enzymatischen Histochemie XXXIII. Die Atmung animaler und vegetativer Keimhälften von Paracentrotus lividus. C. R. Trav. Lab. Carlsberg, Ser. Chim. 23, 257-288.
HÖRSTADIUS, S., 1928. Über die Determination des Keimes bei Echinodermen. Acta Zool. Stockholm 9, 1-191.
HÖRSTADIUS, S., 1935. Über die Determination im Verlaufe der Eiachse bei Seeigeln. Pubbl. Sta. Zool. Napoli 14, 251-479.
HÖRSTADIUS, S., 1939. The mechanics of sea urchin development studied by operative methods. Biol. Rev. 14, 132-179.
HÖRSTADIUS, S., 1952. Induction and inhibition of reduction gradients by the micromeres in the sea urchin egg. J. Exp. Zool. 120, 421-436.
HÖRSTADIUS, S., 1955. Reduction gradients in animalized and vegetalized sea urchin eggs. J. Exp. Zool. 129, 249-256.
HÖRSTADIUS, S., 1959. The effect of sugars on differentiation of larvae of Psammechinus miliaris. J. Exp. Zool. 142, 141-158.
HÖRSTADIUS, S., 1961. The effect of sugars on development of sea urchin larvae. In: Symposium on germ cells and development. Institut international d'embryologie and Fondazione A. Baselli (1960), pp. 384-386. Pavia: Successori Fusi Di Ripa E Figli.
HÖRSTADIUS, S., 1963. Vegetalization of sea urchin larvae by chloramphenicol. Develop. Biol. 7, 144-151.
HÖRSTADIUS, S., JOSEFSSON, L., RUNNSTRÖM, J., 1967. Morphogenetic agents from unfertilized eggs of the sea urchin Paracentrotus lividus. Develop. Biol. 16, 189-202.
IMMERS, J., 1956a. Changes in acid mucopolysaccharides attending the fertilization and development of the sea urchin. Arkiv. f. Zool. (2)9, 367-375.
IMMERS, J., 1956b. Cytological feature of the development of the egg of Paracentrotus lividus reared in artificial seawater devoid of sulphate ions. Exp. Cell Res. 10, 546-548.

IMMERS, J., 1961. Comparative study of the localization of incorporated ^{14}C-labelled amino acids and $^{35}SO_4$ in the sea urchin ovary, egg and embryo. Exp. Cell Res. 24, 356-378.
IMMERS, J., 1962. Investigations on macromolecular sulfated polysaccharides in sea urchin development. Uppsala: Almkvist and Wiksells Boktryckeri AB.
IMMERS, J., MARKMAN, B., RUNNSTRÖM, J., 1967. Nuclear changes in the course of development of the sea urchin studied by means of Hale staining. Exp. Cell Res. 47, 425-442.
IMMERS, J., MARKMAN, B., RUNNSTRÖM, J., 1969. Interactions between proteins and nucleic acids and their changes in the differentiation of sea urchin egg. In: Annales d'Embryologie et de Morphogenèse Suppl. 1, 269.
IMMERS, J., RUNNSTRÖM, J., 1960. Release of respiratory control by 2,4-dinitrophenol in different stages of sea urchin development. Develop. Biol. 2, 90-104.
IMMERS, J., RUNNSTRÖM, J., 1965. Further studies on the effects of deprivation of sulphate on the early development of the sea urchin Paracentrotus lividus. J. Embryol. Exp. Morphol. 14, 289-385.
IMMERS, J., RUNNSTRÖM, J., 1971. In press.
JAGENDORF-ELFVIN, M., GUSTAFSON, T., 1971. In preparation.
KINOSHITA, S., 1969. Periodical release of heparine-like polysaccharides within cytoplasm during cleavage of sea urchin egg. Exp. Cell Res. 56, 39-43.
KINOSHITA, S., YAZAKI, I., 1967. The behaviour and localization of intracellular relaxing system during cleavage in the sea urchin egg. Exp. Cell Res. 47, 449-458.
LALLIER, R., 1958. Analyse expérimentale de la différenciation embryonaire chez les échinodermes. Experientia 14, 309-315.
LALLIER, R., 1962. Les effects du chloramphenicol sur le développement de l'oef d'oursin. J. Embryol. Exp. Morphol. 10, 568-574.
LALLIER, R., 1964. Biochemical aspects of animalization and vegetalization in the sea urchin embryo. In: Advances in morphogenesis 3 (M. Abercrombie, J. Brachet, eds.), pp. 147-191. New York: Academic Press.
LENICQUE, P., HÖRSTADIUS, S., GUSTAFSON, T., 1953. Change of distribution of mitochondria in animal halves of sea urchin eggs by the action of micromeres. Exp. Cell Res. 5, 400-403.
KINNANDER, H., GUSTAFSON, T., 1960. Further studies on the cellular basis of gastrulation in the sea urchin larva. Exp. Cell Res. 19, 278-290.
MARKMAN, B., 1957. Studies on nuclear activity in the differentiation of the sea urchin embryo. Exp. Cell Res. 12, 424-426.
MARKMAN, B., 1961a. Regional differences in isotopic labelling of nucleic acid and protein in early sea urchin development. An autoradiographic study. Exp. Cell Res. 23, 118-129.
MARKMAN, B., 1961b. Differences in isotopic labelling of nucleic acid and protein in early sea urchin development. Exp. Cell Res. 23, 197-200.
MARKMAN, B., 1961c. Differences in isotopic labelling of nucleic acid and protein in sea urchin embryos developing from animal and vegetal halves. Exp. Cell Res. 25, 224-226.
MARKMAN, B., 1967. Isotopic labelling of nucleic acid in sea urchin embryos developing from animal and vegetal halves in relation to protein and nucleic acid content. Exp. Cell Res. 46, 1-18.
MARSLAND, D., LANDAU, J., 1954. The mechanisms of cytokinesis: Temperature-pressure studies on the cortical gel system in various marine eggs. J. Exp. Zool. 125, 507-539.
MAZIA, D., RUBY, A., 1967. A chamber for the prolonged observation of cells under compression. Exp. Cell Res. 45, 505-506.

METSCHNIKOFF, E., 1884. Embryologische Mitteilungen über Echinodermen. Zool. Anz. 7, 43-47 and 62-65.
MONNE, L., HÅRDE, S., 1951. On the formation of the blastocoel and similar embryonic cavities. Arkiv. f. Zool. (2) 1, 463-469.
MONROY, A., MAGGIO, R., 1964. Biochemical studies on the early development of the sea urchin. In: Advances in morphogenesis 3 (M. Abercrombie, J. Brachet, eds.), pp. 95-140. New York: Academic Press.
MOORE, A.R., BURT, A.S., 1939. On the locus and nature of the forces causing gastrulation in the embryos of Dendraster excentricus. J. Exp. Zool. 82, 159-171.
MORI, S., TAKASHIMA, Y., YANO, K., KOJO, T., HASHIMOTO, K., FUJITANI, N., 1958. Studies on mitochondria in growing and developing sea urchin eggs. Bull. Exp. Biol. (Japan) 8, 65-
MOTOMURA, J., 1960. Secretion of mucosubstance in the gastrula of the sea urchin embryo. Bull. biol. Stn. Asamushi 10, 165-169.
NIEUWKOOP, P.D., 1969. The formation of the mesoderm in urodelan amphibians. I. Induction by the endoderm. Wilhelm Roux' Arch. Entwicklungsmech. Organismen 162, 341-373.
OKAZAKI, K., 1960. Skeleton formation of sea urchin larvae. II. Organic matrix of the spicule. Embryologia 5, 283-320.
OKAZAKI, K., 1965. Skeleton formation of sea urchin larvae. V. Continuous observation of the process of matrix formation. Exp. Cell Res. 40, 585-596.
OKAZAKI, K., FUKUSHI, T., DAN, K., 1962. Cyto-embryological studies of sea urchins. IV. Correlation between the shape of the ectodermal cells and the arrangement of the primary mesenchyme cells in sea urchin larvae. Acta Embryol. Morphol. Exp. 5, 17-31.
PATERSON, M.C., 1957. Animal-vegetal balance in amphibian development. J. Exp. Zool. 143, 183-205.
PETHICA, B.A., 1961. The physical chemistry of cell adhesion. Exp. Cell Res. Suppl. 8, 123-140.
RAPPAPORT, R., 1969. Division of isolated furrows and furrow fragments in invertebrate eggs. Exp. Cell Res. 56, 87-91.
REHBUN, L.I., 1964. Saltatory particle movements in cells. In: Primary motile systems in cell biology (R.D. Allen, N. Kamiya, eds.), pp. 503-526. New York: Academic Press.
RUNNSTRÖM, J., 1961. The role of nuclear metabolism in the determination of the sea urchin egg. Pathol. Biol. Semaine Hop. 9, 781-785.
RUNNSTRÖM, J., 1967. Structural changes of the chromatin in early development of sea urchins. Exp. Cell Res. 48, 691-694.
RUNNSTRÖM, J., HÖRSTADIUS, S., IMMERS, J., FUDGE-MASTRANGELO, M., 1964. An analysis of the role of sulphate in the embryonic differentiation of the sea urchin (Paracentrotus lividus). Rev. Suisse Zool. 71, 21-54.
RUNNSTRÖM, J., IMMERS, J., 1970. Heteromorphic budding in lithium-treated sea urchin embryos. Exp. Cell Res. 62, 228-238.
SEMON, R., 1888. Die Entwicklung der Synapta digitata und ihre Bedeutung für die Phylogenie der Echinodermen. Z. Naturwiss. 175, 175-309.
SHAVER, J.R., 1955. The distribution of mitochondria in sea urchin embryos. Experientia 11, 351-353.
SHAVER, J.R., 1956. Mitochondrial populations during development of the sea urchin. Exp. Cell Res. 11, 549-559.
TILNEY, L.G., GIBBINS, J.R., 1969. Microtubules in the formation and development of the primary mesenchyme in Arbacia punctulata. II. An experimental analysis of their role in development and maintenance of cell shape. J. Cell Biol. 41, 227-250.
UBISCH, L., VON, 1932. Untersuchungen über Formbildung. III. Ein vorwiegend spekulativer Beitrag zur Frage der Entstehung und systema-

tischen Bedeutung der Seeigelplutei. Wilhelm Roux' Arch. Entwicklungsmech. Organismen 127, 216-250.
UBISCH, L., VON, 1937. Die normale Skelettbildung bei Echinocyamus pusillus und Psammechinus miliaris und die Bedeutung dieser Vorgänge für die Analyse der Skelette von Keimblatt-Chimären. Z. Wiss. Zool. 149, 402-476.
UBISCH, L., VON, 1950. Die Entwicklung der Echiniden. Verh. K. Ned. Akad. Wet. Afdel. Naturk., Sect. II, 47, 1-50.
UBISCH, L., VON, 1957. Über Seeigelmerogone, II. Pubbl. Sta. Zool. Napoli 30, 279-308.
VACQUIER, V.D., 1968. The connections of blastomeres of sea urchin embryos by filopodia. Exp. Cell Res. 52, 571-581.
VACQUIER, V.D., MAZIA, D., 1968a. Twinning of sand dollar embryos by means of dithiothreitol. The structural basis of blastomere interactions. Exp. Cell Res. 52, 209-219.
VACQUIER, V.D., MAZIA, D., 1968b. Twinning of sea urchin embryos by treatment with dithiothreitol. Role of cell surface interactions and of the hyaline layer. Exp. Cell Res. 52, 459-468.
WEISS, L., 1967. The cell periphery, metastasis, and other contact phenomena. Amsterdam: North Holland Publishing Company.
WILLMER, E.N., 1970. Cytology and evolution, 2nd ed. New York: Academic Press.
WOLPERT, L., 1960. The mechanics and mechanisms of cleavage. In: International Review of Cytology (G.H. Bourne, J.F. Danielli, eds.), pp. 163-216. New York: Academic Press.
WOLPERT, L., 1966. The mechanical properties of the membrane of the sea urchin egg during cleavage. Exp. Cell Res. 41, 385-396.
WOLPERT, L., GUSTAFSON, T., 1961a. Studies on the cellular basis of morphogenesis of the sea urchin embryo. Development of the skeletal pattern. Exp. Cell Res. 25, 311-325.
WOLPERT, L., GUSTAFSON, T., 1961b. Studies on the cellular basis of morphogenesis of the sea urchin embryo. The formation of the blastula. Exp. Cell Res. 25, 374-382.
WOLPERT, L., MERCER, E.H., 1963. An electron microscopy study of the development of the blastula of the sea urchin embryo and its radial polarity. Exp. Cell Res. 30, 280-300.

9. Egg and Embryo Ultrastructure

L. Schmekel

The study of egg and embryo ultrastructure began 30 years ago when HARVEY and T.F. ANDERSON (1943) wrote about the spermatozoon and the fertilization membrane. Investigation continued slowly (DAN, 1952; LANSING et al., 1952; MCCULLOCH, 1952) becoming quite intensive in the last 10 years. Of these recently published articles only a few can be listed in an introducing survey, most of them will be found in the text. Extensive work is done by AFZELIUS, 1955-1972; DAN, 1954-1970; MERCER, 1958, 1962; HARRIS, 1961-1968; WOLPERT, 1961, 1963; LÖNNING, 1963, 1967; BAXANDALL, 1964, 1966; KESSEL, 1964-1968; E. ANDERSON, 1966-1970; RUNNSTRÖM, 1966-1969; TILNEY, 1966-1969; INOUE, 1967-1971; MILLONIG, 1967, 1968; W.A. ANDERSON, 1968, 1969; GIBBINS, 1969; LONGO, 1968-1970.

The cell organelles of echinoderms are very similar. Results of ultrastructural analysis of ophiuroid and asteroid eggs are therefore included here. In each instance the normal morphology not affected by experiments is followed by a summary of the results of experimental work.

9.1 The Mature Egg

The diameter of the mature female gamete ready to be shed varies from species to species (oocytes and oogenesis compare: MCCULLOCH, 1952b; AFZELIUS, 1955a, 1956; KESSEL, 1964, 1968; LÖNNING, 1964; VERHEY and MOYER, 1965, 1967; E. ANDERSON, 1966a, 1968; MILLONIG, 1967; BAL et al., 1968; MILLONIG, BOSCO, and GIAMBERTONE, 1968; TSUKAHAWA and SUGIYAMA, 1969; BAL, 1970). In *Arbacia lixula*, for example, the diameter of the egg is 70 µm while that of the nucleus is about 10 µm. The mature unfertilized egg is surrounded by a thick jelly layer which in many species is difficult to preserve. It is often not sufficiently electron-opaque to be detected by electron microscopy, although its presence may be inferred from the distance between the eggs lying in the embedding medium (BAXANDALL et al., 1964a, 1964b). In other species, particularly in some of the irregular sea urchins, the jelly coat is readily visible (LÖNNING, 1967c). It has a fine fibrous structure and clearly defined edges (LÖNNING, 1967a, 1967c; ITO, 1969).

The egg cell is surrounded by a plasma membrane 60 to 65 Å thick (Fig. 9.1). Immediately outside this is the vitelline layer or primary coat, also called the vitelline membrane, primary envelope, or diffuse layer which is an extracellular layer composed of fuzzy material (AFZELIUS, 1956; MOTOMURA, 1960; BALINSKY, 1961; ENDO, 1961; WOLPERT and MERCER, 1961; BAXANDALL, PERLMANN, and AFZELIUS, 1964a, 1964b; PASTEELS, 1965; RUNNSTRÖM, 1966; LÖNNING, 1967a, 1967c; E. ANDERSON, 1968; ITO, 1969). The *Paracentrotus* vitelline layer separated from the plasma membrane after fertilization is about 30 Å thick. In *Arbacia*

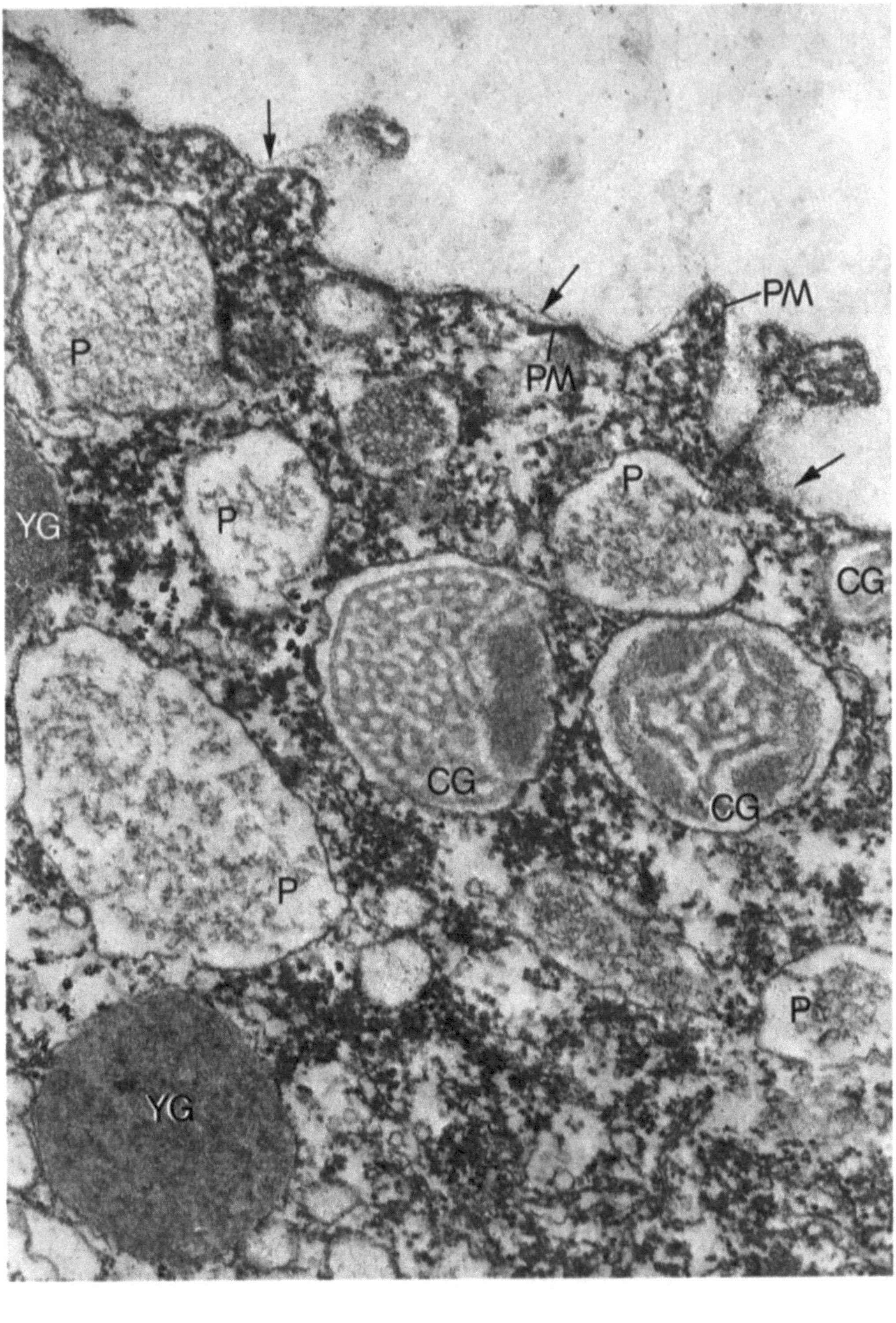
P
PM
PM
P
YG
P
CG
CG
CG
P
P
YG

punctulata and *Strongylocentrotus purpuratus* it is thinner and in *Asterias forbesi* much thicker (E. ANDERSON, 1968). The jelly and vitelline layers correspond together, to what BENNETT has called glykocalyx (BENNETT, 1963; ITO, 1969). From the egg surface microvilli protrude, the length and number of which vary from species to species (HILLER, LANSING, and ROSENTHAL, 1952; AFZELIUS, 1956a, LÖNNING, 1964, 1967c) (Fig. 9.1, 9.6).

The cortical bodies or particles are close to the plasma membrane - but not in direct contact with it (see p. 56) - on the periphery of the egg cytoplasm (WOLPERT and MERCER, 1961; TAKASHIMA, TAKASHIMA, and NODA, 1966; RUNNSTRÖM, 1966; LÖNNING, 1967a, 1967c; MILLONIG and HUNTER, 1968; E. ANDERSON, 1970). Cortical bodies are covered by a unit membrane and often contain 2 structurally different parts the organisation of which is species-specific (MCCULLOCH, 1952b; AFZELIUS, 1956a; MOTOMURA, 1960; RUNNSTRÖM, 1966; LÖNNING, 1967a, 1967c, 1967d; E. ANDERSON, 1968). The cortical bodies of closely related species are of similar structure. Their overall appearance and some genus differences are illustrated (Fig. 9.1, 9.2, 9.6): A peripheral component forms hemispheres in *Sphaerechinus* and the central material forms a network (AFZELIUS, 1956a). Similar marginal hemispheres occur in *Paracentrotus*, but the inner material is arranged in globular lamellae. While the ultrastructure of mature cortical bodies of different species varies, their transformations during egg maturation follow parallel courses in all species examined (AFZELIUS, 1956a; MCCULLOCH, 1952b). The Golgi's complex is believed to be involved in the synthesis and concentration of the precursors of the cortical bodies (E. ANDERSON, 1966a, 1968). During egg maturation the cortical granules migrate to the cortex (MCCULLOCH, 1952b; AFZELIUS, 1956a).

Occasionally seen among the cortical bodies at the egg periphery, yolk granules are usually distributed throughout the cytoplasm (Fig. 9.1, 9.2). Mature yolk granules are dense and spherical and are covered by a membrane. Oocyte growth suggests that the Golgi's complex may influence yolk granule formation during the early stages (AFZELIUS, 1956a; KESSEL, 1966, 1968a). While most echinochrome pigment bodies lie in the cortex, a few may lie deeper in the cytoplasm. They are covered by a membrane and show little internal structure (TILNEY and MARSLAND, 1969).

The organelles of the central cytoplasm (Fig. 9.2) include rough and smooth endoplasmic reticulum, mitochondria, Golgi's complexes, yolk granules, oil droplets, various vacuoles, and annulate lamellae. Mitochondria are often clustered around lipid droplets and dense granules (E. ANDERSON, 1968). Vacuoles vary in size and composition. Those in *Arbacia punctulata*, for example, range from 0.1 μm to 1.0 μm in diameter (TILNEY and MARSLAND, 1969). Some vacuoles are multivesicular bodies and others include short rods (LÖNNING, 1963). The mature egg contains only a few Golgi elements, each of relatively small diameter. In *Echinus esculentus*, *Psammechinus miliaris*, and *Strongylocentrotus droebachiensis*, for example, the Golgi elements consist of stacks of approximately 7 parallel rows of aggregated, flattened vesicles, at the edges of which smaller vesicles generally occur (AFZELIUS, 1956b).

Fig. 9.1. Cortical region of mature unfertilized *Sphaerechinus granularis* egg. Vitelline layer (↑) is outside plasmalemma (PM). On egg cytoplasm periphery are vesicles with cortical granules (CG), pigment bodies (P), and yolk granules (YG). x 57,000

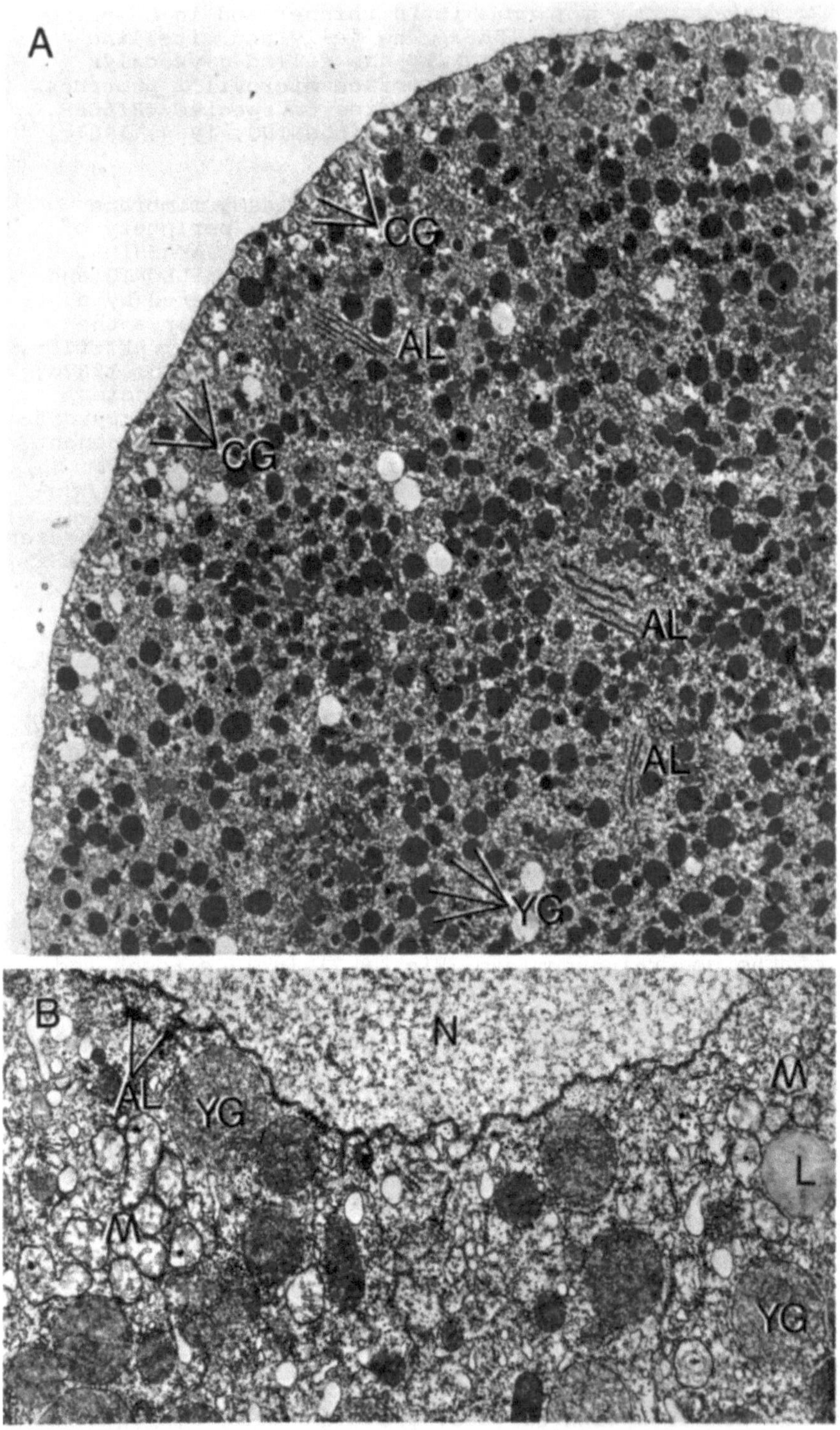
A
CG
AL
CG
AL
AL
YG
B
N
AL
YG
M
M
L
YG

Possibly derived from the nuclear envelope, stacks of annulate lamellae are scattered throughout the cytoplasm (MERRIAM, 1958a, 1959; KESSEL, 1964, 1966, 1968b; HARRIS, 1967a, 1967b; BAL et al., 1968)(Fig. 9.2). The annuli are similar to nuclear pores. Associated with annulate lamellae are osmiophilic granules and fibrils. In *Ophioderma* the granules are 40-90 Å in diameter and in *Dendraster* 150 Å (MERRIAM, 1959). *Ophioderma* fibrils are about 30 Å in diameter (KESSEL, 1968b). It is possible that annulate lamellae influence protein synthesis (KESSEL, 1968b). In *Ophioderma* the osmiophilic granules and fibrils associated with the annulate lamellae have a similar arrangement - and the fibrils a comparable size range - like the material in some large nucleoli. The largest annulate lamellae granules are about the size of the smallest nucleoli granules. The annulate lamellae are often in contact with lamellae and whorls of rough endoplasmic reticulum (PASTEELS et al., 1958; KESSEL, 1966; BAL et al., 1968). The accumulation of osmiophilic granules in the annulate lamellae causes the interlamellar space to widen into areas that have been referred to as heavy bodies (AFZELIUS, 1957). The osmiophilic granules seem not to be glycogen particles but more likely ribosomes or the precursors of ribosomes (MERRIAM, 1958a; HARRIS, 1967a, 1967b; KESSEL, 1968b).

In unfertilized or artificially activated eggs, centrioles have not been observed until the aster forms (DIRKSEN, 1961; SACHS and E. ANDERSON, 1970).

Slightly irregular in shape the female pronucleus is covered by a perforated nuclear envelope (AFZELIUS, 1955a; MILLONIG, BOSCO, and GIAMBERTONE, 1968)(Fig. 9.2B). Near the female pronucleus is an area containing many ribosomes, smooth endoplasmic reticulum, annulate lamellae, but only a few mitochondria, and yolk granules. The nucleoplasm contains fibrillar chromatin and a variable number of nucleolus-like bodies at the periphery of the pronucleus. They adhere to the inner surface of the nuclear membrane. Their shape varies from nearly spherical to lenticular, and they often occur in rounded, cup-like protrusions of the nucleus, from which they are not dislodged by centrifugation. Their number varies, but generally each nucleus contains 1 large nucleolus-like body and as many as 10 smaller ones (HARRIS, 1967a). In the pronuclear envelope overlying the nucleolus-like bodies, pores are of larger diameter and perinuclear space is wider than elsewhere in the pronuclear envelope (LONGO and E. ANDERSON, 1968). In *Arbacia punctulata* each nucleolus-like body is made up of compact, fine-textured material and granules. The nucleoli-like bodies stain purple with methyl green-pyronine, indicating the presence of RNA (LONDO and E. ANDERSON, 1968). The mature egg of *Arbacia lixula* contains only a few agranular nucleoli. They consist of loose fibrils less than 80 Å in diameter. During cleavage the nucleoli remain fibrillar without ribosome-like granules. From the blastula stage on, when cell division almost stops, nucleoli have a few granules. The granules multiply until the gastrula stage while the number of fibrils decreases (MILLONIG, 1967a, 1967b; MILLONIG, BOSCO, and GIAMBERTONE, 1968).

◀ Fig. 9.2. (A) Mature unfertilized *Arbacia lixula* egg. Cortical granules (CG) are visible only at egg periphery. Yolk granules (YG) and annulate lamellae (AL) are distributed through cytoplasm. x 4,140. (B) *Paracentrotus lividus* egg 5 min after fertilization. Female nucleus (N) is surrounded by annulate lamellae (AL), mitochondria (M), lipid droplets (L), and yolk granules (YG). x 15,400

In studies of unfertilized, centrifuged eggs (MCCULLOCH, 1952a; AFZELIUS, 1957; MERRIAM, 1958b; GROSS et al., 1960; MERCER and WOLPERT, 1962) those of *Psammechinus miliaris* centrifuged at 4,400 g for 5 minutes show no shift of cortical granules (WOLPERT and MERCER, 1961). Mitochondria collect at the heavy pole, yolk granules and nucleus at the light. "The separation is fairly complete (only very occasional yolk granules or mitochondria are out of place) and leaves centrally within the egg a broad band (clear in the light microscope) which contains a population of small particulates not greatly influenced by the centrifugation. These same small particulates (dense particles either free or attached to membranes and small empty vesicles (0.1 µm-0.2 µm) are to be seen to a lesser degree between the mitochondria and between the more closely packed yolk granules at the light pole." (MERCER and WOLPERT, 1962).

Eggs of *Arbacia punctulata* centrifuged at 9,000 g for 10-15 minutes do not stratify in uniform layers of lipid droplets, pronucleus, clear zone, mitochondria, yolk, and pigment (E. ANDERSON, 1970). With 2 exceptions, each region may contain elements of another: lipid droplets occupying the extreme centripetal end and pigment bodies occupying the extreme centrifugal end are not seen in other areas. The most heterogeneous region, the clear zone, is made up of annulate lamellae, heavy bodies, Golgi's complexes, and vacuoles containing rods. In mature and immature eggs, centrifugation does not dislodge peripheral cortical granules from the cortical ooplasm. But when mature eggs are treated with 3% urethane for 5 minutes before centrifugation, the cortical bodies leave the periphery and form a stratum. Nothing was found in the cortical ooplasm before or after centrifugation or in the cortex after urethane treatment and centrifugation that would seem to hold the cortical bodies in the peripheral ooplasm. Further centrifugation of the stratified eggs produces nucleated and non-nucleated halves and then quarters (E. ANDERSON, 1970).

During the initial stages of fertilization, the egg surface is particularly important in determining the specificity of sperm-egg interaction. In electron microscopic studies of unfertilized *Paracentrotus lividus* eggs' response to globulin fractions of antisera against isolated homologous jelly-coat substance or homologous homogenates of jelly-free eggs, anti-jelly gamma globulin caused precipitation of the jelly layer (BAXANDALL, PERLMANN, and AFZELIUS, 1964a, 1964b). The precipitation density varied from egg to egg in proportion to gamma globulin concentration. Species-specific jelly-substance agglutination of adjacent eggs was frequent with higher gamma globulin concentrations. Anti-egg gamma globulins from anti-serum against total homogenates or their heat-stable fractions of jelly-free eggs caused structural changes that were mostly stages of parthenogenetic activation induced from those involved in jelly precipitation. Surface changes included formation of papillae, membrane blisters, hyaline layer, and activation membrane; release of cell-surface material; and cortical granulae breakdown. These changes varied with gamma globulin concentration and female reactivity and did not appear in the controls. Intracellular aster formation verified them as parthenogenetic activation and not the result of immune lysis (BAXANDALL, PERLMANN, and AFZELIUS, 1964a, 1964b).

The same authors tried to localize specific antigenic sites using ferritin-labelled antibody and found the jelly layer to be immunologically heterogeneous and to include one component that interacted with anti-jelly gamma globulin and one with anti-egg gamma globulin. Rich in heat-stable and heat-labile egg antigens, the vitelline layer may influence fertilization specificity and parthenogenetic activation and may possibly be analogous to a basement membrane. The plasma membrane reaction with anti-egg gamma globulin showed few antigenic sites. Anti-egg gamma globulin caused specific cortical granule labelling possibly related to egg permeability (BAXANDALL, PERLMANN, and AFZELIUS, 1964a, 1964b).

The effect of partial protein extraction on the structure of *Arbacia* eggs has been described (KANE, 1960). The DNA from sea urchin sperm and embryo nuclei forms strands 21-25 Å wide and over 100 μm long. It is not affected by proteases but is rapidly degraded by DNase and shearing (SOLARI, 1967). The DNA from whole eggs and mitochondria is mostly in the form of closed circular filaments. It takes the form of twisted and open circles ranging from 3.75 μm to 4.83 μm in perimeter (PIKO, TYLER, and VINOGRAD); see page 632.

9.2 The Mature Spermatozoon

The mature spermatozoon has a fairly simple morphology that varies from species to species (AFZELIUS, 1955b; ROTHSCHILD, 1956; BERNSTEIN, 1962; LONGO and E. ANDERSON, 1969)(Figs. 9.3, 9.4). The conical head is pointed in front and flat in back. The enclosed nucleus has one depression in front where the acrosome is inserted and another in back for the tail. The nuclear membrane is double and the nucleoplasm extremely dense and homogeneous. The acrosome (Fig. 9.3) consists of a vesicle delimited by an uninterrupted membrane and containing a fine granular material. In most papers the vesicle is called acrosomal vesicle and its content acrosomal granule. Posterior to the acrosomal vesicle in the subacrosomal region lies the granular periacrosomal substance (DAN, OHORI, and KUSHIDA, 1964; DAN, KUSHIDA, and FUJITA, 1966; LONGO and E. ANDERSON, 1969; DAN, 1970). In *Lytechinus variegatus* a small apical vesicle is sandwiched between the acrosomal vesicle and the sperm membrane (FRANKLIN, 1965). The main part of the middle piece is the ring-shaped mitochondrion that adheres to the flattened posterior surface of the head and presumably develops by fusion of the randomly dispersed spermatid mitochondria (LONGO and E. ANDERSON, 1969). The proximal part of the tail penetrates the central space of the mitochondrion, and an amorphous mass frequently lies at its distal border. The tail has the structure of a cilium: a central pair of filaments surrounded by an outer ring of 9 paired filaments. The central pair terminates in the middle piece while the outer ones form a thickened ring or curved disc - the proximal centriole - lying in the posterior depression of the head. The sperm membrane is continuous over acrosome, middle piece, nucleus, and tail. It is covered all over by a coat of acid mucopolysaccharide or mucoprotein (W.A. ANDERSON, 1968b-d). The ultrastructure of the plasma membrane was studied by means of the surface replica method and the cell surface - including the flagellum - was found to be an assembly of irregularly bent filaments about 40 Å wide (INOUE, BUDAY, and COUSINEAU, 1970b).

Where the mammalian spermatozoon may obtain metabolic energy from exogenous polysaccharides, during its brief but active life span in seawater the sea urchin spermatozoon depends entirely on metabolizable endogenous reserves, probably from the oxidation of phospholipids located mainly in the middle piece. In some species glycogen-like material implies its use for energy production and the presence of glycolytic enzymes (W.A. ANDERSON, 1968c). In the mitochondria of *Paracentrotus lividus* spermatozoa, for example, dense aggregates of granular material ranging from 400 to 1600 Å in diameter are intense-

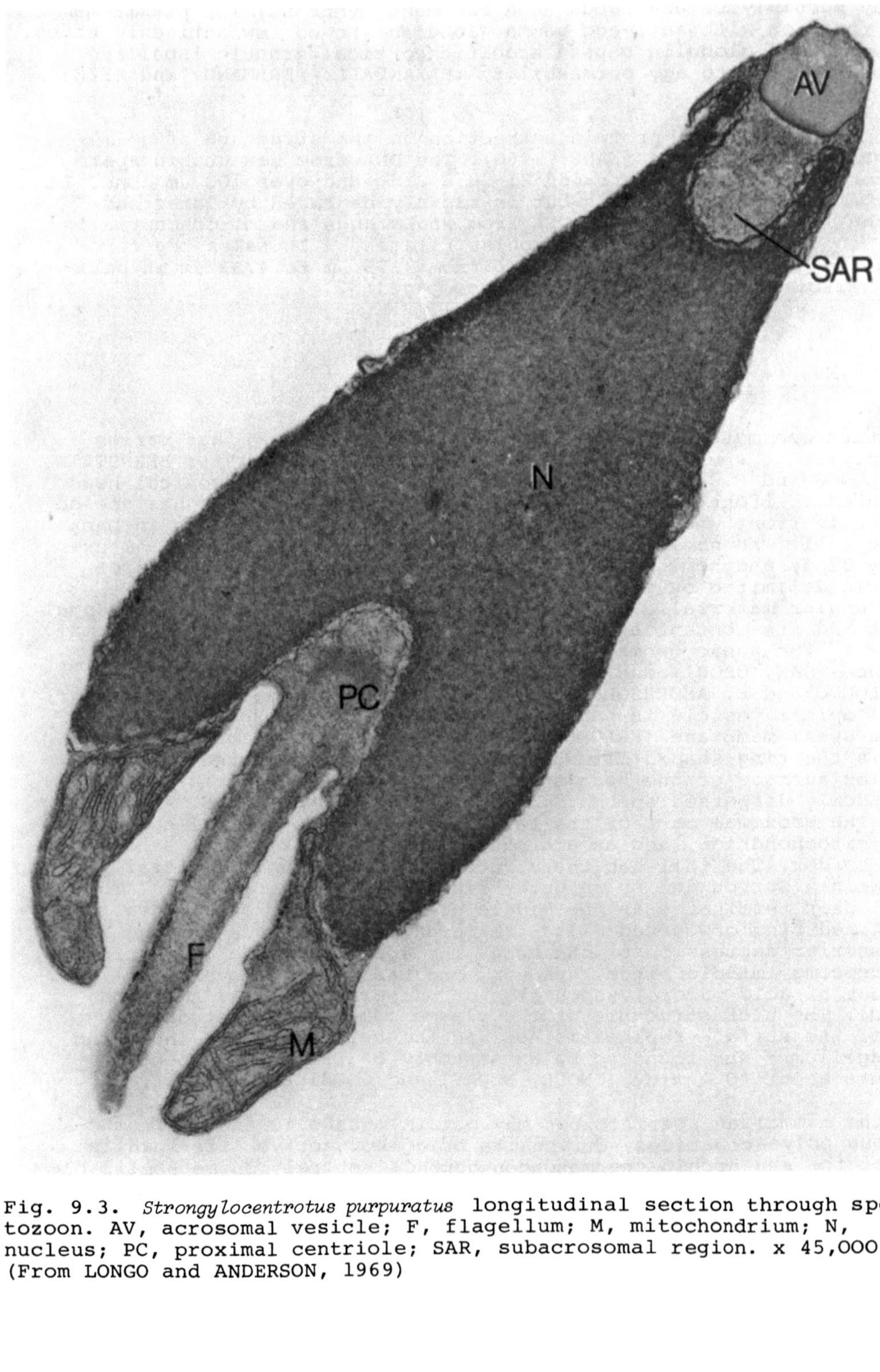

Fig. 9.3. *Strongylocentrotus purpuratus* longitudinal section through spermatozoon. AV, acrosomal vesicle; F, flagellum; M, mitochondrium; N, nucleus; PC, proximal centriole; SAR, subacrosomal region. x 45,000. (From LONGO and ANDERSON, 1969)

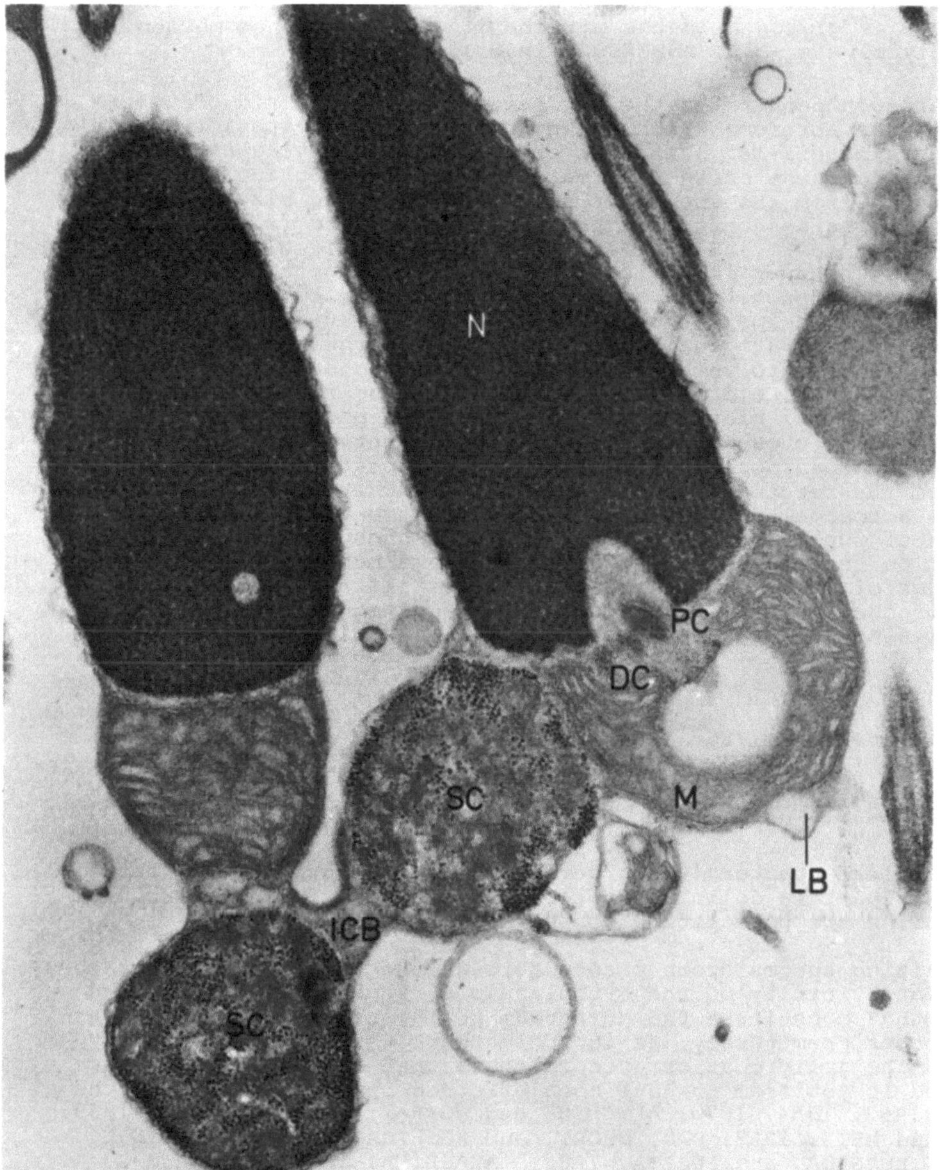

Fig. 9.4. *Strongylocentrotus purpuratus* section through group of spermatids. DC, distal centriole; ICB, intercellular bridges; M, mitochondrium; N, nucleus; PC, proximal centriole; SC, spermatid cytoplasm to be discarded; LB, lipid body. x 51,000. (From LONGO and ANDERSON, 1969)

ly stained with lead citrate, enhanced by lead stains after oxidation in H_2O_2 or periodic acid solutions, extracted with α-amylase or saliva, and composed of 40 Å subunits thought to consist of glycogen particles. Glucose-6-phosphatase and adenosine triphosphatase are active in the middle piece of sea urchin spermatozoa. The enzyme reaction product lies next to the cristae and on the mitochondrial envelope. The presence of glycogen in the mitochondria suggest that metabolism involves glycolysis (W.A. ANDERSON, 1968c).

The spermatozoon penetrates the egg barriers with the aid of lysins released by the acrosome which is known to contain a species-specific jelly-dispersing enzyme. Sulfatase or a jelly-dissolving agent is thought to be present on the spermatozoon surface. During fertilization an enzyme from the sperm acrosomal membrane and plasma membrane probably facilitates contact, adherence, fusion, and fragmentation of the combined sperm and egg membranes. By means of ultrastructural cytochemistry, nonspecific acid and alkaline phosphatases were localized in intact and activated acrosomes, mitochondria, and plasma membrane of sea urchin spermatozoa (W.A. ANDERSON, 1968d). Hydrolases in the middle piece are believed to function in the metabolism of spermatozoa or in autolysis of spermatozoan mitochondria after fertilization. ATPase and acid and alkaline phosphatase end-products appear on the periacrosomal plasmalemma, but glucose-6-phosphatase is absent. Acid and alkaline phosphatase activity is present in the acrosome. During spermatozoan penetration, the acid phosphatase end-product is absent from the reacted acrosome, but enzyme activity continues at the tip of the acrosomal filament. A clear compartment adjacent to it may be an area of ooplasmic hydrolysis. The acid phosphatase end-product is present on the apposed sperm and egg plasma membranes during the early stages of sperm penetration. After fertilization, the enzyme reaction product appears on the plasmalemma of egg microvilli and in the cavities formed by the reacted cortical granules (W.A. ANDERSON, 1968d).

9.3 Fertilization

Since a complete chapter (V) is devoted to the description of fertilization this part is delimited to certain aspects of fine structural changes during the penetration of sperm.

The fertilizing spermatozoon passes through the jelly coat and reaches the immediate vicinity of the cytoplasmic surface where changes are initiated that constitute the acrosomal reaction. The acrosomal reaction may occur prematurely, as in spermatozoa that have only partly penetrated the jelly coat or even in those that are near it but have not entered it yet seem to have been affected by substances from the egg (DAN, 1952, 1954, 1970; AFZELIUS and MURRAY, 1957; COLLIER, 1959; FRANKLIN and METZ, 1962; DAN, OHORI, and KUSHIDA, 1964; FRANKLIN, 1965; DAN, KUSHIDA, and FUJITA, 1966; TAKASHIMA, TAKASHIMA, and Y. NODA, 1966; AUSTIN, 1968; RUNNSTRÖM, 1969).

The membrane of the acrosomal vesicle everts to form a finger-like projection, around the base of which acrosomal granule material squeezed out of the sperm forms a ring. The vesicle membrane and the sperm membrane fuse, and the inner surface of the acrosomal vesicle becomes the new cell surface at the sperm apex. The granular periacrosomal matter from behind the acrosomal granule forms long fibrillar elements. In *Arbacia punctulata* the new sperm apex is pressed into the egg surface 10 to 20 seconds after insemination and 20 to 30 seconds after insemination fusion of sperm and egg-cell membranes normally occurs (FRANKLIN, 1965; LÖNNING, 1967a-c).

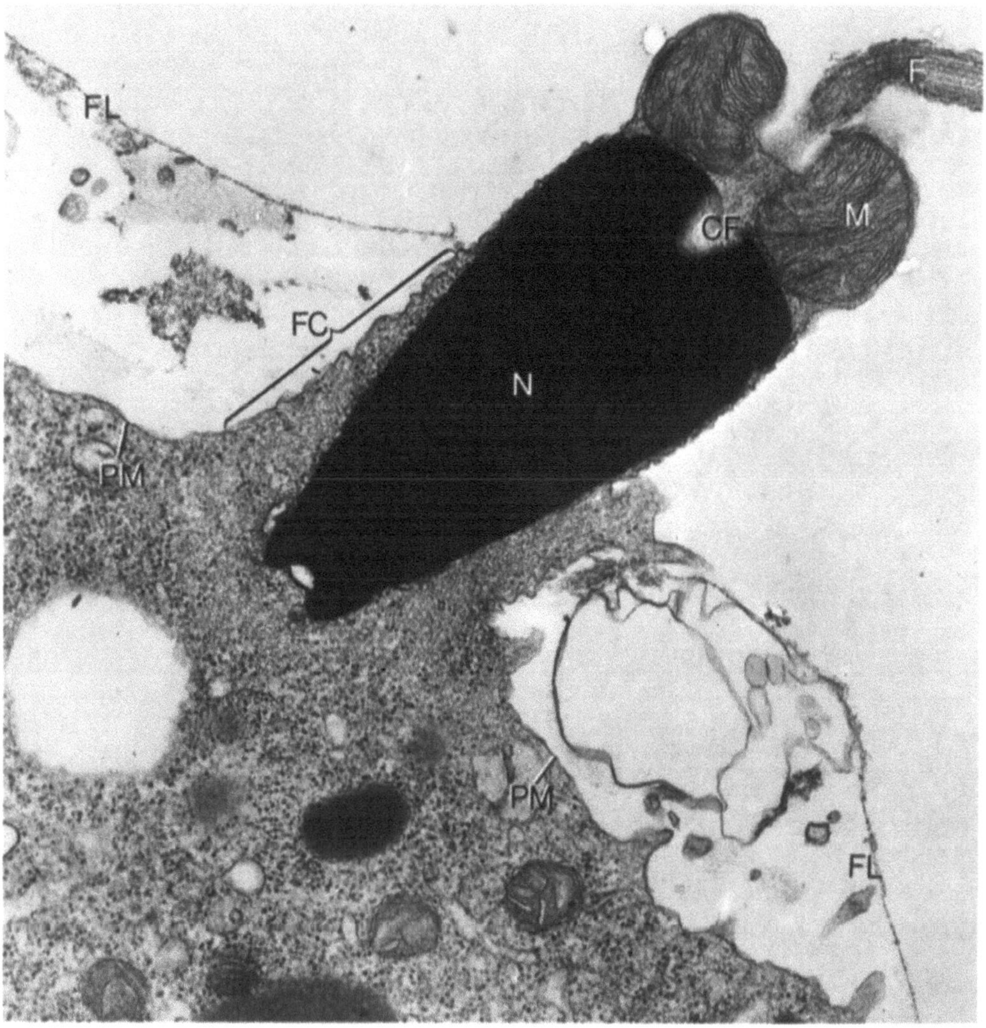

Fig. 9.5. Early stage of spermatozoon incorporation in *Arbacia punctulata*. CF, centriolar fossa; F, flagellum; FC, fertilization cone; FL, fertilization layer; M, sperm mitochondrium; N, sperm nucleus; PM, egg plasmalemma. x 43,000. (From LONGO and ANDERSON, 1969)

The first visible response of the mature egg - the cortical reaction - is the elevation of the vitelline layer from the cell surface accompanied by disruption of cortical granules and the production of the fertilization cone (AFZELIUS, 1956a; WOLPERT and MERCER, 1961; ENDO, 1961; PASTEELS, 1965a, 1965b; E. ANDERSON, 1966a, 1968; RUNNSTRÖM, 1966, 1969; LÖNNING, 1967a, 1967b)(Figs. 9.5-9.7). The vitelline layer now called the fertilization membrane or activation calyx becomes subsequently thicker and hardens into 2 layers (see p. 279). The contents of the cortical bodies empties into the perivitelline space between the egg surface and the fertilization membrane and contributes to the fertilization membrane and the hyaline layer (AFZELIUS, 1956a; ENDO, 1961; BAXANDALL, 1966; RUNNSTRÖM, 1966; LÖNNING, 1967a, 1967b). As

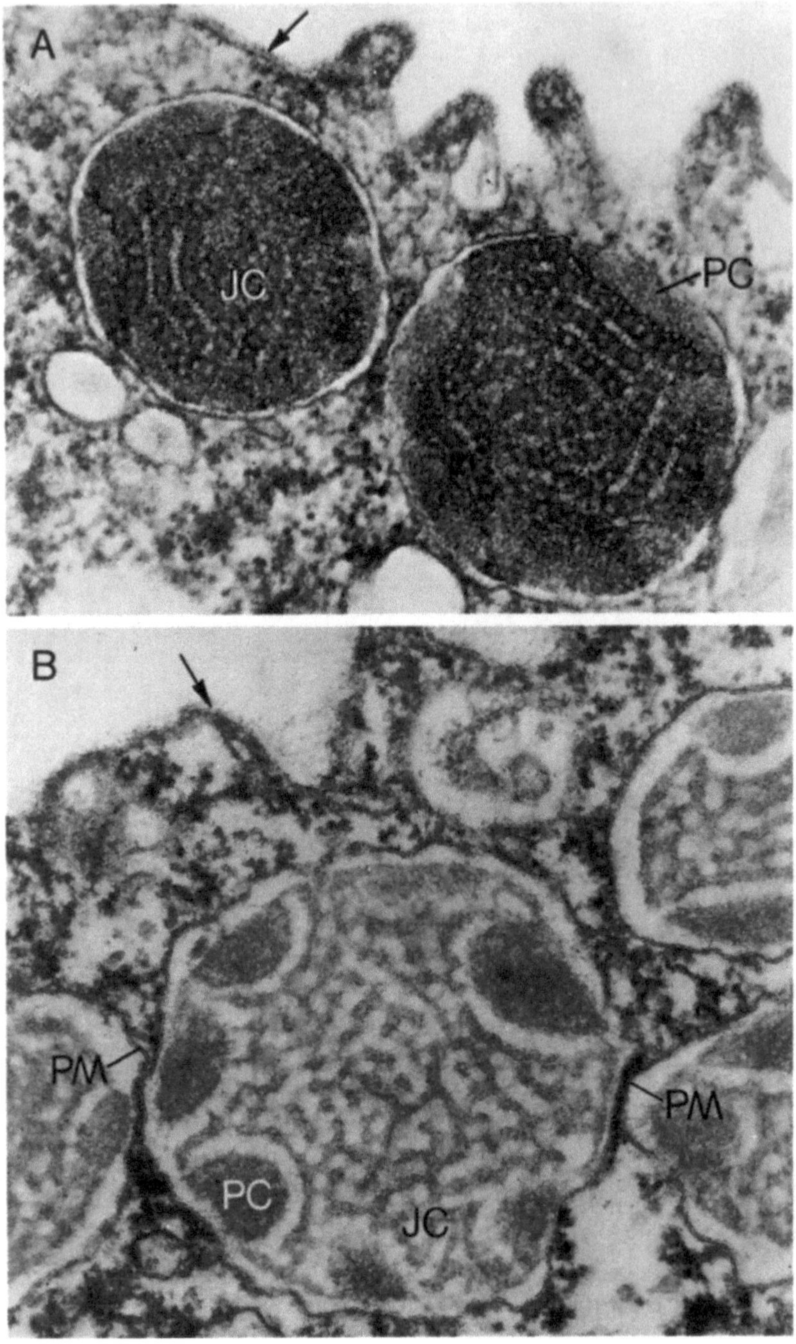

Fig. 9.6. (A) Cortical bodies in unfertilized *Paracentrotus lividus* egg. x 72,000. (B) Cortical bodies in unfertilized *Sphaerechinus granularis* egg. x 75,000. ↑, vitelline layer; IC, inner component; PC, peripheral component; PM, cortical vesicle membrane

already mentioned the cortical granules are found in the peripheral ooplasm of the unfertilized egg immediately beneath the oolemma (Fig. 9.6). The membrane encompassing the granules was in the past said to be in contact with the plasmalemma (ENDO, 1961; WOLPERT and MERCER, 1961), but is recently found 80-200 Å apart (PASTEELS, 1965; E. ANDERSON, 1968; MILLONIG and HUNTER, 1968). Shortly after insemination (after 20-40 seconds in *Paracentrotus lividus*) the apposing membranes fuse, undergo vesiculation, and release the granules between the egg surface and the vitelline membrane (Figs. 9.7A, 9.7B). Some of the vesicles adhere to the inner surface of the fertilization membrane. The major portion of the membrane limiting the cortical granule becomes part of the plasmalemma of the egg (Fig. 9.7). Therefore the zygote plasmalemma is mosaic of original oolemma, cortical bodies' membrane, and sperm membrane (AFZELIUS, 1956a; ENDO, 1961; E. ANDERSON, 1966a, 1966b). After insemination the egg surface is fringed with caves (Fig. 9.7B) where the old cortical granules were. Although they are usually released during the cortical reaction, cortical granules occasionally remain in the blastomeres until gastrula (E. ANDERSON, 1968).

About 7 or 8 minutes after the initial release of the cortical granules into the perivitelline space in *Arbacia punctulata* a stratum of fine filaments is formed known as the hyaline layer presumably formed from part of the cortical granules and condenses into 2 well-defined layers particularly evident after hatching (AFZELIUS, 1956a; WOLPERT, and MERCER, 1963; E. ANDERSON, 1968)(Fig. 9.7B). Associated with the plasmalemma are vesicular bodies with rod-like structures. The vesicular membrane seems to fuse with the egg membrane, releasing the rod-like structures into the perivitelline space. Some of them become part of the hyaline layer (E. ANDERSON, 1968).

As studied in *Strongylocentrotus purpuratus* for 30 minutes after fertilization a great many vesicles about 1 μm in diameter and scattered throughout the cytoplasm migrate to the cell surface, much the way pigment in *Arbacia punctulata* does (HARRIS, 1968). During vesicle migration, while the cortex is indistinguishable from other egg cytoplasm, bundles of fibrils 40-50 Å in diameter spread down into the cortical region and into the cytoplasmic projections that remain after fertilization. In cells in which mercaptoethanol blocks mitosis spindle microtubules are disoriented but not destroyed while cortical fibers are unaffected (HARRIS, 1968).

A fully formed fertilization membrane appears 4 minutes after egg activation and lasts until hatching (ITO, REVEL, and GOODENOUGH, 1967). Dense and about 500 Å thick, it consists of an amorphous central region about 200 Å wide and of 2 rows of circular structures 150 Å in diameter. The circular structures may represent tubular elements (ITO, REVEL, and GOODENOUGH, 1967).

In ultrastructural studies of the fertilization membrane of *Strongylocentrotus purpuratus*, the surface replica method showed the surface consisting of crossed-grid network of 110-155 Å spacings, grid axes crossing at an angle of 74° (INOUE et al., 1967)(Figs. 9.8-9.10). Studies of normal fertilization membrane development and that influenced by 10^{-3}% cystine indicate that cortical granule material attaches in crystals to both sides of the vitelline membrane (INOUE et al., 1970a, 1971). The peripheral layers on either side of the fully developed fertilization membrane seem to be monolayers of flat, parallel fibrils about 160 Å wide and 75 Å thick sandwiched between 2 40-50 Å thick layers of parallel, 40 Å thick filaments about 100 Å apart. The angle between the fibril axes and the 40 Å filaments is

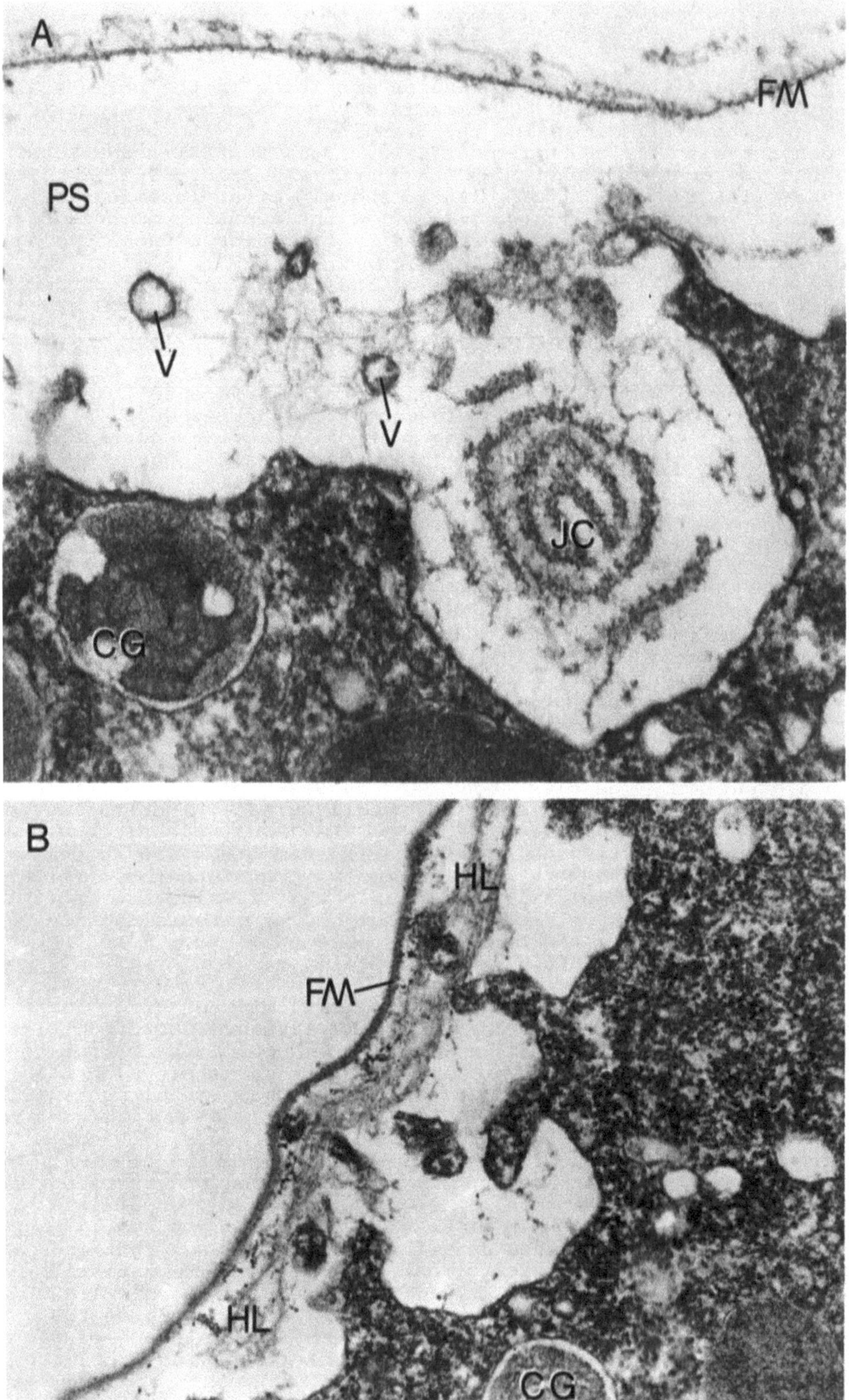
A
FM
PS
V
V
JC
CG
B
HL
FM
HL
CG

◄ Fig. 9.7. (A) *Paracentrotus lividus* 60 sec after fertilization. Left cortical granule (CG) has not yet opened. Right cortical granule releases inner component (IC) into perivitelline space (PS). FM, fertilization membrane; V, vesicles. x 64,000. (B) *Paracentrotus lividus* surface 5 min after fertilization is fringed with caves where old cortical granules were. Fertilization membrane (FM) has thickened and hyaline layer (HL) has formed. x 44,000

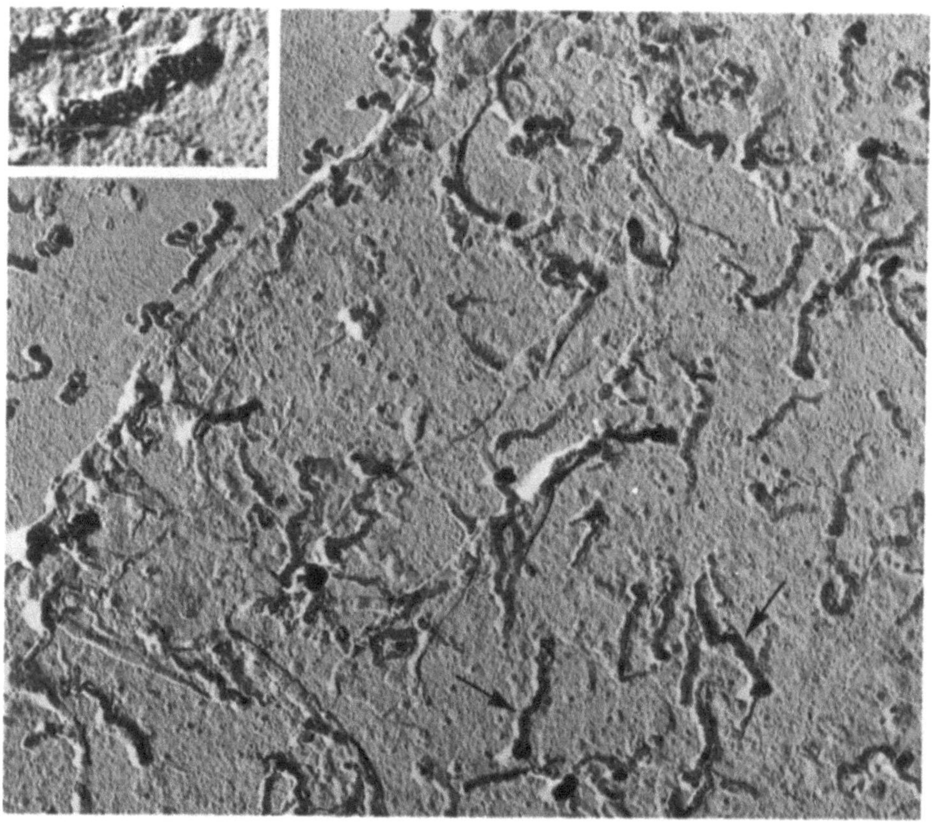

Fig. 9.8. *Strongylocentrotus purpuratus* fertilization membrane 30-60 sec after fertilization, mounted on carbon-coated grid and shadowed with platinum at 30° angle. Arrows indicate narrow electron-dense strips attached to membrane. x 15,700. Inset: enlargement of narrow strip. x 41,000. (From INOUE and HARDY, 1971)

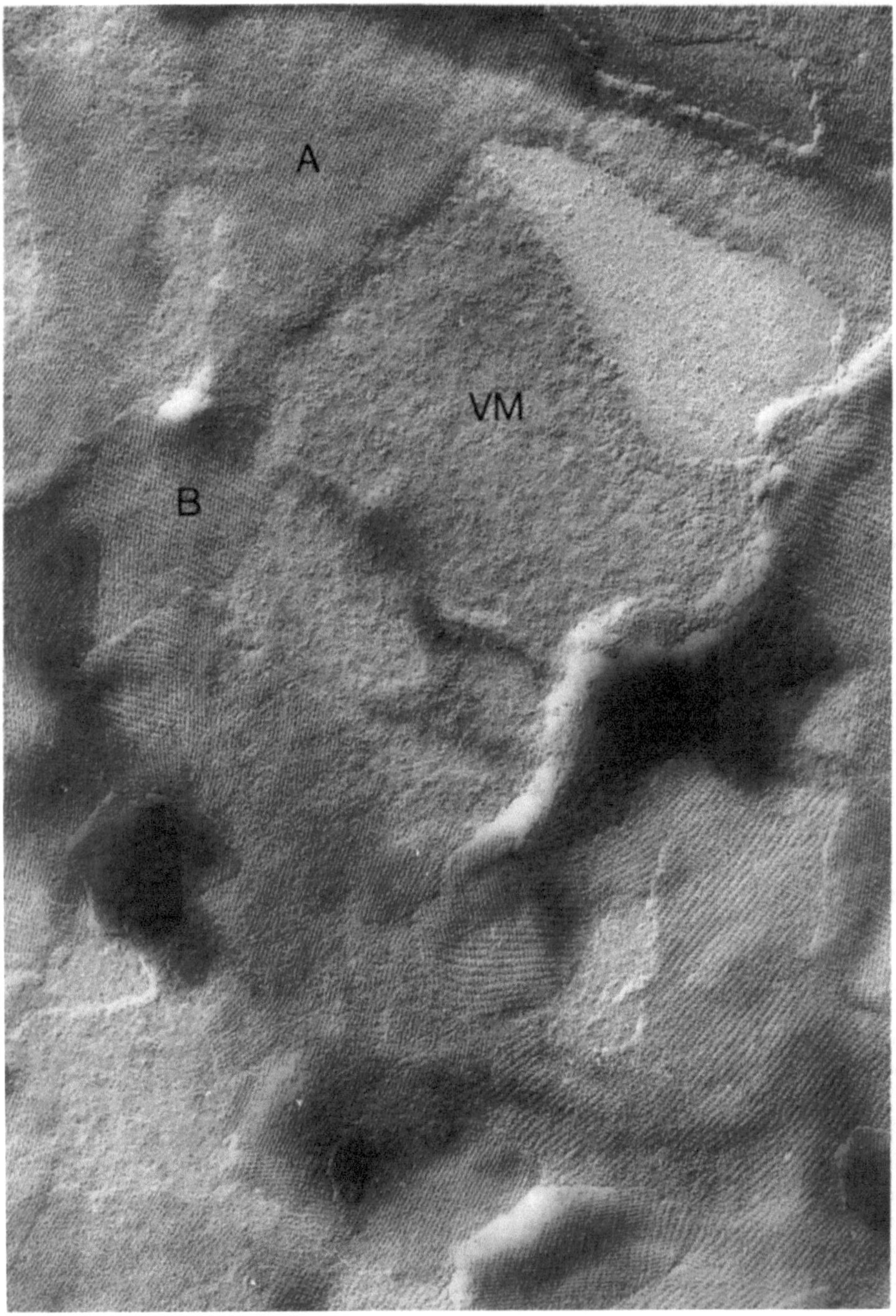

Fig. 9.9. *Strongylocentrotus purpuratus* fertilization membrane 1 - 1 1/2 min after fertilization, just before full development shows an area of crystalline appearing and a smooth flat area (VM). In the former area angle of intersection between 2 sets of parallel lines is slightly different from one part (A) to another (B). x 61,900. (From INOUE and HARDY, 1971)

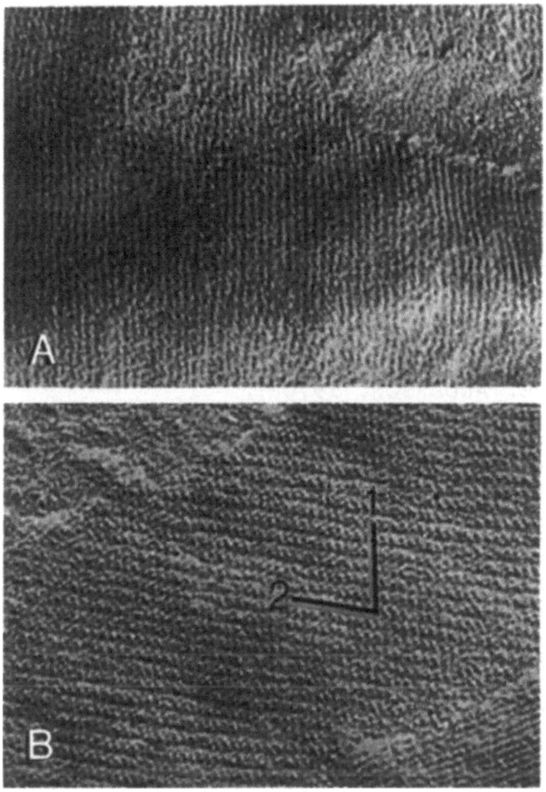

Fig. 9.10. *Strongylocentrotus purpuratus*. Enlargement of Fig. 9.9. (A) Parallel arrangement of filaments 40-50 Å in diameter. (B) Crossed grid network; solid lines indicate directions of 2 sets of parallel lines spaced about 100 Å (1) and 160 Å (2) apart. x 122,000. (From INOUE and HARDY, 1971)

always limited to 70°-80°, indicating the presence of binding-site loci along the fibrils. The peripheral layers are inserted by material of a thickness of about 200 Å, which originally came from the vitelline membrane. Altogether the fertilization membrane is about 500-600 Å thick (INOUE et al., 1970a, 1971).

After prolonged exposure to anti-egg gamma globulin, unfertilized *Paracentrotus lividus* eggs under light microscopy showed no changes, but their fertilizability was greatly depressed (BAXANDALL, 1966). Added sperm fertilized about 40% of the eggs. Electron microscopy revealed that unfertilized eggs were covered by a surface layer 400 Å thick while fertilized eggs were not. Compared with untreated eggs, antibody-treated, fertilized eggs showed delayed cortical reaction (BAXANDALL, 1966).

In the ultrastructure of eggs treated with nicotine, trypsin, soy bean trypsin inhibitor, and chloral hydrate - substances known to induce polyspermy - hyaline layer formation was incomplete, implying that the hyaline layer prevents the penetration of supernumerary spermatozoa (LÖNNING, 1965, 1967b). Eggs treated 10-30 minutes with nicotine concentrations 1/10,000 - 1/1,000 volume/volume showed no cortical reaction. Eggs of *Echinus esculentus*, *Paracentrotus lividus*, and *Psammechinus microtuberculatus* showed cortical swelling and indistinct granules. Changes occurred in the mitochondria and yolk granules, and fertilization membrane and hyaline layer formation was inhibited (LÖNNING, 1965, 1967b).

In *Arbacia punctulata* eggs made polyspermic by varying concentrations of nicotine, cortical response was morphologically the same as that observed in control preparations (LONGO and E. ANDERSON, 1970a). The cortical response was not slower and calyx activation and hyaline layer formation were not different in nicotine-treated eggs. But these events were complete before parallel events occurred in untreated eggs. Nicotine seems to promote greater gamete adhesion, allowing more sperm than normal to fuse with the egg. After they are incorporated into the egg, the spermatozoa change structurally as male pronuclei normally do in monospermic eggs (LONGO and E. ANDERSON, 1970a).

Eggs of *Arbacia punctulata* treated with 3% urethane for 30 seconds and then treated with 0.3% urethane and inseminated are also polyspermic and fail to show normal cortical reaction. On insemination the vitelline layer does not separate or else rises only slightly above the oolemma. 1 to 6 minutes later, nearly all the cortical granules, still intact, are dislodged from the plasmalemma. From 6 minutes after insemination until 2-cell stage, cortical granules are sometimes released at random along the zygote surface. Hyaline layer thickness seems to depend on the number of cortical granules released. Vitelline layer separation from the egg surface without cortical granule breakdown suggests that the 2 events may not be causally related. Although eggs treated with urethane are polyspermic and show variations in cortical breakdown, separation of vitelline layer, and fusion of male pronuclei remains essentially the same as in monospermic eggs. Fertilization events involving the pronuclei thus seem independent of typical cortical response and cortical granule release (LONGO and E. ANDERSON, 1970b).

In eggs activated by a hypotonic and hypertonic medium, cortical granule breakdown and diffuse layer elevation begin but are usually incomplete. The resulting membrane is low and thin often (LÖNNING, 1967d; RUNNSTRÖM, 1968). Eggs of *Arbacia punctulata* artificially activated with hypertonic seawater lack the wave-like cortical granule release seen in inseminated eggs (SACHS and E. ANDERSON, 1970). The perivitelline space formed as a result of cortical reaction is smaller, although fusion and vesiculation are the same as in inseminated eggs. 85 minutes after activation but before nuclear division, annulate lamellae, smooth endoplasmic reticulum, and microtubules organize around a centriole. A streak stage occurs prior to nuclear envelope breakdown. The streak is a central core of annulate lamellae encompassed by endoplasmic reticulum and vesicular components. Chromatin condensation is followed by formation of the mitotic apparatus. Although absent from the mature egg, centrioles are present after activation and before the first nuclear division in the 4-cell embryo, the multicellular embryo, and the blastula (SACHS and E. ANDERSON, 1970).

The 2 meiotic divisions of the sea urchin egg are complete before fertilization. Not until maturity - that is in eggs which have completed the 2 meiotic divisions - does the egg cortex respond properly to the activating stimulus of an attached spermatozoon. Mature eggs may be fertilized only by the sperm of the same species or of closely related species, but oocytes may be penetrated by spermatozoa of unrelated animal groups. From the addition of spermatozoa of the mussel *Cyprina islandica* and the nemertine *Malacobdella grossa* to oocytes and mature eggs of *Psammechinus miliaris*, no spermatozoa attachment to the surface of the mature egg resulted, although spermatozoa of both species were found attached to the oocyte surface (AFZELIUS, 1972a, 1972b). Egg-sperm interaction was different in the 2 species and different from the interaction between sea urchin sperm and sea urchin oocyte. The nemertine sperm was incorporated in the oocyte cortex by phagocytosis. Thin protrusions from the sea urchin oocyte partly surrounded the mussel sperm and extended along most of the sperm head. Following sperm entry into the oocyte cytoplasm, a fibrillar mass - evidently derived from the sperm nucleus - and microtubules in the form of an aster developed in the cytoplasm below the reception cone. The center of the aster may contain 1 or 2 centrioles, and one of them may have a striated rootlet. It is thus possible for a sperm centriole to initiate aster and rootlet formation in the cytoplasm of a foreign species (AFZELIUS, 1972b).

9.4 Pronuclear Development and Fusion

In *Arbacia punctulata* the ooplasm beside the gamete fusion forms a slender protrusion called the fertilization cone or entrance cone (LONGO and E. ANDERSON, 1968)(Fig. 9.5). It contains only ribosomes and filaments, and in its base, vesicles with rod-like elements. On reaching the fertilization cone base, the spermatozoon rotates 180° so that its apex is pointing towards the egg surface. At the same time the nonperforated nuclear envelope vesiculates along its lateral margin and the tightly packed sperm chromatin is dispersed. The nuclear membrane in the anterior and posterior regions of the sperm nucleus remains intact and the sperm mitochondrion and flagellum morphologically unchanged. The male pronucleus becomes heart-shaped and, during aster formation, spherical. Dissociation of the proximal centriole from the sperm flagellum is the first indication that sperm aster development has begun. Composed of ooplasmic, lamellar structures and microtubular bundles, after sperm rotation the aster lies in front of the male pronucleus (LONGO and E. ANDERSON, 1968). The female pronucleus surface facing the advancing male pronucleus becomes highly convoluted. Zygote nuclear formation begins when the inner and outer membrane of the pronuclear envelopes fuse into a small internuclear bridge and a continuous, perforated zygote nuclear envelope. In time the zygote nucleus becomes spheroid. The sperm mitochondrium and flagellum migrate with the sperm aster, keeping their same relative positions throughout pronuclear fusion. In the center of the egg, maternal mitochondria surround the male mitochondrium which, after first cleavage division, seems to undergo progressive degradation and resorption (W.A. ANDERSON, 1968a).

An electron microscopic autoradiographic study of DNA synthesis in the nuclei and cytoplasm of *Paracentrotus lividus* shows that 15-30 minutes after fertilization thymidine-^{3}H starts to be incorporated into the male and female pronuclei of monospermic embryos, indicating that the pronuclei S phase occurs before karyogamy (W.A. ANDERSON, 1969). All differentiated male pronuclei in polyspermic embryos synthesize DNA simultaneously. After Tris-$MgSO_4$, DNase, and trichloroacetic acid treatment, pronuclei lose their radioactivity. Tritiated thymidine is incorporated into the mitochondria and yolk platelets of fertilized sea urchin embryos. Mitochondrial synthesis of DNA begins after fertilization, apparently independent of the pronuclei S phase (W.A. ANDERSON, 1969).

9.5 Mitosis

The interphase nucleus is rather irregularly shaped, and its membrane has regularly spaced pores. Nearby are centrioles, usually in a region without other cytoplasmic organelles. In *Strongylocentrotus purpuratus*, the centrioles are short cylinders about 0.15 µm in diameter and 0.33 µm long. They are made up of a circular array of approximately 9 rods of tubules, and often fine fibers about 150 Å in diameter are associated with them (HARRIS, 1961).

One of the first characteristic features of the formation of the mitotic apparatus is the growth of a mass of radially orientated membranous structures, vesicles, and 15 µm filaments around the centrioles (KUROSUMI, 1958; HARRIS, 1961, 1962)(Fig. 9.11). Yolk and mitochondria appear to be pushed from the center of this aster, while 15 µm filaments radiate out towards the periphery of the cell. As these astral areas grow, the chromosomes condense and the nuclear membrane breaks up into small fragments. In cells, where the nuclear membrane has not yet broken down, no spindle filaments can be seen within the nucleus (HARRIS, 1965). Immediately following the first visible breach in the nuclear membrane, numerous spindle filaments with a diameter of 15 to 20 µm appear within the nuclear region. The filaments become attached to the kinetochores of the chromosomes and side by side form the central spindle. The spindle region contains besides the microtubules all visible components of the ground cytoplasm and more particularly the dense ribonucleoprotein particles, but it excludes all cytoplasmic particulates larger than 40-50 µm (GROSS et al., 1958). Where the microtubules reach the centrosphere region of the aster they suddenly loose the parallel orientation. This centrosphere region, with its disarrayed filaments grows steadily in size from early prophase until it reaches rather huge proportions in late anaphase or in telophase. Till now there has not been observed any direct connection of chromosomal filaments to the centrioles, neither in the first division nor in later embryos, where the aster is reduced or absent. As cleavage progresses, the size of cells and asters decreases. Beyond the 32-cell stage, one often finds figures in which there is little or no aster structure at all. It is possible that asters are a specialization of certain large cells which must separate great masses of cytoplasm during cytokinesis (HARRIS, 1961, 1962, 1965).

As the chromosomes approach the aster at anaphase, their surface is covered with condensed elements of the endoplasmic reticulum, and a double membrane forms which already has pores or annuli. These chromosomes later fuse together and form the nucleus (HARRIS, 1961)(Fig. 9.14A).

During furrowing, the inner furrow region develops a dense layer of cytoplasm called the contractile ring (MERCER and WOLPERT, 1958; WEINSTEIN and HERBST, 1964; GOODENOUGH et al., 1968; TILNEY and MARSLAND, 1969)(Fig. 9.12A). In *Arbacia punctulata* it is amorphous, about 0.1 - 0.2 µm thick, and is made up of amorphous material in which are embedded numerous fine filaments. A filament layer is not found in the cortex away from the furrow region. (About 10 minutes after fertilization similar bundles of fibers 40-50 Å in diameter have been shown to extend from the plasmalemma into the cortical region and out into the projections that remain after cortical granule breakdown! HARRIS, 1968; see p. 279.) Individual contractile ring filaments in *Arbacia punctulata* are 35-60 Å in diameter and occasionally appear hollow (SCHROEDER, 1969, 1972). The contractile ring lasts from about 20

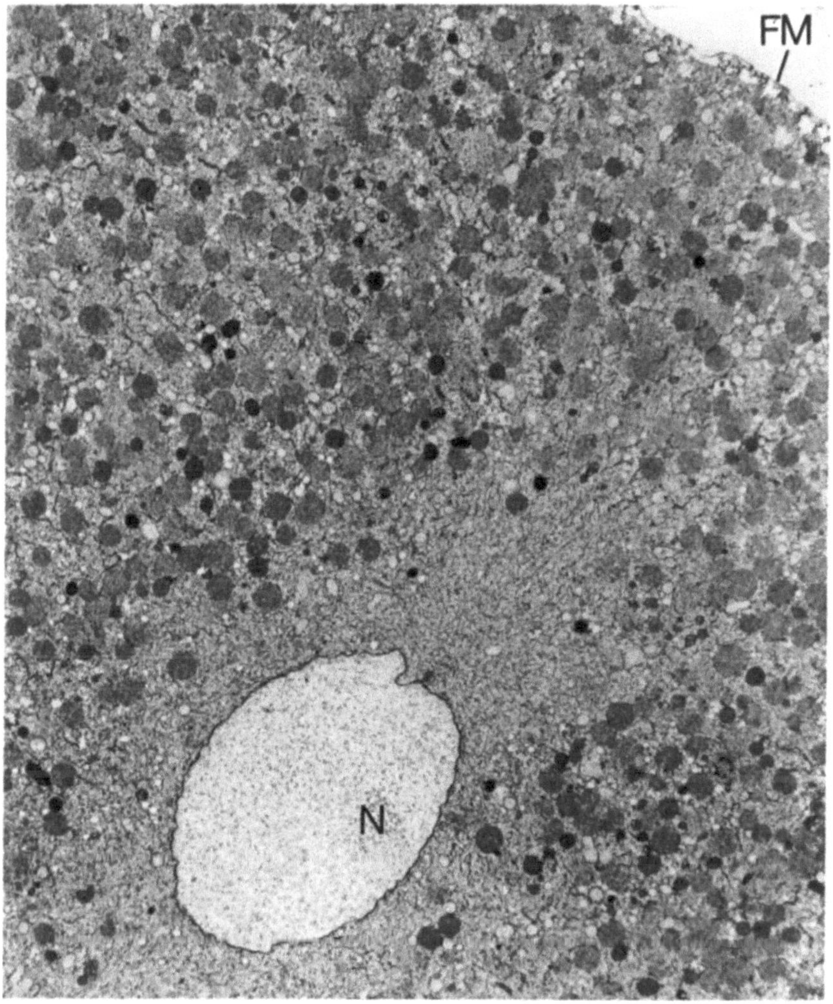

Fig. 9.11. *Paracentrotus lividus* 30 min after fertilization. Nuclear membrane has not yet broken down, but yolk and mitochondria seem to be pushed from center of egg where radially orientated membranous structures are seen. FM, fertilization membrane; N, nucleus. x 4,500.

seconds after anaphase to the end of furrowing activity (for 6 to 7 minutes at 20°C). It is closely associated with the plasma membrane at all times and is probably assembled there. It is about 8 μm wide and 0.2 μm thick throughout cleavage. Its volume decreases, suggesting a contraction-related disassembly of contractile ring filaments rather than a sliding filament mechanism in the strict sense. Cytochalasin B ($>10^{-6}$ M) arrests cleavage within 60 seconds, by which time contractile ring filaments are no longer visible ultrastructurally and the furrow may have receded. Karyokinesis is unaffected. Simultaneous disruption of furrowing activity and the contractile ring largely confirms the vital role of the latter as the organelle of cell cleavage. The contractile ring of *Arbacia punctulata* disappears before

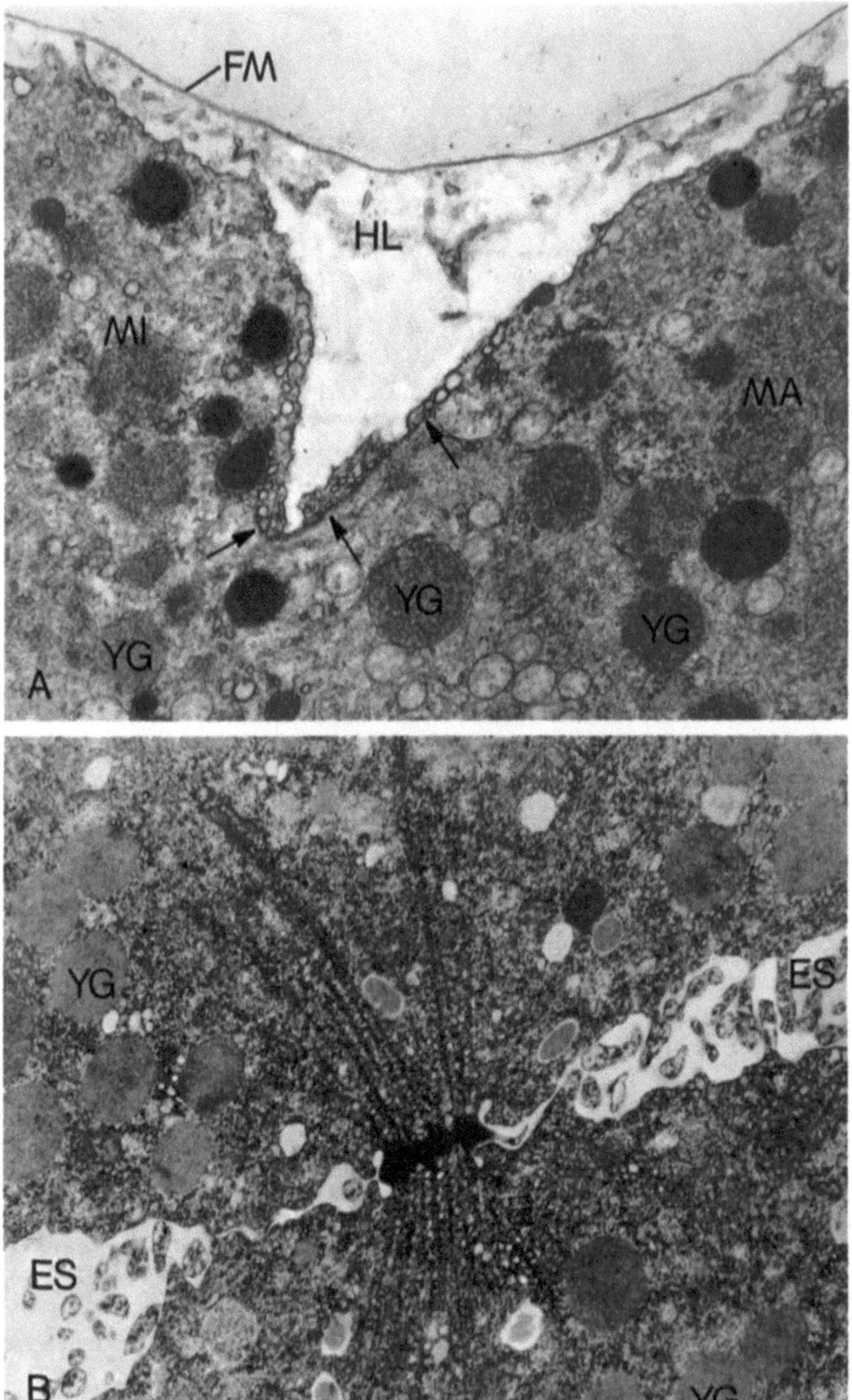
FM
HL
MI
MA
YG
YG
YG
A
YG
ES
ES
B
YG

◀ Fig. 9.12. (A) *Paracentrotus lividus*, fourth division. Dense filament layer (contractile ring) in inner furrow region between macromeres (MA) and micromeres (MI). FM, fertilization membrane; HL, hyaline layer; YG, yolk granules. x 16,800. (B) *Sphaerechinus granularis*, fourth division. Midbody connecting 2 mesomeres. ES, extracellular space; YG, yolk granules. x 15,400

---

the midbody forms (SCHROEDER, 1972)(Fig. 9.12A). A well-developed *Sphaerechinus* midbody is shown (Fig. 9.12B). The last steps of furrowing and new cell membrane production are not yet fully understood.

Annulate lamellae disappear at the time of nuclear membrane breakdown, reappear at the end of first cleavage, and vanish again before second cleavage division (VERHEY and MOYER, 1967; TILNEY and MARSLAND, 1969(Fig. 9.13).

9.6 From 2-Cell Stage to Gastrula

At the end of the first division, *Dendraster excentricus* blastomeres are spherical and have limited surface contact, but 6 to 10 minutes later, as the result of post-division adhesion, they are pressed together nearly into hemispheres and have extensive surface contact. The electron microscope shows that at points of very close contact the plasma membrane of 2 blastomeres touch each other directly like in "tight junctions". Other areas are 25-50 Å apart (VACQUIER and MAZIA, 1968a).

At the 16-cell stage mesomeres, macromeres, and micromeres differ clearly in size. The micromeres have a prolonged interphase and show an extreme RNA synthesis. RNA which is synthesized in the micromeres is assumed to determine the differentiation of the macromeres (CZIHAK, 1965; CZIHAK and HÖRSTADIUS, 1969). While the plasma membranes (Figs. 9.14-9.16) between mesomeres and mesomeres, and mesomeres and macromeres are completely closed during interphase, the boundary between macromeres and micromeres consists of a vesicle plate, which in sections is to be seen as a row of little vesicles (MERCER and WOLPERT, 1958; AGRELL, 1966; HAGSTRÖM and LÖNNING, 1965, 1969; SCHMEKEL, 1969) (Figs. 9.14B, 9.15). Between the vesicles are cytoplasmic bridges that may be used to transport bigger particles rapidly from micromeres to macromeres at the time of the maximum RNA synthesis in the micromeres. A vesicle plate too is the boundary between large and small micromeres in the 32-cell stage (Fig. 9.14B). Such a vesicle plate is found in many dividing cells but normally not persistently in interphase (Fig. 9.16A). While in *Paracentrotus lividus* (AGRELL, 1966) no syncytial connections between cells were seen once the cell membrane was established, combined time-lapse and electron microscopic study in several other species showed that macromeres and micromeres are at first separated by a well defined cell membrane that dissolves into vacuoles and then reforms (HAGSTRÖM and LÖNNING, 1969).

During early embryogenesis, mitosis divides the spherical egg into a blastula the epithelium of which is 1 layer thick and which has a large blastocoel. In some species, cells are consistently close together, but in most, from 2-cell stage to gastrula, they are fairly far apart except at their apicolateral margins. There septate desmosomes develop (Figs. 9.19, 9.20)(in *Paracentrotus miliaris*, 5 1/2 hours after fertilization). At the same time on the inner surface of the cell fibrous

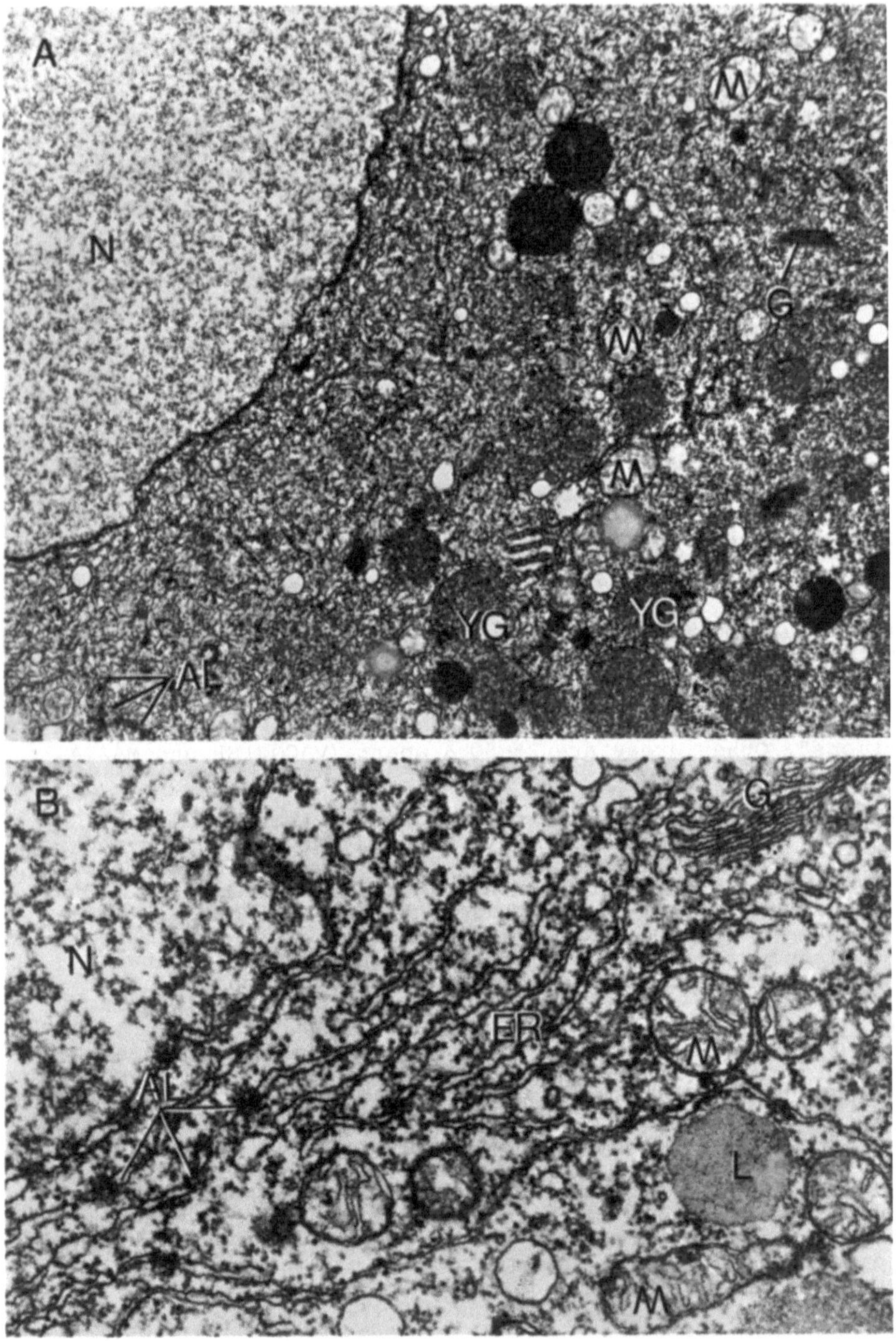

Fig. 9.13. *Paracentrotus lividus* 30 min after fertilization. Annulate lamellae (AL) do not disappear before nuclear membrane breakdown and are continuous with endoplasmic reticulum (ER). G, Golgi's apparatus; L, lipid droplet; M, mitochondrium; N, nucleus; YG, yolk granules. A, x 15,400; B x 44,000

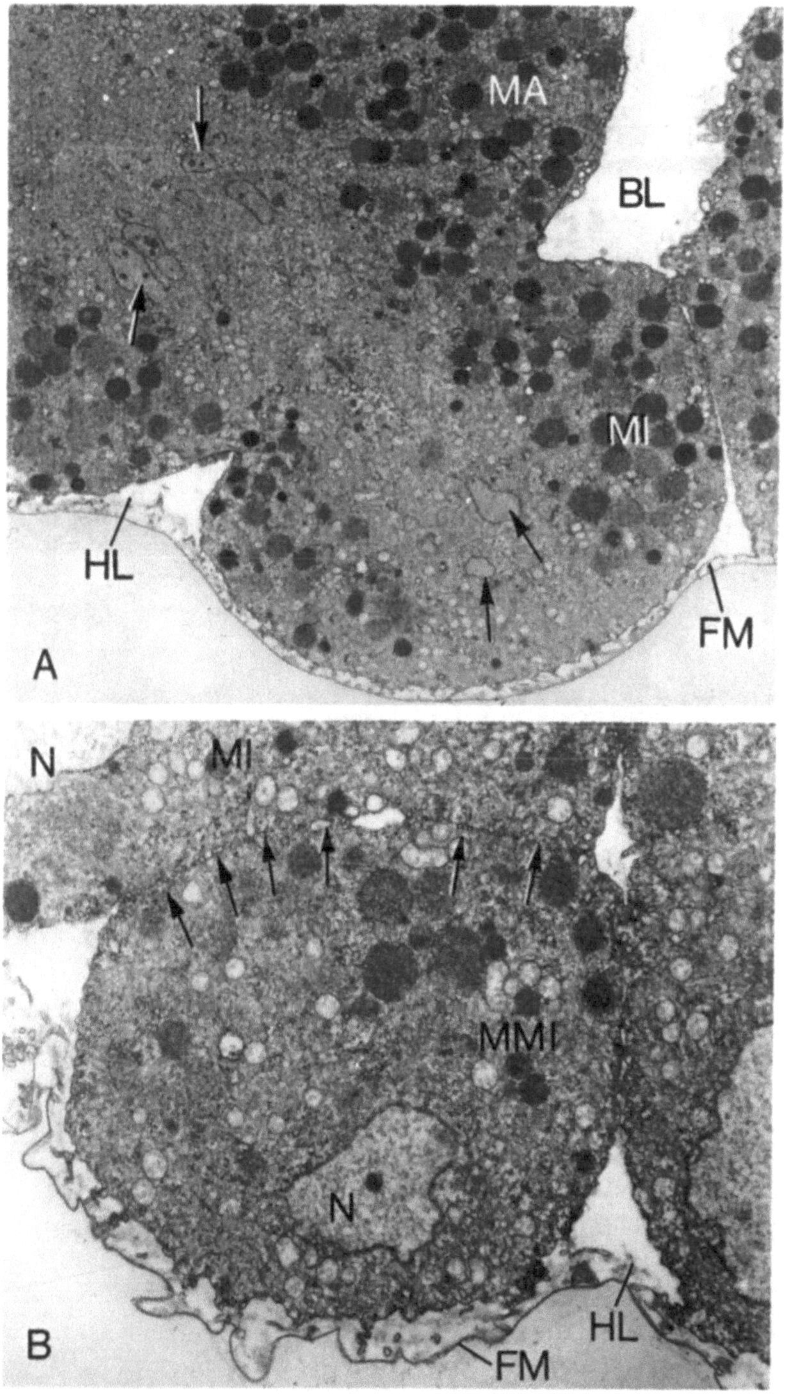

Fig. 9.14. (A) *Paracentrotus lividus* at beginning of 16-cell stage. Section through macromere (MA) and micromere (MI). Each chromosome is surrounded by a vesicle (↑)("Kernbläschen"). BL, blastocoel; FM, fertilization membrane; HL, hyaline layer. x 3,910. (B) *Paracentrotus lividus* at interphase of 32-cell stage. Large (MI) and small (MMI) micromeres are separated by vesicle plate (↑). FM, fertilization membrane; HL, hyaline layer; N, nucleus. x 10,000

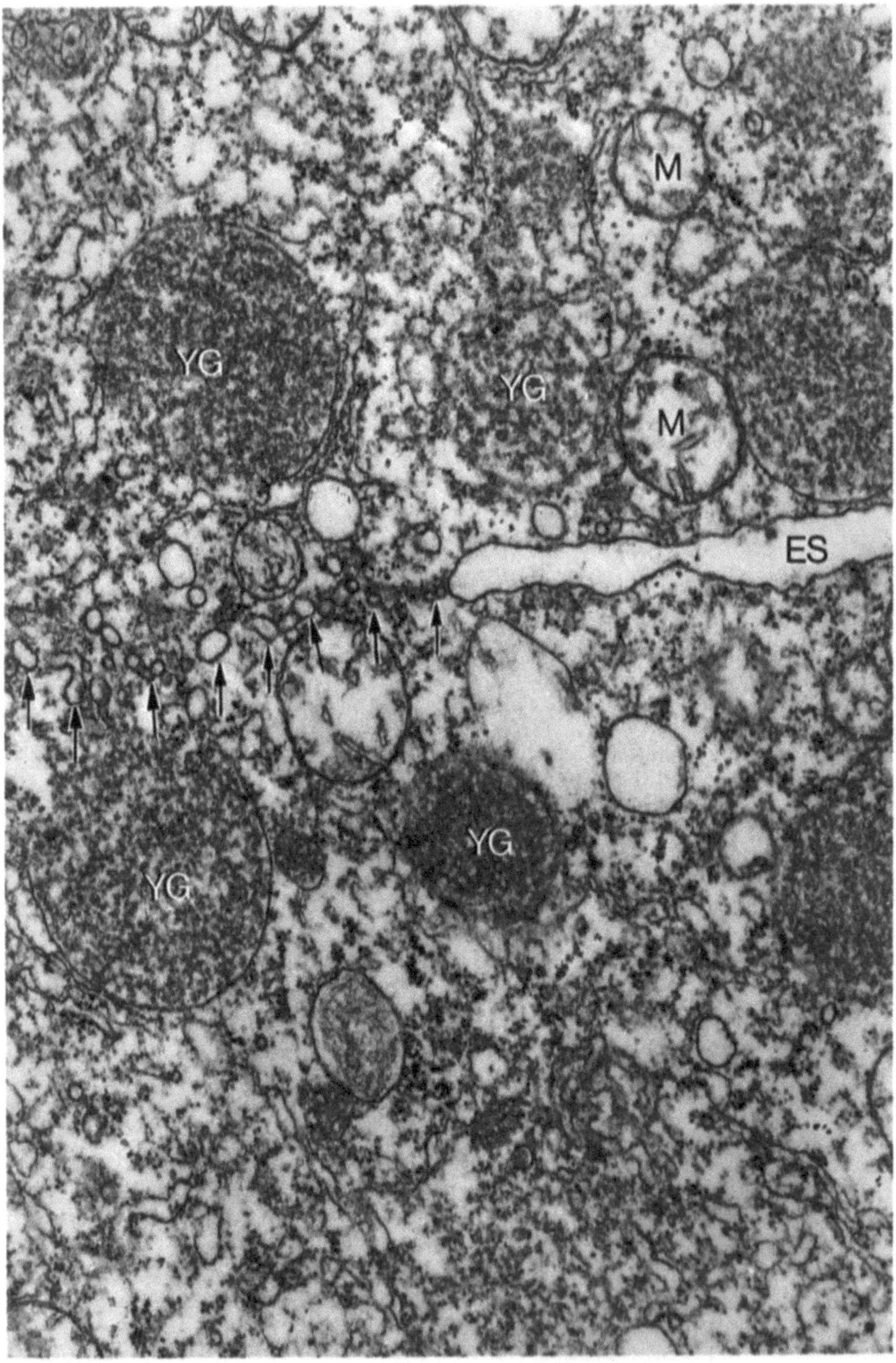
M
M
YG
YG
ES
YG
YG

◀ Fig. 9.15. *Paracentrotus lividus*. Boundary between macromere and micromere is formed by vesicles (↑) during interphase of 16-cell stage. ES, extracellular space; M, mitochondrium; YG, yolk granules. x 46,800

---

material forms a basement layer (Figs. 9.19, 9.20). In the early embryo, surface interdigitations are rare except where 3 or more cells come together (BALINSKY, 1959; WOLPERT and MERCER, 1963; MILLONIG and GIUDICE, 1967; GIUDICE and MUTOLO, 1970).

At the blastula stage, blastomeres are cone- or cube-shaped and show marked structural polarity. In *Paracentrotus lividus* the nucleus is near the inner surface of the cell, the Golgi apparatus is at the apical end, and most of the rough endoplasmic reticulum is at the base. In *Arbacia punctulata*, however, the nucleus is at the apical end of the cell (GIBBINS et al., 1969). The cytoplasm of cells at the blastula and gastrula stages is similar to that of unfertilized eggs. Yolk granules and mitochondria are distributed at random in each cell. The Golgi's complex is much more elaborated, and there are no heavy bodies. Each blastomere has 1 cilium. The blastula cells of *Arbacia punctulata* contain a great many microtubules that converge and make contact with 3 cytoplasmic foci or satellites associated with the basal body of the cilium. At 0°C, the microtubules disassemble while the satellites remain distinct. On warming, the microtubules reappear, 1 end of each seems in contact with its respective satellite. It is possible, therefore, that the assembly of microtubules is initiated at the satellites (TILNEY and GODDARD, 1970).

Before gastrulation, the cells that become the primary mesenchyme lose their cilia, their septate desmosomes disintegrate, and evaginations of the basal surface develop ("basal lobe"; GIBBINS, TILNEY, and PORTER, 1969). These lobes become long, slender protrusions orientated towards the oral field, where the stomodeum will develop (CZIHAK and MEYER, 1964). Later primary mesenchymal cells migrate into the blastocoel where they become spherical (WOLPERT and GUSTAFSON, 1967), each containing a centriole from which microtubules radiate in all directions. They develop slender pseudopods, that contain large numbers of microtubules. The pseudopods of adjacent cells fuse into a "cable" of cytoplasm, a true syncytium, in which the larval skeleton later will be deposited. The cell body containing the nucleus and much of the cytoplasm remains distinct from adjacent cell bodies. Thin stalks connect the cell bodies and the cytoplasmic cable (GIBBINS, TILNEY, and PORTER, 1969; HAGSTRÖM and LÖNNING, 1969)(Figs. 9.16B, 9.17). To test the hypothesis that cytoplasmic microtubules are involved in cell shape development during formation and differentiation of primary mesenchyme, embryos were treated with 2 types of agents that affect microtubules: colchicine and hydrostatic pressure which cause microtubules to disassemble, and D_2O which tends to stabilize them. In sea urchin gastrulae treated with colchicine and hydrostatic pressure, development of primary mesenchyme ceases, microtubules disappear, and cells tend to become spherical. Treatment with D_2O stops development, microtubules appear frozen, and cells remain asymmetric. These agents appear to block development primarily by interfering with the sequential disassembly and reassembly of microtubules into new patterns. Microtubules therefore appear to influence the development of cell form but do little to maintain cell shape in differentiated tissue (TILNEY and GIBBINS, 1966, 1969a).

In cultures of isolated *Paracentrotus* micromeres, the precursors of the primary mesenchyme give rise to aggregates resembling morulae, in which the cells form syncytia much like those of the normal primary

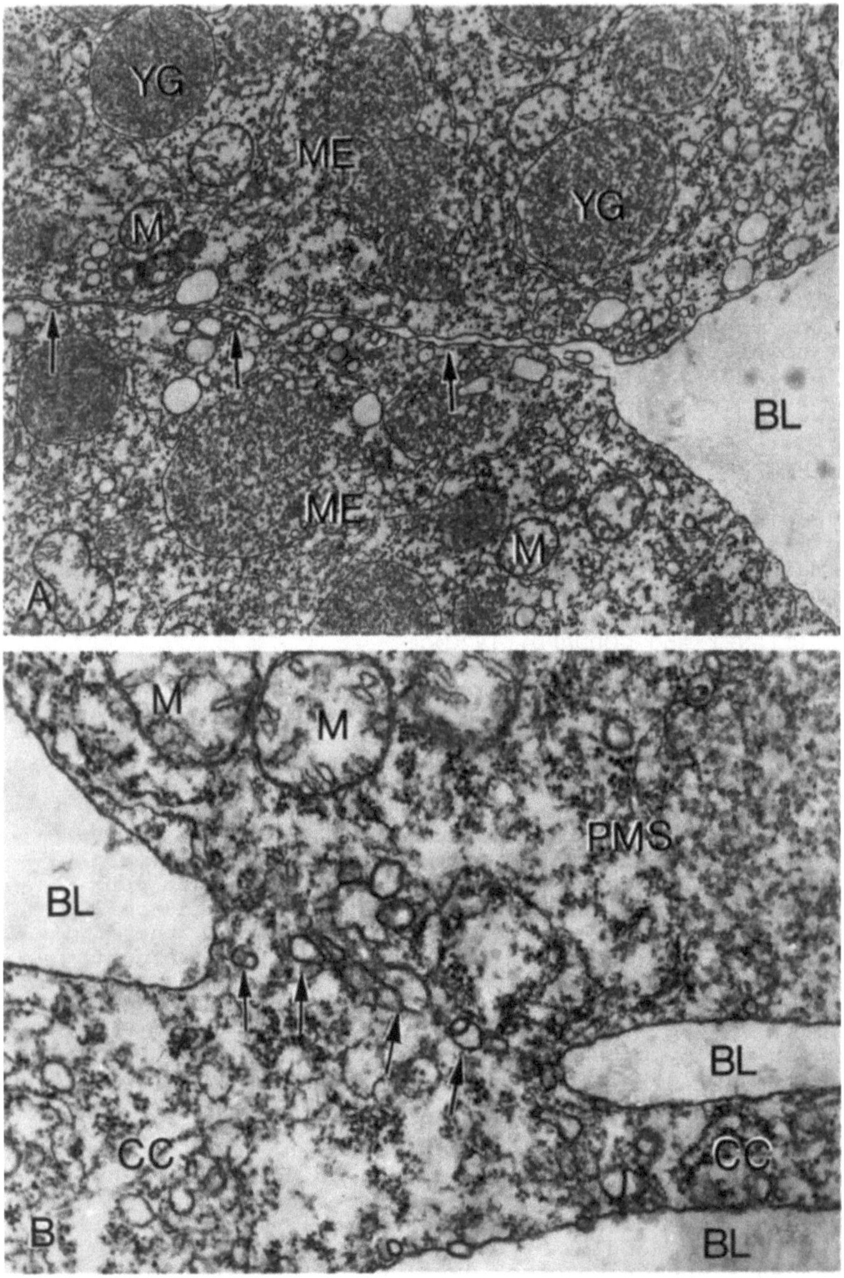

Fig. 9.16. (A) *Paracentrotus lividus*. During interphase of 16-cell stage, membranes (↑) between 2 mesomeres (ME) are closed. BL, blastocoel; M, mitochondrium; YG, yolk granules. x 22,000. (B) *Paracentrotus lividus* 16 hr after fertilization. Primary mesenchymal cells (PMS) have formed cytoplasmic cable (CC). Stalks connecting cell bodies to cable, contain vesicle rows (↑). BL, blastocoel; M, mitochondrium. x 44,000

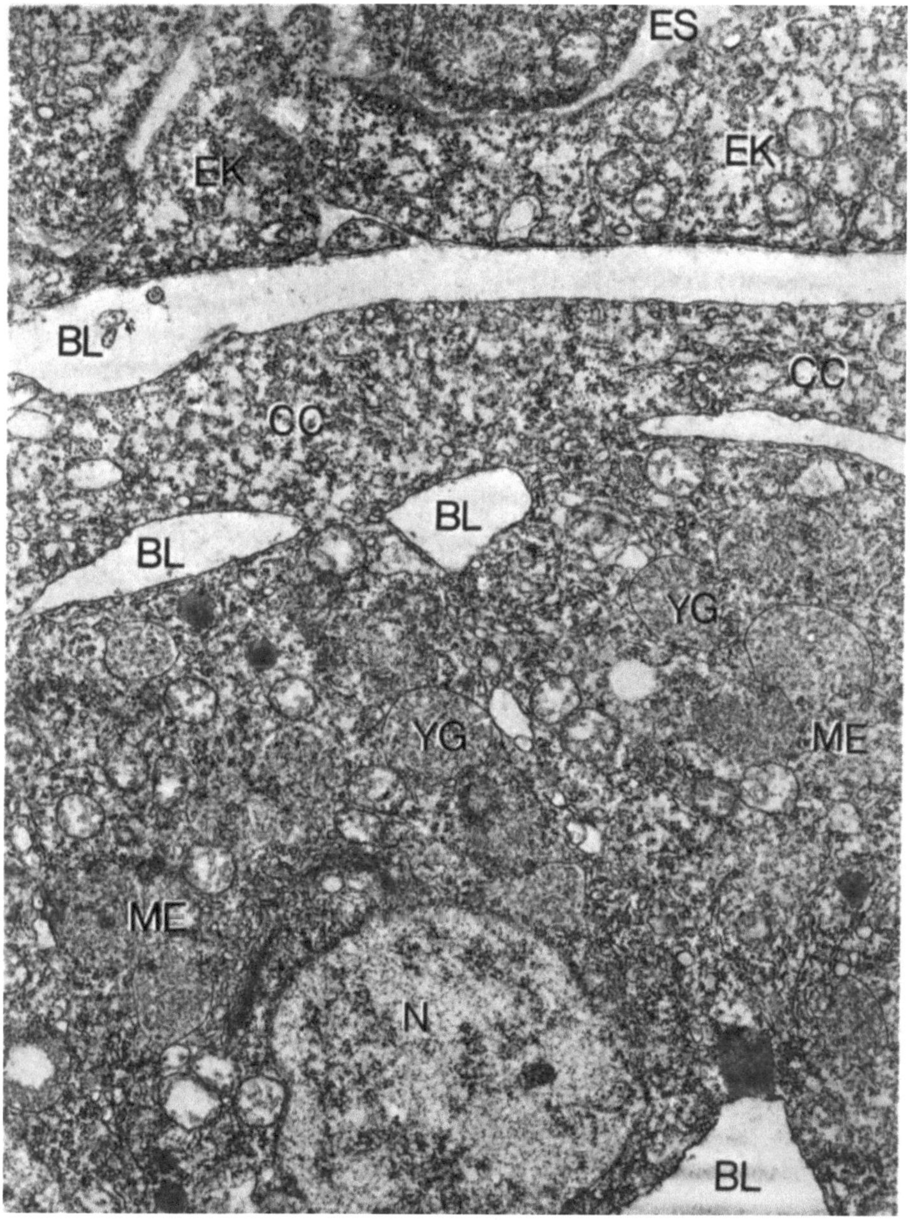

Fig. 9.17. *Paracentrotus lividus* gastrula 16 hr after fertilization. Syncytium of primary mesodermal cells (ME) forms cytoplasmic cable (CC). Stalks connect cell bodies to cable. BL, blastocoel; EK, ectodermal cells; ES, extracellular space; N, nucleus; YG, yolk granules. x 19,000

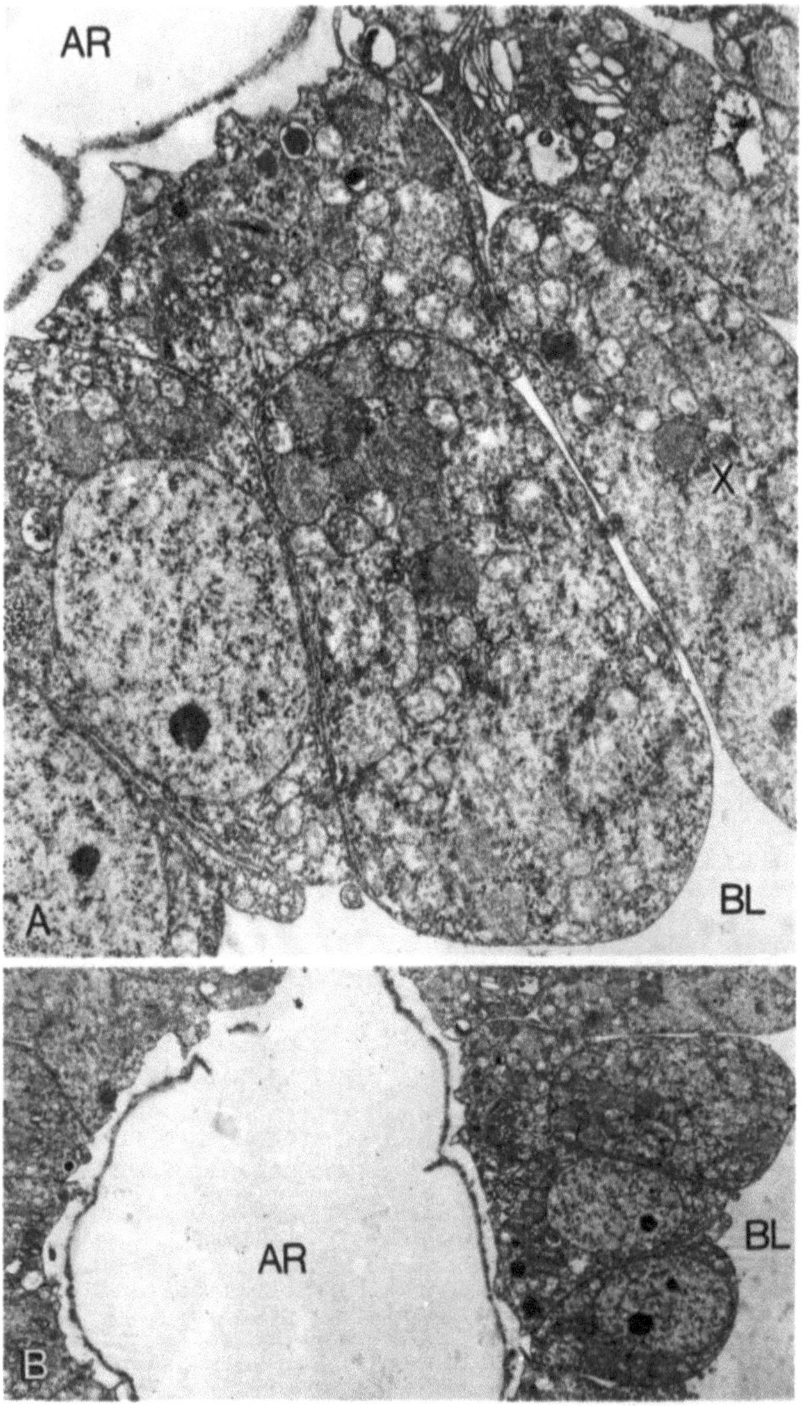

Fig. 9.18. *Paracentrotus lividus* gastrula 16 hr after fertilization. One entodermal cell (X) migrates into blastocoel and forms secondary mesenchyma. AR, archenteron; BL, blastocoel. A, x 12,500; B, x 4,500

mesenchyme of *Paracentrotus*. The plasma membrane of neighboring cells is no longer continuous, and rows of small vesicles occur between cells (PUCCI-MINAFRA, BOSCO, and GIAMBERTONE, 1968).

Toward the end of archenteron invagination, the secondary mesenchyme is derived from the archenteron roof (Fig. 9.18). Soon after their formation the secondary mesenchymal cells send out long filopods to explore the blastocoel wall. The filopods are straight, bristle-like cytoplasmic extensions. Microtubules are common in the cell body and at the base of the filopods but rare at the filopodal tip which is filled with 50 Å filaments. Embryos were treated with colchicine and hydrostatic pressure which stop archenteron movement and cause the microtubules to disintegrate. Hydrostatic pressure causes the disappearance of the filaments as well. The small number of microtubules in the filopods and the fact that in no other system there is any evidence for contraction of these elements suggest that they do not function in the contraction process but that they are probably involved in the formation of filopods. The 50-Å filaments on the other hand may well be responsible for filopodal contraction (TILNEY and GIBBINS, 1969b).

Cilia begin to develop at the blastula stage. Each blastomere first bears 1 cilium. There are cilia which look perfectly normal (9 + 2 aspect: *Arbacia punctulata*; TILNEY and GODDARD, 1970) while others have all characteristics of the choanocytes found in the ciliary bands of *Asterias brachiolaria* larvae and in the ambulacral ampullae and coelomic epithelia of adult echinoderms (NORREWANG and WINGSTRAND, 1970; TILNEY and GODDARD, 1970)(Figs. 9.19, 9.20, 9.21; own observations): The flagellum is situated on the elevated apical cell surface surrounded by a very regular collar of cylindrical microvilli. A string of mucus-like substance covers the inner edge of the microvilli. The flagellum has the typical 9 plus 2 filament pattern. Peripheral filaments end in a basal body situated just below the apical cell surface, while the central filaments end fairly high in the free shaft of the flagellum where a transverse disc of dense matter is developed. A second centriole normally occurs at right angles to the flagellar centriole. The base of the flagellum is markedly constricted below the transvers disc. A cross-striated, ciliary rootlet extends from the basal body into the cytoplasm.

In early gastrulae the only axial differences in the ultrastructure of cytoplasm are a closer packing of the cortical bodies in the cells that will become the entoderm and regional differences in mitochondrial size (BERG and LONG, 1964; RUNNSTRÖM, 1969). While in the 16-cell stage, mitochondrial size is uniform in all regions in the mesenchymal blastula, early and late gastrula, mitochondria are larger in invaginating vegetal and primary mesenchymal cells than in the ectoderm. Calculated from profile dimensions of electron micrographs of embryos fixed in buffered osmium solution, the average size of animal-cell mitochondria is 73% that of vegetal-cell mitochondria. Most sections reveal more cristae in the vegetal-cell mitochondria. Besides there seems to be during early development an overall increase in the number of cristae per mitochondrion. In animal-half embryos, cells converted into entoderm as the result of treatment with LiCl contain significantly larger mitochondria (BERG and LONG, 1964; *Lytechinus anamesus*). Electron microscopic studies of mitochondrial distribution in early gastrulae of *Lytechinus anamesus*, *Strongylocentrotus purpuratus*, and *Dendraster excentricus* revealed no difference in the number of mitochondria. Calculations of mitochondrial densities showed uniform distribution in animal, equatorial, vegetal, and primary mesenchymal cells (BERG, TAYLOR, and HUMPHREY, 1962).

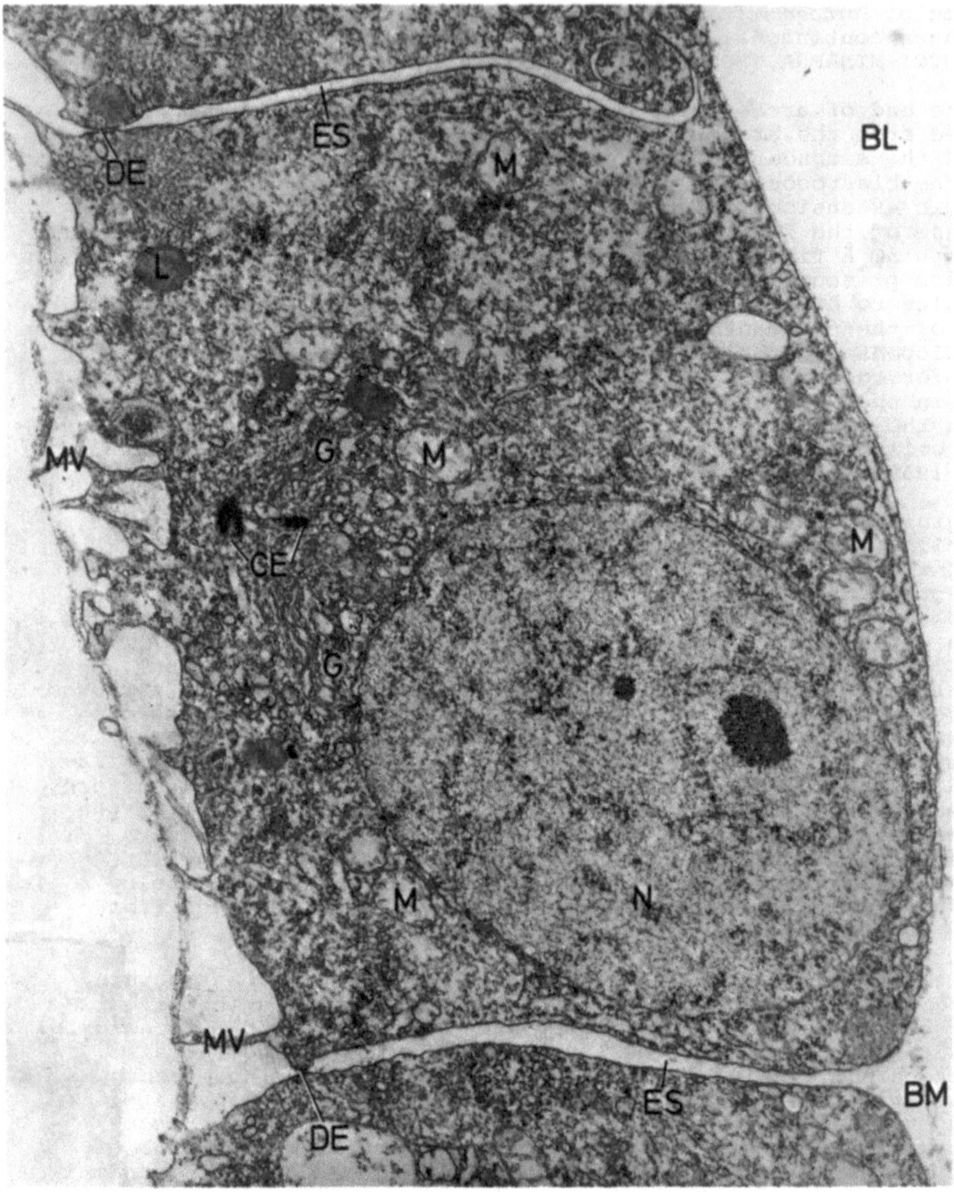

Fig. 9.19. *Paracentrotus lividus* gastrula 16 hr after fertilization. Ectodermal cell has no cilium but typically in right angles arranged centrioles (CE) situated apically. BL, blastocoel; BM, basement layer; DE, septate desmosomes; ES, extracellular space; G, Golgi apparatus; L, lipid droplets; M, mitochondrium; MV, microvilli; N, nucleus. x 24,500

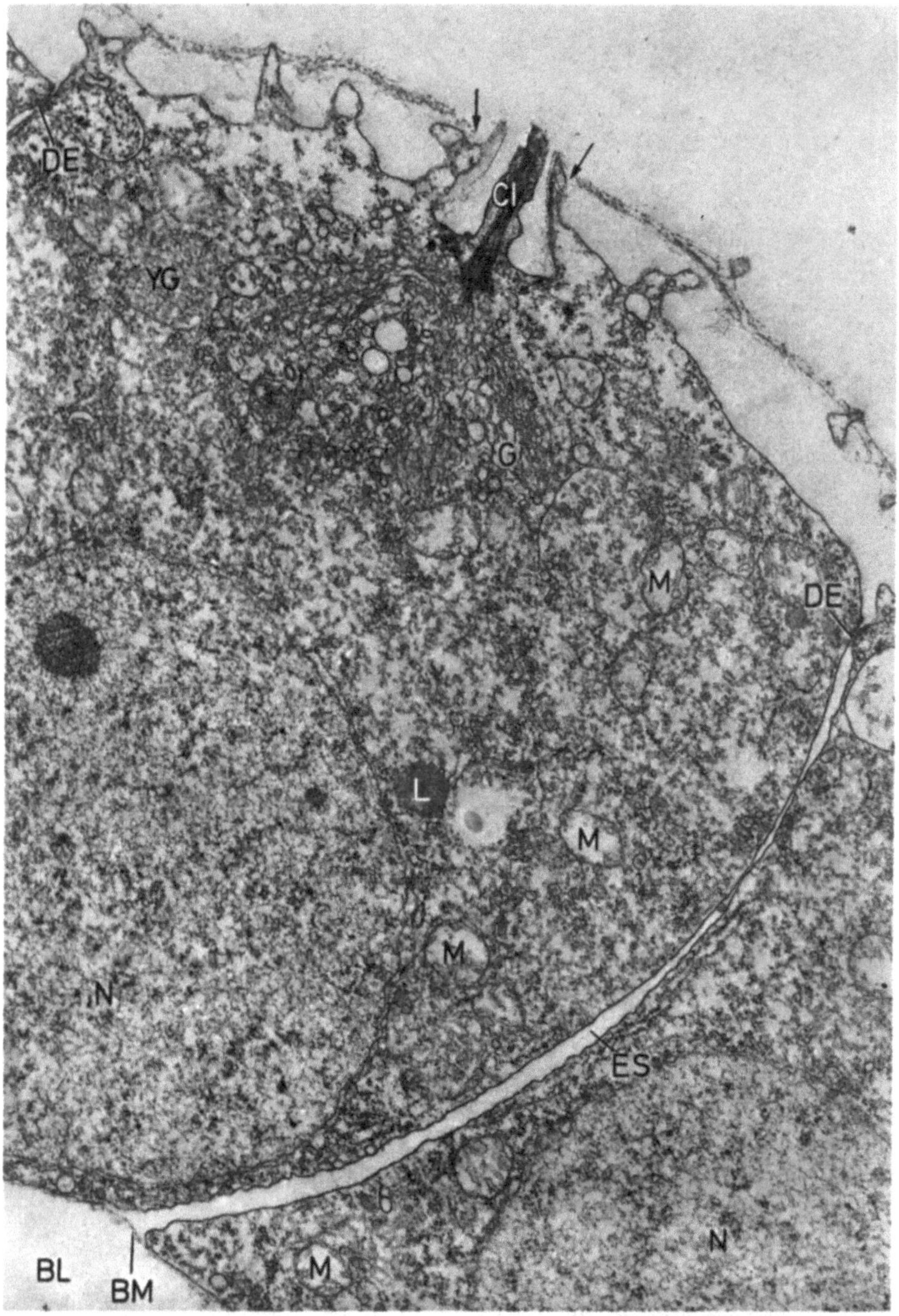

Fig. 9.20. *Paracentrotus lividus* gastrula 16 hr after fertilization. Ectodermal cell is developing single choanocytic cilium (CI) surrounded by microvilli (↑). BL, blastocoel; BM, basement membrane; DE, septate desmosome; ES, extracellular space; G, Golgi apparatus; L, lipid droplet; M, mitochondrium; N, nucleus; YG, yolk granules. x 26,000

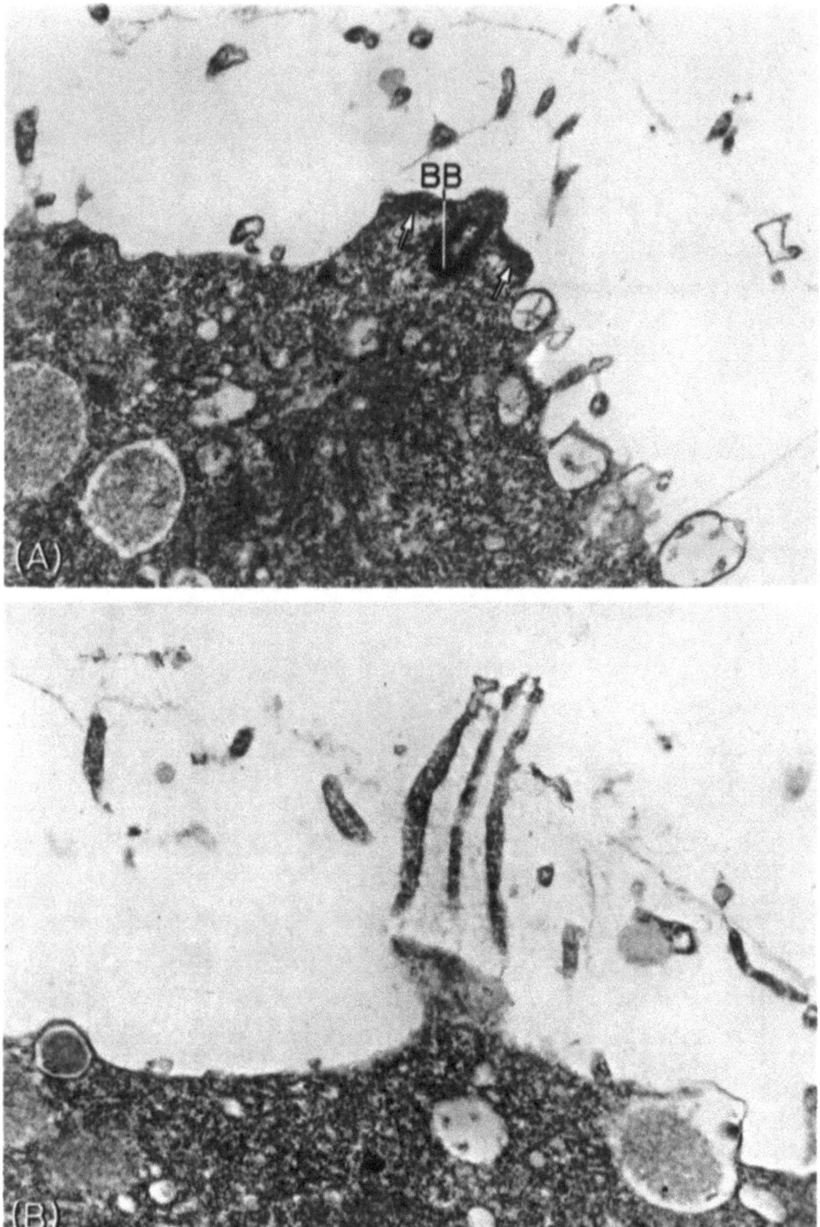

Fig. 9.21. *Paracentrotus lividus* gastrula 21 hr after fertilization. Choanocytic cilium has developed on elevated apical surface of cell. (A) Part of collar of microvilli is cut horizontally. Mucus-like thread is attached to inner edge of microvilli and transverse disc of dense matter (↑). BB, basal body. (B) 3 microvilli are cut tangentially. Cilium base is constricted below transverse disc. A, x 24,200; B, x 26,400

The relationship between structural modifications of the nucleolus and the pattern of nuclear RNA synthesis in *Arbacia punctulata* embryos has been studied by electron microscopic autoradiography. During the cleavage and blastula stages, interphase nuclei contain many dense, fibrillar bodies that do not incorporate uridine although the rest of the nucleus becomes labelled. From late blastula to early gastrula when nuclear RNA synthesis increases significantly, 1 or 2 fibrous nucleoli appear among the chromatin strands. The nucleolus seems to initiate 5-^{3}H-uridine incorporation. During gastrulation, the nucleolus acquires particulate components and becomes positive to cytochemical tests for RNA (KARASAKI, 1966, 1968).

Studies of the twinning of regular and irregular sea urchins showed that in *Dendraster excentricus* embryos dithiothreitol (DTT) destroys blastomere ability to initiate the post-division adhesion process and reverses it once it has formed (VACQUIER and MAZIA, 1968a, 1968b). Treatment with DTT during or after the first cleavage causes twin embryos to develop. In the furrow region, microvilli - projecting from the cell surface between blastomeres produced during exposure to DTT or fixed after cytokinesis before post-division adhesion - serve to pull the blastomeres close together. The retractile ability of the microvilli is destroyed by DTT, post-division adhesion does not occur, and each blastomere develops as a twin although physically apposed to its sister (VACQUIER and MAZIA, 1968a). In difference to the sand dollar, twinning of regular sea urchin does not occur by simple treatment with DTT. But in sea urchin eggs grown in seawater containing pronase, the fertilization membrane is digested and the hyaline layer does not form. When these denuded eggs are exposed to DTT during the first cleavage and returned to normal seawater, the blastomeres separate and each develops a twin (VACQUIER and MAZIA, 1968b).

Studies of normal *Paracentrotus* embryos and lethal hybrids (*Paracentrotus*/*Arbacia*) show that in hybrid blastulae at the time of development arrest, ergastoplasm is atypical containing numerous smooth vesicles scattered throughout it (GEUSKENS, 1968a).

References

AFZELIUS, B.A., 1955a. The ultrastructure of the nuclear envelope of the sea urchin oocyte studied with the electron microscope. Exp. Cell Res. 8, 147-158.

AFZELIUS, B.A., 1955b. The fine structure of the sea urchin spermatozoon as revealed by the electron microscope. Z. Zellforsch. 42, 134-148.

AFZELIUS, B.A., 1956a. The ultrastructure of the cortical granules and their products in the sea urchin egg as studied with the electron microscope. Exp. Cell Res. 10, 257-285.

AFZELIUS, B.A., 1956b. Electron microscopy of Golgi elements in sea urchin eggs. Exp. Cell Res. 11, 67-85.

AFZELIUS, B.A., 1956c. The acrosomal reaction of the sea urchin spermatozoon. Proc. Stockholm Conf. Electron Micr., 167.

AFZELIUS, B.A., 1957. Electron microscopy of the basophilic structures of the sea urchin egg. Z. Zellforsch. 45, 660-675.

AFZELIUS, B.A., 1972a. Reactions of the sea urchin oocyte to foreign spermatozoa. Exp. Cell Res. 72, 25-33.

AFZELIUS, B.A., 1972b. Ultrastructure of species - foreign spermatozoa after penetrating the sea urchin oocyte. Acta Embryol. Exp., 123-133.

AFZELIUS, B.A., MURRAY, A., 1957. The acrosome reaction of spermatozoa during fertilization or treatment with egg water. Exp. Cell Res. 12, 325-337.
AGRELL, J.P.S., 1966. Membrane formation and membrane contacts during the development of the sea urchin embryo. Z. Zellforsch. 72, 12-21.
ANDERSON, E., 1966a. The origin of cortical granules and their participation in the fertilization phenomenon in echinoderms. J. Cell Biol. 31, 5A.
ANDERSON, E., 1966b. The tripartite nature of the oolemma of fertilized echinoderm eggs. J. Cell Biol. 31, 136A.
ANDERSON, E., 1968. Oocyte differentiation in the sea urchin, Arbacia punctulata, with particular reference to the origin of cortical granules and their participation in the cortical reaction. J. Cell Biol. 37, 514-539.
ANDERSON, E., 1970. A cytological study of the centrifuged whole, half, and quarter eggs of the sea urchin Arbacia punctulata. J. Cell Biol. 47, 711-733.
ANDERSON, W.A., 1968a. Structure and fate of the paternal mitochondrion during early embryogenesis of Paracentrotus lividus. J. Ultrastr. Res. 24, 311-321.
ANDERSON, W.A., 1968b. Cytochemistry of sea urchin gametes. II. Ruthenium red staining of gamete membranes of sea urchins. J. Ultrastr. Res. 24, 322-333.
ANDERSON, W.A., 1968c. Cytochemistry of sea urchin gametes. I. Intramitochondrial localization of glycogen, glucose-6-phosphatase, and adenosin triphosphatase activity in spermatozoa of Paracentrotus lividus. J. Ultrastr. Res. 24, 398-411.
ANDERSON, W.A., 1968d. Cytochemistry of sea urchin gametes. III. Acid and alkaline phosphatase activity of spermatozoa and fertilization. J. Ultrastr. Res. 25, 1-14.
ANDERSON, W.A., 1969. Nuclear and cytoplasmic DNA synthesis during early embryogenesis of Paracentrotus lividus. J. Ultrastr. Res. 26, 95-110.
AUSTIN, C.R., 1968. Ultrastructure of fertilization, p. 101. New York: Holt, Rivehart and Winston.
BAL, A.K., 1970. Ultrastructural changes in the accessory-cells and the oocyte surface of the sea urchin Strongylocentrotus droebachiensis during vitellogenesis. Z. Zellforsch. 111, 1-14.
BAL, A.K., JUBINVILLE, F., COUSINEAU, G.H., INOUE, S., 1968. Origin and fate of annulate lamellae in Arbacia punctulata eggs. J. Ultrastr. Res. 25, 15-28.
BALINSKY, B.I., 1959. An electron microscopic investigation of the mechanisms of adhesion of the cells in a sea urchin blastula and gastrula. Exp. Cell Res. 16, 429-433.
BALINSKY, B.I., 1961. The role of cortical granules in the formation of the fertilization membrane on the surface membrane of fertilized sea urchin eggs. Symposium on germ cells and development, 205-219. Pavia: A. Baselli.
BAXANDALL, J., 1966. The surface reactions associated with fertilization of the sea urchin egg as studied by immunoelectron microscopy. J. Ultrastr. Res. 16, 158-180.
BAXANDALL, J., PERLMANN, P., AFZELIUS, B.A., 1964a. Immunoelectron microscope analysis of the surface layers of the unfertilized sea urchin egg. J. Cell Biol. 23, 609-628.
BAXANDALL, J., PERLMANN, P., AFZELIUS, B.A., 1964b. Immunoelectron microscope analysis of the surface layers of the unfertilized sea urchin egg. J. Cell Biol. 23, 629-650.
BERG, W.E., LONG, N.D., 1964. Regional differences of mitochondria size in the sea urchin embryo. Exp. Cell Res. 33, 422-437.

BERG, W.E., TAYLOR, D.A., HUMPHREYS, W.J., 1962. Distribution of mitochondria in echinoderm embryos as determined by electron microscopy. Devel. Biol. 4, 165-176.
BERNSTEIN, M.H., 1962. Normal and reactive morphology of sea urchin spermatozoa. Exp. Cell Res. 27, 197-209.
COLLIER, J.R., 1959. The effect of homologous fertilizin on the sperm of Strongylocentrotus purpuratus. Acta Embryol. Morphol. Exp. 2, 163-170.
CZIHAK, G., 1965. Entwicklungsphysiologische Untersuchungen an Echiniden (Ribonucleinsäure-Synthese in den Mikromeren und Entodermdifferenzierung. Ein Beitrag zum Problem der Induktion). Arch. f. Entw. Mech. 156, 504-524.
CZIHAK, G., HÖRSTADIUS, S., 1969. Transplantation RNS-markierter Mikromeren des Seeigelkeimes (Versuche zur Klärung eines Induktionsvorganges). Zool. Anz. Suppl. 33, Verh. Zool. Ges. 1969, 138-141.
CZIHAK, G., MEYER, G.F., 1964. Differentiation of the oral field of the sea urchin embryo. Nature 201, 315.
DAN, J.C., 1952. Studies on the acrosome. I. Reaction to egg-water and other stimuli. Biol. Bull. 103, 54-66.
DAN, J.C., 1954a. Studies on the acrosome. II. Acrosome reaction in starfish spermatozoa. Biol. Bull. 107, 203-218.
DAN, J.C., 1954b. Studies on the acrosome. III. Effect of calcium deficiency. Biol. Bull. 107, 335-349.
DAN, J.C., 1970. Morphogenetic aspects of acrosome formation and reaction. Adv. in Morphogenesis 8, 1-39.
DAN, J.C., KUSHIDA, H., FUJITA, K., 1966. Fine structure of the sea urchin acrosome. VI. Int. Congr. Electronenmicr., 645.
DAN, J.C., OHORI, Y., KUSHIDA, H., 1964. Studies on the acrosome. VII. Formation of the acrosomal process in sea urchin spermatozoa. J. Ultrastr. Res. 11, 508-524.
DIRKSEN, E.R., 1961. The presence of centrioles in artificially activated sea urchin eggs. J. Biophys. Biochem. Cytol. 11, 244-247.
ENDO, Y., 1961. Changes in the cortical layers of sea urchin eggs at fertilization as studied with the electron microscope. Exp. Cell Res. 25, 383-397.
FRANKLIN, L.E., 1965. Morphology of gamete membrane fusion and of sperm entry into oocytes of the sea urchin. J. Cell Biol. 25, 81-112.
FRANKLIN, L.E., METZ, C.B., 1962. Electron microscope study of sperm entry into sea urchin oocytes. Biol. Bull. 123, 473.
GEUSKENS, M., 1968a. Etude ultrastructurale des embryons normaux et des hybrides létaux entre echinodermes. Exp. Cell Res. 49, 477-487.
GEUSKENS, M., 1968b. Etude au microscope électronique de l'action de l'hydroxyurée sur les embryons. I. Embryons d'oursins. Exp. Cell Res. 52, 608-620.
GIBBINS, J.R., TILNEY, L.G., PORTER, K.R., 1969. Microtubules in the formation and development of the primary mesenchyme in Arbacia punctulata. I. The distribution of microtubules. J. Cell Biol. 41, 201-226.
GIUDICE, G., MUTOLO, V., 1970. Reaggregation of dissociated cells of sea urchin embryos. Adv. in Morphogenesis 8.
GOODENOUGH, D.A., ITO, S., REVEL, J.P., 1968. Electron microscopy of early cleavage stages in Arbacia punctulata. Biol. Bull. 135, 420-421.
GROSS, P.R., PHILPOTT, D.E., NASS, S.J., 1958. The fine structure of the mitotic spindle in sea urchin eggs. J. Ultrastr. Res. 2, 55-72.
GROSS, P.R., PHILPOTT, D.E., NASS, S., 1960. Electron microscopy of the centrifuged sea urchin egg, with a note on the structure of the ground cytoplasm. J. Biophys. Biochem. Cytol. 7, 135-142.
HAGSTRÖM, B.E., LÖNNING, S., 1965. Studies of cleavage and development of isolated sea urchin blastomeres. Sarsia 18, 1-9.

HAGSTRÖM, B.E., LÖNNING, S., 1967. Experimental studies of Strongylocentrotus droebachiensis and S. pallidus. Sarsia 29, 165-176.
HAGSTRÖM, B.E., LÖNNING, S., 1969. Time-lapse and electron microscopic studies of sea urchin micromeres. Protoplasma Wien 68, 271-288.
HARRIS, P.J., 1961. Electron microscope study of mitosis in sea urchin blastomeres. J. Cell Biol. 11, 419-431.
HARRIS, P.J., 1962. Some structural and functional aspects of the mitotic apparatus in sea urchin embryos. J. Cell Biol. 14, 475-487.
HARRIS, P.J., 1965. Some observations concerning metakinesis in sea urchin eggs. J. Cell Biol. 25 (2), 73-77.
HARRIS, P.J., 1967a. Nucleolus-like bodies in sea urchin eggs. Am. Zool. 7, 753.
HARRIS, P.J., 1967b. Structural changes following fertilization in the sea urchin egg. Exp. Cell Res. 48, 569-581.
HARRIS, P.J., 1968. Cortical fibers in fertilized eggs of the sea urchin Strongylocentrotus purpuratus. Exp. Cell Res. 52, 677-681.
HARVEY, E.B., ANDERSON, T.F., 1943. The spermatozoon and fertilization membrane of Arbacia punctulata as shown by the electron microscope. Biol. Bull. 85, 151-156.
HILLIER, J., LANSING, A.J., ROSENTHAL, T.B., 1952. Electron microscopy of some marine egg surfaces. Biol. Bull. 103, 293.
INOUE, S., HARDY, J.P., 1971. Fine structure of the fertilization membranes of sea urchin embryos. Exp. Cell Res. 68, 259-272.
INOUE, S., BUDAY, A., COUSINEAU, G.H., 1970a. Developing fertilization membranes. Exp. Cell Res. 59, 343-346.
INOUE, S., BUDAY, A., COUSINEAU, G.H., 1970b. Observations of sea urchin spermatozoa with the surface replica method. Exp. Cell Res. 61, 285-289.
INOUE, S., HARDY, J.P., COUSINEAU, G.H., BAL, A.K., 1967. Fertilization membranes structure analysis with the surface replica method. Exp. Cell Res. 48, 248-251.
ITO, S., 1969. Structure and function of the glykocalyx. Federation Proc. 28, 12-25.
ITO, S., REVEL, J.P., GOODENOUGH, D.A., 1967. Observations on the fine structure of the fertilization membrane of Arbacia punctulata. Biol. Bull. 133, 471.
KANE, R.E., 1960. The effect of partial protein extraction on the structure of the egg of the sea urchin Arbacia punctulata. J. Biophys. Biochem. Cytol. 7, 21-25.
KARASAKI, S., 1966. The nucleolus and RNA synthesis in sea urchin embryos. J. Cell Biol. 31, 56A.
KARASAKI, S., 1968. The ultrastructure and RNA metabolism of nucleoli in early sea urchin embryos. Exp. Cell Res. 52, 13-26.
KESSEL, R.G., 1964. Electron microscope studies on oocytes of an echinoderm, Thyone briareus, with special reference to the origin and structure of the annulate lamellae. J. Ultrastr. Res. 10, 498-514.
KESSEL, R.G., 1966. Some observations on the ultrastructure of the oocyte of Thyone briareus with special reference to the relationship of the Golgi complex and endoplasmatic reticulum in the formation of yolk. J. Ultrastr. Res. 16, 305-319.
KESSEL, R.G., 1968a. An electron microscope study of differentiation and growth in oocytes of Ophioderma panamensis. J. Ultrastr. Res. 22, 63-89.
KESSEL, R.G., 1968b. Fine structure of annulate lamellae. J. Cell Biol. 36, 658-664.
KUROSUMI, K., 1958. Electron microscope studies on mitosis in sea urchin blastomeres. Protoplasma 49, 116-139.
LANSING, A.I., HILLIER, J., ROSENTHAL, T.B., 1952. Electron microscopy of some marine egg inclusions. Biol. Bull. 103, 294.

LÖNNING, S., 1963. Electron microscopical studies of oocytes and eggs from the sea urchin Brissopsis lyrifera. Sarsia 11, 11-16.
LÖNNING, S., 1964. Studies of the ultrastructure of sea urchin oocytes and changes induced by insemination. Sarsia 15, 9-15.
LÖNNING, S., 1965. Electron microscopic studies of the block to polyspermy. The influence of nicotine. Sarsia 18, 17-22.
LÖNNING, S., 1967a. Studies of the ultrastructure of sea urchin eggs and the changes induced at insemination. Sarsia 30, 31-48.
LÖNNING, S., 1967b. Electron microscopic studies of the block to polyspermy. Sarsia 30, 107-116.
LÖNNING, S., 1967c. Experimental and electron microscopic studies of sea urchin oocytes and eggs and the changes following insemination. Arbok Univers. Bergen, Mat. Nat. Ser. 8, 1-20.
LÖNNING, S., 1967d. Studies of the ultrastructure of sea urchin eggs subjected to hypotonic and hypertonic medium. Arbok Univers. Bergen, Mat. Nat. Ser. 5, 1-8.
LONGO, F.J., ANDERSON, E., 1968. The fine structure of pronuclear development and fusion in the sea urchin Arbacia punctulata. J. Cell Biol. 39, 339-368.
LONGO, F.J., ANDERSON, E., 1969. Sperm differentiation in the sea urchins Arbacia punctulata and Strongylocentrotus purpuratus. J. Ultrastr. Res. 27, 486-509.
LONGO, F.J., ANDERSON, E., 1970a. The effects of nicotine on fertilization in the sea urchin Arbacia punctulata. J. Cell Biol. 46, 308-325.
LONGO, F.J., ANDERSON, E., 1970b. A cytological study of the relation of the cortical reaction to subsequent events of fertilization in urethane-treated eggs of the sea urchin, Arbacia punctulata. J. Cell Biol. 47, 646-665.
MCCULLOCH, D., 1952a. Fibrous structures in the ground cytoplasm of the Arbacia egg. J. Exp. Zool. 119, 47-59.
MCCULLOCH, D., 1952b. Note on the origin of the cortical granules in Arbacia punctulata eggs. Exp. Cell Res. 3, 605-607.
MERCER, E.H., WOLPERT, L., 1958. Electron microscopy of cleaving sea urchin eggs. Exp. Cell Res. 14, 629-632.
MERCER, E.H., WOLPERT, L., 1962. An electron microscope study of the cortex of the sea urchin (Psammechinus miliaris) egg. Exp. Cell Res. 27, 1-13.
MERRIAM, R.W., 1958a. The nuclear envelope as possible agent in specific synthetic events in the cytoplasm of sand dollar eggs. Biol. Bull. 115, 329.
MERRIAM, R.W., 1958b. The contribution of lower oxides of osmium to the density of biological specimens in electron microscopy. J. Biophys. Biochem. Cytol. 4, 579-582.
MERRIAM, R.W., 1959. The origin and fate of annulate lamellae in maturing sand dollar eggs. J. Biophys. Biochem. Cytol. 5, 117-121.
MILLONIG, G., 1967a. The structural changes of the nucleolus during oogenesis and embryogenesis of Arbacia lixula. J. Cell Biol. 35, 177A.
MILLONIG, G., 1967b. The morphological changes of the nucleoli during oogenesis and embryogenesis. Accad. Naz. Lincei 104, 113-115.
MILLONIG, G., BOSCO, M., GIAMBERTONE, L., 1968. Fine structure analysis of oogenesis in sea urchins. J. Exp. Zool. 169, 293-314.
MILLONIG, G., GIUDICE, G., 1967. Electron microscopic study of the reaggregation of cells dissociated from sea urchin embryos. Dev. Biol. 15, 91-101.
MILLONIG, G., HUNTER, A., 1968. Fine structure analysis of the early cortical response to fertilization in sea urchins. IV. Reg. Congr. Electr. Micr. 2, 319-320, Roma.
MITCHISON, J.M., 1953. The thickness of the sea urchin fertilization membrane. Exp. Cell Res. 5, 536-538.

MOTOMURA, J., 1960. On the structure of the cortical granule in the sea urchin egg. Sci. Rep. Tôhoku Univ. Biol. 26, 367-374.
MOYER, F.H., VERHEY, C.A., 1965. Electron microscopy of amino acid incorporating fractions isolated from sea urchin eggs by density gradient centrifugation. Am. Zool. 5, 199.
NØRREVANG, A., WINGSTRAND, K.G., 1970. On the occurrence and structure of choanocyte-like cells in some echinoderms. Acta Zool. 51, 249-270.
PASTEELS, J.J., 1965a. Etude au microscope électronique de la réaction corticale. J. Embryol. Exp. Morphol. 13, 327-339.
PASTEELS, J.J., 1965b. Aspects structuraux de la fécondation vus au microscope électronique. Arch. Biol. Liège 76, 463-509.
PASTEELS, J.J., CASTIAUX, P., VANDERMURSSCHE, G., 1958. Cytochemical localizations and ultrastructure in the fertilized unsegmented egg of Paracentrotus lividus. J. Biophys. Biochem. Cytol. 4, 575-577.
PIKO, L., TYLER, A., VINOGRAD, J., 1967. Amount, location, priming capacity, circularity, and other properties of cytoplasmic DNA in sea urchin eggs. Biol. Bull. 132, 68-90.
PUCCI-MINAFRA, J., BOSCO, M., GIAMBERTONE, L., 1968. Preliminary observations on the isolated micromeres from sea urchin embryos. Exp. Cell Res. 53, 177-183.
ROTHSCHILD, LORD, 1956. Sea urchin spermatozoa. Endeavour 15, 79-86.
ROTHSCHILD, LORD, 1957. The membrane capacitance of the sea urchin egg. J. Biophys. Biochem. Cytol. 3, 103-109.
RUNNSTRÖM, J., 1966. The vitelline membrane and cortical particles in sea urchin eggs and their function in maturation and fertilization. Adv. in Morphogenesis 5, 221-325.
RUNNSTRÖM, J., 1967. Structural changes of the chromatin in early development of sea urchins. Exp. Cell Res. 48, 691-694.
RUNNSTRÖM, J., 1968. The initiation of the development of the egg with special reference to sea urchins. Atti Acc. Naz. dei Lincei 104, 165-178.
RUNNSTRÖM, J., 1969. The appearance of a type of cortical vesicles subsequent to fertilization of the sea urchin egg, their character and possible function. Wilhelm Roux' Arch. Entwicklungsmech. Organismen 162, 254-267.
SACHS, M.J., ANDERSON, E., 1970. A cytological study of artificial parthenogenesis in the sea urchin Arbacia punctulata. Cell Biol. 47, 140-158.
SCHMEKEL, L., 1969. Elektronenmikroskopie der Makromeren-Mikromerengrenze des Seeigelkeimes. Zool. Anz. Suppl. 33, Verh. Zool. Ges., 141-144.
SCHROEDER, T.E., 1969. The role of "contractile ring" filaments in dividing Arbacia eggs. Biol. Bull. 137, 413-414.
SCHROEDER, T.E., 1972. The contractile ring. II. Determining its brief existence, volumetric changes, and vital role in cleaving Arbacia eggs. J. Cell Biol. 53, 419-434.
SOLARI, A.J., 1967. Electron microscopy of native DNA in sea urchin cells. J. Ultrastr. Res. 17, 421-438.
TAKASHIMA, R., TAKASHIMA, Y., NODA, Y., 1966. Ultrastructure of the cortical granules and the cortical reaction of sea urchin eggs. VI. Int. Congr. Electronmicr., 647.
THALER, M.M., COX, M.C.L., VILLEE, C.A., 1969. Isolation of nuclei from sea urchin eggs and embryos. J. Cell Biol. 42, 846-850.
TILNEY, L.G., GIBBINS, J.R., 1966. The relation of microtubules to form differentiation of primary mesenchyme cells in Arbacia embryos. J. Cell Biol. 31, 118A.
TILNEY, L.G., GIBBINS, J.R., 1969a. Microtubules in the formation and development of the primary mesenchyme in Arbacia punctulata. J. Cell Biol. 41, 227-250.

TILNEY, L.G., GIBBINS, J.R., 1969b. Microtubules and filaments in the filopodia of the secondary mesenchyme cells of Arbacia punctulata and Echinarachnius parma. J. Cell Sci. 5, 195-210.
TILNEY, L.G., GODDARD, J., 1970. Nucleating sites for the assembly of cytoplasmic microtubules in the ectodermal cells of blastulae of Arbacia punctulata. J. Cell Biol. 46, 564-575.
TILNEY, L.G., MARSLAND, D., 1969. A fine structural analysis of cleavage induction and furrowing in the eggs of Arbacia punctulata. J. Cell Biol. 42, 170-184.
TSUKAHAVA, J., SUGIYAMA, M., 1969. Ultrastructural changes in the surface of the oocyte during oogenesis of the sea urchin, Hemicentrotus pulcherrimus. Embryologia 10, 343-355.
VACQUIER, V.D., 1968. The connection of blastomeres of sea urchin embryos by filopodia. Exp. Cell Res. 52, 571-581.
VACQUIER, V.D., MAZIA, D., 1968a. Twinning of sand dollar embryos by means of dithiothreitol. Exp. Cell Res. 52, 209-219.
VACQUIER, V.D., MAZIA, D., 1968b. Twinning of sea urchin embryos by treatment with dithiothreitol. Exp. Cell Res. 52, 459-468.
VERHEY, C.A., MOYER, F.H., 1965. A comparison of fine structure changes during oogenesis of regulative and mosaic eggs. Am. Zool. 5, 199 (Abstract).
VERHEY, C.A., MOYER, F.H., 1967. Fine structure changes during sea urchin oogenesis. J. Exp. Zool. 164, 195-226.
WEINSTEIN, R.S., HERBST, R.R., 1964. Electron microscopy of cleavage furrows in sea urchin blastomeres. J. Cell Biol. 23, 101A.
WOLPERT, L., GUSTAFSON, T., 1967. Cell movement and cell contact in sea urchin morphogenesis. Endeavour 26, 85-90.
WOLPERT, L., MERCER, E.H., 1961. An electron microscope study of fertilization of the sea urchin egg Psammechinus miliaris. Exp. Cell Res. 22, 45-55.
WOLPERT, L., MERCER, E.H., 1963. An electron microscope study of the development of the blastula of the sea urchin embryo and its radial polarity. Exp. Cell Res. 30, 280-300.

10. The Biophysics of Cleavage and Cleavage of Geometrically Altered Cells

R. Rappaport

Most of the quantitative and experimentally obtained information about the cleavage mechanism in animal cells comes from studies of echinoderm eggs. No other cells offer the comparable advantages of abundance, size, shape, transparency and the absolute, simple control of division time by fertilization. Their division mechanisms seem similar to those of other eggs, and their appearance differs primarily in the relatively greater size of their asters. Egg cleavage requires energy and force, but useful data on how they are involved are scarce. Investigations have centered on identifying the physical mechanism and determining the role of the visible cell parts.

10.1 General

10.1.1 The Mitotic Apparatus

The cleavage furrow appears during telophase in the plane of the metaphase plate. Although the mitotic apparatus seems a likely participant in furrowing, its removal, destruction or displacement while the cell is still spherical (HIRAMOTO, 1956, 1964, 1965; BEAMS and EVANS, 1940) fails to prevent cleavage when operations are performed between metaphase and the end of anaphase. Although orderly arrays of vacuoles and mucosubstance in normal eggs have been reported in the equatorial plane (MOTOMURA, 1950, 1966, 1967), disruption of the cytoplasmic organization in the equatorial plane does not hinder furrow completion, once started (RAPPAPORT, 1966). The essential role of the mitotic apparatus in furrow formation is usually completed by late metaphase or anaphase (HIRAMOTO, 1956, 1971) when it reaches maximum size (WILSON, 1895; DAN, J.C., 1948) although it can be manipulated to form new furrows after the time of normal division (RAPPAPORT and EBSTEIN, 1965). When the mitotic apparatus is shifted at the beginning of metaphase or before, the cleavage furrow develops in normal relation to the mitotic apparatus in its new position. This is usually thought to imply that the role of the mitotic apparatus in cleavage is confined to establishing the division mechanism in the egg surface or cortex (SWANN and MITCHISON, 1958).

10.1.2 The Egg Surface

The egg surface can accomplish division only if it undergoes regional functional differentiation during each division cycle, and this differentiation is apparently the result of the influence of the mitotic apparatus. Regional differences in surface resistance to rupture during the cleavage cycle have also been shown (JUST, 1922; CHAMBERS, 1938). It is possible that the mitotic apparatus affects a restricted portion of the surface with resulting alteration in behavior and physical properties which precipitate division, and that unaffected surface retains the physical properties and behavior characteristic of the surface before functional differentiation. Experiments involving cleavage pattern alteration have dealt mainly with how and where the physical mechanism forms and what parts of the mitotic apparatus are involved in its formation, while biophysical studies have tried mostly to identify the physical mechanisms.

10.2 Cleavage Theories

3 cleavage theories have been thoroughly scrutinized recently, and most experiments and measurements were designed to test or support them.

10.2.1 Equatorial Constriction

The possibility that cleavage may be accomplished by active constriction in the equatorial plane has been recognized by observers of the living process for many years (ZIEGLER, 1903). LEWIS (1942) proposed that cells are constricted by a gel layer in the median equatorial plane which exerts increased tension as it thickens or becomes more viscous. MARSLAND (1970) postulates that in its initial deformation, the spherical cell becomes elongated in the direction of the spindle axis, apparently as the result of contraction that begins in a broad band of the cortical gel. The surface layers in the polar areas weaken and when the furrow becomes visible, the band of contractile cortex narrows noticeably. As the furrow deepens, the cortical gel with its contractile powers is incorporated in the dividing cell wall. In the final stages of furrowing, cleavage is completed when the region of active contraction shifts from the furrow through to the walls at the side.

10.2.2 Polar Expansion

According to the polar expansion theory, active expansion at the polar surfaces of the cell spreads toward the equator, pushing the furrow passively inward. This possibility was based on a hypothetical effect of molecular disorientation on the looped-protein type of cell membrane structure, and its initiation at the poles of the cell was correlated with a reported pattern of spindle and aster disorientation during mitosis (SWANN and MITCHISON, 1958).

10.2.3 Astral Relaxation

This theory holds that cell membrane tension is uniform before cleavage, that cleavage is initiated by membrane relaxation in the polar regions which expand in area, allowing the furrow region to constrict and divide the egg in two; and that asters or astral centers determine before cleavage the membrane areas that will relax. It combines the polar expansion of the expanding membrane theory with the furrow contraction of the equatorial constriction theory. The crucial point is that tension in the furrow region is greater than at the poles, allowing the furrow region to contract (WOLPERT, 1960, 1966). KINOSHITA (1968, 1969) and KINOSHITA and YAZAKI (1967) have also proposed a polar relaxation theory in which equatorial surface contractility is not altered during furrow formation.

10.3 Biophysics of Cleavage

If the changes in physical properties and behavior of eggs before and during cleavage are assumed to result from the physical activity of the division process, their study should reveal important information about the mechanisms underlying that process. This assumption was the rationale for many biophysical studies.

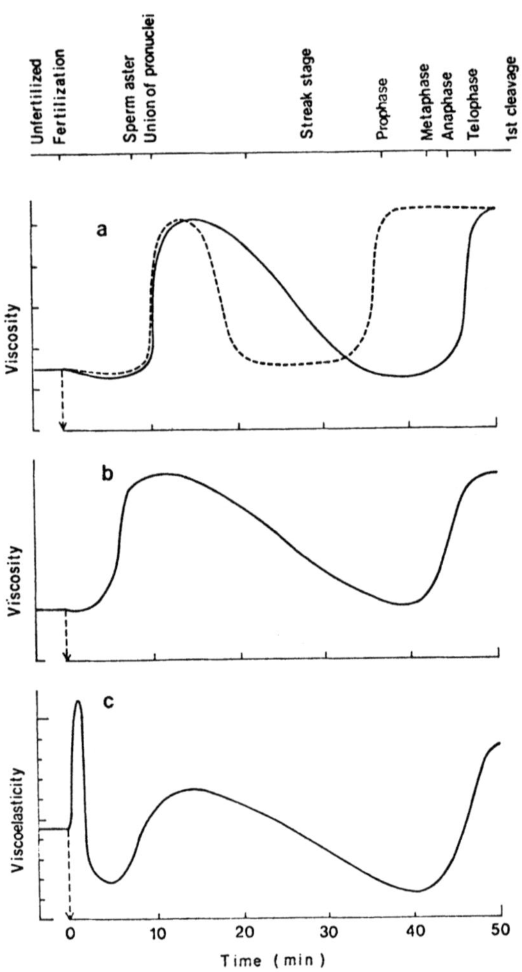

Fig. 10.1a-c. Changes in endoplasmic viscosity from fertilization to onset of cleavage. (a) Change in protoplasmic viscosity determined by HEILBRUNN (1920b) by centrifuge method (solid line) and change determined by CHAMBERS (1919) by microdissection (from HEILBRUNN, 1928). (b) Changes in protoplasmic viscosity determined by centrifuge method (from FRY and PARKS, 1934). (c) Changes in endoplasmic consistency determined by magnetic particle method (from HIRAMOTO, 1969). Intervals from fertilization to onset of cleavage are same throughout. Cytological stages (HARVEY, 1956) referred to are shown at top of figure (from HIRAMOTO, 1970)

10.3.1 Physical Properties that Change during the Cell Cycle

1. Endoplasmic viscosity and viscoelasticity. Fluctuations in endoplasmic viscosity were estimated by centrifugation in *Arbacia punctulata* (HEILBRUNN, 1920a; FRY and PARKS, 1934) and by microdissection in *Echinarachnius parma* (CHAMBERS, 1917). In *Temnopleurus toreumaticus* and *Clypeaster japonicus*, viscoelastic changes were determined by a magnetic particle method, the basis of which is to introduce into the protoplasm a spherical iron particle and to move it by means of external magnetic fields of controlled magnitude and duration (HIRAMOTO, 1968, 1969). From the character and magnitude of the movement, the size of the particle, and the magnitude of the applied force, it is possible to determine the viscoelastic properties of the protoplasm surrounding the particle.

These methods have shown similar changes in protoplasm consistency from fertilization to cleavage. Consistency generally rises after fertilization, peaks when the sperm aster reaches full size, then declines until metaphase, and rises again until the onset of cleavage (Fig. 10.1). This fact, plus the regional consistency differences of protoplasm in the fertilized egg strongly suggests that protoplasm consistency is closely related to astral structure (HIRAMOTO, 1970). Since neither endoplasm nor the mitotic apparatus physically participates in cytokinesis, these changes are probably not directly involved in the division mechanism.

2. Cortical gel strength. Granules lodged in the cortex of sea urchin eggs are more resistant to stratification by centrifugation than endoplasmic granules. In some cases they can be moved only by a combination of great centrifugal force and strong hydrostatic pressure. Before and during cleavage, cortical granules in the equatorial zone of *Arbacia pustulosa* are more firmly fixed than those in the rest of the cell (MARSLAND, 1939). In the echinoderm egg generally, cortical granule stratification shortly before cleavage may require 10 times the centrifugal force necessary to obtain the same degree of stratification in the unfertilized egg (MARSLAND, 1951).

The centrifugal method of measuring gel strength (MARSLAND, 1956) of parts of the protoplasm at various temperatures and pressures is based on the assumption that Stokes' law applied to protoplasmic systems provides a useful approximation. Resistance to visible granule displacement through a gel cannot be regarded as an index of viscosity in a strict physical sense, but assuming that uniform centrifugal force is used, and that the density differential between the granules and the surrounding medium is not altered by the experimental treatment, the centrifuge time required to produce standard granule displacement through the gelated cytoplasm has been taken as an index of relative consistency, stiffness, or strength of that part of the protoplasm.

A relationship between gel strength and cell division (MARSLAND, 1950) is suggested by investigations in which gelation reactions are instrumental to the development of mechanical energy in many kinds of living cells. MARSLAND (1950) proposed that gelated parts of the protoplasm develop an inherent contractility, apparently equivalent to the syneresis of inanimate gels. Periodic determinations of gel strength after fertilization show that the highest peaks come before cleavage (Fig. 10.2).

HIRAMOTO (1970) pointed out that these changes in gel strength are not accompanied by stiffness changes of the egg or extensibility by

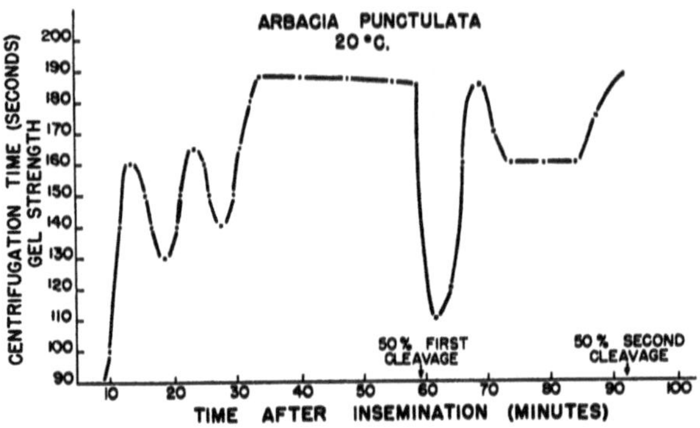

Fig. 10.2. Cortical sol-gel relations following insemination. Pressure-centrifuge measurements of structural state of plasma-gel layer of *Arbacia* egg were made at high pressure (8,000 lb/sq in) and great force (41,000 *G*) (from ZIMMERMAN, LANDAU, and MARSLAND, 1957)

centrifugal force. He considered it hardly possible to compare gel strength with other physical properties of the cell surface because the physical significance of gel strength measurements is by no means clear. He suggested that they may be an indication of some physical property of the cortical cytoplasm.

3. Resistance to deformation. In eggs flattened by the weight of a glass plate, the degree of flattening is not constant. It is maximal between divisions, but just before cell division it diminishes sharply so that the glass plate is raised. It increases again after each division (DANIELLI, 1952).

Egg stiffness or mechanical resistance to deformation depends on the viscoelastic properties of such structures at the cell-surface as the cortex, cell membrane and vitelline membrane and on pressure within

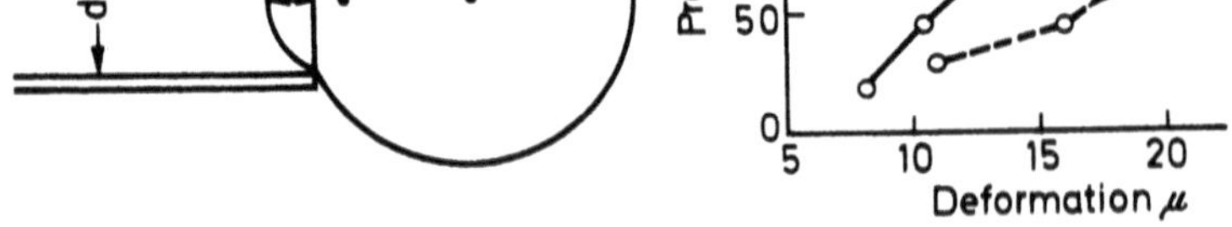

Fig. 10.3 a and b. Cell elastimeter. (a) Egg deformation by suction; *d*, internal diameter of micropipet; *R*, radius of curvature of egg surface outside pipet; *r*, radius of curvature of egg surface inside pipet; *x*, cell surface deformation. (b) Pressure-deformation curve obtained by increasing negative hydrostatic pressure (solid line) and decreasing pressure (broken line) (from MITCHISON and SWANN, 1954; HIRAMOTO, 1970)

the egg. Judging from the degree of endoplasmic deformation and from its rheological properties determined by the magnetic particle method, for instance, endoplasmic stiffness does not seem to have much to do with egg stiffness overall (HIRAMOTO, 1970).

a) Elastimeter measurements. In *Psammechinus miliaris*, *Psammechinus microtuberculatus*, *Paracentrotus lividus*, *Arbacia lixula* and *Sphaerechinus granularis* resistance to deformation was determined by applying locally negative pressure to the egg surface (Fig. 10.3). Resistance increased rapidly after metaphase and reached a peak just before cleavage, after which it fell sharply (Fig. 10.4)(MITCHISON and SWANN, 1954, 1955). These changes were attributed to alterations of the elastic modulus of a relatively thick egg cell membrane.

WOLPERT (1960) disagrees with the MITCHISON-SWANN idea of the sea urchin egg as a thick-walled sphere. Although he believes that, in objects of the range of thickness tested by them, stiffness increases linearly with wall thickness and resistance is directly proportional to wall thickness, he considers the sea urchin egg to be thin-walled and therefore flexure of the membrane may be disregarded.

In other elastimeter experiments on model rubber spheres, when pressure was plotted against deformation, the shape of the curve depended not only on wall thickness and internal pressure, but on the degree of surface slip around the edge of the pipet. The resistance to deformation was considered to be due mostly to tangential stress (HIRAMOTO, 1963a).

b) Compression measurements. The surface forces of *A. punctulata* eggs were calculated from the force required to compress them between parallel plates and from the size of their compressed forms (COLE, 1932). Later, the compression method was used to measure, at intervals from before fertilization to just before cleavage, the surface forces in eggs of *Hemicentrotus pulcherrimus*, *Pseudocentrotus depressus*, *Mespilia globulus* and *C. japonicus* (HIRAMOTO, 1963b). Stiffness changes determined by a compression force of 2.5×10^{-3} dyne for 30 seconds showed a peak immediately before cleavage (Fig. 10.4b). The results of these calculations are shown (Table 10.1). Resistance to flattening by centrifugation was also found to be maximal just before cleavage in *A. punctulata* and *Lytechinus variegatus* (HIRAMOTO, 1967).

Table 10.1. Surface forces and elasticity values of fertilized sea urchin eggs.[a] (From HIRAMOTO, 1970)

	Unfertilized	Sperm Aster Early	Sperm Aster Late	Streak	Diaster (Late anaphase)
Surface force (dyne/cm)	0.03		0.15	0.15	0.2
Elasticity	0.68		0.53	1.85	20
Value (dyne/cm)	1.65	0.75	0.80	0.61	10.1

[a]Fertilization membranes and hyaline layers were removed in calcium-free seawater.

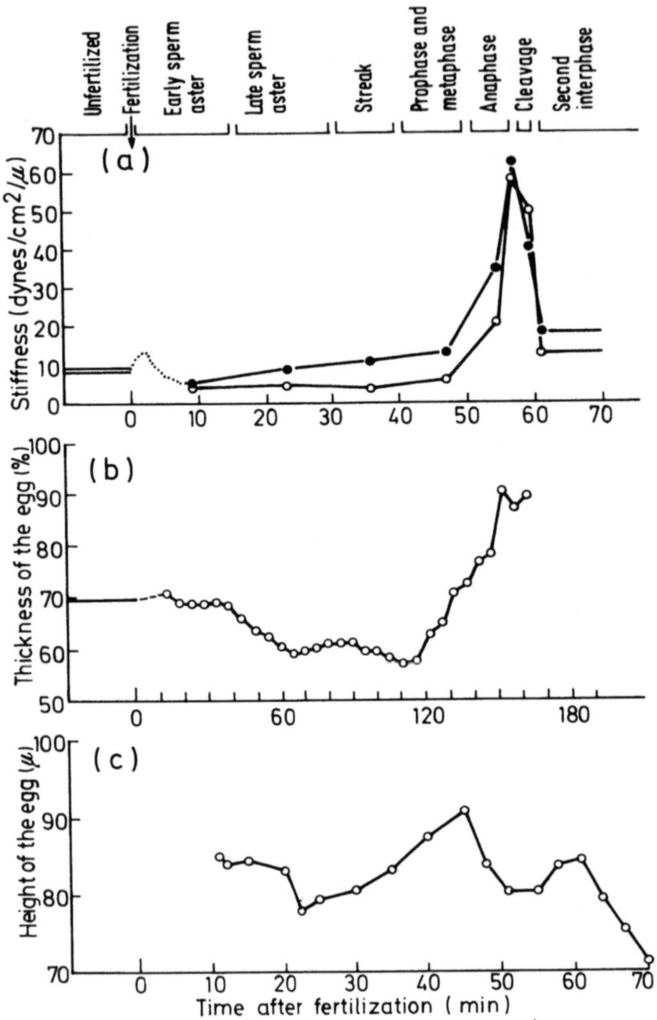

Fig. 10.4a-c. Changes in egg stiffness from fertilization to cleavage. (a) Measured with cell elastimeter (from MITCHISON and SWANN, 1955). (b) Changes in egg thickness following definite force (from HIRAMOTO, 1963b). (c) Changes in degree of egg flattening by centrifugal force of 35 *g* (from HIRAMOTO, 1967). Intervals from fertilization to onset of cleavage are same throughout (from HIRAMOTO, 1970)

c) Birefringence. Studies of variations in the cortical layer of fertilized *P. miliaris* eggs showed that an inconstant cycle of birefringence appears 15-20 minutes after fertilization or parthenogenetic activation, and another cycle, more intense and constant, appears at the end of metaphase up to telophase of the first cleavage or, in parthenogenetic eggs, to monaster expansion. As soon as the first cleavage is completed, cortical birefringence disappears but returns during ana-telophase of the second mitotic cycle (MONROY and MONTALENTI, 1947). Birefringence cycles were maintained in the presence of 0.0002 *M* colchicine which prevented formation of visible asters.

In *P. miliaris* and *P. lividus*, cortical birefringence decrease associated with cleavage begins at the cell poles and appears later at the equator (MITCHISON and SWANN, 1952). In a study of birefringence changes in the mitotic apparatus of *P. miliaris* eggs, decrease in birefringence starting at anaphase in the region around the chromosomes and moving with them to the spindle poles was reported. Astral birefringence decrease begins at the astral centers (SWANN, 1951). Decreased cortical birefringence at the poles may possibly be initiated by the same mechanism that operates in the mitotic apparatus. Early polar decrease in birefringence was considered consistent with the expanding membrane theory of cleavage (SWANN and MITCHISON, 1958).

The relationship between cleavage mechanism and changes in physical properties that occur during the mitotic cycle remains unclear. It was generally thought to be one of cause and effect, but it is possible that protoplasmic structural changes at the surface that accompany mitotic cycles may not be directly controlled by the mitotic apparatus. Changes in sulphhydryl content, cortical birefringence and stiffness occur even when mitotic apparatus formation is suppressed by colchicine treatment or irradiation of the egg with ultraviolet rays (IKEDA, 1965; MONROY and MONTALENTI, 1947; SWANN and MITCHISON, 1953). Furrow-polar differentiation, however, is directly controlled by the mitotic apparatus. The site and direction of the strong contractility at the furrow surface during cleavage are probably determined by a message from the mitotic apparatus before cleavage begins (HIRAMOTO, 1968).

10.3.2 Cleavage Furrow Force

Furrow surface can achieve cytokinesis only if tension at the equator is greater than anywhere else on the cell surface. Tension at the non-furrow surface has been calculated, but it is difficult to study furrow surface properties the same way.

1. Direct measurement. Tension exerted by cleavage furrows in isometric contraction was measured by flexible glass needles with previously calibrated bending characteristics (RAPPAPORT, 1967). The needle was inserted through the polar surface and the furrow plane so that the advancing furrow tip bent the needle until the needle's resistance to deformation equalled the force exerted by the furrow (Fig. 10.5). In *Astriclypeus manni* the maximum tension exerted was $3.04 \times 10^{-3} \pm 0.95 \times 10^{-3}$ dyne and in *P. depressus* $2.00 \times 10^{-3} \pm 0.43 \times 10^{-3}$ dyne. Taking as an average value 2.5×10^{-3} dyne per sea urchin

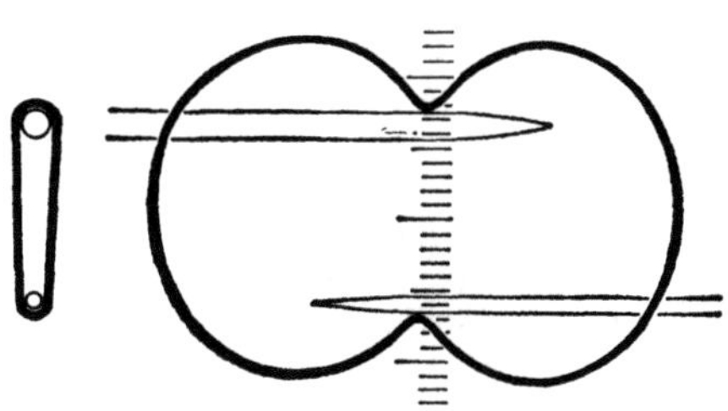

Fig. 10.5. Direct measurement of furrow tension in cleaving eggs. (Left) Cell and needles arranged for measurement. Upper calibrated needle is deflected downward during cleavage while lower holding needle does not move. (Right) Section through and parallel to cleavage plane during isometric contraction. Needle diameter is exaggerated (from RAPPAPORT, 1967)

furrow and assuming an actively contracting, electron-dense or fibrous layer 0.1 µm thick and 5 µm wide, a tension of 2.5×10^5 dynes/sq cm would be required to produce the tension measured (MERCER and WOLPERT, 1958; WOLPERT, 1966; SCHROEDER, 1969). This is close to the 2.45×10^5 dynes/sq cm exerted by actomyosin threads in isometric contraction (PORTZEHL, 1951). A layer 10 µm wide, as studies of *Arbacia* suggest (SCHROEDER, 1969), would exert a force of 1.25×10^5 dynes/sq cm or 1.25 dynes/cm. This type of measurement reveals the maximum force of the furrow, not the force actually used in division. Furrows exerting tension of 1.3×10^{-5} dyne (or 1.3 dynes/cm assuming a band 5 µm wide) continued division after the needles were removed, indicating that the force required for division need be no greater.

2. Indirect measurement. Intracellular pressure in dividing sea urchin eggs was measured after applying a known force to part of the egg surface (HIRAMOTO, 1968). It was calculated from the formula

$$P = F/r^2,$$

where P is intracellular pressure, F equals the force applied, and r is the distance between spindle axis and the point of the cell surface at which tangent is perpendicular to axis (Figs. 10.6, 10.7). Using this information and the curvatures of the normal cell surface during cleavage, tension was calculated by the formulas:

$$T_2 = P_o r_2/2,$$

$$T_2 = P_o r_2\ (2 - r_2/r_1)/2$$

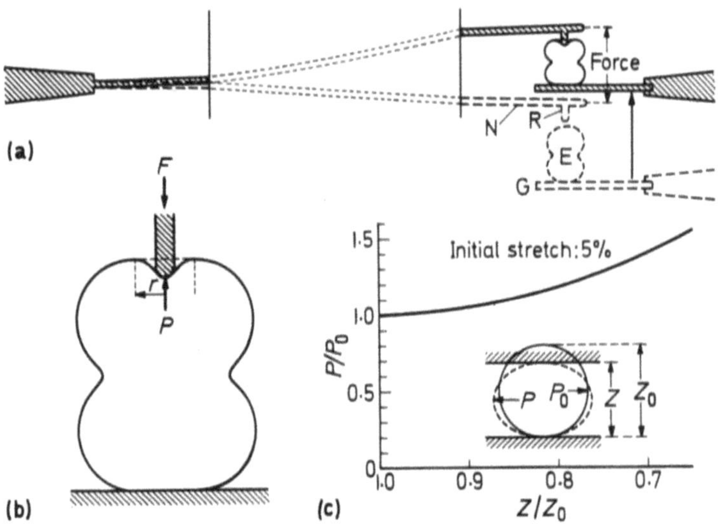

Fig. 10.6a-c. Determination of intracellular pressure in dividing cell. (a) Egg compression and determination of force applied; E, egg; N, glass needle; R, glass rod fixed to tip of needle N; G, glass plate. (b) Compressed egg, dimensions from which intracellular pressure is calculated; F, applied force; P, intracellular pressure ($P = F/\pi\ r^2$); r, distance between spindle axis and cell surface point at which tangent is perpendicular to axis. (c) Relationship between deformation and internal pressure of rubber balls filled with water and compressed between 2 plates after ball surface was stretched 5% by internal pressure through tube. Ball volume remained unchanged during compression; P/P_o, relative internal pressure; Z/Z_o, relative thickness of ball (from HIRAMOTO, 1968)

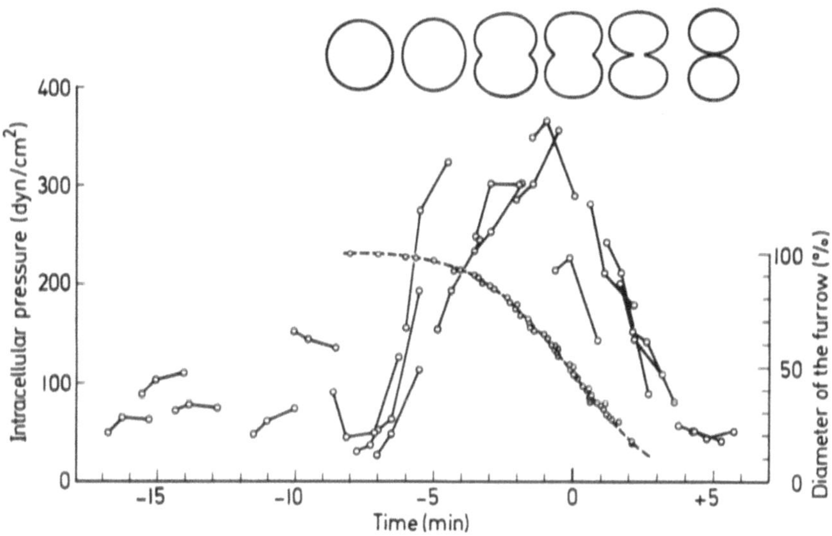

Fig. 10.7. Intracellular pressure of *Temnopleurus toreumaticus* eggs from which fertilization membrane and hyaline layer were removed at 24°C. Pressure was determined at various stages before and during cleavage. 3 connected circles indicate measurements taken in same egg 30 sec, 1 min, and 2 min after constant force of 10^{-3} dyne was applied. Small circles indicate egg furrow diameters (from HIRAMOTO, 1968)

where T_1 is tension at the surface in the longitudinal direction (pole to pole), T_2 is tension at the surface in the latitudinal direction (at right angles to T_1), P_o is intracellular pressure, and r_1 and r_2 are the principal radii of the cell surface curvature. The maximum tension in the latitudinal direction at the furrow surface by this method is about 1.5 dynes/cm, close to the minimum values obtained by direct measurement (HIRAMOTO, 1968)(Fig. 10.8).

Tension at the surface was also calculated using Young's modulus of the surface as determined by elastimeter measurements of stiffness (WOLPERT, 1966). A pre-cleavage stiffness of 3.5 dynes/sq cm/μm gives an *Et* value (Young's modulus times membrane thickness) of 0.64 dynes. Radial strain can be calculated using results from the studies on rubber balls (MITCHISON and SWANN, 1954). The relationship between radial strain *e* and stiffness for R = 50, d = 50, t = 1.6 is

$$S_e = 54_e + 9$$

where S_e is stiffness corresponding to radial strain *e* and S_e = 9 when there is no radial strain. Thus,

$$\frac{S_e}{S_{e = o}} = 6e + 1.$$

In a 12-fold stiffness increase as the result of increased tension alone,

$$S_e/S_{e = o} = 12 \text{ and } e = 1.8,$$

membrane tension is 1.8 x 0.64 = 1.2 dynes/cm. This corresponds with the value of the tension calculated above. Where Young's modulus or

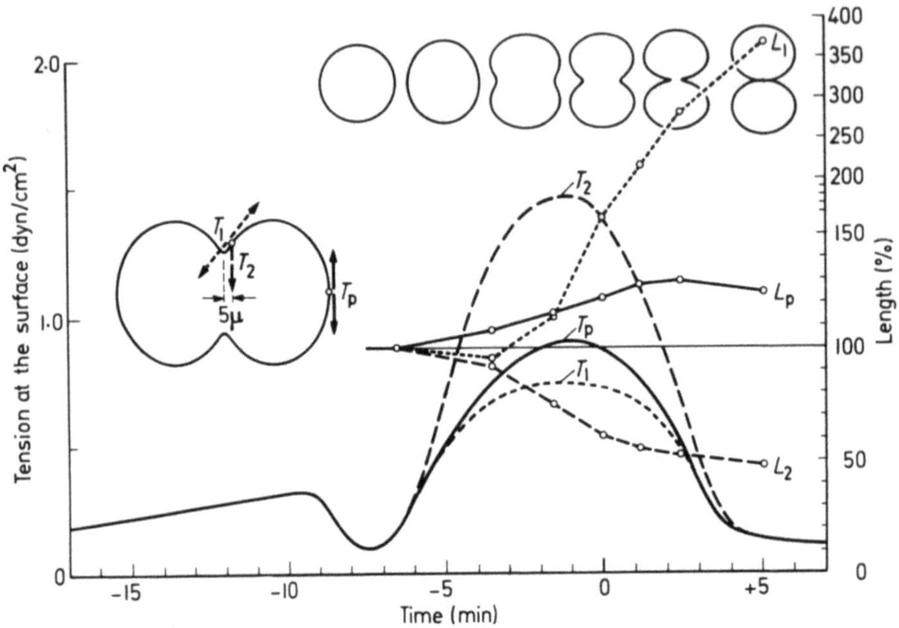

Fig. 10.8. Fluctuations in tension at the surface and surface configuration in *Clypeaster japonicus* eggs; T_1, tension at the surface in longitudinal direction in furrow region (cell surface 5 μm from equatorial plane); T_2, tension at surface in latitudinal direction in furrow region; T_p, tension at surface in polar region; L_1, percentage of linear surface change in longitudinal direction in furrow region; L_2, percentage of linear surface change in latitudinal direction in furrow region; L_p, percentage of linear surface change in polar region. Tension at surface was calculated from intracellular pressure method shown in Fig. 10.7. Surface curvatures of dividing cell was determined in *Temnopleurus toreumaticus* eggs from which fertilization membrane and hyaline layer were removed (from HIRAMOTO, 1968)

thickness increases 4-fold at cleavage surface, Et is 4 times as large as before cleavage and

$$S_e/S_{e=o} = 3.$$

Thus, e = 0.33 and tension is 0.33 x 4 x 0.64 = 0.8 dynes/cm. Even when Et increases 4-fold during cleavage, increase in membrane tension must be about 1 dyne/cm (WOLPERT, 1966). Other methods for calculating the force of cleavage in the sea urchin egg have been described (DAN, K., 1963).

10.3.3 Energy Requirements

If most of the work during egg cleavage is done in the contracting furrow in a region 5 μm wide, and if the furrow contracts with a constant force of 1 dyne/cm, the work done is equivalent to a force of 5×10^{-4} dynes moving $2\pi \times 50 \times 10^{-4}$ cm, or 1.6×10^{-5} ergs. Another possibility is to estimate the work required to stretch the egg surface 27% to provide increased surface area during telophase. This requires each surface element to be extended by a strain e;

$$e^2 + 2e = 0.27, \text{ or}$$
$$e = 0.13.$$

The strain energy is

$$e^2 \times E \times A \times t, \text{ or}$$
$$e^2 \times A \times Et,$$

where A is surface area, E is Young's modulus, and t is thickness. If Et is 2.6 dynes, the strain energy for an egg 100 μm in diameter is 1.4×10^{-5} ergs. Both calculations suggest a value on the order of 10^{-5} ergs, which is very small compared to total egg energy production by the eggs, whose heat production during cleavage is about 2 ergs (WOLPERT, 1966).

10.3.4 Experiments

Theoretical cleavage mechanisms can also be tested by experimentation. According to the polar expansion theory, the force accomplishing division originates at the poles outside the furrow region and cannot operate under tension at the egg surface. But furrows and furrow fragments isolated from the polar and sub-polar regions still cleave, and whole eggs divide under great tensile stress from attached glass weights (RAPPAPORT, 1960, 1969b).

The astral relaxation and equatorial constriction theories both require greater surface tension at the equator than at the poles, but the astral relaxation theory holds that it decreases at the poles, while the equatorial constriction theory holds that it increases at the equator. In late anaphase, surface tension is 0.2 dyne/cm but active furrow tension is 1-2 dynes/cm, suggesting that tension increases at the equator (HIRAMOTO, 1963a).

10.4 Cleavage Pattern Alteration

When sea urchin eggs are distorted or the mitotic apparatus is mechanically relocated during or before metaphase, the division process may be altered and the cleavage plane shifted. These techniques have been useful in studying both embryonic development and cytokinesis.

10.4.1 Nuclear Redistribution in Embryonic Cytoplasm

In disproof of the theory of qualitative division during early embryonic development, flattened sea urchin eggs were caused to segment into flat plates of 8, 16, 32 and even larger numbers of cells (Fig. 10.9). If not flattened too much, the plates, released from pressure, soon divided horizontally, parallel to the direction of elongation, forming flattened blastulae that became rounded, gastrulated and in some cases produced normal larvae (DRIESCH, in WILSON, 1928).

In flattened eggs, the deformed cell mass does not regroup its materials. The nuclei are forced into new and abnormal positions

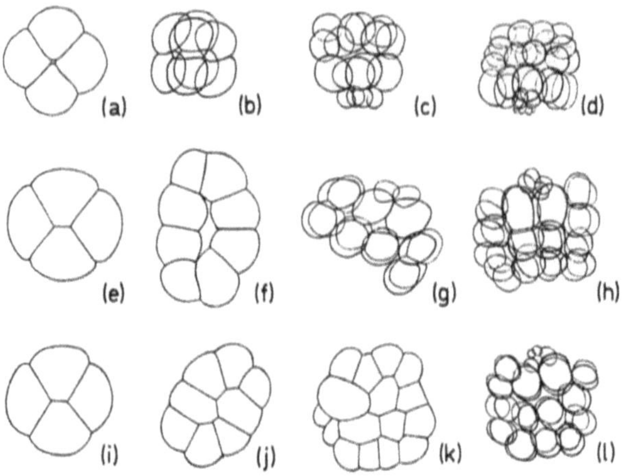

Fig. 10.9a-l. Normal cleavage stages of sea urchin egg, a-d; e, f, cleavage under pressure; g, h, cleavage after pressure is removed; i, j, k, cleavage under pressure; l, cleavage after pressure is removed (from MORGAN, 1927)

but development is still normal. It was clear, therefore, that blastomere specification cannot be due to specific nuclear differences produced by a fixed order of qualitative nuclear divisions but must be due to conditions in the cytoplasm (WILSON, 1928).

Hertwig's modification of Sach's rules of cleavage apply to cells in which the normal geometric relationship between mitotic apparatus and cell surface is experimentally altered. These rules state that the nucleus and the mitotic figure tend to be centered in their sphere of influence: the cytoplasmic mass in which they lie. The spindle axis usually parallels the longest axis of the cytoplasmic mass and is cut transversely by division (WILSON, 1928).

In *P. lividus* eggs on which elongated cylindrical form was imposed at fertilization, cleavage plane and mitotic apparatus orientation were similarly related. The mitotic apparatus is parallel to the longest dimension. The first and second cleavage furrows divide the cells at right angles to the imposed long axis, regardless of the original polarity of the egg. The third cleavage is either parallel or perpendicular to the long axis, depending on the cell dimension ratio (LINDAHL, 1932)(Fig. 10.10).

The growing mitotic apparatus is able to shift into a position offering the least mechanical resistance (Fig. 10.11a), although in many naturally occurring cases this does not account for its position. The force that orients the mitotic apparatus probably orients the cleavage furrow also, but what that force is remains unexplained.

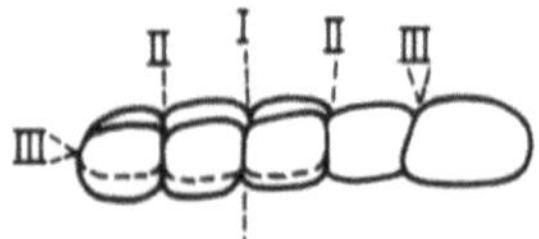

Fig. 10.10. Cleavage of elongated egg. I, II, and III indicate first, second, and third furrows (from LINDAHL, 1932)

10.4.2 The Mitotic Apparatus in Furrow Establishment

The mitotic apparatus takes part in cytokinesis before furrowing begins. It enables the cell surface to divide without the help of subsurface structures, as shown by the fact that isolated furrows and furrow fragments can complete cleavage (RAPPAPORT, 1969b).

Experiments show that a furrow may form on any part of the cell surface, suggesting that, prior to mitotic apparatus action, surface physical properties and receptivity to mitotic apparatus influence are uniform. Location of the division mechanism in the surface implies that the physical and behavioral properties of the surface become different from one area to another. The mitotic apparatus seems to induce the surface changes and alter either the poles or the equator. It is assumed that the changes precipitate cytokinesis and that regions not affected by the mitotic apparatus retain the former properties of the entire surface.

At one time or another, spindles, chromosomes, and asters have been considered essential for furrow formation. But asters acting alone can induce normal furrows in echinoderm eggs. At second cleavage of a torus-shaped egg, furrows form between the polar surfaces of asters of 2 different mitotic apparatuses in surface regions never exposed to spindle or chromosomes (RAPPAPORT, 1961)(Fig. 10.11b). Parts of the mitotic apparatus of *C. japonicus* were removed at various stages before the onset of second cleavage. If they were removed from cells with monasters or cells without asters in the early stages of mitosis, the cleavage furrow usually failed to form. But if parts were removed at metaphase or later, cleavage furrow formed. It seems clear that nuclear materials and spindle are not essential to cleavage furrow formation and complete cytoplasmic division in echinoderm eggs (HIRAMOTO, 1971). Sperm asters and cytasters can also elicit furrows, and the extent of furrowing is correlated with the degree of development of the asters and their geometrical relation to the surface (WILSON, 1901; GRAY, 1931; SUGIYAMA, 1951; KOBAYASHI, 1962; WOLPERT, 1963; RAPPAPORT, 1969a).

It is possible that asters may induce furrow formation by altering or stimulating either the equatorial or polar surfaces. One theory holds that polar surfaces are altered by telophase chromosomes or asters, while the equatorial surface outside the reach of astral stimulation is unaffected (SWANN and MITCHISON, 1958; WOLPERT, 1960) (Fig. 10.12). Polar surface modification precipitates division, and the furrow is comparatively passive.

Others propose that furrow formation results from equatorial surface stimulation (ZIEGLER, 1903; RAPPAPORT, 1961, 1965). In one version of this theory a hypothetical furrow organizing substance is transported by a fountain streaming pattern from the centrosomal region toward the poles and then to the cortical equatorial region. Furrows appear where the fountain streamings converge (CORNMAN and CORNMAN, 1951; HENLEY and COSTELLO, 1965)(Fig. 10.12). Fountain streaming patterns were described in *Rhabditis* and *Crepidula* (SPEK, 1918; CONKLIN, 1902, 1938). Equatorial surface stimulation could also result from an additive effect arising from confluent influence of the asters (RAPPAPORT, 1961, 1965)(Fig. 10.12).

If furrow formation occurred only where astral stimulation could not reach, furrows should appear in any part of the surface positioned farther from the asters than the equatorial surface, but they do not. In the first cleavage of torus-shaped cells, furrowing does not occur between the astral poles (RAPPAPORT, 1961(Fig. 10.11a). Neither does it occur close to the poles of eggs of *E. parma* or *Dendraster excentricus* attenuated by glass weights (RAPPAPORT, 1960)(Fig. 10.11c) or to *P. depressus* with multiple perforations (DAN, 1943; RAPPAPORT, 1968)(Fig. 10.11j).

And if furrow formation occurs only beyond the reach of astral stimulation, equal distance from all parts of the surface to the asters would be expected to block cleavage by preventing functional surface

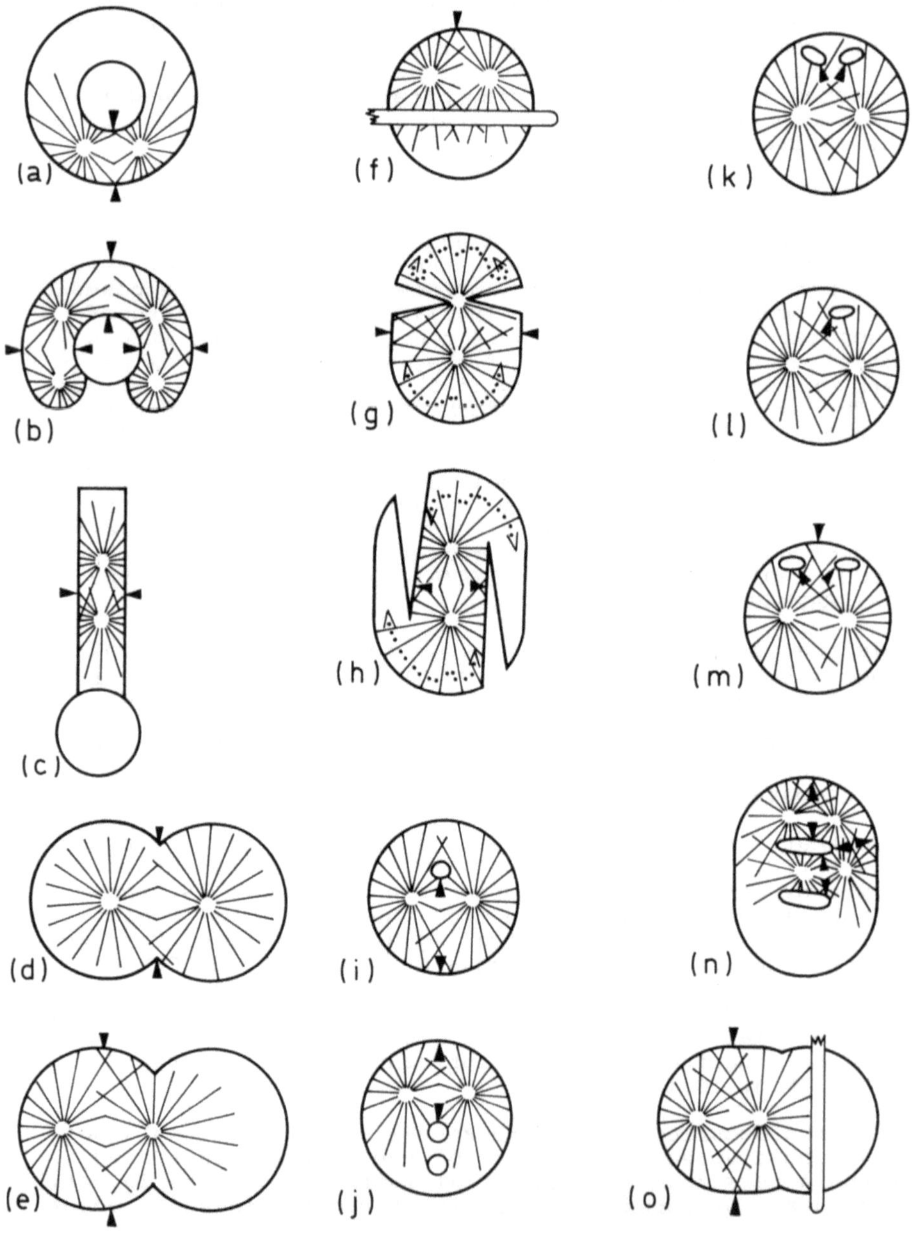

differentiation. But even when distance is equal as the result of artificial constriction in the equatorial plane of *E. parma* (RAPPAPORT, 1964)(Fig. 10.11d), furrow formation and function are normal when the mitotic apparatus straddles the constriction. If the mitotic apparatus lies on one side of the constriction, a furrow forms only in the cell region containing elements of both asters (Fig. 10.11e). None of these results have been explained by the polar stimulation theory.

Evidence indicates that furrow formation depends on a positive relationship between the mitotic apparatus and the surface. When the mitotic apparatus of *E. parma* is pushed to an excentric position with the edge of a needle, cleavage begins on the side nearest the mitotic apparatus and spreads unilaterally across the cell (RAPPAPORT and CONRAD, 1963) (Fig. 10.11f). The dependency involves the relation of asters and the equatorial surface; it involves the distance between asters, and the distance between the pair of asters and surface, as shown in studies of flattened, polyspermic *H. pulcherrimus* eggs.

Asters unusually far apart have little effect on the distant equatorial surface (Fig. 10.13), while reduced spindle-to-surface distance helps to compensate for extreme interastral distance. When spindle-to-surface distance and the interastral distance are 35 μm or greater, furrowing is rare, but when spindle-to-surface distance is 20 μm or less and interastral distance remains 35 μm to 45 μm, furrowing occurs (RAPPAPORT, 1969a)(Fig. 10.13B).

◀ Fig. 10.11a-o. Furrowing following mechanical distortion. Cell is shown at anaphase and subsequent cleavage plane falls between the apices of opposed pairs of black triangles. (a) First cleavage of torus-shaped egg. (b) Second cleavage of torus-shaped egg. Central circle represents glass sphere (from RAPPAPORT, 1961). (c) Cleavage of axially loaded, attenuated cell. Circle represents a glass sphere (from RAPPAPORT, 1965). (d) Furrow orientation after artificial constriction straddled by mitotic apparatus (from RAPPAPORT, 1964). (e) Furrow orientation after artificial constriction. Most of mitotic apparatus lies on one side of constriction (from RAPPAPORT, 1964). (f) Unilateral furrow formation after side of needle displaces mitotic apparatus. (g) Furrowing after deep notches are made at right angles to spindle axis. Dotted lines represent hypothetical fountain streaming currents (from RAPPAPORT, 1970). (h) Furrowing after deep notches are made parallel to spindle axis. Dotted lines represent hypothetical fountain streaming currents (from RAPPAPORT, 1970). (i) Furrow formation in flattened egg involving perforation in equatorial plane (from DAN, 1943). (j) Furrow formation in flattened egg involving 2 perforations in equatorial plane (from DAN, 1943; RAPPAPORT, 1968). (k) Furrow formation in flattened egg with 2 perforations straddling and close to equatorial plane (from DAN, 1943; RAPPAPORT, 1968). (l) Furrow formation involving one perforation close to equatorial plane (from RAPPAPORT, 1968). (m) Furrow formation involving a pair of perforations straddling equatorial plane and lying some distance from it (from RAPPAPORT, 1968). (n) Furrowing formation involving 2 perforations and 2 mitotic apparatuses (from RAPPAPORT, 1968). (o) Second furrow forms after mitotic apparatus is displaced. Indentations next to displacing needle indicate first furrow (from RAPPAPORT and EBSTEIN, 1965)

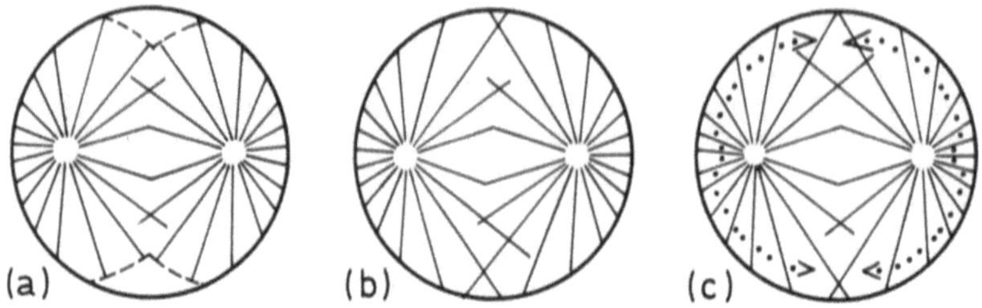

Fig. 10.12a-c. Hypothetical cleavage stimulus patterns. (a) Polar stimulation. Broken line indicates limit of aster influence (from WOLPERT, 1960). (b) Equatorial surface is stimulated by both asters simultaneously (from RAPPAPORT, 1971). (c) Equatorial surface is stimulated when cytoplasmic fountain streams of furrow organizing substance converge below (from RAPPAPORT, 1970)

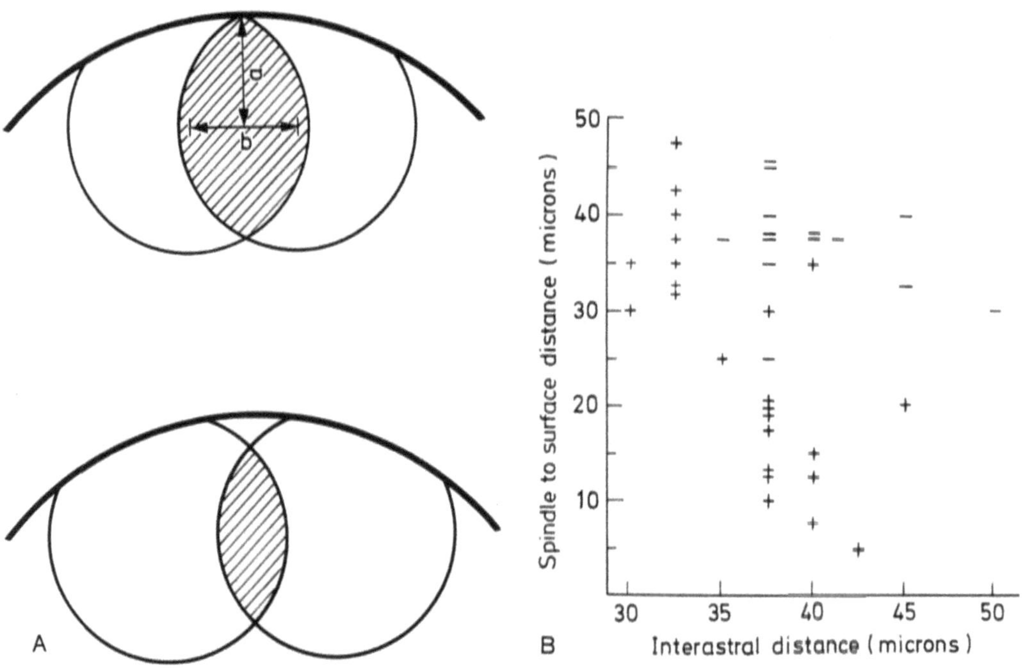

Fig. 10.13 A and B. (A) Relationship between asters and equatorial surface when furrow formation results from joint astral action. Top: Area affected by both asters reaches surface. Bottom: Asters are moved apart. Astral diameter and distance from center of aster to equatorial surface remain unchanged. a, Spindle-to-surface distance measured parallel to flattened surfaces; b, interastral distance. (B) In perforated and unperforated cells, furrowing almost invariably fails when spindle-to-surface and interastral distances are 35 µm or more. But when spindle-to-surface distance is 20 µm or less, furrowing occurs even though interastral distance remains 35 µm. Plus sign indicates furrow has formed adjacent to asters; minus sign that it has not (from RAPPAPORT, 1969b)

Because the polar stimulation theory allows for no interaction between asters and equator, altered spindle-to-surface distance should not affect furrow formation. For a furrow to form, both asters are necessary; in experiments on *P. depressus*, no furrow forms when a block is placed between one aster and the equatorial surface (RAPPAPORT, 1968)(Fig. 10.11l)

10.4.3 The Stimulus Mechanism

Experimental modifications of cell shape and ultrastructural studies of furrow surface and cortex (SCHROEDER, 1969; TILNEY and MARSLAND, 1969, *Arbacia punctulata*) strongly suggest that furrows form when equatorial surface changes as the result of local stimulation (RAPPAPORT, 1971). Asters may alter the equatorial surface by some addition, or removal of a substance, or by introducing a change that propagates (RAPPAPORT, 1965). Mechanical alteration of cell form reveals little of its chemical nature, but something of the vehicle that conveys the stimulus.

The possibility that cytoplasmic fountain streaming induces furrow formation was studied by altering egg shape and disrupting the hypothetical flow pattern. Whatever prevents the confluence of the polar streams would be expected to block cleavage by preventing accumulation of cleavage substance beneath the equatorial surface. This mechanism would hardly explain torus-shaped egg cleavage, however, and deep notches cut parallel and at right angles to the spindle fail to block cleavage but would substantially alter a fountain streaming pattern (Fig. 10.11a, b, g, h). In flattened cells perforated adjacent to the spindle, furrows form only on the surface of the perforation closest to the spindle, even though the region of confluence of a fountain streaming pattern would not be confined to that area (RAPPAPORT, 1970)(Fig. 10.11i). It seems unlikely, therefore, that a fountain streaming pattern of cytoplasmic currents is necessary to furrow formation.

When a perforation or block is placed in the equatorial plane between the mitotic apparatus and the cell margin, no furrow forms in the margin, even though the marginal surface lies within the effective spindle-to-surface range (DAN, 1943; RAPPAPORT, 1968). The walls of perforations in flattened eggs develop cylindrical surfaces in which furrows may occur. When a perforation is made near the spindle, a furrow forms on the side closest to the spindle, although the entire perforation surface may lie within the effective spindle-to-surface range (Fig. 10.11i, j). In the case of 2 perforations in the equatorial plane, a furrow forms on the proximal surface of the perforation nearest the mitotic apparatus (DAN, 1943; RAPPAPORT, 1968)(Fig. 10.11j).

While perforations in the equatorial plane block furrowing in the cell margin, those in the polar and subpolar regions have no effect. Perforations adjacent to the equatorial zone block marginal furrowing even though the equatorial surface is not disturbed and the margin lies within the effective spindle-to-surface range (Fig. 10.11k). When perforations far apart straddle the equatorial plane, furrows form on the cell margin as well as the surfaces of the perforations nearest the equator (Fig. 10.11m).

Perforations also affect the distribution of visible astral rays. Although rays extend to the cell periphery in the rest of the egg, a fan-shaped region distal to a perforation is devoid of rays, like the shadow of a screen in front of a light source (DAN, 1943). Extensive studies have shown that furrows form only on surfaces that can be intersected by straight lines drawn from both astral centers. A block put between one astral center and the margin prevents furrowing (Fig. 10.11l).

Perforations have been shown to interfere with furrow formation only when they are in or near the equatorial plane. This lends support to the idea that furrows are caused by joint astral action. The asters may not serve the same function in later development when they are relatively smaller (RAPPAPORT, 1971). And the mechanism responsible for the delayed, irregular cleavage after astral removal and repeated subthreshold stimulation with activating reagents is not yet understood (LORCH, 1952; KOJIMA, 1969).

When flattened *H. pulcherrimus* eggs that contain 2 mitotic apparatuses are perforated in several places, multiple furrows develop. These separate furrows form close together under the influence of different astral pairs (Fig. 10.11n). In some cases 3 discrete furrows form in 12 μm of a cell margin (RAPPAPORT, 1968). Such a highly restricted locus of furrow formation would be unlikely if cytoplasmic currents or a freely diffusable substance were involved.

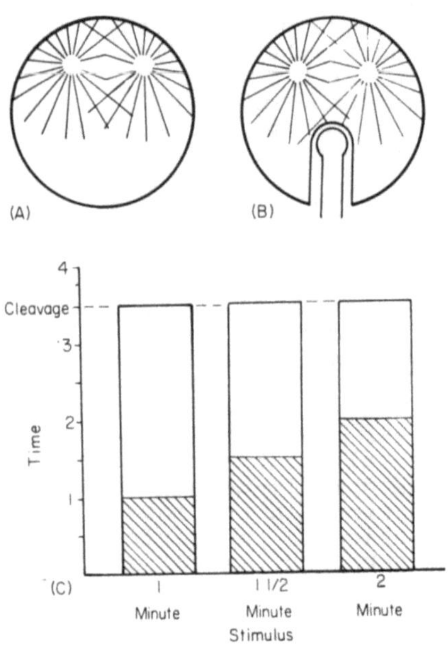

Fig. 10.14A-C. How to determine duration of stimulus and latent period. (A) Mitotic apparatus is pushed to excentric position, creating unilateral furrow. (B) When more distant surface is pushed between asters and held there for more than 1 minute but less than 3 1/2 minutes, cell returns to (A) shape when instrument is removed. Furrow later appears in more distant surface. (C) In experiments combining pressure displacement or mitotic apparatus and surface displacement, cleavage begins 3 1/2 minutes after start of stimulus period (shaded area). Longer stimulus does not significantly alter time of furrow appearance (from RAPPAPORT and EBSTEIN, 1965)

Experiments on sea urchin and sand dollar eggs with excentric nuclei indicate that the extent to which the mitotic apparatus can affect the surface is limited by distance (ZIEGLER, 1903; RAPPAPORT and CONRAD, 1963). When surface beyond the effective range of the mitotic apparatus is pushed closer to it, furrows form. Moving the surface in and out of the effective range makes it possible to estimate the length of time required for irreversible alteration of the surface (RAPPAPORT and EBSTEIN, 1965)(Fig. 10.14a, b). By this method, the minimum time required for furrow formation is 1 minute, followed by a 2 1/2 minute latent period, making a total of 3 1/2 minutes between surface exposure to the mitotic apparatus and start of furrowing. This period cannot be shortened by increasing the exposure time (Fig. 10.14c). During the latent period, the molecular reorganization required for active furrowing occurs.

Multiple furrows form when the mitotic apparatus is pushed back and forth in the egg (Fig. 10.11o). The same mitotic apparatus can establish furrows several times, with a time interval about 2 minutes between mitotic apparatus relocation and furrowing activity (RAPPAPORT and EBSTEIN, 1965).

The rate at which the cleavage stimulus moves from the mitotic apparatus to the cell surface was calculated from measurements made on flattened, fertilized *E. parma* eggs with excentric nuclei. The time interval between the appearance of furrows in the closer and more distant equatorial margins is proportional to the difference in the distance between the axis of the mitotic apparatus and the 2 surfaces. The calculated rate of stimulus movement is 6.3 ± 1.8 μm/minute, a rate slower than that of free diffusion and the usually studied examples of cytoplasmic streaming. It is about the same as that reported for microtubule growth (RAPPAPORT, 1972).

10.4.4 The Cleavage Furrow

It became apparent only recently that the changes that precipitate cleavage occur on the equatorial surface, and there is little information about the properties of the furrow surface. Cortical thickness in the equatorial region is known to exceed that of the rest of the egg (*H. pulcherrimus*, HIRAMOTO, 1957), and ultrastructural studies in *A. punctulata* have shown under the furrow surface a dense layer of circumferentially arranged 40-70 Å diameter microfilaments (WEINSTEIN and HEBERT, 1964; SCHROEDER, 1969; TILNEY and MARSLAND, 1969). After treatment with Cytochalasin B, the microfilaments disappear in 1 minute, leading to regression in shallow furrows and suspended development in deeper ones (SCHROEDER, 1969).

Modification of the normal division process shows that the echinoderm furrow is able to propagate itself on the cell surface. If the mitotic apparatus is mechanically displaced to an extremely excentric position, a unilateral furrow forms, similar to that of coelenterate eggs (RAPPAPORT and CONRAD, 1963)(Fig. 10.15). Experimental behavior of echinoderm unilateral furrows resembles that of coelenterate eggs (YATSU, 1910, 1912). A unilaterally cleaving egg, bisected so that the furrow tip is separated from the uncleaved portion, shows no furrowing activity in the initially uncleaved portion, indicating lack of interaction between the uncleaved part and the mitotic apparatus (Fig. 10.15b). But when the uncleaved portion includes the furrow tip, furrowing continues (Fig. 10.15d). The advancing furrow tip can modify the adjacent surface, and direct interaction between mitotic apparatus and surface is not always necessary. But in the

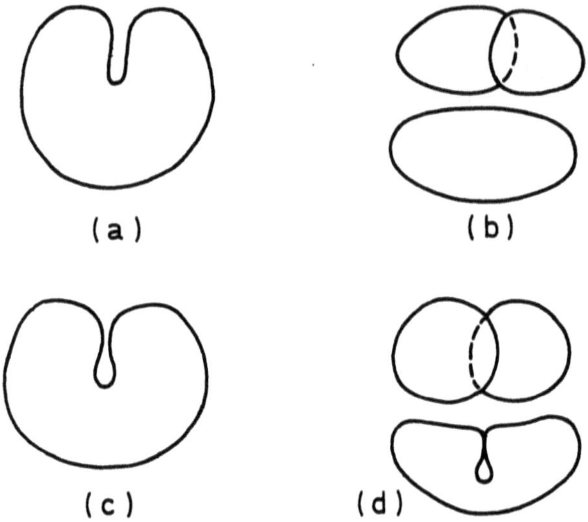

Fig. 10.15a-d. Unilateral furrow. (a, b) Surgically isolated from furrow, lower part of cell shows no independent furrowing activity. (c, d) When furrow tip is included, active, complete furrow forms

normal sea urchin egg in which the mitotic apparatus is comparatively large and the furrow symmetrical, the entire furrow probably results from direct influence of the mitotic apparatus. Parts of the furrow retain the constricting activity characteristic of the whole, and isolated bits of furrow are able to complete furrowing in apparently normal fashion (SCOTT, 1960; RAPPAPORT, 1969b).

10.5 Summary

Direct physical measurement, calculation, and experimental modification of cell shape give us the following concept of the cleavage process in sea urchin eggs. During late metaphase and anaphase, mitotic apparatus asters induce functional differentiation in the egg surface and cortex. Greater tension at the equatorial surface than at the polar surface, appearing to result from interaction between asters and equatorial surface, leads to cleavage. Ultrastructural changes in the equatorial area take place, and that area acquires the capacity to cleave independently of the rest of the egg.

Before and during cleavage, the physical properties of the cell change. Direct measurement and calculation indicate that tension in the active cleavage furrow is 1-2 dynes/cm. Cortical gel strength and resistance to deformation increase, and predictable changes in cortical birefringence occur. How these events relate to the cleavage mechanism is not yet understood.

References

BEAMS, H.W., EVANS, T.C., 1940. Some effects of colchicine upon the first cleavage in Arbacia punctulata. Biol. Bull. 79, 188-198.
CHAMBERS, R., 1917. Microdissection studies. II. The cell aster: a reversible gelating phenomenon. J. Exp. Zool. 23, 483-505.
CHAMBERS, R., 1919. Changes in protoplasmic viscosity and their relation to cell division. J. Gen. Physiol. 2, 49-68.
CHAMBERS, R., 1938. Structural and kinetic aspects of cell division. J. Cell Comp. Physiol. 12, 149-165.
CHAMBERS, R., CHAMBERS, E.L., 1961. Explorations into the nature of the living cell. Cambridge: Harvard University Press.
COLE, K.S., 1932. Surface forces of the Arbacia egg. J. Cell Comp. Physiol. 1, 1-9.
CONKLIN, E.G., 1902. Karyokinesis and cytokinesis in the maturation, fertilization and cleavage of Crepidula and other Gasteropoda. J. Acad. Nat. Sci. of Phila., Ser. II, 12, 1-121.
CONKLIN, E.G., 1938. Disorientations of development in Crepidula plana produced by low temperatures. Proc. Amer. Phil. Soc. 79, 179-211.
CORNMAN, I., CORNMAN, M.E., 1951. The action of podophyllin and its fractions on marine eggs. Ann. N.Y. Acad. Sci. 51, 1443-1481.
DAN, J.C., 1948. On the mechanism of astral cleavage. Physiol. Zool. 21, 191-218.
DAN, K., 1943. Behavior of the cell surface during cleavage. V. Perforation experiment. J. Fac. Sci. Tokyo Imp. Univ. 6, 297-321.
DAN, K., 1963. Force of cleavage of the dividing sea urchin egg. In: Cell growth and cell division (R.J.C. Harris, ed.), 261-276. New York: Academic Press.
DANIELLI, J.F., 1952. Division of the flattened egg. Nature 170, 496.
FRY, H.J., PARKS, M.E., 1934. Studies of the mitotic figure. IV. Mitotic changes and viscosity changes in eggs of Arbacia, Cumingia, and Nereis. Protoplasma 21, 473-499.
GRAY, J., 1931. A textbook of experimental cytology, 203. New York: Macmillan.
HARVEY, E.B., 1956. The American Arbacia and other sea urchins. Princeton: Princeton University Press.
HEILBRUNN, L.V., 1920a. An experimental study of cell-division. J. Exp. Zool. 30, 211-237.
HEILBRUNN, L.V., 1920b. The physical effects of anesthetics upon living protoplasm. Biol. Bull. 39, 307-315.
HEILBRUNN, L.V., 1928. The colloid chemistry of protoplasm. Berlin: Borntrager.
HENLEY, C., COSTELLO, D.P., 1965. The cytological effects of podophyllin and podophyllotoxin on the fertilized eggs of Chaetopterus. Biol. Bull. 128, 369-391.
HIRAMOTO, Y., 1956. Cell division without mitotic apparatus. Exp. Cell Res. 11, 630-636.
HIRAMOTO, Y., 1957. The thickness of the cortex and the refractive index of the protoplasm in sea urchin eggs. Embryologia 3, 361-374.
HIRAMOTO, Y., 1958. A quantitative description of protoplasmic movement during cleavage in the sea urchin egg. J. Exp. Biol. 35, 407-424.
HIRAMOTO, Y., 1963a. Mechanical properties of sea urchin eggs. I. Surface force and elastic modules of the cell membrane. Exp. Cell Res. 32, 59-75.
HIRAMOTO, Y., 1963b. Mechanical properties of sea urchin eggs. II. Changes in mechanical properties from fertilization to cleavage. Exp. Cell Res. 32, 76-88.
HIRAMOTO, Y., 1964. Further studies on the cell division without mitotic apparatus in sea urchin eggs. Biol. Bull. 127, 357-358.

HIRAMOTO, Y., 1965. Further studies on cell division without mitotic apparatus in sea urchin eggs. J. Cell Biol. 25, 161-167.
HIRAMOTO, Y., 1967. Observations and measurements of sea urchin eggs with a centrifuge microscope. J. Cell Physiol. 69, 219-230.
HIRAMOTO, Y., 1968. The mechanics and mechanism of cleavage in the sea urchin egg. Symp. Soc. Exp. Biol. 22, 311-327.
HIRAMOTO, Y., 1969. Mechanical properties of the protoplasm of the sea urchin egg. II. Fertilized egg. Exp. Cell Res. 56, 209-218.
HIRAMOTO, Y., 1970. Rheological properties of sea urchin eggs. Biorheology 6, 201-234.
HIRAMOTO, Y., 1971. Analysis of cleavage stimulus by means of micromanipulation of sea urchin eggs. Exp. Cell Res. 68, 291-298.
IKEDA, M., 1965. Behavior of sulphydryl groups of sea urchin eggs under the blockage of cell division by ultraviolet and heat shock. Exp. Cell Res. 40, 282-291.
JUST, E.E., 1922. Studies on cell division. I. The effect of dilute sea water on the fertilized egg of Echinarachnius parma during the cleavage cycle. Am. J. Physiol. 61, 505-515.
KINOSHITA, S., 1968. Relative deficiency of intracellular relaxing system observed in presumptive furrow region in induced cleavage in the centrifuged sea urchin egg. Exp. Cell Res. 51, 395-405.
KINOSHITA, S., 1969. Periodical release of heparin-like polysaccharide within cytoplasm during cleavage of sea urchin egg. Exp. Cell Res. 56, 39-43.
KINOSHITA, S., YAZAKI, I., 1967. The behavior and localization of intracellular relaxing system during cleavage in the sea urchin egg. Exp. Cell Res. 47, 449-458.
KOBAYASHI, N., 1962. Cleavage of the sea urchin egg recovering from the cleavage-blocking effect of demecolcine. Embryologia 7, 68-80.
KOJIMA, M.K., 1969. Induction of nuclear changes and cleavage by repeated insufficient stimulations with activating reagents in the sea urchin egg. Embryologia 10, 334-342.
LEWIS, W.H., 1942. The relation of viscosity changes of protoplasm to ameboid locomotion and cell division. In: The structure of protoplasm (W. Seifriz, ed.), 163-197. Ames: Iowa State College Press.
LINDAHL, P.E., 1932. Zur experimentellen Analyse der Dorsoventralachse beim Seeigelkeim. I. Versuche mit gestreckten Eiern. Wilhelm Roux' Arch. Entwicklungsmech. Organismen 127, 300-322.
LORCH, I.J., 1952. Enucleation of sea urchin blastomeres with or without removal of asters. Quart. J. Micros. Sci. 93, 475-486.
MARSLAND, D., 1939. The mechanism of cell division. Hydrostatic pressure effects upon dividing egg cells. J. Cell Comp. Physiol. 13, 15-22.
MARSLAND, D., 1950. The mechanisms of cell division; temperature-pressure experiments on the cleaving eggs of Arbacia punctulata. J. Cell Comp. Physiol. 36, 205-227.
MARSLAND, D., 1951. The action of hydrostatic pressure on cell division. Ann. N.Y. Acad. Sci. 51, Art. 8, 1327-1335.
MARSLAND, D., 1956. Protoplasmic contractility in relation to gel structure: Temperature pressure experiments on cytokinesis and amoeboid movement. Internat. Rev. Cytol. 5, 199-227.
MARSLAND, D., 1970. Pressure-temperature studies on the mechanisms of cell division. In: High pressure effects on cellular processes (A.M. Zimmerman, ed.), 259-312. New York: Academic Press.
MERCER, E.H., WOLPERT, L., 1958. Electron microscopy of cleaving sea urchin eggs. Exp. Cell Res. 14, 629-632.
MITCHISON, J.M., SWANN, M.M., 1952. Optical changes in the membranes of the sea urchin at fertilization, mitosis and cleavage. J. Exp. Biol. 29, 357-362.
MITCHISON, J.M., SWANN, M.M., 1954. The mechanical properties of

the cell surface. I. The cell elastimeter. J. Exp. Biol. 31, 443-472.
MITCHISON, J.M., SWANN, M.M., 1955. The mechanical properties of the cell surface. III. The sea urchin egg from fertilization to cleavage. J. Exp. Biol. 32, 734-750.
MONROY, A., MONTALENTI, G., 1947. Variations of the submicroscopic structure of the cortical layer of fertilized and parthenogenetic sea urchin eggs. Biol. Bull. 92, 151-161.
MORGAN, T.H., 1927. Experimental Embryology, 200. New York: Columbia University Press.
MOTOMURA, I., 1950. Studies of cleavage. V. The role of vacuoles in the cleavage plane formation in sea urchin's eggs. Sci. Rep. Tôhoku Univ., 4th Ser., 18, 255-261.
MOTOMURA, I., 1966. Secretion of a mucosubstance in the cleaving egg of the sea urchin. Acta Embryol. Morphol. Exp. 9, 56-60.
MOTOMURA, I., 1967. Formation of diastema in the cleaving egg of the sea urchin. Sci. Rep. Tôhoku Univ., 4th Ser., 33, 135-142.
PORTZEHL, H., 1951. Muskelkontraction und Modellkontraction. Z. Naturforsch. 66, 355-361.
RAPPAPORT, R., 1960. Cleavage of sand dollar eggs under constant tensile stress. J. Exp. Zool. 144, 225-231.
RAPPAPORT, R., 1961. Experiments concerning the cleavage stimulus in sand dollar eggs. J. Exp. Zool. 148, 81-89.
RAPPAPORT, R., 1964. Geometrical relations of the cleavage stimulus in constricted sand dollar eggs. J. Exp. Zool. 155, 225-230.
RAPPAPORT, R., 1965. Geometrical relations of the cleavage stimulus in invertebrate eggs. J. Theoret. Biol. 9, 51-66.
RAPPAPORT, R., 1966. Experiments concerning the cleavage furrow in invertebrate eggs. J. Exp. Zool. 161, 1-8.
RAPPAPORT, R., 1967. Cell division: direct measurement of maximum tension exerted by furrow of echinoderm eggs. Science 156, 1241-1243.
RAPPAPORT, R., 1968. Geometrical relations of the cleavage stimulus in flattened, perforated sea urchin eggs. Embryologia 10, 115-130.
RAPPAPORT, R., 1969a. Aster-equatorial surface relations and furrow establishment. J. Exp. Zool. 171, 59-68.
RAPPAPORT, R., 1969b. Division of isolated furrows and furrow fragments in invertebrate eggs. Exp. Cell Res. 56, 87-91.
RAPPAPORT, R., 1970. An experimental analysis of the role of fountain streaming in furrow establishment. Develop. Growth Diff. 12, 31-40.
RAPPAPORT, R., 1971. Cytokinesis in animal cells. Int. Rev. Cytol. 31, 169-213.
RAPPAPORT, R., 1972. On the rate of movement of the cleavage stimulus in sand dollar eggs. J. Exp. Zool. 183, 115-120.
RAPPAPORT, R., CONRAD, G.W., 1963. An experimental analysis of unilateral cleavage in invertebrate eggs. J. Exp. Zool. 153, 99-112.
RAPPAPORT, R., EBSTEIN, R.P., 1965. Duration of stimulus and latent period preceding furrow formation in sand dollar eggs. J. Exp. Zool. 158, 373-382.
SCHROEDER, T.E., 1969. The role of "contractile ring" filaments in dividing Arbacia egg. Biol. Bull. 137, 413-414.
SCOTT, A.C., 1960. Furrowing in flattened sea urchin eggs. Biol. Bull. 119, 246-259.
SPEK, J., 1918. Oberflächenspannungsdifferenzen als eine Ursache der Zellteilung. Arch. f. Entw. Mech. 44, 5-113.
SUGIYAMA, M., 1951. Refertilization of the fertilized eggs of the sea urchin. Biol. Bull. 101, 335-343.
SWANN, M.M., 1951. Protoplasmic structure and mitosis II. The nature and cause of birefringence changes in the sea urchin egg at anaphase. J. Exp. Biol. 28, 434-441.
SWANN, M.M., MITCHISON, J.M., 1953. Cleavage of sea urchin eggs in colchicine. J. Exp. Biol. 30, 506-514.

SWANN, M.M., MITCHISON, J.M., 1958. The mechanism of cleavage in animal cells. Biol. Rev. 33, 103-135.
TILNEY, L.G., MARSLAND, D., 1969. A fine structural analysis of cleavage induction and furrowing in the eggs of Arbacia punctulata. J. Cell Biol. 42, 170-184.
WEINSTEIN, R.S., HEBERT, R.B., 1964. Electron microscopy of cleavage furrows in sea urchin blastomeres. J. Cell Biol. 23, 101A.
WILSON, E.B., 1895. An atlas of fertilization and karyokinesis of the ovum. New York: Macmillan & Co.
WILSON, E.B., 1901. Experimental studies in cytology. II. Some phenomena of fertilization and cell division in etherized eggs. Arch. f. Entw. Mech. 13, 353-373.
WILSON, E.B., 1928. The cell in development and heredity, 1059. New York: Macmillan & Co.
WOLPERT, L., 1960. The mechanics and mechanism of cleavage. Intern. Rev. Cytol. 10, 163-216.
WOLPERT, L., 1963. Some problems of cleavage in relation to the cell membrane. In: Cell growth and cell division (R.J.C. Harris, ed.), 277-298. New York: Academic Press.
WOLPERT, L., 1966. The mechanical properties of the membrane of the sea urchin egg during cleavage. Exp. Cell Res. 41, 385-396.
YATSU, N., 1910. An experimental study on the cleavage of the ctenophore egg. Proc. 7th Internat. Zool. Congr., 1-4.
YATSU, N., 1912. Observations and experiments on the ctenophore egg. Jour. Coll. Sci. Tokyo 32, Art. 3, 1-21.
ZIEGLER, H.E., 1903. Experimentelle Studien über Zellteilung. IV. Die Zellteilung der Furchungszellen bei Beroë und Echinus. Arch. f. Entw. Mech. 16, 155-175.
ZIMMERMAN, A.M., LANDAU, J.V., MARSLAND, D., 1957. Cell division: A pressure-temperature analysis of the effects of sulfhydryl reagents on the cortical plasmagel structure and furrowing strength of dividing eggs (Arbacia and Chaetopterus). J. Cell Comp. Physiol. 49, 395-435.

11. Centrifugation and Alteration of Cleavage Pattern

B. Brandriff and R. Hinegardner

Centrifugation has been used as a tool in experimental embryology since the turn of the century. By 1910 a number of studies had been published on the results of centrifuging both mosaic and regulative eggs, including the eggs of several species of sea urchin. The object of these early studies was to determine the kind and degree of developmental disturbance caused by redistributing the contents of the egg; to search for morphogenetic substances; to examine egg organization and polarity; and to study cleavage mechanisms.

Centrifugation has continued to be a useful tool in research on sea urchin development. Based on early experimental results, some studies were made in the 1930s on the relation between the future oral field of the pluteus and the centripetal and centrifugal poles of the egg. HARVEY extensively investigated the physical properties, developmental potential, and cleavage patterns in whole, half and quarter eggs obtained by centrifugation. The nature of the cortex and the cortical reaction were also studied. Recently, gene expression in early development has been examined by observing biosynthetic activities in nucleate and anucleate egg fragments obtained with centrifugation.

11.1 Techniques

Current workers use, with slight modifications, the method described by HARVEY (1956) for the red pigmented eggs of *Arbacia punctulata*. Stratified whole, half, and quarter eggs can be obtained. One part of egg and seawater suspension is layered over 2 parts of an isosmotic 0.85 *M* sucrose solution (29.1 G sucrose made up to 100 ml with distilled water). By gently tilting the centrifuge tube several times or stirring it with a narrow rod, a density gradient is established through partial mixing at the seawater and sucrose interface. Centrifugation at 3,000 g for 2 minutes stratifies the eggs, and 10,000 g for 4 minutes causes the stratified eggs to become dumb-bell shaped and then to break into a nucleate white half, located centripetally, and an anucleate red half located centrifugally in the tube.

If the white halves are recentrifuged in a slightly less dense sucrose solution at 10,000 g for 45 minutes, they fragment into clear quarters and mitochondrial quarters. Recentrifuging the red halves in the full-strength sucrose solution at 10,000 g for 45 minutes causes them to break into yolk quarters and pigment quarters (Fig. 11.1).

Current modifications primarily involve changes in the density of the sucrose solution, which may vary from 0.75 *M* to 1.0 *M* or above, and the length of time and centrifugal force used (CRAIG and PIATIGORSKI, 1971). In addition, some workers use step gradients of 3 or 4 layers for better separation of the various egg fragments (TYLER, 1966).

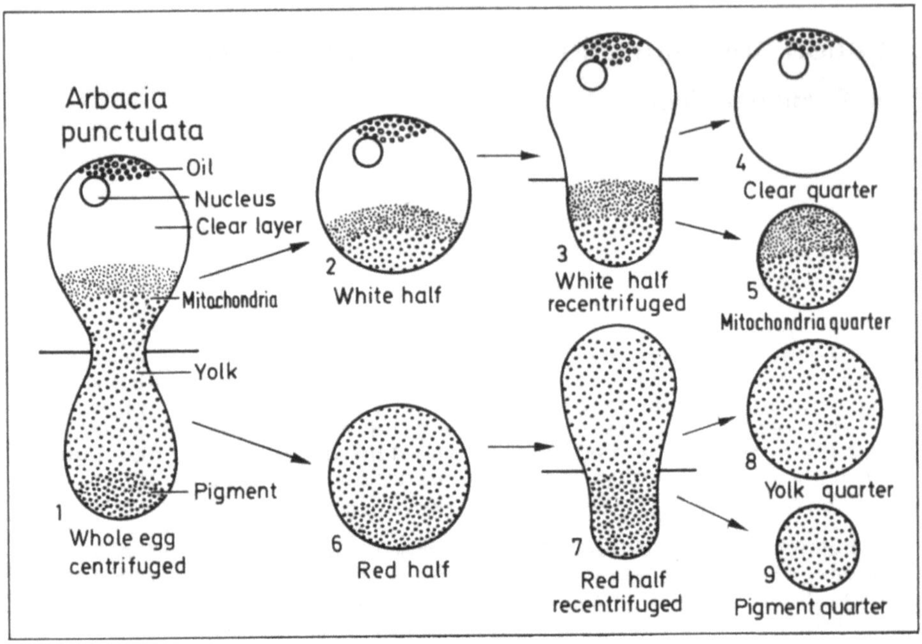

Fig. 11.1. Stratified whole egg of *Arbacia punctulata*, halves and quarters. Clear area at centripetal pole of No. 7 is due to further packing of granules with longer centrifuging. (From HARVEY, 1956)

10 to 15 minutes after the eggs are removed from the centrifuge, the cytoplasmic constituents begin to redistribute, starting with the mitochondrial layer in *A. punctulata*. Similarly, unfertilized stretched eggs return to spherical shape within a short time, although if they are fertilized, they retain their dumb-bell shape into the blastula stage or later. Before the eggs can be fertilized they must be resuspended in seawater since the sucrose solution inhibits fertilization, possibly because of the lack of Ca^{++} ions.

The effects of hypo- and hypertonic seawater, temperature, pressure, and other variables on stratification and fragmenting of eggs are summarized in HARVEY (1956). An inverse relationship exists between temperature and time of centrifuging required for fragmentation (COSTELLO, 1938). Eggs with the vitelline membrane removed fragment at centrifugal gravity forces 25% lower than those required for normal eggs. A decrease in fragmentation takes place upon the addition of uranium ions to the medium (SOLANO and MAZIA, 1953).

11.2 Morphological Changes

11.2.1 Unfertilized Eggs

Most eggs can be stratified into 4 or 5 layers (Fig. 11.1): an oil cap at the centripetal end; a clear layer containing the nucleus; a mitochondrial layer; a yellowish layer of yolk granules; and if the egg is pigmented a pigment layer at the centrifugal end. After the eggs fragment, usually in the region of the mitochondrial zone, the centripetal halves contain the oil cap, clear layer with nucleus, much of the mitochondrial layer, and some yolk. The centrifugal halves are stratified into the yolk and pigment layers, and sometimes part of the mitochondrial layer. On recentrifuging, the centripetal halves break into clear quarters consisting of the oil cap, clear layer and nucleus, and mitochondrial quarters containing the mitochondrial layer and some yolk. The centrifugal halves fragment into yolk quarters, and pigment quarters made up of the pigment layer and some yolk (HARVEY, 1956).

The properties of stratified eggs of several other species of sea urchin are illustrated and tabulated in Figs. 11.2, 11.3 and Table

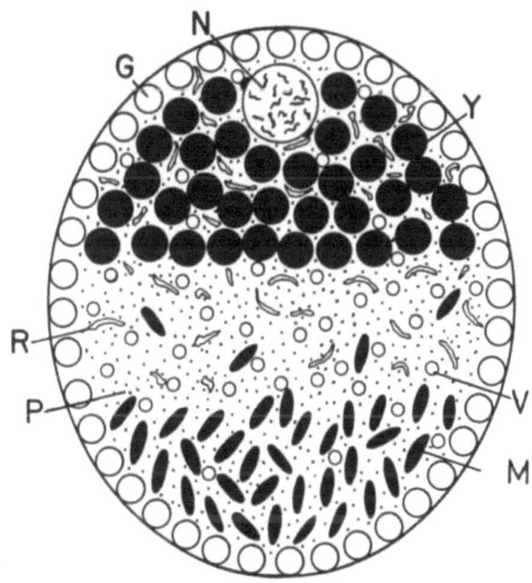

Fig. 11.2. Centrifuged, unfertilized egg of *Psammechinus miliaris*. Nucleus (N) is near centripetal pole. (G) cortical granules; (M) mitochondria; (P) dense ribosomes; (R) reticulum; (V) small vesicles; (Y) yolk granules. (From MERCER and WOLPERT, 1962)

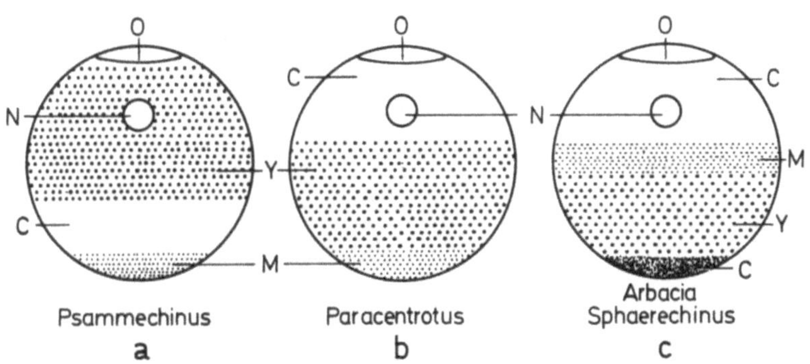

Fig. 11.3a-c. Centrifuged, unfertilized egg of (a) *Psammechinus microtuberculatus*; (b) *Paracentrotus lividus*; (c) *Sphaerechinus granularis* and *Arbacia lixula*. Oil cap (O) marks centripetal pole. (C) clear layer; (M) mitochondria; (N) nucleus; (Y) yolk. Clear layer at centrifugal pole of *Sphaerechinus* is where pigment layer is in *Arbacia*. (From CALLAN, 1949)

Table 11.1. Stratification of sea-urchin eggs. (From HARVEY, 1956)

Species	Layer 1	Layer 2	Layer 3	Layer 4	Layer 5	Layer 6
Echinarachnius parma	Oil	(Clear)[a]	Yolk	Clear	Mitochondria	Neutral-red-staining granules
Echinocardium cordatum	Yolk	Clear	Mitochondria			
Echinometra lucunter	Oil	Clear	Mitochondria	Yolk		
Lytechinus variegatus	Oil	Yolk	Clear	Mitochondria	(Clear)	
Mellita quinquiesperforata	Oil	Clear	Mitochondria	Yolk		
Strongylocentrotus droebachiensis	Oil	(Clear)	Yolk	Clear	Mitochondria	
Strongylocentrotus franciscanus	Oil	(Clear)	Yolk	Clear	Mitochondria	(Clear)
Strongylocentrotus purpuratus	Oil	(Clear)	Yolk	Mitochondria	Clear	
Tripneustes esculentus	Oil	(Clear)	Yolk	(Clear)	Mitochondria	Clear

[a]Layers in parentheses are not always present.

11.1. Information on vital staining, density, osmotic behavior and electrical properties is summarized in HARVEY (1956).

The cortical granules are not dislodged from the periphery of the egg by centrifugation (GROSS, PHILPOTT and NASS, 1960). This has been one reason for ascribing special properties to the egg cortex and what may be a gel-like layer underneath it. Such a layer has been demonstrated to form in *Xenopus* eggs shortly after fertilization (HEBARD and HAROLD, 1967). Since the cortex is believed to be the seat of special morphogenetic properties, the structure of this outer layer would appear to be of particular significance. In *Psammechinus miliaris* the membranes of the cortical granules are fused with the egg membrane and therefore hold the granules in place (MERCER and WOLPERT, 1962). In *A. punctulata* eggs no evidence of such membrane fusion is found, nor of any other structures that might account for the behavior of the cortical granules. However, if centrifuging is preceded by urethane treatment the cortical granules become dislodged and stratify (ANDERSON, 1970).

11.2.2 Fertilized Eggs

Eggs centrifuged after fertilization are more resistant to the effects of high centrifugal force. They stratify less well and the various layers are not distinct, most likely because of the increase in cytoplasmic viscosity known to take place after fertilization. In eggs fertilized while still dumb-bell shaped, the fertilization membrane follows the same outline. If the fertilization membrane is removed, fertilized eggs can be made to fragment like unfertilized eggs, except during a short post-fertilization period when they tend to break into many small pieces (HARVEY, 1956).

In *Arbacia* eggs, one of the intermediate layers disappears from view under the light microscope immediately upon breakdown of the cortical granules. This layer is described as located between the clear and yolk layers, and very likely corresponds to the mitochondrial layer (MOSER, 1939). In this egg, the pigment granules which migrate to the periphery after fertilization cannot be dislodged from that position if the eggs are centrifuged 30 minutes post fertilization. Instead, the granules remain embedded in the apparently strongly gelated cortex (TILNEY and MARSLAND, 1969). Studies of the surface properties and cytoplasmic viscosity of *A. punctulata* and *Lytechinus pictus* eggs with the centrifuge microscope reveal variability in resistance to deformation during the first cleavage cycle. 2 points of least deformation, one just before and one during first cleavage, coincide with 2 peaks in the magnitude of the surface force (HIRAMOTO, 1966).

11.3 Electron Microscope Studies

Electron microscope studies of whole centrifuged eggs of *Arbacia punctulata* have revealed that the layers in the stratified egg are far from homogeneous (ANDERSON, 1970). Each contains in varying quantities most of the cytoplasmic inclusions present in the normal egg. Only the lipid droplets of the oil cap and the pigment bodies are confined to their respective layers and never found anywhere else.

In addition to the lipid droplets, the oil cap contains mitochondria, ribosomes, and some rough endoplasmic reticulum. The clear layer is made up of heterogeneous elements such as rough endoplasmic reticulum, annulate lamellae, heavy bodies, Golgi complexes, and rod-containing vacuoles. Along with the yolk bodies of the yolk layer, there are Golgi complexes, mitochondria, rod-containing vacuoles, and endoplasmic reticulum. The pigment layer, in addition to the pigment bodies, contains mitochondria as well as yolk bodies.

Arbacia eggs fragment into centripetal halves slightly larger than the centrifugal halves and with the exception of the lipid, pigment, and nucleus, there is no exclusive segregation of one kind of inclusion or another into either half. Aside from a few subsurface cisternae in the break region no changes caused by the stretching are detectable in the cell membrane.

Recentrifuging the centripetal halves results in clear quarters containing lipid droplets, the nucleus, Golgi complexes, mitochondria, rod-containing vacuoles, annulate lamellae, and cortical granules which remain attached to the egg membrane in all quarters. In addition to many mitochondria, the mitochondrial quarters include a certain amount of yolk. Recentrifuging the centrifugal halves yields yolk quarters with yolk bodies, some mitochondria, and endoplasmic reticulum. Besides pigment bodies, the pigment quarter holds some yolk bodies, mitochondria, and rod-containing vacuoles.

The heavy bodies, which concentrate in the clear layer, have been postulated to be the site of stored, inactive informational molecules for embryogenesis. In a study employing cytochemical techniques these organelles have been demonstrated to contain RNA (CONWAY, 1971).

11.4 Developmental Patterns

11.4.1 Whole Eggs

Whole centrifuged *Arbacia* eggs, whether spherical or stretched, usually develop into normal plutei, except that the pigment will be localized rather than diffuse (HARVEY, 1956). The ventral area or the area around the mouth and extending into the arms may contain most of the pigment granules, or they may be concentrated in one lateral half of the pluteus, or in the gut (LYON, 1907; MORGAN and LYON, 1907).

11.4.2 Halves

Fertilized centripetal halves of *A. punctulata* are sometimes able to develop into plutei that are half-sized and unpigmented but otherwise apparently normal. Often, however, they develop into permanent blastulae. There is no constant ratio of normal to abnormal development and different batches of eggs vary considerably. Centripetal halves activated parthenogenetically also are capable of becoming plutei.

Even though they have no nucleus, centrifugal halves nevertheless may be fertilized. The sperm nucleus and its aster become visible in the egg cytoplasm 15-20 minutes after the formation of the fertilization membrane. The sperm nucleus swells and after about an hour the nuclear membrane breaks down as in normal prophase. Sometimes fairly normal looking blastulae develop, but more frequently multinucleate structures without cell boundaries arise. Very occasionally hatching occurs with a few plutei developing. Parthenogenetically activated centrifugal halves may have a clear area that superficially looks like a nucleus, with a monaster or diaster arising in its vicinity. Cleavages take place but often without the appearance of cell boundaries, with the eventual formation of structures containing about 500 cells that can survive for up to 1 month (HARVEY, 1956).

11.4.3 Quarters

Clear quarters, which contain the nucleus, cleave much more slowly than normal eggs, with a blastula hatching after some 20 hours. A very few gastrulae and plutei develop. Many of the clear quarters remain no more than clusters of cells, often with greatly enlarged nuclei. Parthenogenetically activated clear quarters undergo early nuclear changes without subsequent development.

Fertilized mitochondrial quarters may develop into swimming blastulae, but no further, often with multinucleate cells without regular cleavage planes separating the cells. In addition, this quarter can sometimes reach the pluteus stage but with irregular skeleton and pigment.

The pigment quarter may produce some normal cleavage stages, which may result in multinucleate structures, with an occasional one being able to swim (HARVEY, 1956).

In another investigation using *Arbacia*, considerably less developmental potential was found to exist in fertilized and artificially activated halves and quarters (ANDERSON, 1970). In this study the only fragment able to produce a pluteus was the centripetal half. All quarters underwent a cortical reaction, but only the nucleated quarter developed into a multicellular structure bearing cilia.

11.4.4 Eggs Centrifuged after Fertilization

Eggs centrifuged within the first minute after fertilization fragment into small pieces. If centrifuged after the first minute and without becoming elongated, they develop normally (HARVEY, 1956). Eggs centrifuged at various intervals through the mitotic stages up to first cleavage yield an unvarying percentage of normal development (SPOONER, 1911).

The developmental potential of halves obtained after fertilization is considerably reduced compared to that of halves obtained before fertilization. Centripetal halves may develop normally to the blastula stage. Thereafter, they either remain permanent blastulae, or occasionally become very abnormal plutei. Centrifugal halves do not cleave except in the event that separation takes place before the male and female nuclei fuse. If this is the case and the male nucleus happens to be in the centrifugal half, it is able to divide (HARVEY, 1956).

11.5 Cleavage Patterns and Micromere Formation

11.5.1 Normal Eggs

In the normal egg the first cleavage plane is meridional, i.e. parallel to the animal-vegetal axis (Fig. 11.4a). The second is meridional but at right angles to the first, while the third is equatorial (Fig. 11.4b). The fourth is meridional in the animal half and subequatorial

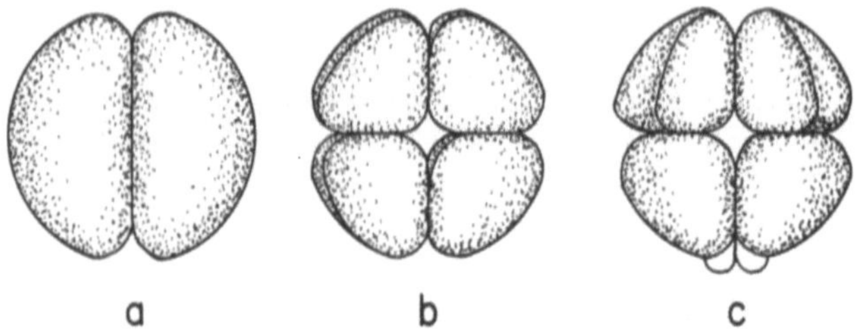

Fig. 11.4a-c. Normal cleavage pattern of sea urchin egg. (a) First cleavage plane is meridional and goes through animal-vegetal axis. (b) 8-cell stage; third cleavage plane is equatorial. (c) 16-cell stage; micromeres are at vegetal pole

in the vegetal half forming 4 micromeres at the vegetal pole (Fig. 11.4c).

11.5.2 Elongated Eggs

In centrifugally elongated eggs the first spindle forms parallel to the long axis of the egg. The first cleavage is therefore parallel to the plane of stratification. Often the egg is divided into 2 unequally sized blastomeres, a small clear (centripetal) one and a larger granular (centrifugal) one. Subsequent cleavages are asynchronous, with the smaller clear blastomere dividing more rapidly (HARVEY, 1956).

11.5.3 Spherical Eggs

In spherical stratified eggs, the first cleavage plane is usually at right angles to the stratification (meridional in terms of stratification) and the second equatorial. The third is again meridional, and as a result the order of the second and third planes seems to be reversed from the normal egg. Sometimes, the first cleavage plane is equatorial, with the second and third meridional. These embryos appear slightly less viable than those with the more common cleavage pattern.

The micromeres frequently form at the centrifugal pole and lose some of their pigment granules, similar to what happens in normal eggs. However, the micromeres may form at the unpigmented pole, or on 1 side between the pigmented and unpigmented halves. The pigment in the pluteus is most often found near the mouth, or mouth and anus.

Since micromeres seem to form in no fixed relation to the stratification, and since the eggs do not orient themselves according to a predetermined axis in the centrifuge but rather fall at random, micromere formation appears to take place independently of the induced axis (LYON, 1907; MORGAN and LYON, 1907).

In normal eggs, the original egg axis can be determined by the location of the micropyle, a funnel-shaped opening in the jelly coat (Fig. 11.5). By use of this marker, it was found that the micromeres in centrifuged eggs form opposite the micropyle at the intersection of 2 cleavage planes, either the first and second, the first and third, or the second and third. Since the crossing points of cleavage planes are not always exactly opposite the micropyle, a certain amount of

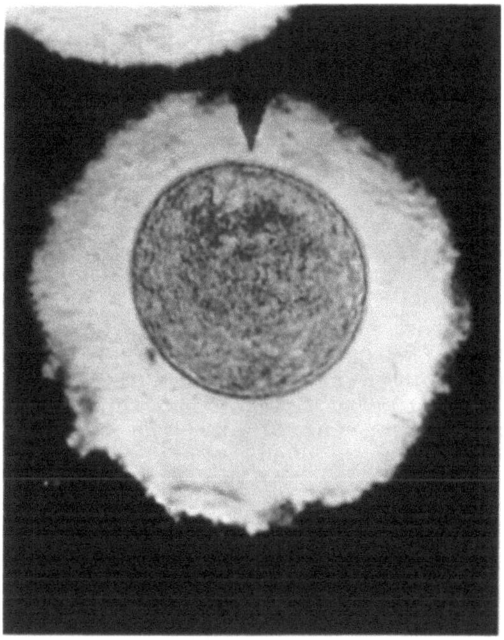

Fig. 11.5a-d. Egg of *Arbacia punctulata* in India ink. Funnel-shaped micropyle passes through jelly coat at top. (Courtesy JULIE R. KORENBERG, 1971)

regulation takes place such that the micromeres arise at the crossing point most nearly opposite the micropyle. Thus, location of the micromeres is determined by the original egg axis and location of the cleavage planes (MORGAN and SPOONER, 1909). The sea urchin thus appears to follow a rather rigid plan of cleaving and exhibits a certain similarity to the egg of *Ilyanassa*, which is categorized as a mosaic egg and and in which the polar lobe forms in the vegetal half even when the yolk normally present there is displaced by centrifuging (CLEMENT, 1968).

Centrifuged eggs gastrulate from the micromere pole as do normal eggs, and development proceeds undisturbed, indicating that gastrulation and differentiation of the embryo take place independently of substances displaced by centrifugation (MORGAN and SPOONER, 1909).

In the eggs of *Temnopleurus* and *Mespilia* granules which can be stained with neutral red are distributed to the centrifugal pole (Fig. 11.6a).

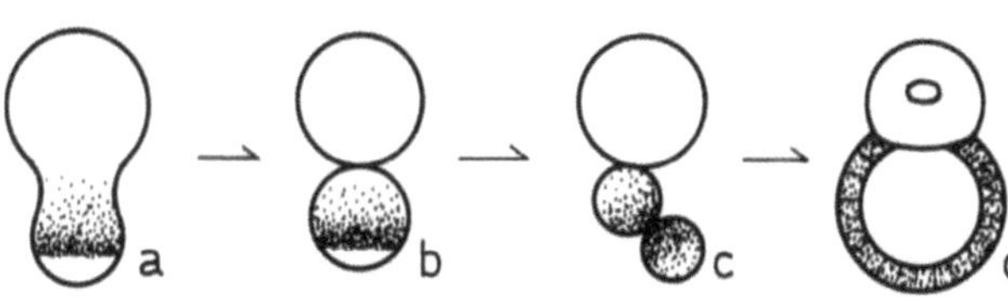

Fig. 11.6a-d. Cleavage pattern in eggs of *Temnopleurus toreumaticus*. Stippling indicates neutral-red staining granules. (a) Centrifuged, stretched egg; (b) first cleavage; (c) second cleavage; (d) blastula with attached, non-dividing blastomere. (From KOJIMA, 1959)

The first cleavage plane which is oriented parallel to the stratification segregates these granules into the centrifugal blastomere (Fig. 11.6b). During subsequent cleavages only the descendants of the centrifugal blastomere divide (Fig. 11.6c), whereas the centripetal blastomere remains uncleaved and attached to the developing blastula (Fig. 11.6d)(KOJIMA, 1959). A similar phenomenon has been reported for the eggs of *Dendraster* which form an uncleaved centripetal lobe that remains attached to the embryo (PEASE, 1939).

11.6 Development of the Dorso-Ventral Axis

After it was reported that in plutei developing from centrifuged *Arbacia* eggs the pigment is concentrated in the ventral area more often than elsewhere (LYON, 1907; MORGAN and LYON, 1907), egg polarity and its relationship to the development of the dorso-ventral axis was investigated using centrifugation (RUNNSTRÖM, 1926; LINDAHL, 1932; PEASE, 1939). With very high frequency, often 90% or above, the ventral side is located opposite the uncleaved centripetal blastomere that forms during development of centrifuged *Dendraster* eggs (PEASE, 1939). With similar frequencies the pigment in *Arbacia* eggs and certain stainable granules in *Psammechinus*, which indicate the centrifugal pole, are located on the ventral side of the pluteus (RUNNSTRÖM, 1926; LINDAHL, 1932). Formation of the ventral side can thus be induced experimentally by centrifugation. Conceivably the entire egg cortex contains a ventral gradient that is immovable by centrifugation and which interacts with some displaced cytoplasmic substrate during development thus inducing formation of the ventral side at the point of greatest interaction between substrate and ventral gradient (PEASE, 1939). Alternately, the granules deposited at the centrifugal pole may have a determining effect (LINDAHL, 1932). It could be that centrifuging as such induces formation of the ventral side (RUNNSTRÖM, 1926).

11.7 Respiration

Respiration in unfertilized centripetal halves of *Arbacia* eggs is about the same as in unfertilized whole eggs, while unfertilized centrifugal halves take up twice as much oxygen. After fertilization the respiratory rate in whole eggs and centripetal halves increases by approximately 2.5 times, and that in centrifugal halves by about 1.2 times. KCN does not affect respiration in unfertilized centripetal halves, but reduces it in unfertilized centrifugal halves. Oxygen uptake is also reduced by KCN in fertilized whole eggs, centripetal and centrifugal halves (SHAPIRO, 1935, 1939, 1940).

11.8 Biochemistry

Enucleate halves obtained by centrifugation have been a useful system to investigate the role of extra-nuclear genes and maternal mRNA in early development. There is a large literature on this subject. Some pertinent references on some aspects are: artificially activated anucleate halves and the incorporation of labelled precursors into protein (BRACHET et al., 1963; DENNY and TYLER, 1964; BURNY et al.,

1965; BALTUS et al., 1965; TYLER, 1966; BRACHET, 1967; CRAIG and PIATIGORSKI, 1971); halves and RNA synthesis (BALTUS et al., 1965; CRAIG, 1970; CHAMBERLAIN, 1968, 1970; SELVIG, GROSS and HUNTER, 1970; CRAIG and PIATIGORSKI, 1971; MAROUN, 1973); polyadenylation (WILT, 1973; WILT and MAZIA, 1974).

References

ANDERSON, E., 1970. A cytological study of the centrifuged whole, half, and quarter eggs of the sea urchin Arbacia punctulata. J. Cell Biol. 47, 711-733.

BALTUS, E., QUERTIER, J., FICQ, A., BRACHET, J., 1965. Biochemical studies of nucleate and anucleate fragments isolated from sea urchin eggs. A comparison between fertilization and parthenogenetic activation. Biochim. Biophys. Acta 95, 408-417.

BRACHET, J., 1967. Protein synthesis in the absence of the nucleus. Nature (London) 213, 650-655.

BRACHET, J., FICQ, A., TENCER, R., 1963. Amino acid incorporation into proteins of nucleate and anucleate fragments of sea urchin eggs: effect of parthenogenetic activation. Exp. Cell Res. 32, 168-170.

BURNY, A., MARBAIX, G., QUERTIER, J., BRACHET, J., 1965. Demonstration of functional polyribosomes in nucleate and anucleate fragments of sea urchin eggs following parthenogenetic activation. Biochim. Biophys. Acta 103, 526-528.

CALLAN, H.G., 1949. Cleavage rate, oxygen consumption and ribose nucleic acid content of sea urchin eggs. Biochim. Biophys. Acta 3, 92-102.

CHAMBERLAIN, J., 1968. Extranuclear RNA synthesis in sea urchin embryos. J. Cell Biol. 39, 23a.

CHAMBERLAIN, J., 1970. RNA synthesis in anucleate egg fragments and normal embryos of the sea urchin, Arbacia punctulata. Biochim. Biophys. Acta 213, 183-193.

CLEMENT, A.C., 1968. Development of the vegetal half of the Ilyanassa egg after removal of most of the yolk by centrifugal force, compared with the development of animal halves of similar visible composition. Develop. Biol. 17, 165-186.

CONWAY, C.M., 1971. Evidence for RNA in the heavy bodies of sea urchin eggs. J. Cell Biol. 51, 889-893.

COSTELLO, D.P., 1938. The effect of temperature on the rate of fragmentation of Arbacia eggs subjected to centrifugal force. J. Cell. Comp. Phys. 11, 301-306.

CRAIG, S.P., 1970. Synthesis of RNA in non-nucleate fragments of sea urchin eggs. J. Mol. Biol. 47, 615-618.

CRAIG, S.P., PIATIGORSKI, J., 1971. Protein synthesis and development in the absence of cytoplasmic RNA synthesis in non-nucleate egg fragments and embryos of sea urchins: effect of ethidium bromide. Develop. Biol. 24, 214-232.

DENNY, P., TYLER, A., 1964. Activation of protein biosynthesis in non-nucleate fragments of sea urchin eggs. Biochem. Biophys. Res. Comm. 14, 245-249.

GROSS, P.R., PHILPOTT, D.E., NASS, S., 1960. Electron microscopy of the centrifuged sea urchin egg, with a note on the structure of the ground cytoplasm. J. Biophys. Biochem. Cytol. 7, 135-142.

HARVEY, E.B., 1956. The American Arbacia and other sea urchins. Princeton Univ. Press, Princeton, 298 pp.

HEBARD, C., HAROLD, R., 1967. The ultrastructure of the cortical cytoplasm in the unfertilized egg and first cleavage zygote of Xenopus laevis. Exp. Cell Res. 46, 553-570.

HIRAMOTO, Y., 1966. Observations and measurements of sea urchin eggs with a centrifuge microscope. J. Cell Phys. 69, 219-230.
KOJIMA, M.K., 1959. Relation between the vitally stained granules and cleavage activity in the sea urchin egg. Embryologia 4, 191-209.
LINDAHL, P.E., 1932. Zur experimentellen Analyse der Determination der Dorsoventralachse beim Seeigelkeim. II. Versuche mit zentrifugierten Eiern. Wilhelm Roux' Arch. Entwicklungsmech. Organismen 127, 323-338.
LYON, E.P., 1907. Results of centrifugalizing eggs. Wilhelm Roux' Arch. Entwicklungsmech. Organismen 23, 151-173.
MAROUN, L.E., 1973. Cytoplasmic localization of DNA dependent RNA polymerase in unfertilized sea urchin eggs. Exp. Cell Res. 79, 459-461.
MERCER, E.H., WOLPERT, L., 1962. An electron microscope study of the cortex of the sea urchin (Psammechinus miliaris) egg. Exp. Cell Res. 27, 1-13.
MORGAN, T.H., LYON, E.P., 1907. The relation of the substances of the egg, separated by a strong centrifugal force, to the location of the embryo. Wilhelm Roux' Arch. Entwicklungsmech. Organismen 24, 147-159.
MORGAN, T.H., SPOONER, G.B., 1909. The polarity of the centrifuged egg. Wilhelm Roux' Arch. Entwicklungsmech. Organismen 28, 104-117.
MOSER, F., 1939. Studies on a cortical layer response to stimulating agents in the Arbacia egg. J. Exp. Zool. 80, 423-445.
PEASE, D.C., 1939. An analysis of the factors of bilateral determination in centrifuged echinoderm embryos. J. Exp. Zool. 80, 225-247.
RUNNSTRÖM, J., 1926. Experimentelle Bestimmung der Dorsoventralachse bei dem Seeigelkeim. Arkiv Zool. 18A, No. 4.
SELVIG, S.E., GROSS, P.R., HUNTER, A.L., 1970. Cytoplasmic synthesis of RNA in the sea urchin embryo. Develop. Biol. 22, 343-365.
SHAPIRO, H., 1935. The respiration of fragments obtained by centrifuging the egg of the sea urchin, Arbacia punctulata. J. Cell. Comp. Physiol. 6, 101-116.
SHAPIRO, H., 1939. Some functional correlatives of cellular metabolism. Cold Spring Harbor Symp. 7, 406-419.
SHAPIRO, H., 1940. Further studies on the metabolism of cell fragments. Biol. Bull. 79, 377.
SOLANO, S.F., MAZIA, D., 1953. The role of the vitelline membrane in the fragmentation of the Arbacia egg by centrifugal force. Biol. Bull. 105, 384.
SPOONER, G.B., 1911. Embryological studies with the centrifuge. J. Exp. Zool. 10, 23-49.
TILNEY, L.G., MARSLAND, D., 1969. A fine structural analysis of cleavage induction and furrowing in the eggs of Arbacia punctulata. J. Cell Biol. 42, 170-184.
TYLER, A., 1966. Incorporation of amino acids into protein by artificially activated non-nucleate fragments of sea urchin eggs. Biol. Bull. 130, 450-461.
WILT, F., 1973. Polyadenylation of maternal RNA of sea urchin eggs after fertilization. Proc. Nat. Acad. Sci. 70, 2345-2349.
WILT, F., MAZIA, D., 1974. The stimulation of cytoplasmic polyadenylation in sea urchin eggs by ammonia. Develop. Biol. 37, 422-424.

12. Irradiation, Chemical Treatment, and Cleavage-Cycle and Cleavage-Pattern Alteration

R. Rustad

When sea-urchin eggs are irradiated or treated with chemical inhibitors, cleavage is usually delayed and mitoses may be morphologically abnormal. This chapter presents examples of the stimulation of premature mitosis, briefly summarizes the general properties of division delay, examines cases of specific inhibition at different mitotic stages, and explores some of the causes of the major division abnormalities. It attempts to provide concise comparisons among the cytologic effects of diverse agents but not to tabulate all of the effects of each agent to which dividing sea-urchin eggs have been exposed. The early literature concerning these problems has been catalogued (HARVEY, 1956), and references to general topics mentioned without literature citation may be found therein. The effects of specific drugs on echinoderm eggs have been summarized (KARNOFSKY and SIMMEL, 1963) and the effects of radiation on sea-urchin eggs reviewed (RUSTAD, 1971). Works on cellular radiation effects are available (ELKIND and WHITMORE, 1967; ALTMAN et al., 1970), and the effects of chemicals on dividing cells have been analyzed extensively (KIHLMAN, 1966). Like that earlier classic treatise on the cell (WILSON, 1924), a more recent analysis of the mitotic cycle is a permanent contribution to our understanding of the problems discussed in this chapter (MAZIA, 1961).

The duration of the first cleavage cycle may be reduced by prefertilization chemical stimulation (KOJIMA, 1967). The first division cycle is normally about twice as long as the others, so the inducement of early mitosis may reflect a stimulation of the relatively quiescent, unfertilized egg into a metabolic state resembling that of the fertilized one.

There are more quantitative data on radiation-induced mitotic delay in the sea-urchin egg than on division delay by any particular chemical treatment. Although their mechanisms remain unknown, the radiation effects seem rather specific. Analyses of changes in radiation sensitivity during the division cycle suggest that ultraviolet radiation affects mainly one mitotic event, while x-radiation affects mainly two events (RUSTAD, 1960a, 1970).

While the effects of some inhibitory chemicals are well understood, the mechanisms of action of others are unknown. The effects of chemicals can be specific or general, as in the colcemid-microtubule interaction that blocks the cells before pronuclear fusion or at metaphase (ZIMMERMAN and ZIMMERMAN, 1967) and respiratory inhibitors that interrupt the mitotic cycle at any point (EPEL, 1963).

Cleavage abnormalities include chromosomal aberrations, division failure, monopolar mitoses, and multipolar cleavage. Unequal cleavage may occur, apparently indicating the formation of micromeres and macromeres at the normal time of the 16-cell stage in eggs in which division is greatly delayed (RUSTAD, 1960b; IKEDA, 1965).

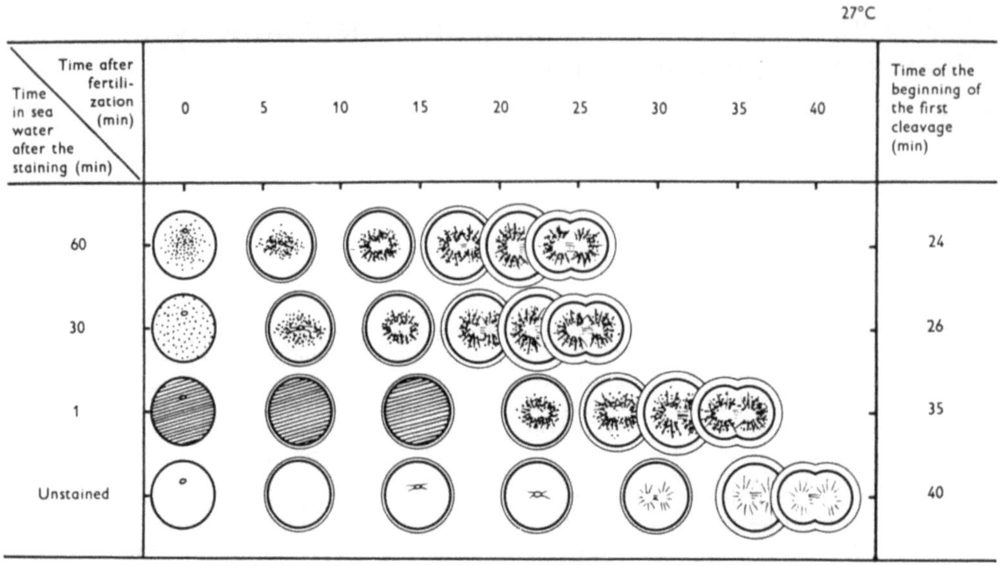

Fig. 12.1. Stimulation of premature nuclear changes and cleavage in *Temnopleurus* eggs. Unfertilized eggs were stained with dilute solution of neutral red in seawater, incubated in seawater for 1, 30, or 60 min, and fertilized. (From KOJIMA, 1960)

12.1 Stimulation of Premature Division

A wide variety of chemical treatments can lead to the stimulation of the fertilization reaction and, in some cases, to parthenogenetic development of the haploid egg. A series of studies indicates that the first post-fertilization mitotic cycle may be shortened by subliminal prefertilization stimulation with dyes, detergents, butyric acid, respiratory inhibitors, or inactivated sperm (Fig. 12.1)(KOJIMA, 1967, 1969a, 1969b). The stimulation of premature division can follow treatments that cause no morphologic changes such as cytaster formation or cortical granule breakdown and trigger no DNA synthesis (UTO and SUGIYAMA, 1969) but stimulate respiration and amino acid incorporation into proteins (NAKASHIMA and SUGIYAMA, 1969; SUGIYAMA et al., 1969; RUSTAD and GREENBERG, 1972). Therefore, it would appear that chemical treatment enhances the metabolic activity of the unfertilized egg so that it may complete some normal post-fertilization activities before fertilization or much sooner than usual after fertilization. Prefertilization stimulation with neutral red, sodium lauryl sulfate, hypertonic or hypotonic seawater, or anoxia may also reduce or eliminate γ-radiation-induced mitotic delay (RUSTAD and GREENBERG, 1972; RUSTAD and FEINGOLD, 1971). The method of KOJIMA, for shortening the mitotic cycle, will likely prove a valuable tool for examining the action of many mitotic inhibitors.

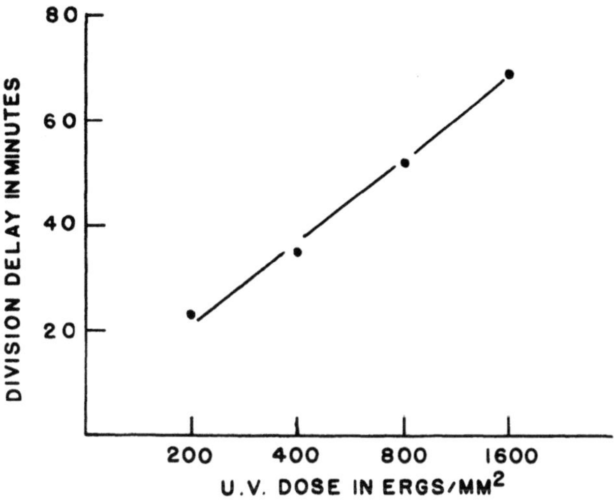

Fig. 12.2. Relationship between mitotic delay and logarithm of ultraviolet radiation dose administered to *Strongylocentrotus* eggs. (From RUSTAD and MAZIA, 1971)

12.2 Division Delay

Exposure of sperm or eggs to ultraviolet or ionizing radiation causes a dose-dependent delay of cleavage time. When sperm are exposed to x-radiation, the amount of mitotic delay is proportional to the logarithm of the dose (HENSHAW, 1940a). Data on the x-radiation sensitivity of eggs are somewhat variable; dose-dependence sometimes appears logarithmic at low doses and linear at high doses (HENSHAW et al., 1933; HENSHAW and FRANCIS, 1936; MIWA et al., 1939, 1941; RAO, 1963). The ultraviolet dose-dependence of eggs kept in the dark is logarithmic (Fig. 12.2)(RUSTAD and MAZIA, 1971) and becomes linear after photoreactivation (WELLS and GIESE, 1950).

There is spontaneous recovery from the damage of x- or γ-radiation in the unfertilized egg but not in the sperm (HENSHAW, 1932; MIWA et al., 1939; FAILLA, 1962a, 1962b). Unfertilized eggs apparently do not exhibit spontaneous dark recovery from ultraviolet radiation damage but can be photoreactivated (WELLS and GIESE, 1950; BLUM et al., 1951; RUSTAD et al., 1966). Sensitivity to x-ray increases early in the mitotic cycle (HENSHAW and COHEN, 1940; FAILLA, 1969; RUSTAD, 1970), and uniform sensitivity to ultraviolet-induced mitotic delay lasts half of a cycle (RUSTAD, 1960a). These sensitivity patterns are consistent with the view that egg damage by ionizing radiation is repaired during the mitotic cycle while ultraviolet damage is not. When the mitotic cycle is interrupted with anoxia, recovery from γ-radiation-induced damage to eggs or sperm continues (FAILLA, 1962a, 1962b). Under these conditions there is recovery from sperm damage resulting from ultraviolet radiation but not from α-particle radiation (RUSTAD and FAILLA, 1971; FAILLA, 1972). The recovery rate from damage by γ-radiation to eggs or sperm seems to depend on cytoplasmic volume (RUSTAD et al., 1971; RUSTAD, 1973), and puromycin, which inhibits protein synthesis, inhibits recovery (RUSTAD and BURCHILL, 1966; RUSTAD et al., 1966; FAILLA and RUSTAD, 1970).

When eggs are cut in half, irradiated, and fertilized with normal sperm, the anucleate eggs exhibit slight but significant sensitivity to γ-radiation and sensitivity to ultraviolet-radiation damage that may be photoreactivated, equivalent to that of whole eggs (RUSTAD et al., 1971, 1972). Sperm and eggs are about equally sensitive to x-radiation (HENSHAW and FRANCIS, 1936), but the sperm are 10 to 15 times more sensitive to ultraviolet radiation (GIESE, 1946). Data on sperm, eggs, and anucleate eggs lead to speculation that the nucleus or some perinuclear structure such as the centriole is most easily damaged but that only cytoplasmic ultraviolet damage is measured in the whole egg in which the nucleus is shielded by 30-50μ of strongly ultraviolet-absorbing cytoplasm.

The nuclear and centriolar targets may be DNA molecules, inasmuch as ultraviolet damage to the sperm may be photoreactivated in the egg (WELLS and GIESE, 1950; BLUM et al., 1951), and photoreactivation usually indicates the enzymatic splitting of ultraviolet-induced pyrimidine dimers in DNA (JAGGER, 1967). Similarly, the cytoplasmic targets for radiation damage leading to mitotic delay may be mitochondrial DNA or free cytoplasmic ribosomal DNA, as ultraviolet-induced cytoplasmic damage may be photoreactivated (RUSTAD et al., 1972).

While the mode of action of some chemical inhibitors of mitosis is obvious, the time of action within the mitotic cycle of the sea-urchin egg is often obscure. The bulk of quantitative information consists of determinations of the concentrations at which various substances delay or block mitosis and a summary of the rhythms of susceptibility to general damage by various agents (HARVEY, 1956; KARNOVSKY and SIMMEL, 1963). In measurements of differential quantitative responses during the mitotic cycle, a biphasic decrease to zero sensitivity to puromycin-induced mitotic delay was found (RUSTAD and BURCHILL, 1966). There are two periods of great susceptibility to colcemid-induced

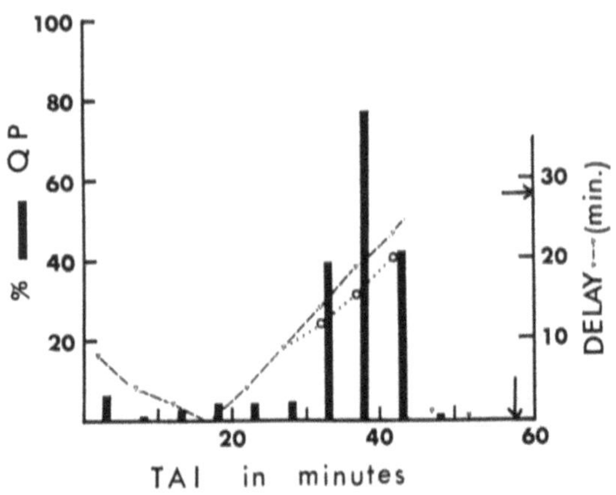

Fig. 12.3. Division delay and early quadripartition induced in *Lytechinus* eggs treated with mercaptoethanol (ME) at various times in cleavage cycle; TAI, time after insemination that eggs were exposed to ME for 20 min. Triangles indicate delay of bipolar division; circles, delay of quadripolar division; solid bars, percentage of quadripartition; arrows, normal first cleavage (58 min after fertilization) and second cleavage (86 min after fertilization). Early quadripartition occurs halfway through second cleavage cycle. (From YUYAMA, 1971)

mitotic delay: one before pronuclear fusion and one before metaphase (ZIMMERMAN and ZIMMERMAN, 1967). Analysis of the effects of β-mercaptoethanol indicates that brief exposures of this -SH reagent may cause uniform or decreasing mitotic delay early in the division cycle and a linearly increasing response in the second half of the cycle (Fig. 12.3)(YUYAMA, 1971).

12.3 Division Abnormalities

Chromosomal abnormalities are known to result from radiation and from the application of compounds that specifically bind to DNA, such as actinomycin D, or interfere with DNA metabolism, such as 5-bromodeoxyuridine (BUdR)(Fig. 12.4)(MAZIA and GONTCHAROFF, 1964; KIEFER et al., 1969, 1971). When sperm are exposed to massive doses of ionizing radiation, the male pronucleus is inactivated and the haploid female pronucleus divides at the normal time (HERTWIG, 1912; RUSTAD, 1969). If pronuclear fusion is mechanically or chemically prevented and only the male pronucleus is associated with the sperm aster, the metaphase spindle will contain only the haploid set of paternal chromosomes. Monopolar mitoses may be induced by a variety of chemical treatments, and karyokinesis may continue after furrow formation has been suppressed, giving rise to multinucleate cells (WILSON, 1924; HARVEY, 1956; MAZIA, 1961).

Multipolar division may occur when multinucleate eggs divide directly into several cells. Such multinucleate eggs may result from chemically enhanced polyspermy, from refertilization after treatment that removes the fertilization membrane and enables the fertilized egg surface to interact again with sperm, or from temporarily inhibiting furrow formation without impairing karyokinesis. Direct multipolar mitoses may result from single nuclei after large doses of ionizing radiation (HENSHAW, 1940c; RUSTAD, 1959a, 1970) or treatment with chemicals such as β-mercaptoethanol (MAZIA et al., 1960; YUYAMA, 1971) and benzimidazole (WENT, 1966).

Multipolar spindles would be expected whenever centriole multiplication proceeded without subsequent cell division or when it was possible for each of the four predivisional centrioles (which normally exist in pairs) to become an autonomous spindle pole. But it seems that mercaptoethanol blocks the synthesis of division centers, possibly centrioles, but allows their separation and may even induce splitting of duplex centrioles (MAZIA et al., 1960; MAZIA, 1961; WENT, 1966; YUYAMA, 1971).

The first three cleavages are normally symmetrical, while the fourth yields 8 mesomeres, 4 macromeres, and 4 micromeres. The first four cleavages usually involve changes in the orientation of the cleavage plane which is determined by the spindle position (HÖRSTADIUS, 1939). When division is delayed by chemicals or by radiation, micromeres may form at the normal time and yield blastomeres of unequal size in the 8-cell, 4-cell, or 2-cell stages (RUSTAD, 1960b; IKEDA, 1965).

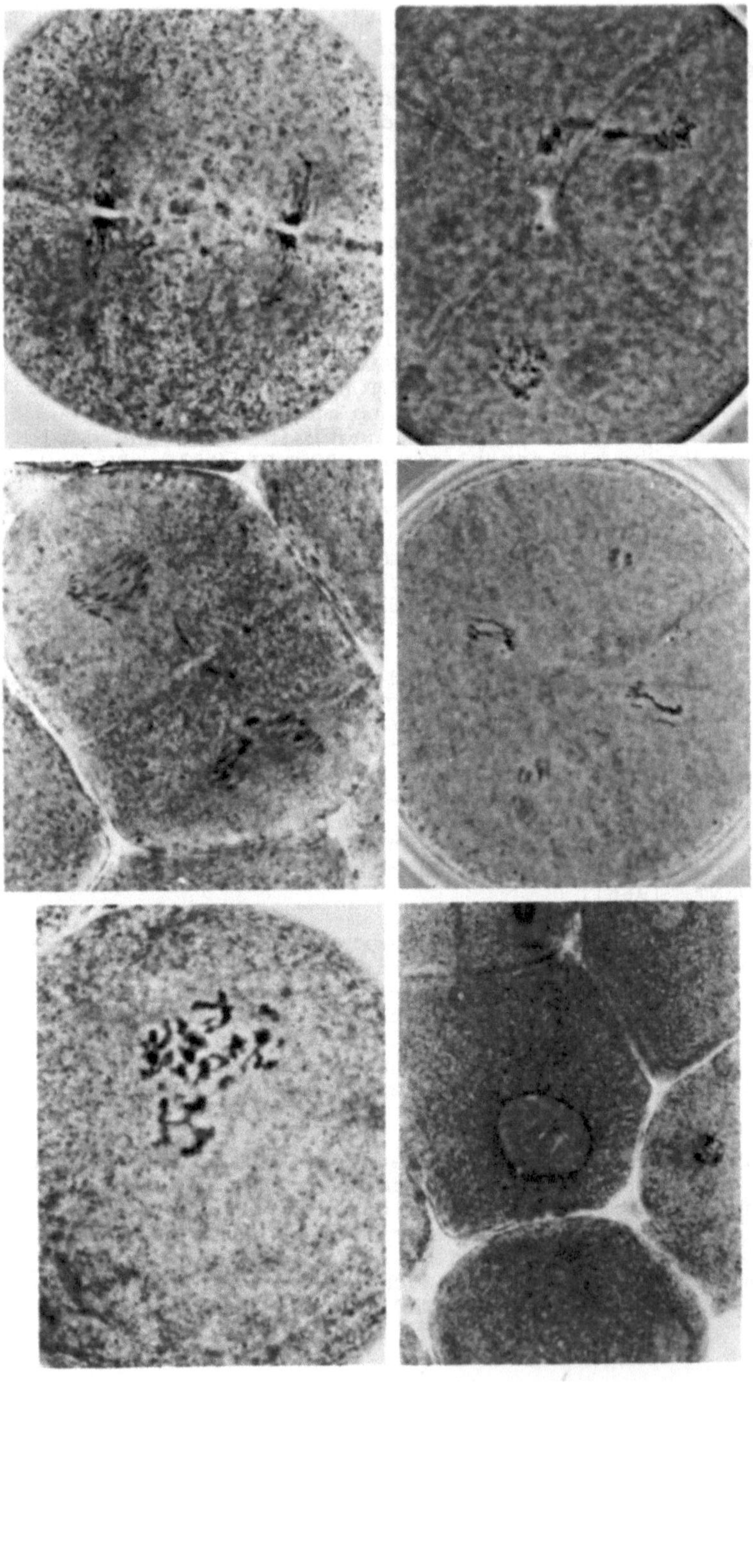

A B C D E F

Fig. 12.4. (A-F) Chromosome abnormalities induced by exposing *Strongylocentrotus* eggs to actinomycin D or (E, F) proflavine. (A) Interphase of 4-cell-stage embryo following bridge formation in both anaphases of previous division. Chromatin in bridges did not decondense. (B) Prophase after bridge formation. Other 2 cells are normal. (C) Anaphase after bridge formation. Chromatin remained at furrow. In lower cell, piece of chromatin was left at equator. (D) Anaphase bridges in 2 cells of 4-cell-stage embryo. (E) Chromosomes exhibiting stickiness and failure to condense. (F) Giant nucleus (center) and normal-sized nuclei (upper center, lower right). (From KIEFER, unpublished)

12.4 Early Post-Fertilization Changes

The first division cycle is exceptionally long. As much as half of it is required to convert the comparatively quiescent, unfertilized egg into a rapidly dividing cell, but the first cycle does not fall neatly into sequential events of fertilization changes and a normal mitotic cycle.

Although parthenogenetic activation and fertilization inhibition by radiation and by chemicals have been studied extensively, attempts seem not to have been made to analyze the activation or inhibition of many of the rapid changes that occur during the first minute after fertilization - such changes as membrane potential, ionic permeability, and coenzyme content (EPEL et al., 1969). Such experiments may be very important to understanding mitotic delay induced by agents with complex actions, for the egg may never reach its normal post-fertilization physiologic state.

12.5 The General Cell Cycle

A diagram of the mitotic cycle is shown (Fig. 12.5). After fertilization, the amino acid incorporation rate increases until shortly before metaphase when it decreases and again increases in the second cycle (Fig. 12.6)(FRY and GROSS, 1970). Nuclear DNA synthesis apparently occurs immediately after pronuclear fusion and immediately after furrow formation; it occupies only about 10% of the first division cycle and 20% of the second (HINEGARDNER et al., 1964). The energy reservoir concept has become relative, because eggs can be blocked at any mitotic stage by carbon monoxide (EPEL, 1963).

12.6 DNA Synthesis

Mitochondrial DNA synthesis may occur continuously in unfertilized and fertilized eggs, because ^{3}H TdR is incorporated into DNase-sensitive material in the cytoplasm (BALTUS et al., 1965; RUSTAD, unpublished). In *Strongylocentrotus* most ^{3}H TdR incorporation occurs within 11 minutes, about 10% of the first mitotic cycle and 20% of the second (HINEGARDNER et al., 1964; ZEITZ et al., 1968, 1970). This incorporation, primarily into nuclear DNA, is initiated after pronuclear fusion in the first cycle and immediately after the first telophase at the

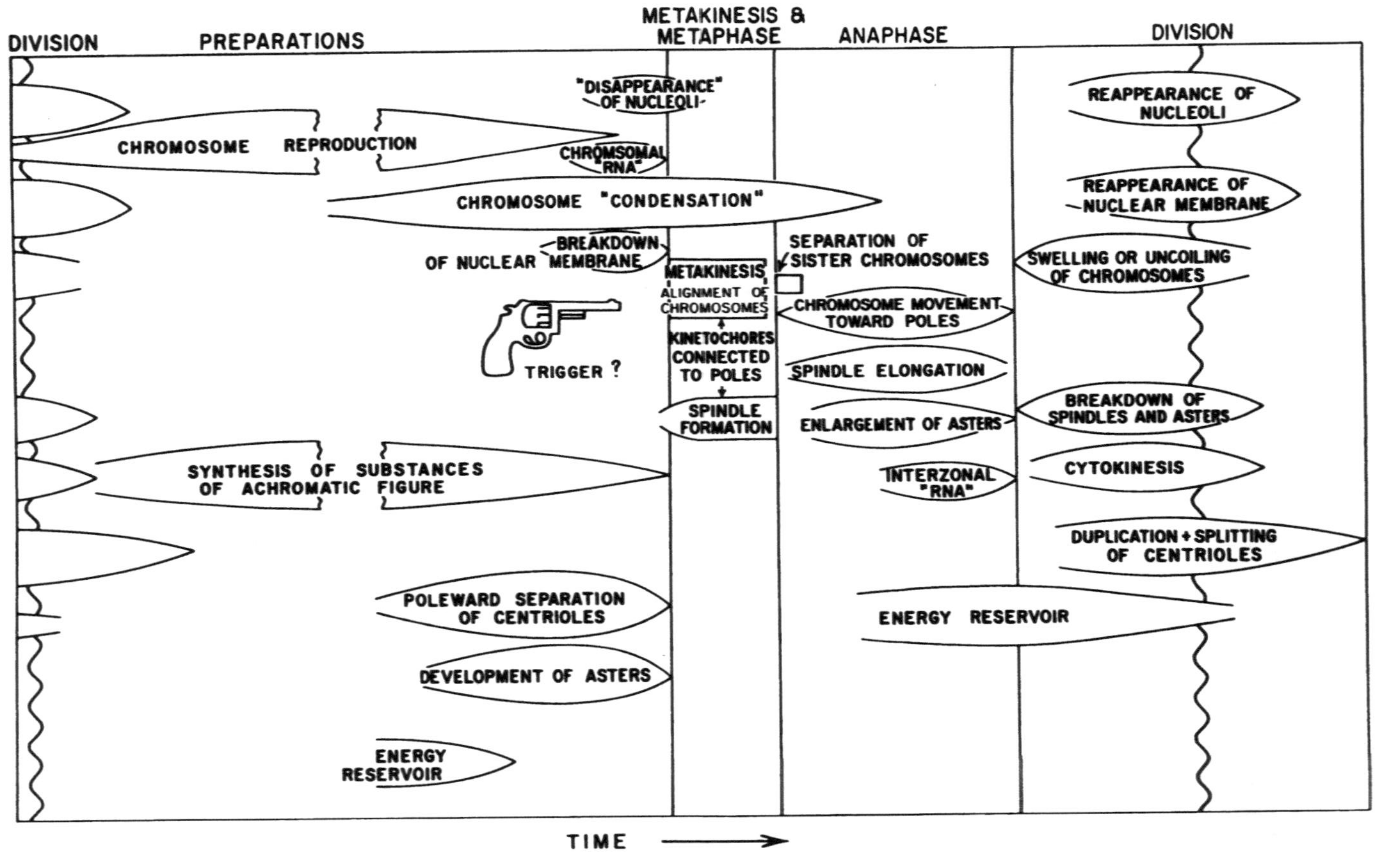

Fig. 12.5. Generalized description of time flow of events in mitotic cycle. (From MAZIA, 1961)

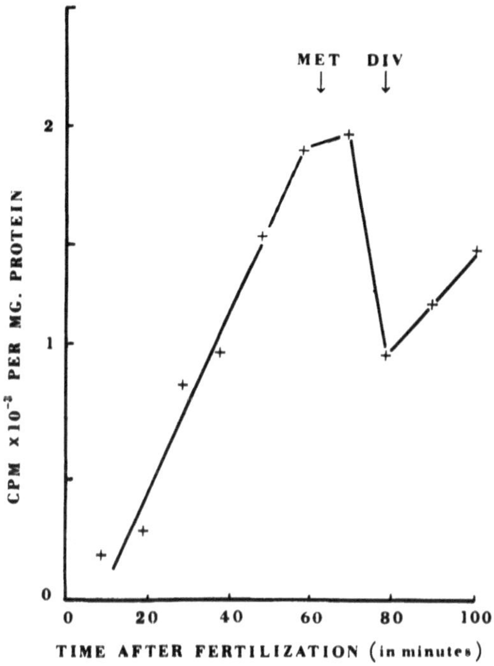

Fig. 12.6. Variations in rate of incorporation of ^{3}H-leucine into protein during division cycle of *Arbacia* eggs. Crosses indicate times at which four-minute exposures to amino acid were begun. Times when half the eggs reached metaphase and division (furrow formation) are indicated. (From PETREK and RUSTAD, unpublished)

beginning of the second cycle. The typical mitotic cycle may thus have no G_1 phase, and the first nuclear division cycle may not begin until the pronuclei fuse. In contrast, DNA synthesis in sand-dollar eggs begins in both pronuclei before fusion (SIMMEL and KARNOFSKY, 1961). There is an unresolved discrepancy between the pattern of incorporation of ^{3}H TdR and ^{3}H BUdR: incorporation of the thymidine analogue begins almost immediately after fertilization (ZEITZ et al., 1970).

Prior x-radiation does not influence the time or rate of DNA synthesis within a mitotic cycle; however, the time of DNA synthesis of the next cycle is postponed until the previous division occurs (RAO and HINEGARDNER, 1965; ZEITZ et al., 1968). In some types of cells, BUdR incorporation reduces the rate of DNA synthesis, but it usually causes no significant mitotic delay in sea-urchin eggs (MAZIA and GONTCHAROFF, 1964; RUSTAD and BURCHILL, unpublished). Division may be delayed by dyes that are known to interact with DNA: acridine orange, ethidium bromide, proflavine, and quinacrine mustard (RUSTAD, 1959c; VAQUIER and BRACHET, 1969; KIEFER et al., 1971). These basic dyes may also interact with other molecules.

12.7 RNA Synthesis

Although some synthesis of nuclear and mitochondrial RNA may occur in the first mitotic cycle, active nuclear RNA synthesis begins in the four micromeres formed at the 16-cell stage (CZIHAK and HÖRSTADIUS, 1970). Exposure of eggs to 8-azaguanine does not delay cleavage, but the incorporation of this analogue into RNA during the 16-cell stage causes gastrulation failure (BAMBERGER et al., 1963; CZIHAK, 1965). Actinomycin D delays cleavage at concentrations that severely inhibit RNA synthesis without noticeably inhibiting total protein synthesis (GROSS, 1967). This delay is in pronuclear fusion in the first mitotic cycle and does not occur in subsequent cycles (WOLSKY and WOLSKY, 1964; CZIHAK, 1968). Actinomycin D causes a number of cleavage abnormalities - including chromosomal bridges (KIEFER et al., 1969, 1971) - because it is bound to the DNA or because it inhibits histone synthesis on newly synthesized messenger RNA (MOAV and NEMER, 1971). Actinomycin D does not seem to influence the length of radiation-induced mitotic delay (RUSTAD and BURCHILL, 1966).

12.8 Protein Synthesis

When actinomycin D blocks RNA synthesis, the rate of protein synthesis is not reduced, which suggests that protein synthesis depends on maternal messenger RNA (GROSS, 1967). Puromycin and cycloheximide reduce the rate of protein synthesis and delay division (HULTIN, 1961; BLACK et al., 1967; WILT et al., 1967). Two different mitotic events may be influenced by a decreased rate of protein synthesis because the sensitivity to the inducement of mitotic delay by puromycin de-

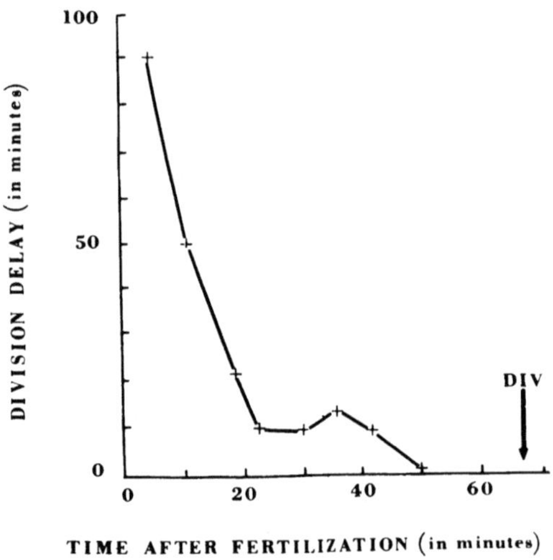

Fig. 12.7. Division delay induced in eggs exposed briefly at various times in mitotic cycle to puromycin which inhibits protein synthesis. There are two periods of sensitivity to puromycin and a refractory period shortly before furrow formation (DIV). (From CVERNA and RUSTAD, unpublished)

creases in two distinct steps (Fig. 12.7). Treatment with puromycin extends the period of sensitivity to irradiation, and vice versa (RUSTAD and BURCHILL, 1966). These observations do not necessarily indicate that irradiation inhibits protein synthesis; the synthesis of recovery proteins may be necessary after irradiation (FAILLA and RUSTAD, 1970; RUSTAD, 1971).

Normal variations in the rate of protein synthesis (Fig. 12.6) are not influenced when colchicine blocks cleavage at metaphase (SOFER et al., 1966; FRY and GROSS, 1970). The normal rhythmic change in the content - not necessarily in the synthesis - of cortical -SH proteins continues when mitosis is delayed by ultraviolet radiation (IKEDA, 1965). Thus, endogenous rhythms controlling cell activities are not associated exclusively with the mitotic cycle.

12.9 Pronuclear Fusion

Neither ionizing radiation nor ultraviolet radiation seems to delay pronuclear fusion (HENSHAW, 1940b; YAMASHITA et al., 1940; MARSHAK, 1949). Pronuclear fusion is prevented by agents such as mercaptoethanol and colcemid that dissociate the microtubules of the sperm aster (BUCHER and MAZIA, 1960; ZIMMERMAN and ZIMMERMAN, 1967). When these agents block syngamy, DNA synthesis takes place at about the normal time in both male and female pronuclei (Fig. 12.8). Mercaptoethanol has been used to block pronuclear fusion in eggs x-irradiated before fertilization to demonstrate a cytoplasmically mediated effect on the nuclear division cycle of the male pronucleus (RAO, 1963).

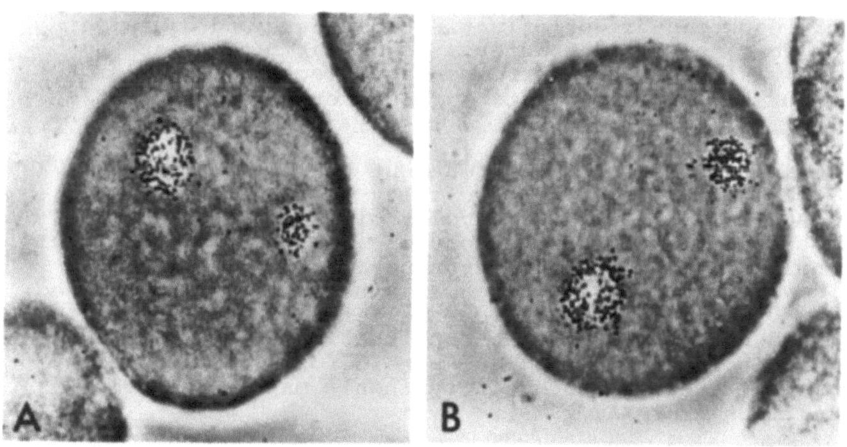

Fig. 12.8.A and B. *Arbacia* eggs showing ^{3}H-thymidine incorporation into both pronuclei following pronuclear fusion inhibition by colcemid. (From ZIMMERMAN and ZIMMERMAN, 1967)

12.10 Duplication and Separation of Centrioles

Evidence suggests that radiation and chemicals such as mercaptoethanol may have rather specific effects on centriolar behavior. It has been hypothesized that radiation-induced mitotic delay arises from damage to the centrioles or to a nuclear trigger for duplication or separation of centrioles (RUSTAD, 1961, 1964). Evidence favoring this hypothesis includes the following: streak stage may be prolonged (HENSHAW, 1940b; YAMASHITA et al., 1940); supernumerary astral development may be delayed; the number of asters may be odd, possibly indicating unequal radiation influence on each of the duplex centrioles (HENSHAW, 1940c; RUSTAD, 1959a); prefertilization recovery from damage leading to mitotic delay and to multipolar mitoses may coincide (RUSTAD, 1959b); radiation-induced delay may occur in that part of the mitotic cycle susceptible to the inducement of monastral blocks by hypotonic seawater (RUSTAD, 1961); and periods of maximum susceptibility to the inducement of mitotic delay and multipolar spindles may coincide (RUSTAD, 1970).

It is possible that mercaptoethanol - and many other chemical agents - may selectively inhibit the duplication but not the separation of division centers - the physiologic equivalent of centrioles. Eggs were placed in a mercaptoethanol solution at various times in the mitotic cycle and the cells washed at the time of the normal second mitosis. Eggs treated early in the first division cycle divided into two cells, while those treated after a critical point divided into four. At the time of the normal third division, mercaptoethanol-treated cells underwent monopolar mitosis. It was suggested that, early in the mitotic cycle, there are two closely associated division centers that are later duplicated. Each aster normally contains two division centers, but mercaptoethanol may permit the separation of those unduplicated, each capable of orienting an aster, and division occurs with two or four monocentral asters. Monopolar mitosis at the time of the third division of the controls may reflect duplication without separation of the division centers (MAZIA et al., 1960; MAZIA, 1961).

As the result of brief mercaptoethanol treatment to obtain a quantitative analysis of the inducement of mitotic delay and the appearance of multipolar mitoses (Fig. 12.3), quadripolar mitoses occurred well in advance of the second mitosis of the controls, possibly indicating that mercaptoethanol induces division center separation. Transition from two to four division centers occurred in anucleate eggs fertilized with normal sperm, confirming the classical belief that centrioles derive from sperm and not from eggs. The fact that puromycin delayed the transition from bipolar to multipolar spindle formation induced by mercaptoethanol indicates that protein synthesis may be essential to centriole duplication (YUYAMA, 1971).

Benzimidazole is an agent that seems to allow duplex division centers to separate. The use of benzimidazole or mercaptoethanol allows counting the number of centrioles present after experimental treatment (WENT, 1966). While γ-radiation delays transition from bipolar to quadripolar division in mercaptoethanol-treated eggs, earlier in the cycle there is a very brief period of susceptibility to multipolar spindle inducement (RUSTAD and YUYAMA, 1966).

12.11 Chromosomal Condensation and Nuclear Membrane Breakdown

Mitosis may be inhibited in prophase by radiation and a number of chemical inhibitors (HARVEY, 1956). X-radiation prolongs the early prophase streak stage when centrioles separate and a later biastral prophase stage preceding chromosomal condensation and nuclear membrane breakdown (HENSHAW, 1940b; YAMASHITA et al., 1940). Low doses of ultraviolet radiation also prolong the streak stage (RUSTAD, unpublished). One of the primary actions of radiation may be to delay chromosomal condensation in many cell types (CARLSON, 1969). X-radiation-induced delay of chromosomal condensation has been found to equal the total mitotic delay of the egg (RAO, 1963).

The relationship among centriolar duplication and separation, chromosomal condensation, and nuclear membrane breakdown is unclear. Various agents may delay nuclear membrane breakdown by delaying some other event that must be completed first. This interdependence makes it difficult to pinpoint mitotic stage delay.

12.12 Metaphase, Anaphase, and Telophase

X-radiation and ultraviolet radiation each significantly prolong chromosomal separation and delay prophase many times longer (HENSHAW, 1940b; YAMASHITA et al., 1940; BLUM et al., 1954). Irradiation or treatment with agents that interact with DNA causes mitotic abnormalities such as anaphase bridges and chromosomal fragmentation (Fig. 12.4).

Substances such as colchicine, colcemid, podophyllotoxin, and mercaptoethanol block cell division at metaphase by disorienting the mitotic apparatus (BEAMS and EVANS, 1940; CORNMAN and CORNMAN, 1951; MAZIA and ZIMMERMAN, 1958; ZIMMERMAN and ZIMMERMAN, 1967). The effects of colcemid and mercaptoethanol are reversible. In other types of cells, C-mitotic agents such as colchicine, podophyllotoxin, and vinblastine have been shown to bind to specific sites on the proteins of the microtubules of the mitotic apparatus (WILSON and FRIEDKIN, 1967; WILSON, 1970). The antimitotic effect may arise from loss of microtubular orientation and from loss of microtubular proteins from the structure of the mitotic apparatus.

Mitosis may be halted in anaphase if the energy source becomes depleted at the proper time or if a spindle-disorienting agent such as mercaptoethanol or colchicine is applied at the right moment (EPEL, 1963). Deuterium oxide also blocks anaphase (GROSS AND SPINDEL, 1960; MARSLAND and ZIMMERMAN, 1965), but there is no evidence that anaphase may be slowed down or halted by specific inhibitors except those that bind to microtubules. As anaphase involves two distinct movements: chromosomal movement toward the poles, and polar separation by spindle elongation, it seems likely that there are agents that selectively inhibit or enhance one of the movements.

Furrow formation may be inhibited if the mitotic apparatus is disoriented before anaphase, after which the mitotic apparatus may be removed microsurgically and furrowing will still occur (HIRAMOTO, 1956). Deuterium oxide enhances furrowing strength as measured by the hydrostatic pressure needed to reverse furrow formation, and deuterium oxide and colchicine may act synergistically or antagonistically in inhibiting cell division (MARSLAND and ASTERITA, 1966; MARSLAND and HECHT, 1968).

In dividing sea-urchin eggs exposed to cytochalasin B, furrows recede and the high concentration of 50-Å microfilaments disappears from the furrow region (SCHROEDER, 1969). As the furrow recedes and the surface area is reduced, cortical pigment granules accumulate in the region where the furrow had been forming (BELANGER and RUSTAD, 1970).

Regional chemical differences between furrow and polar regions are indicated by their susceptibility to lytic agents. Alkalis, for instance, cause local swelling or blistering at the furrows, while wasp venom and detergents cause local cytolysis beginning at the spindle poles (KOJIMA, 1954, 1957).

Acknowledgement

These studies were supported by United States Atomic Energy Commission Contract W-31-109-ENG-78 (Report No. COO-78-268). The author is grateful to the library of the Marine Biological Laboratory, Woods Hole, Massachusetts, Miss A.M. BELANGER, Dr. D.G. CASTON, Dr. R.P. DAVIS, Dr. P.M. FAILLA, Dr. O.F. NYGAARD, and Dr. N.L. OLEINICK.

References

ALTMAN, K.I., GERBER, G.B., OKADA, S., 1970. Radiation Biochemistry, Vol. I. New York: Academic Press.

BALTUS, E., QUERTIER, J., FICQ, A., BRACHET, J., 1965. Biochemical studies of nucleate and anucleate fragments isolated from sea-urchin eggs. A comparison between fertilization and parthenogenetic activation. Biochim. Biophys. Acta, 95, 408-417.

BAMBERGER, J.W., MARTIN, W.E., STEARNS, L.W., JOLLEY, W.B., 1963. Effect of 8-azaguanine on cleavage and nucleic acid metabolism in sea urchin, Strongylocentrotus purpuratus, embryos. Exp. Cell Res., 31, 266-274.

BEAMS, H.W., EVANS, T.C., 1940. Some effects of colchicine upon the first cleavage in Arbacia punctulata. Biol. Bull., 79, 188-198.

BELANGER, A.M., RUSTAD, R.C., 1970. Movement of echinochrome granules during the early development of Arbacia eggs. Biol. Bull., 139, 415 (Abstract).

BLACK, R.E., BAPTIST, E., PILAND, J., 1967. Puromycin and cycloheximide inhibition of thymidine incorporation into DNA of cleaving sea urchin eggs. Exp. Cell Res., 48, 431-439.

BLUM, H.F., ROBINSON, J.C., LOOS, G.M., 1951. The loci of action of ultraviolet and x-radiation and of photorecovery in the egg and sperm of the sea urchin Arbacia punctulata. J. Gen. Physiol., 35, 323-324.

BLUM, H.F., KAUZMANN, E.F., CHAPMAN, G.B., 1954. Ultraviolet light and the mitotic cycle in the sea-urchin egg. J. Gen. Physiol., 37, 325-334.

BUCHER, N.L.R., MAZIA, D., 1960. Deoxyribonucleic acid synthesis in dividing eggs of the sea urchin, Strongylocentrotus purpuratus. J. Biophys. Biochem. Cytol., 7, 651-655.

CARLSON, J.G., 1969. X-ray-induced prophase delay and reversion of selected cells in certain avian and mammalian tissues in culture. Radiat. Res., 37, 15-30.

CORNMAN, I., CORNMAN, M.E., 1957. The action of podophyllin and its fractions on marine eggs. Ann. N.Y. Acad. Sci., 51, 1443-1487.
CZIHAK, G., 1965. Entwicklungsphysiologische Untersuchungen an Echiniden. Ribonucleinsäure-Synthese in den Mikromeren und Entodermdifferenzierung. Ein Beitrag zum Problem der Induktion. Wilhelm Roux' Arch. Entwicklungsmech. Organ., 156, 504-524.
CZIHAK, G., HÖRSTADIUS, S., 1970. Transplantation of RNA-labeled micromeres into animal halves of sea urchin embryos. A contribution to the problem of embryonic induction. Develop. Biol., 22, 15-30.
ELKIND, M.K., WHITMORE, G.F., 1967. The Radiobiology of Cultured Mammalian Cells. New York: Gordon and Breach.
EPEL, D., 1963. The effects of carbon monoxide inhibition on ATP level and rate of mitosis in the sea urchin egg. J. Cell Biol., 17, 315-319.
EPEL, D., PRESSMAN, B.C., ELSAESSER, S., WEAVER, A.M., 1969. The program of structural and metabolic changes following fertilization of sea-urchin eggs. In: The Cell Cycle (G. Padilla, G.L. Whitson, I.L. Cameron, eds.), 279-298. New York: Academic Press.
FAILLA, P.M., 1962a. In vivo and in vitro recovery of irradiated gametes of Arbacia punctulata. Radiat. Res., 17, 767-773.
FAILLA, P.M., 1962b. Recovery from radiation-induced delay of cleavage in gametes of Arbacia punctulata. Science, 138, 1341-1342.
FAILLA, P.M., 1969. Recovery and modification of radiation-induced division delay in developing sea-urchin eggs. Radiology, 93, 643-648.
FAILLA, P.M., 1972. Manuscript in preparation.
FAILLA, P.M. RUSTAD, R.C., 1970. Protein synthesis and recovery from γ-ray-induced mitotic delay in fertilized sea-urchin eggs. Int. J. Radiat. Biol., 17, 385-388.
FRY, B.J., GROSS, P.R., 1970. Patterns and rates of protein synthesis in sea urchin embryos. I. Uptake and incorporation of amino acids during the first cleavage cycle. Develop. Biol., 21, 105-124.
GIESE, A.C., 1946. Comparative sensitivity of sperm and eggs to ultraviolet radiation. Biol. Bull., 91, 81-87.
GIUDICE, G., 1973. Developmental biology of the sea urchin embryo. New York: Academic Press.
GROSS, P.R., 1967. Current Topics in Developmental Biology, Vol. II, 1-43. New York: Academic Press.
GROSS, P.R., SPINDEL, W., 1960. Heavy water inhibition of cell division: An approach to mechanism. Ann. N.Y. Acad. Sci., 90, 500-522.
HARVEY, E.B., 1956. The American Arbacia and Other Sea Urchins. Princeton: Princeton University Press.
HENSHAW, P.S., 1932. Studies of the effect of Roentgen rays on the time of the first cleavage in some marine invertebrate eggs. I. Recovery from Roentgen-ray effects in Arbacia eggs. Am. J. Roentgenol. Rad. Ther., 27, 890-898.
HENSHAW, P.S., 1940a. Further studies on the action of Roentgen rays on the gametes of Arbacia punctulata. I. Delay in cell division caused by exposure to sperm to Roentgen rays. Am. J. Roentgenol. Rad. Ther., 43, 899-906.
HENSHAW, P.S., 1940b. Further studies on the action of Roentgen rays on the gametes of Arbacia punctulata. II. Modification of the mitotic time schedule in the eggs by exposure of the gametes to Roentgen rays. Am. J. Roentgenol. Rad. Ther., 43, 907-912.
HENSHAW, P.S., 1940c. Further studies on the action of Roentgen rays on the gametes of Arbacia punctulata. VI. Production of multipolar cleavage in the eggs by exposure of the gametes to Roentgen rays. Am. J. Roentgenol. Rad. Ther., 43, 923-933.
HENSHAW, P.S., COHEN, I., 1940. Further studies on the action of Roentgen rays on the gametes of Arbacia punctulata. IV. Changes

in radiosensitivity during the first cleavage cycle. Am. J. Roentgenol. Rad. Ther., 43, 917-920.
HENSHAW, P.S., FRANCIS, D.S., 1936. The effects of x-rays on cleavage in Arbacia eggs: evidence of nuclear control of division rate. Biol. Bull., 70, 28-35.
HENSHAW, P.S., HENSHAW, C.T., FRANCIS, D.S., 1933. The effect of Roentgen rays on the time of the first cleavage in marine invertebrate eggs. Radiology, 28, 533-541.
HERTWIG, R., 1912. Das Schicksal des mit Radium bestrahlten Spermachromatins im Seeigelei. Arch. f. mikr. Anat., 79, 201-241.
HINEGARDNER, R.T., RAO, B., FELDMAN, D.E., 1964. The DNA synthetic period during early development of the sea-urchin egg. Exp. Cell Res., 36, 53-61.
HIRAMOTO, Y., 1956. Cell division without mitotic apparatus in sea-urchin eggs. Exp. Cell Res., 11, 630-636.
HÖRSTADIUS, S., 1939. The mechanics in sea urchin development studied by operative methods. Biol. Rev., 14, 132-179.
HULTIN, T., 1961. The effect of puromycin on protein metabolism and cell division in fertilized sea-urchin eggs. Experientia, 17, 410-411.
IKEDA, M., 1965. Behavior of sulfhydryl groups of sea-urchin eggs under the blockage of cell division by UV and heat shock. Exp. Cell Res., 40, 282-291.
JAGGER, J., 1967. Introduction to Research in Ultraviolet Photobiology. New Jersey: Prentice-Hall, Inc.
KARNOVSKY, D.A., SIMMEL, E.B., 1963. Effects of growth-inhibiting chemicals on the sand dollar embryo, Echinorachnius parma. Prog. Exp. Tumor Res., 3, 254-295.
KIEFER, B.I., ENTALIS, C.F., INFANTE, A.A., 1969. Mitotic abnormalities in sea urchin embryos exposed to dactinomycin. Proc. Nat. Acad. Sci. U.S., 64, 857-862.
KIEFER, B.I., INFANTE, A.A., BINKEN, N., 1971. Manuscript in preparation.
KIHLMAN, B.A., 1966. Actions of Chemicals on Dividing Cells. Englewood Cliffs, N.J.: Prentice Hall, Inc.
KOJIMA, M.K., 1954. On the nature of the egg surface during cleavage of the sea-urchin egg. Embryologia, 2, 33-41.
KOJIMA, M.K., 1957. On the regional difference in the nature of the cortex of the sea-urchin egg during cleavage. Embryologia, 3, 279-293.
KOJIMA, M.K., 1960. Acceleration of the cleavage of sea-urchin eggs by vital staining with neutral red. Exp. Cell Res., 20, 565-573.
KOJIMA, M.K., 1967. Acceleration of the cleavage in sea-urchin eggs by insufficient treatments with activating reagents before fertilization. Embryologia, 10, 75-82.
KOJIMA, M.K., 1969a. Acceleration of cleavage in sea-urchin eggs by repeated insufficient stimulations with various activating reagents and spermatozoa. Embryologia, 10, 323-333.
KOJIMA, M.K., 1969b. Induction of nuclear changes and cleavage by repeated insufficient stimulations with activating reagents in the sea-urchin egg. Embryologia, 10, 334-342.
MARSHAK, A., 1949. Recovery from ultraviolet light-induced delay in cleavage of Arbacia eggs by irradiation with visible light. Biol. Bull., 97, 315-322.
MARSLAND, D., ASTERITA, H., 1966. Counteraction of the anti-mitotic effects of D_2O in the dividing eggs of Arbacia punctulata. A temperature-pressure analysis. Exp. Cell Res., 42, 316-327.
MARSLAND, D., HECHT, R., 1968. Cell division: Combined anti-mitotic effects of colchicine and heavy water on first cleavage in the eggs of Arbacia punctulata. Exp. Cell Res., 51, 602-608.
MARSLAND, D., ZIMMERMAN, A.M., 1965. Structural stabilization of the mitotic apparatus by heavy water, in the cleaving eggs of Arbacia punctulata. Exp. Cell Res., 38, 306-313.

MAZIA, D., 1961. In: The Cell (J. Brachet, A.E. Mirsky, eds.), Vol. 3, 77-412. New York: Academic Press.
MAZIA, D., GONTCHAROFF, M., 1969. The mitotic behavior of chromosomes in echinoderm eggs following incorporation of bromodeoxyuridine. Exp. Cell Res., 35, 14-25.
MAZIA, D., HARRIS, P.G., BIBRING, T., 1960. The multiplicity of the mitotic centers and the time-course of their duplication and separation. J. Biophys. Biochem. Cytol., 7, 1-20.
MAZIA, D., ZIMMERMAN, A.M., 1958. SH compound in mitosis. II. The effect of β-mercaptoethanol on the structure of the mitotic apparatus in sea-urchin eggs. Exp. Cell Res., 15, 138-153.
MITCHISON, J.M., 1971. The Biology of the Cell Cycle. Cambridge University Press.
MIWA, M., YAMASHITA, H., MORI, K., 1939. The action of ionizing rays on sea urchins. I. The effects of Roentgen, gamma, and beta rays upon the unfertilized eggs and sperms. Gann, 33, 1-12.
MIWA, M., YAMASHITA, H., MORI, K., 1941. The sperm of sea urchin as a biological test object in Roentgen dosimetry. Gann, 35, 127-133.
MOAV, B., NEMER, M., 1971. Histone synthesis. Assignment to a special class of polyribosomes in sea urchin embryos. Biochem., 10, 881-888.
NAKASHIMA, S.K., SUGIYAMA, M., 1969. Incorporation of amino acids into sea-urchin eggs stimulated insufficiently with an activating reagent. Develop. Growth & Differentiation, 11, 115-122.
RAO, B., 1963. Analysis of x-ray-induced mitotic delay in sea-urchin eggs, Ph.D. Thesis. Berkeley, Calif.: University of California.
RAO, B., HINEGARDNER, R.T., 1965. Analysis of DNA synthesis and x-ray-induced mitotic delay in sea-urchin eggs. Rad. Res., 26, 534-537.
RUSTAD, R.C., 1959a. Induction of multipolar spindles by single x-irradiated sperm. Experientia, 15, 323-324.
RUSTAD, R.C., 1959b. Further observations relating radiation-induced mitotic delay to centriole damage. Biol. Bull., 117, 437 (Abstract).
RUSTAD, R.C., 1959c. The inhbition of mitosis in the sea-urchin egg by acridine orange. Biol. Bull., 117, 437-438 (Abstract).
RUSTAD, R.C., 1960a. Changes in the sensitivity to ultraviolet-induced mitotic delay during the cell division cycle of the sea-urchin egg. Exp. Cell Res., 21, 596-602.
RUSTAD, R.C., 1960b. Dissociation of the mitotic time schedule from the micromere "clock" with x-rays. Acta Embryo. Morph. Exp., 3, 155-158.
RUSTAD, R.C., 1961. The centriole hypothesis of radiation-induced mitotic delay. Path. et Biol., 9, 493-494.
RUSTAD, R.C., 1964. UV-induced mitotic delay in the sea-urchin egg. Photochem. Photobiol., 3, 529-538.
RUSTAD, R.C., 1969. The independence of the mitotic rate of sea urchin eggs from ploidy and cytoplasmic volume. Biophys. J., A-186 (abstract).
RUSTAD, R.C., 1970. Variations in the sensitivity to x-ray-induced mitotic delay during the cell division cycle of the sea-urchin egg. Rad. Res., 42, 498-512.
RUSTAD, R.C., 1971. In: Developmental Aspects of the Cell Cycle (I. Cameron, G. Padilla, A. Zimmerman, eds.), 127-159. New York: Academic Press.
RUSTAD, R.C., 1972. In: Concepts in Radiation Cell Biology (G. Whitson, ed.), 153-181. New York: Academic Press.
RUSTAD, R.C., 1973. Nuclear-cytoplasmic relations in the mitosis of sea-urchin eggs. IV. Modification of γ-radiation-induced damage in whole eggs and in haploid and diploid half-eggs. Rad. Res. 54, 328-332.
RUSTAD, R.C., BURCHILL, B.R., 1966. Radiation-induced mitotic delay in sea-urchin eggs treated with puromycin and actinomycin D. Rad. Res., 29, 203-210.

RUSTAD, R.C., FAILLA, P.M., 1971. Post-fertilization dark recovery of UV-irradiated Arbacia sperm. Biol. Bull., 141, 400-401 (Abstract).
RUSTAD, R.C., FEINGOLD, R., 1971. Unpublished experiments.
RUSTAD, R.C., GREENBERG, M.A., 1972. Prefertilization stimulation of sea-urchin eggs can "cure" radiation-induced mitotic delay. Exp. Cell Res. 75, 285-288.
RUSTAD, R.C., MAZIA, D., 1971. The relationship of UV dose to the delay and loss of synchrony of mitosis in the sea-urchin egg. (Manuscript in preparation).
RUSTAD, R.C., McGURN, E., YUYAMA, S., RUSTAD, L.C., 1966. Recovery from γ-ray and UV radiation damage in unfertilized sea-urchin eggs. Rad. Res., 27, 543 (Abstract).
RUSTAD, R.C., YUYAMA, S., 1966. Unpublished experiments.
RUSTAD, R.C., YUYAMA, S., RUSTAD, L.C., 1971. Nuclear-cytoplasmic relations in the mitosis of sea-urchin eggs. III. γ-ray-induced damage to whole eggs and nucleate and anucleate half-eggs. J. Cell Biol., 49, 906-912.
RUSTAD, R.C., YUYAMA, S., RUSTAD, L.C., 1972. Manuscript in preparation.
SCHROEDER, T.E., 1969. The role of "contractile ring" filaments in dividing Arbacia egg. Biol. Bull., 137, 413 (Abstract).
SIMMEL, E.B., KARNOFSKY, D., 1961. Observations on the uptake of tritiated thymidine in the pronuclei of fertilized sand dollar embryos. J. Biophys. Biochem. Cytol., 10, 59-65.
SOFER, W.H., GEORGE, J.H., IVENSON, R.M., 1966. Rate of protein synthesis: Regulation during first division cycle. Science, 153, 1644.
SUGIYAMA, M., ISHIKAWA, M., KOJIMA, M.K., 1969. Enhancement of respiration in the sea-urchin egg by a subthreshold stimulation with an activating reagent. Embryologia, 10, 318-322.
UTO, N., SUGIYAMA, M., 1969. Incorporation of thymidine-^{3}H into the activated sea-urchin egg with special reference to subthreshold stimulation with an activating reagent. Develop. Growth & Differentiation, 11, 123-129.
VACQUIER, V.D., BRACHET, J., 1969. Chromosomal abnormalities resulting from ethidium bromide treatment. Nature, 222, 193-195.
WELLS, P.H., GIESE, A.C., 1950. Photoreactivation of ultraviolet light injury in gametes of the sea urchin Strongylocentrotus purpuratus. Biol. Bull., 99, 163-172.
WENT, H.A., 1966. An indirect method to assay for mitotic centers in sand dollar (Denraster excentricus) eggs. J. Cell Biol., 30, 555-562.
WILSON, E.B., 1924. The Cell in Development and Heredity. New York: The Macmillan Co.
WILSON, L., 1970. Properties of colchicine binding protein from chick embryo brain. Interactions with vinca alkaloids and podophyllotoxin. Biochem., 9, 4999-5007.
WILSON, L., FRIEDKIN, M., 1967. The Biochemical Events of Mitosis. II. The in vivo and in vitro binding of colchicine in grasshopper embryos and its possible relation to inhibition of mitosis. Biochem., 6, 3126-3135.
WILT, H., SAKAI, H., MAZIA, D., 1967. Old and new protein in the formation of the mitotic apparatus in cleaving sea-urchin eggs. J. Mol. Biol., 27, 1-7.
WOLSKY, A., WOLSKY, M., 1964. An analysis of the actinomycin induced cleavage delay in sea-urchin eggs. Z. Zellforsch., 63, 960-964.
YAMASHITA, H., MORI, K., MIWA, M., 1940. The action of ionizing rays on sea urchin. V. The mitotic observations on the effects of Roentgen rays upon the unfertilized eggs and sperms. Gann, 34, 239-245.

YUYAMA, S., 1971. Delay and quadripartition in sea-urchin eggs induced by short exposure to 2-mercaptoethanol. Biol. Bull., 140, 339-351.
ZEITZ, L., FERGUSON, R., GARFINKEL, E., 1968. Radiation-induced effects on DNA synthesis in developing sea-urchin eggs. Rad. Res., 34, 200-208.
ZEITZ, L., GARFINKEL, E., FERGUSON, R., 1970. Determination of the first S-period in sea-urchin eggs by the uptake of labeled precursors. Difference found with different precursors. Biophys. Soc. Abstract, 10, 141a.
ZIMMERMAN, A.M., ZIMMERMAN, S., 1967. Actions of colcimid in sea-urchin eggs. J. Cell Biol., 34, 483-488.

13. Isolation and Transplantation Experiments

S. Hörstadius

The story of experimental embryology begins with attempts to study the development of single early cleavage cells. Blastomeres of ascidian eggs were killed by CHABRY (1887) and ROUX (1888) killed blastomeres of amphibian eggs by pricking them with a needle and the behaviour of their living partners was followed. ROUX has been called the father of experimental embryology (Entwicklungsmechanik). A few years later, DRIESCH (1891) and FIEDLER (1891) isolated blastomeres of 2-cell stages of sea urchin eggs by shaking. Later, it was discovered by HERBST that isolation is facilitated by lack of calcium ions in seawater which loosens the contact between cleavage cells (1900). A flow of papers in the decades at the turn of the century by Theodor BOVERI, Hans DRIESCH, Curt HERBST, Thomas Hunt MORGAN, and others, gave the sea urchin egg a position as one of the foremost objects for developmental physiology, a position held ever since.

Before starting experimental work it is important to acquire an exact knowledge of the normal development. It is necessary to keep in mind, the position of the axes, and the fate of the different layers in the egg (i.e. the prospective significance of the layers, and the position of the presumptive organs). A short recapitulation of important points is given here.

The egg axis is often called the animal-vegetal axis, referring to the work of VON BAER (1828) on the germ layers of the chick and the differentiation of the material in different parts of the egg. Animal cells give rise to ectoderm which differentiates not only to skin but also to the neural system which is a purely animal characteristic. Vegetal cells form the greater part of the digestive tract on which the further growth of the larva depends.

In the sea urchin ovary, the oocyte is attached to the wall by its vegetal pole (JENKINSON, 1911b; LINDAHL, 1932a). The polar bodies are formed at the animal pole. The first 2 furrows are meridional and the third cleavage furrow divides the egg equatorially into animal and vegetal halves (Fig. 13.1C). The fourth division is unequal, producing 8 animal cells of equal size (mesomeres) and 4 large and 4 small vegetal cells (macromeres, micromeres), thereby identifying the egg axis and enabling the operator to work with well defined meridional, and animal and vegetal fragments (Fig. 13.1D). In the 32-cell stage, the mesomeres have divided into 2 8-celled rings, called animal one and animal two (AN_1, AN_2) (Fig. 13.1E), and in the 64-cell stage, the macromeres divide similarly into vegetal one and vegetal two layers (VEG_1, VEG_2) (Fig. 13.1F) (HÖRSTADIUS, 1931, 1935). The micromeres divide off 4 smaller micromeres in the 32-cell stage. This cleavage of the micromeres is somewhat delayed in relation to the other blastomeres (THEEL, 1892; HAGSTRÖM and LÖNNING, 1965).

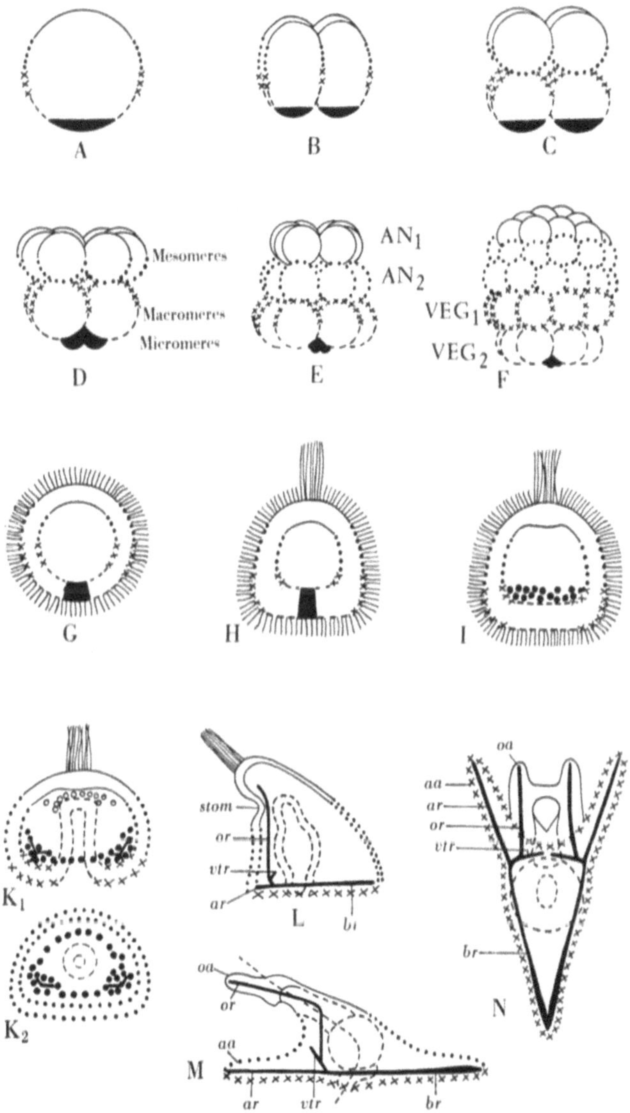

Fig. 13.1. Diagram of the normal development of *Paracentrotus*. Indication of layers: AN_1, continuous lines; AN_2, dotted; VEG_1, crosses; VEG_2, broken lines; micromeres black. A, uncleaved egg; B, 4-cell stage; C, 8-cell stage; D, 16-cell stage; E, 32-cell stage; F, 64-cell stage; G, young blastula; H, later blastula, with animal tuft, before the formation of the primary mesenchyme; I, blastula after the formation of the primary mesenchyme; K_1, gastrula: secondary mesenchyme and the 2 triradiate spicules formed; K_2, transverse optical section of the same gastrula, bilateral symmetry established; L, the prism stage, stomodaeum invaginating; M, pluteus larva from the left side, the broken line indicates the position of the egg axis; N, pluteus from the anal side. aa, anal arm; ar, anal rod; br, body rod, in the apex; oa, oral arm; or, oral rod; stom, stomodaeum; vtr, ventral transverse rod. (HÖRSTADIUS, 1935, 1939)

The larval constitution is expressed by formulae referring to the normal 16-cell stage: 8 + 4 + 4 (8 meso-, 4 macro- and 4 micromeres).

Thus a larva from which the AN_1-ring has been removed is characterized as AN_2 + 4 + 4, a larva 8 + VEG_1 + 0 is devoid of both VEG_2 and the micromeres (HÖRSTADIUS, 1931, 1935).

It has recently been shown that minute parts of cell membranes break down between micromeres and between micromeres and macromeres (HAGSTRÖM and LÖNNING, 1969). This communication may be of great importance for induction processes dealt with in this chapter. It also explains the observation, that the blastomeres of vegetal halves of 32-cell stages, after treatment with calcium-free (Ca-free) seawater, show less tendency to fall apart than other cleavage cells.

The blastula (Fig. 13.1G) soon acquires active cilia. At the animal pole a group of long immovable cilia, stereocilia, grows out, forming the animal tuft (Fig. 13.1H,I). The vegetal side of the originally spherical blastula becomes somewhat flattened. From there the primary mesenchyme cells, derived from the micromeres, migrate into the blastocoel (Fig. 13.1I). These cells ring the base of the archenteron, fuse to syncytia, and give rise to the skeleton. The invaginating archenteron comprises only the VEG_2-material (Fig. 13.1K,L). From the tip of the archenteron, the secondary mesenchyme and coelom are budded off (Fig. 13.1K_1) whereupon the tip fuses with an invagination, the stomodaeum (Fig. 13.1L), which forms the mouth (Fig. 13.1M) and is derived from AN_1-material. The first sign of the ventral side is a flattening of a part of the gastrula wall (Fig. 13.1K_2). While most of the ectoderm differentiates as thin, squamous epithelium, the ventral side becomes encircled by a ciliated band of columnar cells, thereby forming the oral field.

The first traces of skeleton are 2 triradiate spicules at the base of the archenteron on the ventral side (Fig. 13.1K_2). They form the oral rods in the oral arms, the anal rods in the anal arms, the body rods in the pointed dorsal side (the apex), and moreover, 2 short ventral transverse rods (Fig. 13.1L-N). In the pluteus larva thus formed, the digestive tract (oesophagus, stomach, intestine) has been bent, thereby bending the egg axis (stippled line in Fig. 13.1M). The animal tuft has disappeared, and the larva is swimming with the ciliated band, which runs up and down the arms. The animal pole is represented by a thickening of the ciliated band between the oral arms, the animal plate (acron).

Some points may be summarized as follows. Along the egg axis, the egg can be divided in cleavage stages into 5 layers. The AN_1, AN_2 and VEG_1 - layers differentiate as ectoderm. The AN_1-layer forms both animal tuft and stomodaeum. The VEG_2-layer invaginates as the archenteron to form secondary mesenchyme, coelom and digestive tract. A vegetal half contains about one third of the presumptive ectoderm (VEG_1). The ventral side of the pluteus includes the anal arms, while the pointed part with the body rods, the apex, is on the dorsal side.

SELENKA (1883) discovered a ring of light reddish pigment in ripe eggs of *Paracentrotus lividus* which forms a girdle below the equator at right angles to the egg axis. BOVERI (1901a, 1901b) observed the ring and used it for estimating how much material invaginates during gastrulation, but mistakingly stated that all of the pigmented area invaginates as archenteron. VON UBISCH (1925a) misinterpreted his vital staining of the vegetal half, like BOVERI thinking that all vegetal cells invaginated. But it was shown that more than half of the pigmented area remains ectoderm. Vital staining, by the method of VOGT (1925) or by staining isolated layers and retransplanting them, which produces sharper stain borders, or staining individual VEG_2-cells or VEG_1 and VEG_2-cells show that only VEG_2-material invaginates (HÖRSTADIUS, 1931, 1935, 1936b). This basic fact was reconfirmed by VON UBISCH (1933) through new vital staining experiments.

The pigment ring, although distinct only in some batches, allows experimental operations along or perpendicular to the egg axis even before the eggs are fertilized.

13.1 Methods

Isolation of blastomeres or parts of later stages is nowadays performed by using glassneedles (SPEMANN, 1906) by hand. The fertilization membrane is first removed a minute or two after fertilization, before it toughens. DRIESCH (1893b) achieved this by shaking the eggs. PLOUGH (1927) recommended sucking up the eggs into a fine capillary pipette with an opening about two thirds the diameter of the egg membrane. HÖRSTADIUS (1935) aspirates the eggs rapidly through a pipette with an orifice about 3 or 4 times the diameter of an egg. LINDAHL and LUNDIN (1948) and MARKMAN (1958) describe devices for gently sucking or pressing the eggs through silk gauze with mesh a bit smaller than the diameters of the eggs. BERG (1965) removed membranes with a 2% protease solution.

Cleavage stages of some species, such as *Psammechinus miliaris*, can be divided with glassneedles without any pretreatment. In other cases, as with *Paracentrotus lividus*, it is preferable to operate in Ca-free seawater (HERBST, 1900). The time needed to loosen the connections between the cells depends partly on the purity of the chemicals used for preparing the artificial seawater (Table 13.1). Blastomeres or groups of blastomeres that remain too long in the solution may fall apart during the operation or when returning them back to normal seawater. Blastomeres of *Arbacia* are not easily separated even in a Ca-free medium.

Table 13.1. Artificial seawater formulas[a]

Solution (ml)		British Channel 35°/oo Saline %	Gulf of Naples 38°/oo Saline %
NaCl	(100)	3.29	3.60
KCl	(2)	4.28	4.64
$CaCl_2$	(3.5)	4.16	4.43
$MgCl_2$	(10.5)	3.49	3.68
$MgSO_4 \cdot 7H_2O$	(3)	19.50	21.25
$NaHCO_3$	(0.1)	5.21	5.21
NaOH	(0.1)	2.60	2.60

[a]For calcium-free seawater, omit $CaCl_2$. (From HÖRSTADIUS, 1935). For other recipes see Chapter 2A.

Fig. 13.2. For transplantation, fragments are placed in a depression, and a small glass ball is placed on top of the upper piece to give the necessary pressure. (HÖRSTADIUS, 1928a, b)

Before transplantation the parts to be fused are isolated in Ca-free seawater and then placed in normal seawater, one on top of the other, in a small excavation made with a somewhat flattened point of a pin, in a thin plate of celluloid, or some other not too soft, inert and transparent material. A small glass ball is put on top of the upper piece for a short time to provide the necessary pressure (Fig. 13.2) (HÖRSTADIUS, 1928b).

The staining of single, small blastomeres was developed by LINDAHL (1932b) and requires using a micropipette filled with Nile blue agar.

Smoking is not allowed where larvae are reared or used for experiments. Seawater rapidly absorbs poisonous substances which preclude normal development.

13.2 Determination of Cleavage

Isolated early blastomeres show partial cleavage as regards the number of meso-, macro- and micromeres (1/2-, 1/4-cleavage, or groups of 1/8 blastomeres, DRIESCH, 1891; FIEDLER, 1891; ZOJA, 1895). Whole eggs may segment as fragments after treatment with hypotonic seawater or after shaking (DRIESCH, 1893a; MORGAN, 1894; ZOJA, 1895)(Fig.13.3B_4-F_3). They may divide as meridional halves or as quarters of whole eggs, and may even cleave as vegetal halves with 4 big and 4 small cells in the 8-cell stage displaying what DRIESCH named "vorzeitige Mikromeren" (premature micromeres). BOVERI (1901b, 1905) assumed that the position of the spindles in different planes (first equatorial, then animal-vegetal) at different times is dependent upon a determination process initiated at fertilization and independent of the cleavage of the cytoplasm. MORGAN (1894, *Arbacia*) and BOVERI (1901a, 1901b, *Paracentrotus*) observed that the cytoplasm of the vegetal pole has a particular structure.

DRIESCH (1896b) obtained fragments of unfertilized eggs by shaking, and after fertilizing them observed both whole and partial cleavage. He attributed the difference to a different capacity for regulation. Meridional and vegetal fragments cleave like whole eggs, while animal fragments yield blastomeres of equal size, as the animal cells normally do (BOVERI, 1901b; HÖRSTADIUS, 1927, 1928a, 1928b). Therefore, in

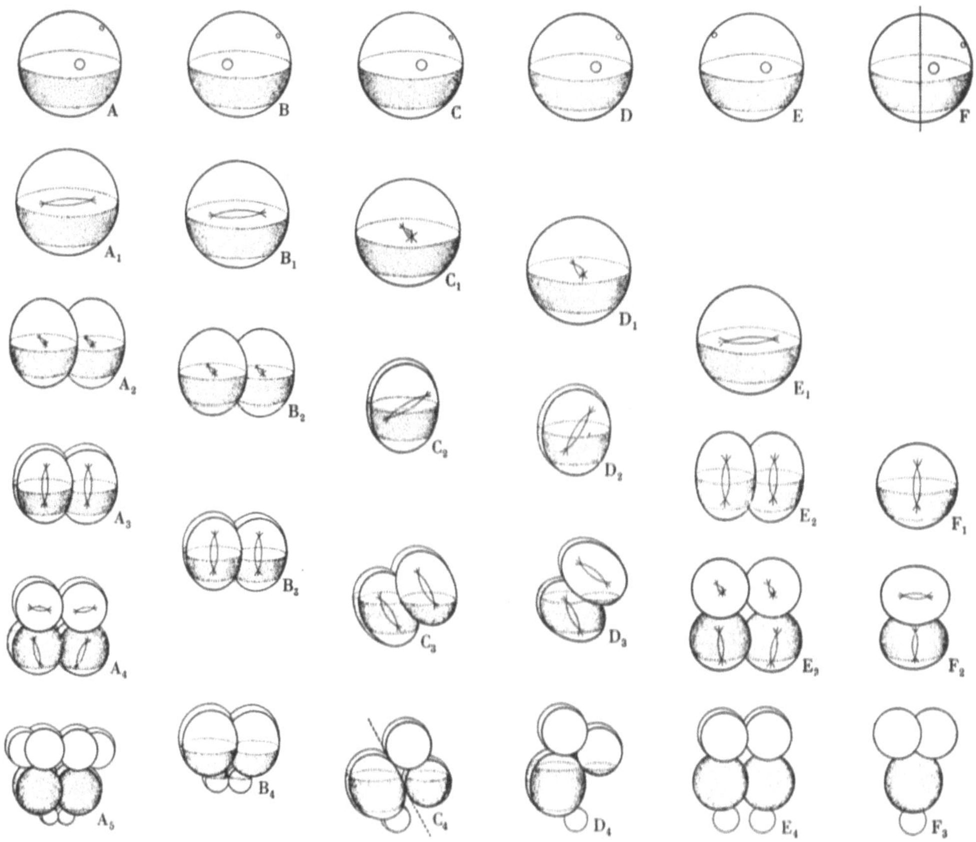

Fig. 13.3. Diagram of cleavage in *Paracentrotus lividus*. A_1-A_5, normal cleavage. B-F_3, cleavage delayed in relation to the process of determination, affecting the position of the cleavage spindles and causing the activation of the micromere cytoplasm. The delay was caused by hypotonic seawater or by shaking. The normal cleavage A-A_5 serves as time scale. All stages oriented with vertical axis. Pigment band stippled. (HÖRSTADIUS, 1927, 1928b)

contrast to DRIESCH, the lack of micromeres in animal fragments was ascribed to the lack of micromere-forming material. The fact that a vegetal half does not form micromeres in its 8-cell stage when the spindles occupy nearly the same position as in whole eggs before the 16-cell stage, indicates that the micromere material is not activated to form small cells until the time of normal micromere formation (HÖRSTADIUS, 1927, 1928b).

Plutei can develop from fragments that have segmented as fragments or as whole eggs, and normal larvae appear after fragment cleavage of whole eggs. A normal cleavage pattern is therefore not a prerequisite for differentiation. Through pressure, atypical placement of nuclei and cleavage furrows was observed and typical differentiation took place just the same (Fig. 13.4). These experiments by DRIESCH (1892, 1893b) and MORGAN (1894) proved that differentiation is not due to distribution of qualitatively different nuclei.

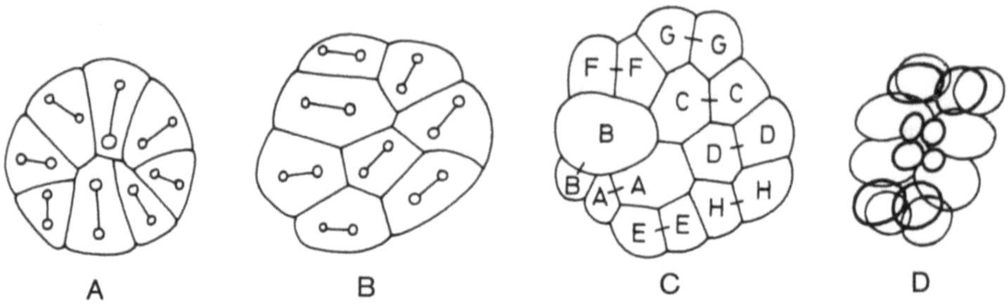

Fig. 13.4. Cleavage of eggs under pressure. A, B, 8-cell stages with all blastomeres in one layer. In A, the spindles in radial position, in B some tangential. C, 16-cell stage in one layer with 2 macromeres and 2 marginal micromeres. D, in another egg, after release of the pressure 2 layers of cells have been formed, with 4 micromeres in the center. (DRIESCH, 1892, 1893b)

In some egg batches, one may encounter unequal cleavage in the 8-cell stage, where the vegetal quartet is smaller than the animal. Such eggs are not of good quality and should not be used for experiments.

13.3 Determination along the Egg Axis

Early isolation experiments with shaking yielded plutei from blastomeres of 2-cell stages and also from 4-cell stages (DRIESCH, 1891, 1892, 1900a, 1906). BOVERI (1908) reported 4 plutei from 1 egg in some cases. Fig. 13.5 illustrates 4 perfect plutei derived from 1 egg. Fragments smaller than 1/8 of an egg can develop into plutei as shown in Fig. 13.6 (HÖRSTADIUS and WOLSKY, 1936). Micromeres, at the formation of the 32-cell stage, cleave off 4 smaller micromeres, and do not form a ring of 8 cells, thereby disallowing the division of the egg into 8 meridional parts of equal constitution. The micromeres were therefore removed before the sectioning into 8 fragments. The 7 small larvae in Fig. 13.6 accordingly represent eighths of 8 + 4 + 0. That harmonic larvae can develop from isolated 1/2- and 1/4-blastomeres and even from smaller parts of an egg, is due to the fact that they

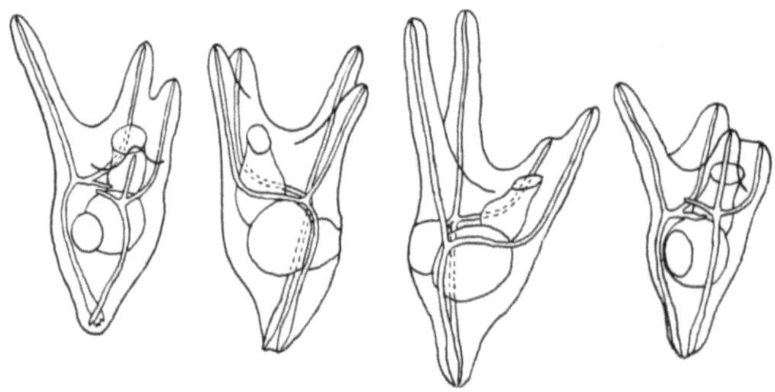

Fig. 13.5. Plutei from the 4 1/4-blastomeres of one egg of *Paracentrotus lividus*. (HÖRSTADIUS and WOLSKY, 1936)

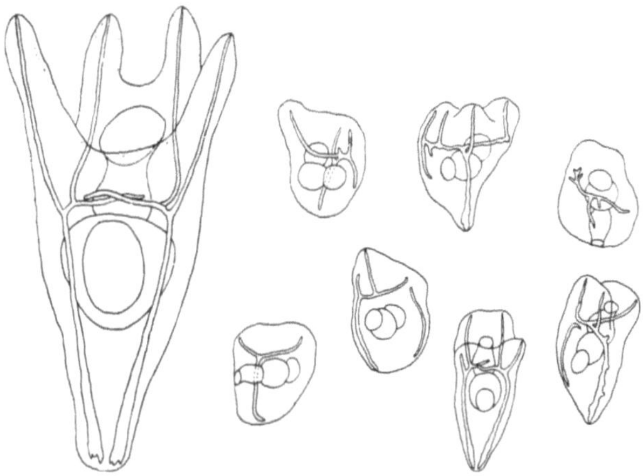

Fig. 13.6. A control pluteus and 7 1/8-larvae. (HÖRSTADIUS and WOLSKY, 1936)

represent meridional fragments which contain material from different strata along the egg axis (with exception of the 1/8-larvae which have been deprived of the most vegetal cells).

BOVERI advanced, in 1901, the hypothesis of a stratification with layers of different physiological qualities along the animal-vegetal axis in the sea urchin egg. He spoke of a gradient (Gefälle, 1910), where the most vegetal material would exert a determining influence upon the rest of the egg. If the most vegetal material is removed, the material next to it would act as a substitute, provided it still had enough vegetal properties. This notion has been proved by a great number of various experiments. A basic fact is, that isolated animal halves do not gastrulate, nor do they form any skeleton (ZOJA, 1895; BOVERI, 1901b, 1902; TERNI, 1914; VON UBISCH, 1925c, 1925d; HÖRSTADIUS, 1928b, 1935).

In a series of papers beginning 1891, DRIESCH, on the other hand, considered all parts of the sea urchin egg to be totipotent, the egg being a "harmonic equipotential system". As support for this idea, he states (1893a, 1900a, 1900b, 1901, 1902) that at least 25% of "animal" halves develop into plutei. Thanks to a mysterious force, entelechy (1901), fragments from any part of the egg should be able to develop into whole larvae. This was the basis for his philosophy of vitalism. As we have seen, micromere formation can often be inhibited by shaking. When DRIESCH isolated groups of 4 cells from 8-cell stages he judged those segmenting without micromeres, as animal halves. But many of them must actually have been meridional and vegetal halves which developed into plutei (HÖRSTADIUS, 1928b). This conclusion was demonstrated by the following experiment. After shaking 4-cell stages, the 4 blastomeres were isolated. The following division yields 1 animal and 1 vegetal cell. These were isolated, and the pairs reared separately. Of the vegetal cells 26% showed equal cleavage but gastrulated (l.c.p.124).

Isolated animal and vegetal halves, and other layers isolated at right angles to the egg axis, do not, as a rule, differentiate in accordance with their prospective significance (cf. Fig. 13.1, 13.7). In some cases, however, animal halves have an animal tuft of normal size and

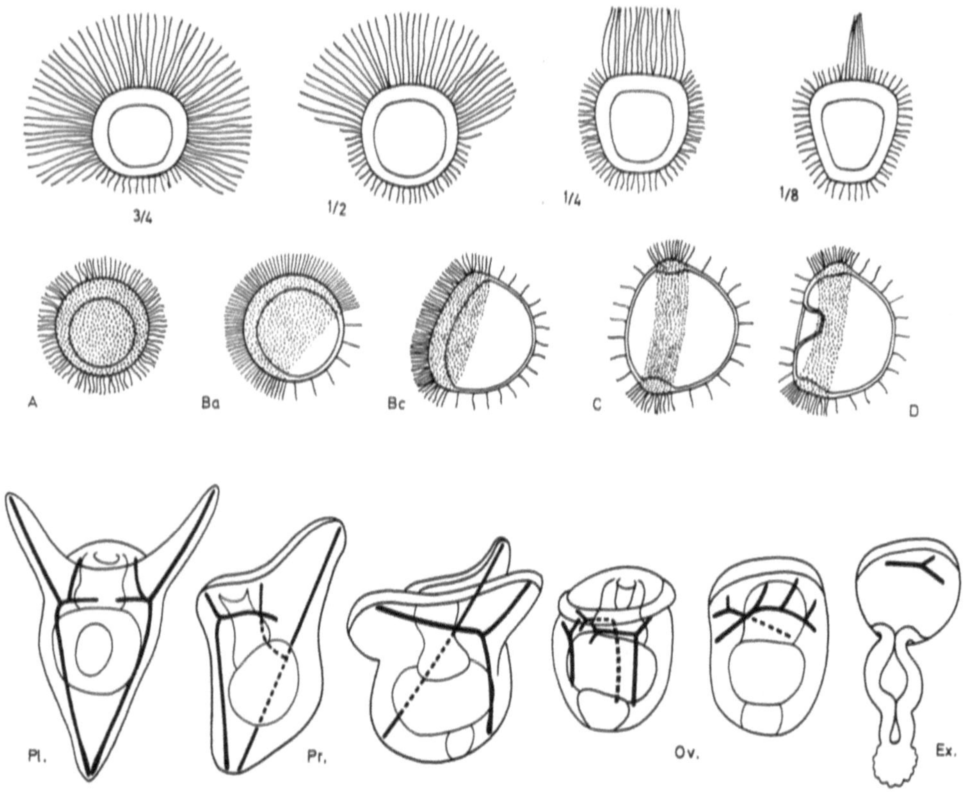

Fig. 13.7. Upper row development of isolated animal halves first day after fertilization and isolation. The young blastulae have more or less enlarged animal tufts, types designated 3/4, 1/2, 1/3, 1/4, 1/6. 1/8 normal tuft. Second row, fully differentiated animal halves, types called A, Ba, Bc, C, D. Third row vegetal halves. Pl. pluteus, Pr. intermediate types between plutei and ovoid larvae, Ov. Ex. exogastrula. The animal and vegetal halves on the left represent animal types, those on the right vegetal type of differentiation. (Partly from HÖRSTADIUS, 1935, 1939)

later form both a ciliated band and a stomodaeum (Fig. 13.7 1/8, D). But in the great majority of cases the animal tuft is more or less enlarged, covering up to about three fourth of the surface of the blastula (Fig. 13.7, 3/4). For classification, the different types are called 3/4, 1/2, 1/3, 1/4, 1/6, 1/8, the last designating a tuft of normal size (HÖRSTADIUS, 1935). Absence of the tuft is marked O. The more extended the tuft is, the less diversified the differentiation. Halves with a much enlarged tuft usually give only equally ciliated blastulae (Fig. 13.7A). Halves with tufts of middle size, yield blastulae with a larger or smaller ciliated field (Fig. 13.7Ba, Bc) or a ciliated band (Fig. 13.7C). One cannot predict exactly from the size of the tuft, the further development, but the tendency is such that a blastula with a tuft of size 3/4 or 1/2 never forms mouth and ciliated band (Fig. 13.7D), nor a 1/8 or 1/6 corresponds to Fig. 13.7A.

Some of the many types of isolated vegetal halves are depicted in Fig. 13.7. About one third of the presumptive ectoderm is included in the half (Fig. 13.1F, K-M). In spite of this unharmonic composition, some halves develop into plutei (Pl) or into larvae more or less like an irregular prism stage (Pr). Most frequent, are ovoid larvae (Ov), sometimes with a mouth and a ciliated band but more often with a ciliated field and no mouth. In such larvae the egg axis is only slightly or not at all bent, and the skeleton consists of a number of irregular spicules in a rather animal position. A deviating type is the exogastrula (Ex).

Based on thorough studies of the action of Li- and K-ions and informed of new results of isolations and transplants, RUNNSTRÖM, in 2 papers (1928a, 1928b) pointed out that there are a number of interactions between animal and vegetal potentials which are responsible for determination. In a paper dated 1929, this hypothesis is clearly stated: "Wir nehmen zur Erklärung der Determination entlang der Eiachse zwei gegeneinander gerichtete und sich gegenseitig beeinflussende Gefälle an ..." BOVERI (1901b) had suggested a vegetal gradient, and CHILD (1916a, 1916b) had described gradients of susceptibility, whereas RUNNSTRÖM's 2 gradients implied that they overlap throughout the whole egg, and are partially hostile to each other. If they are properly balanced, they interact mutually and a typical differentiation ensues. Many experimental results are understandable when viewed in the light of this conception.

In animal halves of type 1/8-D there are sufficient vegetal properties present to restrict the extension of the tuft and to allow the formation of mouth and ciliated band. In the other types, the vegetal tendencies have been more or less suppressed. In vegetal halves, we also witness differences in gradient strengths. The ovoid larvae without mouths are perhaps those which correspond most to the prospective significance of the material. In exogastrulae, the endoderm is usually enlarged, a differentiation more vegetal than in the ovoid larvae. (Exogastrulation can also occur without changing the ratio of ectoderm to endoderm.) In plutei, the ectoderm is comparatively better developed than in ovoid larvae, as mouth and ciliated band are fully developed and the skeleton is typically arranged and arms grow out. As demonstrated by HERBST (1892, 1893, 1896) and RUNNSTRÖM (1917, 1928a, 1928b, 1929), the interaction between ectoderm and primary mesenchyme is necessary to form a normal skeleton.

Using the terms from LINDAHL (1933), we can say that plutei of vegetal halves are animalized, and exogastrulae vegetalized, in relation to ovoid larvae.

The different types of halves do not occur randomly, and it rarely happens that the halves represent only one type. On the other hand, one seldom finds all types in fairly equal numbers. Instead, the eggs in a batch usually belong to a few types, classifiable as more or less animal, or vegetal, or intermediate (Fig. 13.34, 13.35).

In eggs of some batches, the third cleavage furrow is not strictly equatorial, but somewhat subequatorial, yielding vegetal blastomeres smaller than the animal ones. It was first thought that halves showing a more vegetal development were derived from such subequatorial eggs (HÖRSTADIUS, 1935), but further experiments have shown that equatorial halves can also differentiate in a more vegetal way, the result depending upon difference in the gradient system. It was also believed, that a strong animal differentiation of animal halves was coupled with a similar development of the vegetal partner, and vice versa, but exceptions to this rule have also been recorded.

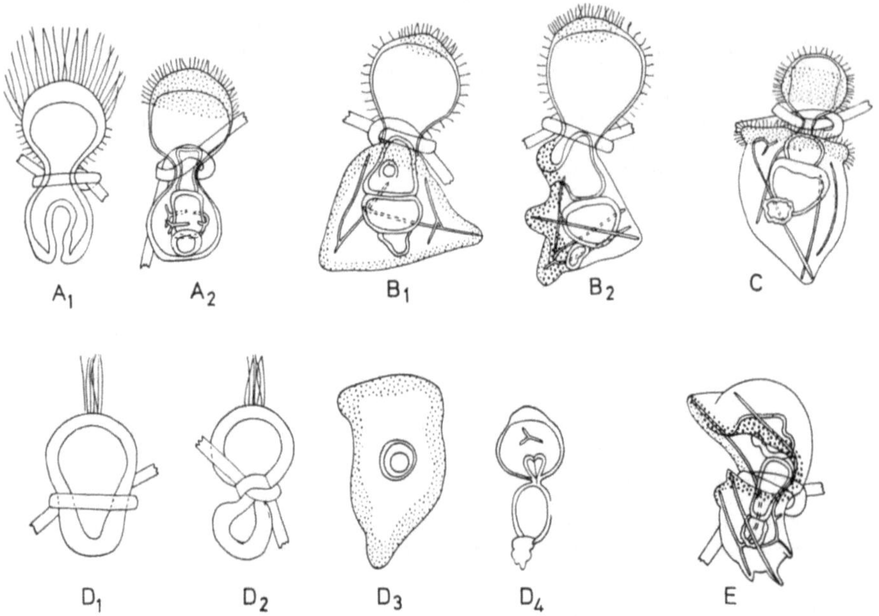

Fig. 13.8. A-C, equatorial constriction in early cleavage stage. A_1, animal tuft enlarged because of reduced communication with the vegetal material. A_2, the animal part has only differentiated as an isolated animal half of type B (Fig. 7). B_1-B_2, a larva seen from the ventral and the left side. The vegetal part has formed a mouth as some isolated vegetal halves can do (Fig. 7). In C the animal region has a ciliated band, cf. type C. The originally loose hair-loop (D_1) has at a late blastula stage been more tightly constricted (D_2). The animal half (D_3, separated from the vegetal, D_4) due to the early open connection with the vegetal material has formed both ciliated band and stomodaeum (type D). In a similar way E has differentiated like a pluteus. (HÖRSTADIUS, 1938)

It is not necessary to isolate animal and vegetal halves to illustrate changes in interactions between the 2 systems. LINDAHL (1932b, 1936) has shown that stretching whole eggs along the egg axis is sufficient to cause an enlargement of the animal plate. The same occurs through early equatorial constriction by means of a silk fibre (HÖRSTADIUS, 1938). Neither ciliated band nor stomodaeum are formed, even if the oesophagus reaches up into the animal half (Fig. 13.8A_1, A_2). In Fig. 13.8B, the most animal third of the egg is partly isolated by the constriction: ciliated band and stomodaeum are lacking. Instead, through a regulatory mechanism, they have been formed in the vegetal component. These cases correspond to differentiation of isolated halves. The necessity of mutual influence is proven by constrictions in later stages. In Fig. 13.8D_1, D_2 the originally loose hair-loop has at a late blastula stage been more tightly constricted. The interactions took place before the constriction, the animal part possessing both ciliated band and stomodaeum (Fig. 13.8D_3) and the vegetal half has exogastrulated, as many isolated halves do (Fig. 13.8D_4; the 2 halves had fallen apart). In Fig. 13.8E all interactions took place before the tight constriction. The result is a pluteus-like larva with a narrow belt and undersized rods.

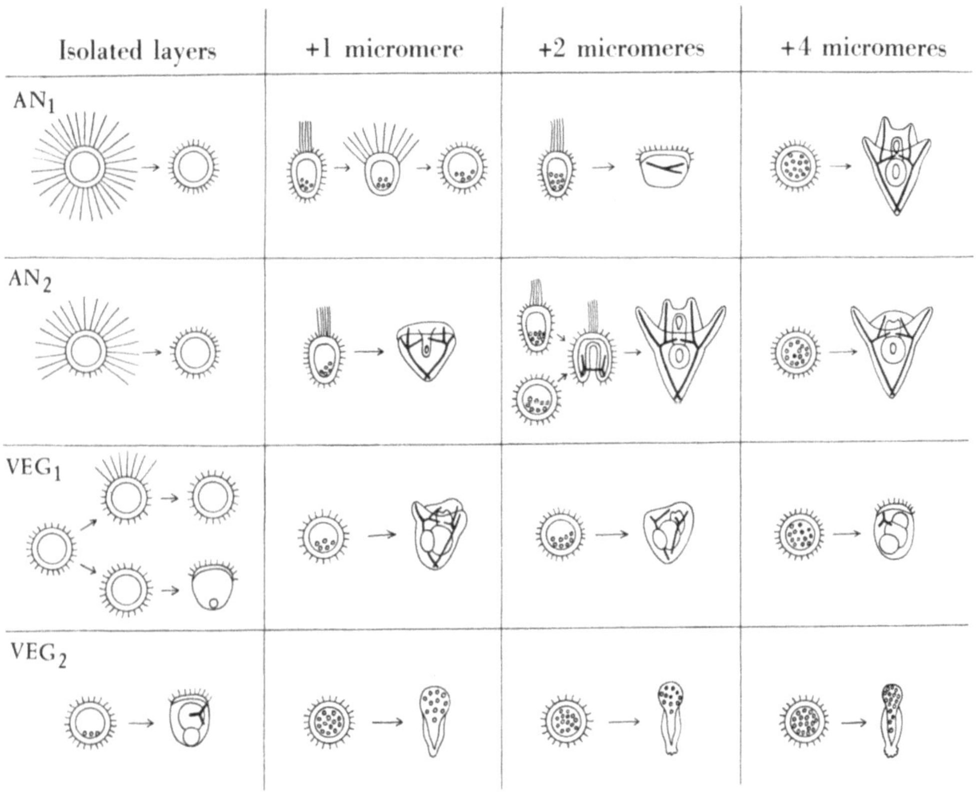

Fig. 13.9. Diagram of the development of the layers AN_1, AN_2, VEG_1 and VEG_2 isolated (left column) and with 1, 2, and 4 implanted micromeres. (HÖRSTADIUS, 1935)

An isolated AN_1-ring develops into a blastula completely covered with stereocilia, a differentiation more animal than in any animal half, and subsequently gives rise to a uniformly ciliated blastula with no further development (Fig. 13.9, left). AN_2 behaves more like an animal half. VEG_1-layers produced 2 lines of differentiation. This material is presumptive ectoderm but lies next to the endoderm-forming cells. In some cases, the development follows an animal, ectodermal line, the tuft is of intermediate size and gastrulation does not occur, but here, the outgrowth of the tuft was delayed. In other cases, no tuft differentiated, and instead, a small invagination, appeared, although VEG_1 normally does not participate in the gastrulation. VEG_1 thus offers an example of a subtle balance between the 2 gradients. VEG_2, the material of the future archenteron, was found to give a small ovoid larva. This is an interesting case of regulation, at both ends. The ectoderm is a more animal structure than expected from the material, and the skeleton has been formed in spite of the absence of the material for the primary mesenchyme. The removal of the animal material AN_1 + AN_2 + VEG_1 on the one side and the most vegetal material, the micromeres, on the other side, has left the VEG_2 with a gradient constitution similar to that of an isolated vegetal half.

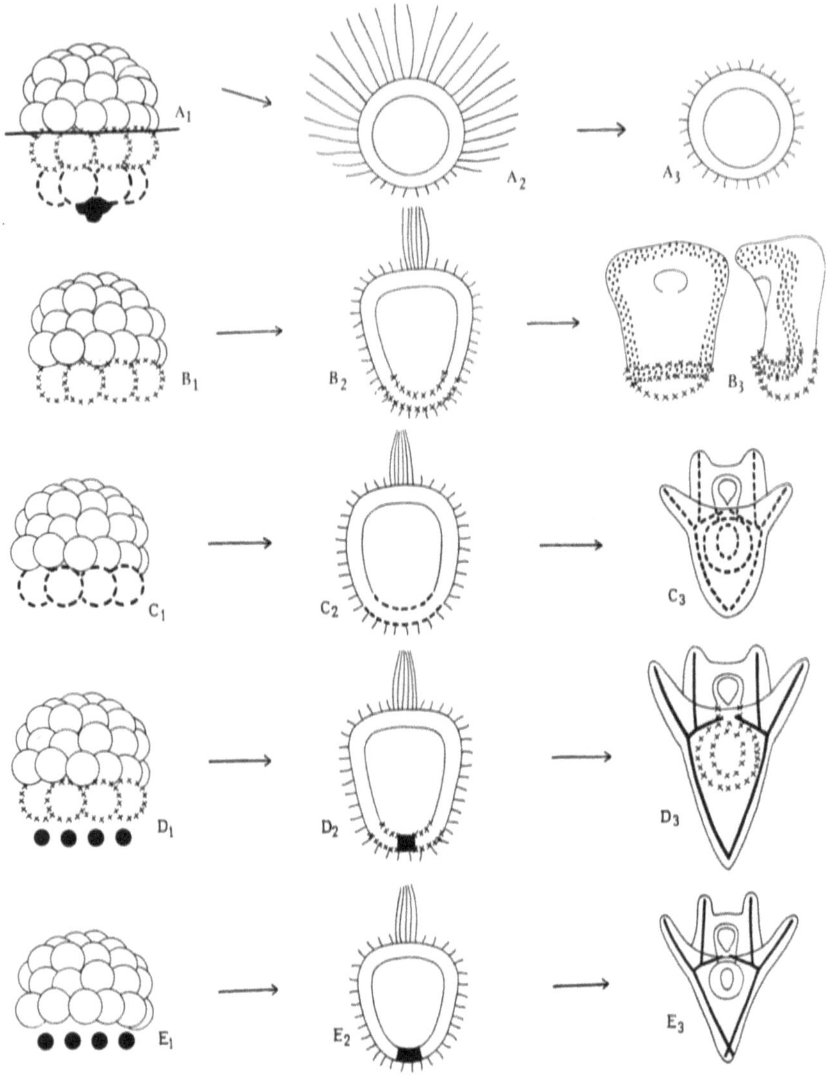

Fig. 13.10. Diagram of the influence of VEG_1, VEG_2 and the micromeres on the differentiation of animal material. AN_1 and AN_2, continuous lines; VEG_1, crosses; VEG_2, broken lines; micromeres black. A, isolated animal half. B, 8 + VEG_1 + 0. C, 8 + VEG_2 + 0. D, 8 + VEG_1 + 4. E, 8 + 0 + 4. (HÖRSTADIUS, 1935, 1939)

Isolated micromeres first form a lump of cells, but at the time when the primary mesenchyme migrates into the blastocoel the cells wander apart (HÖRSTADIUS, 1935). The cleavage pattern of isolated single micromeres showed the temporary occurrence of a syncytium with blunt pseudopodia, which eventually divided into more than 20 cells spread out on the bottom (HAGSTRÖM and LÖNNING, 1969). Isolated micromeres are evidently too vegetal to be able to regulate in animal direction.

Culturing isolated micromeres in a medium consisting of 1-2% horse serum in seawater has, however, shown that without any contact with other cells they are capable of skeleton formation (OKAZAKI, 1971). 1972).

Fragments with $AN_1 + AN_2 + VEG_1$ (Fig. 13.10B_1) contain the entire presumptive ectoderm. They do not gastrulate, but contrary to isolated animal halves (A), they possess in VEG_1 enough vegetal properties to inhibit an extension of the animal tuft and to differentiate both ciliated band and stomodaeum (Fig. 13.10B_2, B_3). The same power is also inherent in VEG_2, as shown when a VEG_2-ring is added to an animal half ($AN_1 + AN_2 + VEG_2$)(Fig. 13.10C_1). A larva of this composition will develop into a pluteus, and the VEG_2-layer will form the archenteron. But it also produces cells of more vegetal character; the skeleton forming mesenchyme (Fig. 13.10C_2, C_3). If the VEG_2-layer and the micromeres are fused with an animal half, a pluteus also forms, the difference being that the skeleton is secreted by proper primary mesenchyme cells.

Of special interest is the implantation of a group of 4 micromeres into the entire presumptive ectoderm ($8 + VEG_1 + 4$)(Fig. 13.10D_1-D_3), or into an animal half (8 + 0 + 4)(Fig. 13.10E_1-E_3). We obtain plutei if the micromeres become attached to the most vegetal part of the host. In these cases, the origin of the organs is different from that of Fig. 13.10C_3 where the archenteron was formed by endoderm material and the skeleton by a strengthened vegetal capacity. In Fig. 13.10D and E, the micromeres not only act on the development of tuft, ciliated band and stomodaeum, and produce spicules, but moreover they have the power of transforming presumptive ectoderm to endoderm, thus inducing an archenteron.

An example of the inducing power of vegetal material is demonstrated by fusing a 16-cell stage meridional half with an animal half which has been vitally stained before transplantation (Fig. 13.11). Part

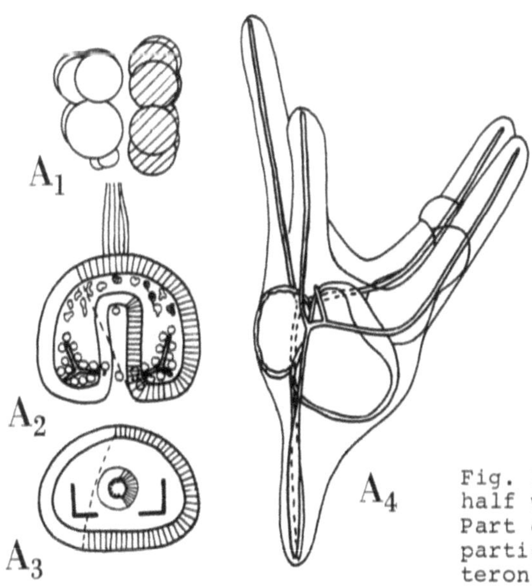

Fig. 13.11. Fusion of a meridional half with an animal half (hatched). Part of the presumptive ectoderm participates in forming the archenteron. (HÖRSTADIUS, 1928b, 1935)

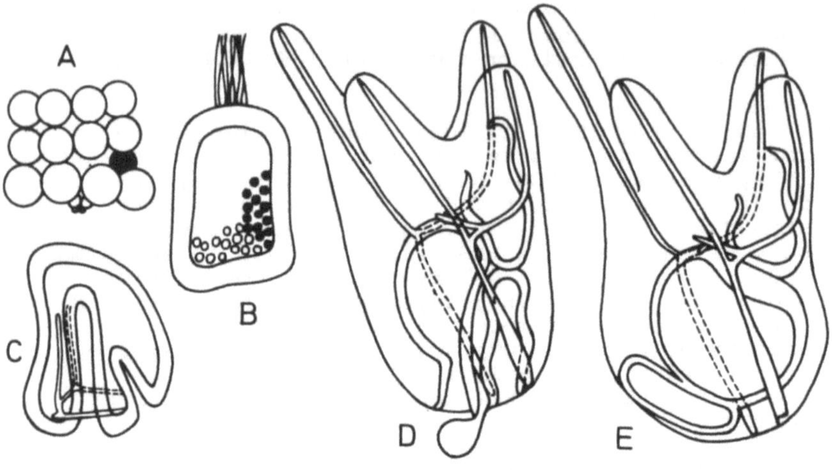

Fig. 13.12. Implantation of 4 micromeres between AN_2 and VEG_1 (A). The micromeres produce primary mesenchyme cells (B), and induce a second, smaller archenteron (C, D) which fuses with the normal archenteron (E). (HÖRSTADIUS, 1935)

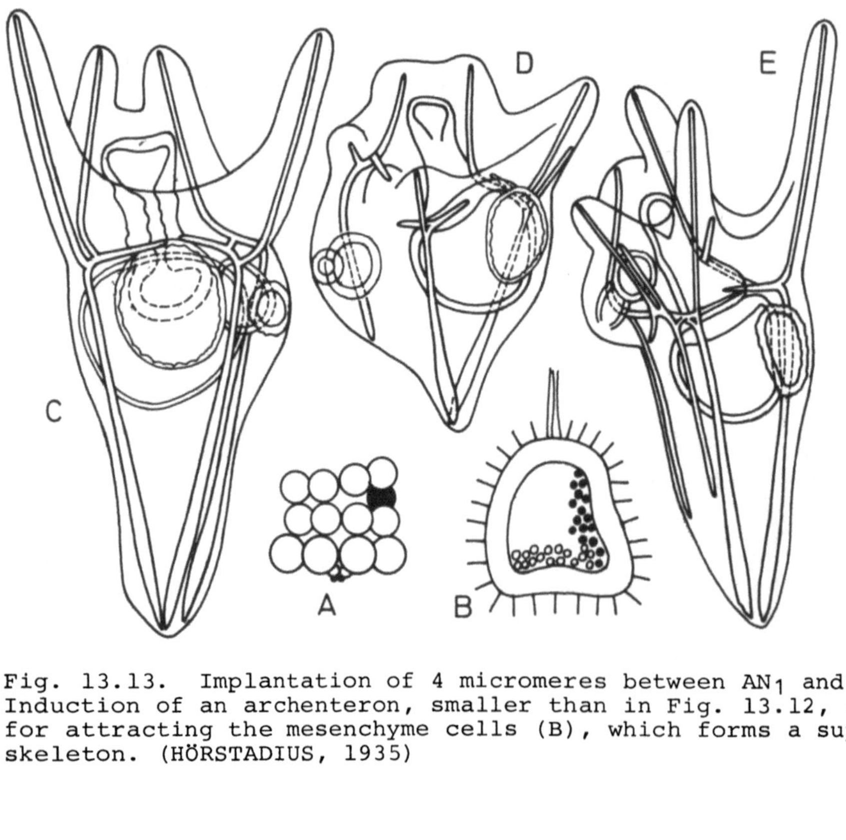

Fig. 13.13. Implantation of 4 micromeres between AN_1 and AN_2 (A). Induction of an archenteron, smaller than in Fig. 13.12, and a center for attracting the mesenchyme cells (B), which forms a supernumerary skeleton. (HÖRSTADIUS, 1935)

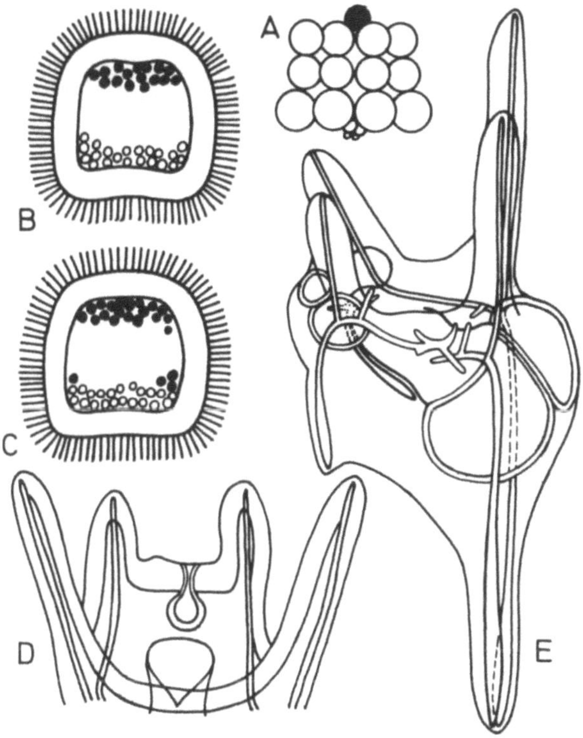

Fig. 13.14. Implantation of 4 micromeres in the animal pole. The resistance against induction is stronger, the archenteron is smaller. (HÖRSTADIUS, 1935)

of the presumptive ectoderm contributes to the archenteron, and the polarity of the halves is somewhat changed.

The gradual increase of the animal gradient towards the animal pole in a whole egg can be illustrated by implanting micromeres at different levels of a whole egg. A group inserted between VEG_1 and AN_2 will induce a fairly large archenteron, but being so close to the normal archenteron, both will soon fuse and a typical pluteus will be the result (Fig. 13.12). Placed between AN_2 and AN_1-layers, the micromeres induce a bipartite archenteron, and a pair of skeletal spicules (Fig. 13.13). At the animal pole, the animal resistance is stronger and the induced archenteron and skeleton appear smaller (Fig. 13.14).

A clear illustration of the gradual changes in intensity of the gradients is given by the implantation of 1, 2 or 4 micromeres in isolated layers AN_1-VEG_2 of eggs of the animal type (Fig. 13.9). 1 micromere in the AN_1-layer first seemed to stop the extension of the tuft, but then the animal forces prevailed. 2 micromeres were strong enough to produce a spicule, and 4 sufficed to form a pluteus. In the AN_2-layer, 1 micromere could restrict tuft extension, develop skeleton and induce a small invagination. 2 micromeres resulted in more uniform plutei than the implantation of 4. The VEG_1-layer offered less resistance to induction, the result with 1 micromere being a fairly typical pluteus, while 4 created a balance as in ovoid vegetal halves. In VEG_2-layers, even 1 micromere was enough to cause exogastru-

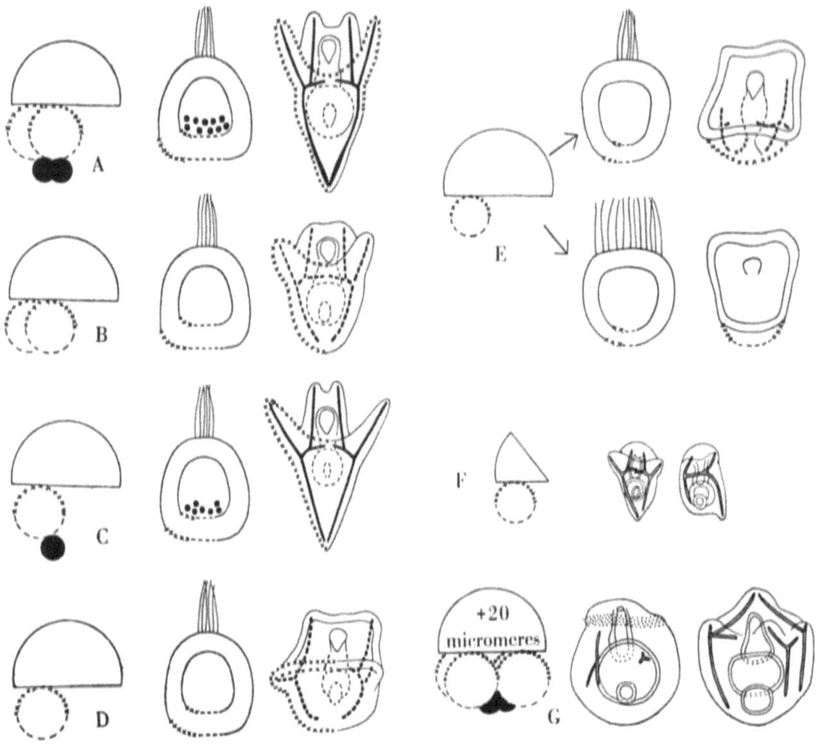

Fig. 13.15. Diagram of gradual diminution of vegetal material (A-E). Animal half = half circle; VEG_1, crosses; VEG_2, broken lines; micromeres black. A, 8 + 2 + 2; B, 8 + 2 + 0; C, 8 + 1 + 1; D, 8 + 1 + 0; E, 8 + 1/2 + 0; F, if the animal material also is diminished, this small larva will differentiate more harmonically, as the animal and vegetal qualities are more typically balanced; G, if a whole egg is filled with micromeres, it will differentiate like a vegetal half. (HÖRSTADIUS, 1935, 1939)

lation. Particularly elucidating, is the fact that the most harmonic larvae are to be found along the diagonal line VEG_2 + 0, VEG_1 + 1, AN_2 + 2, and AN_1 + 4.

The strong inducing forces of the micromeres, as shown in 8 + VEG_1 + 4, 8 + 0 + 4, might give the impression that vegetal forces can always bring about a harmonic differentiation. But, this is possible only within certain quantitative limits, as already shown in Fig. 13.9. Some further experiments are proving this. A gradual reduction of vegetal material (8 + 2 + 2, 8 + 2 + 0, 8 + 1 + 1, 8 + 1 + 0), may still give plutei, although with a smaller digestive tract (Fig. 13.15 A-D). But with 8 + 1/2 + 0 (Fig. 13.15E) the VEG_2-material is only in some cases able to assert itself, in inhibiting the extension of the tuft, and inducing gastrulation and skeleton formation, while in other cases the vegetal forces can be suppressed, the result being a blastula with enlarged tuft, ciliated band and stomodaeum. The necessity for a balance between animal and vegetal systems is evident when a half macromere instead is left together with a proportionally small amount of animal cells, the result being a dwarf pluteus (Fig. 13.15F). If a whole egg is filled with micromeres it differentiates as a vegetal half (Fig. 13.15G)(HÖRSTADIUS, 1935).

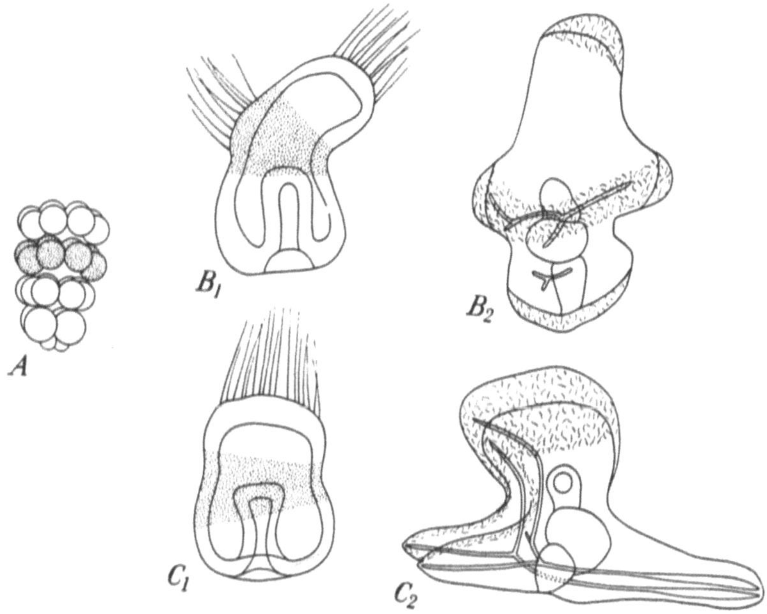

Fig. 13.16. If 2 animal halves are added to a whole egg (A, the half next to the egg vitally stained, stippled) the animal tuft is more or less extended or also double (C_1, B_1). A pluteus may develop, although with a considerably enlarged oral lobe (C_2). In other cases no reorganization to a single gradient system occurred (B_1, B_2). (HÖRSTADIUS, 1950)

By adding one animal half on top of the mesomeres of whole 16-cell stages, a pluteus was obtained which could be mistaken for normal. But on closer inspection it was revealed that the oral lobe and arms as well as the mouth were enlarged. With 2 rings of mesomeres, the vegetal system did not suffice at all to produce larvae of typical proportions (Fig. 13.16)(HÖRSTADIUS, 1950). In some cases, a regulation took place to only one, although enlarged apical tuft, and to a kind of pluteus, but this one had an enormous oral lobe with ciliated field instead of arms (C_1, C_2). In other cases, however, several tufts appeared, and the larva showed a radial symmetry with only a small archenteron, thus showing no regulation towards a typical pluteus (B_1, B_2).

When the ectoderm material borders directly on the most vegetal cells, the differentiation is more harmonic than expected from the material. But if isolated vegetal halves followed the principle of harmonizing, a part of the VEG_2-cells ought to give ectoderm to restore the proportions. But this is evidently not the case. The ecto-endoderm border in vegetal halves did not change or else it moved unexpectedly in the animal direction, producing a larger archenteron than in normal gastrulae (HÖRSTADIUS, 1935, 1936b). As animal forces increase in intensity in animal fragments the vegetal forces seem to spread in vegetal fragments, thus an advance when the opposite gradient has been weakened, and this leads to a less harmonic development. But when strong vegetal material is brought into close contact with presumptive ectoderm, we saw that induction can lead to surprisingly harmonic development (Fig. 13.9-14).

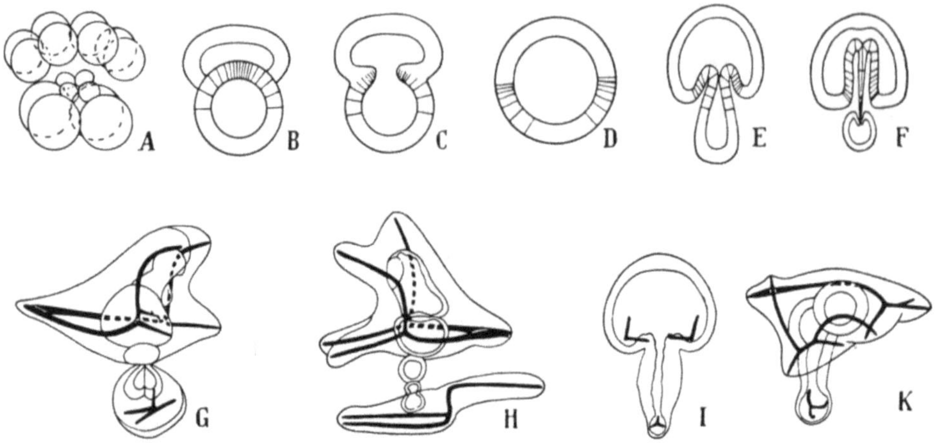

Fig. 13.17. Reversal of the vegetal half (A). The animal half is sitting as a cap on the vegetal half (B). A blastula is formed, with the most vegetal material close to the equator (C, D). Here the invagination begins (E). A small part of the vegetal half remains outside the gastrula (F). In this vesicle a small invagination as well as skeleton formation can occur (G, H) which implies a small animal-vegetal axis reversed to the larger one. I, K, treatment with lithium reveals endodermization along both axes. (HÖRSTADIUS, 1928b, 1936)

On the other hand, we must remember the dominating influence of the animal forces in some fragments, e.g. ciliated band and stomodaeum inhibited in animal halves, skeleton formation and induction inhibited in AN_1 + 1. Another example is a larva with the vegetal half transplanted upside down, Fig. 13.17A. Both halves want to close, forming a blastula in the normal way, which results in the animal half sitting as a cap on the vegetal (Fig. 13.17B). The latter will open, and a blastula is formed with the most vegetal material immediately below the equator (Fig. 13.17C, D). Here the invagination begins (Fig. 13.17E). The greater part invaginates into the animal half but a small vesicle remains outside the blastopore, mostly differentiating as ectoderm (VEG_1, Fig. 13.17F). But also in this diminutive blastula, a small invagination occurs (Fig. 13.17G-H). The larger ectoderm, the animal half, has thus attracted the major part of the presumptive endoderm although its invagination is proceeding in a direction opposite to its prospective fate. Only a small part of the VEG_2-material has invaginated into its own VEG_1. This is a good example of the influence of animal material upon vegetal.

A true reversal of the polarity of the egg axis can in some cases be obtained by implanting a group of micromeres in the animal pole of an animal half (HÖRSTADIUS, 1935). When inserted in the most vegetal part of an animal half the result is a small pluteus (8 + 0 + 4, Fig. 13.10E). When placed in the animal pole of a whole egg the micromeres are confronted with a stronger resistance, being able to induce only a small archenteron (Fig. 13.14). In an animal half the vegetal forces lack the support of the gradient in the vegetal part of the egg. We would now find it probable to obtain induction of a fairly large archenteron. But the regulation in many cases, took a different course. To determine if the micromeres actually remained at the pole, they were stained with Nile blue sulphate and the pole was stained with neutral red (Fig. 13.18A_2). Although isolated animal halves never

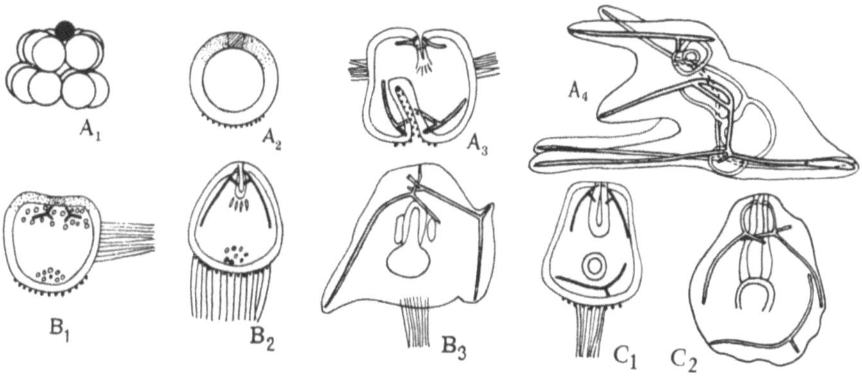

Fig. 13.18. A_1, implantation of 4 micromeres into the animal pole of an animal half. A_2, the site of implantation could be checked as the micromeres had been stained with Nile blue (hatched) and the animal pole with neutral red (stippled). A_3, A_4, larvae with 2 archenterons, 1 induced, animal, and 1 vegetal, and 2 pairs of skeleton. B_1-B_3, C_1-C_2, reversal of the polarity. Explanation in text. (HÖRSTADIUS, 1935)

gastrulate, it was astonishing to find in some cases a larger archenteron invaginated from the vegetal part of the half and only a smaller one from the pole (Fig. 13.18A_3). Both had, in a typical way, 2 triradiate spicules at their base. Apparently the micromeres had created a new vegetal center in the animal pole and weakened the animal gradient in the whole half so that the vegetal properties of the original equatorial region were about the same strength relative to the animal ones, as normally occurs in the VEG_2-material. If this interpretation is correct, we should expect to find the strongest animal gradient pushed aside somewhere between the 2 new vegetal centers, and this was actually the case. The stereocilia formed a belt in proper distance from the centers (Fig. 13.18A_3) and the larva developed to a pluteus with a smaller archenteron and skeleton in the oral lobe (Fig. 13.18A_4). The larva in Fig. 13.18B_1-B_3 gives us another picture of the labile balance between the gradients. Only 1 small archenteron invaginated, at the animal pole, where spicules also appeared. Vegetal forces at the other end were indicated by a concentration of mesenchyme cells and by compelling the tuft to grow out between the 2 centers. But the new system was not yet stabilized. The new center at the animal pole became dominant and alone formed the archenteron and skeleton, thereby pushing the tuft down to the vegetal side. Fig. 13.18C_1 and C_2 give an example of a nearly complete dominance of the new center at the animal pole. From the beginning the tuft is located at the opposite side and the vegetal properties have only been able to assert themselves by forming a skeletal spicule. The egg axis has been reversed.

When one vegetal half is transplanted on top of another vegetal half (Fig. 13.19), it shows a strange migration. Within a day or two it has reoriented itself from the animal pole of the partner until the 2 endodermal regions are able to fuse (Fig. 13.19A_4). When 3 vegetal halves are placed on top of each other, the upper 2 move in the same way down along the side of their neighbour until endodermal contact is reached between the 3 (HÖRSTADIUS, 1928b, 1950).

While plutei can develop from meridional fragments, it is not possible to obtain 2 plutei from 1 egg after a transverse cut. The animal frag-

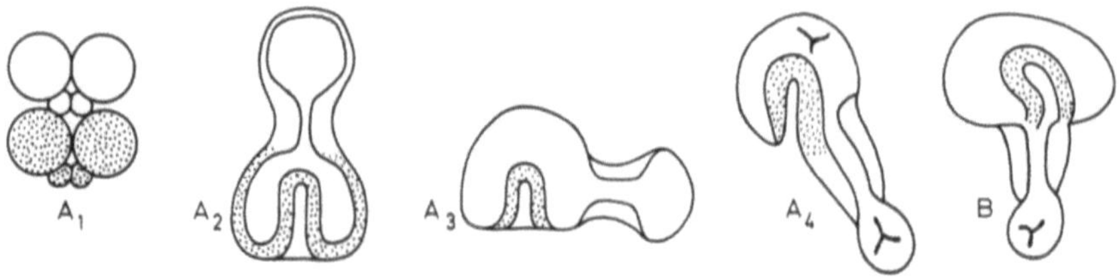

Fig. 13.19. A_1, a vegetal half placed on top of another vegetal half. A_2, the ecto- and endodermal regions recognizable. A_3, migration of the upper partner along the side of the lower towards its vegetal part. A_4, B, 2 larvae in which the endodermal regions have fused. (HÖRSTADIUS, 1928b, 1949)

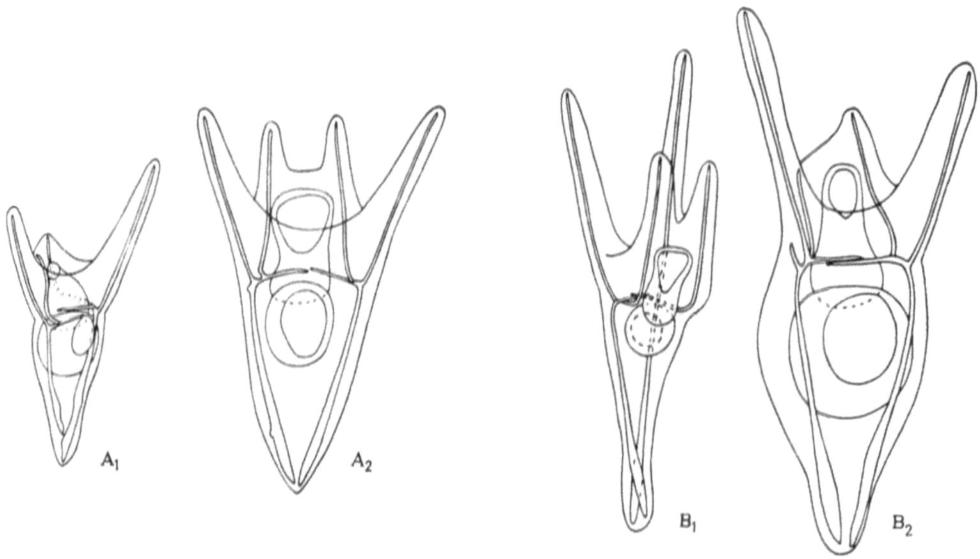

Fig. 13.20. 2 plutei can only be obtained from 1 egg after transverse cuts by isolating the middle part and fusing the animal and vegetal fragments. A_1, A_2, AN_1 + O + 4 and AN_2 + 4 + O. B_1, B_2, 8 + O + 4 and O + 4 + O. (HÖRSTADIUS, 1936b)

ment has to be as large as 8 + 4 + O to gastrulate and to form a skeleton. But 2 plutei can easily be produced after 2 transverse sections, if the middle fragment is left alone and the polar fragments brought together: AN_1 + O + 4 and AN_2 + 4 + O, respectively 8 + O + 4 and O + 4 + O (Fig. 13.20)(HÖRSTADIUS, 1936b). This gives a further support to the double gradient hypothesis.

CHILD (1936) described differential reduction of methylene blue and Janus green in the early development of sea urchin and starfish eggs. A weak staining with Janus green gives a light greyish blue colour which, at reduction, changes into light red, and after further reduction, the colour vanishes completely, thereby providing a stepwise control.

In a normal late blastula, reduction begins at the vegetal pole and proceeds in animal direction. Half way, a new reduction center appears at the animal pole (Fig. 13.21A_1, A_2). The last region to change colour is a ring about half way between the animal pole and the equator (Fig. 13.21A_3). The ectoderm of a young gastrula shows the same

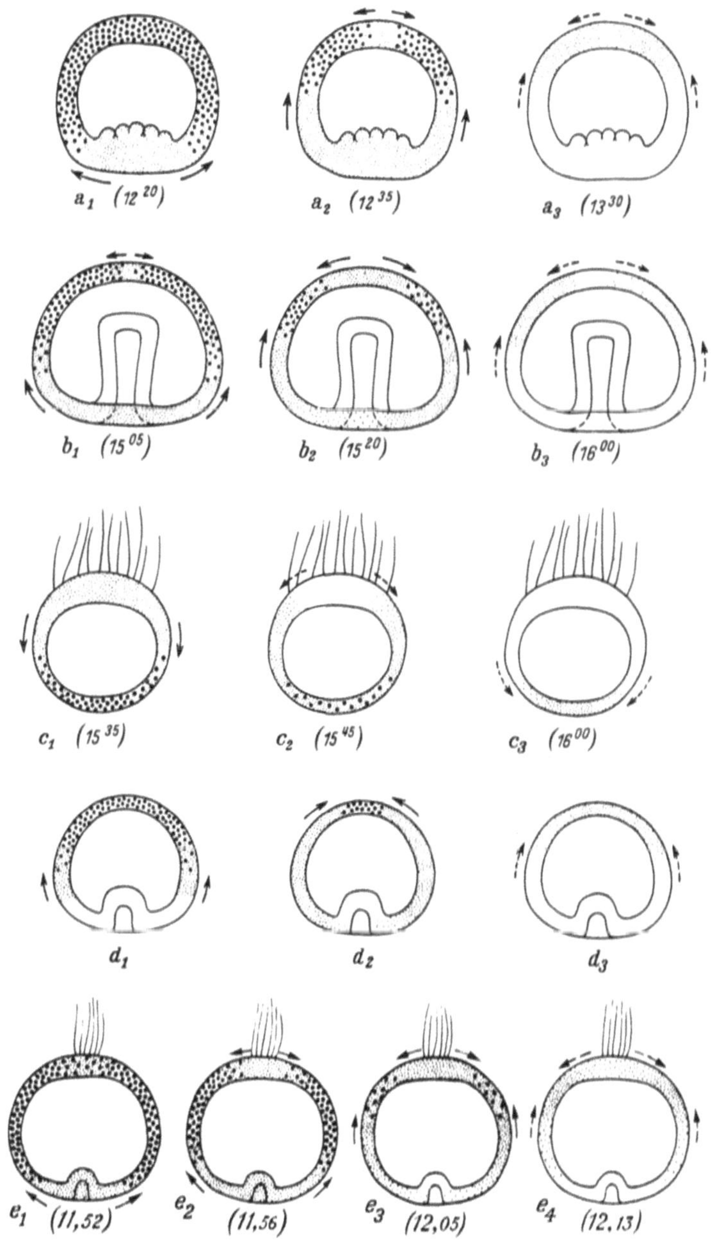

Fig. 13.21. Change of color of janus green showing reduction gradients. Large dots, blue; small dots, red; no dots, the color has faded. Figures indicate time of observations. a_1-a_3, late blastula; b_1-b_3, gastrula; c_1-c_3, animal; d_1-d_3, vegetal half of the same egg; e_1-e_4, implantation of 4 micromeres into an animal half. This normally results in a pluteus (Fig. 13.10.E). The reduction pattern is now the same as in a whole egg, a_1-b_3. (HÖRSTADIUS, 1952, diagram partly from KÜHN, 1955)

picture (Fig. 13.21B_1-B_3). In animal halves which never gastrulate there is logically no vegetal reduction center. Instead, reduction begins at the animal pole and continues around the blastula (Fig. 13.21C_1-C_3). In vegetal halves the reduction direction is reversed (Fig. 13.21D_1-D_3). Sometimes, however, we observe a delayed animal center. This is in keeping with the fact that some isolated halves possess so much of the animal qualities that they can differentiate as plutei (Fig. 13.7Pl).

Micromeres implanted laterally in a whole egg induce a vegetal center with archenteron and skeletal spicules (Fig. 13.12,13). Similar experiments with Janus green reveal the induction of a reduction center. In animal halves there is no vegetal reduction center (Fig. 13.21C_1-C_3). When 4 micromeres are implanted in an isolated animal half (8 + O + 4) the system is changed into a complete counterpart of a normal larva. The reduction now begins vegetally, a second center starts later at the animal pole, and the last region to be reduced is in the same region as in whole eggs (Fig. 13.21E_1-E_4) (HÖRSTADIUS, 1952).

HERBST (1892, 1893) discovered that lithium ions vegetalize sea urchin eggs. Several methods of animalization have also been described, lack of SO_4-ions (HERBST, 1897), with SCN- or J-ions in Ca-free seawater (LINDAHL, 1936), iodosobenzoic acid (RUNNSTRÖM and KRISZAT, 1952), proteolytic enzymes (HÖRSTADIUS, 1953a). Staining of Li-vegetalized whole eggs with Janus green showed that the vegetal reduction gradient was dominant. No animal center was visible. Eggs more or less animalized by trypsin showed an animal center domination, but as animalization was not complete, both mesenchyme and a small archenteron formed, and a delayed and weak vegetal reduction became visible (HÖRSTADIUS, 1955).

The effect of lithium has been used in many ways to study interaction and metabolism in the gradient system. VON UBISCH (1925c, 1925d) found that isolated animal halves treated with lithium could develop into plutei. HÖRSTADIUS (1936b) tested the most vegetal material of such halves to see if it had acquired the same inducing power as micromeres. The vegetal side of a Li-animal half was locally stained. This material was in the blastula stage isolated and added to a normal animal half on its vegetal side (Fig. 13.22D, E, F, H, I, M, P, O). This artificially created vegetal material restricted the extension of the apical tuft, and induced archenteron and skeleton formation, thereby resulting in a typical pluteus differentiation. To ascertain that it was not the effect of residual ions, but the effect of vegetalized material, the most animal material of the Li-half was removed and attached to another animal half, which differentiated into the typical animal blastula with large tuft (Fig. 13.22F, G, K, N, Q, R).

Cleavage stages with a reversed vegetal half (Fig. 13.17A) were transferred to seawater with 5% isotonic LiCl. The endodermization did not affect the larva as a whole. Instead, we can follow the effect working along the 2 axes, thus in opposite directions (Fig. 13.17I, K). The VEG_1-ectoderm vesicle is strongly diminished, and the endoderm increases also in the animal half. In some cases the VEG_1 was completely endodermized and the larva could hardly be distinguished from a typical exogastrula (HÖRSTADIUS, 1936b).

Although the cytoplasm of sea urchin eggs can be stratified by centrifugation, larvae can develop into plutei, regardless at which angle the stratification has to the egg axis (LYON, 1906a, 1906b; MORGAN

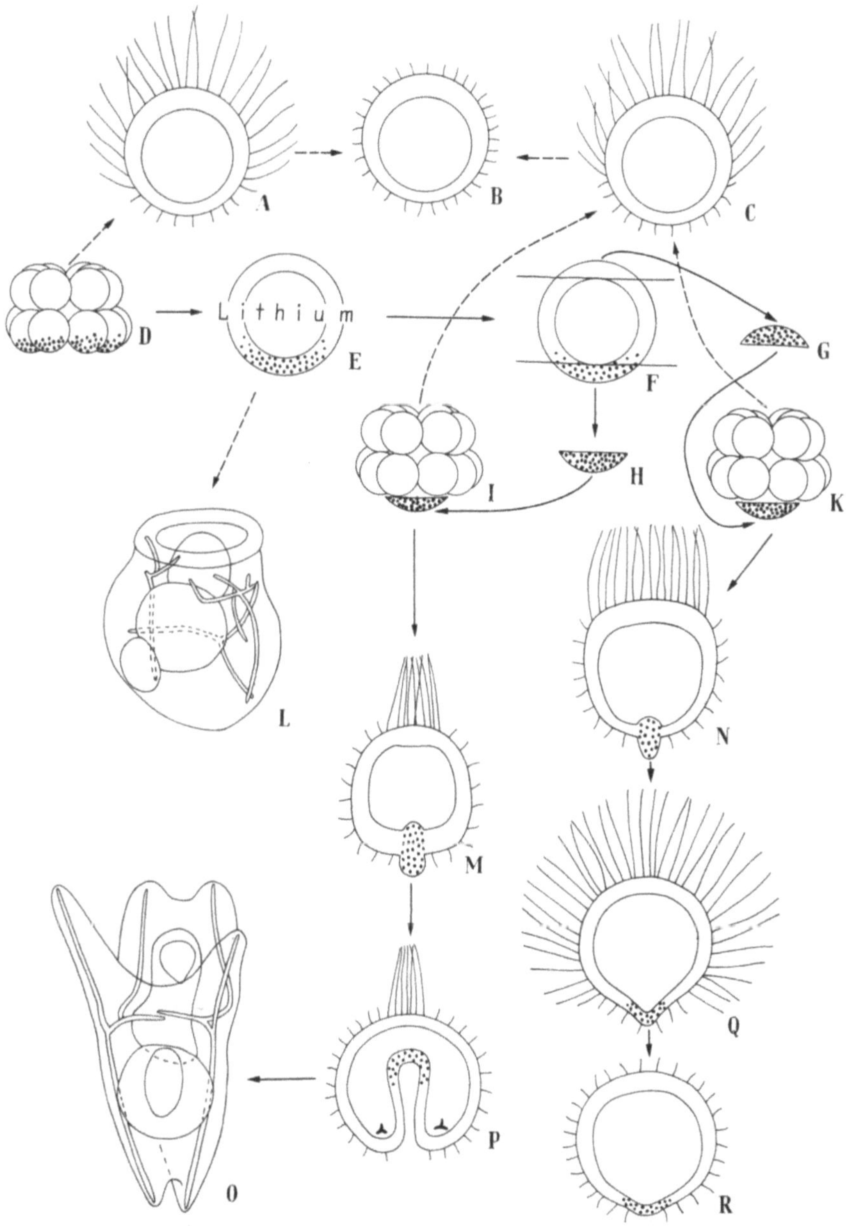

Fig. 13.22. Diagram of development of a pluteus from an animal half to which had been added the most vegetal part of another, with lithium treated animal half. Explanation in text. (HÖRSTADIUS, 1936b)

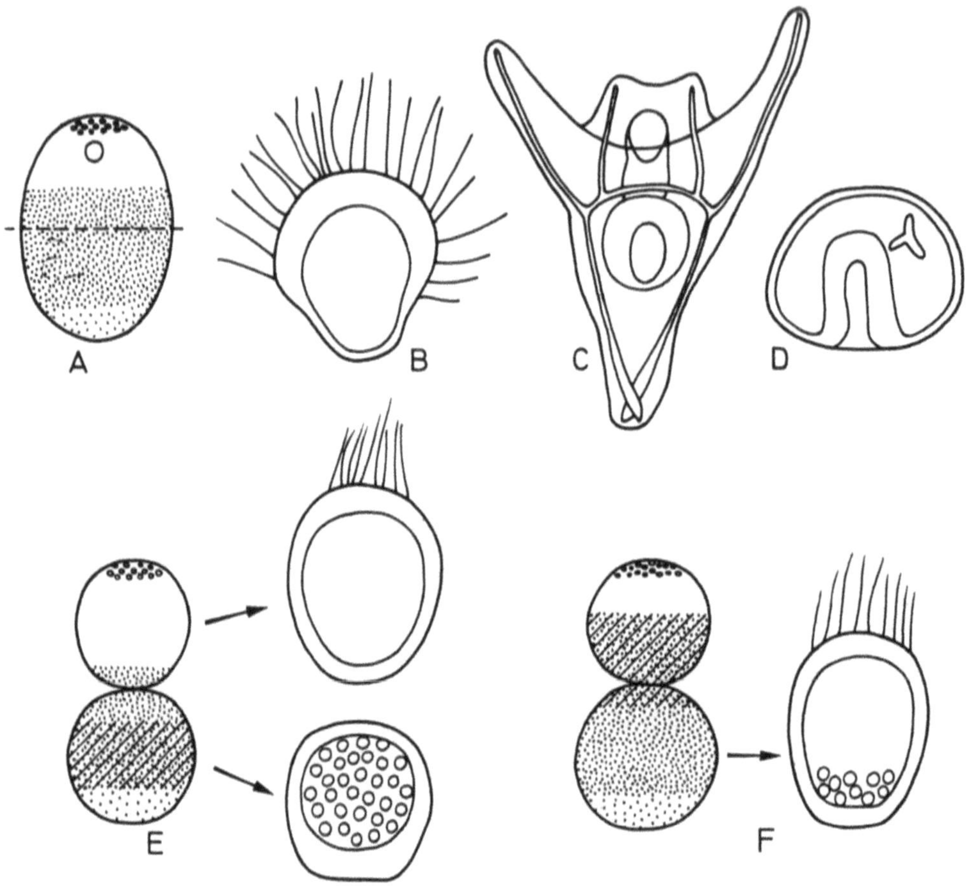

Fig. 13.23. Development of fragments of centrifuged, stratified eggs. A, unfertilized egg, broken line indicates plane of separation by glass needle. B, animal blastula with enlarged animal tuft. C, pluteus. D, larva of vegetal type. E, development of light animal and heavy vegetal half of the same egg to larvae of animal or vegetal type respectively. F, heavy animal half with enlarged animal tuft. Pigment band obliquely hatched in E and F. (HÖRSTADIUS, 1953)

and LYON, 1907; MORGAN and SPOONER, 1909). Fragments of stratified eggs differentiated in conformity with the origin of the material in relation to the egg axis, irrespective of strata represented in the fragment (HÖRSTADIUS, 1953b). Fig. 13.23A illustrates an unfertilized stratified egg. The nucleus lies in the lighter half, and the stippled line indicates the plane of separation into 2 halves which can develop after fertilization. Both light and heavy fragments can yield plutei (C), blastulae of animal type (B), gastrulae with large archenteron, thus of vegetal type (D), or intermediate forms.

In E the pigment band is visible in the heavy fragment, with stratification along the egg axis. The light fragment differentiates like an animal half, the heavy fragment in vegetal direction. In F, where the egg axis is reversed to the stratification the light, pigmented vegetal fragment did not develop, whereas the larger, heavy fragment developed an enlarged tuft and the vegetal wall did not gastrulate. Lithium was tested on centrifuged fragments which may represent animal, vegetal, meridional or intermediate fragments. Although many of them did not contain some substances, which normally are spread more evenly throughout the egg, they were all strongly affected by lithium.

The methods used in this chapter allow a study of how determination proceeds with time, and 1 example was given in the constriction experiments illustrated in Fig. 13.8.

That determination gradually proceeds during cleavage was demonstrated by the differentiation of animal halves isolated in 8- to 64-cell stages from the same batch of eggs (HÖRSTADIUS, 1965a). Advances in determination were also noted by isolating halves every second hour up to beginning gastrulation (16 hrs)(HÖRSTADIUS, 1936a). In animal halves, the percentage of types with enlarged tuft decreases to zero, while the types of fully differentiated halves with ciliated band and stomodaeum increase. Ovoid larvae from vegetal halves gradually disappear, with a concomitant increase in more pluteus-like larvae. Operations at late blastula or beginning gastrula result in plutei well developed, except for a total lack of the most animal parts: the oral lobe, oral arms, and mouth.

Animal halves were isolated from the 16- and 32-cell stages and a group of 4 micromeres were implanted either immediately, (4 hrs after fertilization), or at 2 hour intervals up to beginning gastrulation (16 hrs)(Fig. 13.24)(HÖRSTADIUS, 1936a). The isolated halves gradually loose the ability to react upon the vegetal influences from the micromeres. The first stage of effect is the reduced size of the digestive tract of the pluteus. In halves with later implantation, the archenteron was extremely small, but the skeleton remained bilateral. The blastula eventually loses the capacity to react by endodermization. A stomodaeum may still be formed, but often only 1 tripartite spicule appears. Later lost, is the faculty to form a mouth; blastulae with spicules develop, and finally, blastulae are obtained in which the skeleton-forming capacity of the micromeres seems to have been inhibited. In another series, the isolations were performed 4-16 hrs after fertilization and micromeres were immediately implanted. These halves, due to a longer period of cellular interactions with the vegetal halves, appeared to be accessible to vegetal influences at later hours. The animal properties, in the sense of suppressing vegetal tendencies, therefore, work faster in isolated animal halves than in halves having for a longer time been a part of the whole gradient system. Similar results were obtained when lithium was used instead of implantation of micromeres.

Animal and vegetal halves of cleavage stages, blastulae, and beginning gastrulae, with age differences from 2 to 17 hours, were fused together. Typical plutei developed in spite of great age difference between halves. An old vegetal half was still able to inhibit animal tuft extension and to induce in the animal material, formation of the ciliated band and stomodaeum. In larvae of great age difference, bilateral symmetry was often abnormal (HÖRSTADIUS, 1950).

The distribution along the animal-vegetal axis of mitochondria vitally stained with Nile blue sulphate was studied in late blastula and gastrula stages of whole eggs, animal and vegetal halves, and animal halves with implanted micromeres (LENICQUE et al., 1953). As shown in Fig. 13.25, the curves in all cases express a gradient. The curve for animal halves is lying above, that for vegetal halves below that of the control whole eggs. Implantation of 4 micromeres shifts the

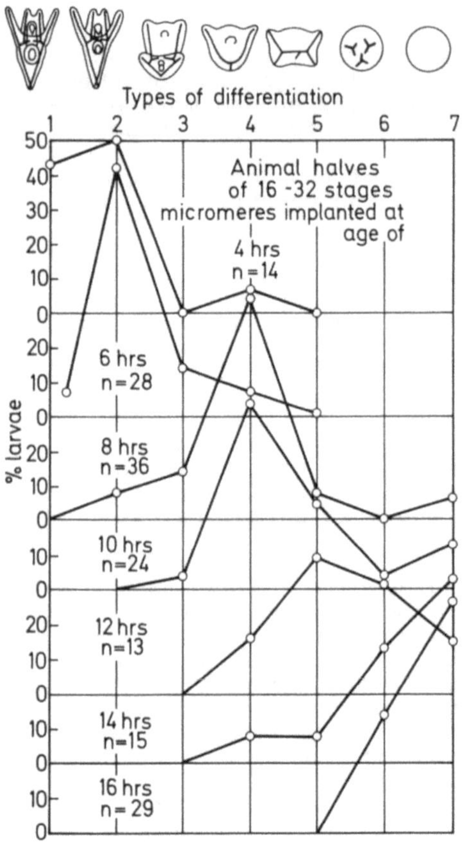

Fig. 13.24. Differentiation of animal halves isolated in 16- or 32-cell stages, in which micromeres were implanted 4-16 hrs after fertilization. (KÜHN, 1955, from figures in HÖRSTADIUS, 1936a)

pattern towards that of normal larvae. With 8 micromeres the level is close to that of vegetal halves. Electron microscopy has been used for identification of mitochondria in different regions of sea urchin gastrulae (BERG et al., 1962). The counts indicate uniform distribution in animal, equatorial, vegetal, and mesenchyme cells. Several authors have described marked changes in activity of mitochondria at the onset of gastrulation (lit. in GUSTAFSON, 1969). It is probable that the differences in stainability recorded in Fig. 13.25 depend on differences in metabolic activity of the mitochondria along the gradients (GUSTAFSON, 1969).

13.4 Dorsoventral Axis Determination

Studies of the determination of bilateral symmetry in the sea urchin egg have yielded conflicting results: that the first cleavage furrow coincides with the future median plane (BOVERI, 1902, 1905, 1908; HERBST, 1907; RUNNSTRÖM, 1914, 1925) or that it coincides with the frontal plane (DRIESCH, 1906, 1908). After one of the first blastomeres was vitally stained, the first furrow seemed to cut through the

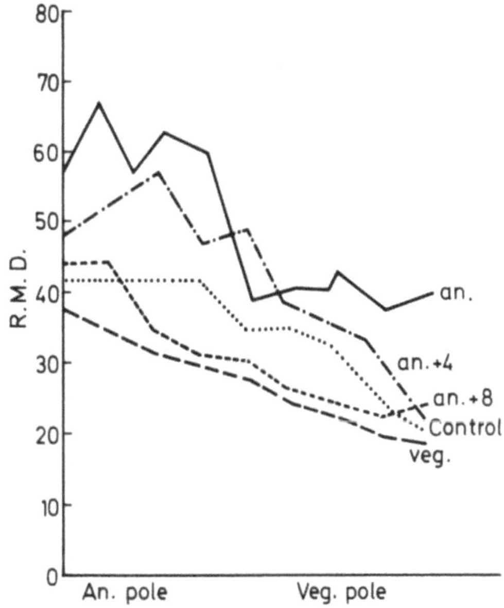

Fig. 13.25. Distribution of vitally stained mitochondria along the animal-vegetal axis in late blastulae-early gastrulae (Control), in isolated animal halves (an) and vegetal halves (veg) as well as in animal halves with 4 (an + 4) or 8 implanted micromeres (an + 8). R.M.D. (relative mitochondrial density). (LENICQUE, HÖRSTADIUS and GUSTAFSON, 1953)

median of the frontal plane, rarely occurring obliquely to it (VON UBISCH, 1925a; RUNNSTRÖM, 1926; HÖRSTADIUS, 1928b). Using small marks of Nile blue stain applied with micropipettes, the first furrow was found to have no fixed position in relation to the future median plane and no particular affinity for either median or frontal plane (HÖRSTADIUS and WOLSKY, 1936).

Investigations of sperm influence on bilateral symmetry led to equally contradictory conclusions that sperm entry and first furrow position coincide (SELENKA, 1878; WILSON and MATTHEWS, 1895; BOVERI, 1901b; RUNNSTRÖM, 1925) or that they do not (GARBOWSKI, 1905; BOVERI in HEFFNER, 1908). When the side opposite to the sperm entry was stained, the first furrow was found to occur in any position (HÖRSTADIUS, 1928b). As the first furrow is unrelated to the sperm entry point and to the median plane, the possibility that the sperm induces bilateral symmetry cannot be excluded.

In case the sperm does not determine bilateral symmetry question arises whether the egg is bilaterally organized before fertilization. An aspect of this kind must exist, because unfertilized eggs, activated by artificial parthenogenesis, differentiate into plutei (LOEB, 1899). Furthermore, several authors assume some special properties of the ventral region in the unfertilized egg (RUNNSTRÖM, 1925; FOERSTER and ÖRSTRÖM, 1933; GUSTAFSON, 1952; MARKMAN, 1963).

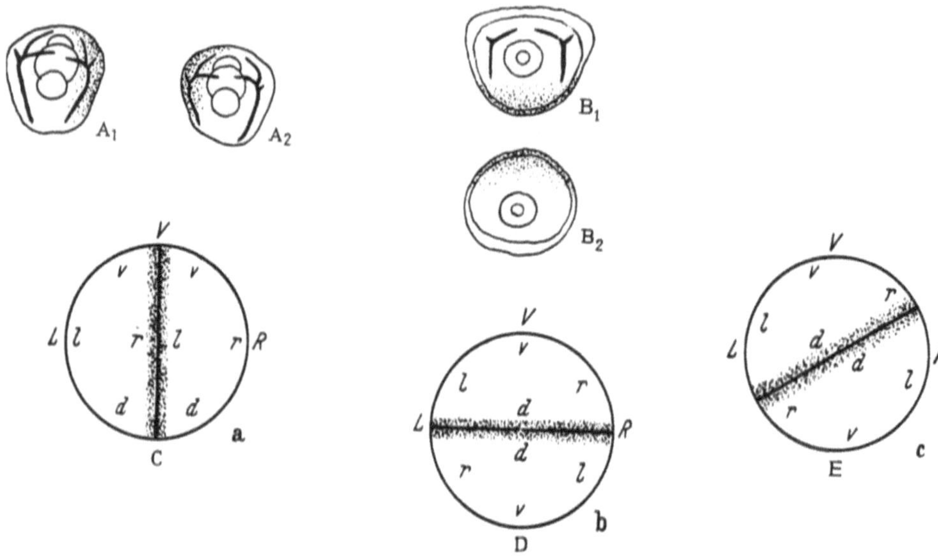

Fig. 13.26. The dorso-ventral axis in meridional halves of the same egg. The cut sides vitally stained. A_1, A_2, left and right halves with delay in differentiation of the cut sides. The axes coincide with the presumptive axis (C). B_1, B_2, ventral and dorsal halves. In the dorsal half, the axis has been reversed, its differentiation of the new ventral side is delayed. The reversal illustrated in the diagram D. E, diagram of cases with an oblique plane of separation in relation to the presumptive organization. (HÖRSTADIUS and WOLSKY, 1936)

The early dorsoventral axis cannot be very stable as quarter-larvae give complete plutei. After local application of metabolism inhibiting substances, the inhibited part of the egg developed the dorsal side (PEASE, 1941, 1942a, 1942b). Many observations have been made on the position of the median plane in larvae from blastomeres or in fused twins in relation to the presumptive dorsoventral axis. The dorso-ventral axis in halves was supposed to rotate 90° (BOVERI in HEFFNER, 1908) or to be reversed in ventral halves but to remain in dorsal halves (DRIESCH, 1908) and not to change in isolated half-blastomeres (RUNNSTRÖM, 1914).

A method was developed where isolated fragments were vitally stained on the cut side to study whether a stable bilaterality already exists in early stages. Skeletal and arm development on the left side of one half-blastomere and the right side of the other was delayed (Fig. 13.26A_1, A_2). As the delayed sides were stained, the halves must correspond to left and right halves of the egg. Other pairs of half-blastomeres were stained in their dorsal parts, and in each case one half formed the ventral side - thick epithelium and skeleton - faster than the other (Fig. 13.26B_1, B_2). The half that formed the ventral side faster (the ventral half) has retained its dorsoventral polarity. The polarity of the dorsal half was reversed and delayed because the new ventral side required more time to form (Fig. 13.26D). In half-blastomere pairs in which the first furrow formed obliquely to the original bilateral symmetry, the results varied (HÖRSTADIUS and WOLSKY, 1936).

Axial reversal in the dorsal half has been explained as follows: In a normal egg the side with the stronger capacity to form a ventral side suppresses any such tendency in other parts of the egg (LINDAHL, 1932b). When the influence of the presumptive ventral center of dorsal halves is suppressed, the point farthest away and least affected is most likely to form another ventral side. Similar results were obtained in meridionally constricted eggs and in meridional halves isolated before fertilization (HÖRSTADIUS, 1938).

Isolated 4-cell-stage blastomeres develop into plutei and in a series of experiments where the blastomeres were isolated from their position about the egg, ventral-side differentiation was sometimes faster in 2 blastomeres than in the other 2. This evidently signified that first 2 furrows had been median and frontal, creating 2 ventral halves and 2 dorsal halves. At other times, 1 larva developed first, then 2 intermediately, and finally the fourth - the results of 1 ventral, 2 lateral, and 1 dorsal blastomere. In many quartets, however, the results varied (HÖRSTADIUS and WOLSKY, 1936).

Early bilateral organization may also be studied by rotating animal and vegetal fragments. After a meridional side of an 8-cell stage is vitally stained, the fragments may be rotated when they are replanted. Rotating 8 + 0 + 0 and 0 + 4 + 4, or 8 + VEG_1 + 0 and VEG_2 + 4 90°-180° results in typical plutei when a round blastula appears after replanting. When, however, reduced surface contact, due to oblique transplantation, occurred, ectodermal irregularities were observed, indicating an early bilateral organization of the presumptive ectoderm. The endoderm seemed to adjust itself after the ectoderm (HÖRSTADIUS and WOLSKY, 1936).

When animal and vegetal halves of different stages are brought together without considering dorsoventral character, bilateral symmetry is often abnormal in larvae in which the stage difference is great. The determination of bilateral symmetry seems to be accomplished while the vegetal material is still able to counteract animal influences (HÖRSTADIUS, 1950).

Further studies on the reorganization within the bilateral symmetry system were made in 2 ways (HÖRSTADIUS, 1957). The one involved exchange of meridional halves between 2 eggs. These halves, that could have represented dorsal, ventral or oblique sections were fused with other 16-cell halves along parallel egg axes (Fig. 13.27). Typical plutei developed, indicating that a reorganization of whole and of parts of the hypothetical ventral centers had occurred, resulting in a dominant ventral center.

The other method yielded giant larvae from 2 whole eggs. 2 32-cell stages were cut meridionally on 1 side, both were flattened and then fused together with the same orientation (Fig. 13.28). The ensuing blastula will thus have an egg axis of normal length but a circumference twice the normal size. Some of the giant plutei are perfect, while in the dorsal region, others have a supernumary spicule together, in some cases, with an extra ciliated band indicating a first sign of a new ventral center formation (Fig. 13.29A-C). In addition to these intermediate types, there are joined twins with a common intestine and stomach, but with 2 mouths and esophagi, 2 oral fields, and 2 pairs of skeletons (Fig. 13.29D). In all these larvae, the second ventral side appears opposite and as far away as possible from the first. Fig. 13.30 is a diagram showing a configurative constitution of the operated 32-cell stages. If the presumptive ventral centers come close together, they may fuse by natural interactions

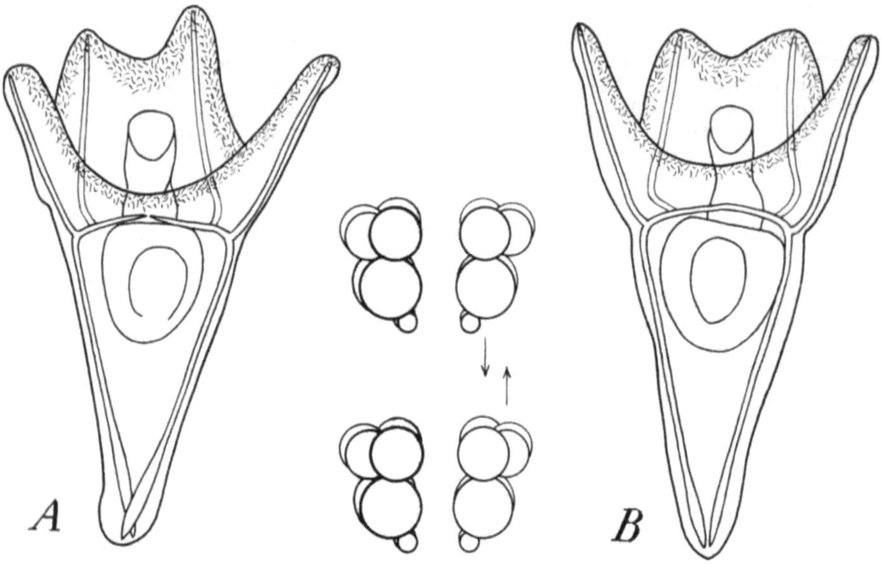

Fig. 13.27. A pair of typical plutei after exchange of meridional halves between 2 eggs. (HÖRSTADIUS, 1957)

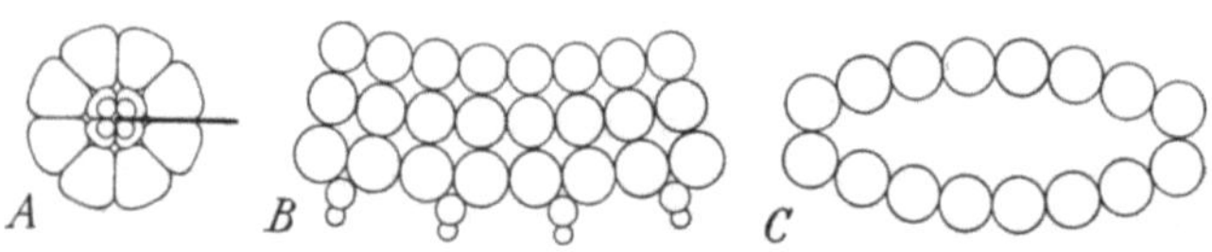

Fig. 13.28. Fusion of 2 32-cell stages to form a giant larva with a single animal-vegetal axis. A, the egg is placed on its animal pole and cut along a meridian from the vegetal to the animal pole (thick line). B, the 32-cell stage flattened to 1 layer. C, 1 flattened egg placed on top of another, as seen from their animal ends. (HÖRSTADIUS, 1957)

or one may dominate and suppress the other. If they are opposite, however, both may persist.

The determination of bilateral symmetry versus time has been studied by isolation at different stages (HÖRSTADIUS, 1936a). Isolated left and right halves of early cleavage stages 4 hours after fertilization yielded plutei only slightly deficient on the cut side (Fig. 13.31A). 10 hour blastulae halves developed only a small spicule and no arms on the cut side, and halves isolated at the beginning of gastrulation eventually looked like plutei cut in two (Fig. 13.31B, C). Of greater interest are the ventral and dorsal halves. Pairs isolated at 10 hours may develop like early isolated halves. The axis is reversed in the dorsal half and its ventral side is delayed (Fig. 13.32A_1, A_2). In other pairs separated 10-12 hours after fertilization, the struggle between 2 ventral centers in the dorsal half is apparent. 2 small spicules develop in a gastrula at the new ventral side opposite a larger spicule (Fig. 13.32A_2), while in another gastrula the new ventral side contains the larger of the pairs of spicules (Fig. 13.32B). In another, the presumptive dorsal side, which should have become the

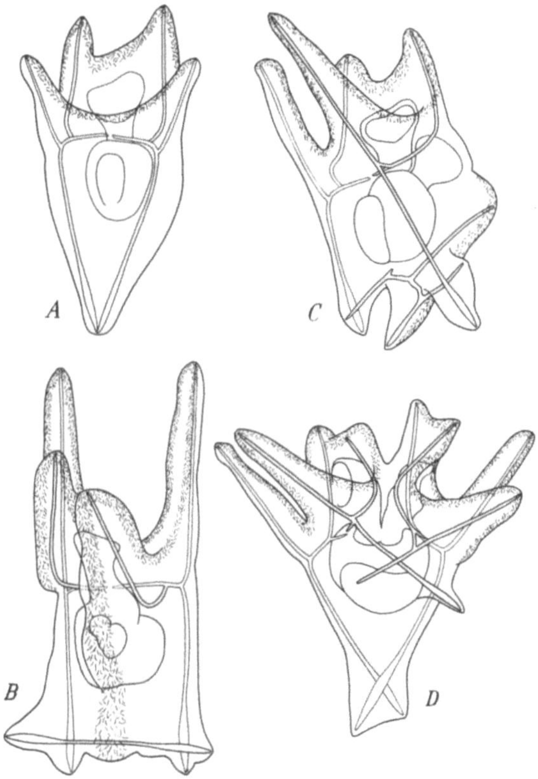

Fig. 13.29. A, giant pluteus. B, C, giant plutei with supernumerary dorsal rod and ciliated band and also with 2 esophagi. D, double larva with 2 oral fields, 2 pairs of skeleton, 2 esophagi, and 2 mouths. (HÖRSTADIUS, 1957)

new ventral side, has the largest skeleton by far, arranged as in a young pluteus, but the ectoderm has thickened on the cut side and has shown its dominance by forming the mouth (Fig. 13.32C, D). Frontally divided early gastrulae isolated 16 hours after fertilization develop as dorsal and ventral parts of plutei (HÖRSTADIUS, 1936a).

13.5 Interactions between Species

Micromeres of one species implanted into a blastocoel of another species result in a larva containing 2 sets of skeleton-forming cells. If the one species had plain skeletal rods, such as *Psammechinus* and the other fenestrated rods, such as *Echinocardium*, the result was all types of intermediate skeletons. It is not clear whether the mixture of different structures was due to a mosaic of mesenchymal cells or to a mixed syncytium. The role of the ectoderm in the growth of long rods was illustrated in chimeras, where an unpaired body-rod, as in the plutei of *Echinocardium*, appeared only with ectoderm of this species (VON UBISCH, 1934, 1939).

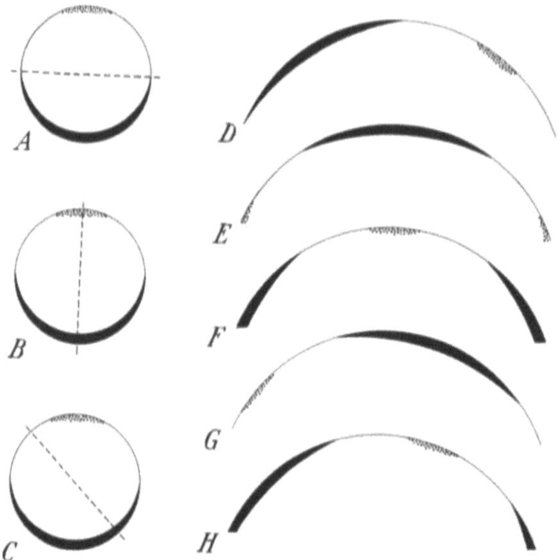

Fig. 13.30. Diagrams of eggs with a supposed dominating ventral center (black) decreasing in a dorsal direction. The least inhibited material (dotted area) will be found at the presumptive dorsal pole. A-C, constitution of dorsal and ventral, left and right, and oblique halves. D-H, constitution of flattened whole cleavage stages cut open along radii corresponding to dotted lines in A-C. (HÖRSTADIUS, 1957)

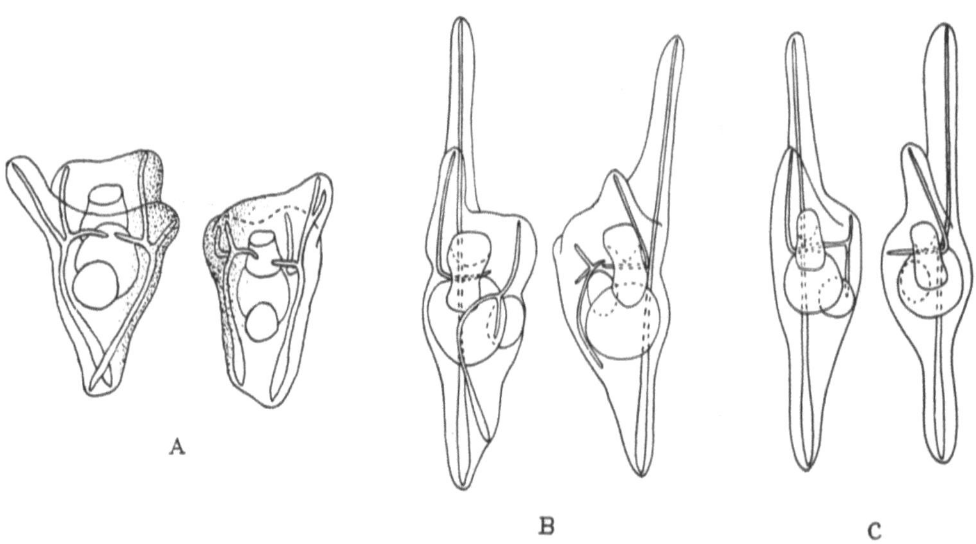

Fig. 13.31. Determination of the bilateral symmetry in left-right halves. A, a pair isolated at 16-cell stage, about 4 hrs after fertilization. B, isolation in blastula stage (10 hrs). C, isolation at the beginning of gastrulation (16 hrs). (HÖRSTADIUS and WOLSKY, 1936; HÖRSTADIUS, 1936a)

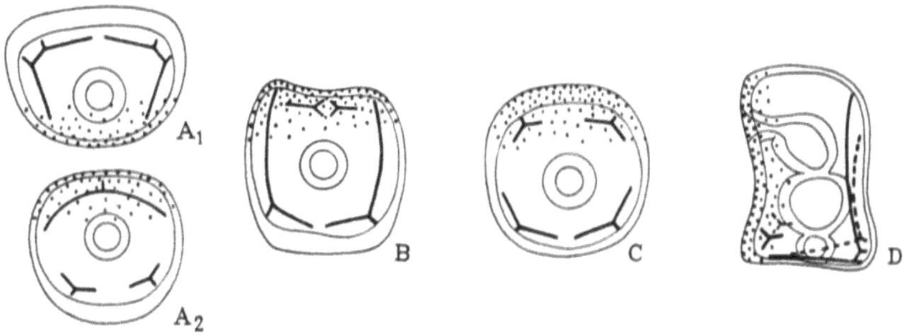

Fig. 13.32. Contrasting influences between the presumptive most ventral part (stained) and the presumptive dorsal side to form ventral side in dorsal halves. A_1 ventral, A_2 dorsal half of a pair isolated 10 hrs after fertilization. B-D, dorsal halves isolated 10 hrs (B) resp. 12 hrs (C, D) after fertilization. (HÖRSTADIUS, 1936a)

The experiments described above by VON UBISCH took place too late to allow the induction of an archenteron. When micromeres of one species were implanted into 8 + VEG_1 + 0, or 8 + 0 + 0 of another species, plutei could be obtained with the induced formation of a typical alimentary canal and a skeleton showing characters of the micromere species. The combination of presumptive ectoderm and micromeres from more closely related species occurred without difficulties as with *Psammechinus*, *Paracentrotus* and *Sphaerechinus*. However, in experiments with the regular *Sphaerechinus granularis* and the irregular *Echinocardium cordatum*, the micromeres were rejected. The margins of the open late cleavage stage could bend down towards the foreign micromeres, seize them and expel them. If the connection between the cells after treatment with Ca-free seawater was rather loose, the blastomeres made an opening between themselves large enough to let the micromeres fall out. These results had also been observed with the implantation into animal halves of a piece of vegetal cytoplasm devoid of nucleus (HÖRSTADIUS, 1928b, 1935, 1936c).

Although in many attempts to raise heterosperm merogones the larvae have died at an early age (BOVERI, 1889, 1914, 1918; TAYLOR and TENNENT, 1924; FRY, 1927; HARVEY, 1933), *Psammechinus microtuberculatus* cytoplasm and a *Paracentrotus lividus* sperm nucleus, as well as the opposite combination, resulted in plutei that swam and fed for 2 weeks and showed species characteristics (HÖRSTADIUS, 1936c).

Germ-layer chimeras were made from one species' normal larval presumptive ectoderm and micromeres from a heterosperm merogone with nuclei of the other species. An archenteron was induced and the larvae developed into plutei (Fig. 13.33). Although all that represented the one species were the nuclei of the skeleton-forming cells (Fig. 13.33A_4, B_1), body-rod clubs in chimeras having *Psammechinus* nuclei had the bent, thickened, irregular shape characteristic of that species. But in the opposite combination they were straight and slender (Fig. 13.33A_5, B_2).

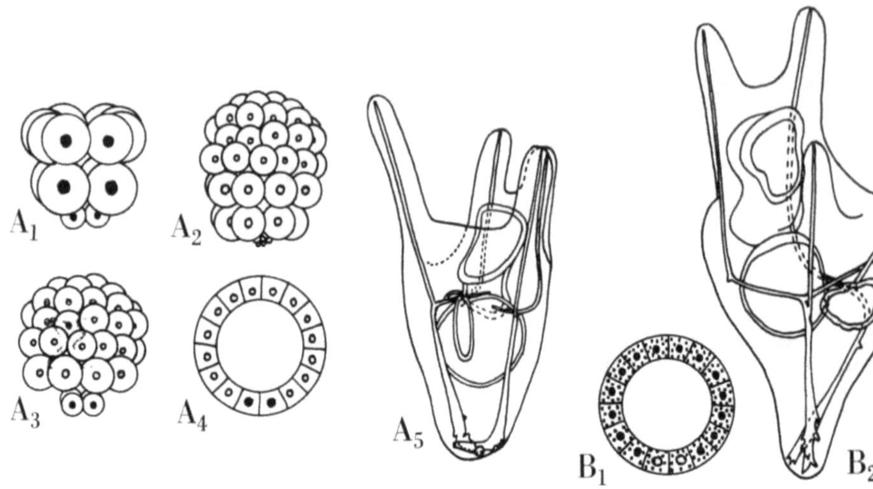

Fig. 13.33. A_1, diagram of a 16-cell stage of a heterosperm merogone *Paracentrotus lividus* cytoplasm (white) + *Psammechinus microtuberculatus* nucleus (black). A_2, a 64-cell stage of *Paracentrotus* (nuclei white). A_3, the presumptive ectoderm (8 + VEG_1) of *Paracentrotus* combined with the micromeres of the merogone. A_4, the blastula consists of *Paracentrotus* cytoplasm and nuclei, except for the nuclei of the micromeres. A_5, these nuclei give the pluteus the appearance of a *Psammechinus* larva with thickened, bent, irregular clubs of the body rods. B_1, the reciprocal combination. B_2, although the larva consists of *Psammechinus* cytoplasm (dotted in B_1) and nuclei, except for the nuclei of the skeleton-forming cells, the pluteus looks like a *Paracentrotus* larva with straight, slender body rods. (HÖRSTADIUS, 1936c)

13.6 Fragments Used for Physiological Tests

Bilateral symmetry in whole-egg larvae can be modified towards radial symmetry. Slighter shifts in the animal or vegetal direction, such as an archenteron of somewhat abnormal volume, are harder to detect. The animal-vegetal balance in whole eggs is too stable to be markedly affected by all active substances, in contrast to animal and vegetal fragments, which are much more sensitive (Fig. 13.34, 13.35).

Animal halves treated with lithium can yield gastrulae and exogastrulae (VON UBISCH, 1925c, 1925d, 1929). Sodium pyruvate, pyocyanin, and glutamine further animalize animal halves, while carbon monoxide animalizes halves and vegetalizes vegetal halves (HÖRSTADIUS and STRÖMBERG, 1940). Sulphate-ion absence has the same double effect on animal and vegetal halves (RUNNSTRÖM et al., 1964). Propanediol phosphate, phosphogluconic acid, and sodium lactate tend to cause animalization (HÖRSTADIUS and GUSTAFSON, 1947). The effects of many antimetabolites and amino acids have been tested (GUSTAFSON and HÖRSTADIUS, 1955, 1956, 1957; HÖRSTADIUS and GUSTAFSON, 1954; MARKMAN, 1963). Trypsin and ficin were found to animalize while dinitrophenol and chloramphenicol act strongly vegetalizing (HÖRSTADIUS, 1953a, 1963, 1965b). The effect of actinomycin on halves was reported by MARKMAN (1963) and MARKMAN and RUNNSTRÖM (1963), and on the potentialities of animal halves in combination with micromeres by GIUDICE and

HÖRSTADIUS (1965). MARKMAN has, in several papers (1961a, 1961b, 1967), used isotopic labelling of nucleic acids and proteins to study regional differences in early development. Incorporation of labelled amino acids and of labelled SO_4 have also been performed (RUNNSTRÖM et al., 1964; BERG, 1965; HÖRSTADIUS et al., 1966). Controlled and released respiration was measured by DE VINCENTIIS et al. (1966).

Implantation of vegetal cytoplasm in animal halves did not have any effect on the differentiation (HÖRSTADIUS, 1935). In whole embryos and half-embryos raised individually in micro-depression slides in the presence of large numbers of segregated micromeres or macromere-micromere mixtures, no animalization or vegetalization was observed (BERG and CHENG, 1962).

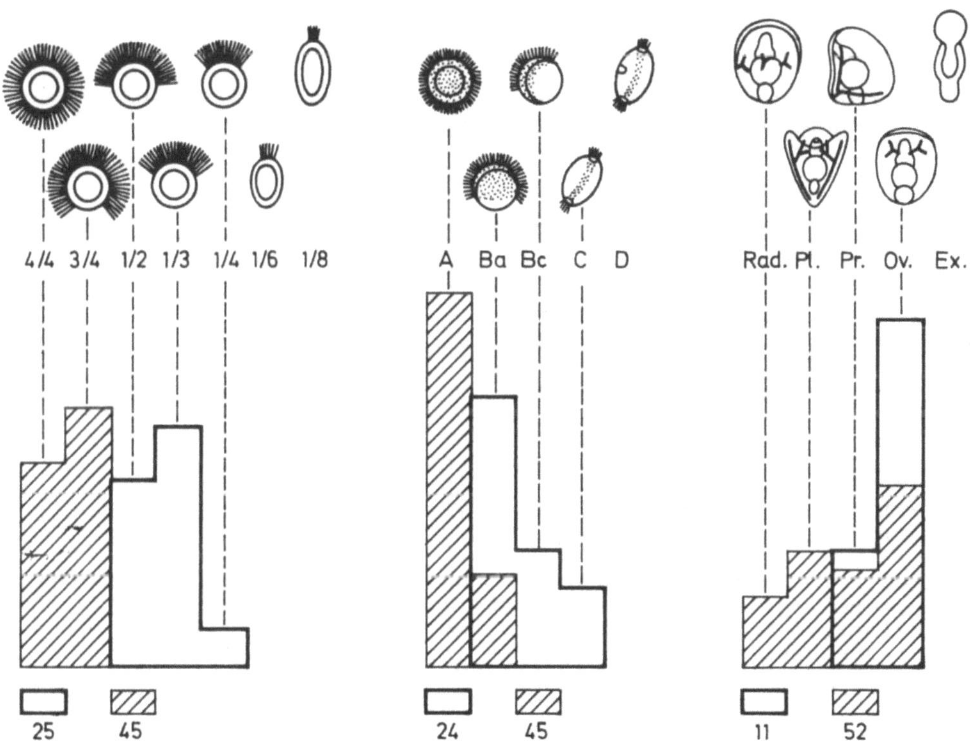

Fig. 13.34. Animalizing activity of a substance isolated from lyophilized unfertilized eggs of *Paracentrotus lividus*. In the block diagrams the columns indicate the percentage distribution of different larval types. The numerals below each diagram state the total number of control halves (heavy lines) and treated halves (hatched columns), concentrations 0.001-0.0005%. The diagram to the left gives the frequency of the types of animal halves (4/4 to 1/8) on the first day after application; the diagram in the middle, of the types of animal halves (A to D) on the second day or later; and the diagram to the right, the frequency of the types of vegetal halves (Radialized (Rad.) to Exogastrulae (Ex.)). As to the types, compare with Fig. 13.7. Types more animal than those found in untreated halves are 4/4 (the whole blastula covered with stereocilia), and Rad. (in early stage animal tuft, later radial symmetry). (JOSEFSSON and HÖRSTADIUS, 1969)

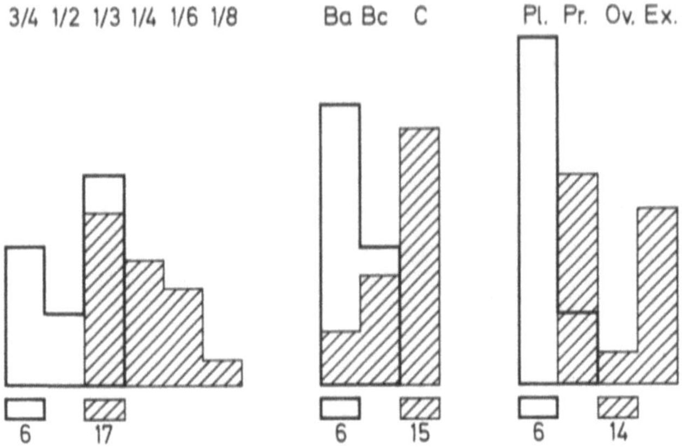

Fig. 13.35. Vegetalizing effect of a substance isolated from unfertilized eggs, added in concentrations from 0.01 to 0.0025%. (JOSEFSSON and HÖRSTADIUS, 1969)

Morphogenetic substances from sea urchin eggs have been studied by separating them on Dowex or Sephadex and testing the embryological effects of the fractions. These effects are seldom apparent in whole eggs but are often obvious in halves (Fig. 13.34, 13.35). The animalization caused by the fractions may result in halves more animal than those raised in normal seawater (HÖRSTADIUS et al., 1967; JOSEFSSON and HÖRSTADIUS, 1969; HÖRSTADIUS and JOSEFSSON, 1972).

Actinomycin D has been shown to remove the animalizing effect trypsin has on vegetal halves and to counteract the vegetalizing effect of lithium on animal halves. This suggests that trypsin and lithium affect the gene transcription that determines animal and vegetal differentiation (DE ANGELIS and RUNNSTRÖM, 1970; MARKMAN and RUNNSTRÖM, 1970).

In unlabelled animal halves in which micromeres, that have incorporated ^{14}C-uridine, are implanted, labelled material leaves the micromeres and becomes located in the cytoplasm of the other cells. Micromeres are known to synthesize RNA which may be involved in the inductive action of micromeres (AGRELL, 1958; CZIHAK and HÖRSTADIUS, 1970). Other studies show no marked difference of protein synthesis rate among isolated micromeres, macromeres, and mesomeres (SPIEGEL and TYLER, 1966).

References

AGRELL, I., 1958. A cytoplasmic production of ribonucleic acid during the cell cycle of the micromeres in the sea urchin embryo. Arkiv Zool. Ser. 2, 11, 435-438.

BERG, W.E., 1965. Rates of protein synthesis in whole and half embryos of the sea urchin. Exp. Cell Res. 40, 469-489.

BERG, W.E., CHENG, A.C., 1962. Tests for diffusible morphogenic substances in the sea urchin embryo. Acta Embryol. Morphol. Exp. 5, 167-171.

BERG, W.E., TAYLOR, D.A., HUMPHREYS, W.J., 1962. Distribution of mitochondria in echinoderm embryos as determined by electron microscopy. Develop. Biol. 4, 165-176.
BOVERI, Th., 1889. Ein geschlechtlich erzeugter Organismus ohne mütterliche Eigenschaften. S.B. Ges. Morph. Physiol. München 5, 73-80.
BOVERI, Th., 1901a. Die Polarität von Ovocyte, Ei und Larve des Strongylocentrotus lividus. Zool. Jb. Abt. Anat. Ontog. 14, 630-653.
BOVERI, Th., 1901b. Über die Polarität des Seeigeleies. Verh. phys.-med. Ges. Würzburg (N.F.) 34, 145-175.
BOVERI, Th., 1902. Über mehrpolige Mitosen als Mittel zur Analyse des Zellkerns. Verh. phys.-med. Ges. Würzburg 35, 67-88.
BOVERI, Th., 1905. Zellenstudien V. Jena Z. Naturwiss. 39, 445.
BOVERI, Th., 1908. Zellenstudien VI. Die Entwicklung dispermer Seeigeleier. Ein Beitrag zur Befruchtungslehre und zur Theorie des Kerns. Jena Z. Naturwiss. 43, 1.
BOVERI, Th., 1910. Die Potenzen der Ascaris-Blastomeren bei abgeänderter Furchung. Festschr. Richard Hertwig 3, 131-214.
BOVERI, Th., 1914. Über die Charaktere von Echiniden-Bastardlarven bei verschiedenem Mengenverhältnis mütterlicher und väterlicher Substanzen. Verh. phys.-med. Ges. Würzburg (N.F.) 43, 117.
BOVERI, Th., 1918. Zwei Fehlerquellen bei Merogonieversuchen und die Entwicklungsfähigkeit merogonischer und partiell-merogonischer Seeigelbastarde. Arch. f. Entw. Mech. 44, 417-471.
CHABRY, L., 1887. Contribution à l'émbryologie normale et tératologique des ascidies simples. J. Anat. Phys. 23, 167-319.
CHILD, C.M., 1916a. Experimental control and modification of larval development in the sea urchin in relation to axial gradients. J. Morphol. 28, 65-133.
CHILD, C.M., 1916b. Axial susceptibility gradients in the early development of the sea urchin. Biol. Bull. (Woods Hole) 30, 391-405.
CHILD, C.M., 1936. Differential reduction of vital dyes in the early development of echinoderms. Wilhelm Roux' Arch. Entwicklungsmech. Organismen 135, 426-456.
CZIHAK, G., HÖRSTADIUS, S., 1970. Transplantation of RNA-labeled micromeres into animal halves of sea urchin embryos. A contribution to the problem of embryonic induction. Develop. Biol. 22, 15-30.
DE ANGELIS, E., RUNNSTRÖM, J., 1970. The effect of temporary treatment of animal half embryos with lithium and the modification of this effect by simultaneous exposure to Actinomycin D. Wilhelm Roux' Arch. Entwicklungsmech. Organismen 164, 236-246.
DE VINCENTIIS, M., HÖRSTADIUS, S., RUNNSTRÖM, J., 1966. Studies on controlled and released respiration in animal and vegetal halves of the embryo of the sea urchin, Paracentrotus lividus. Exp. Cell Res. 41, 535-544.
DRIESCH, H., 1891. Entwicklungsmechanische Studien. I. Der Wert der beiden ersten Furchungszellen in der Echinodermenentwicklung. Experimentelle Erzeugung von Teil- und Doppelbildungen. Z. wiss. Zool. 53, 160-184.
DRIESCH, H., 1892. Entwicklungsmechanische Studien III-VI. Z. wiss. Zool. 55, 1-62.
DRIESCH, H., 1893a. Entwicklungsmechanische Studien VII-X. Mitth. Zool. Stat. Neapel 11, 221-253.
DRIESCH, H., 1893b. Zur Verlagerung der Blastomeren des Seeigeleies. Anat. Anz. 8, 348-357.
DRIESCH, H., 1896a. Die taktische Reizbarkeit der Mesenchymzellen von Echinus microtuberculatus. Arch. f. Entw. Mech. 3, 362-380.
DRIESCH, H., 1896b. Betrachtungen über die Organisation des Eies und ihre Genese. Arch. f. Entw. Mech. 4, 75-125.
DRIESCH, H., 1900a. Die isolierten Blastomeren des Echinidenkeimes. Arch. f. Entw. Mech. 10, 361-410.

DRIESCH, H., 1900b. Studien über das Regulationsvermögen der Organismen. IV. Arch. f. Entw. Mech. 10, 411-434.
DRIESCH, H., 1901. Die organischen Regulationen. Leipzig.
DRIESCH, H., 1902. Neue Ergänzungen zur Entwicklungsphysiologie des Echinidenkeimes. Arch. f. Entw. Mech. 14, 500-531.
DRIESCH, H., 1906. Studien zur Entwicklungsphysiologie der Bilateralität. Arch. f. Entw. Mech. 21, 756-791.
DRIESCH, H., 1908. Zur Theorie der organischen Symmetrie. Arch. f. Entw. Mech. 26, 130-145.
FIEDLER, K., 1891. Entwicklungsmechanische Studien an Echinodermeneiern. Festschr. Nägeli u. Kölliker, Zürich, 189-196.
FOERSTER, M., ÖRSTRÖM, Å., 1933. Observations sur la prédétermination de la partie ventrale dans l'oeuf d'oursin. Trav. Stat. biol. Roscoff, 11, 63-83.
FRY, H.J., 1927. The cross-fertilization of enucleated Echinarachnius eggs by Arbacia sperm. Biol. Bull. (Woods Hole) 53, 173.
GARBOWSKI, M.T., 1905. Über die Polarität des Seeigeleies. Bull. int. Acad. Sci. Cracovie, 599-635.
GIUDICE, G., HÖRSTADIUS, S., 1965. Effect of actinomycin D on the segregation of animal and vegetal potentialities in the sea urchin egg. Exp. Cell Res. 39, 117-120.
GUSTAFSON, Tr., 1952. Studies on the determination of the oral side of the sea urchin egg. II. The dorso-ventral structure of the unfertilized egg. Arkiv Zool. Ser. 2,3 Nr. 19, 273-282.
GUSTAFSON, Tr., 1969. Fertilization and development. Chem. Zool. 3, 149-206.
GUSTAFSON, Tr., HÖRSTADIUS, S., 1955. Vegetalization and animalization in the sea urchin egg induced by antimetabolites. Exp. Cell Res. Suppl. 3, 170-180.
GUSTAFSON, Tr., HÖRSTADIUS, S., 1956. 2-Thio-5-methyl-cytosine, an animalizing agent. Zool. Anz. 156, 102-106.
GUSTAFSON, Tr., HÖRSTADIUS, S., 1957. Changes in determination of the sea urchin egg induced by amino acids. Pubbl. Sta. Zool. Napoli 29, 407-424.
HAGSTRÖM, B., LÖNNING, S., 1965. Studies of cleavage and development of isolated sea urchin blastomeres. Sarsia 18, 1-9.
HAGSTRÖM, B., LÖNNING, S., 1969. Time-lapse and electron microscopic studies of sea urchin micromeres. Protoplasma 68, 271-288.
HARVEY, E.B., 1933. Development of the parts of sea-urchin eggs separated by centrifugal force. Biol. Bull. (Woods Hole) 64, 125-148.
HARVEY, E.B., 1956. The American Arbacia and other sea urchins. Princeton Univ. Press, 1-298.
HEFFNER, B., 1908. Über experimentell erzeugte Mehrfachbildungen des Skeletts bei Echinodermenlarven. Arch. f. Entw. Mech. 26, 1-46.
HERBST, C., 1892. Experimentelle Untersuchungen über den Einfluß der veränderten chemischen Zusammensetzung des umgebenden Mediums auf die Entwicklung der Tiere. I. Z. wiss. Zool. 55, 446-518.
HERBST, C., 1893. Experimentelle Untersuchungen über den Einfluß der veränderten chemischen Zusammensetzung des umgebenden Mediums auf die Entwicklung der Tiere. II. Weiteres über die Wirkung der Lithiumsalze und ihre theoretische Bedeutung. Mitth. Zool. Stat. Neapel 11, 136-220.
HERBST, C., 1896. D:o III-VI. Arch. f. Entw. Mech. 2, 455-516.
HERBST, C., 1897. Über die zur Entwicklung der Seeigellarven notwendigen anorganischen Stoffe, ihre Rolle und ihre Vertretbarkeit. I. Arch. f. Entw. Mech. 5, 649-793.
HERBST, C., 1900. Über das Auseinandergehen von Furchungs- und Gewebezellen in kalkfreiem Medium. Arch. f. Entw. Mech. 9, 424-463.

HERBST, C., 1904. Über die zur Entwicklung der Seeigellarven notwendigen anorganischen Stoffe, ihre Rolle und ihre Vertretbarkeit. III. Arch. f. Entw. Mech. 17, 306.
HERBST, C., 1907. Vererbungsstudien. V. Arch. f. Entw. Mech. 24, 185-238.
HÖRSTADIUS, S., 1927. Studien über die Determination bei Paracentrotus lividus Lk. Wilhelm Roux' Arch. Entwicklungsmech. Organismen 112, 239-246.
HÖRSTADIUS, S., 1928a. Transplantationsversuche am Keim von Paracentrotus lividus Lk. Wilhelm Roux' Arch. Entwicklungsmech. Organismen 113, 312-322.
HÖRSTADIUS, S., 1928b. Über die Determination des Keimes bei Echinodermen. Acta Zool. (Stockholm) 9, 1-191.
HÖRSTADIUS, S., 1931. Über die Potenzverteilung im Verlaufe der Eiachse. Arkiv Zool. 23, 1-6.
HÖRSTADIUS, S., 1935. Über die Determination im Verlaufe der Eiachse bei Seeigeln. Pubbl. Sta. Zool. Napoli 14, 251-479.
HÖRSTADIUS, S., 1936a. Über die zeitliche Determination im Keim von Paracentrotus lividus Lk. Wilhelm Roux' Arch. Entwicklungsmech. Organismen 135, 1-39.
HÖRSTADIUS, S., 1936b. Weitere Studien über die Determination im Verlaufe der Eiachse bei Seeigeln. Wilhelm Roux' Arch. Entwicklungsmech. Organismen 135, 40-68.
HÖRSTADIUS, S., 1936c. Studien über heterosperme Seeigelmerogone nebst Bemerkungen über einige Keimblattchimären. Mém. Mus. Hist. nat. Belg. 2me sér. fasc. 3, 801-880.
HÖRSTADIUS, S., 1938. Schnürungsversuche an Seeigelkeimen. Wilhelm Roux' Arch. Entwicklungsmech. Organismen 138, 197-258.
HÖRSTADIUS, S., 1939. The mechanics of sea urchin development, studied by operative methods. Biol. Rev. 14, 132-179.
HÖRSTADIUS, S., 1950. Transplantation experiments to elucidate interactions and regulations within the gradient system of the developing sea urchin egg. J. Exp. Zool. 113, 245-276.
HÖRSTADIUS, S., 1952. Induction and inhibition of reduction gradients by the micromeres in the sea urchin egg. J. Exp. Zool. 120, 421-436.
HÖRSTADIUS, S., 1953a. Vegetalization of the sea urchin egg by dinitrophenol and animalization by trypsin and ficin. J. Embryol. Exp. Morphol. 1, 327-348.
HÖRSTADIUS, S., 1953b. The effect of lithium ions on centrifuged eggs of Paracentrotus lividus. Pubbl. Sta. Zool. Napoli 24, 45-60.
HÖRSTADIUS, S., 1955. Reduction gradients in animalized and vegetalized sea urchin eggs. J. Exp. Zool. 129, 249-256.
HÖRSTADIUS, S., 1957. On the regulation of bilateral symmetry in plutei with exchanged meridional halves and in giant plutei. J. Embryol. Exp. Morphol. 5, 60-73.
HÖRSTADIUS, S., 1963. Vegetalization of sea urchin larvae by chloramphenicol. Develop. Biol. 7, 144-151.
HÖRSTADIUS, S., 1965a. Über die fortschreitende Determination in Furchungsstadien von Seeigeleiern. Z. Naturforsch. 20b, 331-333.
HÖRSTADIUS, S., 1965b. Über die animalisierende Wirkung von Trypsin auf Seeigelkeime. Zool. Jb. Abt. Physiol. 71, 241-244.
HÖRSTADIUS, S., 1973. Experimental Embryology of Echinoderms. Oxford: Clarendon Press, 1-192.
HÖRSTADIUS, S., GUSTAFSON, Tr., 1947. Change of determination in the sea urchin egg through the action of propanediol phosphate, phosphogluconic acid, and lactate. Zool. Bidr. Uppsala 25, 571-581.
HÖRSTADIUS, S., GUSTAFSON, Tr., 1954. The effect of 3 antimetabolites on sea urchin development. J. Embryol. Exp. Morphol. 2, 216-226.
HÖRSTADIUS, S., IMMERS, J., RUNNSTRÖM, J., 1966. The incorporation of $^{35}SO_4$ in whole embryos and meridional, animal and vegetal halves of the sea urchin Paracentrotus lividus. Exp. Cell Res. 43, 444-450.

HÖRSTADIUS, S., JOSEFSSON, L., 1972. Morphogenetic substances from sea urchin eggs. Isolation of animalizing substances from developing eggs of Paracentrotus lividus. Acta Embryol. Exp. 1972, 7-23.
HÖRSTADIUS, S., JOSEFSSON, L., RUNNSTRÖM, J., 1967. Morphogenetic agents from unfertilized eggs of the sea urchin Paracentrotus lividus. Develop. Biol. 16, 189-202.
HÖRSTADIUS, S., STRÖMBERG, St., 1940. Untersuchungen über Umdeterminierung von Fragmenten des Seeigeleies durch chemische Agentien. Wilhelm Roux' Arch. Entwicklungsmech. Organismen 140, 409-462.
HÖRSTADIUS, S., WOLSKY, A., 1936. Studien über die Determination der Bilateralsymmetrie des jungen Seeigelkeimes. Wilhelm Roux' Arch. Entwicklungsmech. Organismen 135, 69-113.
JENKINSON, J.W., 1911a. On the development of isolated pieces of the gastrulae of the sea urchin Strongylocentrotus lividus. Arch. f. Entw. Mech. 32, 269.
JENKINSON, J.W., 1911b. On the origin of the polar and bilateral structure of the egg of the sea urchin. Arch. f. Entw. Mech. 32, 699-716.
JOSEFSSON, L., HÖRSTADIUS, S., 1969. Morphogenetic substances from sea urchin eggs. Isolation of animalizing and vegetalizing substances from unfertilized eggs of Paracentrotus lividus. Develop. Biol. 20, 481-500.
KÜHN, A., 1955. Vorlesungen über Entwicklungsphysiologie. Berlin, Göttingen, Heidelberg, 1-506.
LENICQUE, P., HÖRSTADIUS, S., GUSTAFSON, Tr., 1953. Change of distribution of mitochondria in animal halves of sea urchin eggs by the action of micromeres. Exp. Cell Res. 5, 400-403.
LINDAHL, P.E., 1932a. Zur Kenntnis des Ovarialeies bei dem Seeigel. Wilhelm Roux' Arch. Entwicklungsmech. Organismen 126, 373-390.
LINDAHL, P.E., 1932b. Zur experimentellen Analyse der Determination der Dorsoventralachse beim Seeigelkeim. I. Versuche mit gestreckten Eiern. Wilhelm Roux' Arch. Entwicklungsmech. Organismen 127, 300-302.
LINDAHL, P.E., 1933. Über "animalisierte" und "vegetativisierte" Seeigellarven. Wilhelm Roux' Arch. Entwicklungsmech. Organismen 128, 661-664.
LINDAHL, P.E., 1936. Zur Kenntnis der physiologischen Grundlagen der Determination im Seeigelkeim. Acta Zool. (Stockholm) 17, 179-365.
LINDAHL, P.E., LUNDIN, J., 1948. Removal of the fertilization membranes from large quantities of sea urchin eggs. Science 108, 481-482.
LOEB, J., 1899. On the nature of the process of fertilization and the artificial production of normal larvae (plutei) from the unfertilized eggs of the sea urchin. Amer. J. Physiol. 3, 135-138.
LYON, E.P., 1906a. Some results of centrifugalizing the eggs of Arbacia. Amer. J. Physiol. 15, 21.
LYON, E.P., 1906b. Results of centrifugalizing eggs. Arch. f. Entw. Mech. 23, 151.
MARKMAN, B., 1958. Studies on the formation of the fertilization membrane in sea urchins. Acta Zool. (Stockholm) 39, 103-115.
MARKMAN, B., 1961a. Regional differences in isotopic labelling of nucleic acid and protein in early sea urchin development. Exp. Cell Res. 23, 118-129.
MARKMAN, B., 1961b. Differences in isotopic labelling of nucleic acid and protein in early sea urchin development. Exp. Cell Res. 23, 197-200.
MARKMAN, B., 1963. Morphogenetic effects of some nucleotide metabolites and antibiotics on early sea urchin development. Arkiv Zool. Ser. 2,16 Nr. 12, 207-217.
MARKMAN, B., 1967. Isotopic labelling of nucleic acids in sea urchin embryos developing from animal and vegetal halves in relation to protein and nucleic acid content. Exp. Cell Res. 46, 1-18.
MARKMAN, B., RUNNSTRÖM, J., 1963. Animal and vegetal halves of sea

urchin larvae subjected to temporary treatment with actinomycin C and mitomycin C. Exp. Cell Res. 31, 615-618.
MARKMAN, B., RUNNSTRÖM, J., 1970. The removal by actinomycin D of the effect of endogenous or exogenous animalizing agents in sea urchin development. Wilhelm Roux' Arch. Entwicklungsmech. Organismen 165, 1-7.
MORGAN, Th.H., 1894. Experimental studies on echinoderm eggs. Anat. Anz. 9, 141-152.
MORGAN, Th.H., LYON, E.P., 1907. The relation of the substances of the egg, separated by a strong centrifugal force, to the location of the embryo. Arch. f. Entw. Mech. 24, 147.
MORGAN, Th.H., SPOONER, G.B., 1909. The polarity of the centrifuged egg. Arch. f. Entw. Mech. 28, 104.
OKAZAKI, K., 1971. Spicule formation in sea urchin larvae; observations in vivo and in vitro. Symp. Cell Biol. (Japan) 22, 163-171.
PEASE, D.C., 1941. Echinoderm bilateral determination in chemical concentration gradients. I. J. exp. Zool. 86, 381-404.
PEASE, D.C., 1942a. Idem II. J. exp. Zool. 89, 329-345.
PEASE, D.C., 1942b. Idem III. J. exp. Zool. 89, 347-356.
PLOUGH, H.H., 1927. Defective pluteus larvae from isolated blastomeres of Arbacia and Echinarachnius. Biol. Bull. (Woods Hole) 52, 373-393.
ROUX, W., 1888. Über die künstliche Hervorbringung halber Embryonen durch Zerstörung einer der beiden ersten Furchungskugeln, sowie über die Nachentwicklung (Postgeneration) der fehlenden Körperhälfte. Virchows Arch. Anat. Phys. 114, 113 und 246.
RUNNSTRÖM, J., 1914. Analytische Studien über die Seeigelentwicklung. I. Arch. f. Entw. Mech. 40, 526-564.
RUNNSTRÖM, J., 1917. Analytische Studien über die Seeigelentwicklung. III. Arch. f. Entw. Mech. 43, 223-328.
RUNNSTRÖM, J., 1925. Experimentelle Bestimmung der Dorso-Ventralachse bei dem Seeigelkeim. Arkiv Zool. 18A, Nr. 4, 1-6.
RUNNSTRÖM, J., 1926. Über die Verteilung der Potenzen der Urdarmbildung bei dem Seeigelkeim. Acta Zool. (Stockholm) 7, 117-121.
RUNNSTRÖM, J., 1928a. Plasmabau und Determination bei dem Ei von Paracentrotus lividus Lk. Wilhelm Roux' Arch. Entwicklungsmech. Organismen 113, 556-581.
RUNNSTRÖM, J., 1928b. Zur experimentellen Analyse der Wirkung des Lithiums auf den Seeigelkeim. Acta Zool. (Stockholm) 9, 365-424.
RUNNSTRÖM, J., 1929. Über Selbstdifferenzierung und Induktion bei dem Seeigelkeim. Wilhelm Roux' Arch. Entwicklungsmech. Organismen 117, 112-145.
RUNNSTRÖM, J., HÖRSTADIUS, S., IMMERS, J., FUDGE-MASTRANGELO, M., 1964. An analysis of the role of sulfate in the embryonic differentiation of the sea urchin (Paracentrotus lividus). Revue Suisse Zool. 71, 21-54.
RUNNSTRÖM, J., KRISZAT, G., 1952. Animalizing action of iodosobensoic acid in the sea urchin development. Exp. Cell Res. 3, 497-499.
SELENKA, E., 1878. Zoologische Studien. I. Befruchtung des Eies von Toxopneustes variegatus. Leipzig.
SELENKA, E., 1883. Die Keimblätter der Echinodermen. Studien über die Entwicklungsgeschichte der Tiere. 1:2. Wiesbaden.
SPEMANN, H., 1906. Über eine neue Methode der embryonalen Transplantation. Verh. Deutsch. Zool. Ges., 195-202.
SPIEGEL, M., TYLER, A., 1966. Protein synthesis in micromeres of the sea urchin egg. Science 151, 1233-1234.
TAYLOR, C.V., TENNENT, D.H., 1924. Preliminary report on the development of egg fragments. Yb. Carnegie Inst. Wash. 23, 201-206.
TERNI, T., 1914. Studio sulle larve atipiche (blastulae permanenti) degli Echinoidi. Analisi della limitata equipotentialità dell'uovo di Echinoide. Mitth. Zool. Stat. Neapel 22, 59-97.

THEEL, H., 1892. On the development of Echinocyamus pusillus (O.F. Müller). Nova Acta R. Soc. Scient. upsal. Ser. III, 15 (6), 1-57.
VOGT, W., 1925. Gestaltungsanalyse am Amphibienkeim mit örtlicher Vitalfärbung. Wilhelm Roux' Arch. Entwicklungsmech. Organismen 106, 542-610.
VON BAER, K.E., 1828. Über Entwicklungsgeschichte der Tiere, Beobachtung und Reflexion. Königsberg.
VON UBISCH, L., 1925a. Entwicklungsphysiologische Studien an Seeigelkeimen. I. Über die Beziehung der ersten Furchungsebene zur Larvensymmetrie, und die prospektive Bedeutung der Eibezirke. Z. wiss. Zool. 124, 361-381.
VON UBISCH, L., 1925b. D:o II. Die Entstehung von Einheitslarven aus verschmolzenen Keimen. Z. wiss. Zool. 124, 457-468.
VON UBISCH, L., 1925c. D:o III. Die normale und durch Lithium beeinflusste Anlage der Primitivorgane bei animalen und vegetativen Halbkeimen von Echinocyamus pusillus. Z. wiss. Zool. 124, 469-486.
VON UBISCH, L., 1925d. Über die Entodermisierung ektodermaler Bezirke des Echinoiden-Keimes und die Reversion dieses Vorganges. Verh. phys.-med. Ges. Würzburg 50, 13-19.
VON UBISCH, L., 1929. Über die Determination der larvalen Organe und der Imaginalanlage bei Seeigeln. Wilhelm Roux' Arch. Entwicklungsmech. Organismen 117, 80-122.
VON UBISCH, L., 1933. Untersuchungen über Formbildung. IV. Wilhelm Roux' Arch. Entwicklungsmech. Organismen 129, 45-67.
VON UBISCH, L., 1934. Untersuchungen über Formbildung. VI. Wilhelm Roux' Arch. Entwicklungsmech. Organismen 131, 95-112.
VON UBISCH, L., 1939. Keimblattchimärenforschung an Seeigellarven. Biol. Rev. 14, 88-103.
WILSON, E.B., MATTHEWS, A.P., 1895. Maturation, fertilization and polarity in the echinoderm egg. J. Morphol. 10, 319-342.
ZOJA, R., 1895. Sullo sviluppo dei blastomeri isolati dalle uova di alcune meduse (e di altri organismi). II. Arch. f. Entw. Mech. 2, 1-37.

14. Blastomere Reaggregation

G. Millonig

The first discovery that sea urchin embryos can be dissociated into single cells dates back to experiments, by which the calcium ion was found to be responsible for the adhesion between blastomeres (HERBST, 1900). In *Psammechinus microtuberculatus* and *Sphaerechinus granularis* it was shown that isolated embryonic cells reaggregate when returned to normal conditions. Fertilized *Psammechinus* eggs were transferred to artificial sea water without calcium (3% NaCl, 0.08% KCl, 0.66% $MgSO_4$, and small amounts of Li_2HPO_4 or $MgHPO_4$ to alkalize the solution) and observed in vivo with the light microscope. Blastomeres divide regularly in this solution but round off and remain isolated from each other within the fertilization membrane. At the end of the cleavage stage the superficial cells do not form an epithelium but they are capable of differentiating cilia. Motility is observed up to 23 hours, and there is no indication of gastrulation.

Similar experiments were performed also on embryos which had the fertilization membrane removed by shaking. Up to about the 100-cell stage, embryos cleave almost regularly in the solution mentioned above but blastomeres are spherical and only loosely connected. Differentiation seems to continue because many cells form cilia and are motile while cells without cilia migrate by amoeboid movement.

Blastomeres survive for 48 hours when kept in a large beaker. Results from experiments with *Sphaerechinus granularis* are similar.

The same calcium-free medium was also tested on normal membrane-free 2-, 4-, 8-, and 32-cell stages. Blastomeres lose reciprocal contact in less than 25 minutes after transfer to the medium; *Psammechinus* blastulae fall apart within 3 minutes and plutei within 1 hour. Blastomeres of the 2-, 4-, or 16-cell stage transferred from the medium to normal sea water produce larvae one-half, one-fourth, or one-sixteenth the normal size while blastomeres of the more advanced stages give rise only to clumps of cells. It was not determined whether these small larvae form an intestine or skeletons.

In embryos grown within the fertilization membrane in the calcium-free medium, cells lose their intercellular connections but do not fall apart. Transferred to normal sea water at the 2-, 4-, and 16-cell stages they develop into normal larvae, but when transferred at the 32- and 100-cell stages, they develop abnormally.

When normal early blastulae are transferred before mesenchyme formation and before hatching to the calcium-free medium, the blastomeres gradually dissociate and after a very few minutes almost completely lose contact with each other. When these embryos are transferred back to sea water, the blastulae are re-formed within 12 minutes and develop normally. When embryos are treated with calcium-free sea water for more than 7 hours, a large number of cells completely lose contact with each other, and the size of the embryos re-formed is considerably reduced because comparatively few cells re-establish contact. These larvae recover, but the resulting embryos are to some degree abnormal, having only rudimentary skeletons and intestines. The superficial cells which completely lose contact with the embryo are not reincorporated.

From these experiments HERBST concluded that the absence of calcium can separate embryonic cells without damage and that embryonic cell contact is not a permanent link.

Histological aspects, as well as genetic and biochemical control mechanisms of differentiation and morphogenesis during early embryonic development were studied in reaggregating systems by GIUDICE (1962a). Blastulae, gastrulae, and plutei of *Arbacia lixula* and *Paracentrotus lividus* were suspended first in calcium- and magnesium-free sea water[1] and then in 0.04 *M* sucrose and 0.001 *M* EDTA and 0.5 *M* citrate buffer pH 7.8. The embryos were gently dissociated manually in a glass homogenizer and the isolated cells were filtered through a layer of gauze, diluted 1:1 with calcium- and magnesium-free sea water, and centrifuged at 1,000 *g* for 12 minutes. The supernatant was then discarded. The separated blastomeres were usually spherical and showed amoeboid movement. Some cells still had a beating cilium. Cell size depended on embryonic stage.

To observe the mode of reaggregation in vivo, cells were transferred to Syracuse dishes filled with sea water where they immediately aggregated into small clumps which continued to grow larger through aggregation of other blastomeres. After 3 to 4 hours some of them were as large as normal embryos and after 12 hours of culture, the clusters became spherical and rotated by means of uniformly distributed, superficial cilia.

In aggregates from dissociated mesenchyme blastulae or young gastrulae, spicules, pigmented cells, and intestine-like structures appeared after 48 hours. Aggregates from late gastrulae formed these structures after as little as 24 hours. Histologic analysis showed that after a few hours several cavities containing an amorphous substance appeared within the aggregates. 2 kinds of cavities could be distinguished: one had an epithelial lining and resembled intestine, while the other was similar to a blastocoel. After a few days some aggregates resembled plutei and had spicules and arm-like structures. They could be kept alive for more than 2 weeks.

When blastomeres of pigmented *Arbacia* and unpigmented *Paracentrotus* mesenchyme blastulae were mixed, there was an initial random aggregation, followed by the formation of distinct small pigmented and unpigmented clusters.

Inasmuch as the time the re-formed larvae require to reach a given degree of differentiation depends on the embryonic stage at which the cells are isolated, the original differentiation seems not to have been lost by the dissociation.

To study reaggregation of cells dissociated from vegetalized and animalized sea urchin embryos, fertilized eggs were raised for 18 hours, i.e. up to the late blastula stage, in 0.066 *M* LiCl or 0.001 *M*

---

[1]Calcium-magnesium-free sea water: 2.37% NaCl, 0.066% KCl, and 0.4% Na_2SO_4, adjusted to pH 8.0 with 4.62% $NaHCO_3$. To correct tonicity to local sea water conditions, NaCl concentration can be increased.

$ZnSO_4$ in sea water (GIUDICE, 1963). Embryos were kept in lithium until they became exogastrulae. Aggregation of the isolated animalized and vegetalized blastomeres occurred in the same way and at the same rate as in blastomeres of normal embryos.

After 24 hours the vegetalized aggregates appeared to be solid spheres with no active motion. Instead of a ciliated ectodermal surface, they were covered by a layer of flat endothelial-like cells, indicating that the vegetalizing effect of the lithium upon the cells was permanent. Intestine, blastocoel, and spicules did not form. Their absence was attributed to the missing inductive effect of ectoderm tissue. Animalized cell aggregates were one-third to one-fourth the size of normal embryos and appeared as hollow spheres having active surface cilia but no spicules. Both animalized and vegetalized aggregates degenerated after a few days.

The experiments suggested that the cells retain their individual changes during isolation and aggregation. Aggregation experiments performed with a mixed population gave inconclusive results.

To investigate reaggregation of blastomeres obtained from 2 different embryonic stages, aggregates of cells dissociated from labeled and unlabeled embryos were analyzed by autoradiography (GIUDICE et al., 1969). Fertilized eggs of *Paracentrotus lividus* were suspended under continuous rotation in sea water with antibiotics and radioactive thymidine: 100 IU penicillin + 54 μg streptomycin + 50 μg 3-sulfonamide-6-methoxypyridazine/ml, and 2.3 μCi/ml thymidine ^3H-methyl with a specific activity of 17.2 Ci/m*M*. At hatching, blastulae were dissociated in 0.44 *M* sucrose and allowed to reaggregate with dissociated unlabeled cells from the young prism stage in sterile sea water containing streptomycin, sulfonamide, 1 m*M* each of uridine and thymidine, at 19°C while stirred by a propeller.

The autoradiographs showed in the pluteus-like aggregates a random distribution of labeled and unlabeled cells. No preferential adhesion seemed to occur among cells of the same embryonic stage when mixed aggregates developed. The specific cell surface characteristics that differentiated in the blastomeres from the blastula to the prism stages did not entirely prevent association between cells of the 2 stages, however, it is possible that the surface differences were lost during the dissociation-reaggregation process.

Reaggregation and the morphology of reaggregated embryos were studied with the light and electron microscope (MILLONIG and GIUDICE, 1967). Mesenchyme blastulae of *Paracentrotus* were washed with calcium-free artificial sea water and dissociated in 0.44 *M* sucrose and 0.001 *M* EDTA in 0.05 *M* tris-citrate buffer pH 8.0 in a Potter homogenizer. Blastomeres were allowed to reaggregate in agitated normal sea water and fixed after 10 minutes, 1 hour, 22 hours, and 3 days. The blastomeres became almost spherical after dissociation (Figs. 14.1 - 14.3). They adhered to each other almost immediately when transferred to normal sea water. After 10 minutes (Fig. 14.4) the contact areas were rather extensive and electron-dense material was observed within the 15 nm intercellular space. After about 30 minutes there appeared solid aggregates of 10 or more cells (Fig. 14.5), which were linked on the surface by penta-laminar connections (Fig. 14.6).

After 7 hours the aggregates were made up of 100 cells or more (Figs. 14.7 - 14.9). The exterior cell-layer had an epithelial appearance; the cells became cylindrical with nuclei in a basal position; microvilli and initial stages of terminal bars appeared; and an extra-

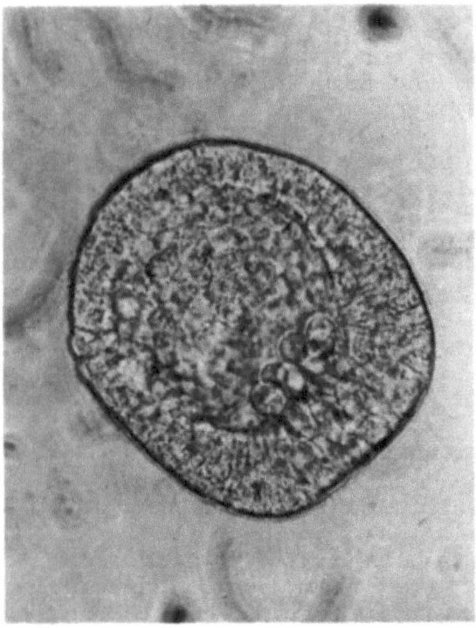

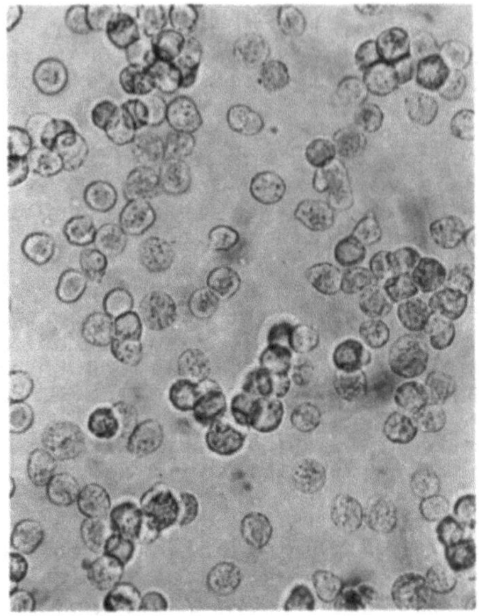

Fig. 14.1. *Paracentrotus lividus* mesenchyme blastula before dissociation. X 800

Fig. 14.2. Suspension of dissociated mesenchyme blastula cells in artificial calcium- and magnesium-free sea water. Cells are almost spherical and nuclei and yolk platelets are recognizable. X 200

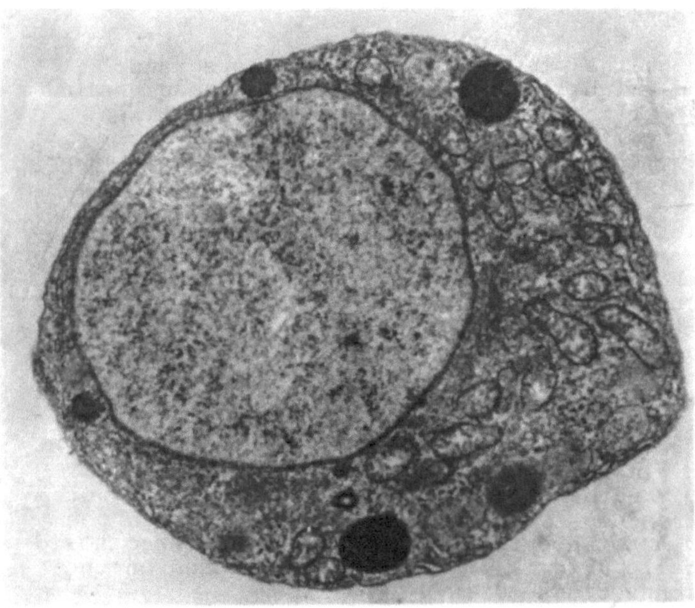

Fig. 14.3. Dissociated mesenchyme blastula cell. Nucleus and the cytoplasmic components - mitochondria, yolk granules, ergastoplasmic lamellae, polyribosomes - appear normal. Fixed with OsO_4, embedded in Epon. X 11,000

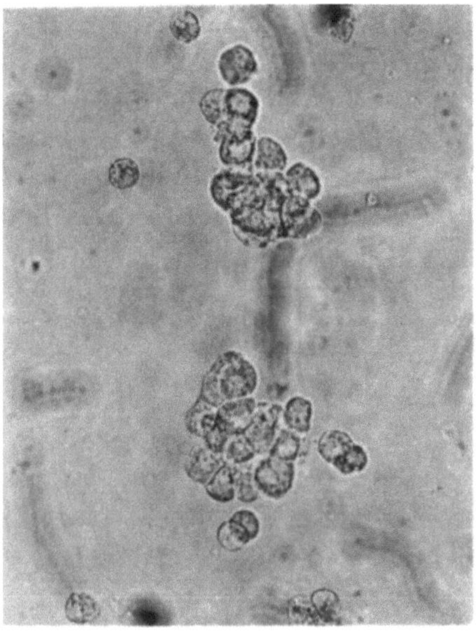

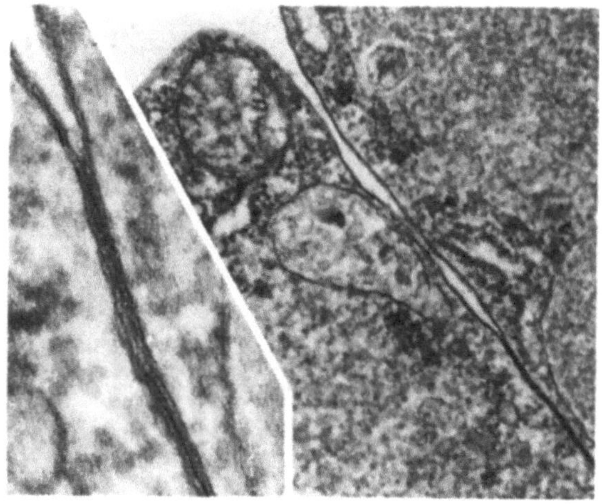

Fig. 14.4. After 10 min of reaggregation in continuously agitated seawater, cells start to form small, irregularly-shaped aggregates. X 200

Fig. 14.6. Contact area between 2 cells near surface of 30-min aggregate. 2 cell membranes in deeper part form pentalaminar connection. At this stage, terminal bars are not present. $KMnO_4$ fixation. X 5,500 and 156,000

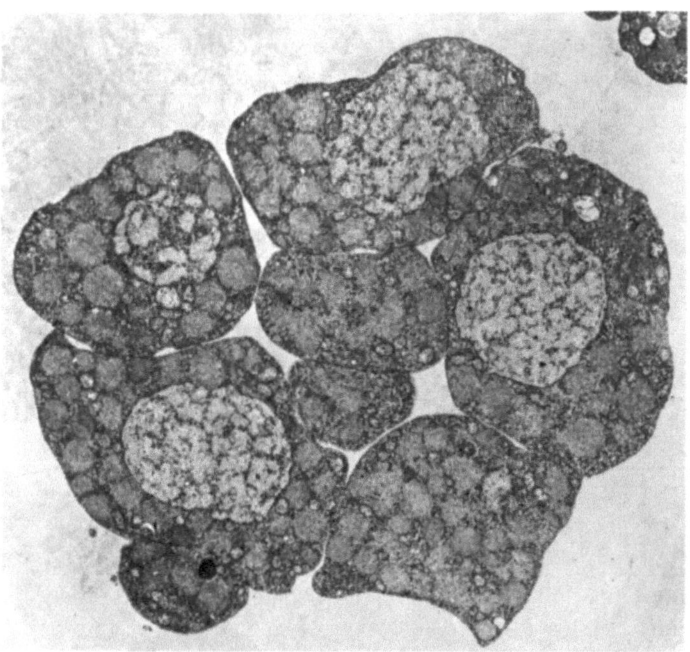

Fig. 14.5. Small group of blastomeres 30 min after reaggregation. In areas of cell contact, blastomeres are flattened by which reciprocal contact increases. X 3,700

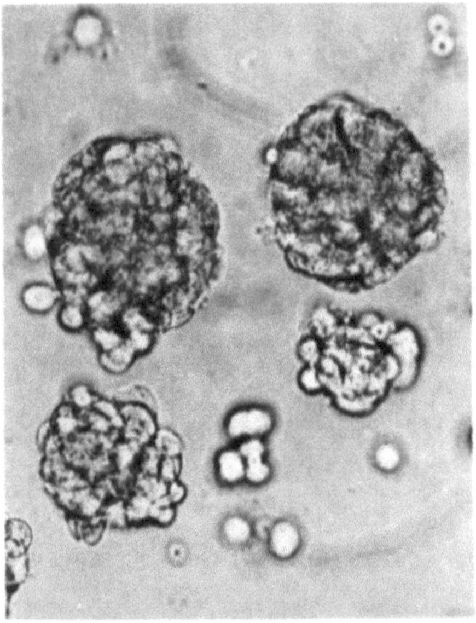

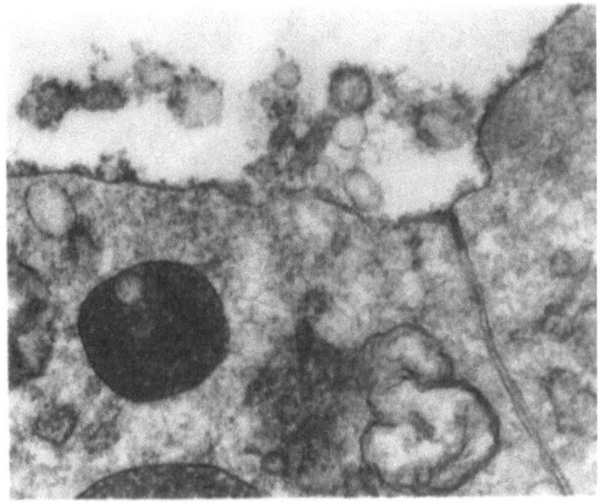

Fig. 14.7. 3 hours later, solid aggregates with cilia are formed, slightly smaller than normal embryos and almost spherical. Up to this stage, single blastomeres may be incorporated and become part of aggregate. X 200

Fig. 14.9. Initial stage of terminal bar formation in 7-hour aggregate. With $KMnO_4$ as fixative fluffy extracellular coat and intercellular substance are observed. X 45,000

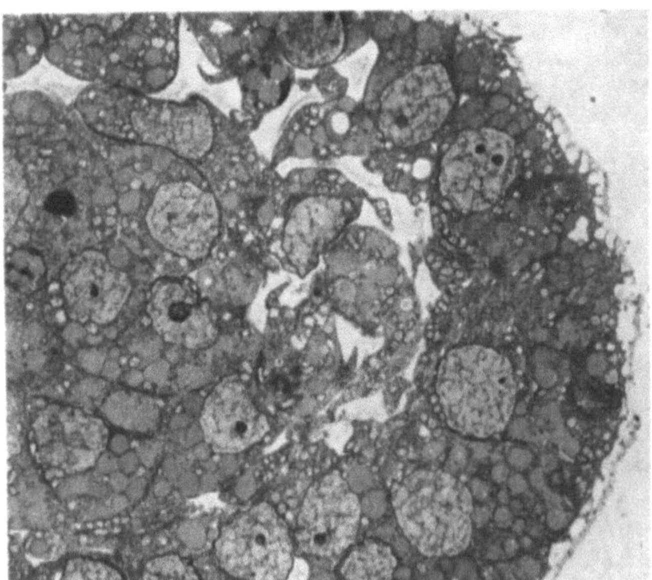

Fig. 14.8. 7-hour aggregate in which the superficial cells have assumed morphological arrangement as in normal blastula. Cells inside are separated partly by intercellular spaces. X 4,600

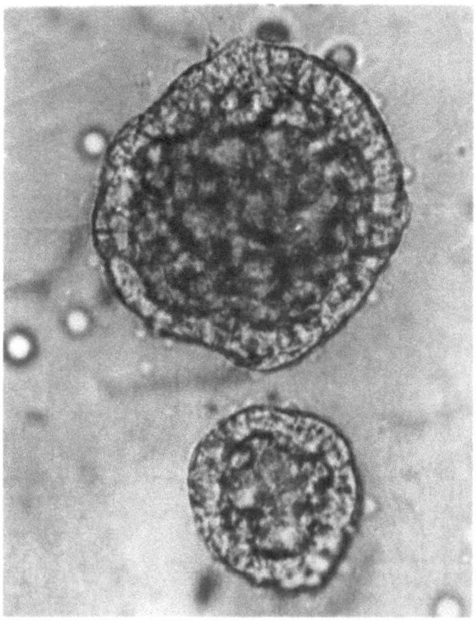

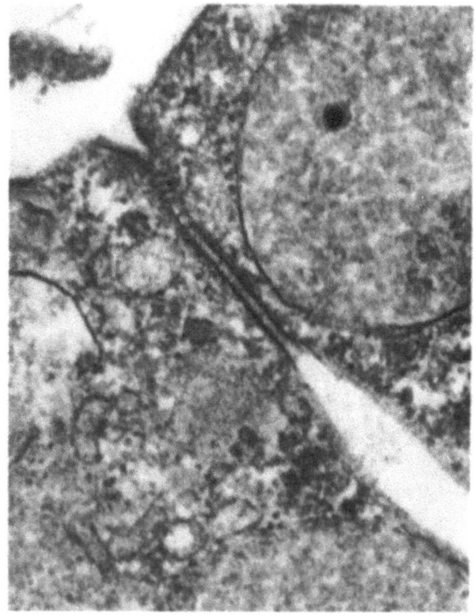

Fig. 14.10. After developing for 9 hours, aggregates resemble blastulae. Their size varies considerably and is determined by the number of aggregated cells. Superficial cells appear as normal blastoderm layer, and ciliary activity gives aggregates uncoordinated rotary movement. Isolated cells, occasionally adhering to surface, are not incorporated into aggregates at this stage. X 200

Fig. 14.12. Terminal bar in 9-hour aggregate. Dense fibrillar bridges connect the 2 cell membranes. X 46,500

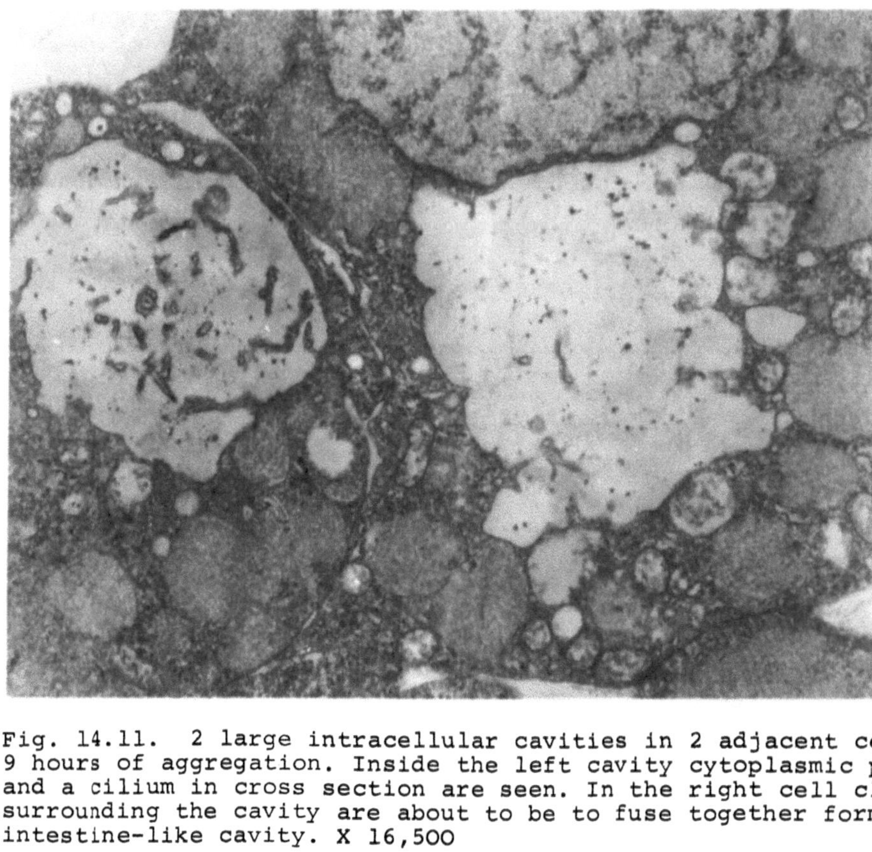

Fig. 14.11. 2 large intracellular cavities in 2 adjacent cells after 9 hours of aggregation. Inside the left cavity cytoplasmic protrusions and a cilium in cross section are seen. In the right cell clear vesicles surrounding the cavity are about to be to fuse together forming an intestine-like cavity. X 16,500

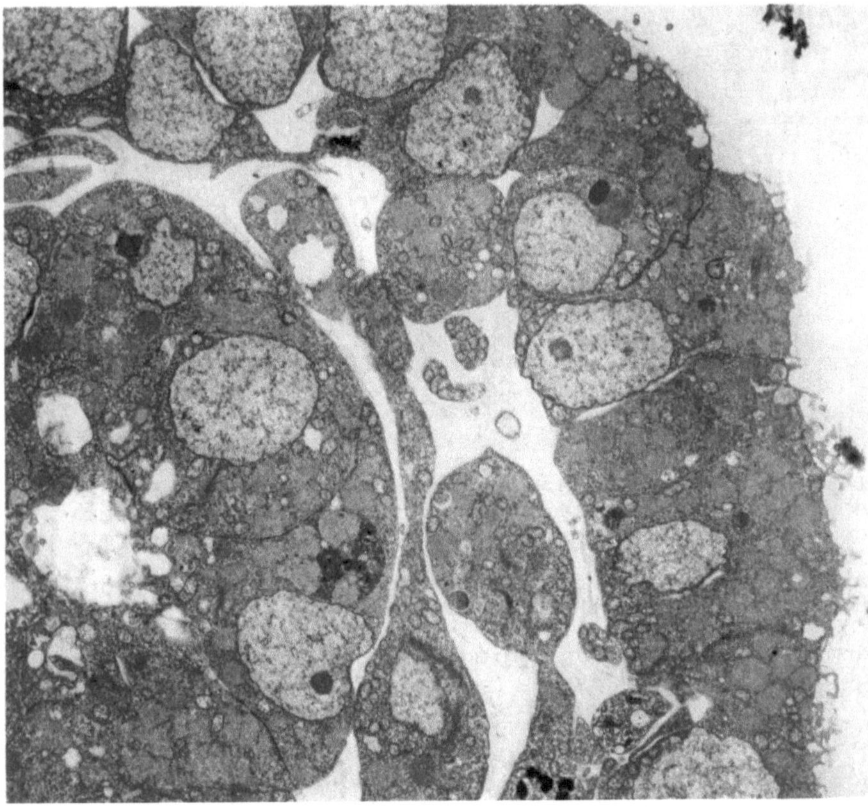

Fig. 14.13. After 22 hours, superficial cells have assumed epithelial arrangement, Nuclei are in basal position, and microvilli are on free surface. At left a circular, intestine-like structure and in the center, a large space resembling a coelomic cavity can be recognized. X 5,400

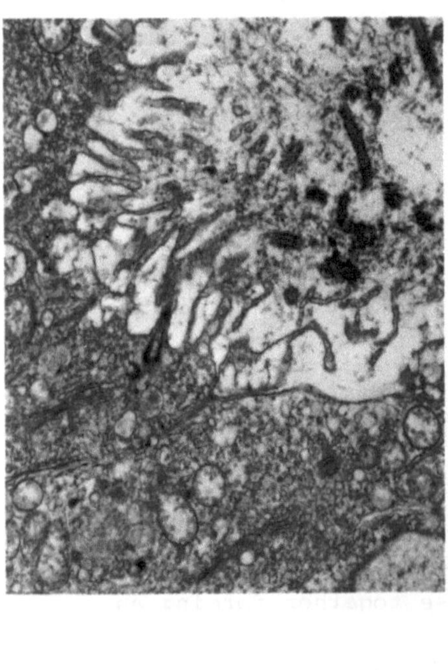

Fig. 14.14. In 3-day aggregate, cells lining intestinal cavity form large number of microvilli, cilia, and Golgi vesicles. Amorphous material is deposited inside cavity. X 14,800

cellular coat similar to the hyaline layer covered the free surface. Inside, an intercellular space resembling a coelomic cavity appeared.

After 9 hours, when the aggregates resembled blastulae (Fig. 14.10), some cells in the center of the aggregates developed one or more intracellular cavities with protoplasmic protrusions, cilia, and an amorphous content (Fig. 14.11). Several vacuoles of one cell seemed to fuse with each other, and also with vacuoles of neighboring cells. Terminal bars with fibrillar bridges sealed off the intercellular spaces towards the exterior (Fig. 14.12).

After 22 hours the cavity, or cavities, became similar to an intestine and lined with cells epithelially arranged (Fig. 14.13). Cilia and microvilli (up to 1.5 µm long) were observed on the free surface (Fig. 14.14). In large aggregates, more than 1 intestine-like cavity was formed. They never had openings to the outside. Another space, similar to a coelomic cavity, developed between the cell group surrounding the intestine-like cavity and the external epithelium. Here, no secretory activity or microvilli or cilia were noticed. One or more spicules formed, which after 2 days were as long but not as regular as in normal plutei.

Aggregates of isolated micromeres of 16-cell stage *Paracentrotus* embryos have been observed by light and electron microscopy (PUCCI-MINAFRA, BOSCO, and GIAMBERTONE, 1968). The membrane-free embryos were dissociated and the micromeres collected on a sucrose gradient by gravity. The cells, transferred to Syracuse dishes filled with sterile sea water, aggregated into small, morula-like clusters. Light microscopy revealed amoeboid movement of the blastomeres and pseudopods, while electron microscopy showed microvilli on the cell surfaces. Rows of vesicles in the cell contact areas of 24 hour aggregates indicated cell-membrane breakdown to form a syncytium. Intestine-like cavities were observed, as in aggregates of blastula blastomeres, but spicules, which are normally produced by the micromere derivative, the primary mesenchyme, were not observed.

The cilia of sea urchin embryos may be removed by brief treatment with hypertonic sea water or with 1 *M* sodium acetate, after which they form again in a few hours (IWAIKAWA, 1967). Mesenchyme blastulae of *Anthocidaris crassispina* and *Pseudocentrotus depressus* were deciliated and immediately dissociated. One batch of blastomeres was kept in agitated sea water at low concentration to prevent aggregation and another at high concentration to induce reaggregation (AMEMIYA, 1971). Light microscopy showed that isolated cells were not able to regenerate cilia, while aggregates formed cilia if the cells were aggregated immediately after dissociation or within 8 hours of isolation. Regenerative growth was apparently suppressed during dissociation and the assembly of subunit proteins for cilia formation seemed to be inhibited inasmuch as precursor proteins of cilia are stored in embryonic cells (AUCLAIR and SIEGEL, 1966). Inhibition of precursor protein biosynthesis should, however, not be ruled out. More information might be obtained from embryos and aggregates during repeated ciliation and deciliation. Although there is evidence that in isolated blastomeres rRNA and enzyme synthesis is triggered almost automatically at a given time of development, even without reaggregation and without clear evidence of cellular differentiation, the cilium should be considered a specialized cell organelle which is formed only in a more complex epigenetic cell association. Thus ciliation would result from a complex stimulation rather than from suppression of inhibitory factors.

Comparison of biochemical events in normal embryos and isolated or reaggregating blastomeres (GIUDICE, 1962b) showed that isolated cells incorporate amino acids at the same relative but about 50% reduced rate as whole embryos; and that osmolarity, pH, and calcium and potassium concentrations influence the incorporation in isolated cells at pH 7.8 in sucrose. It also showed that amino acid incorporation is influenced by adenosine triphosphate and phosphoenolpyruvate and that homogenates from different embryologic stages have different rates of protein synthesis. For these experiments mesenchyme blastulae, early gastrulae, midgastrulae, and late gastrulae of *Paracentrotus l.* were used and amino acid incorporation was measured in whole embryos, dissociated cells, and in homogenates of whole embryos. Embryos were suspended in 5 ml of sea water containing 0.5 μCi L-leucine ^{14}C at 20°C for 20 minutes. To prepare isolated blastomeres embryos were transferred to calcium-free sea water (3% NaCl, 0.07% KCl, and 0.4% Na_2SO_4 adjusted with 4.62% $NaHCO_3$ to pH 7.8) and then to 0.05 *M* citrate buffer pH 7.8 containing 0.4 *M* sucrose and 0.001 *M* EDTA. The cells were dissociated manually in a Potter glass homogenizer in 0.035 *M* tris buffer pH 7.8 containing 0.2 *M* KCl and 0.005 *M* $MgCl_2$ and then incubated in 5 ml of the same medium containing 0.5 μCi L-leucine ^{14}C at 20°C for 20 minutes. 1 milliliter of packed embryos was homogenized in a 2 ml glass homogenizer with a motor-driven plunger and diluted with about 3 ml of a medium containing 0.27 *M* sucrose in 0.035 *M* tris buffer pH 7.8 with 0.2 *M* KCl, 0.005 *M* $MgCl_2$, 0.001 *M* adenosinetriphosphate, and 0.01 *M* phosphoenolpyruvate. To 0.8 ml of this homogenate was added 0.2 ml of 0.5 μCi L-leucine ^{14}C (specific activity 7.7 mCi/mmole).

Incorporation rates were determined for isolated cells in the above solution with different molarities of sucrose: 0.3, 0.4, 0.5, and 0.6 *M*. Cells had the highest incorporation rate in 0.4 *M* sucrose (3380, 4431, 3945, 3502 counts/min/mg protein, resp.). In calcium- and magnesium-free sea water, leucine incorporation was 2.4 times greater than in normal sea water. This increase of incorporation has been subsequently found to be due to changes in cell permeability, rather than to a direct effect on metabolism since vital dyes readily penetrate isolated cells but have difficulty penetrating whole embryos and since leucine incorporation is stimulated by brief pretreatment in calcium-free sea water, while a 6-hour pretreatment reduces it to about half that of the controls (MUTOLO et al. in GIUDICE and MUTOLO, 1970).

Dissociated cells were kept up to 6 hours either in normal sea water or in calcium-free sea water before adding amino acids. To correct for permeability variations, data were calculated as percent of total ^3H-leucine uptake incorporated into protein after 20 minutes of exposure. After preincubation in normal sea water, for 1, 3, and 6 hours, leucine incorporation was about 38%, 42%, and 38%, respectively, while after preincubation in calcium-free sea water it was about 41%, 37%, and 22%. Following normal sea water preincubation for a few minutes it was only 32%, as compared with 40% after brief preincubation in calcium-free sea water.

Comparing incorporation at different pH values, specific activity (counts/min/mg protein), in isolated blastomeres at pH 6.5 was 195; at pH 7.0, 234; and at pH 7.8, 1818 (GIUDICE, 1962b). When 0.4 *M* sucrose in 0.035 *M* tris buffer pH 7.8 and 0.0005 *M* $MgCl_2$, was tested with 0.1, 0.2, and 0.4 *M* KCl, activities were 968, 1334, and 2205, respectively. L-leucine incorporation was more than twice as great in cells and embryos of midgastrulae as in mesenchyme blastulae but was incorporated into the cells at a 50% lower rate. Later, at the gastrula stage, incorporation decreased more in embryos than in cells, until the rate was the same in both. Since in a following paper (GIUDICE and MUTOLO, 1970) it is reported that EDTA, and probably also citrate, which was used for dissociation, lowers the incorporation in cells, the findings need further confirmation.

Storage of dissociated cells in the cold for several hours did not impair their capacity for incorporation but a 30-minute heat treatment at 35°C reduced the capacity to 5%. Also, in the cell debris of homogenized embryos incorporation increased until the gastrula stage, but was 100 times lower. It depended entirely on adenosine-triphosphate and phosphoenolpyruvate, whereas in isolated cells these substances reduced incorporation by 60%.

It is concluded that the mechanism of L-leucine incorporation is evidently the same in isolated cells as in embryos and that the blastocoel fluid does not play a significant role. The changes in activity during development are due to physiological cellular changes, while the decline in synthesis after the midgastrula stage remains unexplained.

To determine whether there is any correlation between aggregation and biochemical cellular activity, protein synthesis and oxygen consumption of reaggregating cells were measured and the effects of puromycin and actinomycin D were studied (GIUDICE, 1965). Cells of *Paracentrotus* mesenchyme blastulae were dissociated in 0.05 *M* citrate - 0.01 *M* tris buffer pH 8.1, 0.44 *M* sucrose, and 0.001 *M* EDTA. Puromycin or actinomycin D (40 μg/ml) was added to the reaggregating cells and protein synthesis was estimated by measuring the rate of ^{14}C L-leucine incorporation (160 μCi/μ*M*; 1.2×10^6 cells/ml) after 10 minutes of incubation. Protein was determined according to LOWRY et al. (1951), and oxygen consumption measured manometrically in a Warburg apparatus.

Reaggregation was observed microscopically while cells were kept in Syracuse dishes at 18°C without stirring. The monolayered cells (about 4×10^3 cells/ml) begin adhering to each other a few minutes after they are transferred to dishes and then form small groups. After about an hour a lattice structure develops, and after 2 hours single cell groups begin ciliary movement. Large, compact clumps form, leaving only a few cells in the open spaces of the lattice and after about 4 hours the lattice breaks up and is replaced by large, irregularly shaped aggregates. 8 hours later they have differentiated an epithelial layer. Intestine-like structures and spicules appear, as in aggregates grown in beakers and stirred continuously. In the presence of puromycin the cells started to reaggregate normally but disaggregated suddenly after about 90 minutes and were completely dissociated after 2 hours. During a period of 3 hours, the normal reaggregating cells showed an increase in L-leucine incorporation but in the presence of puromycin incorporation drops rapidly to less than 20% after 90 minutes and to less than 5% after 2 hours.

Actinomycin D causes a steady decline of L-leucine incorporation after the first hour until, after 8 hours, it is less than 5%. The aggregates develop normally, and begin to form an epithelial layer after 6 hours, when synthesis is about 20%, but then fall apart into single cells or into clear clusters and develop no further. Cells that disaggregate as the result of puromycin or actinomycin D appear normal in vivo and show no evidence of cytolysis. Puromycin seems to have an irreversible effect on aggregation, while aggregates surviving after actinomycin D treatment continue to differentiate in sea water, forming intestine-like structures and pigment cells, but not skeletons.

When fertilized eggs are treated for 3 hours with L-ethionine (0.0015 *M*) the differentiation of primary mesenchyme is impeded and gastrulation inhibited (BOSCO and MONROY, 1960). After 14 hours, when the embryos were at the mesenchyme blastula stage, the cells were dissociated and allowed to reaggregate. Untreated embryos were used for comparison. After 20-minutes isolation, amino acid incorporation dropped 60% and after 2 hours, 90%. In treated whole embryos amino acid incorporation was reduced only 30-40%. Dissociated cells from L-ethionine-treated embryos aggregated and formed a lattice that did not break up later into single aggregates as happens normally but dissociated completely into cells having irregular shapes. When dissociated blastula cells were transferred to sea water with L-ethionine, amino acid incorporation immediately dropped to 10% and reaggregation was delayed but not inhibited. In experiments in which the aggregates dissociated again or formed an outer epithelial layer, amino acid incorporation was reduced to 30%.

For 3 ml of intact embryos at the blastula stage (3 mg protein per ml), oxygen consumption was 50 µl after 1 hour and 200 µl after 4 hours (GIUDICE, 1965). Isolated cells that reaggregated in Warburg flasks consumed only about 5% less oxygen. When puromycin was present during reaggregation, oxygen consumption dropped to 43% after 2 1/2 hours and amino acid incorporation to 7%; the aggregates then dissociated. Sodium azide, dinitrophenol, actinomycin D, and L-ethionine had similar effects. Low temperature (0°C) strongly inhibited protein synthesis, and some aggregation did take place but only as far as lattice formation.

If the medium does not contain potassium or calcium, aggregation is inhibited whereas lack of magnesium delays aggregation which only proceeds to lattice formation. Without potassium ions amino acid incorporation and respiration are inhibited whereas in absence of calcium ions amino acid incorporation is stimulated (Table 14.1).

The experiments do not contradict the hypothesis that aggregation is mediated by a protein which is continuously synthesized by the cells, and since inhibitors of protein synthesis also influence respiration, cellular aggregation may be inhibited when overall cellular metabolism is affected.

To determine whether the physiological onset of rRNA synthesis at late gastrula is suppressed by dissociation and reaggregation, *Paracentrotus* embryos and reaggregating cells at the midgastrula stage were labeled with 0.5 µCi/ml of carrier-free ^{32}P for 5 hours, after which rRNA was extracted and analyzed by sucrose gradient centrifugation (GIUDICE, MUTOLO, and MOSCONA, 1967). The 28S peak was taken as an rRNA marker. At the end of the labeling period, the embryos had reached the late gastrula stage and irregular aggregates were formed by the cells. The pattern of rRNA synthesis was similar in both, although lower in the aggregates than in the embryos, indicating that the increase in synthesis, which normally starts at late gastrula, occurred even in aggregates that did not reach morphological differentiation.

In another experiment, dissociated cells of blastulae were cultured until the control embryos had reached the midgastrula stage. Embryos and aggregates were then exposed for 5 hours to ^{32}P. A radioactive 28S peak was observed in the normal gastrulae but not in the blastula-like aggregates.

It was therefore concluded that cell interaction is an important factor for the initiation of rRNA synthesis during embryonic development and

Table 14.1. Amino acid incorporation and cell aggregation in *Paracentrotus* embryos under inhibitory conditions. (From GIUDICE, 1965)

Medium	Amino acid incorporation (cpm/mg protein)[a]	Aggregation
Sea water	815	Normal
Sea water + 5 x 10^{-3} *M* sodium azide	37	Delayed and arrested at lattice stage
Sea water + 5 x 10^{-4} *M* 2,4-dinitrophenol	63	Totally inhibited
Sea water + 1 x 10^{-4} *M* 2,4-dinitrophenol	155	Totally inhibited
Sea water + 1 x 10^{-5} *M* 2,4-dinitrophenol	776	Normal
Sea water at $0^{o}C$	20	Strongly delayed and arrested at lattice stage
Ca^{++}-free sea water	1,425	Totally inhibited
Ca^{++}- and Mg^{++}-free sea water	908	Totally inhibited
Mg^{++}-free sea water	347	Delayed and arrested at lattice stage
K^{+}-free sea water	11	Totally inhibited

[a]Amino acid incorporation into proteins was measured after 20 minutes of incubation with 0.5 μCi of ^{14}C-L-leucine (U) 7.7 μCi/μmole at $22^{o}C$ (except in sample at $0^{o}C$) immediately after cells were resuspended in medium.

that cells become committed to this synthesis shortly before the gastrula stage.

In a subsequent paper these conclusions were questioned since it was found that the permeability to inorganic phosphorus is much lower in dissociated blastula cells than in gastrula cells (GIUDICE and MUTOLO, 1970). In the following experiment, therefore, isolated blastomeres were obtained from previously labeled embryos (SCONZO et al., 1970).

Fertilized *Paracentrotus* eggs were continuously labeled for 7 hours until the blastula stage when one part was transferred to sea water to develop further and the other part dissociated. One batch of these cells was kept isolated in diluted sea water while another was used to form aggregates. Embryos were mechanically dissociated in 0.01 *M* tris-HCl buffer pH 8.0 and 0.5 *M* sucrose and labeled with ^{32}P (EDTA and citrate were not used because they reduce amino acid incorporation). Acid-insoluble RNA and the acid-soluble nucleotide pool were measured for RNA synthesis and DNA was measured for mitotic activity.

In all 3 samples, specific activity of the acid-soluble nucleotide pool remained almost constant up to 22 hours after fertilization, and no nucleotide leakage was observed after dissociation. In hatching blastulae (10 hours) and midgastrulae (17 hours) the amount of 28S rRNA was nearly the same as in cells kept dissociated or in re-aggregating cells. The rRNA synthesis was slightly reduced in aggregates and in nonaggregating cells of early blastulae (7 hours). In cells isolated for 24 hours, DNA remained constant and thymidine was not incorporated.

Inasmuch as rRNA synthesis occurred in both reaggregating and non-reaggregating embryonic cells after the advanced blastula stage, cell contact and mitotic division did not seem to be significantly involved. The rather slow DNA increase during reaggregation suggested reduced mitotic activity (GIUDICE and MUTOLO, 1970). In normal developing embryos 94% of the cells divide once from blastula to pluteus stage (25 hours), while only 70% divide in the aggregates. After fertilization the egg begins a number of complex synthesizing activities indicated by increased oxygen consumption and rRNA synthesis. They result from a specific genetic function during this period of differentiation, but at what level the stimulus is initiated remains obscure.

As had been shown, the level of deoxycytidylate (dCMP)-aminohydrolase in embryos decreases gradually from fertilization to the pluteus stage (SCARANO and MAGGIO, 1959), a similar enzyme variation in blastomeres after dissociation was investigated (SCARANO et al., 1964).

Enzyme levels were studied in unfertilized *Paracentrotus* eggs and embryos at the 32-cell, blastula, gastrula, and pluteus stages and in dissociated cells of these embryos. To obtain dissociated cells from 32-cell-stage embryos, the fertilization membrane was removed first by forcing the eggs through a syringe immediately after fertilization. The embryos were then dissociated with a Potter homogenizer in calcium-free sea water and the blastomeres were grown in calcium-free sea water with 1.5 ml saturated sulfadiazine/100 ml. Enzyme assays were made on the cells dissociated from blastulae and gastrulae shortly after isolation and again when the control embryos had reached the gastrula stage (24 hours) or pluteus stage (50 hours).

Embryos and blastomeres were homogenized in 0.01 *M* phosphate buffer pH 7.3 and centrifuged at 18.000 *g* for 15 minutes at 0°C. The assays were made with the supernatant and the following incubation mixture: 9.2 m*M* ^{14}C-dCMP (100,000 cpm/μ*M*), 0.125 m*M* dCTP, and 10 m*M* phosphate buffer pH 7.3 at 30°C for 0, 5, and 10 minutes. The mixtures were acidified with PCA to a concentration of 5% for 10 minutes at 0°C and centrifuged. An aliquot of the supernatant was neutralized with 0.933 *M* K_2CO_3 in 0.1 *M* tris buffer pH 7.0. The extracted dCMP and dUMP were separated by thin-layer chromatography on DEAE cellulose in 0.02 *N* HCl and dUMP standard solution added as carrier. The spots of dUMP were counted in a liquid scintillation counter. Enzyme activity was calculated as micromoles of dCMP deaminated every 10 minutes per milligram of protein.

The calculated enzyme activity gradually dropped from fertilization to the pluteus stage (52 hours) from 0.35 μ*M* to 0.05 μ*M*. In blastomeres dissociated from blastulae and gastrulae and kept disaggregated in calcium-free sea water, activity remained at the same high level for at least 30 hours. It has been concluded that at the level of regulation of enzyme content in cells, intercellular interactions are of importance for differentiation. This opinion, however, was questioned by GIUDICE and MUTOLO (1970) since a severe impairment of cell metabolism results when isolated cells are kept for long periods without calcium. In fact, blastomeres which are kept for 6 hours in calcium-free sea water are unable to reaggregate. It would also be important to show whether or not the enzyme content remains high in blastomeres kept isolated and to examine reaggregating cells of different embryonic stages to obtain positive proof of the effect of cell adhesion.

Alkaline phosphatase activity is extremely low during early development but increases sharply at the late pluteus stage and rises continuously during the next 40 hours (PFOHL, 1965). To determine whether dissociation and reaggregation affect the gene activity that initiates enzyme synthesis, PFOHL and GIUDICE (1967) dissociated *Arbacia punctulata* embryos at different stages of development and homogenized the reaggregates. A butanol extract was used to determine alkaline phosphatase activity at pH 9.4, and *p*-nitrophenylphosphate was used as substrate. Alkaline phosphatase activity was calculated as micromoles of nitrophenol liberated per hour per milligram of protein.

When plutei are dissociated shortly before the onset of enzyme activity, and the cells reaggregated, alkaline phosphatase is synthesized almost at the same time and at the same rate in embryos and cells, although it takes 10 hours to form compact aggregates. In aggregates from mesenchyme blastula cells, however, enzyme activity remains extremely low until after 75 hours (60 hours after dissociation), activity becomes pronounced (Fig. 14.15).

Evidently a commitment to initiate enzyme synthesis is made in the embryo after the mesenchyme blastula stage and before the pluteus stage since subsequent dissociation into single cells does not inhibit the pattern of enzyme activity. This is not the case when dissociation is performed before the blastula stage; enzyme activity is resumed after the formation of pluteus-like aggregates in which correct cell interaction has been established.

Current dissociation methods (GIUDICE and MUTOLO, 1970) call for embryos to be washed twice in artificial calcium-free sea water buffered at pH 8 with 0.01 *M* tris buffer, and subsequent suspension in 5 volumes of 0.01 *M* tris-HCl buffer pH 8 and 0.5 *M* sucrose. Dissociation is performed in a glass homogenizer using slow strokes of a Teflon plunger with periodic microscopical examination. The suspension is diluted 1:1 with calcium-free sea water and centrifuged 2 minutes at low speed (about 1,000 rpm). The supernatant (mainly cell debris) is discarded. All solutions are cooled and the tubes kept in an ice bath during the procedures. The yield is about 75% of the cellular proteins, with 25% being lost in the nonsedimented or broken cell portion. Cells may be suspended in a small amount of calcium-free sea water and kept in cold storage for over 3 hours. To reaggregate, the blastomeres are transferred to Millipore-filtered sea water containing 10,000 IU penicillin + 5 mg streptomycin + 5 mg sulfadiazine/ 100 ml to inhibit bacterial growth in experiments longer than 12 hours. For continuous observation, the suspension may be placed in small Petri or Syracuse dishes without agitation. The newly formed embryos, however, will be irregularly shaped.

To obtain large quantities of uniform aggregates, the cells are kept in 100 - 2,000 ml beakers continuously stirred by a plastic propeller (30 rpm). The best results are obtained with a cell concentration of 1.5×10^6/ml. At higher speeds or at higher cell concentration, larger aggregates are formed. The optimal temperature is 18°-20°C.

Small aggregates from mesenchyme blastulae remain spherical and form only few calcareous skeletons and intestine-like structures (Fig. 14.16). Large aggregates (up to 1 mm in diameter) are almost spherical and have several protruding spicules resembling the arms of normal plutei and a number of intestine-like structures (Fig. 14.17). Aggregates generated from cells of later stages (such as gastrulae or early prisms) have spicules but not arm-like protrusions, while aggregates from pluteus cells have no spicules.

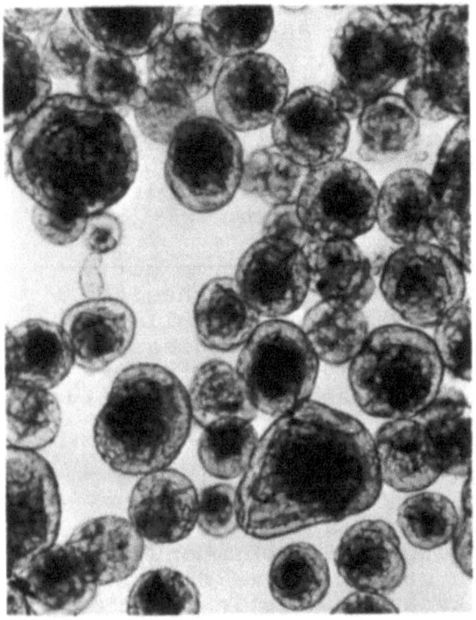

◀ Fig. 14.15. Histochemical Gomori reaction for alkaline phosphatases on 49-hour aggregates of mesenchyme blastula cells. All aggregates, regardless of size, contain positive reaction product. X 200

Fig. 14.16

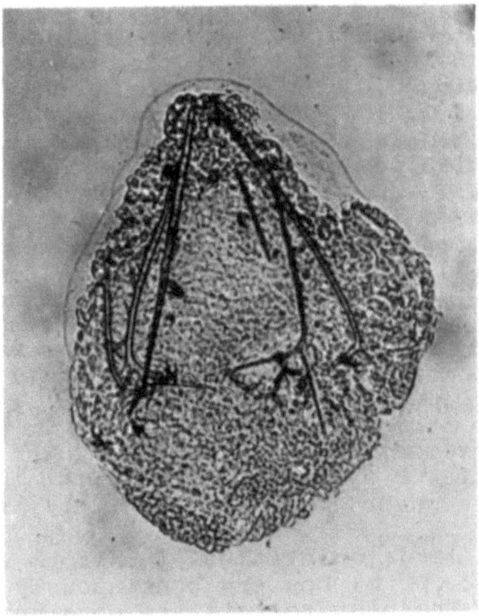

Fig. 14.17

Fig. 14.16. 49-hour aggregates, slightly compressed by the coverslip with rod-like, calcareous skeletons. X 200

Fig. 14.17. 5-day-old pluteus-like aggregate, slightly compressed by coverslip. Several rods similar to spicules are seen. (From GIUDICE and MUTOLO, 1970)

References

AMEMIYA, S., 1971. Relationship between cilia formation and cell association in sea urchin embryos. Exp. Cell Res. 64, 227-230.

AUCLAIR, W., SIEGEL, B.W., 1966. Cilia regeneration in the sea urchin embryo. Evidence for a pool of ciliary proteins. Science 154, 913-915.

BOSCO, M., MONROY, A., 1960. Inhibition of the differentiation of the primary mesenchyme in the sea urchin embryo caused by ethionine. Acta Embryol. Morphol. Exp. 3, 53-64.

GIUDICE, G., 1962a. Restitution of whole larvae from disaggregated sea urchin embryos. Develop. Biol. 5, 402-411.

GIUDICE, G., 1962b. Amino acid incorporation into the proteins of isolated cells and total homogenates of sea urchin embryos. Arch. Biochem. Biophys. 99, 447-450.

GIUDICE, G., 1963. Aggregation of cells isolated from vegetalized and animalized sea urchin embryos. Experientia 19, 83.

GIUDICE, G., 1965. The mechanism of aggregation of embryonic sea urchin cells; a biochemical approach. Develop. Biol. 12, 233-247.

GIUDICE, G., MUTOLO, V., 1970. Reaggregation of dissociated cells of sea urchin embryos. Adv. in Morphog. 8, 115-158.

GIUDICE, G., MUTOLO, V., DONATUTI, G., BOSCO, M., 1969. Reaggregation of mixtures of cells from different developmental stages of sea urchin embryos. Exp. Cell Res. 54, 279-281.

GIUDICE, G., MUTOLO, V., MOSCONA, A.A., 1967. The role of cell interactions in the control of RNA synthesis. Biochim. Biophys. Acta 138, 607-610.

GIUDICE, G., VITTORELLI, M.L., MONROY, A., 1962. Investigations on the protein metabolism during early development of the sea urchin. Acta Embryol. Morphol. Exp. 5, 113-122.

HERBST, C., 1900. Über das Auseinandergehen von Furchungs- und Gewebezellen in kalkfreiem Medium. Arch. f. Entw. Mech. 9, 424-463.

IWAIKAWA, Y., 1967. Regeneration of cilia in the sea urchin embryo. Embryologia 9, 287-294.

LOWRY, O.H., ROSENBROUGH, N.J., FARR, A.L., RANDALL, R.J., 1951. Protein measurements with the Folin phenol reagent. J. Biol. Chem. 193, 265-275.

MILLONIG, G., GIUDICE, G., 1967. Electron microscopic study of the reaggregation of cells dissociated from sea urchin embryos. Develop. biol. 15, 91-101.

PFOHL, R.J., 1965. Changes in alkaline phosphatase during the early development of the sea urchin, Arbacia punctulata. Exp. Cell Res. 39, 496-503.

PFOHL, R.J., GIUDICE, G., 1967. The role of cell interaction in the control of enzyme activity during embryogenesis. Biochim. Biophys. Acta 142, 263-266.

PUCCI-MINAFRA, I., BOSCO, M., GIAMBERTONE, L., 1968. Preliminary observations on the isolated micromeres from sea urchin embryos. Exp. Cell Res. 53, 177-183.

SCARANO, E., MAGGIO, R., 1959. The enzymatic deamination of 5'-deoxycytidylic acid and of 5-methyl-5'-deoxycytidylic acid in the developing sea urchin embryo. Exp. Cell Res. 18, 333-346.

SCARANO, E., DE PETROCELLIS, B., AUGUSTI-TOCCO, G., 1964. Deoxycytidilate aminohydrolase content in disaggregated cells from sea urchin embryos. Exp. Cell Res. 36, 211-213.

SCONZO, G., PIRRONE, A.M., MUTOLO, V., GIUDICE, G., 1970. Synthesis of ribosomal RNA in disaggregated cells of sea urchin embryos. Biochim. Biophys. Acta 199, 441-446.

15. Morphology and Biochemistry of Diploid and Androgenetic Haploid (Merogonic) Hybrids

P. S. Chen and F. Baltzer †

Sea urchin hybrids have often been used to study nucleo-cytoplasmic interaction during embryonic development. In diploid hybrids the sperm nucleus of one species is introduced into the egg of another species, while in haploid merogonic hybrids anucleate egg fragments are cross-fertilized. In the former the egg contains 2 pronuclei of different species, and in the latter a male pronucleus is enclosed by foreign cytoplasm without a second haploid nucleus. Both combinations often result in disharmonic interaction between nucleus and cytoplasm, leading sooner or later to developmental arrest and death of the embryo. It is evident that analysis of the cytologic behavior and morphogenetic patterns of these hybrids provide us with valuable information about the basic mechanism of embryonic development.

As recent advances in genetics and molecular biology have clearly demonstrated that development is under genic control, it is understandable that emphasis was placed on the nucleus which is the site of chromosomal genes, but morphogenetic importance of cytoplasm cannot be neglected. Various observations have indicated, in fact, that gene activity is controlled by cytoplasm. Owing to their inherent difficulties, sea urchins have contributed very little to modern genetics. Repeated attempts have been made to study inheritance in echinoid hybrids by utilizing different species which differ in rates of development and in body form and skeletal structure of the pluteus larva. The work so far has been limited and the results fragmentary, but they nevertheless add to our understanding of the control mechanism in embryonic differentiation.

Earlier studies on sea urchin hybrids and merogones were largely limited to mapping cytologic and morphologic characteristics, but recent analyses at the molecular level have been carried out with the help of refined biochemical techniques.

Instead of trying to cover all the possible hybrid combinations, in this paper we will select examples to illustrate the types of developmental and biochemical problems involved in the study of sea urchin hybrids. Unlike homospermic combinations, crosses between 2 species, genera, or orders may not be easy. As such difficulties have escaped the attention of many investigators, we will discuss briefly some of the problems involved in hybrid fertilization.

---

†Prof. F. Baltzer died on March 18, 1974, shortly after his 90th birthday.

15.1 Classification

Echinoderm classification used for hybrid combinations discussed here is as follows:

Phylum Echinodermata
Class Echinoidea
Subclass Euechinoidea
Superorder Echinacea
Order Stirodonta
- Family Arbaciidae
  Arbacia lixula L.

Order Camarodonta (Echinoida)
- Family Temnopleuridae
  Genocidaris maculata Mortensen
- Family Toxopneustidae
  Sphaerechinus granularis Lamarck
  Lytechinus variegatus Lamarck
  Tripneustes esculentus Leske
- Family Echinidae
  Psammechinus microtuberculatus Blainville
  Paracentrotus lividus Lamarck
- Family Strongylocentrotidae
  Strongylocentrotus purpuratus Stimpson
  Strongylocentrotus franciscanus Agassiz
  Strongylocentrotus droebachiensis Müller

Superorder Gnathostomata
Order Clypeasteroidea
- Family Scutellidae
  Dendraster excentricus Eschscholtz

Experience with hybrids has shown that there is no positive correlation between the developmental potency and the taxonomic relationship of the parental species. Even between the same 2 species, the cross in one direction may be highly viable while in the reciprocal cross the paternal chromosomes are partly eliminated and the hybrid embryo invariably becomes lethal. It is the task of embryologists to clarify the causal relationship of these profound differences and especially the physiological-biochemical basis of incompatibility in the lethal crosses.

15.2 Hybrid Fertilization Techniques

Compared to homospermic insemination, the rate of cross-fertilization is usually low and varies greatly according to the parental species used. For example, crosses between *P. lividus* ♂ and *Ps. microtuberculatus* or *S. granularis* ♀ are easy, while other crosses, especially those using *Arbacia* sperm, are more difficult. Various methods have been suggested to overcome the block against heterospermic insemination and to increase the rate of hybridization: high concentration of the sperm or egg suspension, aging the gametes in seawater, hyperalkalinity, treating the male gametes with egg water, and removing the jelly coat or vitelline membrane (RUNNSTRÖM, 1952; HARVEY, 1956). Our own experience has shown that although these procedures are useful in some hybrid combinations, they may have no effect in others. If no special precaution is taken, pretreatment may harm the hybrid embryos and cause experimental results to be misinterpreted.

Still very little is known about the nature of the block to heterologous fertilization. The jelly coat, the vitelline membrane, and the property of the egg surface are considered the main barriers. Accordingly, removal of the jelly coat by brief treatment in seawater acidified to pH 5.5 to 5.8 was suggested (HAGSTRÖM, 1956). In the hybrid combination *A. lixula* ♀ x *P. lividus* ♂, we first removed the jelly coat by treating the egg for less than 1 minute with acidified seawater, after which we treated the jelly-free eggs for a few minutes with alkaline seawater at about pH 9. Insemination was carried out with a rather concentrated sperm suspension that contained glycine (0.05%). The hybridization rate obtained was about 1%-5%, which is still very low compared to that of homospermic fertilization.

To facilitate cross-fertilization, the use of trypsin, known to act on the vitelline membrane and inhibit formation of the fertilization membrane, was suggested (RUNNSTRÖM et al., 1943; HULTIN, 1948a, 1948b). But its effect varies greatly in different hybrid combinations, the concentration used for increasing cross-fertilization rate is in the range of 0.2%-0.5%, much higher than what is needed to prevent fertilization membrane formation. This indicates that the action of trypsin is complex and probably affects egg components besides the vitelline membrane (BOHUS-JENSEN, 1953). Even more important, cleavage and later embryonic development may be abnormal (HAGSTRÖM, 1959). Without the fertilization membrane, fertilized eggs usually clump together, and subsequent development is not uniform (BROOKBANK, 1970). For the hybrid combination *P. lividus* ♀ x *S. granularis* ♂ we treated eggs without jelly coat very briefly with a low concentration of trypsin (0.001%). We washed the eggs in seawater and cross-fertilized them with a sperm suspension containing 0.1% glycine. This procedure yielded a fairly high rate of hybridization, but inhibited later development compared to the controls. While useful in some crosses, pretreatment with trypsin should be used with caution.

The use of hyperalkalinity for heterologous insemination was suggested by LOEB (1904) and ELSTER (1935). In the combination *P. lividus* ♀ x *A. lixula* ♂ we pretreated the egg 5 to 6 minutes in seawater containing 2.5% of 0.1 *N* NaOH (BALTZER and BERNHARD, 1955). While the method was simple, the rate of fertilization was low (12%-42%). Hyperalkalinity may not improve cross-fertilization in sperm concentrations below 5×10^7/ml (HAGSTRÖM, 1959).

Pretreatment with calcium-free seawater also improves cross-fertilization. When *Hemicentrotus pulcherrimus* eggs were immersed in that solution for 180 minutes and returned to normal seawater in which they were inseminated with *Strongylocentrotus intermedius* sperm, cross-fertilization values were significantly higher. This method may also be used in intraclass and interorder crossings such as those between *Scaphechinus (Echarachnius) mirabilis* and *Strongylocentrotus nudus*.

Or cross-fertilization may be improved by treatment with pronase. Only a few eggs were fertilized in the reciprocal crosses between *Glyptocidaris crenularis* and *H. pulcherrimus*. If the eggs were pretreated with 0.1% seawater solution of pronase P for 120 minutes and inseminated with the heterologous sperm in normal seawater, many more hybrids developed (OKADA, 1950).

Whatever method is chosen to increase hybridization rate, pretreatment of gametes must be carried out under carefully controlled conditions.

15.3 Early Development of Parental Species

15.3.1 *Paracentrotus lividus*

The unfertilized egg of *P. lividus* has a diameter of 95 µm and a volume of 440 x 103 µm^3 (CHEN, 1958; WHITELEY and BALTZER, 1958). The diploid chromosome number is 36 (BALTZER, 1909). The chromosomes at the metaphase of the first cleavage mitosis are shown in Fig. 15.1. At 18°-19°C the 16-cell stage is reached 3 hours and 45 minutes after insemination. The 10-hour blastula develops a rather wide blastocoel the diameter of which is about three-fifths that of the embryo. At 20-21 hours the archenteron begins to form when the vegetative area invaginates (Fig. 15.2). In the 29-hour-old prism, the oral rods, body rods, and transverse rods are well developed but the anal rods are rudimentary (Fig. 15.3). The pluteus at 40-50 hours has a conical

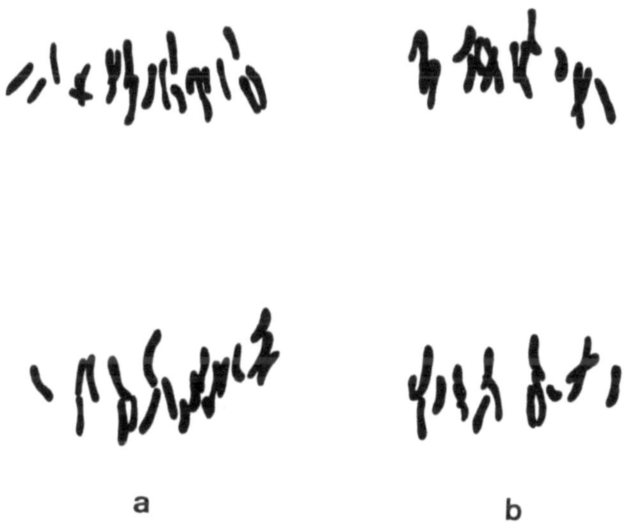

Fig. 15.1 a and b. Number of chromosomes in first cleavage mitosis of *Paracentrotus lividus*. (a) Upper: 20. Lower: 19. (b) Upper: 14. Lower: 15. x 3,393. (From BALTZER, 1909)

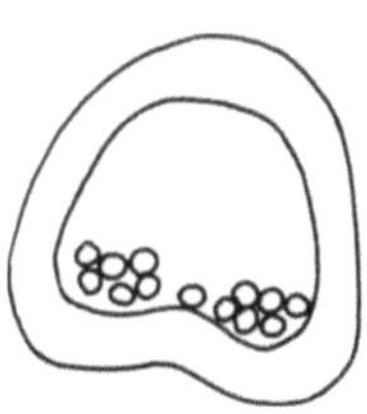

Fig. 15.2. Early gastrula of *Paracentrotus lividus* (20-21 hr) showing first invagination of vegetal area. x 192. (From WHITELEY and BALTZER, 1958)

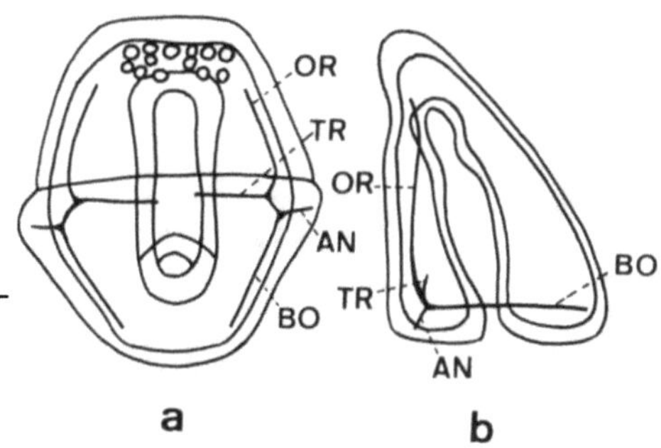

Fig. 15.3 a and b. Prism stage of *Paracentrotus lividus* (29 hr or over). (a) Seen from presumptive frontal side. (b) Lateral view from left. AN, anal rod; OR, oral rod; BO, body rod; TR, transverse rod. x 200. (From WHITELEY and BALTZER, 1958)

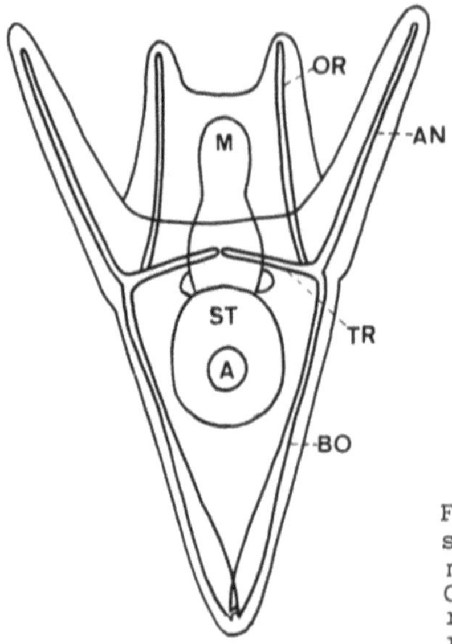

Fig. 15.4. Pluteus of *Paracentrotus lividus* shortly before feeding (45-50 hr). M, mouth; ST, stomach; A, anus; AN, anal rod; OR, oral rod; BO, body rod; TR, transverse rod. x 204. (From WHITELEY and BALTZER, 1958)

body, and its skeleton consists of one pair each of simple anal, oral, and transverse rods (Fig. 15.4). The club-shaped body rods are unjoined and in later pluteus development remains unchanged up to 20-21 days except that their apical ends thicken and become spiny (HÖRSTADIUS, 1936). As the rod tip grows medially and apically, the body rods bend toward each other when the larva is 27 days old.

15.3.2 *Psammechinus microtuberculatus*

In *Ps. microtuberculatus* there are 36 rod-shaped chromosomes, 4 of which are very long (Fig. 15.5)(BALTZER, 1909). The rate of development is similar to that of *P. lividus*. At about 18°C the mesenchyme blastula stage occurs after 14-17 hours, the late gastrula stage after 27 hours, and the pluteus stage after 40-50 hours. In early stages the embryo develops a spaceous blastocoel and appears larger than in *P. lividus* (Fig. 15.6a, b). The plutei of both species are similar in body form and skeleton, the simple type with thickened body rods (Fig. 15.6c). On the 4th day the apical tips of the body rods of *Psammechinus* plutei broaden transversely and in the next 4 or 5 days each develops a pair

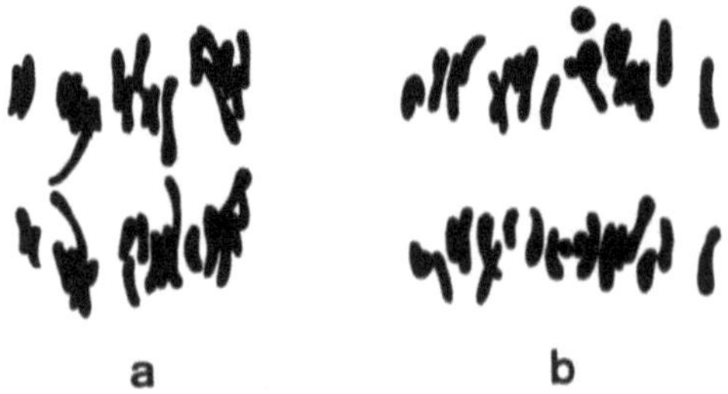

Fig. 15.5 a and b. 18 chromosomes in each view of first cleavage mitosis of *Psammechinus microtuberculatus*. x 3,654. (From BALTZER, 1909)

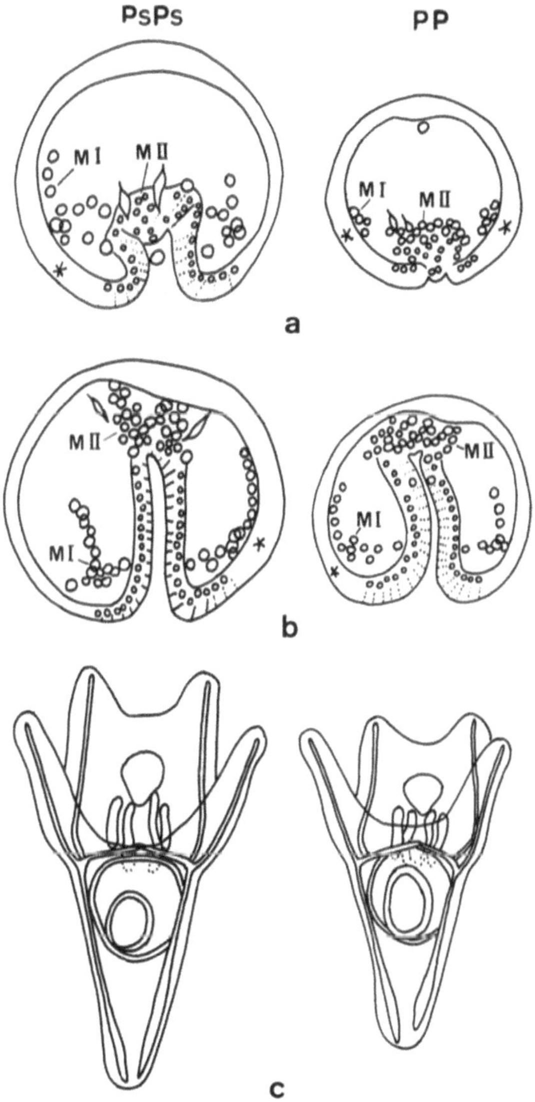

Fig. 15.6a-c. Parallel gastrula and pluteus stages of *Psammechinus microtuberculatus* (PsPs) and *Paracentrotus lividus* (PP). (a) Early gastrula 20 hr, 30 min; and 21 hr, 40 min, respectively. (b) Late gastrulae 21 hr, 30 min; and 35 hr, respectively. x 325. (c) Plutei 3 days old. x 156. (From HÖRSTADIUS, 1936; BALTZER et al., 1954)

of forked processes that meet medially and form a ring. By contrast, the body rods of *Paracentrotus* plutei retain their club shape for at least 3 to 4 weeks. The different times these species require to form skeletons is important to the study of hereditary species-specific characteristics of hybrids (HÖRSTADIUS, 1936).

15.3.3 *Sphaerechinus granularis*

Unfertilized eggs of *S. granularis* and *P. lividus* are about the same size and contain similar amounts of nitrogen (CHEN, 1958). Of the 40 rod-shaped chromosomes in the diploid set, 2 maternal and 2 paternal chromosomes are very long (Fig. 15.7). As the eggs are almost colorless, the early stages are easy to follow. At about 18°C, *S. granularis* eggs hatch in 15 hours and *P. lividus* eggs in 12. In *S. granularis* gastrulation begins at about 30 hours and is complete at 45, while in *P. lividus* it begins at 19-20 hours and ends at 29. At the blastula and gastrula stages, because of blastocoel size, the embryo of *S. granularis* is larger than that of *P. lividus* and at the end of gastrulation is about twice as large. The ectoderm of *S. granularis* contains vacuoles, while the homogenous entoderm does not (Fig. 15.8). At 28 1/2 hours the

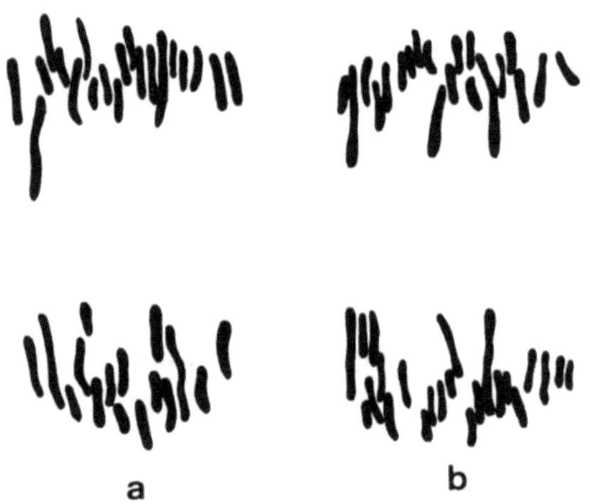

Fig. 15.7 a and b. Number of chromosomes in first cleavage mitosis of *Sphaerechinus granularis*. (a) Upper: 20. Lower: 16. (b) Upper: 20. Lower: 23. x 3,387. (From BALTZER, 1910)

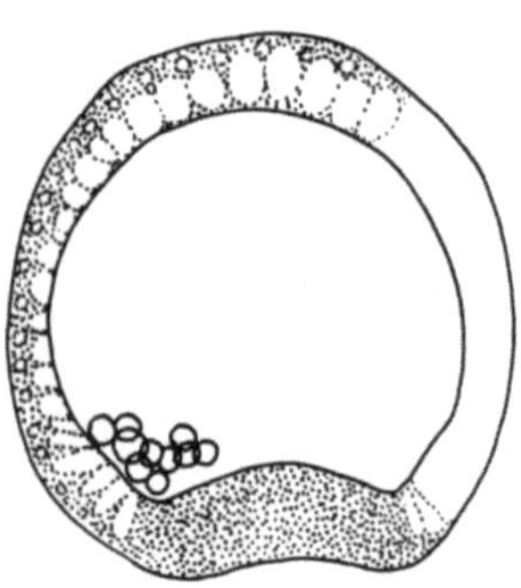

Fig. 15.8. Lateral view of living blastula of *Sphaerechinus granularis* (28 1/2 hr) showing detailed left half of primary mesenchyme ring. Boundary between vacuolized ectoderm and non-vacuolized entoderm is near mesenchyme ring. x 256. (From BALTZER et al., 1959)

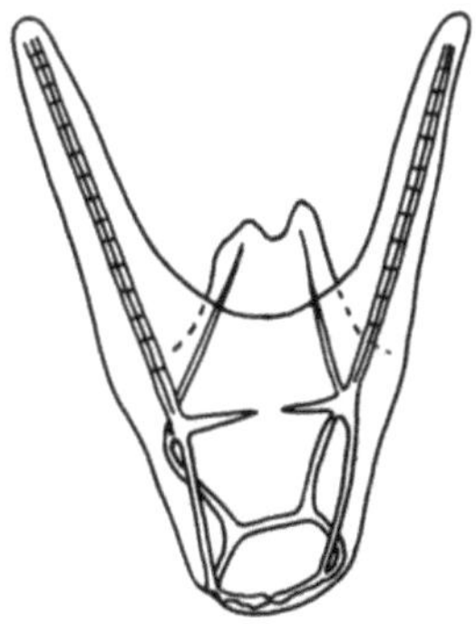

Fig. 15.9. Pluteus of *Sphaerechinus granularis* 10 days old. (From VON UBISCH, 1957)

boundary between the vacuolized ectoderm and nonvacuolized entoderm is near the mesenchymal ring. *Paracentrotus* ectoderm contains no vacuoles.

The prism and pluteus stages of *S. granularis* and *P. lividus* differ significantly in body form, skeleton, and chromatophore pattern. The 10-day-old *S. granularis* pluteus is broad and appears round and flat at the apical end (Fig. 15.9). Each anal arm contains 3 parallel rods having transverse connections. The body rods and recurrent rods form a rectangular skeletal framework in the apical region. The pair of fenestrated anal rods in the unusually long anal arms are easily distinguished from those of *P. lividus*.

15.3.4 *Arbacia lixula*

The volume of the unfertilized *A. lixula* egg (260 x 10^3 μm^3) is 59% that of the *P. lividus* egg, and its diameter is 80 µm (CHEN, 1958; WHITELEY and BALTZER, 1958). Nitrogen content of *A. lixula* eggs is 8.5 $\mu g/10^3$ and of *P. lividus* eggs, 13.3 $\mu g/10^3$. In *A. lixula* the diploid set contains 40 small, dotted chromosomes (Fig. 15.10). Because of red-brown cytoplasmic pigment, the embryo is opaque and early stages are hard to identify. According to our experience cultivation of *A. lixula* to the pluteus stage is difficult but may be improved by adding Versene (EDTA, 10^{-4} *M*) to the culture medium (BERNHARD, 1957).

Embryos of *A. lixula* develop more slowly than those of *P. lividus*. In the blastula 12 or 13 hours old, the diameter of the blastocoel at most is only one-fifth that of the whole embryo (Fig. 15.11a). At the prism stage, reached after about 34 hours, 3 short anal rods develop, the first species-specific characteristic (Fig. 15.11b). After about 42 hours in the early pluteus, anal and oral lobes are still rudimentary (WHITELEY and BALTZER, 1958). The pair of body rods bend toward each other to form the apical arch. Each anal arm contains 3 parallel rods, the tips of which unite to form the first transverse connections. After 55 hours the body of the pluteus differs considerably from that of the *P. lividus* (Fig. 15.11c). The anal arms lengthen, but the oral lobe remains stubby and the apical end blunt. The anal rods form a triple-barred complex, and a second transverse connection gives the apical arch a ring shape in which additional knobs appear. At the base of the oral rods, short, apical branches develop.

Fig. 15.10. 20 chromosomes in each view of first cleavage mitosis of *Arbacia lixula*. x 3,387. (From BALTZER, 1910)

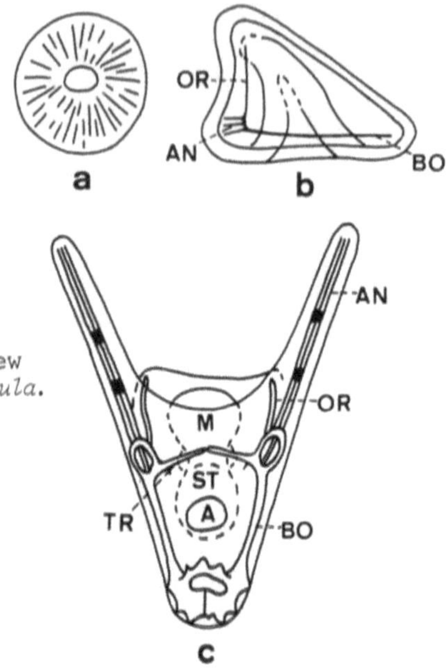

Fig. 15.11a-c. Representative development stages of *Arbacia lixula*. (a) Blastula (12-13 hr). (b) Prism (34 hr or over); lateral view from left. (c) Pluteus shortly before feeding (55 hr or over); ventral view. A, anus; AN, anal rod; BO, body rod; M, mouth; ST, stomach; TR, transverse rod. x 195. (From WHITELEY and BALTZER, 1958)

15.3.5 *Genocidaris maculata*

Among sea urchin species, *G. maculata* is unique: its diploid set contains only 16 rod-shaped chromosomes of varying lengths (VON UBISCH, 1959). Egg diameter is 87/113 μm, and at 18° the 16-cell stage occurs about 4 3/4 hours after insemination. Cells of the cleavage stages are loosely connected by cytoplasmic bridges. The blastula/gastrula stage occurs after about 24 hours. In the early pluteus, each anal arm develops 3 short, parallel rods that become the complex, fenestrated, anal rods in larvae 6 to 10 days old (Fig. 15.12). The body rods are smooth at first but develop spines. Their distal tips thicken and form the processes that become the ring in the apical region.

15.4 Developmental Capacity and Morphologic Characteristics of Hybrids

Experience has shown that, depending on the parental species used, the morphogenetic capacity of hybrids varies greatly, from combinations fully viable to highly lethal. Typical examples of diploid and androgenetic haploid hybrids are listed in Table 15.1. The combinations are

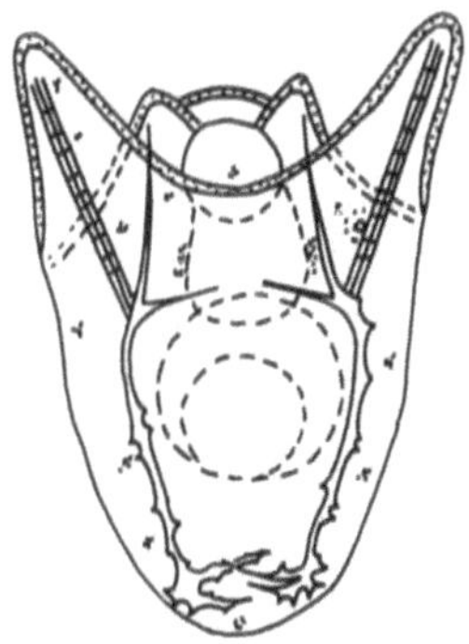

Fig. 15.12. Pluteus of *Genocidaris maculata* (10 days old). x 117. (From VON UBISCH, 1959)

interordinal, interfamiliar, intergeneric, or interspecific. But there are no clear-cut criteria for evaluating hybrid development potency. Even in the same combination there is a large variation in the morphogenetic capacity among individual hybrid embryos.

15.4.1 *Ps. microtuberculatus* ♀ x *P. lividus* ♂ (PsP) and *P. lividus* ♀ x *Ps. microtuberculatus* ♂ (PPs)[1]

In these optimal hybrid combinations, the fertilization rate is almost 100%, and nearly all the hybrid embryos develop normally. Species-specific body rods become first discernible in the 4-day-old pluteus. In both reciprocal crosses, young plutei are of maternal size. Body rods of 8- to 26-day-old hybrids are of the intermediate type, suggesting that both maternal and paternal genomes influence larval skeletal formation (HÖRSTADIUS, 1936).

In studies of androgenetic haploid hybrids of *Ps. microtuberculatus* and *P. lividus*, the nucleus of an unfertilized egg, as little cytoplasm as possible, was cut off with a fine glass needle and the anucleate egg fragment cross-fertilized (HÖRSTADIUS, 1936). Homospermic merogones served as controls. In contrast to the diploid hybrids, only a few heterospermic merogones could be raised to plutei. With *P. lividus* cytoplasm and *Ps. microtuberculatus* nucleus, larvae developed body rods characteristic of the paternal species, indicating nuclear control.

---

[1]The following abbreviations are used for parental controls: PP, *Paracentrotus lividus*; PsPs, *Psammechinus microtuberculatus*; SS, *Sphaerechinus granularis*; AA, *Arbacia lixula*; GG, *Genocidaris maculata*; SpSp, *Strongylocentrotus purpuratus*; SfSf, *Strongylocentrotus franciscanus*; SdSd, *Strongylocentrotus droebachiensis*; LL, *Lytechinus variegatus*; TT, *Tripneustes esculentus*; DD, *Dendraster excentricus*.

In the abbreviations used for hybrids, in each instance the first indicates the maternal species: PsP, PPs, *Ps. microtuberculatus* x *P. lividus*; SP, PS, *S. granularis* x *P. lividus*; PA, AP, *P. lividus* x *A. lixula*; SPs, PsS, *S. granularis* x *Ps. microtuberculatus*; SpSf, SfSp, *St. purpuratus* x *St. franciscanus*; SpSd, SdSp, *St. purpuratus* x *St. droebachiensis*; SpD, DSp, *St. purpuratus* x *D. excentricus*; LT, TL, *L. variegatus* x *T. esculentus*; GP, *G. maculata* x *P. lividus*; GPs, *G. maculata* x *Ps. microtuberculatus*; AS, *A. lixula* x *S. granularis*; LA, *L. variegatus* x *A. punctulata*.

Table 15.1. Combinations, cytologic behavior and developmental potency of diploid and androgenetic haploid hybrids

Combination		Diploid chromosome number		Diploid Hybrid				Androgenetic haploid hybrid	Author
Female	Male			Cleavage	Blastula/ gastrula	Pluteus	Skeleton		
Type A									
Paracentrotus lividus	*Psammechinus microtuberculatus*	18	18	Normal	Normal	Numerous	Intermediate	Pluteus; paternal	HÖRSTADIUS, 1936
Ps. microtuberculatus	*P. lividus*	18	18	Normal	Normal	Numerous	Intermediate	Pluteus; paternal	BOVERI, 1918; HÖRSTADIUS, 1936
Strongylocentrotus purpuratus	*Strongylocentrotus franciscanus*	18	18(?)	Normal	Normal	Numerous	Intermediate	-	MOORE, 1943
Type B									
Sphaerechinus granularis	*P. lividus* or *Ps. microtuberculatus*	20	18	Normal	Normal	Numerous	Intermediate	Abnormal gastrulae, rare plutei	BOVERI, 1889,1895; BALTZER, 1910; v.UBISCH, 1953, 1954, 1955, 1957
St. purpuratus	*Dendraster excentricus*	18	18	Normal	Normal	Numerous	Intermediate	-	MOORE, 1957; BROOKBANK, 1970
Genocidaris maculata	*P. lividus* or *Ps. microtuberculatus*	8	18	Normal	Normal	Numerous	Intermediate	Gastrula; abnormal plutei	v.UBISCH, 1959
P. lividus or *Ps. microtuberculatus*	*G. maculata*	18	8	Normal	Normal	Numerous	Intermediate	Blastula	v.UBISCH, 1959
Lytechinus variegatus	*Tripneustes esculentus*	18	(?) 16-17	Normal	Normal	Numerous	Intermediate	Blastula	TENNENT, 1910, 1912, 1914; TAYLOR and TENNENT, 1924; BADMANN and BROOKBANK, 1970

Type C									
P. lividus or *Ps. microtuberculatus*	*Arbacia lixula*	18	20	Normal	Lethal	Rare	Intermediate or maternal	-	BALTZER, 1910; BALTZER et al., 1954; BALTZER and BERNHARD, 1955; WHITELEY and BALTZER, 1958; FICQ and BRACHET, 1963; DENIS and BRACHET, 1969; BRACHET and HULIN, 1970
A. lixula	*S. granularis*	20	18	Normal	Lethal	Rare	Intermediate or maternal	-	BALTZER et al., 1961
Type D									
P. lividus or *Ps. microtuberculatus*	*S. granularis*	18	20	Elimination of paternal chromosomes	Lethal	Very rare	Maternal	Early blastula	BALTZER, 1910; BOVERI, 1918; HÖRSTADIUS, 1936; BALTZER and CHEN, 1960
A. lixula	*S. granularis*	20	20	Elimination of paternal chromosomes	Lethal	Very rare	Maternal (?)	-	BALTZER, 1910
T. esculentus	*L. variegatus*	(?) 16-17	18	Elimination of paternal chromosomes	(?) Lethal	(?) Frequent	Maternal and intermediate		TENNENT, 1912a, 1912b

In the reciprocal combination (*Ps. microtuberculatus* cytoplasm and *P. lividus* nucleus) most of the merogones followed the paternal pattern, although 22 of 66 plutei had body rods of the *Psammechinus* type. Whether this indicated cytoplasmic control of skeleton formation or maternal characteristics that were artefacts resulting from external causes such as operation procedure and water temperature was unclear.

In the hybrid *St. purpuratus* ♀ x *St. franciscanus* ♂ (SpSf), the 2 parental species differ in body form although both pluteus skeletons consist of simple rods (MOORE, 1943). The recurrent rod is present only in *St. franciscanus*, the body rod of which is spiny and curved at the apical end; the body rod of *St. purpuratus* is smooth, club-shaped, and straight. Quantitative measurements showed that in the SpSf hybrid pluteus shape is maternal but skeletal form intermediate. The recurrent rod and spiny apical body rod are less pronounced than in the paternal species. The PsP and SpSf combinations have remarkably similar parental and hybrid characteristics.

15.4.2 *S. granularis* ♀ x *P. lividus* ♂ (SP), and *S. granularis* ♀ x *Ps. microtuberculatus* ♂ (SPs)

In these hybrid combinations, the parental species differ sharply in body form and larval skeleton. Whereas in the *Sphaerechinus* pluteus each anal rod forms a fenestrated, triple-barred complex and body rods in the apical region form a 4-cornered frame, *Paracentrotus* and *Psammechinus* plutei have simple rod-type skeletons (p. 428). Thus maternal and paternal characteristics of hybrid larvae are easy to distinguish. The hybrid fertilization rate is relatively high and embryos are easily raised to older pluteus stages.

The SP hybrids initially have a normal maternal cleavage pattern and retain all 38 chromosomes: 20 from *Sphaerechinus* and 18 from *Paracentrotus* (Fig. 15.13). They form a large blastocoel and develop into rather uniform plutei having intermediate characteristics. The anal arm of the early pluteus contains 2 or 3 parallel rods as in *S. granu-*

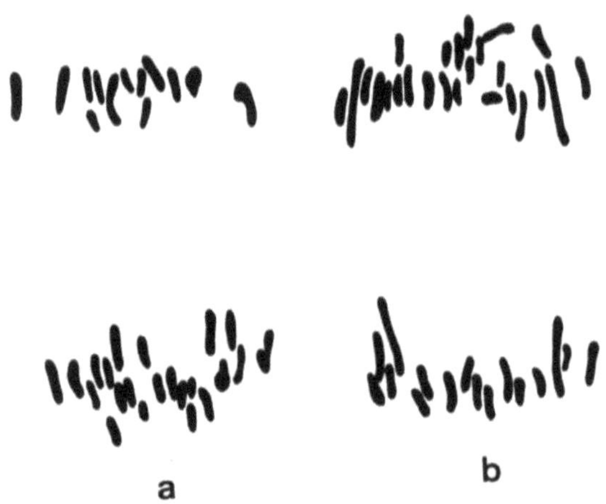

Fig. 15.13 a and b. Number of chromosomes of hybrid *Sphaerechinus granularis* ♀ x *Paracentrotus lividus* ♂; metaphase at 2-cell stage. (a) Upper: 13. Lower: 22. (b) Upper: 25. Lower: 16. x 3,225. (From BALTZER, 1910)

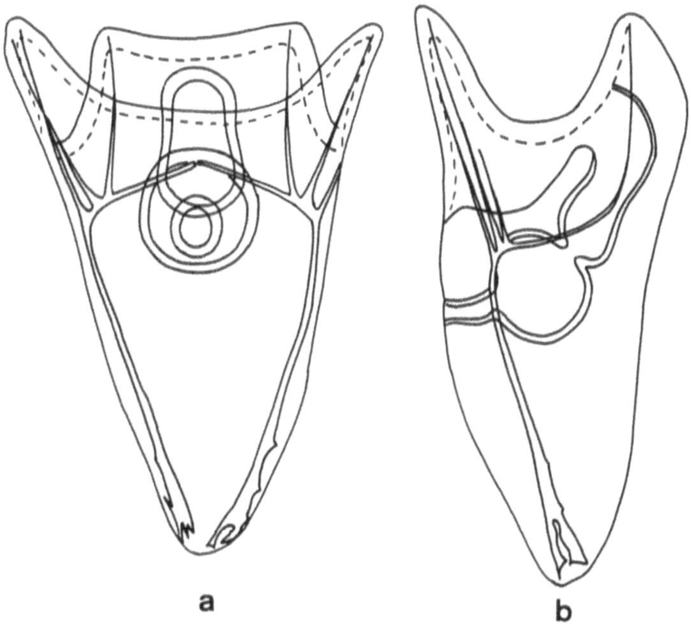

Fig. 15.14 a and b. Plutei of hybrid *Sphaerechinus granularis* ♀ x *Psammechinus microtuberculatus* ♂. (a) Young larva; ventral view. (b) Older larva; lateral view. x 310. (From BOVERI, 1895)

laris but without the typical transverse connections (Fig. 15.14). The body rods have a horn-like structure in the apical region where the skeleton is intermediate between the complex apical framework of *Sphaerechinus* and the simple, club-shaped thickening of *Paracentrotus* and *Psammechinus*.

The androgenetic haploid hybrid (*S. granularis* cytoplasm and nucleus of *P. lividus* or *Ps. microtuberculatus*) does not develop as far as the diploid combination. Most merogones die soon after the onset of gastrulation if not earlier (BOVERI, 1918; HÖRSTADIUS, 1936). When the glass needle procedure was used to produce anucleate egg fragments for cross-fertilization, very few of the plutei survived until metamorphosis. The oldest merogones lived for 46 days. The external larval form was at first paternal but became increasingly maternal, probably due to cytoplasmic influence. Because the hybrid has only the paternal genome, the larval skeleton and such structures as the ciliary epaulettes are all paternal (VON UBISCH, 1954).

In the hybrid combination of *St. purpuratus* ♀ x *D. excentricus* ♂ (SpD), as in the SP combination, the larval skeletons of the parental species differ morphologically. The *Strongylocentrotus* pluteus has a skeleton of simple rods and no recurrent rods, while the *Dendraster* pluteus has fenestrated anal rods and recurrent rods that join the apical region to form a reticulum. The maternal rate of hybrid development from cleavage to mesenchyme blastula becomes slower at gastrulation. The paternal characteristic in the hybrid is a rudimentary fenestration of anal rods and formation of recurrent rods in the pluteus. Thus the intermediate skeletal pattern of SpD parallels that of SP. Developmental rate and skeletal formation in the reciprocal cross are similar (MOORE, 1957; BROOKBANK, 1970; BADMAN and BROOKBANK, 1970).

In SP and SpD hybrids, paternal chromosomes are not eliminated. Plutei of SpD are smaller than those of the parental species and opaque (MOORE, 1957). It is possible that reduced larval size is due to loss of cellular material and that opacity is due to cell degeneration characteristic of the lethal hybrid.

As mentioned previously (p. 428), the skeletons of *Paracentrotus* and *Psammechinus* plutei are of the simple rod type, and the complex skeleton of the *Genocidaris* pluteus has fenestrated anal rods and horn-like structures on the body rods in the apical region. Cross-fertilization of *G. maculata* eggs with sperm of *P. lividus* (GP) or *Ps. microtuberculatus* (GPs) produces healthy plutei having intermediate skeletal characteristics, while the reciprocal crosses produce malformed plutei that nevertheless have the skeletal characteristics of both parents, somewhat like the cross between *S. granularis* and *P. lividus* (VON UBISCH, 1959). While the SP hybrid develops normally, the reciprocal combination (PS) is lethal, and the paternal chromosomes are soon eliminated (p. 441). *G. maculata* has a diploid number of 16 chromosomes, while *P. lividus* and *Ps. microtuberculatus* each have 36. But there is no information about the chromosomal behavior in their hybrids. That the hybrid embryos of the reciprocal crosses behave so differently suggests that their egg cytoplasm contains species-specific differences.

The merogones of *Genocidaris* cytoplasm and *Psammechinus* nucleus, and of the opposite combination, die at the gastrula stage. Those of *Genocidaris* cytoplasm and *Paracentrotus* nucleus may develop into abnormal plutei having skeletons characteristic of the paternal species.

In the hybrid *L. variegatus* ♀ x *T. esculentus* ♂ (LT), as in the SP hybrid, the larval skeleton of the parental species is different: the *L. variegatus* pluteus has simple rods, while the *T. esculentus* pluteus has fenestrated anal rods and a quadrangular framework in the apical region. More slowly than the maternal species, the LT hybrid develops to the pluteus stage. Its intermediate skeleton contains double anal rods without transverse connections (TENNENT, 1910, 1912a, 1912b, 1914; BADMAN and BROOKBANK, 1970).

L. variegatus has a diploid set of 36 chromosomes and *T. esculentus* 32 to 34. The LT hybrid retains all the parental chromosomes, but in the reciprocal cross (*T. esculentus* ♀ x *L. variegatus* ♂, TL) chromosomes are eliminated (TENNENT, 1912a, 1912b; MCCLUNG, 1939). If the TL combination also eliminates most paternal chromosomes at cleavage, like PS it must also be lethal. In the haploid combination of *L. variegatus* cytoplasm and *T. esculentus* nucleus, of 49 merogones 15 began to cleave, and 1 reached the blastula stage without gastrulating (TAYLOR and TENNENT, 1924).

15.4.3 *P. lividus* (or *P. microtuberculatus*) ♀ x *A. lixula* ♂ (PA)

The cross-fertilization of *P. lividus* eggs with *A. lixula* sperm gives rise to a typical lethal hybrid in which egg cytoplasm and nucleus interact disharmonically during morphogenesis. The rate of successful cross-fertilization is low: 12% to 42%. None of the 38 chromosomes (18 from *Paracentrotus*, 20 from *Arbacia*) is eliminated at cleavage. The rod-shaped maternal chromosomes are easily distinguished from the smaller dotted paternal chromosomes (Fig. 15.15). Normal cleavage and blastula formation occur at the maternal rate. Primary mesenchyme cell migration is normal but is usually unilateral (Fig. 15.16a). Development is delayed at the onset of gastrulation and in most embryos becomes arrested when the archenteron is half-developed.

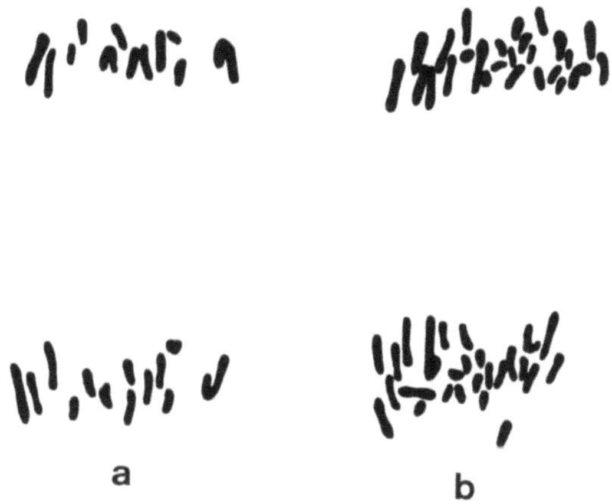

Fig. 15.15

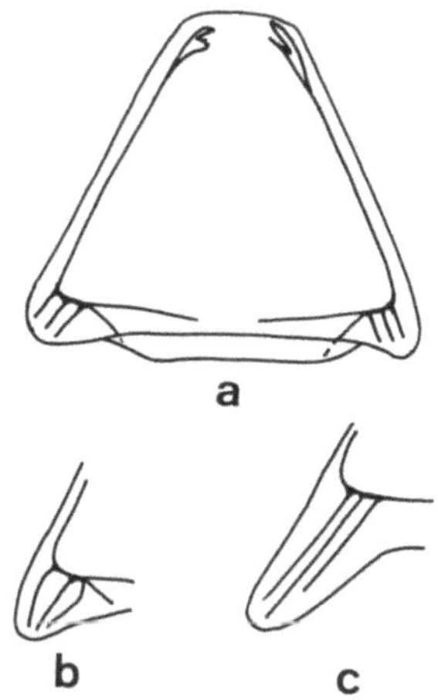

Fig. 15.17

Fig. 15.15 a and b. Number of chromosomes of hybrid *Paracentrotus lividus* ♀ x *Arbacia lixula* ♂ at metaphase of first cleavage mitosis. (a) Upper: 12. Lower: 13. (b) Upper: 26. Lower: 26. x 3,387. (From BALTZER, 1910)

Fig. 15.17a-c. Intermediate skeletons of PA hybrid showing varying degrees of paternal influence. (a) Short, parallel anal roots and forked apical ends of body rods. (b) Anal roots with convergent distal ends. (c) Long, parallel anal rods without transverse connections. (From BALTZER and BERNHARD, 1955)

In the inhibited gastrula the blastocoel gradually fills with degenerating material, and many body-wall cells contain pyknotic nuclei (Fig. 15.16b-d). The embryos swim vigorously on the second day and die on the third or fourth. In 1% of the cases at most, embryos develop to the prism or pluteus stage with maternal or intermediate skeletal characteristics. The intermediate type has 3 rudimentary, parallel rods in the anal arm, and the apical ends of the body rods are fork-shaped, indicating paternal influence on skeletal formation (Fig. 15.17).

To overcome lethality, large populations of PA embryos were raised under controlled conditions (BALTZER et al., 1954; BALTZER and BERNHARD, 1955; WHITELEY and BALTZER, 1958). About 14 hours after fertilization, the actively swimming hybrids were separated from unfertilized and abnormally cleaved eggs and transferred into seawater to which 10^{-4} *M* of EDTA was added. They were kept there at 18°-19°C for at least 3 days. Under the same condition, parental species' development was optimal.

10 cultures, each derived from 1 female and 1 male, were set up (Table 15.2), from which a number of conclusions were drawn: The PA hybrid's failure to develop is not due to culturing conditions but to incompatibility in the nucleo-cytoplasmic system. The degree of lethality differs significantly in the progeny of different parental pairs. In some cultures all embryos die at gastrulation, while in others some embryos develop to the pluteus stage. In different cultures heredity occurs in different ways. The pluteus skeleton may be purely maternal or may have paternal characteristics.

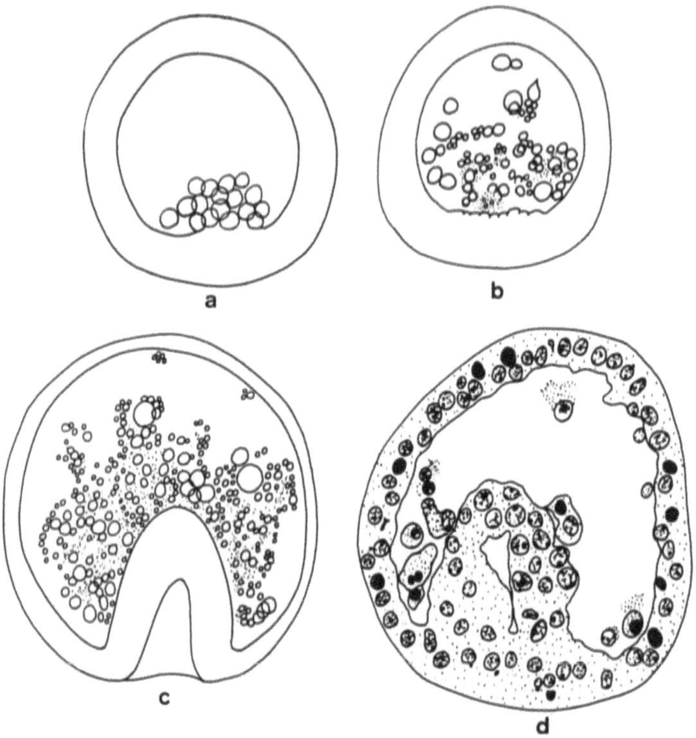

Fig. 15.16a-d. Representative development stages of hybrid *Paracentrotus lividus* ♀x *Arbacia lixula* ♂.(a) Mesenchyme blastula before degeneration (22 hr). (b) Early gastrula after onset of degeneration (26 1/2 hr). (c) Late, inhibited gastrula (47 hr); blastocoel filled with degenerating material. x 297. (d) Hybrid gastrula with pyknotic nuclei (30 hr). x 562. (From BALTZER, 1910)

Although it had been shown that hybrid nuclei retain their chromosomes from cleavage to blastula, nuclei at the pluteus stage were found to be much smaller than nuclei of maternal species, suggesting the possibility that paternal chromosomes were eliminated during later development at the retardation period. But nuclei of PA plutei were later found to be of diploid size, and staining by the Feulgen method did not reveal chromosome elimination (BALTZER et al., 1954; BALTZER and CHEN, 1965). However, estimation of chromosome number from nuclear size was not accurate, as chromosomes of the 2 parental species are not the same size. Autoradiographic study of the PA hybrid embryos showed extranuclear radioactivity that could possibly indicate chromosome elimination (FICQ and BRACHET, 1963). Feulgen-positive granules and filaments in PA blastula cytoplasm soon after hatching may have indicated fragments of eliminated chromosomes, suggesting that lethality in the hybrid embryo is caused by aneuploidy resulting from abnormal mitotic division (BRACHET and HULIN, 1970). To determine if extranuclear chromosomal material is always of paternal origin, merogonic combinations of *Paracentrotus* cytoplasm and *Arbacia* nucleus need to be examined.

Table 15.2. Skeleton formation in progeny of different PA crosses. (From BALTZER and BERNHARD, 1955)

Experiment	Culture	Number of blastulae	Skeleton formation
51 P_1 A_1	Go	6,500	Begins at 65 hours; over 100 prisms and plutei; intermediate skeleton shows anal roots and forked ends of body rods; rare maternal skeleton
	VGo	7,800	
P_1 A_2	Go, VGo	2,100 each	Very rare intermediate skeleton; 1 maternal skeleton
52 P_2 A_3	VGo	1,280	No prism despite normal invagination; irregular, scattered, skeletal fragments
53 P_2 A_3 P_2 A_4 P_2 A_5	Go, VGo	14,000	At 40-50 hours, 104 fully grown plutei and pure maternal skeleton; continuing development; very rare intermediate skeleton
P_5 A_{3-5}	Go, VGo	5,000	Arrested at midgastrula stage; very rare maternal skeleton
54 P_2 A_7	Go, Sil	4,100	Skeleton-rich but rarely distinctly intermediate; 1 maternal skeleton
P_2 A_8	Go, Sil	1,200	Poorer skeletal formation than in P_2 A_7
P_4 A_{7-9}	Go, VGo, Sil	50,000	

Go, Gulf water; VGo, Gulf water and Versene; Sil, silicate.

At the ultrastructural level the formation of a typical ergastoplasm and the appearance of nuclear membrane projections with attached ribosomes are delayed in the arrested PA blastulae. There are numerous small vesicles dispersed throughout the cytoplasm (GEUSKENS, 1968).

In a study of the reciprocal cross *A. lixula* ♀ x *P. lividus* ♂ (AP), analysis of both mitosis during early cleavage and nuclear size of the hybrid blastula showed no elimination of paternal chromosomes. The AP embryos developed at the maternal rate until late blastula and became inhibited at the onset of gastrulation. Although most embryos stop developing at the gastrula stage, occasional hybrid plutei with maternal or intermediate skeletons were seen (BALTZER et al., 1961).

15.4.4 *P. lividus* ♀ x *S. granularis* ♂ (PS), *Ps. microtuberculatus* ♀ x *S. granularis* ♂ (PsS), and *A. lixula* ♀ x *S. granularis* ♂ (AS)

In these hybrid combinations nearly all the paternal chromosomes are eliminated during the first cleavages. In the cross *P. lividus* ♀ x *S. granularis* ♂, of the 20 *Sphaerechinus* chromosomes, 16 or 17 are removed during the first cleavage mitoses (Fig. 15.18) (BALTZER, 1910; BALTZER and CHEN, 1960). In later development the hybrid nuclei retain only 21 or 22 chromosomes: 18 of the maternal species and 3 or 4 of the

Fig. 15.18 a and b. Number of single elements of chromosome pattern in first cleavage mitosis of hybrid *Paracentrotus lividus* ♀ x *Sphaerechinus granularis* ♂. Unseparated double elements at equatorial spindle region are eliminated. (a) Upper: 10. Lower: 11. (b) Upper: 12. Lower: 10. x 2,902. (From BALTZER, 1910)

paternal. Different form and size of chromosomes of the 2 parental species make it possible to identify those eliminated as being of paternal origin.

The *Sphaerechinus* nuclear complement forms daughter chromosomes that, unable to separate from each other, remain in the equatorial spindle region and are lost in later mitoses. At the 4- to 8- and 32- to 64-cell stages, eliminated chromosomes, that may exceed the original number of 32 to 34, are easily identified outside the mitotic plate (Fig. 15.19)(BALTZER and CHEN, 1960). In hybrid aged 8 hours, the blastula wall contains numerous mitoses with reduced chromosomal set and interphase nuclei of reduced size. Large nuclear vesicles probably derived from the eliminated chromosomes are also visible (Fig. 15.20). Eliminated nuclear material and adjacent cytoplasm are soon extruded into the blastocoel.

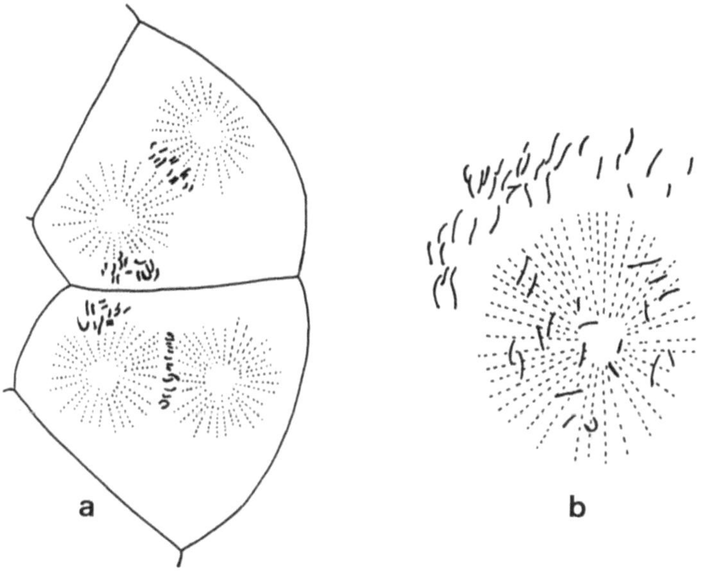

Fig. 15.19 a and b. Mitotic patterns at 32- to 64-cell stage of hybrid *Paracentrotus lividus* ♀ x *Sphaerechinus granularis* ♂ (7 hr) at 13°C. (a) 2 cells at metaphase; eliminated chromosomes have accumulated near inner centrospheres. (b) Equatorial plate viewed from spindle pole; 20 chromosomes; at upper left, 34 eliminated chromosomes are visible outside polar region. x 2,437. (From BALTZER and CHEN, 1960)

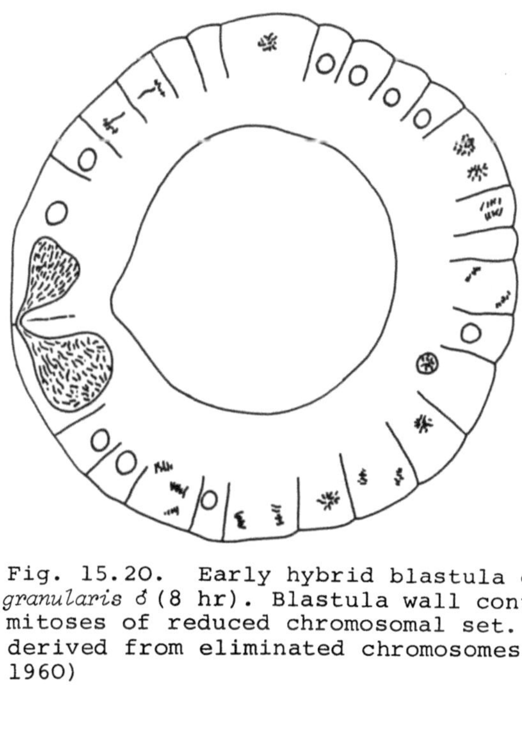

Fig. 15.20. Early hybrid blastula of *Paracentrotus lividus* ♀ x *Sphaerechinus granularis* ♂ (8 hr). Blastula wall contains nuclei of reduced size and mitoses of reduced chromosomal set. At left 2 giant nuclear vesicles derived from eliminated chromosomes. x 870. (From BALTZER and CHEN, 1960)

Early cleavage of the PS hybrid occurs at the maternal rate. At 8 to 14 hours, while the embryos are still in the fertilization membrane, the blastocoel fills with degenerating material, and primary mesenchyme cell migration may occur without the customary bilateral distribution. Most hybrid development stops at the stereoblastula or stereogastrula stage, and very few hybrids become plutei. The skeletons of these rare larvae are entirely maternal.

The behavior of the combination of *S. granularis* ♂ x *Ps. microtuberculatus* ♀ (PsS) or *A. lixula* ♀ (AS) is similar. The PsS hybrid has 38 chromosomes (18 from *Psammechinus*, 20 from *Sphaerechinus*), and the AS hybrid has 40 (20 from *Arbacia*, 20 from *Sphaerechinus*). In PsS 16 paternal chromosomes are eliminated at early cleavage, and in AS 18 are eliminated. The opacity of *Arbacia* eggs makes it hard to identify eliminated elements and early developmental stages. In all 3 crosses it would be interesting to know if the same paternal chromosomes are always eliminated.

In these hybrids the cause of lethality is probably not the same as in the PA hybrid. In these elimination occurs when development begins and involves only the paternal chromosomes. But in the PA embryo mitotic abnormality starts only as late as the hatching blastula.

15.5 Biochemical Properties of Hybrids

The morphogenetic capacity of hybrid combinations varies from normal development to extreme lethality. It is important to know how these variations are related to physiologic and biochemical properties, but this has been studied very little. Our discussion, therefore, will be limited to respiration, amino acid metabolism, and synthesis of macromolecules such as proteins and nucleic acids. The results of some of these studies have been recently reviewed by GIUDICE (1973).

15.5.1 Respiratory Rate

As an optimal utilization of reserve materials in the developing egg is a prerequisite for embryogenesis, arrested development in the lethal hybrid may be related to respiratory metabolic block. The PP respiratory rate increases linearly from cleavage to early blastula, reaches a plateau at hatching, and rises again at gastrulation (Fig. 15.21a). It is highest in fully grown plutei and declines when larvae need feeding. The AA respiratory pattern from mesenchyme blastula to pluteus is similar, although the overall values are about half as great as those of PP, probably because *Arbacia* eggs are smaller and develop more slowly (WHITELEY and BALTZER, 1958).

Initially the PA respiratory rate is the same as in the maternal species, but when development slows down at 8 to 18 hours, the respiratory rate also decreases. During the maintenance period it is about the same as the mean of the 2 parental species and then declines when degeneration begins in the last phase of hybrid development.

In the reciprocal combination AP, the rate of oxygen uptake of newly hatched AA embryos is only half that of PP, but it increases in later stages (Fig. 15.21b). The respiratory rate of the 19-hour-old mesenchyme blastula or hybrid AP is similar to that of the maternal species, but at 24 to 30 hours when gastrulation is inhibited, it drops to

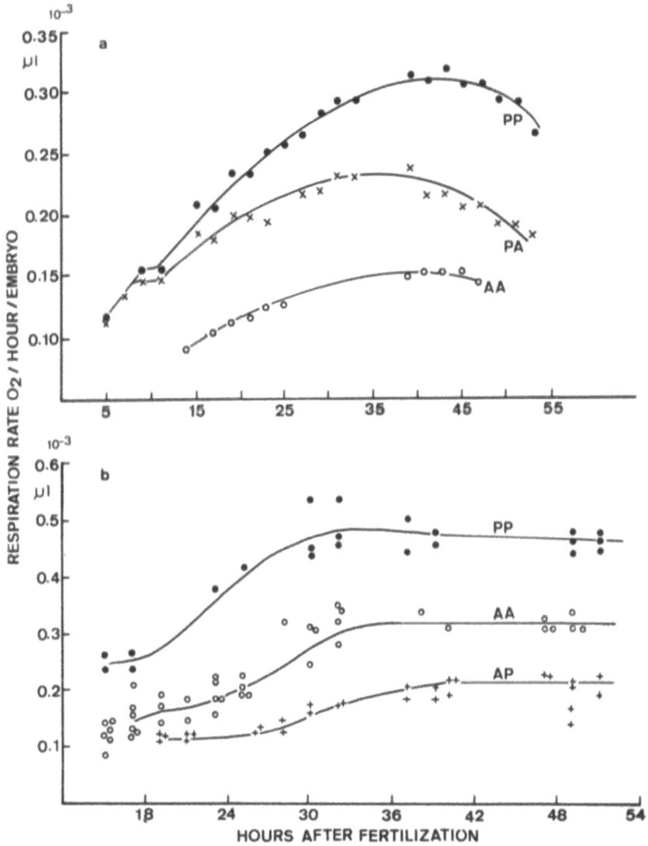

Fig. 15.21. Respiratory rate changes during embryonic development of the hybrids PA (a) and AP (b) and corresponding parental controls PP and AA. (From WHITELEY and BALTZER, 1958; BALTZER et al., 1961)

61% of the AA respiratory rate. Oxygen uptake in AP is lower than in both parental species but intermediate in PA (BALTZER et al., 1961).

It is possible that in early stages respiratory behavior results from the positive genic influence of both parents and that the 2 parental genomes normally regulate the energy metabolism of the developing hybrid. The PA respiratory pattern has been shown to be similar to the mean pattern estimated for PP and AA. But oxygen uptake of the reciprocal combination AP is lower than in both parental species.

It is also possible that the hybrid respiratory rate is low because of morphogenetic inhibition. During the early period when development is normal, oxygen uptake is determined by the maternal genome, but when embryonic retardation begins, respiration is reduced as the result of disharmonious interaction between P cytoplasm and A nucleus or between A cytoplasm and P nucleus. Energy metabolism is normally higher in PP than in AA, and if the respiratory rate in both hybrids is reduced to a similar extent, it follows that PA values are intermediate between the 2 parents, while AP values are lower than both parental ones.

But what causes respiratory depression in hybrid embryos? It may be a problem of substrate supply and/or enzyme synthesis and activation. PP and AA require similar amounts of oxygen during gastrulation, while PA requires more (WHITELEY and BALTZER, 1958). It is possible that oxygen is used less efficiently in hybrids, but evidence to support this is lacking.

In the embryo of the anuran lethal hybrid *Rana esculenta* ♀ and *Rana temporaria* ♂, when normal $NADP^+$/NADPH equilibrium is not maintained, the amount of ribose reduced to deoxyribose, which is NADPH-dependent, falls below normal. This may possibly explain why the lethal hybrid cannot utilize uridine (by reduction of ribose to deoxyribose) for DNA synthesis (BRACHET et al., 1962). But this is apparently not true of the sea urchin hybrid because the NADPH content was found to be higher in PA than in the maternal control (DENIS, 1968).

The fact that the rate of oxygen uptake in the developing sea urchin embryo may be elevated by adding 2,4-dinitrophenol suggests that potential respiratory capacity may exceed energy demand (IMMERS and RUNNSTRÖM, 1960). Blocked hybrid development is probably not the result of insufficient oxidative phosphorylation. The presence of oxidative enzymes and an operative tricarboxylic acid cycle has been demonstrated in sea urchin eggs and embryos many times (CHEN, 1967). Respiration and mitochondrial activity in egg cytoplasm increase at cleavage, suggesting that de novo synthesis of respiratory enzymes occurs during embryogenesis (GUSTAFSON, 1952, 1965; GUSTAFSON and LENICQUE, 1952). Studies of lethal sea urchin hybrid energy metabolism at subcellular and molecular levels would help to clarify the situation.

15.5.2 Amino Acid and Protein Metabolism

Sea urchin eggs and embryos are known to be rich in free amino acids that function as precursors in protein synthesis and play an important role in regulating osmosis (KAVANAU, 1953, 1954; MONROY and MAGGIO, 1966). In microorganisms free amino acids are essential to catabolite repression of gene activity in enzyme synthesis (PAIGEN, 1963). And some amino acids affect embryo morphogenesis. Phenylalanine, for example, causes amphibian epidermal cells to resemble melanoblasts, and leucine accumulates during amphibian muscle cell differentiation (WILDE, 1955a, 1955b, 1956; DEUCHAR, 1961). In sea urchin embryogenesis, glutamine and lysine have an animalizing effect while other amino acids have a vegetalizing influence, and lithium treatment interferes with the amino acid pool (GUSTAFSON and HJELTE, 1951; GUSTAFSON and HÖRSTADIUS, 1957). Glutamine is a major, free, ninhydrin-positive component of sea urchin eggs. Its synthesis is known to be ATP-dependent (CHEN and BACHMANN-DIEM, 1964). It is important to know how free amino acids and their derivatives behave in hybrid embryos as compared with the parental species.

Paper and column chromatography have shown at least 23 amino acids in the eggs and embryos of *P. lividus*, *A. lixula*, *S. granularis*, *Ps. microtuberculatus*, *G. maculata*, and *E. cordatum* (Table 15.3)(CHEN, 1958, 1962; CHEN and BALTZER, 1958). Except in *A. lixula*, glycine has the highest concentration; it is about 23% of the total nitrogen of the unfertilized *Paracentrotus* egg (BERG, 1950; GUSTAFSON and HJELTE, 1951; KAVANAU, 1953, 1954a, 1954b). Low in glycine content, *Arbacia* has another concentrated ninhydrin-positive component, the paper chromatogram position of which is very close to that of alpha-aminobutyric acid

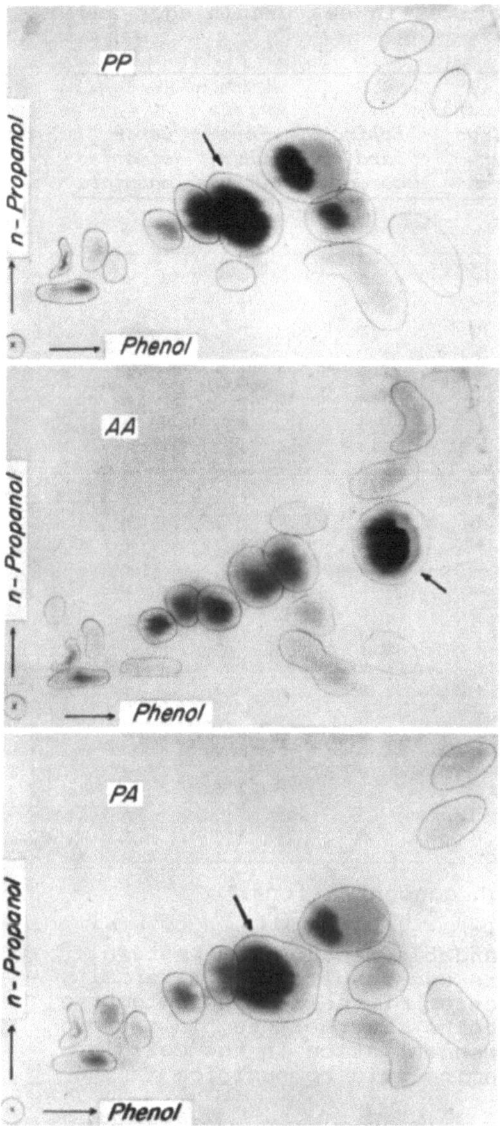

Fig. 15.22. 2-dimensional chromatograms of free amino acids and derivatives in parental species PP (unfertilized eggs) and AA (early plutei) and hybrid PA (late blastula). Pattern does not change qualitatively. Maternal pattern of PA blastula is clear. Arrows indicate components of greatest concentration; glycine in PP and PA and one unknown substance in AA. (From CHEN, 1967)

(Fig. 15.22). Free ninhydrin-positive components in *Arbacia* are only one-seventh to one-fifth as much as in other sea urchin species (Table 15.4). Unfertilized *Echinocardium* eggs have high concentrations of valine and leucine, and *Paracentrotus* eggs contain peptides (GUSTAFSON and HJELTE, 1951; CHEN, 1958). Many peptides consist of common amino acids: aspartic acid, glutamic acid, and glycine in particular (Table 15.5).

It appears that patterns of free ninhydrin-positive compounds are entirely maternal, as in lethal hybrid PA and in the fully viable combinations SP and *E. cordatum* ♀ x *P. lividus* ♂, though the larval skeletons of the last 2 show both maternal and paternal characteristics

Table 15.3. Free amino acids and derivatives in sea urchin eggs and embryos. (From CHEN, 1958, 1962)

	Paracentrotus lividus	*Arbacia lixula*	*Sphaerechinus granularis*	*Echinocardium cordatum*	*Psammechinus microtuberculatus*	*Genocidaris maculata*
α-Alanine	+	+	+	+	+	+
β-Alanine	+	-	-	+	-	-
Arginine	+	+	+	+	+	+
Aspartic acid	+	+	+	+	+	+
Asparagine	+	+	+	-	-	-
Cysteine	-	-	+	-	-	-
Cystine	+	+	+	+	+	+
Glutamic acid	+	+	+	+	+	+
Glutamine	+	+	+	+	+	+
Glycine	+[a]	+	+[a]	+[a]	+[a]	+[a]
Histidine	+	+	+	+	-	-
Isoleucine	+	+	+	+	+	+
Leucine	+	+	+	+[a]	+	+
Lysine	+	+	+	+	+	+
Methionine	+	-	-	+	+	+
Ornithine	+	+	-	-	-	-
Phenylalanine	+	+	+	-	-	-
Proline	+	+	+	+	+	-
Serine	+	+	+	+	+	+
Threonine	+	+	+	+	-	-
Tyrosine	-	+	+	-	-	-
Valine	+	+	+	+[a]	+	+
"Fast"-aminobutyric acid	-	+[a]	-	+(?)	-	-
Peptides	+	+	+	+	+	+
Unknown	+	+	+	+	+	+

[a]Substances that occur in especially high concentrations.

(BALTZER et al., 1958). This is understandable because the pattern of the free amino acid pool in the mature egg results from maternal influence during oogenesis, and genic action of the developing embryo consists mainly of synthesizing new proteins. Furthermore, in microorganisms and other animals amino acid concentration in the cellular pool has been shown to be unrelated to amino acid composition in the protein molecule (CHEN, 1971).

As the major biochemical event underlying embryonic differentiation is protein synthesis, detailed analysis of the influence of maternal and paternal genomes on protein pattern in developing hybrid embryos would be of great interest. Past studies have investigated species-specific and hybrid-specific proteins, direct gene products, and the times the products appear during embryogenesis. Unfertilized sea urchin eggs are known to contain mRNA in an inactive state. Protein synthesis is triggered by fertilization and is even necessary to cleavage. Labeled precursors are actively incorporated into protein fractions from midblastula to gastrulation, and in *Paracentrotus* embryos an antigen of new specificity appears at the mesenchymal blastula stage. In early development, protein synthesis seems to utilize existing RNA, and the messenger-type RNA then synthesized is not used until after the blastula stage (PERLMANN and GUSTAFSON, 1948; PERLMANN, 1953;

Table 15.4. Size, total nitrogen, and total free ninhydrin-positive substances of sea urchin eggs. (From CHEN, 1958)

Species	Egg diameter (mm)	Volume relation (mm^3/egg)	Total nitrogen (μg/10^3 eggs)	Total free ninhydrin-reacting substances	
				Extinction/ 10^3 eggs	Extinction/ μg nitrogen
Arbacia lixula	0.080	512×10^{-6}	8.50	0.080	0.0094
Paracentrotus lividus	0.095	857×10^{-6}	13.32	0.390	0.0293
Sphaerechinus granularis	0.095	857×10^{-6}	14.51	0.410	0.0282
Echinocardium cordatum	0.122	1816×10^{-6}	17.24	0.549	0.0318

Table 15.5. Peptides that occur in relatively high concentrations in sea urchin eggs. (From CHEN, 1962)

Species	Peptide	Fraction	Amino acid composition[a]
Paracentrotus lividus	P 17 a,b	21- 30	Asp,Glu,Gly
	P 22	59- 71	Asp,Glu,Gly
	P 26 a,b	110-117	Asp,Glu,Gly,Ser
	P 27	118-120	Asp,Glu,Gly
	P 28	121-124	Asp,Glu,Gly
	P 29 a,b	125-131	Asp,Gly
	P 30	132-133	Asp,Glu,Gly,Ala
	P 32 a,b,c	136-141	Asp,Glu,Gly,Val,Leu
	P 33	142-153	Asp,Glu,Gly
Arbacia lixula	P 1 a,b	5- 15	Asp,Glu,Ala
	P 4 a,b,c,d	14- 16	Glu,Gly,Ala
	P 7	69- 79	Asp,Glu
	P 8	80- 89	Asp,Glu,Gly,Ala
	P 9-11	90- 94	Asp,Glu,Gly,Ala,Cys
	P 12 a,b,c	91- 99	Glu,Ala
	P 15	108-110	Asp,Glu,Gly,Ala,Val,Leu
	P 17 a,b	122-127	Asp,Gly
	P 18-19	128-132	Asp,Glu,Gly,Ala,Val,Leu,Ser
Sphaerechinus granularis	P 2-4	7- 13	Asp,Glu,Gly
	P 5	9- 20	Asp,Glu,Gly
	P 6	21- 30	Glu,Gly
	P 8	34- 45	Asp,Glu,Gly
	P 9-10	81- 83	Asp,Glu,Gly
	P 15-16	90- 94	Asp,Glu,Gly,Ala,Val
	P 17	92- 96	Asp,Glu,Gly,Ala
	P 18	94-101	Asp,Glu,Ala
	P 25	126-138	Asp,Glu,Gly,Ser,Val,Leu

[a]Composition indicates only those amino acids found in acid hydrolysates and does not show amino acid sequence or frequency in peptide chain.

HULTIN, 1961; MARKMAN, 1961a, 1961b; GROSS and COUSINEAU, 1963; NEMER and BARD, 1963; BERG, 1965; SPIRIN and NEMER, 1965; TYLER and TYLER, 1966; MONROY and GROSS, 1967).

In lethal hybrid PA, a paternal antigen was identified as early as the mesenchyme blastula stage, the only positive evidence of paternal genomic influence before development was inhibited (HARDING et al., 1954). The viable hybrids LT and SpD showed neither paternal nor hybrid-specific antigens, although the skeletons of hybrid plutei have clearly paternal characteristics (BADMAN and BROOKBANK, 1970). It seems that the zygote nucleus was not actively involved in synthesizing any new class of antigenically detectable protein in these hybrids, and even in the 4 parental species no antigens of new specificity appeared during early embryogenesis. Previous positive results may have been due to changes in existing egg protein solubility and may not have indicated de novo protein synthesis at the genomic level. In fact, proteins synthesized during early embryogenesis are thought to be of the same type. The distribution pattern may change during development and a few new types may later appear (SPIEGEL et al., 1965; TERMAN and GROSS, 1965). It is also possible that existing protein molecules are only rearranged during early embryonic differentiation. It is still necessary to study other hybrid combinations, particularly those that show distinct morphologic characteristics in the larvae of both parental species.

15.5.3 Enzyme Activity

It is clear that the level of enzyme activity is closely related to the morphogenetic process. During growth and differentiation, enzymes involved in maintenance, substrate transport, and protein formation predominate, while in later embryogenesis enzymes affecting functional differentiation appear (MOOG, 1965). Enzyme studies reveal the function of individual components and imply morphogenetic significance.

Fertilization of sea urchin eggs, for instance, is associated with the activation of proteolytic enzymes probably used to remove the protein repressor that keeps the mRNA-ribosome complex inactive prior to insemination (LUNDBLAD, 1949, 1950; MONROY et al., 1965). Enzymes such as alkaline phosphtase, apyrase, glutaminase, and succinodehydrogenase are low and rather constant at the pregastrula stage but increase rapidly at gastrulation when respiration increases, protein synthesis intensifies, and differentiation becomes visible (GUSTAFSON and HASSELBERG, 1950, 1951; GUSTAFSON, 1965).

Little is known about the level of enzyme activity in the hybrid embryo. In the cross *St. purpuratus* ♀ x *D. excentricus* ♂ (SpD), alkaline phosphatase activity at the prism stage is intermediate between low maternal and high paternal values (Fig. 15.23). As the result of introducing the *Dendraster* genome into the *Strongylocentrotus* egg, enzyme activity is elevated (FLICKINGER, 1957). In the reciprocal cross *D. excentricus* ♀ x *St. purpuratus* ♂ (DSp), only maternal forms of esterases, malate dehydrogenases, and hatching enzymes specific to the maternal membrane are synthesized (OZAKI, 1965; OZAKI and WHITELEY, 1970; WHITELEY and WHITELEY, 1972). In the reciprocal hybrids between *St. purpuratus* and *St. franciscanus*, hatching enzymes are also entirely maternal (BARRETT and ANGELLO, 1969). But in the reciprocal combinations between *Allocentrotus fragilis* and *St. purpuratus* both activity and electrophoretic mobility of aryl sulfatase have been reported to be intermediate between the parental enzymes (FEDECKA-BRUNNER and EPEL, 1969; FEDECKA-BRUNNER et al., 1971).

Pigment formation involves synthesis of the corresponding enzymes. The eggs of many sea urchin species contain only carotenoid pigments, and naphthaquinone pigments (echinochromes and spinochromes) form at the onset of gastrulation. In the combination *L. variegatus* ♀ x *A. punctulata* ♂ (LA), larval pigmentation is maternal. Hybrid pigment patterns and quantity are similar to those of the parthenogenetically activated embryo, suggesting that the maternal nucleus may direct pigment synthesis. The paternal genome does not seem essential to pigmentation in hybrid larvae (YOUNG, 1958).

In the reciprocal hybrids of *St. purpuratus* and *St. franciscanus*, echinochrome synthesis begins at the same time it starts in the maternal species. The rate of pigment accumulation is intermediate in the combination *St. purpuratus* ♀ x *St. franciscanus* ♂ (SpSf) but maternal in the reciprocal cross (SfSp), indicating that the egg cytoplasm of the 2 species differs in the supply of enzyme synthesis precursors or else indicating other limiting factors in the process of pigment formation (CHAFFEE and MAZIA, 1963).

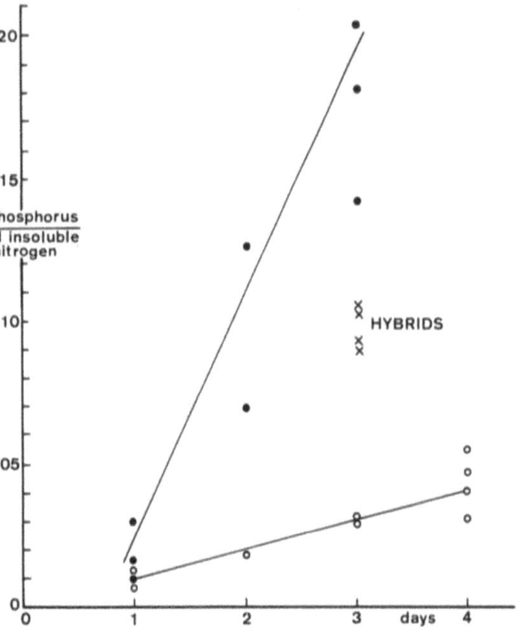

Fig. 15.23. Alkaline phosphatase changes in parental species SpSp (o) and DD (●) and hybrid SpD (X). Abscissa indicates age in days; ordinate, micrograms of phosphorus released over 2 hour period per micrograms of acid-insoluble nitrogen. (From FLICKINGER, 1957)

The species *St. purpuratus* and *St. droebachiensis* produce quite different pigments during development, making it possible to assess maternal and paternal influences on hybrid pigmentation. The cross *St. purpuratus* ♀ x *St. droebachiensis* ♂ (SpSd) is more difficult than the reciprocal combination SdSp because of the low rate of hybrid fertilization. However, both hybrids develop normally, and well-differentiated plutei are obtained at 120 hours. *Purpuratus* larvae synthesize 2 naphthaquinone pigments, A and B, the absorption spectra of which are identical to those of adult echinochromes (Fig. 15.24a). *Droebachiensis* larvae produce 2 major pigments, X and Y, the absorption maximum of which, 420 nm, differs from that of typical adult echinochromes. Pigment Y is a napthaquinone derivate despite its atypical echinochrome spectrum, while pigment X, at 280 nm, shows an absorption band in the ultraviolet range (Fig. 15.24b). Thus, the 5-day-old plutei of the 2 parental species synthesize different kinds of pigments (GRIFFITHS, 1965).

The 2 reciprocal hybrids have the same genomic makeup but different cytoplasm. The plutei of hybrids SpSd and SdSp contain both pigments A and B (Fig. 15.24c and d). In SpSd the *St. purpuratus* genome occurs in the cytoplasm of its own species, and the synthesis of pigments A and B is not surprising. But in SdSp *St. purpuratus* genes have to utilize precursors available in *St. droebachiensis* cytoplasm to produce pigments typical of the paternal species. Larval pigments X and Y are absent from both hybrids because these 2 compounds are thought to be intermediates in the biosynthesis of adult echinochromes. In the presence of *St. purpuratus* chromosomes they are converted to pigments A and B. Small amounts of pigments C and D occur in both hybrid plutei but not in any detectable amount in the parental species. (In

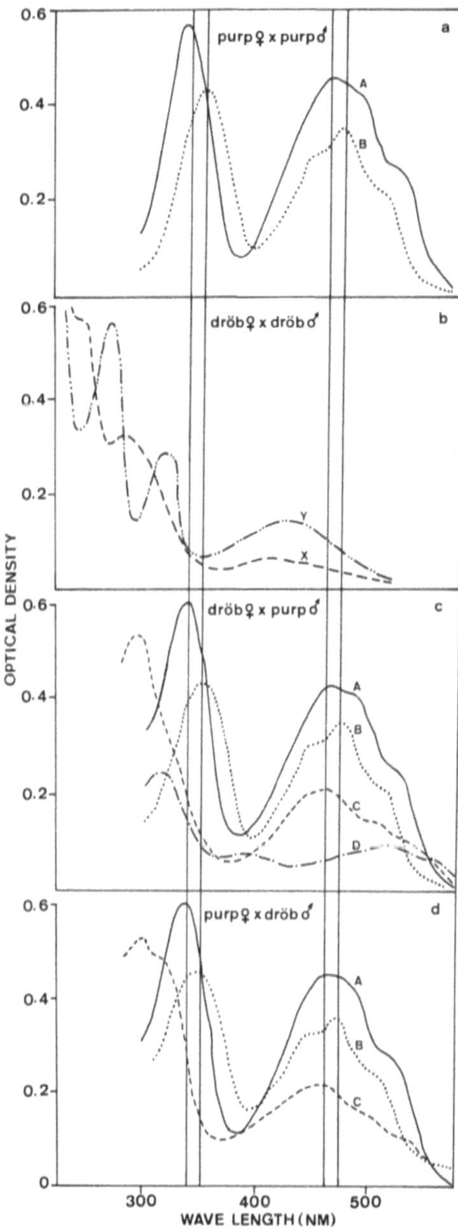

Fig. 15.24a-d. Absorption spectra of naphthaquinone pigments (A-D, X, Y) in pluteus larvae of parental species SpSp (a) and SdSd (b) and reciprocal hybrid combinations SdSp (c) and SpSd (d). Vertical lines through maxima of SpSp spectrum to show comparison. (From GRIFFITHS, 1965)

Fig. 15.24D the pigment D absorption curve is not shown because its concentration is too low to yield enough material for purification.) These pigments are believed to be intermediate that occur later than X and Y on the biosynthetic pathway of adult pigments or to be abnormal pigments, the synthesis of which is caused by hybrid enzymes (GRIFFITHS, 1965).

In both hybrids pigment synthesis begins at the same time as in *St. purpuratus* - about 50 hours after fertilization. In the 2 reciprocal crosses, the synthesis rates are alike, but both are lower than in *St. purpuratus*. It seems that cytoplasm does not affect the type of pigment hybrids produce or the time synthesis begins or the rate at which it proceeds (GRIFFITHS, 1965).

Echinochrome pigments that normally appear at gastrulation are not synthesized in hybrid DSp (WHITELEY and WHITELEY, 1972). Hybrid LT plutei raised to 90 hours form no visible echinchrome pigment although pigment granules appear in embryos of the *Lytechinus* maternal species at about 18 hours and in embryos of the *Tripneustes* paternal species at about 20 hours. As mentioned previously, the pigment patterns of *Lytechinus* embryos produced by artificial activation and those normally fertilized are the same, suggesting that a haploid set of maternal chromosomes is sufficient to support echinochrome synthesis in the hybrid (p. 451). It is possible that the paternal nucleus interferes with the normal biosynthesis process of the *Lytechinus* embryo by converting intermediates to colorless compounds or interacts with the maternal genome to form colorless hybrid products (BADMAN and BROOKBANK, 1970).

15.5.4 DNA and RNA Synthesis

Comparatively little is known about DNA and RNA synthesis in sea urchin hybrids. In lethal combinations it is possible that lethality is linked to blocks in DNA and RNA synthesis. Paternal chromosomes in various crosses are eliminated during embryogenesis, and degeneration sets in when cell differentiation increases. Analysis of nucleic acid patterns in hybrid embryos will help us to understand the molecular bases of these phenomena.

Total DNA of hybrid PA and reciprocal cross AP during morphogenetic inhibition is intermediate between the parental species (WHITELEY and BALTZER, 1958; BALTZER and CHEN, 1960; CHEN et al., 1960; BALTZER et al., 1961). When PP embryos reach the late blastula or early gastrula stage about 20 hours after fertilization, the DNA content in PA falls to 82%-85% of the maternal level, indicating in the hybrid embryo re-

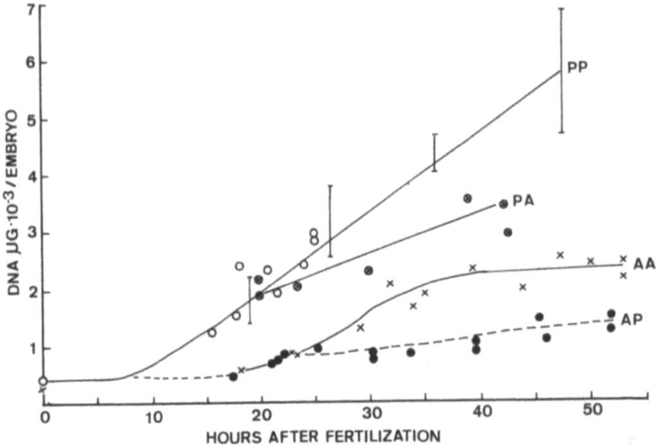

Fig. 15.25. Total DNA content changes during embryonic development of parental species PP and AA and hybrids PA and AP. (From BALTZER et al., 1961)

duced DNA synthesis that may be detected at the first sign of morphogenetic inhibition (Fig. 15.25). At the early pluteus stage, DNA synthesis in PA is about 71% of the maternal species; at the late pluteus stage it is about 52%.

At about 25 hours total DNA in AP is the same as in AA. When gastrulation begins, the DNA level rises in the maternal species but less in AP. When AA embryos are fully grown plutei, most hybrids are arrested at the early gastrula stage. During this maintenance period, DNA synthesis in AP increases slightly, to about half of what it is in AA. While the DNA content of PA is intermediate, that of AP is much lower than in both parental species.

Reduced DNA patterns may indicate that hybrid nuclei have less DNA than nuclei in maternal controls or that hybrid embryos have fewer nuclei than maternal controls. From the mesenchyme blastula 15- to 17-hours-old to the 2-day-old pluteus, the number of nuclei increases 4 times in PP and twice in AA (Table 15.6)(CHEN and BALTZER, 1964). When gastrulation begins and hybrid development becomes arrested, the number of nuclei in PA and AP is lower than in the maternal species and increases only slightly in older hybrids. Reduced DNA synthesis in hybrids is at least in part due to depressed mitotic rate.

^{3}H-thymidine incorporation into DNA in PP is initially 4 times higher than in AA and at later stages 2 to 3 times higher (Table 15.7)(BALTZER and CHEN, 1965). ^{3}H-thymidine incorporation in PA is intermediate at 17-30 hours but falls far below normal at 36-47 hours. This decline, plus the slow increase in number of nuclei, indicates almost no mitotic division. As DNA here was pulse-labeled, these values cannot be compared directly with earlier data on total DNA.

Autoradiography has shown that AA and PA nuclei from cleavage to early gastrula incorporate ^{3}H-thymidine faster than PP nuclei. Counting the silver granules over the labeled nuclei showed even higher values in the hybrid than in both parental controls (FICQ and BRACHET, 1963).

Microspectrophotometric measurement of Fuelgen-stained preparations later showed equal DNA in nuclei of PP and AA at the blastula and gastrula stages but consistently more DNA in nuclei of PA. Nuclear DNA continued to increase in PA from the blastula stage to embryos as old as 42 hours, indicating accumulated polyploid and aneuploid nuclei resulting from abnormal mitoses. Fuelgen-positive grains and filaments in PA embryo cytoplasm were thought to be eliminated chromosomal material (BRACHET and HULIN, 1970). Molecular hybridization has shown that chromosomes eliminated during hybrid development are of paternal origin and that the PA embryo synthesizes up to twice as much DNA from *Paracentrotus* as from *Arbacia* (Fig. 15.26a). The number of chromosomes of the 1-day-old hybrid may be as little as 30% of those of PP (DENIS and BRACHET, 1969a). Earlier findings had shown greater ^{3}H-thymidine incorporation and Fuelgen stainability in nuclei of PA than of PP (FICQ and BRACHET, 1963; BRACHET and HULIN, 1970). DNA-DNA annealing experiments have since shown equal amounts of both parental DNAs in the blocked hybrid DSp (WHITELEY and WHITELEY, 1972).

Our earlier data on total DNA and ^{3}H-thymidine incorporation were obtained by using PCA extracts from whole embryos, giving no direct information about DNA and ^{3}H-thymidine incorporation in individual nuclei. Microscopy and autoradiography were therefore used in further studies. Based on total ^{3}H-thymidine uptake in each embryo, the total number of nuclei, and the percentage of labeled nuclei, the ^{3}H-thymidine incorporation rate in a single replicating nucleus of a given

Table 15.6. Average number of nuclei at developmental stages of normal and hybrid sea urchin embryos. Symbols for developmental stages are explained in WHITELEY and BALTZER, 1958. Hours after fertilization indicate time of fixation for countings and RNA determinations but do not show precise limits of developmental stages. (From CHEN and BALTZER, 1964)

Developmental stage	Age after fertilization at 18°C (hours)	Number of countings	Average Number of nuclei	Developmental stage	Age after fertilization at 18°C (hours)	Number of countings	Average Number of nuclei
PP				AA			
Bl with immigrating My	15	4	391	Bl with immigrating My	- 17	2	318
Ga j1/2-j4/5	22-27	5	700	Ga j1/3-j2/3	24-27	4	390
Pri-Pl-1	38	4	761	Ga j2/3-Pri	38-40	2	605
Pl-1	41	5	950	Pri-Pl-1	48	3	685
Pl-2	48	2	1429	Pl-1-2	66-72	2	1050
				Pl-2	78	2	1473
PA				AP			
Bl with immigrating My	17-19	7	492	Bl with immigrating My	22	2	307
Ga j1/4-j1/2	24-26	6	533	Ga j0-j2/3	26-33	5	357
Ga j1/2-j4/5	48	2	641	Ga j1/2-j2/3	38	2	304
Ga j1/3-j4/5	61-65	2	982	Ga j0-j1/3	45	3	402
SS				PS			
Bl; thick wall; before My moves in	21	1	414	Bl; chromosome fragments eliminated in blastocoel	21	1	339
Bl with immigrating My	24	2	431	Bl with degraded material and	25-25 1/2	3	411
Ga with fully invaginated gut	41	1	669	BlMy-Ga j0	41	2	517
	47 1/4	2	692		49	3	670

Table 15.7. H^3-thymidine incorporation into DNA at developmental stages of normal and hybrid sea urchin embryos. (From BALTZER and CHEN, 1965)

Developmental stage	Age after fertilization at 18°C (hours)	Number of determinations	Radio-activity (cpm/100 embryos)
PP (6 series)			
Morula	4 hr,30 min- 5 hr,30 min	4	122.4
MyBl	14 hr,40 min-17 hr,20 min	10	120.5
Ga J 1/2-3/4	21 hr,30 min-22 hr, 5 min	4	112.9
Ga J 4/5-Pri	24 hr,10 min-29 hr,45 min	10	155.1
Pl-1	35 hr,35 min-41 hr,25 min	8	115.4
Pl-2	45 hr -48 hr,20 min	8	99.4
AA (3 series)			
MyBl	17 hr -17 hr,20 min	6	30.0
Ga J 1/2-2/3	24 hr,10 min-28 hr	6	39.6
Ga J 2/3-Pri	35 hr,55 min-41 hr,25 min	4	49.8
Pl-1 (2)	45 hr -52 hr,25 min	8	52.7
SS (2 series)			
Morula	6 hr	2	60.4
Bl (before hatching)	14 hr,40 min	2	66.8
Bl (after hatching)	21 hr,30 min-22 hr, 5 min	4	75.9
MyBl	27 hr,50 min	2	72.5
Ga J 1/3-2/3	36 hr	2	74.5
Ga J 4/5-Pri	45 hr	2	92.1
PA (3 series)			
MyBl	17 hr -17 hr,40 min	6	61.3
Ga J 0-1/2	24 hr,15 min-29 hr,45 min	6	62.2
Ga J 1/4-3/4	35 hr,40 min-42 hr, 5 min	6	23.2
Ga J 1/4-3/4	45 hr, 5 min-47 hr,10 min	6	26.4
PS (3 series)			
Morula (chromosome fragments eliminated)	4 hr,30 min- 6 hr, 5 min	4	60.1
Stereoblastula	14 hr,40 min-16 hr	4	46.8
Stereoblastula/GaJ0	21 hr,30 min-22 hr, 5 min	4	40.5
Stereoblastula/GaJ0	27 hr,50 min	2	22.3
Stereoblastula/GaJ0	38 hr,55 min	2	15.6
Stereoblastula/GaJ0	45 hr	2	18.6

type of embryo at a given stage of development was calculated (Table 15.8) (BALTZER et al., 1967). Development in AA is slower than in PP, and at the mesenchyme blastula and gastrula stages PP has twice as many nuclei as AA. At 17-18 hours and 26-29 hours, ^3H-thymidine incorporation in individual PP nuclei is 0.768 cpm x 10^{-2} and in AA 0.427 x 10^{-2}. Replicating nuclei of PP incorporate about 79% more ^3H-thymidine than those of AA. ^3H-thymidine incorporation in PA is initially lower than in the parental species and later becomes intermediate. At neither of the stages studied did it exceed ^3H-thymidine incorporation in the maternal controls.

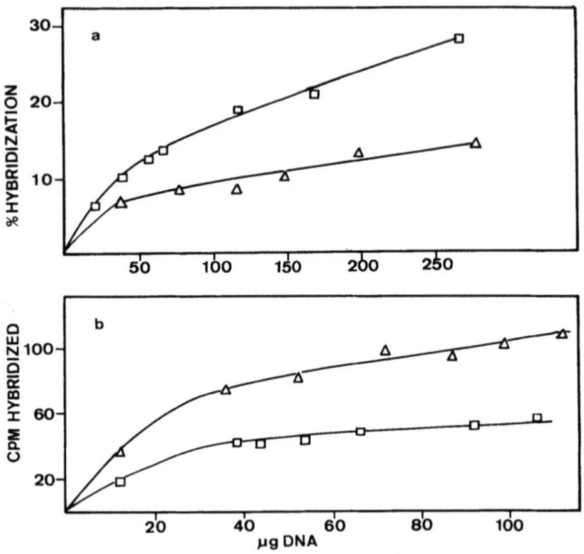

Fig. 15.26 a and b. Analysis of DNA and RNA synthesis in lethal cross PA by molecular hybridization. (a) DNA fragments (250 mµg, 2,500 cpm) labeled from third to 24th hour of development of PA hybridized with increasing amounts of DNA-agar from PP (□) and AA (△). (b) RNA (5 µg, 8,000 cpm) labeled from 24th to 27th hour of development of PA hybridized with increasing amounts of DNA from AA (△) and PP (□). (From DENIS and BRACHET, 1969a, 1969b)

Another Fuelgen-microspectrophotometric study of the DNA content of sperm and blastula nuclei showed in PP nuclei about 1.36 times more DNA than in AA nuclei (VILLIGER et al., 1970). The total number of chromosomes does not explain this difference, as PP has a diploid set of 36 and AA 40. Pool size and permeability of the precursors, and time and duration of isotopic labeling may be involved. But essentially these findings point up the enormous variability in lethal hybrid development. Healthy embryos must be collected individually for each study. And unless the developmental state is rigorously controlled, the use of heterogeneous materials may cause experimental results to be misinterpreted.

Examining newly hatched PA blastulae by the Fuelgen method revealed eliminated chromosomes, blocked mitoses, and nuclei of unequal size and stainability. During the subsequent 3 1/2 hours, nuclear pyknosis occurred and cells degenerated, suggesting that the primary cause of PA lethality may possibly be aneuploidy resulting from mitotic interference (BRACHET and HULIN, 1970).

Study of the RNA pattern provides information about the transcriptive activity of maternal and paternal genomes and about capacity for protein synthesis. Although DNA synthesis is reduced in hybrid PA, total RNA content remains strikingly constant during embryonic development (CHEN et al., 1960). Most RNA in the sea urchin embryo is stable rRNA. mRNA synthesis normally begins soon after fertilization, although the new mRNA is not used until after the blastula stage. Precursor incorporation into tRNA occurs at the onset of development. Even with rRNA, old ribosomes stored in egg cytoplasm decay gradually and then more

Table 15.8. Total ^{3}H-thymidine uptake, total number of nuclei, percentage of labeled nuclei per embryo, and incorporation rate per single nucleus in lethal hybrid PA and parental controls PP and AA. (From BALTZER et al., 1967)

Combination	Age	Stage	cpm/embryo x 10^{-2}	Mean of total number of nuclei per embryo	Labeled nuclei (%)	Average number of labeled nuclei per embryo (abs.)	cpm/replicating nucleus x 10^{-2}
PP	18 hr, 25 min	MyBl/GaJ0	120.5	468.3 ± 136.4	33.5	156.88	0.768
AA	17 hr, 20 min-18 hr, 25 min	MyBl	30.0	240.3 ± 43.0	29.1	69.92	0.429
PA	16 hr, 30 min-18 hr, 50 min	MyBl	61.3	499.8 ± 86.8	40.8	203.91	0.305
PP	26 hr, 30 min-29 hr, 45 min	GaJ4/5	155.1	795.1 ± 195.5	29.5	234.55	0.661
AA	26 hr, 20 min-28 hr	GaJ 1/2	39.6	376.8 ± 63.8	27.6	103.99	0.380
PA	26 hr, 30 min	GaJ0-1/2	62.2	496.0 ± 134.2	27.4	135.90	0.458

MyBl: blastula with mesenchymal cells; GaJ 1/2, 4/5: gastrula; entoderman invagination completed to 1/2-4/5.

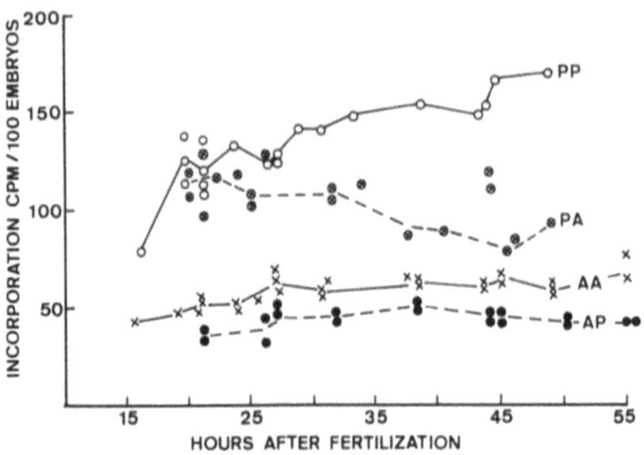

Fig. 15.27. Incorporation of ^{14}C-8-adenine into RNA during embryonic development of parental species PP and AA and hybrids PA and AP. (From CHEN and BALTZER, 1964)

rapidly after the mesenchymal blastula stage. Synthesis of new ribosomes is detectable at midgastrula and becomes very rapid thereafter (NEMER, 1963; GLISIN and GLISIN, 1964; GROSS et al., 1965; NEMER and INFANTE, 1965; SPIRIN and NEMER, 1965).

Experiments with isotopic labeling have shown differential patterns of RNA turnover. Developing embryos of the hybrid and parental species were labeled with ^{14}C-adenine and the incorporation rate measured from the tagged RNA prepared by differential extraction. The patterns of ^{14}C-adenine incorporation into RNA in PP and AA are remarkably similar to those of DNA synthesis (Fig. 15.27). In PA the rate of ^{14}C-adenine incorporation is comparable to the maternal level when development is first retarded, but then it falls off. ^{14}C-adenine incorporation in AP does not decrease during development but remains consistently lower than ^{14}C-adenine incorporation in maternal controls. In both lethal hybrids examined, the overall rate of RNA metabolism was reduced (CHEN and BALTZER, 1962, 1964). It is possible that the labeling, at least at early stages, reflects mainly terminal exchange of tRNA (GIUDICE, 1973).

Based on the analysis of base composition, all types of RNA - soluble, ribosomal, and DNA-like RNA - are synthesized in the PA hybrid. Until the late blastula stage at 18-20 hours, ^{3}H-uridine incorporation into the total RNA occurs in PA at about the same rate as in PP. Then it declines as development becomes arrested at 24 hours, much like ^{14}C-adenine incorporation into total RNA in PA. "Heavy" RNA (rRNA plus dRNA) synthesized in PA is less stable than in normal controls. Although synthesized, rRNA seems not to be assembled into ribosome but to accumulate in the nuclei of arrested hybrid gastrulae. Annealing experiments showed that RNA synthesized by the inhibited PA gastrula hybridized up to twice as well with paternal (*Arbacia*) DNA as with maternal (*Paracentrotus*) DNA (DENIS and BRACHET, 1969a, 1969b).

But if the hybrid eliminates chromosomes of the paternal species, why does the embryo synthesize more RNA of the paternal type? It may be that the parental species have a different number of active loci. More cistrons in the paternal genome than in the maternal genome, coding for hybrid RNA, would result in better hybridization of RNA from the PA gastrula with paternal DNA than with maternal DNA. Or it may be that the hybrid favors transcription of the paternal genome. The latter seems more likely because the number of active genes of the paternal and maternal types is apparently the same, and it may be that stage-specific transcription of the paternal but not the maternal genome is lost in the hybrid. Paternal genes normally transcribed during development are much more active in a foreign cytoplasm other than that of their own species. And paternal genes normally active only in the gastrula stage are repressed in cytoplasm other than their own. But in the PA hybrid cytoplasmic regulation of gene activity needs further proof.

DNA synthesis in hybrids PS and SP is of interest because they develop so differently. In PS most of the paternal chromosomes are eliminated at early cleavage, and development becomes blocked at late blastula to early gastrula. SP does not eliminate chromosomes and develops normally to well-differentiated plutei. Total DNA in PS and PP is the same until about 5 hours after fertilization (Fig. 15.28)(BALTZER et al., 1961). DNA synthesis does not begin until about the 32-cell stage, and the eliminated paternal chromosomes probably replicate further within this period. At 11 1/2 - 12 hours the DNA content of the hatching PS blastula falls to 64% that of PP. DNA synthesis in PS is thus inhibited considerably earlier than in PA and AP. When PS development becomes completely arrested at the stereoblastula or stereogastrula stage, DNA content is only about half of maternal controls. As the rate of DNA synthesis is about the same in both parental species and about half of the 38 chromosomes are eliminated at cleavage, theoretically the PS hybrid should have a DNA content about 58% of the maternal level.

The difference in DNA content of hybrid and maternal control becomes even more pronounced later when PP embryos have become well-developed plutei. Counting the number of nuclei in the different types of embryos shows a very low hybrid mitotic rate (Table 15.6). At 30-49 hours the average number of nuclei in PS is much lower than in the maternal species but close to the number in the paternal species. An increasingly inhibited mitotic rate may account for this. All maternal chromosomes and the 3 or 4 paternal chromosomes retained in the hybrid nuclei

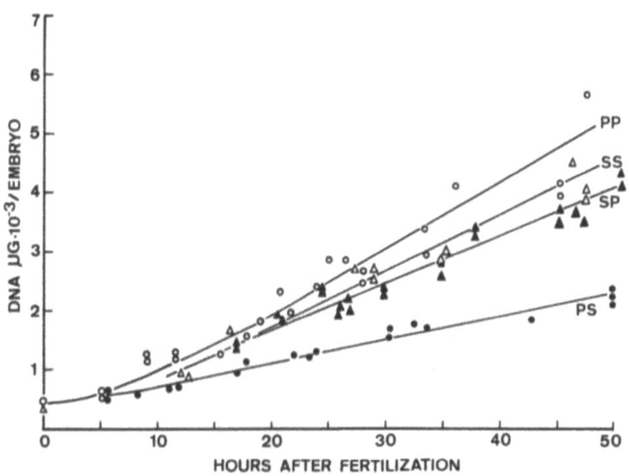

Fig. 15.28. Total DNA content changes during embryonic development of parental species PP and SS and hybrids PS and SP. (From BALTZER et al., 1961)

still have normal DNA replication, at least in the maintenance period. As each *Sphaerechinus* sperm nucleus contains about 1 1/2 times as much DNA as *Paracentrotus* (VILLIGER et al., 1970), the few remaining paterna chromosomes in PS probably contribute more to DNA content than has beer recognized.

The DNA content of the reciprocal combination SP is similar to that of the maternal control but consistently lower than that of the paternal species. SP is fully viable and develops at the maternal rate. Total RNA in hybrid PS remains at the maternal level throughout development (CHEN et al., 1960). It would be interesting to compare RNA turnover and synthesis in PS and SP. In PS the paternal genome is reduced to one-fifth its normal size, while in SP neither of the parental genomes is reduced.

In the reciprocal crosses between *Strongylocentrotus* (SpSp) and *Dendraster* (DD), DD embryos develop sooner than SpSp embryos. Hybrids SpD and DSp develop first at the maternal rate but are delayed at gastrulation. In both hybrids embryo labeling and autoradiographic analysis showed precocial replication of *Dendraster* chromosomes before pronuclear fusion. At early cleavage stages this caused heavier labeling of SpD nuclei than maternal control nuclei. Total radioactivity at the S_1 and S_2 phases showed greater ^{14}C-thymidine incorporation in hybrid embryos than maternal controls. But in the 16-hour-old blastula the ^{14}C-thymidine incorporation rate in SpD fell far below that in SpSp, probably due to gradually decreasing mitotic rate in the hybrid embryo. Precocial replication of *Dendraster* chromosomes is believed to occur because the pronucleus has just enough DNA polymerase for the S_1 phase and has to synthesize new enzyme for later S periods. Increased reduction of the DNA synthesis rate at later stages may be related to a negative feedback mechanism regulated by the amount of DNA per unit volume of cytoplasm. More frequent chromosomal replication and greater precursor incorporation at early cleavage and subsequent decrease of mitotic rate are reflected by the fact that DNA content in the hybrid is higher and then lower than DNA content of the maternal control embryo. And studies on total DNA in SPD and SpSp bear this out. Precocial replication and differential reduction of DNA synthesis may possibly cause heteroploidy and chromosomal imbalance resulting in variable morphologic characteristics in the developing hybrid (BROOKBANK, 1970; BROOKBANK and CUMMINS, 1972).

The hybrid combination *D. excentricus* ♀ x *St. purpuratus* ♂ (DSp) is similar to PA in that its development stops when gastrulation begins, although it has been known to give rise to plutei having intermediate larval skeletons. Cytologic observation indicates that early mitoses are normal and that paternal chromatin is not lost. At 10°C in embryos 70 hours old, morphogenesis is totally inhibited but cells are morphologically normal and show no signs of degeneration (WHITELEY and WHITELEY, 1972).

DNA from hybrid DSp, labeled with ^3H-thymidine, was annealed to DNA from *D. excentricus* or *St. purpuratus* sperm. The proportional binding to *St. purpuratus* DNA in relation to total annealing to both parental DNAs was 43.6% and to *D. excentricus* DNA 56.4%, indicating that the 2 parental genomes are replicated in the hybrid embryo and that both parental DNAs are synthesized in nearly equal amounts. When hybrid DNA was annealed to DNA of each parental species, identical melting curves with values of 72°C or 72.5°C were obtained as in homologous annealing (Fig. 15.29). Thus, DNA sequences replicated in the hybrid are qualitatively normal for both species (WHITELEY and WHITELEY, 1972).

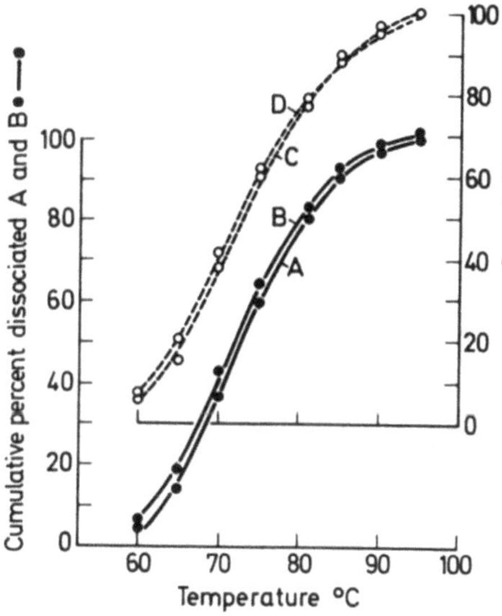

Fig. 15.29

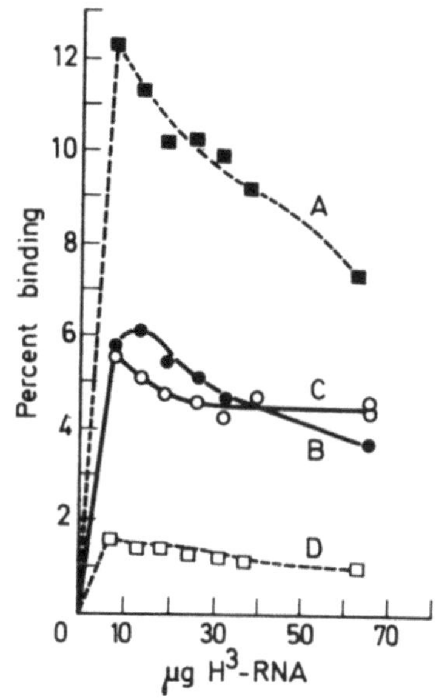

Fig. 15.30

Fig. 15.29. Profiles of thermal dissociation of DNA-DNA duplexes made by annealing ^{3}H-DNA from DSp hybrid to *excentricus* DNA (curves A and B) and *purpuratus* DNA (curves C and D). (From WHITELEY and WHITELEY, 1972)

Fig. 15.30. Increasing amounts of ^{3}H-RNA from the DSp hybrid and DD embryos were annealed to D- and Sp-DNA. (A) Homologous annealing of ^{3}H-RNA from DD embryos to D-DNA. (B, C) Annealing of ^{3}H-RNA from DSp hybrid to D- and Sp-DNA, respectively. (D) Heterologous annealing of ^{3}H-RNA from DD embryos to Sp-DNA. (From WHITELEY and WHITELEY, 1972)

In segments of reiterated genomes, *D. excentricus* and *St. purpuratus* show conservatism and the transcription of these reiterated genomes contributes to much of the RNA in the unfertilized egg. New synthesis of species-specific RNA begins shortly before gastrulation. To examine the type of RNA synthesized in the hybrid and the rate of synthesis, RNA from DSp and DD embryos of the same age was labeled with ^{3}H-uridine for 1 hour. Specific activity in DSp was 947 dpm/μg RNA and in DD 1031 dpm/μg RNA. This suggests that the rate of RNA synthesis in hybrid and maternal control is the same. Autoradiography shows the same cell pattern of RNA synthesis in both types of embryos (WHITELEY et al., 1970; WHITELEY and WHITELEY, 1972).

Homologous binding of DD RNA to DD DNA has a maximum value of 12.3%, while heterologous binding of DD RNA to SpSp DNA has a maximum value of only 1.6%. The corresponding values for hybrid RNA to maternal and paternal DNA are 6.1% and 5.5%, respectively (Fig. 15.30). When RNA bound to DNA is calculated from equal specific activity of RNA and expressed as the increasing ratio of RNA to DNA in the incubation medium, the sum of affinity of DSp RNA to each parental DNA approaches

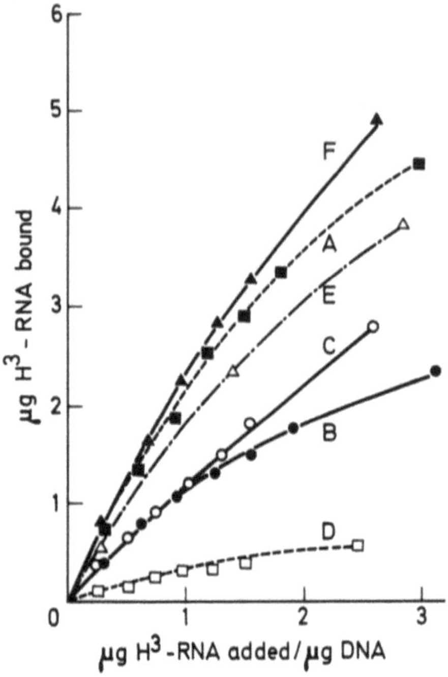

Fig. 15.31

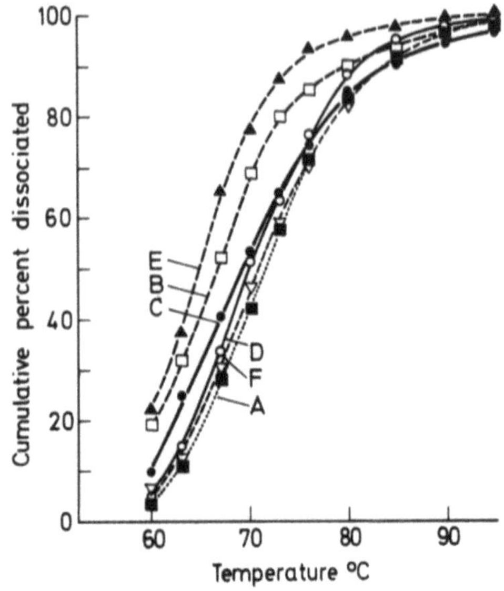

Fig. 15.32

Fig. 15.31. Ratios of RNA to DNA increased as increasing amounts of ^{3}H-RNA from DD and DSp embryos were annealed to both parental DNAs. Curves A, B, C, and D indicate data calculated from Fig. 15.30 on basis of equal specific activity of RNAs. Curve E shows annealing of increasing amounts of ^{3}H-RNA from DSp hybrid in presence of 10.5 μg of Sp-DNA. Arithmetic sum of B and C is shown by curve F. (From WHITELEY and WHITELEY, 1972)

Fig. 15.32. Profiles of thermal dissociation of RNA-DNA duplexes made by annealing ^{3}H-RNA from DD, DSp, and SpSp embryos to D- and Sp-DNA. (A) ^{3}H-RNA from DD bound to D-DNA (T_m = 71.5°C). (B) ^{3}H-RNA from DD bound to Sp-DNA (T_m = 66.5°C). (C) ^{3}H-RNA from DSp bound to D-DNA (T_m = 69.5°C). (D) ^{3}H-RNA from DSp bound to Sp-DNA (T_m = 70°C). (E) ^{3}H-RNA from SpSp bound to D-DNA (T_m = 65°C). (F) ^{3}H-RNA from SpSp bound to Sp-DNA (T_m = 71°C). (From WHITELEY and WHITELEY, 1972)

that of homologous binding (Fig. 15.31). This suggests that the amount of RNA in the hybrid and in the normal DD embryo is the same (WHITELEY and WHITELEY, 1972).

In homologous annealings DD shows values of 71.5°C and SpSp 71°C, while in heterologous annealings DD shows values of 65°C and SpSp 66.5°C (Fig. 15.32). When hybrid RNA is bound to parental DNA, DD shows values of 69.5°C and SpSp 70°C. RNA molecules synthesized in the DSp hybrid thus seem to have complete base-pair matching when they form duplexes with DNA from both parental species (WHITELEY and WHITELEY, 1972).

Almost as well as unlabeled RNA from the parental species, that from the DSp hybrid competes with pulse-labeled RNA in forming RNA-DNA duplexes (Fig. 15.33). In the reiterated part of the genome, the DSp

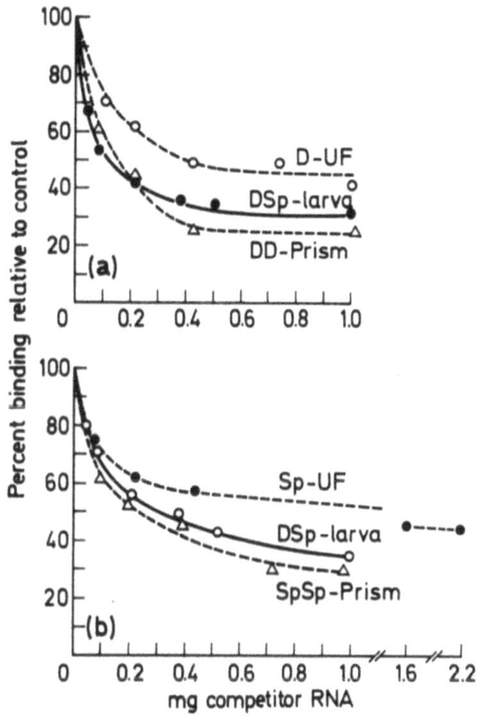

Fig. 15.33 a and b. Competition of unlabeled RNAs from maternal (a) and paternal (b) unfertilized eggs (UF) and prism embryos and hybrid in in vitro annealing of pulse-labeled ^{3}H-RNA from DSp to D- and Sp-DNA, respectively. (From WHITELEY and WHITELEY, 1972)

hybrid, as much as the parental species, synthesizes and stores transcription products of conservative and species-specific genes (WHITELEY and WHITELEY, 1972).

In the DSp hybrid DNA replication and transcription of species-specific genes are surprisingly normal in spite of morphogenetic block and other altered biochemical properties. It is possible that the morphogenetic block may be due to certain defects in the replication of only the unique part of the genome or in the expression of few regulatory genes in early development which are difficult to be detected. Other possible causes are abnormal transport of nuclear RNA to the cytoplasm, or failure of peptide chains to be released from the site of synthesis.

15.6 Conclusions

The development of sea urchin hybrids is clearly complex. Broadly speaking, it begins at the maternal rate and is delayed at gastrulation. The difference between the development rates of the hybrid and the maternal control becomes greater as development proceeds. Depending on the parental species, hybrid viability varies from highly lethal to fully viable. When the hybrid embryo can be raised to a well-developed pluteus larva, the influence of the paternal genome is usually visible in the skeleton and the body form.

Studies of the mechanism underlying the interaction between the nucleus of one species and the cytoplasm of another have been concerned primarily with DNA and RNA synthesis, particularly in lethal hybrids. In DNA synthesis, chromosomes may be eliminated, chromosomal DNA may be asynchronously replicated, and mitotic rate may be differentially reduced. In RNA synthesis, nucleocytoplasmic interaction may interfere with transcription activity or lead to the production of unstable RNA.

Sea urchin embryos are very sensitive to culturing conditions: temperature, salinity, shaking, crowding, and contamination. In hybridization this is especially true when the fertilization rate is low and hybrid viability reduced. Unless the culture dishes are tended carefully, larvae usually fail to develop and in morphologic observation or biochemical analysis lead to erroneous results. The developmental capacity and morphologic characteristics of hybrid embryos nearly always show a wide range of variability that must be taken into account in evaluating the experimental results.

In the early stages pluteus larvae differ only in body form and skeleton, severely restricting genetic analysis. And because of technical difficulties, data on reciprocal crosses are often unobtainable. It is almost impossible to raise the hybrid to sexual maturation, so that heredity cannot be studied in the second generation.

Common in interspecific hybridization, lethality is difficult to interpret because differences between lethal and viable combinations are not clear-cut. Viability in hybrids is nearly always less than in parental controls and hybrid development is usually delayed. But even in such typical lethal combinations as PA and PS some well-developed pluteus larvae that show the characteristics of both parents may still develop. The frequency of these larvae varies, however, from cross to cross, and many intermediate forms occur. If lethality is really due to positive influence of the parental genomes and not to culture conditions, then its variability may indicate the genetic makeup of the individuals used for cross-fertilization. In spite of such complexities, future work at the cellular and molecular level will make it possible to pinpoint the primary cause of lethality in the hybrid embryo.

Acknowledgement

Our investigation was supported by the Schweizerische Nationalfonds zur Förderung der wissenschaftlichen Forschung and the Stiftung Dr. J. De Giacomi der SNG. We are grateful to the Zoological Station at Naples where our experiments were carried out.

References

BADMAN, W.W., BROOKBANK, J.W., 1970. Serological studies of 2 hybrid sea urchins. Develop. Biol. 21, 243-256.

BALTZER, F., 1909. Die Chromosomen von Strongylocentrotus lividus and Echinus microtuberculatus. Arch. Zellforsch. 2, 549-632.

BALTZER, F., 1910. Über die Beziehung zwischen Chromatin und der Entwicklung und Vererbungsrichtung bei Echinodermusbastarden. Arch. Zellforsch. 5, 497-621.

BALTZER, F., BERNHARD, M., 1955. Weitere Beobachtungen über Letalität und Vererbungsrichtung beim Seeigelbastard Paracentrotus ♀ x Arbacia♂. Exp. Cell Res. Suppl. 3, 16-26.

BALTZER, F., CHEN, P.S., 1960. Über das zytologische Verhalten und die Synthese der Nukleinsäuren bei den Seeigelbastarden Paracentrotus ♀ x Arbacia ♂ und Paracentrotus ♀ x Sphaerechinus♂. Rev. Suisse Zool. 67, 183-194.
BALTZER, F., CHEN, P.S., 1965. A study of DNA synthesis in sea urchin hybrids by the incorporation of H^3-thymidine. Experientia 21, 194-196.
BALTZER, F., CHEN, P.S., TARDENT, P., 1961. Embryonalentwicklung, DNS-Synthese und Respiration des Bastards Arbacia ♀ x Paracentrotus♂, mit Vergleichen zu anderen Seeigelbastarden. Arch. Julius Klaus-Stift. Vererbungsforsch. Sozialanthropol. Rassenhyg. 36, 126-135.
BALTZER, F., CHEN, P.S., WHITELEY, A.H., 1958. Biochemical studies on sea urchin hybrids. Exp. Cell Res. Suppl. 6, 192-209.
BALTZER, F., HARDING, C., LEHMAN, H.E., BOPP, P., 1954. Über die Entwicklungshemmungen der Seeigelbastarde Paracentrotus♀ x Arbacia♂ und Psammechinus ♀ x Arbacia♂. Rev. Suisse Zool. 61, 402-416.
BALTZER, F., TARDENT, P., CHEN, P.S., 1967. About the DNA-synthesis during the early development of Paracentrotus lividus, Arbacia lixula and their hybrids. Experientia 23, 777-779.
BARRETT, D., ANGELLO, G.M., 1969. Maternal characteristics of hatching enzymes in hybrid sea urchin embryos. Exp. Cell Res. 57, 159-166.
BERG, W.E., 1950. Free amino acids in sea urchin eggs and embryos. Proc. Soc. exp. Biol. and Med. 75, 30-32.
BERG, W.E., 1965. Rates of protein synthesis in whole and half embryos of the sea urchin. Exp. Cell Res. 40, 469-489.
BERNHARD, M., 1957. Die Kultur von Seeigellarven (Arbacia lixula L.) in künstlichem und natürlichen Meerwasser mit Hilfe von Ionenaustauschsubstanzen und Komplexbildern. Pubbl. Sta. Zool. Napoli 29, 80.
BOELL, E.J., 1955. Energy exchange and enzyme development during embryogenesis. In: Analysis of Development (B.H. Willier, P.A. Weiss und V. Hamburger, eds), pp. 520-555. Philadelphia and London: Saunders.
BOHUS-JENSEN, A.A., 1953. The effect of trypsin on the cross fertility of sea urchin eggs. Exp. Cell Res. 5, 325-328.
BOREI, H., 1948. Respiration of oocytes, unfertilized eggs and fertilized eggs from Psammechinus and Asterias. Biol. Bull. 95, 124-150.
BOVERI, TH., 1889. Ein geschlechtlich erzeugter Organismus ohne mütterliche Eigenschaften. Sitzber. Ges. Morph. Physiol. München 5, 73-80.
BOVERI, TH., 1895. Über die Befruchtungs- und Entwicklungsfähigkeit kernloser Seeigeleier und über die Möglichkeit ihrer Bastardierung. Wilhelm Roux' Arch. Entwicklungsmech. Organismen 2, 394-443.
BOVERI, TH., 1901. Die Polarität von Oocyte, Ei und Larve des Strongylocentrotus lividus. Zool. Jahrb. Anat. Ont. 14, 630-653.
BOVERI, TH., 1918. 2 Fehlerquellen bei Merogonieversuchen und die Entwicklungsfähigkeit merogonischer und partiell-merogonischer Seeigelbastarde. Wilhelm Roux' Arch. Entwicklungsmech. Organismen 44, 417-471.
BRACHET, J., BIELIAVSKY, N., TENCER, R., 1962. Nouvelle observation et hypothèses sur la lètalitê chez les hybrides. Bull. Acad. roy. Belg. 48, 255-277.
BRACHET, J., HULIN, N., 1970. Observations sur les acides dèsoxyribonucleiques des hybrides lètaux entre oursins. Exp. Cell Res. 60, 393-400.
BROOKBANK, J.W., 1970. DNA synthesis and development in reciprocal interordinal hybrids of a sea urchin and a sand dollar. Develop. Biol. 21, 29-47.
BROOKBANK, J.W., CUMMINS, D.E., 1972. Microspectrophotometry of nuclear DNA during the early development of a sea urchin, a sand dollar, and their interordinal hybrids. Develop. Biol. 29, 234-240.

CHAFFEE, R.R., MAZIA, D., 1963. Echinochrome synthesis in hybrid sea urchin embryos. Develop. Biol. 7, 502-512.
CHEN, P.S., 1958. Further studies on free amino acids and peptides in eggs and embryos of different sea urchin species and hybrids. Experientia 14, 369-371.
CHEN, P.S., 1962. Trennung der freien Aminosäuren und Peptide von Seeigeleiern mittels Ionenaustauschchromatographie. Rev. Suisse Zool. 69, 288-296.
CHEN, P.S., 1967. Biochemistry of nuclei-cytoplasmic interactions in morphogenesis. In: The Biochemistry of Animal Development (R. Weber, ed.), Vol. II, pp. 115-191. New York and London: Academic Press.
CHEN, P.S., 1971. Biochemical Aspects of Insect Development. Basel: Karger Verlag.
CHEN, P.S., BACHMANN-DIEM, C., 1964. Studies on the transamination reactions in the larval fat body of Drosophila melanogaster. J. Insect Physiol. 10, 819-829.
CHEN, P.S., BALTZER, F., 1958. Species-specific differences in free amino-acids and peptides in sea urchin eggs and embryos (pure species and hybrids). Nature 181, 98-100.
CHEN, P.S., BALTZER, F., 1962. Experiments concerning the incorporation of labeled adenine into ribonucleic acid in normal sea urchin embryos and in the hybrid Paracentrotus ♀ x Arbacia♂. Experientia 18, 522-524.
CHEN, P.S., BALTZER, F., 1964. Further morphological and biochemical studies on normal and hybrid embryos of sea urchins. Experientia 20, 236-240.
CHEN, P.S., BALTZER, F., ZELLER, CH., 1960. Changes in nucleic acids in early amphibian and sea urchin embryos (pure species, merogonic and hybrid combinations). In: Symposium Germ Cells and Development (S. Ranzi, ed.), pp. 506-523. Milan: Inst. Intern. Embryol. and Fondazione A. Baseli.
CHEN, P.S., LEVENBOOK, L., 1966. Studies on the haemolymph proteins of the blowfly Phormia regina. II. Synthesis and breakdown as revealed by isotopic labeling. J. Insect Physiol. 12, 1611-1627.
DENIS, S., 1968. Changes in the level of triphosphopyridine nucleotides during development of sea urchin eggs (normal and lethal hybrids). Biochim. Biophys. Acta 157, 212-214.
DENIS, H., BRACHET, J., 1969a. Gene expression in interspecific hybrids. I. DNA synthesis in the lethal cross Arbacia lixula ♂ x Paracentrotus lividus ♀. Proc. Nat. Acad. Sci. 62, 194-201.
DENIS, H., BRACHET, J., 1969b. Gene expressions in interspecific hybrids. II. RNA synthesis in the lethal cross Arbacia lixula ♂ x Paracentrotus lividus ♀. Proc. Nat. Acad. Sci. 62, 438-445.
DENIS, H., BRACHET, J., 1970. Expression du génome chez les hybrides interspecifiques. Fidélité de la transcription dans le croisement létal Arbacia lixula ♂x Paracentrotus lividus♀. Eur. J. Biochem. 13, 86-93.
DEUCHAR, E.M., 1961. Amino-acid activation in embryonic tissues of Xenopus laevis. I. Increased 32 p exchange between pyrophosphate and adenosine triphosphate in the presence of added l-leucine. Exp. Cell Res. 25, 364-373.
ELSTER, H.J., 1935. Experimentelle Beiträge zur Kenntnis der Physiologie der Befruchtung bei Echinoiden. Wilhelm Roux' Arch. Entwicklungsmech. Organismen 133, 1.
FEDECKA-BRUNNER, B., ANDERSON, M., EPEL, D., 1971. Control of enzyme synthesis in early sea urchin development. Aryl sulfatase activity in normal and hybrid embryos. Develop. Biol. 25, 655-671.
FEDECKA-BRUNNER, B., EPEL, D., 1969. Nuclear control of "lysosomal" aryl sulfatase activity in sea urchin embryos. J. Cell Biol. 43, 35a.

FICQ, A., BRACHET, J., 1963. Métabolisme des acides nucleiques et des protéines chez les embryos normaux et les hybrides létaux entre echinodermes. Exp. Cell Res. 32, 90-108.
FLICKINGER, R.A., 1957. Evidence from sea urchin-sand dollar hybrid embryos for a nuclear control of alkaline phosphatase activity. Biol. Bull. 112, 21-27.
GEUSKENS, M., 1968. Etude ultrastructurale des embryos normaux et des hybrides létaux entre echinodermes. Exp. Cell Res. 49, 477-487.
GIUDICE, G., 1973. Developmental Biology of the Sea Urchin Embryo. New York and London: Academic Press.
GLIŠIN, V.R., GLIŠIN, M.V., 1964. Ribonucleic acid metabolism following fertilization in sea urchin eggs. Proc. Nat. Acad. Sci. 52, 1548-1553.
GRIFFITHS, M., 1965. A study of the synthesis of naphthaquinone pigments in the larvae of 2 species of sea urchins and their reciprocal hybrids. Develop. Biol. 11, 433-447.
GROSS, P.R., COUSINEAU, Q.H., 1963. Synthesis of spindle associated proteins in early cleavage. J. Cell Biol. 19, 260-265.
GROSS, P.R., KRAEMER, K., MALKIN, L.J., 1965. Base composition of RNA synthesized during cleavage of the sea urchin embryo. Biochem. Biophys. Res. Comm. 18, 569-575.
GUSTAFSON, T., 1952. Nitrogen metabolism, enzymic activity, and mitochondrial distribution in relation to differentiation in the sea urchin egg. From the Wenner-Gren Inst. Exp. Biol., Uppsala.
GUSTAFSON, T., 1965. Morphogenetic significance of biochemical patterns in sea urchin embryos. In: The Biochemistry of Animal Development (R. Weber, ed.), Vol. I, pp. 139-202. New York and London: Academic Press.
GUSTAFSON, T., HASSELBERG, I., 1950. Alkaline phosphatase activity in developing sea urchin eggs. Exp. Cell Res. 1, 371-375.
GUSTAFSON, T., HASSELBERG, I., 1951. Studies on enzymes in the developing sea urchin egg. Exp. Cell Res. 2, 642-672.
GUSTAFSON, T., HJELTE, M.B., 1951. The amino acid metabolism of the developing sea urchin egg. Exp. Cell Res. 11, 474-490.
GUSTAFSON, T., HÖRSTADIUS, S., 1957. Changes in the determination of the sea urchin egg induced by amino acids. Pubbl. Sta. Zool. Napoli 29, 407-424.
GUSTAFSON, T., LENICQUE, P., 1952. Studies on mitochondria in the developing sea urchin egg. Exp. Cell Res. 3, 251-274.
HAGSTRÖM, B.E., 1956. Studies on the fertilization of jelly-free sea urchin eggs. Exp. Cell Res. 10, 24-28.
HAGSTRÖM, B.E., 1959. Experiments on hybridization of sea urchins. Arkiv Zool., Ser. 2, 12, 127-135.
HARDING, C.V., HARDING, D., 1952a. Cross fertilization with the sperm of Arbacia lixula. Exp. Cell Res. 3, 475-484.
HARDING, C.V., HARDING, D., 1952b. The hybridization of Echinocardium cordatum and Psammechinus miliaris. Arkiv Zool. 4, 91-93.
HARDING, C.V., HARDING, D., PERLMANN, P., 1954. Antigens in sea urchin hybrid embryos. Exp. Cell Res. 6, 202-210.
HARVEY, E.B., 1956. The American Arbacia and other sea urchins. New Jersey: Princeton University Press.
HÖRSTADIUS, S., 1936. Studien über heterosperme Seeigelmerogone nebst Bemerkungen über einige Keimblattchimäsen. Mém. Mus. R. d'Hist. Nat. Bruxelles, sér. 2, fasc. 3, 801-880.
HULTIN, T., 1948a. Species specificity in fertilization reaction. The role of the vistelline membrane of sea urchin eggs in species specificity. Arkiv Zool. 40A, no. 12, 1-9.
HULTIN, T., 1948b. Species specificity in fertilization reaction. II. Influence of certain factors on the cross-fertilization capacity of Arbacia lixula (L.). Arkiv Zool. 40A, no. 20, 1-8.

HULTIN, T., 1961. The effect of puromycin on protein metabolism and cell division in fertilized sea urchin eggs. Experientia 17, 410-411.
IMMERS, J., RUNNSTRÖM, J., 1960. Release of respiratory control by 2,4-dinitrophenol in different stages of sea urchin development. Develop. Biol. 2, 90-104.
KAESTNER, A., 1963. Lehrbuch der speziellen Zoologie. (Teil I. Wirbellose), pp. 1160-1299. Jena: Gustav Fischer.
KAVANAU, J.L., 1953. Metabolism of free amino acids, peptides and proteins in early sea urchin development. J. exp. Zool. 122, 285-337.
KAVANAU, J.L., 1954a. Amino acid metabolism in developing sea urchin embryos. Exp. Cell Res. 6, 563-566.
KAVANAU, J.L., 1954b. Amino acid metabolism in the early development of the sea urchin Paracentrotus lividus. Exp. Cell Res. 7, 530-557.
LINDAHL, P.E., 1939. Zur Kenntnis der Entwicklungsphysiologie des Seeigeleies. Z. vergl. Physiol. 27, 233-250.
LOEB, J., 1904. Über Befruchtung, künstliche Parthenogenese und Cytolyse des Seeigeleies. Pflüger's Arch. ges. Physiol. 103, 257-265.
LUNDBLAD, G., 1949. Proteolytic activity in eggs and sperms from sea urchins. Nature 163, 643.
LUNDBLAD, G., 1950. Proteolytic activity in sea urchin gametes. Exp. Cell Res. 1, 264-271.
MARKMAN, B., 1961a. Regional differences in isotopic labeling of nucleic acid and protein in early sea urchin development. Exp. Cell Res. 23, 118-129.
MARKMAN, B., 1961b. Differences in isotopic labeling of nucleic acid and protein in sea urchin embryos developing from animal and vegetal egg halves. Exp. Cell Res. 25, 224-227.
MCCLUNG, C.E., 1939. Chromosome numbers in animals. Tab. Biol. 18, 34.
MONROY, A., GROSS, P.R., 1967. The control of gene action during echinoderm embryogenesis. Exp. Biol. Med. 1, 37-51.
MONROY, A., MAGGIO, R., 1966. Amino acid metabolism in the developing embryos. In: Physiology of Echinodermata (R.A. Boolootian, ed.), pp. 743-756. New York, London and Sidney: John Wiley and Sons.
MONROY, A., MAGGIO, R., RINALDI, A.M., 1965. Experimentally induced activation of the ribosomes of the unfertilized sea urchin egg. Proc. Nat. Acad. Sci. 54, 107-111.
MOOG, F., 1965. Enzyme development in relation to functional differentiation. In: The Biochemistry of Animal Development (R. Weber, ed.), Vol. I, pp. 307-365. New York and London: Academic Press.
MOORE, A.R., 1943. Maternal and paternal inheritance in the plutei of hybrids of the sea urchins Strongylocentrotus purpuratus and Strongylocentrotus franciscanus. J. exp. Zool. 94, 211-228.
MOORE, A.R., 1957. Biparental inheritance in an interordinal cross of sea urchin and sand dollar. J. exp. Zool. 135, 75-79.
NEMER, M., 1963. Old and new RNA in the embryogenesis of the purple sea urchin. Proc. Nat. Acad. Sci. 50, 230-235.
NEMER, M., BARD, S.G., 1963. Polypeptide synthesis in sea urchin embryogenesis: An examination with synthetic polyribonucleotides. Science 140, 664-666.
NEMER, M., INFANTE, A.A., 1965. Messenger RNA in early sea urchin embryos: size classes. Science 150, 217-221.
OZAKI, H., 1965. Differentiation of esterases in the development of echinoderms and their hybrids. Ph.D. Thesis, University of Washington. (Cited in WHITELEY and WHITELEY, 1972).
OZAKI, H., WHITELEY, A.H., 1970. L-malate dehydrogenase in the development of the sea urchin Strongylocentrotus purpuratus. Develop. Biol. 21, 196-215.
PAIGEN, K., 1963. Changes in inducibility of galactochinase and galactocidase during inhibition of growth in Escherichia coli. Biochem. Biophys. Acta 77, 318-328.

PERLMANN, P., 1953. Soluble antigens in sea urchin gametes and developmental stages. Exp. Cell Res. 5, 394-399.
PERLMANN, P., GUSTAFSON, T., 1948. Antigens in the egg and early developmental stages of the sea urchin. Experientia 40, 481-483.
RUNNSTRÖM, J., 1952. The cell surface in relation to fertilization. Symp. Soc. exp. Biol. 6, 39-88.
RUNNSTRÖM, J., MONNE, L., BROMAN, L., 1943. On some properties of the surface layers in the sea urchin egg and their changes upon activation. Arkiv Zool. 35A, no. 3, 1-32.
SPIEGEL, M., OZAKI, H., TYLER, A., 1965. Electrophoretic examination of soluble proteins synthesized in early sea urchin development. Biochem. Biophys. Res. Comm. 21, 135-140.
SPIRIN, A.S., NEMER, M., 1965. Messenger RNA in sea urchin embryos: cytoplasmic particles. Science 150, 214-217.
TAYLOR, C.V., TENNENT, D.H., 1924. Preliminary report on the development of egg fragments. Carnegie Inst. Year Book Nr. 23.
TENNENT, D.H., 1912b. Studies in cytology I. A further study of the ters in Echinoderm hybrids. Wilhelm Roux' Arch. Entwicklungsmech. Organismen 29, 1-14.
TENNENT, D.H., 1912a. The behavior of the chromosomes in cross-fertilized echinoid eggs. J. Morph. 23, 17-29.
TENNENT, D.H., 1912b. Studies in cytology I. A further study of the chromosomes of Toxopneustes variegatus. II. The behavior of the chromosomes in Arbacia-Toxopneustes crosses. J. exp. Zool. 12, 391-411.
TENNENT, D.H., 1914. The early influence of the spermatozoan upon the characters of echinoid larvae. Carnegie Inst. Wash. 5, 127-138.
TERMAN, S.A., GROSS, P.R., 1965. Translation-level control of protein synthesis during early development. Biochem. Biophys. Res. Comm. 21, 595-600.
TYLER, A., TYLER, B.S., 1966. Physiology of fertilization and early development. In: Physiology of Echinodermata (R.S. Boolootian, ed.), pp. 683-741. New York, London and Sydney: Interscience Publishers, John Wiley and Sons.
UBISCH, VON, L., 1953. Über Seeigelmerogone. Experientia 9, 294.
UBISCH, VON, L., 1954. Über Seeigelmerogone. Pubbl. Sta. Zool. Napoli 25, 246-340.
UBISCH, VON, L., 1955. Über Seeigelbastarde. Exp. Cell Res. Suppl. 3, 358-365.
UBISCH, VON, L., 1957a. Merogone und Bastarde von Seeigeln. Nova Acta Leopoldina 19, 5-12.
UBISCH, VON, L., 1957b. Über Seeigelmerogone II. Pubbl. Sta. Zool. Napoli 30, 279-308.
UBISCH, VON, L., 1959. Die Entwicklung von Genocidaris maculata und Sphaerechinus granularis, sowie Bastarde und Merogone von Genocidaris. Pubbl. Sta. Zool. Napoli 31, 159-208.
VILLIGER, M., CZIHAK, G., TARDENT, P., BALTZER, F., 1970. Feulgen microspectrophotometry of spermatozoa and blastula nuclei of different sea urchin species. Exp. Cell Res. 60, 119-126.
WHITELEY, A.H., BALTZER, F., 1958. Development, respiratory rate and content of desoxyribonucleic acid in the hybrid Paracentrotus ♀ x Arbacia ♂. Pubbl. Sta. Zool. Napoli 30, 402-457.
WHITELEY, H.R., MCCARTHY, B.J., WHITELEY, A.H., 1970. Conservatism of base sequences in RNA for early development of echinoderms. Develop. Biol. 21, 216-242.
WHITELEY, A.H., WHITELEY, H.R., 1972. The replication and expression of maternal and paternal genomes in a blocked echinoid hybrid. Develop. Biol. 29, 183-198.
WILDE, C.E., 1955a. The urodele neuroepithelium. I. The differentiation in vitro of the cranial neural crest. J. exp. Zool. 130, 573-596.

WILDE, C.E., 1955b. The urodele neuroepithelium. II. The relationship between phenylalanine metabolism and the differentiation of neural crest cells. J. Morph. 97, 313-344.
WILDE, C.E., 1956. The urodele neuroepithelium. III. The presentation of phenylalanine to the neural crest by archenteron roof mesoderm. J. exp. Zool. 133, 409-440.
YOUNG, R.S., 1958. Development of pigment in the larvae of the sea urchin, Lytechinus variegatus. Biol. Bull. 114, 394-403.

16. Animalization and Vegetalization

R. Lallier

16.1 Introduction

In the sea urchin larva the proportions between ectodermic and entomesodermic structures are established at the beginning of development. If one changes the proportions between ectoderm and entomesoderm, the subsequent differentiation results in the formation of 2 larval types with characteristic morphology, very different from the normal pluteus larva. These phenomena have received the names animalization and vegetalization. Their realization involves a change in the orientation and differentiation of certain larval areas.

We shall examine, successively, the techniques used to provoke these phenomena, their effects on metabolism and certain bio-physical properties of eggs.

The chemical techniques took their origin from the research of HERBST on the effects of lithium and sulphocyanide ions. From this starting point they have expanded considerably with the discovery of new substances capable of orientating the development of larvae towards animalization or vegetalization. Recently, substances have been isolated from eggs that have proven to be very active animalizing and vegetalizing agents.

The metabolism of transformed larvae was studied, and compared with that of normal larvae. This research was first concerned with respiratory metabolism, at a time when the analysis of this aspect was advancing, thanks to the improvements in measuring gaseous exchange, and to the progress in our knowledge of the different stages of carbohydrate degradation. Subsequently, the technical and theoretical enrichment resulting from research on the metabolism of nucleic acids and proteins in microorganisms permitted the examination of these phenomena to be undertaken on the sea urchin.

The interpretation of the sea urchin larva's differentiation lies in the concept of the double gradient. The analysis of this organism's differentiation therefore involves the elucidation of the nature of these gradients, the mechanisms which govern their formation and development and lastly the interaction of the gradients and the way in which they intervene in differentiation. Research on the mode of action of the vegetalizing and animalizing agents, be it a question of natural agents present in eggs, or chemical substances exercising an effect on differentiation, offers a way of approaching the analysis of these phenomena.

Recent reviews on these problems have been presented by CZIHAK (1962), LALLIER (1964a) and GUSTAFSON (1965, 1969).

16.2 Definitions

The modifications in development of the sea urchin egg can be classed into 3 categories: inhibition, exogastrulation and lastly animalization and vegetalization.

Inhibition corresponds to the stopping of development at a more or less early stage, according to the nature and conditions of the experimental interference.

Exogastrulation corresponds to the exovagination of entomesodermic structures without the modification of proportions between the ectodermic and entomesodermic structures.

On the contrary, in animalization as in vegetalization, these proportions are changed. Ectodermic structures are enhanced at the expense of entomesodermic structures in animalization, and vice versa in vegetalization. A part of the structures normally destined to form ectoderm or entomesoderm becomes changed under the influence of experimental treatment, and this has an effect on the orientation of the larva's differentiation. Fig. 16.1a$_1$-a$_3$ shows a series of larvae, animalized to varying degrees while Fig. 16.1b$_1$-b$_4$ shows different degrees of vegetalization. Fig. 16.1a$_4$ represents a type of larva with a peculiar structure. These larvae have a radial symmetry, i.e. their archenteron is orientated axially according to the animal-vegetal axis, and the spicules are arranged in a crown around the archenteron base. Radialization, corresponding to a mild type of animalization, is provoked by animalizing agents in weaker concentrations than those necessary for complete animalization. Finally, we

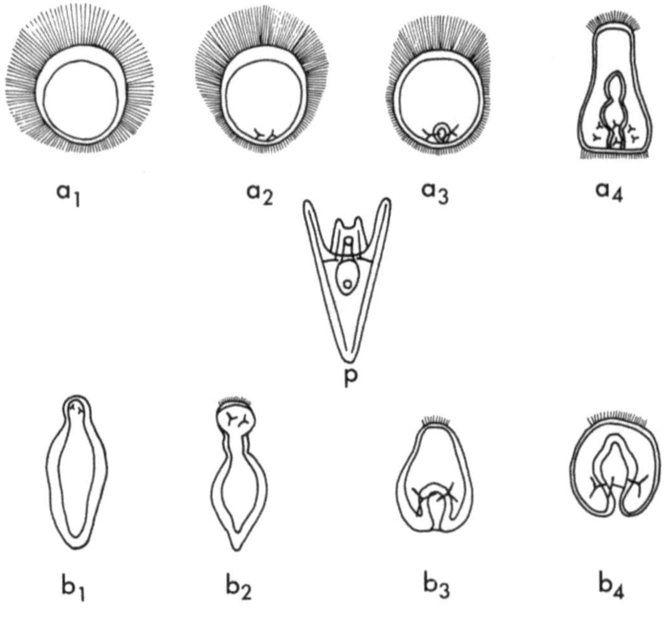

Fig. 16.1. a$_1$-a$_3$: Animalized larvae arranged in order of decreasing animalization. a$_4$: Radialized larva. b$_1$-b$_4$: Vegetalized larvae arranged in order of decreasing vegetalization. p, normal pluteus larva

shall point out that animalization and vegetalization generally result from experimental interference. Up to the present time, only 1 example of spontaneous animalization has been reported (HÖRSTADIUS and GUSTAFSON, 1956).

16.3 Experimental Methods

16.3.1 Operative Methods

Inaugurated by DRIESCH (1891) and BOVERI (1901) and developed by RUNNSTRÖM (1914), VON UBISCH (1925a, 1925b) and HÖRSTADIUS (1935, 1939), this method permitted the analysis of the first stages of differentiation of the sea urchin egg. Technically, it is based upon the isolation of blastomeres and their reassociation in varying combinations. In Chapter 13 of this book HÖRSTADIUS explains the principles of this method and its results. These have led to the sea urchin egg being described as a bipolar system with an animal pole corresponding to the position where the polar bodies are formed and a vegetal pole localized at the point of attachment of the oocyte in the ovary. The animal-vegetal axis passes through these 2 poles. The first 2 cleavage planes are parallel to this axis and result in blastomeres of the same prospective value in view of subsequent differentiation. The cleavage planes perpendicular to this axis mark the limit of blastomeres with different prospective value. The blastomeres situated nearest the animal pole form ectodermic structures, and if isolated, they will result in animalized larvae. In contrast, the blastomeres situated near the vegetal pole will form mainly entomesodermic structures, and these groups of blastomeres, when isolated, will give vegetalized larvae. The double gradient concept was conceived by RUNNSTRÖM (1928a, 1929) to interpretate these data. According to this theory, 2 qualitatively different metabolic gradients overlap one another. One of the gradients would have its maximum at the animal pole, the other at the vegetal pole. The interaction of these 2 gradients would control the normal development of the larvae. Animalization and vegetalization would each correspond to the predominance of one type of gradient over the other.

The theoretical aspects of the gradient concept have been examined by APTER (1966), and WOLPERT (1969).

16.3.2 Chemical Methods

Eggs are treated during the cleavage stage, i.e. from fertilization to the blastula stage. Treatment with chemical agents after this stage provokes non-specific inhibition of development. The cleavage stage corresponds to the embryonic determination period during which differentiation tendencies are progressively stabilized. After this period, the determination of these tendencies is irreversible and can no longer be experimentally changed. Animalization and vegetalization are never complete. The extreme animal and vegetal polar regions always retain some of their original characteristics which indicates a very early determination.

16.3.3 Combination of Operative and Chemical Methods

This method was used by HÖRSTADIUS (1936) to show that blastomeres vegetalized by Li possess physiological properties analogous to those of blastomeres originating from the most vegetal region of the larva. After isolation, the animal and vegetal halves have been used to test the activity of chemical agents, and the animal halves in particular have proven to be more susceptible to experimental deviations of differentiation than whole eggs (HÖRSTADIUS, 1953).

We shall now examine the main chemical agents exercising an effect on differentiation, and we will expose some of their properties which can appear useful for the study of their mode of action.

16.4 Vegetalizing Agents

16.4.1 Lithium

Lithium was the first known vegetalizing agent (HERBST, 1892) and at the present time is the most studied from the biochemical point of view.

The lithium ion has the smallest diameter of all the ions of alkaline metals. It is also the most strongly hydrated and occupies an extreme in the HOFMEISTER's (1888) lyotropic serie. This serie classifies ions according to their effect on the precipitation and hydration of colloids and macromolecules. Lithium's precipitating action on cytoplasmic colloids was moreover put forward to explain its vegetalizing effects (RUNNSTRÖM, 1928b; LINDAHL, 1936). In fact, through some of its chemical properties, in particular the formation of insoluble sulphates and phosphates, Li approximates alkaline earth metals. In a recent review, DIAMOND and WRIGHT (1969) have examined the physiological basis of ion selectivity in some biological systems. It is the lithium ion which is active, therefore it is used in the form of an ionized soluble salt, generally chloride. Increasing the lithium concentration and the treatment time result in a stronger degree of vegetalization. The unfertilized egg does not appear susceptible to the vegetalizing effects of lithium, and 2 phases of maximum susceptibility exist during development. The first is at the beginning of cleavage (HÖRSTADIUS, 1936; LINDAHL, 1936; OTOMO, 1961). The second phase appears at the advanced gastrula stage where lithium provokes a non-specific inhibition of development (BÄCKSTRÖM and GUSTAFSON, 1953). We further point out that lithium reduces the cleavage rate (HÄGSTROM and LÖNNING, 1967). Although the penetration of lithium into sea urchin eggs has been shown by spectrophotometry (RANZI and FALKENHEIM, 1937), ELBERS (1959), through electron microscope studies, has suggested that lithium does not penetrate into cells but acts on the surfaces. The absence of ultrastructural alteration however, does not appear sufficient to infer the absence of penetration of this ion. From data obtained from different tissues, in particular muscle, it appears that lithium is capable of cell penetration. Research on this problem is made difficult in the sea urchin because the half-life of radioactive lithium isotopes is very short. In addition, according to LOEWENSTEIN (1968), lithium can break down intercellular communications. We shall see further on by studying the effects of proteolytic enzymes that a lithium effect on the cell surface appears probable. At the present time, lithium is the only metal known to exercise a vegetalizing action. Amongst the other alkaline metals, only potassium and to a lesser degree rubidium, counteract the vegetalizing effects (LOEB, 1920; RUNNSTRÖM, 1928a; LALLIER, 1960a). Caesium and other monovalent ions inhibit development without counteracting lithium effects (LALLIER, 1960a). The protective action of potassium ions is particularly interesting. These ions play an important role in carbohydrate catabolism (PRESSMAN and LARDY, 1952; GAMBLE, 1957), in RNA polymerase activity (HURWITZ, 1963) and in the aminoacyl - tRNA-ribosome complex formation (LUBIN, 1963; SPYRIDES, 1964). The effects of lithium, which inhibit the formation of this complex are partially reversed by potassium (SUZUKA and KAJI, 1968). Lithium also reduces the acceptance of amino acids by tRNA in bacteria (SMITH, 1969). Although the toxity of lithium ions has been known for a long time, the mechanism behind this is not known. We shall point out in this respect that, in addition to their aptitude for modifying potassium ion distribution, lithium also appears capable of interfering with the distribution of sodium, as has been observed in nerve tissue (COPPEN, 1965). Lithium ions also disturbe the distribution of biogenic amines (KATZ et al., 1968). Intracellular anionic polymeres carrying phosphate and carboxyl groups show an affinity for alkaline ions. This affinity decreases in the following order: lithium > sodium > potassium, for different polyanions (BREGMAN, 1953; GREGOR et al., 1956). It seems then that an extensive spectrum of research relating to various organisms or biochemical systems would be advantageous in order to analyze the mode of action of Li ions, and so that embryologists, especially interested in the morphological effects of these ions, could extract information useful in the orientation of their research.

16.4.2 Isoniazid (Isonicotinic Acid Hydrazide)

Isoniazid induces the vegetalization of sea urchin eggs. It enhances the vegetalizing effects of weak lithium concentrations and reduces the animalizing effects of zinc (LALLIER, 1966a). Isoniazid inhibits diphosphopyridine nucleotidase activity in the sea urchin (FUJI and OHNISHI, 1962). Other inhibitors of this enzyme, like nicotinamide and 3-acetylpyridine appear, however, lacking in vegetalizing activity. Isoniazide's affinity for pyridoxal phosphate may be responsible for the inhibition of different enzymatic systems, the activity of which depends on this coenzyme. Isoniazid also inhibits diamino-oxydases (ZELLER et al., 1952), yet other diamine oxidase inhibitors have proven incapable of inducing vegetalization. Research on several isoniazid analogs shows that vegetalizing activity depends on molecular structure and in comparison, nicotinic acid hydrazide (Fig. 16.2b) is far less active than isoniazid (Fig. 16.2a). Picolinic (Fig. 16.2c) and benzoic acid hydrazides (Fig. 16.2d) are inactive (LALLIER, 1966a). A similar situation arises when the bacteriostatic activity of isoniazid is compared with that of its homologues (ALBERT,

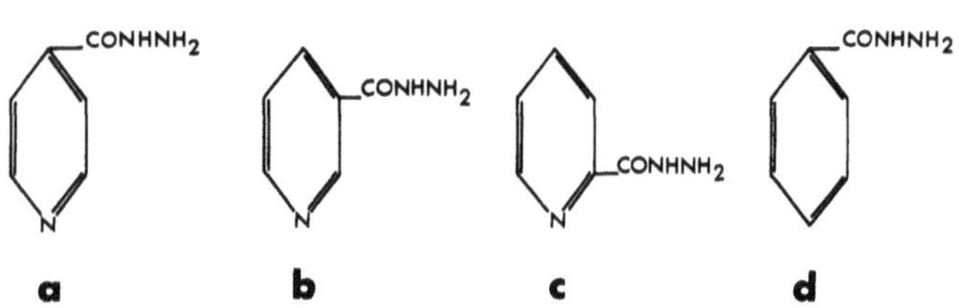

Fig. 16.2a-d. (a) Isoniazid. (b) Nicotinic acid hydrazide. (c) Picolinic acid hydrazide. (d) Benzoic acid hydrazide

1956; MCMILLAN et al., 1953). Recent research has shown that isoniazid disturbs the metabolism of bacterial lipids (WINDER, 1964). In the sea urchin, Marchi's reaction, carried out by SWANK's and DAVENPORT's method (1935), appears more intense in animalized larvae than in vegetalized larvae (LALLIER, 1966a), which suggests a modification in lipid structure during sea urchin differentiation. The ultrastructure of liver cells, particularly mitochondria, appears altered by isoniazid (KENDREY et al., 1967). The possibility of an action of isoniazid on mitochondria in the sea urchin egg is interesting in that chloramphenicol, an inhibitor of mitochondrial protein synthesis, also induces vegetalization.

16.4.3 Chloramphenicol

Chloramphenicol induces the vegetalization in whole eggs and isolated animal and vegetal halves (LALLIER, 1961, 1962a; HÖRSTADIUS, 1963). These effects are enhanced by lithium ions and depressed by animalizing agents like zinc or Evans blue. Only D-threo-chloramphenicol (Fig. 16.3a) is vegetalizing. The stereoisomer L-threo is inactive (Fig. 16.3b). The substitution of the nitro group in the D-threo-chloramphenicol molecule by a thiomethyl group (D-Win 5094)(Fig. 16.3c) greatly reduces vegetalizing activity. The replacing of this nitro group by an aminosulphonyl group (D-AMP-3)(Fig. 16.3d) or a methylsulphonyl (thiophenicol) group (Fig. 16.3e) completely suppresses animalizing activity. Thus it appears that the D-threo configuration is essential and that the presence of the nitro group is important for vegetalizing activity (LALLIER, 1966b, 1968a). D-threo-chloramphenicol inhibits protein synthesis in bacteria, and the D-threo configuration is essential for this activity. A substitution applying to the aromatic part of the chloramphenicol molecule would influence its interaction with the protein synthesizing systems in bacteria (TELESNINA et al., 1967). This data suggests that chloramphenicol's vegetalizing activity is linked with its action on protein synthesis. In the sea urchin, chloramphenicol inhibits polyribosome formation as well as the incorporation of valine, leucine (HULTIN, 1964a, 1964b) and a protein hydrolysate (BLACK, 1967). The formation of the hatching enzyme is also inhibited (YASUMASU, 1963). It has been shown that protein synthesis in mitochondria is inhibited by chloramphenicol whereas the cytoplasmic ribosomes are not susceptible to its action (MAGER, 1960; KROON, 1963; LAMB et al., 1968). It is known that mitochondria, in addition to containing protein synthesizing systems, contain DNA (Fig.20.2), and thus appear capable of synthesizing some proteins characteristic of mitochondria, and perhaps other cell membrane structures, too (WAGNER, 1969). Recent studies from PIKO and TYLER (1969) suggest that the sea urchin egg mitochondria synthesize proteins which, in part, are transported into the cytoplasm.

The vegetalizing effects of chloramphenicol could result from a differential action on protein synthesis; the intracellular target being perhaps the mitochondrial ribosomes. The development of mitochondria undergoes a particularly interesting evolution in embryonic development, the importance of which has been shown by GUSTAFSON. Puromycin, which also inhibits protein synthesis and development (HULTIN, 1961), does not vegetalize (LALLIER, 1962b). D-threo-chloramphenicol and the more active L-threo form are decouplers of oxidative phosphorylations (HANSON and KRUEGER, 1966). The fact that L-threo-chloramphenicol is not vegetalizing seems to indicate that the decoupling of oxidative phosphorylations is not responsible for vegetalization.

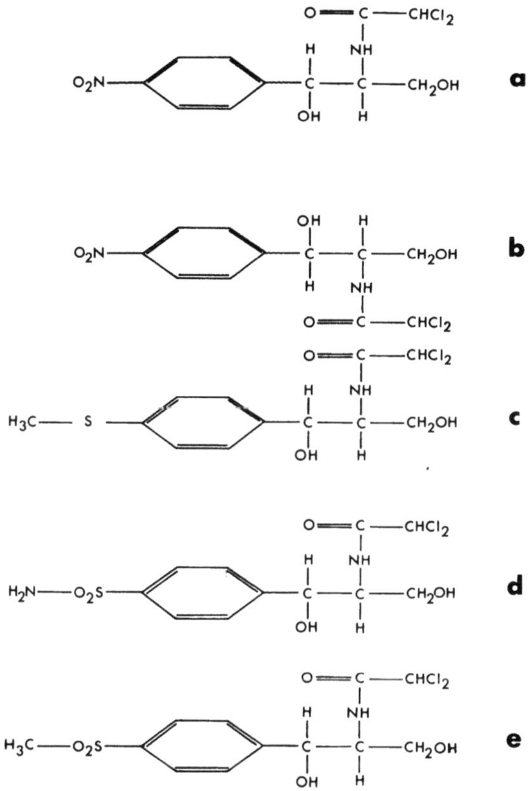

Fig. 16.3a-e. (a) D-threo chloramphenicol. (b) L-threo-chloramphenicol. (c) D-Win 5094. (d) D-AMP-e. (e) Thiophenicol

16.4.4 2-Mercapto-1-(β-4-Pyridethyl)Benzimidazole (MPB)

MPB (Fig. 16.4), an antiviral agent, is capable of vegetalizing whole sea urchin eggs (LALLIER, 1969a). It interferes with viral growth and the nucleic acid metabolism of the host cells (BUCKNALL, 1967). MPB inhibits uridine and thymidine incorporation and according to SKEHEL et al. (1967) and BUCKNALL and CARTER (1967), it might inhibit the phosphorylation of nucleosides. Like actinomycin, MPB would preferentially inhibit the synthesis of RNA governed by DNA (WALTERS et al., 1967). In the sea urchin, the effects of MPB on morphogenesis are very different from those of actinomycin where the latter is not vegetalizing. DNA added to the culture medium does not protect eggs from the effects of MPB. Cytidine, adenosine, guanosine, inosine, and uridine as well as thymidine and uridine-5-tri-phosphate do not protect eggs from the vegetalizing effects of MPB. But varying concentrations of zinc ions reduce or completely suppress the effects of MPB. This protection is perhaps due to the combination of zinc ions with MPB (LALLIER, 1969a). In other organisms, MPB appears to have a specific reversible effect on the nucleoside uptake (NAKATA and BADER, 1969) and it possibly also blocks phospholipid synthesis (FRIEDMAN and PASTAN, 1970). Recent investigations from CRAIG (1973) showed that MPB affected protein synthesis by inhibiting the rate of initiation and not the level of messengers in mouse L-cells.

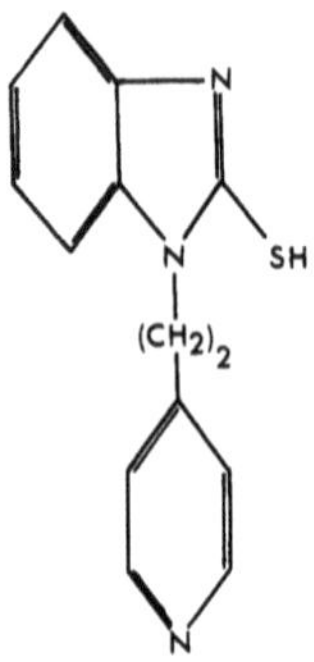

Fig. 16.4. 2-Mercapto-1-(β-4-pyridethyl) benzimidazole (MPB)

16.5 Animalizing Agents

16.5.1 Thiocyanate and Iodide

The animalizing effects of these agents have been studied by HERBST (1896), RUNNSTRÖM (1928b, 1967a) and LINDAHL (1936).

The anions thiocyanate and iodide are large in size, weakly hydrated, and besides their differences in electrical charge, they exercise a dispersing influence on colloids, properties which distinguish them from lithium ions. This dispersing action on cytoplasmic colloids has also been suggested to explain their animalizing effects (RUNNSTRÖM, 1928b; LINDAHL, 1936). Thiocyanate ions are, however, very irregular in their effects on morphogenesis (LINDAHL, 1936; LINDAHL et al., 1951). The discovery of agents more active and consistent in their effects has replaced the use of these ions as animalizing agents. Lastly, 2 other anions, the perchlorate ions, very similar to thiocyanate, and the persulphate ions are also animalizing (LALLIER, 1956a, 1957a).

16.5.2 Sulphate Deprivation

In addition to respiration depression, blastulae cultivated in sulphate-free seawater develop into animalized or radialized larvae (HERBST, 1904; LINDAHL, 1936; RUNNSTRÖM et al., 1964). The sensitivity of the entomesoderm, however, does not appear before the blastula stage. According to LINDAHL (1936), sulphate ions would serve in the detoxification of phenolic derivatives produced by the catabolism of proteins. The apparent susceptibility of the entomesoderm to the lack of sulphates would be due to the preponderance of protein catabolism in the vegetal region. In eggs and larvae, $^{35}SO_4$ is incorporated into substances identifiable as mucopolysaccharides (IMMERS, 1961, 1962). The $^{35}SO_4$ incorporation rate is the same in the isolated animal and vegetal halves (HÖRSTADIUS et al., 1966). The incorporation of labelled amino acids, thymidine, and tritiated uridine is depressed (IMMERS and RUNNSTRÖM, 1965). The nuclei of the animal and vegetal parts appear to react differently to Hale staining and IMMERS (1956) interprets these differences by suggesting that phosphate groups of nuclear RNA exist in a free state in the ectodermic cell nuclei and would be combined with proteins in the nuclei of entomesodermic cells. Where sulphates are absent mucopolysaccharide formation is completely suppressed, and it seems as if these substances combined with proteins have a role in the motility of mesenchymatous cells (IMMERS, 1962). According to RUNNSTRÖM et al. (1964), the polysaccharide sulphate protein compounds may exercise a protecting influence vis-a-vis some hydrolytic enzymes. When they are absent, the activity of these enzymes, by impairing the protein structure, would modify the anabolic processes, particularly in the vegetal regions most sensitive to the lack of sulphate. GEZELIUS (1969) has suggested the role of sulphate in the transport of RNA from the nucleus to the cytoplasm.

16.5.3 Heavy Metals

Different heavy metals favor ectodermization (WATERMAN, 1937) and treating eggs for several minutes with $HgCl_2$ induces animalization of larvae (LINDAHL, 1936). Zinc has proven particularly active, and weak concentrations induce radialization of *Paracentrotus lividus* eggs (LALLIER, 1955a) and sand dollar eggs (RULON, 1955). These phenomena are suppressed in the presence of glutathion which combines with zinc ions (RULON, 1955) or ethylenediamine tetraacetic acid, which chelates zinc ions (LALLIER, 1955a). The developmental stages which are sensitive to the animalizing effects of zinc, also correspond to the stages sensitive to lithium. The effects of zinc are, moreover, counteracted by lithium and enhanced in the presence of another animalizing agent like Evans blue (LALLIER, 1959a). Zinc detection with dithizon shows accumulation in the mesenchyme and entoderm of gastrula, and in the pluteus, zinc appears localized in the pigment cells (LALLIER, 1956b). Fixation of ^{35}Zn by the larvae increases during development (TIMOURIAN, 1968); it rapidly augments immediately after fertilization, then becomes stabilized until hatching and, after a temporary decrease, increases rapidly until the pluteus stage (OZRETIC and KRAJNOVIC, 1968). The incorporation of ^{14}C-valine into zinc treated larvae has been studied by TIMOURIAN (1968), but the results obtained are inconclusive.

The effects of copper, iron and manganese have also been studied, and with valencies of 2, these metals show characteristics in common with those of zinc. Nickel and cobalt induce radialization of sand dollar larvae (RULON, 1953a, 1953b) and *Paracentrotus lividus* larvae are radialized by nickel, but cobalt disturbs gastrulation and inhibits subsequent larval development (LALLIER, 1956b). Iron and manganese inhibit or retard development according to the concentrations used but do not alter the orientation of differentiation (LALLIER, 1956b; TIMOURIAN, 1968).

Of all the metals investigated, zinc is the most active on embryonic determination. Its mode of action is not known but due to its affinity for -SH groups, a zinc action on the enzymes whose activity depends on -SH groups has been suggested (RULON, 1955). Nevertheless, specific agents of -SH groups like p-chloromercuribenzoid acid, do not exhibit animalizing activity (LALLIER, 1956a). Moreover, cadmium, which has an affinity for -SH groups stronger than that of zinc, is a less effective animalizing agent (LALLIER, 1955a). Other groups show an affinity for zinc; the COOH groups (PERKINS, 1961), the amine groups and especially imidazole, present in histidine (GURD and GOODMAN, 1952). It is possible that the fixation of zinc ions into proteins influences their structure, and consequently their biochemical properties, as well as their affinity for other proteins or non-protein molecules, are altered. Recent experiments by PIRRONE et al. (1970) have shown that zinc ions strongly inhibit the initiation of ribosomal RNA synthesis. In contrast, lithium does not seem clearly to affect the synthesis of this RNA. Lastly, we will recall that zinc is a normal constituent of cells. It has an essential role in the activity of several enzymes and is present in the midpiece of spermatozoids and intervenes in their mobility. Zinc also accumulates in the nucleoli and could be involved in the mitotic processes (FUJII, 1954). ERDMAN (1959) showed that zinc modifies the contractile properties of muscle fibers. Further on, we shall see that other animalizing agents are also capable of reacting with contractile macromolecules. In this way, animalizing agents may interfere with cell mobility and influence the formation of mesenchyme and entoderm. Zinc interferes with the biosynthesis of nucleic acids (FUJIOKA and LIEBERMAN, 1964; WESER et al., 1969) and appears capable of modifying the template activity of chromatin (KROEGER, 1963; HUANG and HUANG, 1969). Zinc appears to play a functional role in DNA polymerase, SLATER et al. (1971) and RNA polymerase, SCRUTTON et al. (1971). Zinc appears as a stabilizer of macromolecules and biological membranes, CHVAPIL (1973). Up to the present time, research carried out with the zinc ion on the sea urchin egg has been essentially morphological. Research on its mode of action is necessary to account for the great activity of this ion on sea urchin differentiation.

16.5.4 Polyanionic and Polycationic Substances

Polyanionic Agents. Animalization and radialization can be induced with polyanionic derivatives. These substances, active at weak concentrations, belong to different chemical categories. Their activity is concomitant with the presence of acid, sulphonic or carboxylic groups (LALLIER, 1955b, 1958a). Numerous stains fulfil these conditions. The comparison of 3 polysulphonated dyes of closely related structure (Fig. 16.5) shows that their animalizing activity diminishes at the same rate as their affinity for proteins: Evans blue > Niagara blue 6B > Niagara blue 4B. The enhancement of the strength of the acid function by the introduction of halogens into the heterocycle of xanthenic dyes enhances their animalizing activity (LALLIER, 1957b). Research on the distribution of an active dye, sky blue chlorazol, indicates a preferential affinity for the vegetal parts of the larvae (LALLIER, 1957c). These substances also exercise other effects such as, the inhibition of fertilization and hatching. Lithium counteracts animalization by these substances but does not alleviate inhibition of hatching. Other polyanionic substances of high molecular weight formed by an aliphatic chain carrying sulphonic or sulphate groups, like polyvinylsulphate (PUS) and polyethylenesulphonate (PES) are not animalizing. They inhibit neither fertilization nor hatching. Suggestions only can be put forward to explain the activity of polyanionic derivatives. It is possible that they act either on the cell surface or on the internal structures. In the first possibility, the fixation of polyanionic derivatives to basic protein groups may modify the cortical structures and interfere with the mechanisms which determine the egg's polarity and its development during cleavage. According to the second possibility, they may interfere with the connections between proteins and nucleic acids. Several observations on different organisms suggest this possibility. PUS is able to become fixed to ribosomes and thereby interferes with protein synthesis (SHINOZAWA et al., 1968). PES appears to stimulate, in vitro, the transcription of RNA by chromatin (CHAMBON et al., 1968). In sea urchin eggs, neither PES nor PUS exercise animalizing effects, and it would be interesting to investigate whether some effects comparable to those observed with PUS and PES can be obtained with animalizing polysulphonic agents. The active agents may influence morphogenesis by combining with morphogenic factors. An example of this type of action is the neutralization, by PUS, of mesodermalization inducing factors in amphibians (TIEDEMANN et al., 1969).

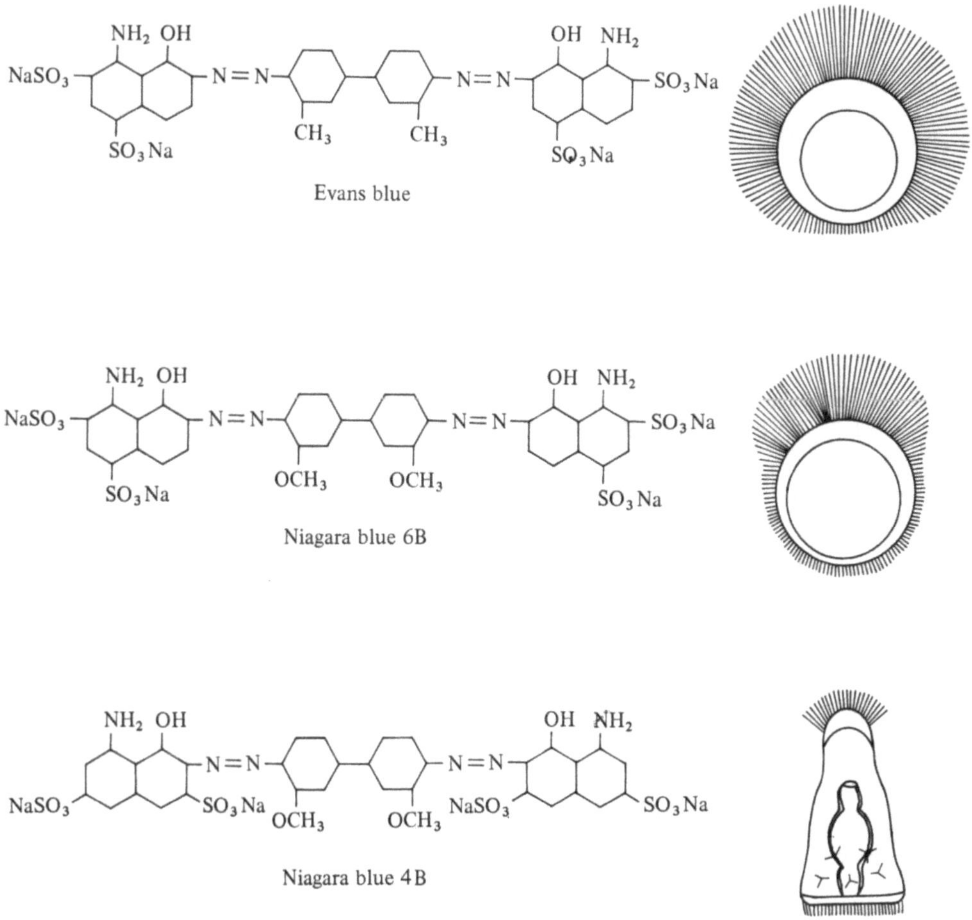

Fig. 16.5. Decreasing animalization resulting from 3 polysulphonated dyes of the same concentration. Evans blue: strongly animalizing; Niagara blue 6B: moderately animalizing; Niagara blue 4B: induces radialization

Polycationic Agents. Aliphatic polyamines provoke animalization and radialization of larvae. The comparison of the effects of several of these polyamines shows that at least 2 groups are necessary for animalizing and radializing activity. The presence of secondary amine functions enhances the activity, and the length of the carbon chain and the spacing of the main functions also play an important role. A relation appears to exist between the animalizing activity of the polyamines and their stabilizing action on DNA (LALLIER, 1963a, 1966c). Some of these polyamines are normal cell constituents; they seem to control nucleic acid synthesis (FOX and WEISS, 1964; RAINA and COHEN, 1966) and protein synthesis (MARTIN and AMES, 1962). In the sea urchin, one of these polyamines, spermidine, stimulates RNA synthesis (BARROS and GIUDICE, 1968).

Synthetic polycationic substances, especially cobalt complexes, are also animalizing and the activity increases with the valency where hexavalent complexes and to a lesser extent trivalent complexes are effective. Non-ionized or monovalent complexes are inactive (LALLIER, 1966d). Animalizing activity exercised by both polyanionic and polycationic agents may result from their action on cellular macromolecular complexes, the association of which would depend on the reciprocal affinity of acid and basic groups.

16.5.5 Sulphydryl Groups

Sulphydryl groups are involved in many cellular processes, and may exist freely or in the form of S-S bridges. The -SH groups are very reactive and may combine with heavy metals, organo-metallic derivatives; they may be alkylized or oxidized. The S-S bridges may be opened by reducing agents.

The effects of most of the agents capable of effecting these operations have been studied in the sea urchin egg. These can be classified into 3 categories: antimitotic action, inhibition of differentiation and animalization. Only the animalizing agents will be examined here. They are: thiomalic acid ($COOH-CH_2-CHSH-COOH$), 2-mercaptopropionyl-glycine ($CH_3-CHSH-CONH-CH_2-COOH$) and N-acetylcysteine (LALLIER, 1952, 1968b). The structural analogy that exists between thiomalic acid and the cysteine part of glutathion led BÄCKSTRÖM (1958a) to suggest that the animalizing effects of thiomalic acid are due to a phenomenon of competition with glutathion. Yet other glutathion analogues like S-methylglutathion and S-N-ethylsuccinimido-glutathion are not animalizing. Moreover, reduced glutathion enhances the animalizing effects of thiomalic acid (LALLIER, 1959b). Cysteine alone inhibits development, and this inhibition is not alleviated by thiomalic acid. The -SH groups of thiomalic acid are necessary for activity. Oxidized thiomalic acid is inactive (LALLIER, 1958b). In comparing the effects of thiomalic acid with those of other sulphydryl agents, it appears that some, like mercaptoethanol (MAZIA and ZIMMERMAN, 1958) thioglycerol and thiosorbitol (LALLIER, 1962c) are antimitotic and act by destroying the fibrous structure of the spindle. Other sulphydryl agents like cysteine, homocysteine and penicillamine inhibit differentiation. Their effects, furthermore, are counteracted by different structurally analogous amino acids (LALLIER, 1965a). These differences may be due to the fact that these antimitotic or inhibitory agents penetrate into the cells, whereas the animalizing sulphydryl agents exercise their action on the cell surface.

Certain substances capable of reacting with -SH groups are animalizing. These are o-iodosobenzoic acid (RUNNSTRÖM and KRISZAT, 1952), lipoic acid (RUNNSTRÖM, 1956) and an organic derivative of mercury, Salyrgan (LALLIER, 1960b). Iodoacetate is weakly animalizing (IMMERS, 1965), but its amide, iodoacetamide, inhibits development (LALLIER, 1962d). This may be due to differences in permeability; iodoacetamide penetrates into the cells whereas iodoacetate act on the cell surface. The stage of maximum sensitivity to o.iodosobenzoic acid also corresponds to that of lithium (BÄCKSTRÖM, 1953), but in contrast, o.iodosobenzoic acid speeds up cleavage and development (HAGSTRÖM, 1963).

The use of Salyrgan labelled with ^{20}Hg shows that this substance reacts with the surface of the unfertilized egg (FERRINI and MANELLI, 1967). Salyrgan inhibits ATPase (MANELLI and FERRINI, 1967), which is present at the surface of the egg (OHNISHI, 1963).

It is known, moreover, that the sea urchin egg surface bears -SH groups and that their number increases during development (WOLFSON, 1966).

However, neither animalization nor vegetalization seem to bring about changes in the distribution of -SH groups detected by cytochemical techniques; these do not appear to be distributed in gradients (BÄCKSTRÖM, 1959a, FENAUX, 1961).

In conclusion we see that agents, capable of combining with -SH groups or reducing the S-S bridges exercise animalizing effects and that they are not all active, may result from differences in accessibility of the cellular targets.

16.5.6 Proteolytic Enzymes

Various proteolytic enzymes are animalizing: trypsin (HÖRSTADIUS, 1949, 1953; MOORE, 1952), chymotrypsin, ficin (HÖRSTADIUS, 1953), papain, chymopapain, and pronase (LALLIER, 1967). Inhibition of enzyme activity through heating, or with an ovo mucoid inhibitor, or in the case of papain by Hg, suppresses animalizing activity. Reciprocally, the reactivation of the enzyme restores its animalizing power. Other non-proteolytic enzymes like lipase and phospholipase C are devoid of animalizing activity (LALLIER, 1969b). Trypsin interferes with the formation of the fertilization membrane (RUNNSTRÖM, 1948) and speeds up the development of the larvae (HAGSTRÖM and LÖNNING, 1963). According to RUNNSTRÖM and IMMERS (1966), trypsin exercises a double action, depressing the activity of vegetal regions and enhancing the activity of animal regions where the Golgi apparatus seems, in fact, to be especially active in the animal regions of larvae treated with trypsin. Lithium counteracts the effects of trypsin (MOORE, 1952) and pronase (LALLIER, 1969b). Proteins extracted from embryos vegetalized by lithium resist more strongly the proteolytic action of trypsin, than proteins extracted from animalized or normal embryos (RANZI et al., 1957; RANZI, 1959). Proteolysis of albumin from bull serum by pronase is considerably slowed down in the presence of a lithium concentration equal to that used for inducing vegetalization (LALLIER, 1969b). In the concentrations used, where the proteolytic enzymes exercise their animalizing action, it seems improbable that they penetrate into the egg. In the unfertilized eggs treated with a concentration of trypsin sufficient to inhibit fertilization, only the exterior membrane seems to be less thick (WOLPERT and MERCER, 1961). It would be interesting to examine the surface of the developing larvae by this technique, during trypsin or pronase treatment. In order to exercise their animalizing effect, proteolytic enzymes have to act during the whole period of cleavage. Under these conditions, it is not only the surface of the fertilized egg but also all of the newly formed surfaces, formed during cleavage, which are subjected to proteolytic action. It is possible that the proteolytic enzymes remove a protein constituent from the surface of the blastomeres. The change resulting in the structure of the cell membranes could modify the distribution of cortical factors responsible for the differentiation of the animal and vegetal regions. According to GUSTAFSON and SÄVHAGEN (1949), some detergents exert radializing effects on larvae; an action on the surface of the cells also seems to be involved in these phenomena.

It is known, that lithium protects proteins from proteolysis. This protecting effect may be due to the action of lithium on protein structure, a change of which produces an increase in resistance to proteolysis. Numerous examples exist of this phenomenon of protection by metallic ions (GORINI and AUDRAIN, 1952; AZARI and FEENEY, 1961).

16.5.7 Natural Morphogenetic Agents

A new approach to the analysis of differentiation in the sea urchin egg has been brilliantly opened up with the isolation of morphogenetic substances from unfertilized eggs (HÖRSTADIUS et al., 1967; JOSEFSSON and HÖRSTADIUS, 1969). It has been possible to isolate several different substances by chromatography on Dowex 50 W-X2 and Sephadex G-25. Their effects have been investigated on isolated animal and vegetal halves and on whole eggs. Some exercise an animalizing activity, others are vegetalizing. These extremely important investigations offer new perspectives for the research on the sea urchin egg's differentiation.

16.6 Research on the Metabolism of Animalized and Vegetalized Larvae

16.6.1 Carbohydrate Catabolism

In the sea urchin larva, carbohydrates are degraded via the Embden Meyerhof pathway, and the tricarboxylic acid cycle. The energy provided by carbohydrate catabolism is accumulated in the form of ATP directly available for use by the cells. The hexose monophosphate shunt, also functional in the degradation of carbohydrates in the sea urchin, contributes to the formation of phosphate ribose, utilizable in the synthesis of nucleic acids.

Lithium, vegetalizing agent, inhibits the increase of oxygen consumption; the percentage of inhibition increases with the degree of vegetalization. Inhibition reaches its maximum at the beginning of segmentation at the time when the egg is most sensitive to the morphogenetic effects of lithium.

Lithium inhibits some glycolytic reactions, like anaerobic reduction of methylene blue by an egg homogenate in the presence of substrates of glycolysis (LINDAHL, 1936). From different observations LINDAHL concluded in favor of a preponderance of a carbohydrate catabolism in the larva's animal regions. Depression of carbohydrate catabolism by lithium would provoke vegetalization and this hypothesis has been experimentally tested. The respiratory quotient increases during development and this points to an increase in the utilization of carbohydrates during development (ÖHMAN, 1940). But, according to LINDAHL and HOLTER (1940), the oxygen consumption of isolated animal and vegetal halves does not show any appreciable differences. It is, however, possible that the respiration of the animal and vegetal halves after their separation does not correspond to the normal respiration of these halves when they are associated in the whole embryo. The normal oxygen consumption (controlled respiration) has been compared with that resulting from the action of an agent decoupling the oxidation of phosphorylation, like dinitrophenol (released respiration), in whole embryos of different types and in isolated animal and vegetal halves. These observations bring to light differences in respiratory control between the animal and vegetal regions. Lithium would cause a decrease in the number of respiratory chains available in the cells (IMMERS and RUNNSTRÖM, 1960; DE VICENTIIS et al., 1966; DE VICENTIIS and RUNNSTRÖM, 1967).

LINDAHL's hypothesis has also been tested by studying the effects of various inhibitors of respiration and glycolysis. To what extent do these inhibitors exercise vegetalizing effects like those of lithium? The following glycolytic inhibitors have been studied: bromoacetic acid (TCHAKHOTINE, 1938), monoiodoacetamide (LINDAHL, 1939), DL-glyceraldehyde, phlorizin (NEEDHAM and NEEDHAM, 1940), 2-deoxy-D-glucose (LALLIER, 1960c). None of these glycolytic inhibitors exercise vegetalizing effects. The effects of several inhibitors of respiration on morphogenesis have also been examined. KCN, CO and partial anaerobiose, under the conditions where they partially inhibit respiration, do not exercise any vegetalizing effects; they are, however, capable of enhancing the vegetalizing effects of lithium (RUNNSTRÖM, 1928a; HÖRSTADIUS and STRÖMBERG, 1940). Sodium azide and KCN induce exogastrulation and vegetalization (CHILD, 1948; RULON, 1950; CZIHAK, 1963). Other inhibitors like malonate and sodium selenite are not vegetalizing (RULON, 1951, 1952a). Stimulation of oxidation by pyocyanine appears to be capable of counteracting lithium effects in some sea urchin species (RUNNSTRÖM, 1935), where in others, pyocyanine and lithium do not act as antagonists (MOORE et al., 1945). From these observations it follows that a depression of carbohydrate catabolism may, under certain conditions, either provoke the vegetalization of larvae, or enhance the effects of lithium. The decrease in ATP synthesis at the time of carbohydrate catabolism inhibition may intervene in vegetalization. 2 observations favor this hypothesis: ATP (KRISZAT and RUNNSTRÖM, 1957), and adenosine and inosine counteract the effects of lithium (LALLIER, 1963b). Dinitrophenol increases oxygen consumption, but ATP synthesis decreases. Dinitrophenol is vegetalizing; its action, however, seems to be different from that of lithium. Dinitrophenol, while active on isolated animal and vegetal halves, induces vegetalization very rarely in whole eggs (HÖRSTADIUS, 1953).

Animalization appears to require oxygen because KCN or partial anaerobiosis suppresses animalization (LINDAHL, 1936). Stimulation of oxidation by pyocyanine enhances the animalizing effects of thiocyanate (RUNNSTRÖM and THÖRNBLOM, 1938). Different substrates of carbohydrate catabolism favor animalization. This is the case with pyruvate (HÖRSTADIUS and STRÖMBERG, 1940), propanediol phosphate and lactate (HÖRSTADIUS and GUSTAFSON, 1947). Glucose has been shown to induce radialization and favor the development of the ectoderm (RULON, 1952b; HÖRSTADIUS, 1959), but the high concentrations used imply the possibility of a mechanism other than the metabolic utilization of glucose as a substrate.

According to HULTIN (1953a) the hexose monophosphate shunt would intervene in embryonic determination, and more especially during animalization. Its activity has been studied in various larval types by BÄCKSTRÖM (1959b, 1963); BÄCKSTRÖM et al. (1960) and accordingly the shunt activity evolves in a comparable way in different types of larvae. After hatching, the shunt activity decreases, remaining however, at a higher level in animalized larvae than in vegetalized larvae, the normal larvae occupying an intermediate position. Phosphogluconic acid, a metabolite of the shunt, is animalizing (HÖRSTADIUS and GUSTAFSON, 1947), while an inhibitor of the shunt, dinitrocreol, is vegetalizing (LINDAHL, 1940). This agent, however, can also affect other points in metabolism, especially by decoupling the oxidation of phosphorylation like dinitrophenol. Shunt inhibitors like p-chloromercuribenzoic acid and N-ethylmaleimide do not alter larval determination (LALLIER, 1956a and unpublished observations). It is possible that the shunt activity is linked particularly to cell division, which would explain its high level during cleavage (BÄCKSTRÖM, 1962). Several metabolic systems may be sensitive to the inhibitors, particularly when these react with the -SH groups. The permeability of the cells and an action limited to the cell surface should imply different interpretations of inhibitor action.

16.6.2 Gradients of Reduction

The degradation of carbohydrate has been visualized as gradients of activity through use of oxidation-reduction indicators.

The reduction capacity of normal and lithium treated eggs has been studied by RANZI and FALKENHEIM (1937) who also examined oxidase activity and -SH group distribution. They conclude that the reducing activity is higher at the animal pole than at the vegetal pole and they interpret these differences as the consequence of inhibition of metabolism at the vegetal pole by micromeres.

Gradients of reduction have also been observed during normal development (CHILD, 1936; HÖRSTADIUS, 1952, 1955). We shall just point out that during normal development, a gradient decreasing from the animal pole to the vegetal pole, then in the opposite direction, vegetal-animal, appear successively and a third gradient appears in the oral area of the late gastrula (CZIHAK, 1961). In entirely animalized larvae or in isolated animal halves, only the animal gradient is observed and the opposite is true for entirely vegetalized larvae or isolated vegetal halves (HÖRSTADIUS, 1955; LALLIER, 1956c).

The gradients of reduction may correspond to a differential distribution of reducing substances or enzymes catalyzing an oxidation-reduction process. Among reducing substances, the -SH containing substances and ascorbic acid are especially interesting. It has not been possible to establish any correlation between -SH group distribution and gradients of reduction (BÄCKSTRÖM, 1958b, 1959a). According to BÄCKSTRÖM (1956, 1957), ascorbic acid distribution seems to follow the vegetal-animal gradient. The differentiating entomesodermic structures seem, in fact, to be richer in ascorbic acid and suggests the participation of ascorbic acid in the differentiation of the vegetal region. We will recall that ascorbic acid is capable of enhancing the vegetalizing effects of lithium ions (RUNNSTRÖM, 1956), whereas an ascorbic acid antagonist, glucoascorbic acid, augments animalization in isolated animal halves (GUSTAFSON and HÖRSTADIUS, 1955).

The reduction of a redox indicator, tetrazolium blue, has been investigated in different types of embryos, in the presence of various substrates and coenzymes DPN and TPN (BÄCKSTRÖM, 1959c, 1959d). These coenzymes participate in numerous oxidation-reduction reactions catalyzed by dehydrogenases. The gradients of reduction observed in these experiments appear to correspond to those described by CHILD and HÖRSTADIUS and they have shown differences in the degree of coenzyme availability during the development. TPN, which intervenes in hexose-monophosphate shunt activity, is available at all stages of development in the different types of larvae. DPN, on the other hand, appears to be available only after hatching and only in those larvae which differentiate the entomesoderm.

Mitochondrial distribution has been studied by GUSTAFSON and LENICQUE (1952, 1955); LENICQUE et al. (1953). These authors observe that, at the beginning of development, mitochondria are distributed in an animal-vegetal gradient. Subsequently, the mitochondria decrease in number at the animal pole, whereas at the vegetal pole, new mitochondria populations appear in successive waves corresponding to the different regions of the entomesoderm. The staining of mitochondria, weak at the beginning of development, gradually increases, suggesting a progressive organization of mitochondria. The formation of mitochondrial enzymes has also been followed and some are less active in lithium treated larvae. On the other hand, non-mitochondrial enzymes do not appear to be influenced by lithium (GUSTAFSON and HASSELBERG, 1950, 1951). From these observations GUSTAFSON has presented the following interpretation: the micromeres which form the most vegetal material of the larva would contain a material inhibiting protein synthesis which interferes with the formation of mitochondria.

The distribution of mitochondria has been studied by other authors with different results. According to SHAVER (1955, 1956, 1957) the number of mitochondria increases at gastrulation and then decreases and no gradient is observed. In the vegetal regions the mitochondria appear larger than in the animal regions and the study of fixed material observed with the light microscope (MORI et al., 1958), or the electron microscope (BERG et al., 1962) has given results analogous to those of SHAVER. Differences relating to the size of the mitochondria do not appear until after the 16-cell stage. The large mitochondria of the vegetal regions would contain more cristae than those of the animal regions and would also, without doubt, be richer in cytochrome oxidase (BERG and LONG, 1964).

16.6.3 Nucleic Acid Metabolism

During development the RNA content of the larvae shows fluctuations which probably reflect successive phases of synthesis and degradation. The RNA content of animalized and vegetalized larvae is higher than that of normal larvae (BÄCKSTRÖM, 1959e). No differences however, are observed between isolated animal and vegetal halves (MARKMAN, 1967).

Incorporation of various nucleotide precursors rapidly increases from the beginning of the blastula stage (HULTIN, 1953b, 1953c, 1957). Autoradiographic studies with ^{14}C adenine show incorporation in the animal regions at the young blastula stage and in the vegetal regions in advanced blastulae (MARKMAN, 1961a, 1961b). At gastrulation, incorporation of ^{14}C uridine increases in the vegetal halves and decreases in the animal halves (MARKMAN, 1967). Uridine is incorporated and utilized for RNA synthesis, and probably for messenger RNA in the micromeres (CZIHAK, 1965; CZIHAK et al., 1967). BERG's recent research (1968a) shows that after hatching, the ^{14}C uracil incorporation rate is greater in the vegetal halves. The differences, however, are not observed with other precursors such as ^{14}C adenine and ^{3}H guanosine.

Cytochemical techniques indicate a preferential accumulation of RNA in the cytoplasm of the more vegetal regions (MARKMAN, 1957).

Another line of approach for studying the role of nucleic acids in embryonic determination is the use of agents capable of interfering with their metabolism: purines, pyrimidines and their analogues and various inhibitors. Synthetic nucleosides, analogues of cytidine, favor the animalization of animal halves, but paradoxically increase the vegetalization of whole eggs by lithium (GUSTAFSON and HÖRSTADIUS, 1955). 2-thio-5-methylcytosine is animalizing (GUSTAFSON and HÖRSTADIUS, 1956); 6-diazo-5-L-norleucine and azaserine interfere with purine metabolism and arrest development at the blastula and young gastrula stages (KARNOFSKY and BEVELANDER, 1958). Development arrest at the blastula stage is also obtained with 5-fluorodeoxyuridine, fluorouracil (KARNOFSKY and BASCH, 1960), and 6-methylpurine (LALLIER, 1965b). Adenine completely reverses the effects of 6-methylpurine. 6-azauracil and 8-azaguanine are radializing (LALLIER, 1962e). These effects are not reversed by adding the corresponding natural metabolites. Nitroorotic acid, azauridine and azaguanine appear to slow development of the entomesoderm (CZIHAK, 1965).

In general, animalization or radialization induced by antimetabolites are not very intense and are irregularly observed; the absence of any reversal of the morphogenetic effect, in the presence of normal metabolites, suggests that the antimetabolic activity of the agents is not responsible for their action on morphogenesis.

Experiments have been made by studying the effects of nucleic acids on development. DNA increases the animalization of isolated animal halves (HÖRSTADIUS et al., 1954) while with whole eggs, this effect is not observed (LALLIER, 1963c; MATEYKO, 1967).

Ribonucleic acid is not animalizing. On the other hand, the transfer ribonucleic acid of yeast (tRNA) induces larval radialization, moreover, it enhances the effects of other animalizing agents. Alkaline hydrolysis of tRNA does not suppress its activity, which seems to be linked to the presence of cytidylic acid which exercises the same animalizing effects as tRNA (LALLIER, 1964b, 1966e).

Actinomycin D inhibits the synthesis of ribonucleic acids governed by DNA. Consequently, it is a very valuable agent for studying the influence of the nucleus on differentiation. Actinomycin retards cleavage (WOLSKY and WOLSKY, 1964) and inhibits differentiation (GROSS and COUSINEAU, 1964). The formation of entomesoderm is repressed by actinomycin (GIUDICE et al., 1968), which prevents also animalization and vegetalization induced with chemical agents (MARKMAN and RUNNSTRÖM, 1963; RUNNSTRÖM and MARKMAN, 1966; DE ANGELIS and RUNNSTRÖM, 1970). According to these authors the activation of genes and the transfer of genetic information would be involved in animalization as in vegetalization. LALLIER (1963c) has shown that actinomycin D depresses the animalizing effects of zinc ions and Evans blue, but enhances the vegetalizing effects of lithium. A possible explanation of this difference is that the development of vegetal structures is partially controlled by preexisting RNA messengers stored in a masked or inactive state in the cells and strong concentrations of lithium ions might activate these messengers.

16.6.4 Protein Metabolism

LINDAHL (1936) suggested that protein catabolism predominates in the vegetal regions. This hypothesis was presented to explain the strong susceptibility of vegetal regions to the lack of sulphate, the latter being necessary for the detoxification of products formed during protein catabolism. To test this hypothesis HOLTER et al. (1938), HOLTER and LINDAHL (1941) measured the peptidase activity in isolated animal and vegetal halves. From their results it appears that the enzyme is uniformly distributed until the pluteus stage.

The larvae's antigenic composition was also examined to note possible differences between animalized and vegetalized larvae. Larvae vegetalized with lithium do not differ in this respect from normal larvae (PERLMANN and GUSTAFSON, 1948). Antigenic differences appear between normal larvae and larvae animalized by o-iodosobenzoic acid (RANZI and CITTERIO, 1956).

Lithium slows down the synthesis of peptides and proteins and reduces the rate of glutamine turnover (GUSTAFSON and HJELTE, 1951). Autoradiographic research on adenine and leucine show first a more pronounced incorporation in the animal regions of the young blastula, and which afterwards predominates in the vegetal regions of the advanced blastula (MARKMAN, 1961a). BOSCO and MONROY (1962) observed that the incorporation of ^{35}S-DL-methionine, ^{14}C-DL-leucine and ^{14}C-DL-alanine is stronger in the ectoderm of the blastula. No difference appears between the different regions of the gastrula. MARKMAN (1961b) observed that the separation of animal and vegetal halves increases ^{14}C-valine incorporation which exceeds that of the whole egg. Furthermore, incorporation in the animal halves is higher than in the vegetal halves. BERG (1965) measured the ^{14}C-valine incorporation rate and found it to be equal in isolated animal and vegetal halves and accordingly (BERG, 1968a) noted that the differences in incorporation rate shown by MARKMAN were not due to a real increase in protein synthesis but resulted from modifications in the rate of amino acid transfer. Regional differences in incorporation rates detected by autoradiography on whole larvae are not observed when isolated animal and vegetal halves are used. HYNES and GROSS (1970) have separated micromeres, macromeres and mesomeres from the 16-cell stage and examined the rates of DNA, RNA and protein synthesis in each cell class. No marked differences were apparent.

During cleavage lithium does not inhibit protein synthesis but after the cleavage period, lithium reduces the protein synthesis rate, measured by ^{14}C-valine incorporation. No differential effect of lithium on protein synthesis rate can be detected in animal and vegetal halves and important qualitative or quantitative differences do not appear in the electrophoretic fractions of normal and vegetalized embryo proteins (BERG, 1968b). O'MELIA (1971, 1972, 1973) has examined by an electrophoretic method the proteins and enzymes from normal, animalized and radialized *Arbacia punctulata* embryos. Animalized larvae protein patterns showed a reduction in band number as compared to the controls. Proteins appearing during normal gastrulation are absent in exogastrula produced by lithium treatment (SPIEGEL et al., 1970).

Most amino acids are inactive or weakly vegetalizing while L-glutamine and L-lysine animalize isolated animal halves (GUSTAFSON and HÖRSTADIUS, 1957). Tyrosine is capable of vegetalizing whole eggs or isolated animal and vegetal halves (FUDGE, 1959). Tyrosine decreases the incorporation of an algal hydrolysat and ^{14}C leucine into proteins. The vegetalizing effect of tyrosine is suppressed in the presence of a full complement of the free amino acids normally present in the egg. According to FUDGE-MASTRANGELO (1966) tyrosine seems to limit the endogenous supply of amino acids used in the protein synthesis of animal structures.

The effects of amino acid analogues have also been studied: canavanine is animalizing; allylglycine, DL-phenyllactic acid, β-2-thionylalanine, D-leucine and DL-norleucine augment the vegetalization of whole eggs in the presence of lithium (GUSTAFSON and HÖRSTADIUS, 1955). Ethionine inhibits the differentiation of primary mesenchyme and provokes the radialization of larvae (BOSCO and MONROY, 1960). Some vegetalized larvae can be obtained by treating whole eggs with 7-azatryptophane (LALLIER, 1962e). Many analogues inhibit development without altering determination, which is the case with cycloserine (LALLIER, 1962f), p-flurophenylalanine (LALLIER, 1962e), proline analogues (LALLIER, 1965c), cysteine (LALLIER, 1965a), asparagine (LALLIER, 1968c). Amino acid analogues generally retard or inhibit development. Animalization or vegetalization, when apparent, are weak. However, a number of these analogues are capable of enhancing the effects of weak concentrations of lithium.

16.7 Biophysical Aspects

Modifications in cell appearance and in the properties of the proteins extracted from embryos are observed during the development of normal, animalized and vegetalized larvae. The normal egg of the sea urchin *Paracentrotus lividus* shows an orange pigmented band that extends beneath the equator with the exception of a zone, free of pigments, located at the vegetal pole (SELENKA, 1883; BOVERI, 1901). At gastrulation this pigmented zone is only partially invaginated (RUNNSTRÖM, 1928a; LINDAHL, 1936). Thiocyanate appears to exercise a dispersing effect on this pigmented band (LINDAHL, 1936). Different observations indicate that cell surface properties are modified during animalization and vegetalization. Animalization is associated with an increase of cellular cohesion (GUSTAFSON and WOLPERT, 1963). Lithium favors the adhesion of cells in the vegetal region of embryos. In extreme cases the fusion of embryos leads to the coalescence of the blastocoels. The appearance of the cytoplasm itself is different according to the regions of the larva. Under the electron microscope, the cytoplasm of the micromeres appears to have a finer, denser structure than that of the other blastomeres (LEHMANN, 1950). With phase-contrast examination, the cytoplasm of animal cells appears darker than that of vegetal cells and according to RUNNSTRÖM (1957) these differences would reflect the increase of synthetic activities in the animal regions. Animalizing and vegetalizing agents are capable of modifying some physical-chemical properties of proteins. The viscosity of fibrillar proteins is increased by lithium and vegetalizing agents, and decreased by animalizing agents. This phenomenon is observed in vivo and in vitro (AROSIO et al., 1949). In contrast, globular proteins show only an increase in viscosity through the use of different animalizing and vegetalizing agents. The effects of animalizing agents on a fibrillar protein, have been studied. The flow birefringence is decreased (RANZI, 1955). Hydroxyl and phenolic groups are augmented (AROSIO and BOSSI, 1954). The immunological properties are altered (AROSIO, 1953). These changes suggest a denaturation process induced by animalizing agents. Lithium, on the other hand, would exercise a stabilizing effect on the structure of fibrillar proteins. RANZI et al. (1957), showed that lithium protects proteins against the effects of animalizing agents and the action of proteolytic enzymes. According to RANZI (1957) animalizing agents would induce the destruction of pre-existing protein structures, with, as a consequence, an activation of metabolic processes. Vegetalizing agents, in contrast, by their stabilizing action on protein structures, would retard these metabolic activation processes.

16.8 Concluding Remarks

The operative method was used to show the capacity of development of different groups of blastomeres and their interaction during the first phases of larval differentiation. These observations are put into concrete form in the double gradient theory.

The chemical method has been considerably developed since HERBST's research and numerous substances have been shown capable of changing the orientation of sea urchin differentiation. The utilization of the most active of them has made it possible to obtain large quantities of modified larvae of the same type, a necessary condition for their biochemical study. The relatively considerable number and variety of animalizing and vegetalizing agents is striking. This situation suggests that these agents act at different stages of the processes involved in differentiation. The knowledge of the chemical composition and properties of these agents, the possibility of comparing the activity of related substances differing only by a single structural characteristic may supply indications about the cellular targets and mode of action. Broadly speaking, the inhibitors, whose mode of action on metabolism is best known, prove to be only slightly active or inactive. Chloramphenicol represents an exception in this respect. On the other hand, with the most active of the agents, it has not been possible so far to show in the sea urchin, any specific action on a specific metabolic system.

Isolation of animalizing and vegetalizing substances from unfertilized eggs constitutes a very important stage in the analysis of differentiation. This new line of research seems to be full of promise for the study of the sea urchin egg's differentiation and its experimental modifications.

The first hypothesis, made on the differentiation of the sea urchin egg, implied the existence of different types of metabolism in the animal and vegetal regions. It has not been possible to demonstrate any obvious difference in metabolism up to now, between animal and vegetal blastomeres. Differentiation, being based on the synthesis of specific proteins - it seems necessary to examine, with the aid of the most sensitive methods available, the proteins of animalized and vegetalized larvae.

The existence of reduction gradients is a particularly interesting phenomenon. Their evolution during development, and their chemically or operatively provoked modifications show that they develop in a direction parallel with the morphogenetic gradients postulated by the double gradient theory. The way in which the gradients are revealed suggests that they reflect, at least for a large part, the mitochondrial activity. These gradients should express the activity of the mitochondrial enzymatic systems which supply, in the form of ATP molecules, the energy necessary for the synthesis of specific proteins. This synthesis would be distributed in gradients corresponding to the morphogenetic gradients. The fact that chloramphenicol, which inhibits protein synthesis governed by the mitochondrial ribosomes, exercises a vegetalizing action, should be incentive to study the role of mitochondria in differentiation.

Cell differentiation implies the synthesis of specific proteins that intervene in metabolism (enzymes) and in the composition of cellular structures (proteins of structure). The knowledge of the processes that ensure the synthesis of proteins and the control of their structural specificity by genes, is essentially the result of research on microorganisms. The processes of transcription and translation of genetic information, shown in bacteria, are also at work in the more complex organisms, and in particular in the sea urchin. In this organism, the pluricellularity and the complexity of the genome enclosed in the nucleus, imply the existence of additional regulating mechanisms. In pluricellular organisms, the genome appears to be equally distributed between all the cells, and differentiation would result from a reversible and sequential activation of the genes. These phenomena would be controlled by the environment of the genome. Suggestions concerning gene activation at the beginning of development have been offered by RUNNSTRÖM (1967b), and CZIHAK (1968). LALLIER (1966f) has examined different possibilities of action of animalizing and vegetalizing agents on the processes of transcription and translation of genetic manifestations.

The control of nuclear activity by the cell medium may be considered in various ways. A possibility to be considered is the control of specific protein synthesis by the product of the synthesis itself, i.e. the proteins themselves. In the case of proteins of structure, for example, their utilization for the establishment of cell structures (membranes, mitochondria, etc.) by associating with other cell constituents of protein, lipid or carbohydrate nature, for example, implicates the adjustment of their synthesis to supply the necessary quantity of specific proteins for the elaboration of these structures. It is possible that the animalizing and vegetalizing agents inhibit the formation of certain cell structures, and therefore indirectly control the synthesis of the specific proteins which have a part in their formation. We will recall that numerous agents which are active on differentiation show a marked affinity for proteins, thereby modifying their structure and physical-chemical properties.

The role of the cell membrane must also be considered. It can act by controlling the exchange of material between the blastomeres. Experiments on isolating and recombining blastomeres indicate that a regulating influence is exercised by contact between the blastomeres, which is compatible with an exchange of material. But in the last resort, differentiation must originate in the blastomeres themselves. From the beginning of its development, the egg shows a heterogeneity of its surface marked by the distribution of pigment. The egg's cleavage and the resulting position of the blastomeres in relation to the animal-vegetal axis imply a differential distribution of the initial surface of the egg between the blastomeres. The cell surface, and perhaps also the underlying cortex, therefore, represent a component of the genome environment which changes very early in the course of development, and in particular during the period when the respective volumes of ecto- and entomesodermic structures are fixed. Some suggestions can be made about the possible role of superficial cell structures. RAVEN (1958), after studying the egg of Limnaea, suggested that information was inscribed as a blue print in the egg cortex. It is also possible to consider that superficial structures indirectly control the manifestation of genetic activity by the reversible fixation of factors responsible for the activation or inactivation of the genes. According to this concept, chemical agents could provoke animalization or vegetalization by modifying the affinity of the cell membrane for some regulating factors of genetic activity.

The effects of animalizing and vegetalizing agents have also been studied in various organisms: amphibians (LEHMANN, 1937; BÄCKSTRÖM and NWAGWU, 1963); prochordates (NIEUWKOOP, 1953); molluscs (RAVEN, 1952), and protista (KÖNIG, 1967). Modifications yielding a change in orientation of differentiation have been obtained in amphibian eggs with lithium ions and chloramphenicol. These observations, in conjunction with sea urchin research, suggest the existence of common underlying mechanisms controlling the orientation of the differentiation at the beginning of development.

Acknowledgement

I wish to thank Mr. P. CHANG for improvement of the English.

References

ALBERT, A., 1956. Mode of action of isoniazid. Nature 177, 525-526.
DE ANGELIS, E., RUNNSTRÖM, J., 1970. The effect of temporary treatment of animal half embryos with lithium and the modification of its effect by simultaneous exposure to actinomycin D. Wilhelm Roux' Arch. Entwicklungsmech. Organismen 164, 236-246.
APTER, M.J., 1966. Cybernetics and development. International series of monographs on pure and applied biology. Oxford: Pergamon Press.
AROSIO, R., 1953. Ricerche sierologiche sull'azione di solfocianuro e ioduro su macromolecole. Rend. ist. lombardo Sci. 86, 873-878.
AROSIO, R., BOSSI, J., 1954. Gruppi chimicamente atti vi dell'actomiosina dopo trattamento con sostanze animalizzanti e vegetativizzanti. Rend. ist. lombardo Sci. 87, 445-451.
AROSIO, R., CITTERIO, P., RANZI, M.E.S., TOSI, L., 1949. Le proteine nella determinazione embrionale degli Echinodermi. Rend. ist. lombardo Sci. 82, 143-178.
AZARI, P.R., FEENEY, R.E., 1961. The resistance of conalbumin and its iron complex to physical and chemical treatment. Arch. Biochem. Biophys. 92, 44-52.
BÄCKSTRÖM, S., 1953. Studies on the animalizing action of iodosobenzoic acid in the sea urchin development. Arkiv Zool. Ser. 2, 4, 485-491.
BÄCKSTRÖM, S., 1956. The total content of ascorbic acid in the developing sea urchin egg. Exp. Cell Res. 11, 322-326.
BÄCKSTRÖM, S., 1957. Content and distribution of ascorbic acid in sea urchin embryos of different developmental trends. Exp. Cell Res. 13, 333-340.
BÄCKSTRÖM, S., 1958a. The inhibitory effect of glutathione on some processes of animalization. Exp. Cell Res. 14, 426-429.
BÄCKSTRÖM, S., 1958b. Glutathione in aging sperm and developing eggs of the sea urchin. Arkiv Zool. Ser. 2, 11, 441-446.
BÄCKSTRÖM, S., 1959a. Studies on sulphydryl-containing substances in sea urchin embryos of various developmental trends. Exp. Cell Res. 16, 165-173.
BÄCKSTRÖM, S., 1959b. Activity of glucose-6-phosphate dehydrogenase in sea urchin embryos of different developmental trends. Exp. Cell Res. 18, 347-356.
BÄCKSTRÖM, S., 1959c. Reduction of blue tetrazolium in developing sea urchin eggs after addition of various substrates and phosphopyridine nucleotides. Exp. Cell Res. 18, 357-363.
BÄCKSTRÖM, S., 1959d. Reducing agents and activities in sea urchin development. Uppsala: Almqvist and Wiksells Boktr.
BÄCKSTRÖM, S., 1959e. Changes in ribonucleic acid content during early sea urchin development. Arkiv Zool. Ser. 2, 12, 339-342.
BÄCKSTRÖM, S., 1962. The hexose monophosphate shunt and mitosis. 3rd Scandinavian Conference on Cell Research.
BÄCKSTRÖM, S., 1963. 6-phosphogluconate dehydrogenase in sea urchin embryos. Exp. Cell Res. 32, 566-569.
BÄCKSTRÖM, S., GUSTAFSON, T., 1953. Lithium sensitivity in the sea urchin in relation to the stage of development. Arkiv Zool. Ser. 2, 6 No. 9, 185-188.
BÄCKSTRÖM, S., HULTIN, K., HULTIN, T., 1960. Pathways of glucose metabolism in early sea urchin development. Exp. Cell Res. 19, 634-636.
BÄCKSTRÖM, S., NWAGWU, M., 1963. Effects of chloramphenicol on early development of Xenopus. Acta Embryol. Morphol. Exp. 6, 333-338.
BARROS, C., GIUDICE, G., 1968. Effect of polyamines on ribosomal RNA synthesis during sea urchin development. Exp. Cell Res. 50, 671-674.
BERG, W.E., 1965. Rates of protein synthesis in whole and half embryos of the sea urchin. Exp. Cell Res. 40, 469-489.

BERG, W.E., 1968a. Rates of protein and nucleic acid synthesis in half embryos of the sea urchin. Exp. Cell Res. 50, 679-683.
BERG, W.E., 1968b. Effect of lithium on the rate of protein synthesis in the sea urchin embryo. Exp. Cell Res. 50, 133-139.
BERG, W.E., LONG, N.D., 1964. Regional differences of mitochondria size in the sea urchin embryo. Exp. Cell Res. 33, 422-437.
BERG, W.E., TAYLOR, D.A., HUMPHREYS, W.J., 1962. Distribution of mitochondria in echinoderm embryos as determined by electron microscopy. Develop. Biol. 4, 165-176.
BLACK, R.E., 1967. Partial activation of intact, unfertilized sea urchin eggs by treatment with polyuridylic acid. Exp. Cell Res. 46, 452-466.
BOSCO, M., MONROY, A., 1960. Inhibition of the differentiation of the primary mesenchyme in the sea urchin embryo caused by ethionine. Acta Embryol. Morphol. Exp. 3, 53-64.
BOSCO, M., MONROY, A., 1962. Differential incorporation of labeled amino acids in the territories of the sea urchin blastula. Experientia 18, 124-125.
BOVERI, TH., 1901. Die Polarität von Ovocyte, Ei und Larve des Strongylocentrotus lividus. Zool. Jahrb. 14, 630-653.
BREGMAN, J.I., 1953. Cation exchange processes. Ann. N.Y. Acad. Sci. 57, 125-143.
BUCKNALL, R.M., 1967. The effects of substituted benzimidazoles on the growth of viruses and the nucleic acid metabolism of host cells. J. gen. Virol. 1, 89-99.
BUCKNALL, R.A., CARTER, S.B., 1967. A reversible inhibitor of nucleic acid synthesis. Nature 213, No. 5081, 1099-1101.
CHAMBON, P., KARON, H., RAMUZ, M., MANDEL, P., 1968. The influence of ionic strength and a polyanion on transcription in vitro. II. Effects on the template efficiency of rat liver chromatin for a purified bacterial RNA polymerase. Biochim. Biophys. Acta 157, 520-531.
CHILD, C.M., 1936. Differential reduction of vital dyes in the early development of Echinoderms. Arch. f. Entw. Mech. 135, 426-493.
CHILD, C.M., 1948. Exogastrulation by sodium azide and other inhibiting conditions in Strongylocentrotus purpuratus. J. Exp. Zool. 107, 1-38.
CHVAPIL, M., 1973. New aspects in the biological role of zinc: a stabilizer of macromolecules and biological membranes. Life Sci. 13, 1041-1049.
COPPEN, A., 1965. Mineral metabolism in affective disorders. Brit. J. Psychiat. 111, 1133-1142.
CRAIG, N., 1973. The effects of inhibitors of RNA and DNA synthesis on protein synthesis and polysome levels in mouse L-cells. J. Cell Physiol. 82, 133-150.
CZIHAK, G., 1961. Ein neuer Gradient in der Pluteusentwicklung. Wilhelm Roux' Arch. Entwicklungsmech. Organismen 153, 353-356.
CZIHAK, G., 1962. Entwicklungsphysiologie der Echinodermen. Forts. d. Zool. 14, 238-267.
CZIHAK, G., 1963. Entwicklungsphysiologische Untersuchungen an Echiniden. (Verteilung und Bedeutung der Cytochromoxydase). Wilhelm Roux' Arch. Entwicklungsmech. Organismen 154, 272-292.
CZIHAK, G., 1965. Entwicklungsphysiologische Untersuchungen an Echiniden. Ribonucleinsäuresynthese in den Mikromeren und Entodermdifferenzierung. Ein Beitrag zum Problem der Induktion. Wilhelm Roux' Arch. Entwicklungsmech. Organismen 156, 504-524.
CZIHAK, G., 1968. Molekularbiologische Aspekte der frühen Embryonalentwicklung von Seeigeln. Eine Übersicht mit neuen Ergebnissen und einer Hypothese zur Erklärung der Regulationsfähigkeit. Pubbl. Sta. Zool. Napoli 36, 321-345.

CZIHAK, G., WITTMANN, H.G., HINDENNACH, I., 1967. Uridineinbau in die Nucleinsäuren von Furchungsstadien der Eier des Seeigels Paracentrotus lividus. Zeit. Naturforsch. 22b, 1176-1182.
DIAMOND, J.M., WRIGHT, E.M., 1969. Biological membranes: the physical basis of ion and non electrolyte selectivity. Ann. Rev. Physiol. 31, 581-646.
DRIESCH, H., 1891. Entwicklungsmechanische Studien. I. Der Wert der beiden ersten Furchungszellen in der Echinodermenentwicklung. Experimentelle Erzeugung von Teil- und Doppelbildung. II. Über die Beziehungen des Lichtes zur ersten Etappe der tierischen Formbildung. Zeit. Wiss. Zool. 53, 160-184.
ELBERS, P.F., 1959. Over de beginoorzaak van het Li-effect in de morphogenese. Een electronenmicroscopisch onderzoek aan eieren van Limnaea stagnalis en Paracentrotus lividus, 59 p. Utrecht: Drukkerij Libertas N.V., Amsterdam: L. van Leer and Co. N.V.
ERDMAN, K.A.P., 1959. Studies on the interaction of zinc and other bivalent metals with contractile element in glycerol-extracted muscle. Acta Soc. Med. Upsaliensis 64, 97-110.
FENAUX, L., 1961. Etude cytochimique des groupes sulfhydriles au cours des modifications de la détermination embryonnaire chez l'oeuf de l'oursin Paracentrotus lividus. Experientia 17, 316-317.
FERRINI, U., MANELLI, H., 1967. Studio quantitativo dei gruppi sulfidrili reattivi con il Mersalyl Hg^{203} nell'uovo vergine ed in alcuni precoci stadi della segmentazione del Riccio di mare (Paracentrotus lividus). Ist. Lombardo R, C.B. 101, 204-211.
FOX, F., WEISS, S.B., 1964. Enzymatic synthesis of ribonucleic acid. II. Properties of the deoxyribonucleic acid-primed reaction with Micrococcus lysodeikticus ribonucleic acid polymerase. J. Biol. Chem. 239, 175-185.
FRIEDMAN, R.M., PASTAN, I., 1970. Mechanism of action of 2-mercapto-1 (β-4-pyridethyl) benzimidazole. A reversible inhibitor of interferon activity. Proc. Nat. Acad. Sci. 65, 104-109.
FUDGE, M.W., 1959. Vegetalization of sea urchin embryos by treatment with tyrosine. Exp. Cell Res. 18, 401-404.
FUDGE-MASTRANGELO, M., 1966. Analysis of the vegetalizing action of tyrosine on the sea urchin embryo. J. exp. Zool. 161, 109-128.
FUJII, T., 1954. Presence of zinc in nucleoli and its possible role in mitosis. Nature 174, 1108-1109.
FUJII, T., OHNISHI, T., 1962. Inhibition of acid production at fertilization by nicotinamide and other inhibitors of DPNase. J. Fac. Sci. Univ. Tokyo Ser. IV, 9, 333-348.
FUJIOKA, M., LIEBERMAN, I., 1964. A Zn^{++} requirement for synthesis of deoxyribonucleic acid by rat liver. J. Biol. Chem. 239, 1164-1167.
GAMBLE, J.L., JR., 1957. Potassium binding and oxidative phosphorylation in mitochondria and mitochondrial fragments. J. Biol. Chem. 228, 955-971.
GEZELIUS, G., 1969. On the effect of sulphate deficiency and of o-iodosobenzoate and the role of sulphate in sea urchin development. Zool. Bidrag 37, 211-221.
GIUDICE, G., MUTOLO, V., DONATUTI, G., 1968. Gene expression in sea urchin development. Wilhelm Roux' Arch. Entwicklungsmech. Organismen 161, 118-128.
GORINI, L., AUDRAIN, L., 1952. Action de quelques métaux bivalents sur la sensibilité de la sérum-albumine à l'action de la trypsine. Biochim. Biophys. Acta 9, 180-192.
GREGOR, H.P., HAMILTON, M.J., OZA, J., BERNSTEIN, F., 1956. Studies on ion exchange resins. XV. Selectivity coefficients of methacrylic acid resins toward alkali metal cations. J. Phys. Chem. 60, 263-267.
GROSS, P.R., COUSINEAU, G.H., 1964. Macromolecule synthesis and the influence of actinomycin on early development. Exp. Cell Res. 33, 368-395.

GURD, F.R.N., GOODMAN, D.S., 1952. Preparation and properties of serum and plasma proteins. 32. The interaction of human serum albumin with Zn ions. J. Amer. Chem. Soc. 74, 670-675.
GUSTAFSON, T., 1965. Morphogenetic significance of biochemical patterns in sea urchin embryos. In: The biochemistry of animal development (R. Weber, ed.), 139-202. New York: Academic Press.
GUSTAFSON, T., 1969. Fertilization and development. In: Chemical Zoology (M. Florkin, B.T. Scheer, eds.), 149-206. Vol. III: Echinodermata, Nematoda and Acantocephala.
GUSTAFSON, T., HASSELBERG, I., 1950. Alkaline phosphatase activity in developing sea urchin eggs. Exp. Cell Res. 1, 371-375.
GUSTAFSON, T., HASSELBERG, I., 1951. Studies on enzymes in the developing sea urchin egg. Exp. Cell Res. 2, 642-672.
GUSTAFSON, T., HJELTE, M.-B., 1951. The amino acid metabolism of the developing sea urchin egg. Exp. Cell Res. 2, 474-490.
GUSTAFSON, T., HÖRSTADIUS, S., 1955. Vegetalization and animalization in the sea urchin egg induced by antimetabolites. Exp. Cell Res. Suppl. 3, 170-180.
GUSTAFSON, T., HÖRSTADIUS, S., 1956. 2-Thio-5-methylcytosine, an animalizing agent. Zool. Anz. 156, 102-106.
GUSTAFSON, T., HÖRSTADIUS, S., 1957. Changes in the determination of the sea urchin egg induced by amino acids. Pubbl. Sta. Zool. Napoli 29, 407-424.
GUSTAFSON, T., LENICQUE, P., 1952. Studies on mitochondria in the developing sea urchin egg. Exp. Cell Res. 3, 251-274.
GUSTAFSON, T., LENICQUE, P., 1955. Studies on mitochondria in early cleavage stages of the sea urchin egg. Exp. Cell Res. 8, 114-117.
GUSTAFSON, T., SÄVHAGEN, R. 1949. Studies on the determination of the oral side of the sea urchin egg. I. The effects of some detergents on the development. Arkiv Zool. 42A, No. 10, 1-6.
GUSTAFSON, T., WOLPERT, L., 1963. The cellular basis of morphogenesis and sea urchin development. Inter. Rev. Cytol. 15, 139-214.
HAGSTRÖM, B.E., LÖNNING, S., 1963. The effect of trypsin on the early development of the sea urchin egg. Arkiv Zool. Ser. 2, 15 No. 26, 377-380.
HAGSTRÖM, B.E., LÖNNING, S., 1967. Cytological and morphological studies of the action of lithium on the development of the sea urchin embryo. Wilhelm Roux' Arch. Entwicklungsmech. Organismen 158, 1-12.
HANSON, J.B., KRUEGER, W.A., 1966. Impairment of oxidative phosphorylation by D-threo and L-threo-chloramphenicol. Nature 211 No. 5055, 1322.
HERBST, C., 1892. Experimentelle Untersuchungen über den Einfluß der veränderten chemischen Zusammensetzung des umgebenden Mediums auf die Entwicklung der Tiere. I. Teil. Versuche an Seeigeleiern. Z. Wiss. Zool. 55, 446-518.
HERBST, C., 1896. Experimentelle Untersuchungen über den Einfluß der veränderten chemischen Zusammensetzung des umgebenden Mediums auf die Entwicklung der Tiere. Arch. f. Entw. Mech. 2, 455-516.
HERBST, C., 1904. Über die zur Entwicklung der Seeigellarven notwendigen anorganischen Stoffe, ihre Rolle und ihre Vertretbarkeit. III. Teil. Die Rolle der notwendigen anorganischen Stoffe. Arch. f. Entw. Mech. 17, 306-520.
HOFMEISTER, F., 1888. 11. Zur Lehre von der Wirkung der Salze. Arch. exp. path. Pharmakol. 24, 247-260.
HOLTER, H., LANZ, H., JR., LINDERSTRØM-LANG, K., 1938. Studies on enzymatic histochemistry. 30. Localization of peptidase during the first cleavages of the egg of the sea urchin Psammechinus miliaris. J. Cell Comp. Physiol. 12, 119-127.
HOLTER, H., LINDAHL, P.E., 1941. The distribution of peptidase in Paracentrotus lividus. J. Cell Comp. Physiol. 17, 235-241.

HÖRSTADIUS, S., 1935. Über die Determination im Verlaufe der Eiachse bei Seeigeln. Pubbl. Sta. Zool. Napoli 14, 251-479.

HÖRSTADIUS, S., 1936. Über die zeitliche Determination im Keim von Paracentrotus lividus. Wilhelm Roux' Arch. Entwicklungsmech. Organismen 135, 1-39.

HÖRSTADIUS, S., 1939. The mechanics of sea urchin development, studied by operative methods. Biol. Rev. 14, 132-179.

HÖRSTADIUS, S., 1949. Experimental researches on the developmental physiology of the sea urchin. Pubbl. Sta. Zool. Napoli Suppl. 21, 131-172.

HÖRSTADIUS, S., 1952. Induction and inhibition of reduction gradients by the micromeres in the sea urchin egg. J. Exp. Zool. 120, 421-436.

HÖRSTADIUS, S., 1953. Vegetalization of the sea-urchin egg by dinitrophenol and animalization by trypsin and ficin. J. Embryol. Exp. Morphol. 1, 327-348.

HÖRSTADIUS, S., 1955. Reduction gradients in animalized and vegetalized sea urchin eggs. J. Exp. Zool. 129, 249-256.

HÖRSTADIUS, S., 1959. The effect of sugars on differentiation of larvae of Psammechinus miliaris. J. Exp. Zool. 142, 141-158.

HÖRSTADIUS, S., 1963. Vegetalization of sea urchin larvae by chloramphenicol. Develop. Biol. 7, 144-151.

HÖRSTADIUS, S., GUSTAFSON, T., 1947. Change of determination in the sea urchin egg through the action of propanediol phosphate, phosphogluconic acid and lactate. Zool. Bidrag Uppsala 25, 571-581.

HÖRSTADIUS, S., GUSTAFSON, T., 1956. A remarkable case of animalization in a batch of eggs of Paracentrotus lividus. J. Embryol. Exp. Morphol. 4, 217-219.

HÖRSTADIUS, S., IMMERS, J., RUNNSTRÖM, J., 1966. The incorporation of $^{35}SO_4^{=}$ in whole embryos and meridional, animal and vegetal halves of the sea urchin Paracentrotus lividus. Exp. Cell Res. 43, 444-450.

HÖRSTADIUS, S., JOSEFSSON, L., RUNNSTRÖM, J., 1967. Morphogenetic agents from unfertilized eggs of the sea urchin Paracentrotus lividus. Develop. Biol. 16, 189-202.

HÖRSTADIUS, S., LORCH, J., CHARGAFF, E., 1954. The effect of deoxyribonucleic acids extracted from sea urchin sperm on the development of sea urchin eggs. Exp. Cell Res. 6, 440-452.

HÖRSTADIUS, S., STRÖMBERG, S., 1940. Untersuchungen über Undeterminierung von Fragmenten des Seeigeleies durch chemische Agentien. Wilhelm Roux' Arch. Entwicklungsmech. Organismen 140, 409-462.

HUANG, R.C.C., HUANG, P.C., 1969. Effect of protein-bound RNA associated with chick embryo chromatin on template specificity of the chromatin. J. Mol. Biol. 39, 365-378.

HULTIN, T., 1953a. Studies on the structural and metabolic background of fertilization and development, 28 p. Stockholm: E. Kihlströms.

HULTIN, T., 1953b. Incorporation of N^{15}-labelled ammonium chloride into pyrimidines and purines during the early sea-urchin development. Arkiv Kemi 5, 267-275.

HULTIN, T., 1953c. Incorporation of C^{14}-labelled carbonate and acetate into sea urchin embryos. Arkiv Kemi 6, 195-200.

HULTIN, T., 1957. The metabolic utilization of C^{14}-formate in sea urchin embryos. Exp. Cell Res. 12, 518-525.

HULTIN, T., 1961. The effect of puromycin on protein metabolism and cell division in fertilized sea urchin eggs. Experientia 17, 410-411.

HULTIN, T., 1964a. On the mechanism of ribosomal activation in newly fertilized sea urchin eggs. Develop. Biol. 10, 305-328.

HULTIN, T., 1964b. Factors influencing polyribosome formation in vivo. Exp. Cell Res. 34, 608-611.

HURWITZ, J., 1963. RNA polymerase. In: Methods in Enzymology, VI (S.P. Colowick, N.O. Kaplan, eds.), 23-27. New York: Academic Press.

HYNES, R.O., GROSS, P.R., 1970. A method for separating cells from early sea urchin embryos. Develop. Biol. 21, 383-402.

IMMERS, J., 1956. Cytological features of the development of the eggs of Paracentrotus lividus reared in artificial sea water devoid of sulphate ions. Exp. Cell Res. 10, 546-548.
IMMERS, J., 1961. Comparative study of the localization of incorporated ^{14}C-labelled amino acids and $^{35}SO_4$, in the sea urchin ovary, egg and embryo. Exp. Cell Res. 24, 356-378.
IMMERS, J., 1962. Investigations on macromolecular sulfated polysaccharides in sea urchin development. A survey of histochemical and autoradiographic studies on sulfated polysaccharides, 19 p. Uppsala: Almqvist and Wiksells.
IMMERS, J., 1965. Monoiodoacetate as an animalizing agent in experiments with embryo of the sea urchin, Paracentrotus lividus. Acta Embryol. Morphol. Exp. 8, 205-212.
IMMERS, J., RUNNSTRÖM, J., 1960. Release of respiratory control by 2,4-dinitrophenol in different stages of sea urchin development. Develop. Biol. 2, 90-104.
IMMERS, J., RUNNSTRÖM, J., 1965. Further studies of the effects of deprivation of sulfate on the early development of the sea urchin Paracentrotus lividus. J. Embryol. Morphol. Exp. 14, 289-305.
JOSEFSSON, L., HÖRSTADIUS, S., 1969. Morphogenetic substances from sea urchin eggs. Isolation of animalizing and vegetalizing substances from unfertilized eggs of Paracentrotus lividus. Develop. Biol. 20, 481-500.
KARNOFSKY, D.A., BASCH, R.S., 1960. Effects of 5-fluoro-deoxyuridine and related halogenated pyrimidines on the sand-dollar embryo. J. Biophys. Biochem. Cytol. 7, 61-71.
KARNOFSKY, D.A., BEVELANDER, G., 1958. Effects of DON (6-Diazo-5-oxo-L-norleucine) and Azaserine on the sand dollar embryo. Proc. Soc. Exp. Biol. N.Y. 97, 32-37.
KATZ, R.I., CHASE, Z.N., KOPIN, I.J., 1968. Evoked release of norepinephrine and serotonin from brain slices: inhibition by lithium. Science 162 No. 3852, 466-467.
KENDREY, G., JUHASZ, J., SZENDE, B., 1967. Ultrastructure of the liver and of liver tumors provoked by isonicotinic acid hydrazide in rats. Morph. Igasz. orv. Szle. 7, 176-186.
KÖNIG, K., 1967. Wirkung von Lithium- und Rhodanid-Ionen auf die polare Differenzierung und die Morphogenese von Stentor coeruleus Ehrbg. Arch. Protistenk. 110, 179-230.
KRISZAT, G., RUNNSTRÖM, J., 1957. Effect of dinitrophenols on the physical properties, developmental capacity and differentiation of sea urchin eggs. Arkiv Zool. 10, 595-606.
KROEGER, H., 1963. Experiments on the extranuclear control of gene activity in dipteran polytene chromosomes. J. Cell Comp. Physiol. 62, 45-59.
KROON, A.M., 1963. Protein synthesis in heart mitochondria. I. Amino acid into the protein of isolated beef-heart mitochondria and fractions derived from them by ionic oscillation. Biochim. Biophys. Acta 72, 391-402.
LALLIER, R., 1952. Recherches sur le problème de la détermination chez les Echinodermes. Experientia 8, 271.
LALLIER, R., 1955a. Effets des ions zinc et cadmium sur le développement de l'oeuf de l'oursin Paracentrotus lividus. Arch. Biol. (Liège) 66, 75-102.
LALLIER, R., 1955b. Animalisation de l'oeuf d'oursin par les colorants azoiques et les blues d'aniline sulfonés. Exp. Cell Res. 9, 232-240.
LALLIER, R., 1956a. Recherches sur la détermination embryonnaire chez les Echinodermes. L'action des oxydants et de l'acide p.chloromercuribenzoique sur le développement de l'oeuf de l'oursin Paracentrotus lividus. Arch. Biol. 67, 181-209.
LALLIER, R., 1956b. Les ions des métaux lourds et le problème de la détermination embryonnaire chez les Echinodermes. J. Embryol. Exp. Morphol. 4, 265-278.

LALLIER, R., 1956c. Réduction du vert Janus B par les embryons de l'oursin Paracentrotus lividus animalisés par les ions zinc. Arch. Biol. 67, 475-483.
LALLIER, R., 1957a. Perchlorate et détermination embryonnaire chez les Echinodermes. Comp. Rend. Soc. Biol. 151, 471-474.
LALLIER, R., 1957b. Colorants acides et détermination embryonnaire chez les Echinodermes. Experientia 13, 362-363.
LALLIER, R., 1957c. Recherches sur l'animalisation de l'oeuf de l'oursin Paracentrotus lividus par les dérivés polysulfoniques. Pubbl. Sta. Zool. Napoli 30, 185-209.
LALLIER, R., 1958a. Recherches sur les relations entre la structure et l'activité animalisante des dérivés organiques acides. Arch. Biol. (Liège) 69, 497-517.
LALLIER, R., 1958b. Effets de substances sulfhydrilées sur la détermination embryonnaire de l'oeuf de l'oursin Paracentrotus lividus. Comp. Rend. 246, 2810-2812.
LALLIER, R., 1959a. Recherches sur l'animalisation de l'oeuf d'oursin par les ions zinc. J. Embryol. Exp. Morphol. 7, 540-548.
LALLIER, R., 1959b. Les groupes sulfhydriles et la détermination embryonnaire de l'oeuf de l'oursin Paracentrotus lividus. Comp. Rend. 248, 2397-2399.
LALLIER, R., 1960a. Recherches sur les effets des ions de métaux alcalins sur le développement embryonnaire de l'oeuf d'oursin. Exp. Cell Res. 21, 556-563.
LALLIER, R., 1960b. Effets d'un dérivé organique du mercure, le salyrgan, sur le développement et la détermination embryonnaire de l'oeuf de l'oursin Paracentrotus lividus. Comp. Rend. 250, 3380-3382.
LALLIER, R., 1960c. Le 2-désoxy-D-glucose et le développment de l'oeuf de l'oursin Paracentrotus lividus. Comp. Rend. 250, 3509-3510.
LALLIER, R., 1961. Les effets du chloramphénicol sur la détermination embryonnaire de l'oeuf de l'oursin Paracentrotus lividus. Comp. Rend. 253, 3060-3062.
LALLIER, R., 1962a. Les effets du chloramphénicol sur le développement de l'oeuf d'oursin. J. Embryol. Exp. Morphol. 10, 563-571.
LALLIER, R., 1962b. Recherches sur le contrôle de la différenciation de l'oeuf d'oursin par des inhibiteurs des synthèses protéiques. Comp. Rend. Soc. Biol. 156, 1249-1252.
LALLIER, R., 1962c. Les effets du thiosorbitol et du thioglycérol sur la structure de l'appareil mitotique de l'oeuf de l'oursin Paracentrotus lividus. J. Cell Biol. 15, 382-384.
LALLIER, R., 1962d. Les groupes sulfhydriles et la morphogenèse de l'oeuf de l'oursin Paracentrotus lividus. Develop. Biol. 5, 218-231.
LALLIER, R., 1962e. Les effets de quelques analogues de purines, de pyrimidines et d'acides aminés sur le développement de l'oeuf d'oursin. Acta Embryol. Morphol. Exp. 5, 179-188.
LALLIER, R., 1962f. Effets de la D-cyclosérine sur la morphogenèse de la larve d'oursin. Comp. Rend. Soc. Biol. 156, 762-765.
LALLIER, R., 1963a. Effets de substances polycationiques sur la détermination embryonnaire de l'oeuf de l'oursin Paracentrotus lividus. Comp. Rend. 256, 5409-5412.
LALLIER, R., 1963b. Effet inhibiteur de nucléosides sur les processus de la végétalisation chez l'oeuf d'oursin. Exp. Cell Res. 29, 119-127.
LALLIER, R., 1963c. Effets de l'actinomycine D sur le développement de l'oeuf de l'oursin Paracentrotus lividus. Comp. Rend. 257, 2159-2162.
LALLIER, R., 1964a. Biochemical aspects of animalization and vegetalization in the sea urchin embryo. In: Advances in Morphogenesis (M. Abercrombie, J. Brachet, eds.). New York: Academic Press.
LALLIER, R., 1964b. Effet d'un acide ribonucléique soluble de levure

sur la morphogénèse de l'oeuf de l'oursin Paracentrotus lividus. Comp. Rend. 259, 2140-2142.
LALLIER, R., 1965a. Effets de substances sulfhydrilées (Cystéine, Homocystéine et Pénicillamine) sur le développement de l'oeuf de l'oursin Paracentrotus lividus et action protectrice des acides aminés. Acta Embryol. Morphol. Exp. 8, 12-19.
LALLIER, R., 1965b. Effets du 6-fluoro-uracile et de la 6-méthyl-purine sur le développment de l'oeuf de l'oursin Paracentrotus lividus. J. Embryol. Exp. Morphol. 14, 181-189.
LALLIER, R., 1965c. Effets de différents analogues de la proline sur le développement de l'oeuf de l'oursin Paracentrotus lividus. Exp. Cell Res. 40, 630-636.
LALLIER, R., 1966a. Effets de l'isoniazide et de substances apparentées sur la différenciation de l'oeuf d'oursin. Life Sci. 5, 1761-1765.
LALLIER, R., 1966b. Effets d'analogues du chloramphénicol sur la différenciation de l'oeuf de l'oursin Paracentrotus lividus. Experientia 22, 724.
LALLIER, R., 1966c. Relations entre la structure des polyamines et leur action sur la differenciation de l'oeuf de l'oursin Paracentrotus lividus. Comp. Rend. 262, 1460-1463.
LALLIER, R., 1966d. Relation entre la valence et l'activité de complexes cationiques sur la différenciation de l'oeuf de l'oursin Paracentrotus lividus. Comp. Rend. 263, 386-388.
LALLIER, R., 1966e. Recherches sur les effets de l'acide ribonucléique soluble sur la différenciation de l'oeuf de l'oursin Paracentrotus lividus. Comp. Rend. Soc. Biol. 160, 269-272.
LALLIER, R., 1966f. Les modifications expérimentales de la morphogenèse chez la larve d'oursin. Essai d'interprétation des effets des agents animalisants et végétalisants. Ann. Biol. V, 313-335.
LALLIER, R., 1967. Modifications de la différenciation de la larve d'oursin Paracentrotus lividus par la pronase. Comp. Rend. 265, 751-753.
LALLIER, R., 1968a. Relations entre la structure du chloramphénicol et ses effets végétalisants sur l'oeuf de l'oursin Paracentrotus lividus. Acta Embryol. Morphol. Exp. 10, 280-287.
LALLIER, R., 1968b. Effets de la 2-mercaptopropionyl-glycine sur la différenciation de l'oeuf d'oursin. Experientia 24, 803-804.
LALLIER, R., 1968c. Effets de la L-asparaginase et d'un analogue de l'asparagine, la L-2-amino-2-carboxy-ethanesulfonamide, sur le développement de l'oeuf d'oursin. Life Sci. 8, 801-804.
LALLIER, R., 1969a. Changement de la différenciation de l'oeuf de l'oursin Paracentrotus lividus par le 2-mercapto-1-(β-4-pyridéthyl) benzimidazole. Comp. Rend. 268, 2592-2594.
LALLIER, R., 1969b. Recherches sur les modifications expérimentales de la différenciation de la larve d'oursin par les enzymes proteolytiques. Comp. Rend. Soc. Biol. 163, 2028-2032.
LAMB, A.J., CLARK-WALKER, G.D., LINNANE, A.W., 1968. The biogenesis of mitochondria. 4. The differentiation of mitochondrial and cytoplasmic protein synthesizing systems in vitro by antibiotics. Biochim. Biophys. Acta 161, 415-427.
LEHMANN, F., 1937. Mesodermisierung des präsumptiven Chordamaterials durch Einwirkung von Lithiumchlorid auf die Gastrula von Triton alpestris. Wilhelm Roux' Arch. Entwicklungsmech. Organismen 136, 112-146.
LEHMANN, F.E., 1950. Elektronenmikroskopische Untersuchungen an den Polplasmen von Tubifex und den Mikromeren von Paracentrotus. Arch. Julius Klaus-Sift. Vererbungsforsch. Sozialanthropol. u. Rassenhyg. 25, 611-619.

LENICQUE, P., HÖRSTADIUS, S., GUSTAFSON, T., 1953. Change of distribution of mitochondria in animal halves of sea urchin eggs by the action of micromeres. Exp. Cell Res. 5, 400-403.
LINDAHL, P.E., 1936. Zur Kenntnis der physiologischen Grundlagen der Determination im Seeigelkeim. Acta Zool. 17, 179-365.
LINDAHL, P.E., 1939. Zur Kenntnis der Entwicklungsphysiologie des Seeigeleies. Z. vergl. Physiol. 27, 233-250.
LINDAHL, P.E., 1940. Neue Beiträge zur physiologischen Grundlage der Vegetativisierung des Seeigelkeimes durch Lithiumionen. Wilhelm Roux' Arch. Entwicklungsmech. Organismen 140, 168-194.
LINDAHL, P.E., HOLTER, H., 1940. Beiträge zur enzymatischen Histochemie. 33. Die Atmung animaler und vegetativer Keimhälften von Paracentrotus lividus. Comp. Rend. Lab. Carlsberg, Ser. Chim. 23, 257-287.
LINDAHL, P.E., SWEDMARK, B., LUNDIN, J., 1951. Some new observations on the animalization of the unfertilized sea urchin egg. Exp. Cell Res. 2, 39-46.
LOEB, J., 1920. Chemical characters and physiological action of the potassium ions. J. Gen. Physiol. 3, 237-245.
LOEWENSTEIN, W.R., 1968. III. Emergence of order in tissue and organs. Communication through cell junctions. Implications in growth control and differentiation. Develop. Biol. 2, 151-183.
LUBIN, M., 1963. A priming reaction in protein synthesis. Biochim. Biophys. Acta 72, 345-348.
MAGER, J., 1960. Chloramphenicol and chlortetracycline inhibition of amino acid incorporation into proteins in a cell-free system from Tetrahymena pyriformis. Biochim. Biophys. Acta 38, 150-152.
MANELLI, H., FERRINI, U., 1967. Sui meccanismi di protezione contro la polispermia da un composto mercuriale (Mersalyl) nell'uovo di riccio di mare (Paracentrotus lividus). Ist. lombardo, R.C. B 101, 191-203.
MARKMAN, B., 1957. Studies on nuclear activity in the differentiation of the sea urchin embryo. Exp. Cell Res. 12, 424-426.
MARKMAN, B., 1961a. Regional differences in isotopic labelling of nucleic acid and protein in early sea urchin development. An autoradiographic study. Exp. Cell Res. 23, 118-129.
MARKMAN, B., 1961b. Difference in isotopic labelling of nucleic acid and protein in sea urchin embryos developing from animal and vegetal egg halves. Exp. Cell Res. 25, 224-227.
MARKMAN, B., 1967. Isotopic labelling of nucleic acids in sea urchin embryos developing from animal and vegetal halves in relation to protein and nucleic acid content. Exp. Cell Res. 46, 1-18.
MARKMAN, B., RUNNSTRÖM, J., 1963. Animal and vegetal halves of sea urchin larvae subjected to temporary treatment with actinomycin C and mitomycin C. Exp. Cell Res. 31, 615-618.
MARTIN, R.G., AMES, B.N., 1962. The effect of polyamines and of poly U size on phenylalanine incorporation. Proc. Nat. Acad. Sci. U.S. 48, 2171-2178.
MATEYKO, G.M., 1967. Developmental modifications in Arbacia punctulata by various metabolic substances. Biol. Bull. 133, 184-228.
MAZIA, D., ZIMMERMAN, A.M., 1958. SH compounds in mitosis. II. The effect of mercaptoethanol on the structure of the mitotic apparatus in sea urchin eggs. Exp. Cell Res. 15, 138-153.
MCMILLAN, F.H., LEONARD, F., MELTZER, R.I., KING, J.A., 1953. Antitubercular substances. II. Substitution products of isonicotinic hydrazide. J. Am. Phar. Assoc. 42, 457-464.
MOORE, A.R., 1952. An animalizing effect of trypsin and its inhibition by lithium in the developing eggs of Strongylocentrotus droebachiensis. J. Exp. Zool. 121, 99-104.

MOORE, A.R., BLISS, H.S., ANDERSON, E.H., 1945. The effects of pyocyanine and of lithium on the development and respiration of the eggs of 2 echinoderms. J. Cell Comp. Physiol. 25, 27-35.
MORI, S., TAKASHIMA, Y., YANO, K., KOJO, T., HASHIMOTO, K., FUJITANI, N., 1958. Studies on mitochondria and developing sea urchin egg. Bull. Exp. Biol. (Japan) 8, 65-70.
NAKATA, Y., BADER, J.P., 1969. Uptake of nucleosides by cells in culture. II. Inhibition by 2-mercapto-1-(β-4-pyridethyl)benzimidazole. Biochim. Biophys. Acta 190, 250-256.
NEEDHAM, J., NEEDHAM, D.M., 1940. A note on the biochemistry of embryonic determination in echinoderms. J. Exp. Biol. 17, 147-152.
NIEUWKOOP, P.D., 1953. The influence of the Li ion on the development of the egg of Ascidia malaca. Pubbl. Sta. Zool. Napoli 24, 101-141.
ÖHMAN, L.O., 1940. Über die Veränderung des respiratorischen Quotienten während der Frühentwicklung des Seeigeleies. Arkiv Zool. 32A, 1-9.
O'MELIA, A.F., 1971. Animalizing effects of Evans blue in embryos of Arbacia punctulata. Exp. Cell Res. 67, 402-406.
O'MELIA, A.F., 1972. Changes in esterase and cholinesterase isozymes in normally developing, animalized and radialized embryos of Arbacia punctulata. Exp. Cell Res. 73, 469-474.
O'MELIA, A.F., 1973. Animalizing ability of Evans blue in embryos of Arbacia punctulata. Exp. Cell Res. 77, 280-284.
OHNISHI, T., 1963. Adenosine triphosphatase activity relating to active transport in the cortex of sea urchin egg. J. Biochem. 53, 238-241.
OTOMO, T., 1961. Stage sensitive to lithium in sea urchin embryo. Sci. Rep. Tohoku Univ. Ser. IV, 27, 183-185.
OZRETIC, B., KRAJNOVIC, M., 1968. The turnover of Zn^{65} during the early embryonal development of the sea urchin Paracentrotus lividus. Lam. Rev. Inter. Oceanogr. med. 11, 101-106.
PERKINS, D.J., 1961. Studies on the interaction of zinc, cadmium and mercuric ions with native and chemically modified human serum albumin. Biochem. J. 80, 668-672.
PERLMANN, P., GUSTAFSON, T., 1948. Antigens in the egg and early developmental stages of the sea urchin. Experientia 4, 481-483.
PIKO, L., TYLER, A., 1969. Properties and function of mitochondrial DNA of sea urchin eggs. Ann. Embryol. et Morphogénèse, Suppl. 1, 266.
PIRRONE, A.M. SCONZO, G., MUTOLO, V., GIUDICE, G., 1970. Effect of chemical animalization and vegetalization on the synthesis of ribosomal RNA in sea urchin embryos. Wilhelm Roux' Arch. Entwicklungsmech. Organismen 164, 222-225.
PRESSMAN, B.C., LARDY, H.A., 1952. Influence of K and other alkalin cations on respiration mitochondria. J. Biol. Chem. 197, 547-555.
RAINA, A., COHEN, S.S., 1966. Polyamines and RNA synthesis in a polyauxotrophic strain of E. coli. Proc. Nat. Acad. Sci. U.S. 55, 1587-1593.
RANZI, S., 1955. Le macromolecule nella determinazione embrionale. Rend. Ist. lombardo Sci. 89, 200-223.
RANZI, S., 1957. Early determination in development under normal and experimental conditions. In: The beginnings of embryonic development, 291-318. Washington: Am. Assoc. Advanc. Science.
RANZI, S., 1959. On the stability of the structures of animalized and vegetalized sea urchin embryos. Biol. Bull. 117, 436-437.
RANZI, S., CITTERIO, P., 1956. Sulle proteine dell'Arbacia lixula nello sviluppo embrionale e postembrionale. Rend. Ist. lombardo Sci. 90, 515-521.
RANZI, S., CITTERIO, P., COPES, M.V., SAMUELLI, C., 1957. Proteine e determinazione embrionale nella Rana, nel Riccio di mare e negli incroci dei Rospi. Acta Embryol. Morphol. Exp. 1, 78-98.

RANZI, S., FALKENHEIM, M., 1937. Ricerche sulle basi fisiologiche della determinazione nell'embrione degli Echinodermi. Pubbl. Sta. Zool. Napoli 16, 436-458.
RAVEN, CHR.P., 1952. Lithium as a tool in the analysis of morphogenesis in Limnaea stagnalis. Experientia 8, 252-257.
RAVEN, CHR.P., 1958. Information versus preformation in embryonic development. Arch. néerl. Zool. 13, Suppl. 1, 185-193.
RULON, O., 1950. The modification of developmental patterns in the sand dollar with sodium azide. Physiol. Zool. 23, 236-247.
RULON, O., 1951. The modification of developmental patterns in the sand dollar by malonic acid. Physiol. Zool. 24, 85-92.
RULON, O., 1952a. The modification of developmental patterns in the sand dollar by sodium selenite. Physiol. Zool. 25, 333-346.
RULON, O., 1952b. The modification of developmental patterns in the sand dollar by glucose. Physiol. Zool. 25, 346-357.
RULON, O., 1953a. The modification of developmental patterns in the sand dollar with cobaltous chloride. Anat. Rec. 117, 522-523.
RULON, O., 1953b. The modification of developmental patterns in the sand dollar with nickelous chloride. Anat. Rec. 177, 615.
RULON, O., 1955. Developmental modifications in the sand dollar caused by zinc chloride and prevented by glutathione. Biol. Bull. 109, 316-327.
RUNNSTRÖM, J., 1914. Analytische Studien über die Seeigelentwicklung. Wilhelm Roux' Arch. Entwicklungsmech. Organismen 40, 520-564.
RUNNSTRÖM, J., 1928a. Zur experimentellen Analyse der Wirkung des Lithiums auf den Seeigelkeim. Acta Zool. 9, 365-424.
RUNNSTRÖM, J., 1928b. Plasmabau und Determination bei dem Ei von Paracentrotus lividus Lk. Wilhelm Roux' Arch. Entwicklungsmech. Organismen 113, 556-581.
RUNNSTRÖM, J., 1929. Über Selbstdifferenzierung und Induktion bei dem Seeigelkeim. Wilhelm Roux' Arch. Entwicklungsmech. Organismen 117, 123-145.
RUNNSTRÖM, J., 1935. An analysis of the action of lithium on sea-urchin development. Biol. Bull. 68, 378-383.
RUNNSTRÖM, J., 1948. On the action of trypsin and chymotrypsin on the unfertilized sea-urchin egg. A study concerning the mechanism of formation of the fertilization membrane. Arkiv Zool. 40A, No. 17, 1-16.
RUNNSTRÖM, J., 1956. The animalizing effect of α-lipoic acid on sea urchin eggs. Exp. Cell Res. 11, 660-664.
RUNNSTRÖM, J., 1957. Cellular structure and behavior under the influence of animal and vegetal factors in sea urchin development. Arkiv Zool. Ser. 2, 10, 523-537.
RUNNSTRÖM, J., 1967a. The animalizing action of pretreatment of sea urchin egg with thiocyanate in calcium-free seawater and its stabilization after fertilization. Arkiv Zool. Ser. 2, 19, 251-263.
RUNNSTRÖM, J., 1967b. The mechanism of control of differentiation in early development of the sea urchin. A tentative discussion. Exp. Biol. Med. 1, 52-62.
RUNNSTRÖM, J., HÖRSTADIUS, S., IMMERS, J., FUDGE-MASTRANGELO, M., 1964. An analysis of the role of sulfate in the embryonic differentiation of the sea urchin (Paracentrotus lividus). Rev. suisse Zool. 71, 21-54.
RUNNSTRÖM, J., IMMERS, J., 1966. The animalizing action of trypsin on embryos of the sea urchin (Psammechinus miliaris, Paracentrotus). A study of interaction in early embryonic differentiation. Arch. Biol. (Liège) 77, 365-410.
RUNNSTRÖM, J., KRISZAT, G., 1952. Animalizing action of iodosobenzoic acid in the sea urchin development. Exp. Cell Res. 3, 497-502.
RUNNSTRÖM, J., MARKMAN, B., 1966. Gene dependency of vegetalization in sea urchin embryos treated with lithium. Biol. Bull. 130, 402-414.

RUNNSTRÖM, J., THÖRNBLOM, D., 1938. The influence of pyocyanine on the animalization of the sea urchin egg induced by calcium free seawater and sulphocyanate. Acta Biol. latvica, 97-115.
SCRUTTON, M.C., WU, C.W., GOLDTHWAIT, D.A., 1971. The presence and possible role of zinc in RNA polymerase obtained from Escherichia coli. Proc. Nat. Acad. Sci. U.S. 68, 2497-2501.
SELENKA, E., 1883. Studien zur Entwicklungsgeschichte der Tiere. II. Die Keimblätter der Echinoderm, 28-61. Wiesbaden: C.W. Kreidel's Verlag.
SHAVER, J.R., 1955. The distribution of mitochondria in sea urchin embryos. Experientia 11, 351-353.
SHAVER, J.R., 1956. Mitochondrial populations during development of the sea urchin. Exp. Cell Res. 11, 548-559.
SHAVER, J.R., 1957. Some observations on cytoplasmic particles in early echinoderm development. Publ. Am. Assoc. Advanc. Sci. 48, 263-290.
SHINOZAWA, T., YAHARA, I., IMAHORI, K., 1968. Interaction of polyvinyl sulfate with ribosomes. J. Mol. Biol. 36, 305-319.
SKEHEL, J.J., HAY, A.J., BURKE, D.C., CARTWRIGHT, L.N., 1967. Effects of actinomycin D and 2-mercapto-1-(β-4-pyridethyl)benzimidazole on the incorporation of (^{3}H)Uridine by chick embryo cells. Biochim. Biophys. Acta 142, 430-439.
SLATER, J.P., MILDVAN, A.S., LOEB, L.A., 1971. Zinc in DNA polymerase. Biochem. Biophys. Res. Comm. 44, 37-43.
SMITH, D.W.E., 1969. The effect of salt solutions on the acceptance of amino acids by transfer ribonucleic acid. J. Biol. Chem. 244, 896-901.
SPIEGEL, M., SPIEGEL, E.S., MELTZER, P.S., 1970. Qualitative changes in the basic protein fraction of developing embryos. Develop. Biol. 21, 73-86.
SPYRIDES, G.J., 1964. The effect of univalent cations on the binding of s-RNA to the template-ribosome complex. Proc. Nat. Acad. Sci. U.S. 51, 1220-1226.
SUZUKA, I., KAJI, A., 1968. Reversible effect of lithium chloride on ribosomes. J. Biol. Chem. 243, 3136-3141.
SWANK, R.L., DAVENPORT, H.A., 1935. Chlorate-osmic-formalin method for staining degenerating myelin. Stain Technol. 10, 87-90.
TCHAKHOTINE, S., 1938. Cancérisation expérimentale des éléments embryonnaires, obtenue sur des larves d'oursins. Comp. Rend. Soc. Biol. 127, 1195-1197.
TELESNINA, G.N., NOVIKOVA, M.A., ZHDANOV, G.L., KOLOSOV, M.N., SHEMYAKIN, M.M., 1967. The effect of chloramphenicol analogs on protein biosynthesis in a cell-free Escherichia coli B system. Experientia 23, 427-428.
TIEDEMANN, H., TIEDEMANN, H., BORN, J., 1969. Polyvinylsulfate: Interaction with complexes of morphogenetic factors and their natural inhibitors. Science 164, 1175-1177.
TIMOURIAN, H., 1968. The effect of zinc on sea urchin morphogenesis. J. Exp. Zool. 169, 121-132.
UBISCH, VON, L., 1925a. Entwicklungsphysiologische Studien an Seeigelkeimen. I. Über die Beziehungen der ersten Furchungsebene zur Larven-Symmetrie und die prospektive Bedeutung der Eibezirke. Z. wiss. Zool. 124, 361-381.
UBISCH, VON, L., 1925b. Entwicklungsphysiologische Studien an Seeigelkeimen. II. Die Entstehung von Einheitslarven aus verschmolzenen Keimen. Z. wiss. Zool. 124, 457-468.
DE VICENTIIS, M., HÖRSTADIUS, S., RUNNSTRÖM, J., 1966. Studies on controlled and released respiration in animal and vegetal halves of the embryo of the sea urchin Paracentrotus lividus. Exp. Cell Res. 41, 535-544.

DE VICENTIIS, M., RUNNSTRÖM, J., 1967. Studies on controlled and released respiration in animalized and vegetalized embryos of the sea urchin Paracentrotus lividus. Exp. Cell Res. 45, 681-689.
WAGNER, R.P., 1969. Genetics and phenogenetics of mitochondria. Science 163, 1026-1031.
WALTERS, S., BURKE, D.C., SKEHEL, J.J., 1967. Interferon production and RNA inhibitors. J. gen. Virol. 1, 349-362.
WATERMAN, A.J., 1937. Effect of salts of heavy metals on development of the sea urchin, Arbacia punctulata. Biol. Bull. 73, 401-420.
WESER, U., SEEBER, S., WARNECKE, P., 1969. Reactivity of Zn^{2+} on nuclear DNA and RNA biosynthesis of regenerating rat liver. Biochim. Biophys. Acta 179, 422-428.
WINDER, F., 1964. Early changes induced by isoniazid in the composition of Mycobacterium tuberculosis. Biochim. Biophys. Acta 82, 210-212.
WOLFSON, N., 1966. Dosage de groupes sulfhydriles à la surface d'embryons vivants de l'oursin Paracentrotus lividus. Comp. Rend. Soc. Biol. 160, 1996-1999.
WOLPERT, L., 1969. Positional informations and the spatial pattern of cellular differentiation. J. Theoret. Biol. 25, 1-47.
WOLPERT, L., MERCER, E.H., 1961. An electron microscope study of fertilization of the sea urchin egg Psammechinus miliaris. Exp. Cell Res. 22, 45-55.
WOLSKY, A., DE ISSEKUTZ WOLSKY, M., 1964. An analysis of the actinomycin induced cleavage delay in sea urchin eggs. Z. Zellforsch. 63, 960-964.
YASUMASU, I., 1963. Inhibition of the hatching enzyme formation during embryonic development of the sea urchin by chloramphenicol, 8-aza-guanine and 5-bromo-uracil. Sci. Rep. Cool. gen. Educ. Univ. Tokyo 13, 241-246.
ZELLER, E.A., BARSKY, J., FOUTS, J.R., KIRCHHEIMER, W.F., VAN ORDEN, L.S., 1952. Influence of isonicotinic acid hydrazide (INH) and 1-isonicotinyl-2-isopropyl hydrazide (IIH) on bacterial and mammalian enzymes. Experientia 8, 349-350.

Addendum

RUNNSTRÖM and IMMERS (1971) have suggested that the vegetalizing action of lithium ions resulted in the inhibition of the diffusion of animalizing substances. Lithium ions affect β-glucanase activity. The activity increased in lithium treated larvae normally invaginated. In exogastrulates the activity of the enzyme is slightly lower than that of the controls (VACQUIER, 1971). In a preliminary study, RUNNSTRÖM et al. (1972) noted a modification of the sedimentation pattern of the ribosomes in lithium treated embryos. Concanavalin A induces animalization. These effects were counteracted by methyl-α-D-mannopyranoside. The role of carbohydrates of the cell membrane in the animalizing effect of Concanavalin A is suggested (LALLIER, 1972). The suppression of the differentiation of entomesodermal parts of the larvae may be obtained by a 3 minutes treatment, 30 seconds after fertilization of the eggs with a detergent. A material located at the surface of the egg appears responsible for the differentiation of entomesodermic structures (LALLIER, 1973a). Various amines and also ammonium ions are capable of inducing vegetalization of larvae. The lowering of the concentration of potassium ions in seawater induces animalization of larvae, confirming thus the former observations from HERBST and RUNNSTRÖM. From these results, LALLIER (1973b) suggested a difference in the intracellular pH between the animal and vegetal parts of the egg. A regional chemical differentiation was detected by means of PAS-staining. A marked reduction of the nuclear stainability appears in the ectodermal area (IMMERS, 1973). TIMOURIAN et al. (1973) studying the intrinsic differences in the blastula cells have suggested that the original animal-vegetal polarity possibly represents a differential gradient of ability to synthetize hyalin, a material intervening in the processes of differentiation and reaggregation. MIZUNO et al. (1974) have examined the cellular distribution of RNA population in the embryos of *Dendraster excentricus* at the 16-cell stage. Their results suggest the presence of stage-specific RNA in micromeres. According to these authors the ooplasmic segregation at the beginning of segmentation could result from an unequal distribution of factors in relation with the specific gene activation. In a recent paper, BERG (1974) studied heterogenous nuclear and cytoplasmic RNA in lithium treated sea urchin embryos. It appears that the nuclear RNA from the lithium treated embryos is less easily extractible with cold phenol. Studying the incorporation of sulphate ions during early sea urchin development, GESELIUS (1974) suggested the sulphate ions are directly involved in the regulation of nucleo-cytoplasmic interactions. FUJIWARA and YASUMASU (1974a, 1974b) studied 3 sorts of morphogenetically active substances in eggs and embryos. One of them cancelled the morphogenetic effects of chloramphenicol. ECKBERG and OZAKI (1972), through observations on embryos animalized with zinc, presented data indicating that the activation of RNA synthesis does not necessarily follow cellular multiplication. In an ensuing article CARROLL et al. (1975) studied the protein synthesis pattern in normal and animalized embryos. At 24 hours of development, the normal and animalized embryos presented the same difference in protein synthesis.

References of the Addendum

BERG, W., 1974. Heterogenous nuclear and cytoplasmic RNA in Li-treated sea urchin embryos. Exp. Cell Res. 88, 439-442.

CARROLL, A.G., ECKBERG, W.R., OZAKI, H., 1975. Comparison of protein synthetic patterns in normal and animalized sea urchin embryos. Exp. Cell Res. 90, 328-332.

ECKBERG, W.R., OZAKI, H., 1972. Temporal pattern of RNA synthesis in animalized sea urchin embryos. Exp. Cell Res. 73, 177-181.

FUJIWARA, A., YASUMASU, I., 1974a. Some observations of abnormal embryos induced by short-period treatment with chloramphenicol during early development of sea urchin. Develop., Growth and Diff. 16, 83-92.

FUJIWARA, A., YASUMASU, I., 1974b. Morphogenetic substances found in the embryos of sea urchin, with special reference to the anti-vegetalizing substances. Develop., Growth and Diff. 16, 93-103.

GEZELIUS, G., 1974. Incorporation of sulphate ions during early sea urchin development. Zoon 2, 105-116.

IMMERS, J., 1973. Nuclear and cytoplasmic changes in early development of lithium treated sea urchin embryos. Acta Embryol. Exp., 205-221.

LALLIER, R., 1972. Effects of Concanavalin A on the development of sea urchin egg. Exp. Cell Res. 72, 157-163.

LALLIER, R., 1973a. Changes in the differentiation of sea urchin larva by action of a detergent upon the unsegmented egg. Experientia 29, 1022.

LALLIER, R., 1973b. Amines et morphogenèse: une nouvelle interprétation du mode d'action des ions lithium sur la différenciation de l'oeuf de l'oursin Paracentrotus lividus. Comp. Rend. 276, 3473-3476.

MIZUNO, S., YANG RIM LEE, WHITELEY, A.H., WHITELEY, H.R., 1974. Cellular distribution of RNA populations in 16-cell stage embryos of the sand dollar, Dendraster excentricus. Develop. Biol. 37, 18-27.

RUNNSTRÖM, J., IMMERS, J., 1971. Treatment with lithium as a tool for the study of animal-vegetal interactions in sea urchin embryos. Wilhelm Roux' Arch. Entwicklungsmech. Organismen 167, 222-242.

RUNNSTRÖM, J., NUZZOLO, C., CITRO, G., 1972. Ribosomes and polyribosomes in sea urchin development, particularly in embryos undergoing animalization or vegetalization. Exp. Cell Res. 72, 252-256.

TIMOURIAN, H., CLOTHIER, G., WATCHMAKER, G., 1973. Reaggregation of sea urchin blastula cells. 1. Intrinsic differences in the blastula cells. Develop. Biol. 31, 252-263.

VACQUIER, V.D., 1971. The effects of glucose and lithium chloride on the appearance of β-1,3 glucanohydrolase activity in sand dollar plutei. Develop. Biol. 26, 11-16.

17. Respiration and Energy Metabolism

T. Yanagisawa

The embryo develops as cells move and multiply, and the molecular and morphologic complexity of cell structures requires energy. One of the reasons for studying sea urchin egg metabolism is to translate the observed biological results of fertilization and embryonic development into quantitative chemical language (ROTHSCHILD, 1951). What are the principal sources of energy in the egg, and what metabolic changes make this energy available for structural maintenance and alteration, cell division and movement, and tissue differentiation?

17.1 Overall Changes in Respiration during Development

Unless sea-urchin eggs have oxygen, cell and nucleus cleavage does not occur (LOEB, 1896). And removal of oxygen or addition of KCN (1/2,000%) keeps eggs from disintegrating after membrane formation is artificially induced. This suggests that the oxidation rate increases when the egg is activated (LOEB, 1906a, 1906b).

Respiration in unfertilized eggs is slow and cells are fragile, making respiration hard to measure. As determined by the WINKLER titration method, the oxygen consumption of *Arbacia pustulosa* eggs in seawater increases 6- to 7-fold after fertilization. Unfertilized eggs consume 0.055 mg of oxygen, while eggs just fertilized and those in the 8-cell and 32-cell stages consume, respectively, 0.303, 0.355, and 0.576 mg of oxygen per hour per 28 mg of nitrogen (WARBURG, 1908). After fertilization oxygen consumption of *Arbacia punctulata* eggs increases 4-fold and of *Strongylocentrotus purpuratus* eggs 4- to 6-fold (LOEB and WASTENEYS, 1911, 1913).

Using the WARBURG manometer, similar results were obtained in the eggs of *Paracentrotus lividus* (Fig. 17.1), European *Arbacia*, *Echinus microtuberculatus*, *Psammechinus miliaris*, *Arbacia punctulata*, and *Dendraster excentricus* (Table 17.1) (WARBURG, 1915; SHEARER, 1922a, 1922b; GRAY, 1927; RUNNSTRÖM, 1930a; TANG, 1931a; WHITAKER, 1933a; GERARD and RUBENSTEIN, 1934; RUBENSTEIN and GERARD, 1934; TYLER and HUMASON, 1937).

In fertilized embryos of *Psammechinus miliaris*, *Paracentrotus lividus*, *Pseudocentrotus depressus*, and *Arbacia lixula* and its hybrid with *Psammechinus miliaris*, respiratory metabolism increases rapidly and exponentially during cleavage and the early blastula stage. It reaches a plateau at the blastula stage, increases at the beginning of the mesenchymal blastula stage, and peaks at the middle pluteus stage (Figs. 17.2-17.4) (GRAY, 1927; EPHRUSSI, 1933; LINDAHL and ÖHMAN, 1938; LINDAHL, 1939b; BOREI, 1948; WHITELEY and BALTZER, 1958; IMMERS and RUNNSTRÖM, 1960; ISONO and YASUMASU, 1968).

Table 17.1. Oxygen consumption of marine ova before and after fertilization; ml O_2 per hr per ml eggs. (From BALLENTINE, 1940)

Species	Unfertilized	Fertilized	Fertilized/ Unfertilized	Temperature (C)	Source
Arbacia punctulata	0.158	0.790	5.0	25°	TANG, 1931a
		0.550		25°	TANG and GERARD, 1932
	0.040	0.200	5.0	21°	WHITAKER, 1933a, 1933b
	0.038	0.148	3.9	24°	RUBENSTEIN and GERARD, 1934
	0.023	0.101	4.4	21°	RUBENSTEIN and GERARD, 1934
	0.149	0.385	2.6	26°	SHAPIRO, 1935
	0.060	0.320	5.2	20°	KORR, 1937
	0.090	0.450	5.0	25°	KORR, 1937
	0.100	0.450	4.5	25°	BALLENTINE, 1940
Arbacia pustulosa	0.028	0.168	6.0	20°	WARBURG, 1908; WHITAKER, 1933b
Echinus miliaris	0.021	0.120	5.7	15°	SHEARER, 1922b
Paracentrotus lividus	0.036	0.216	6.0	23°	WARBURG, 1915; WHITAKER, 1933b
		0.216		21°	RUNNSTRÖM, 1930a
Asterias	0.080	0.080	1.0	23°	TANG, 1931b; WHITAKER, 1933b
Sabellaria alveolata	0.183	0.205	1.1	20°	FAURE-FREMIET, 1922; WHITAKER, 1933b
Chaetopterus variopedatus	0.240	0.130	0.54	21°	WHITAKER, 1933a, 1933b
Chaetopterus pergamentaceus	0.160	0.130	0.81	25°	BALLENTINE, 1940
Nereis limbata	0.110	0.154	1.4	21°	WHITAKER, 1931c, 1933b[a]
Cumingia tellinoides	0.310	0.140	0.45	21°	WHITAKER, 1931b, 1933b
Ostrea virginica	0.340	0.460	1.4	25°	BALLENTINE, 1940
Mactra lateralis	0.230	0.430	1.8	25°	BALLENTINE, 1940
Fundulus heteroclitus	0.0015	0.025	16.7	20°	BOYD, 1928
Fucus vesiculosus	0.520	0.970	1.9	21°	WHITAKER, 1931a, 1933b

[a]BARRON (1932) found no change on fertilization.

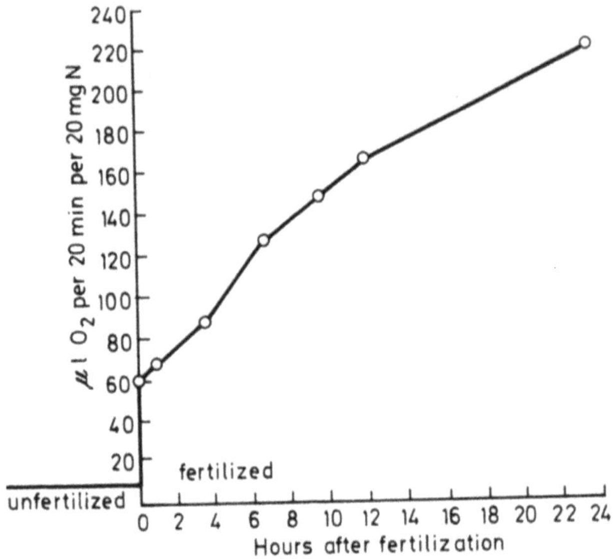

Fig. 17.1. Oxygen consumption during development of *Paracentrotus lividus* eggs at 16°C. By manometric measurement at 23°C. (From WARBURG, 1915)

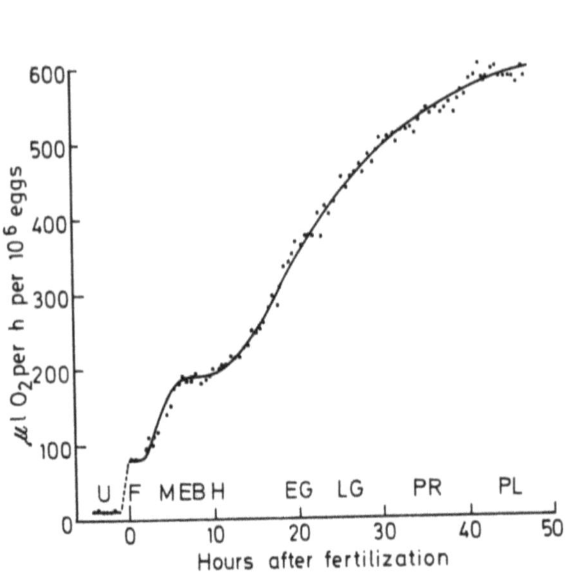

Fig. 17.3. Oxygen consumption during development of *Pseudocentrotus depressus* eggs at 20°C as measured by WARBURG manometer. U, unfertilized; F, fertilization; M, cleavage; EB, early blastula; H, hatching; EG, early gastrula; LG, late gastrula; PR, prism; PL, pluteus. (From ISONO and YASUMASU, 1968)

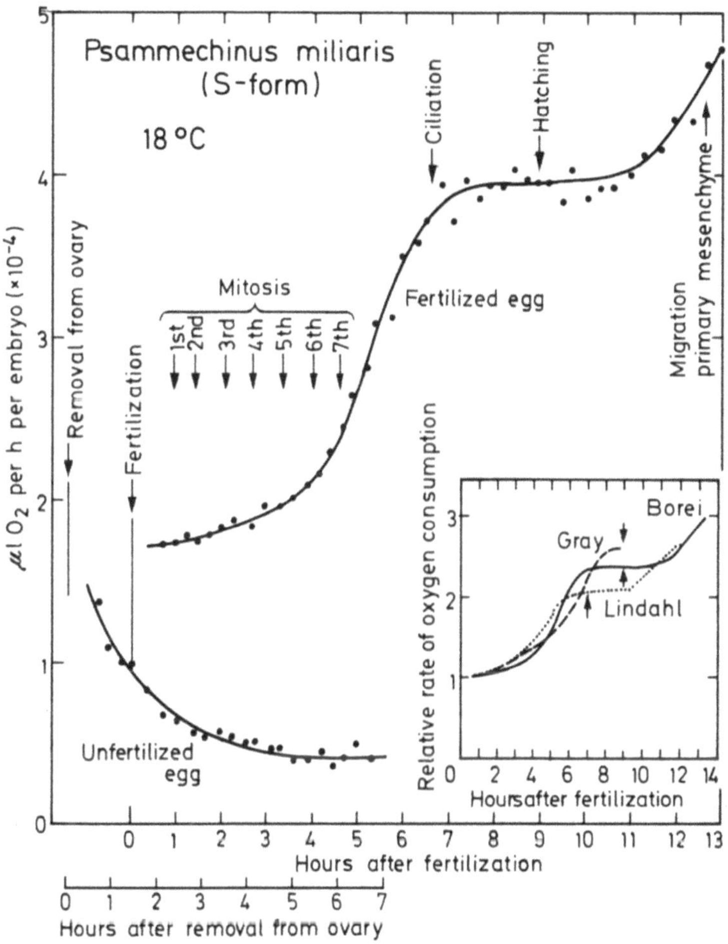

Fig. 17.2. Oxygen consumption of *Psammechinus* eggs before and after fertilization was measured by Cartesian divers. Measurement of respiration was started 40 min after eggs were removed from ovaries, the minimum time required to prepare diver respirometer. (From BOREI, 1948). Other results (GRAY, 1927, *Psammechinus* eggs, 17°C; and LINDAHL, 1939, *Paracentrotus lividus* eggs, 22°C) by WARBURG manometer are compared with results of this investigation, 18°C, and plotted on a relative scale. Extrapolated values at 30 min after fertilization are 1.0

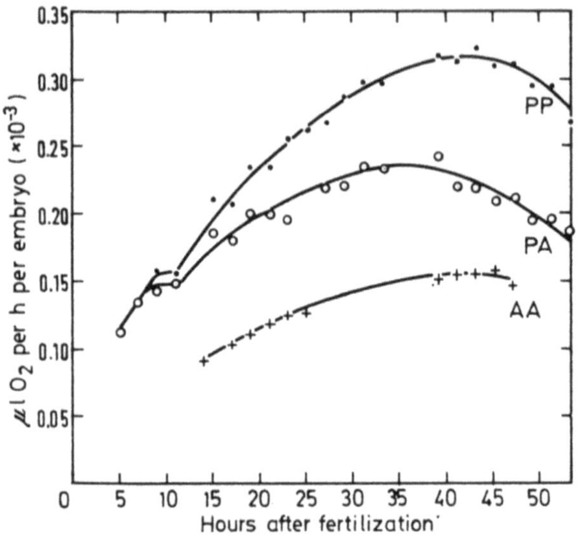

Fig. 17.4. Oxygen consumption of eggs of *Paracentrotus lividus* (PP), *Arbacia lixula* (AA), and their hybrid (*Paracentrotus* female and *Arbacia* male, PA) as measured by Cartesian diver at 18°C. (From WHITELEY and BALTZER, 1958)

17.2 Increased Respiration after Fertilization

After fertilization the rate of oxygen consumption increases rapidly in sea urchins. But in other organisms there is practically no change at fertilization (*Sabellaria* and *Nereis*) or respiration is reduced (*Cumingia* and *Chaetopterus*) (Table 17.1) (WHITAKER, 1931b, 1931c). In eggs of *Chaetopterus*, *Cumingia*, *Nereis*, and *Arbacia*, absolute rates of oxygen consumption before fertilization range from 0.4 to 3.1 and after fertilization range from 1.3 to 2 μl per hour per 10 mm^3 eggs at 21°C (Fig. 17.5) (WHITAKER, 1933b).

Respiration greatly increases only in eggs fertilized after maturation division is complete (Table 17.2). Special importance has been devoted to the stages of maturation divisions in eggs at the time of fertilization (BALLENTINE, 1940; CLELAND, 1950a). Oxygen uptake increases at fertilization when the nucleus is quiescent but not when it is in the "kinetic" state (ROTHSCHILD, 1956). In *Ciona intestinalis* eggs, increased respiration was detected by the WARBURG manometer but not by the Cartesian diver (TYLER and HUMASON, 1937; HOLTER and ZEUTHEN, 1944). The effect of fertilization on respiration in *Urechis caupo* depends on how long the animals are kept in aquarium tanks (TYLER and HUMASON, 1937). When temperature is increased, respiration rate in *Arbacia punctulata* eggs before and after fertilization is more nearly equal. And in those eggs that may survive at 32° and 36°C, fertilization causes no increase in oxygen uptake (RUBENSTEIN and GERARD, 1934; KORR, 1937).

Respiration in unfertilized *Arbacia punctulata* and *Strongylocentrotus purpuratus* eggs is faster in the early season (January to March) than in the late season (March to July) (TANG, 1931a; GONSE, 1960), while in

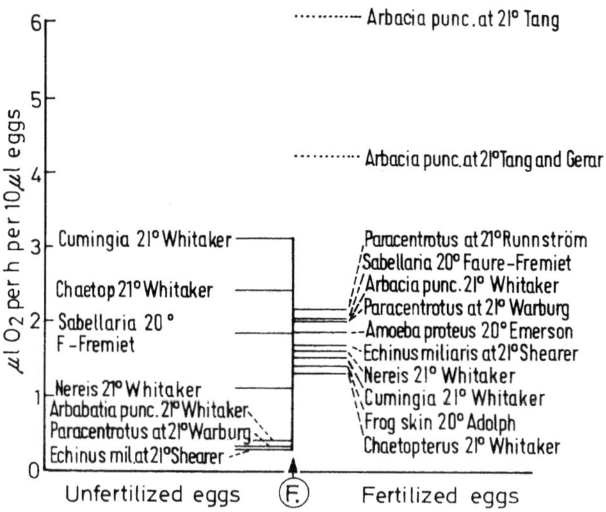

Fig. 17.5. Absolute rates of oxygen consumption per unit volume of fertilized and unfertilized eggs of several marine animals. Temperature correction has been made; value would probably be lower if measurements were confined to first hour after fertilization. (From WHITAKER, 1933b)

fertilized *Paracentrotus lividus* eggs it is slower in the early season (January to early April) than in the late season (April to May) (IMMERS and RUNNSTRÖM, 1960).

The Cartesian diver method shows that respiration in ripe *Psammechinus miliaris* eggs decreases rapidly when eggs are removed from the ovary and placed in seawater (Fig. 17.2). From a rate sometimes higher than newly fertilized eggs, after several hours respiration drops to a fairly low and constant level. Respiration increase at fertilization may depend on at what point on the slope of the decreasing respiration curve fertilization occurs. In natural spawning, slightly increased respiration may be due to early fertilization (BOREI, 1948, 1949).

After unfertilized *Arbacia punctulata* eggs are removed from the female, oxygen uptake remains fairly constant for several days if there is no bacterial growth (Fig. 17.6) (TYLER et al., 1938). Even immediately after shedding, oxygen uptake is slow in unfertilized eggs of *Pseudocentrotus depressus*, *Hemicentrotus pulcherrimus*, and *Anthocidaris crassispina*. It remains constant for over 3 hours and increases sharply on fertilization, even under natural conditions (Fig. 17.7) (YASUMASU and NAKANO, 1963). At 20°C oxygen uptake in unfertilized *Hemicentrotus* and *Pseudocentrotus* eggs is nearly constant from a few minutes after spawning until more than 1 hour (OHNISHI and SUGIYAMA, 1963). The rate of oxygen consumption in unfertilized eggs of *Spisula solidissima*, *Asterias forbesi*, and *Arbacia punctulata* also remains constant (HORWITZ, 1965).

In primary oocytes of *Paracentrotus lividus*, respiration rate is slightly higher than in newly fertilized eggs and much higher than in mature, unfertilized eggs (Fig. 17.8) (LINDAHL and HOLTER, 1941). And in the eggs of the freshwater fish *Oryzias latipes*, the same phenomenon has been observed: oocyte respiration increases with cell growth (Fig.

Table 17.2. Ratio of oxygen uptake in fertilized and unfertilized eggs. (From ROTHSCHILD, 1956, and MONROY, 1965)

Stage of maturation	Species	Fertilized/ Unfertilized	Source
Germinal vesicle	*Nereis succinea*	1.3	BARRON, 1932
	Mactra lateralis	1.8	BALLENTINE, 1940
	Urechis caupo	1.2	TYLER and HUMASON, 1937
	Nereis limbata	1.35=1.45	WHITAKER, 1931c
	Marthasterias glacialis	1.0	TANG, 1931b; WHITAKER, 1933b
Metaphase of first maturation division	*Cumingia tellinoides*	0.45	WHITAKER, 1931b
	Chaetopterus variopedatus	0.53	WHITAKER, 1933a
	Marthasterias glacialis	1.0	BOREI, 1948
	Saxostrea commercialis	1.0	CLELAND, 1950a
	Sabellaria alveolata	1.1	FAURE-FREMIET, 1922
	Ciona intestinalis	1.0	HOLTER and ZEUTHEN, 1944
		1.5	TYLER and HUMASON, 1937
		2.0	LENTINI, 1961
	Phallusia mamillata	2.0	LENTINI, 1961; MINGANT, 1957
Metaphase of second maturation division	*Rana temporaria*	1.0	ZEUTHEN, 1944
	Bufo bufo	1.0	STEFANELLI, 1938
	Fundulus heteroclitus	1.0	PHILIPS, 1940
	Oryzias latipes	1.0	NAKANO, 1953
Matured	*Fucus vesiculosus*	1.9	WHITAKER, 1931a
	Strongylocentrotus purpuratus	3.7	TYLER and HUMASON, 1937
		6.2	GONSE, 1960
	Psammechinus miliaris	3.6	BOREI, 1948
	Paracentrotus lividus	4.7	BROCK et al., 1938
	Arbacia punctulata	4.5	BALLENTINE, 1940

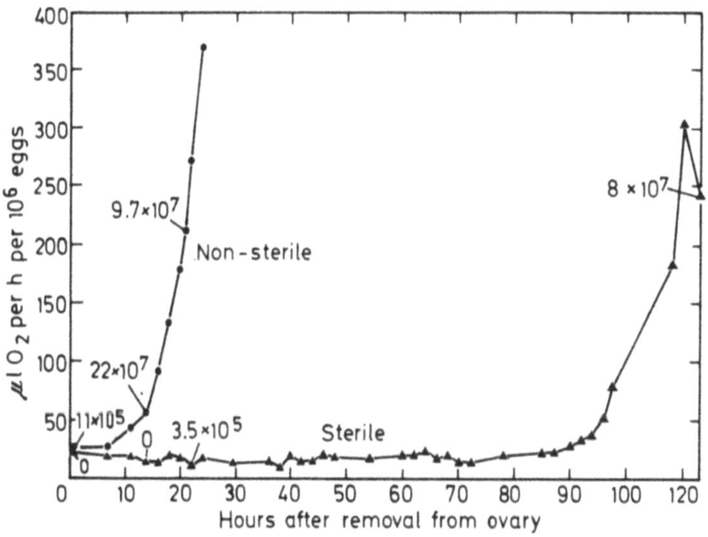

Fig. 17.6. Oxygen consumption of unfertilized *Arbacia* eggs under sterile and nonsterile conditions. Figures on graph indicate bacteria per ml of egg suspension. (From TYLER et al., 1938)

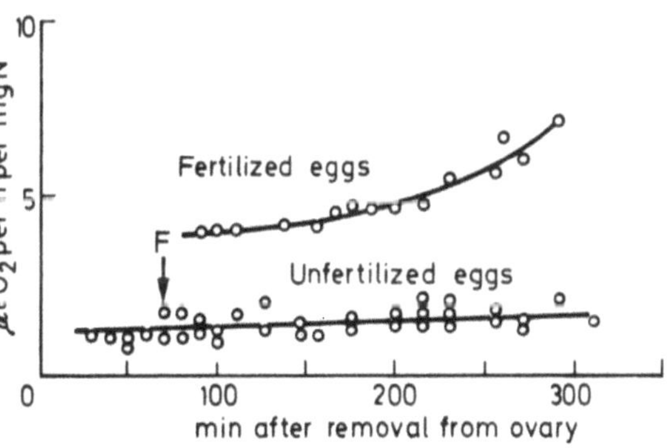

Fig. 17.7. Oxygen consumption of *Hemicentrotus* eggs at 10°C before and after fertilization. Diver measurement of respiration was started 40 min after eggs were removed from ovaries. F, spermatozoa were added. (From YASUMASU and NAKANO, 1963)

17.9). Respiration is more intense in oocytes than in ripe, unfertilized eggs. With maturation division proceeding, it decreases gradually to a low and constant level. The respiration of unfertilized eggs does not decline upon removal from the ovary. Fertilization causes a gradual increase in oxygen consumption (NAKANO, 1953).

Respiration in unfertilized eggs of *Strongylocentrotus purpuratus* is twice as fast in the coelomic fluid of freshly collected sea urchins (or in coelomic fluid containing ovarian homogenate, sucrose, or glucose) as in normal seawater. Thus, the fast respiration of the oocyte in the ovary may be due to an adequate supply of nutrients (GONSE, 1960).

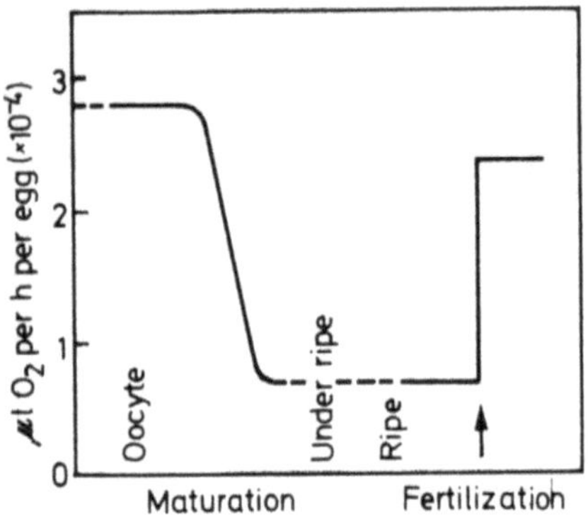

Fig. 17.8. Changes in oxygen consumption during maturation and after fertilization of *Paracentrotus lividus* eggs as measured by Cartesian diver respirometer. (From LINDAHL and HOLTER, 1941)

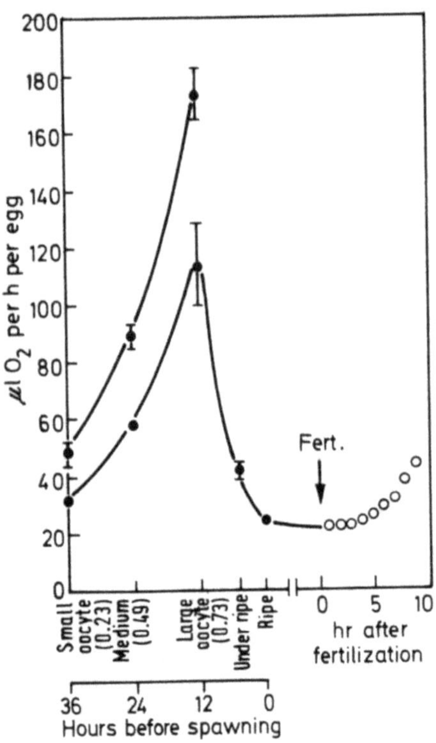

Fig. 17.9. Oxygen consumption during maturation and after fertilization of eggs of freshwater fish *Oryzias latipes*. Upper curve: oocyte with follicle cells. Lower curve: respiration rate of small and medium-size oocyte was calculated as two-thirds that of larger oocyte. Volume is indicated in µl in parentheses. (From NAKANO, 1953)

17.2.1 Temperature Coefficient of Respiratory Rate

As a resting cell, the unfertilized sea-urchin egg has simple maintenance metabolism. But in the fertilized egg, metabolism is also needed for growth and differentiation. These 3 processes have conceivably different temperature coefficients. In *Arbacia* the temperature coefficient was found to be very low in unfertilized eggs but normal in fertilized eggs, while in *Paracentrotus lividus* Q_{10} values for oxygen consumption appeared to be the same in unfertilized eggs and gastrulae at 14.9^{o} - $22.4^{o}C$ (LOEB and WASTENEYS, 1911; EPHRUSSI, 1933). In *Arbacia punctulata* Q_{10} values were found to be much higher in unfertilized eggs than in fertilized eggs. In unfertilized eggs the average value at 13°-$30^{o}C$ was 4.1 and in fertilized eggs, 1.8 (RUBENSTEIN and GERARD, 1934; KORR, 1937). In careful experiments, temperature coefficients of respiration rates of fertilized and unfertilized eggs of *Urechis*, *Strongylocentrotus*, *Ciona*, and *Dendraster* did not differ significantly over most of the temperature range (5°-$22^{o}C$) in which development is possible (TYLER and HUMASON, 1937). These results were principally confirmed by the Cartesian diver technique in the *Strongylocentrotus-Psammechinus*- group with equal Q_{10} values before and after fertilization, while in the *Arbacia*-group they are higher before fertilization (BOREI and LYBLING, 1949).

17.2.2 Oxygen Pressure and Consumption

At 50 mm Hg, oxygen consumption in fertilized *Arbacia punctulata* eggs is the same as in air (Fig. 17.10). Below 70-80 mm Hg, oxygen consumption in unfertilized eggs declines as pressure is reduced and

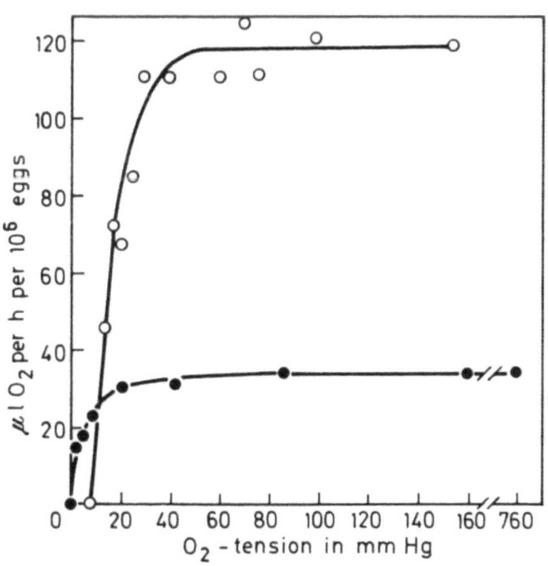

Fig. 17.10. Relationship between oxygen tension and oxygen consumption in fertilized (o) and unfertilized (●) eggs of *Arbacia punctulata*. Oxygen consumption was measured by WARBURG manometer. Gas mixtures were prepared by mixing nitrogen and air or pure oxygen. Oxygen tension in air was 155-160 mm Hg. (From TANG, 1931a, and TANG and GERARD, 1932)

at 20 mm Hg falls very rapidly. In fertilized eggs, however, at 35 mm Hg, oxygen consumption is reduced by 5% and then declines nearly linearly until at 4 mm Hg it is almost negligible. Below that pressure, fertilized eggs do not cleave, and at 4-40 mm Hg cleavage is slower than at higher pressure (TANG, 1931a; TANG and GERARD, 1932).

17.2.3 Respiratory Quotient

The respiratory quotient (R.Q.) is determined by dividing the volume of carbon dioxide produced, by the volume of oxygen consumed and allows some conclusions upon the main substrates for respiration. In unfertilized eggs, it was found to be 1.06 in *Arbacia pustulosa*, 1.1 - 1.3 in *Psammechinus miliaris*, 0.92 in *Sphaerechinus granularis*, and 1.01 in *Anthocidaris crassispina* (ASHBEL, 1930; BOREI, 1933; HERK, 1933; LASER and ROTHSCHILD, 1939; ISONO, 1963a). This suggests that carbohydrates are the main source of oxidative metabolism.

In fertilized eggs the respiratory quotient was found to be 0.85 or 0.95 in *Arbacia punctulata*, 0.92 in *Psammechinus microtuberculatus*, 0.73 or 0.90 in *Paracentrotus lividus* and 0.90 in *Psammechinus miliaris* (WARBURG, 1915; SHEARER, 1922b; ASHBEL, 1930; RUNNSTRÖM, 1932; BOREI, 1933; EPHRUSSI, 1933; HERK, 1933; ÖHMAN, 1940; HUTCHENS et al., 1942a). Low respiratory quotient indicates lipid participation during early development.

Discrepancies in respiratory quotients as they show up between the different authors mentioned above may be due to a number of factors. One manometric flask of eggs absorbing the carbon dioxide produced compares with another flask not absorbing it. Respiration is faster when the bicarbonate content of seawater is normal than when it is low following absorption of carbon dioxide from the gas phase. Carbon dioxide retention must be measured using mixtures with known carbon dioxide pressure in the gas phase. This is troublesome and necessitates the use of mixtures with known tensions of carbon dioxide in the gas phase. Unfertilized eggs prepared for manometric measurement may have been out of the ovary for over an hour, have low metabolism, and are easily damaged by manometric shaking. It is also possible that some of the carbon dioxide produced may be used in carbon dioxide fixation (HULTIN and WESSEL, 1952; HULTIN, 1953; ROTHSCHILD, 1956).

Table 17.3. Respiratory quotients of *Anthocidaris crassispina* eggs and embryos cultured at 24°-26°C. (From ISONO, 1963a)

Stage	Unfertilized	Fertilized	Morula	Swimming blastula	Gastrula	Prism
Hours after fertilization	0	1-2	5-6	10-11	16-17.5	24-27
Oxygen uptake[a]	11.6	61	118	209	295	406
CO_2 output[a]	11.7	62	117	186	200	321
Respiratory Quotient	1.01	1.01	0.98	0.90	0.68	0.79

[a]Determined at 20°C by WARBURG manometer and DICKENS-ŠIMER flask (microliter/10^6 embryos/hour).

The respiratory quotient in unfertilized *Anthocidaris* eggs and embryos is estimated as almost 1 at the morula stage, 0.90 at the swimming blastula stage, 0.68 at the gastrula stage, and 0.79 at the prism larva stage (Table 17.3). This indicates that until the hatching stage, the main substrate of respiration is carbohydrate and thereafter it is lipid (Fig. 17.21)(ISONO, 1963a).

17.2.4 Effect of Phenol Derivatives on Respiration

In *Arbacia punctulata*, oxidation-reduction dyes such as o-cresol indophenol, methylene blue, neutral red, dimethyl-p-phenylene diamine, tetramethyl-p-phenylene diamine, and pyocyanine stimulate respiration most strongly in the range of 5×10^{-4} to 5×10^{-3} *M* and inhibit cleavage irreversibly. Substituted phenols stimulate respiration most strongly and initially block cleavage reversibly at concentrations as low as 10^{-6} *M*. The cleavage block caused by nitrophenol derivatives is fully reversible up to concentrations 100 to 500 times as strong as those required for initial cleavage block. Eggs develop normally when they are returned to normal seawater (CLOWES and KRAHL, 1936).

Among the mononitrophenols, p-nitrophenol stimulates respiration and suppresses division much more than m-nitrophenol. o-Nitrophenol stimulates oxygen consumption and blocks division only at very strong and permanently injurious concentrations. 2,4-Dinitrophenol (DNP) has a stronger effect than 2,6-dinitrophenol, but upon introducing a third nitro group to form 2,4,6-trinitrophenol (picric acid) activity is lost entirely (Fig. 17.11)(CLOWES and KRAHL, 1936).

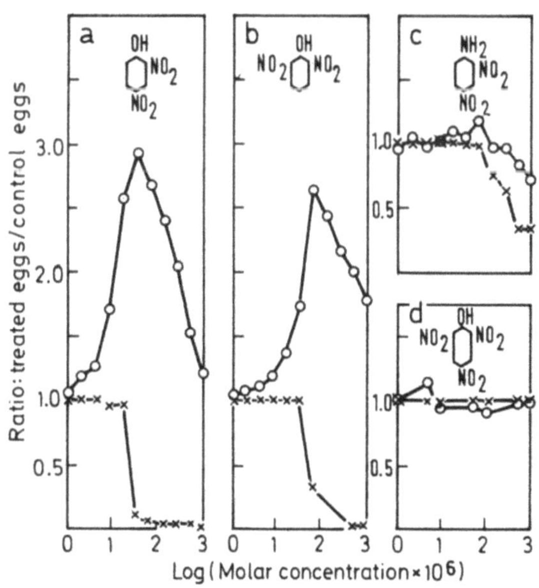

Fig. 17.11a-d. Fertilized *Arbacia* egg oxygen consumption stimulated (o) and cell division blocked (x) by different concentrations of 2,4-dinitrophenol (a), 2,6-dinitrophenol (b), 2,4-dinitroaniline (c), and 2,4,6-trinitrophenol (d) at 20°C. Reagent was added 15 min after fertilization. (From CLOWES and KRAHL, 1936)

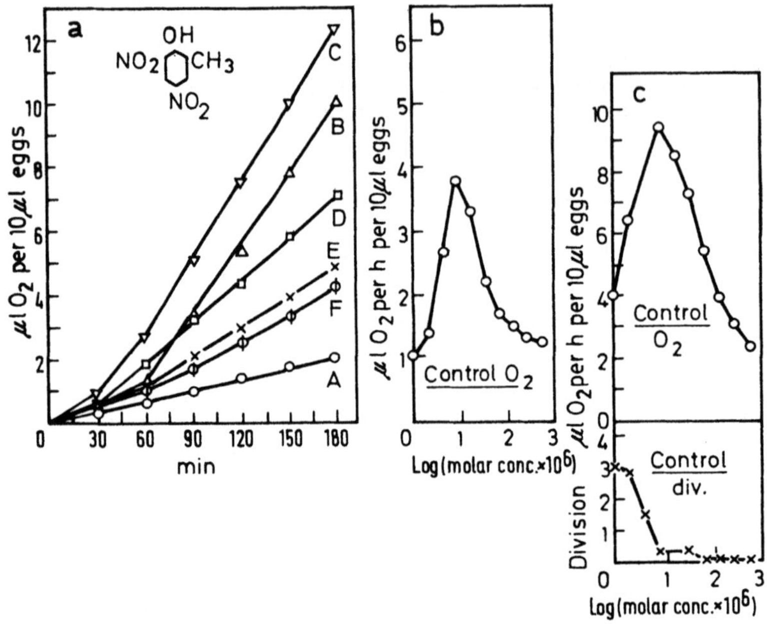

Fig. 17.12a-c. Effect of 2,4-dinitro-o-cresol on *Arbacia* egg respiration and cleavage at 20°C. (a) Oxygen consumption of unfertilized eggs in concentrations of DNC: A, control; B, 4 x 10^{-6} *M*; C, 8 x 10^{-6} *M*; D, 3.2 x 10^{-5} *M*; E, 1.28 x 10^{-4} *M*; F, 5.12 x 10^{-4} *M*. (b) Respiratory rate of unfertilized eggs in concentrations of DNC. (c) Respiratory rate of fertilized eggs in concentrations of DNC. Fertilized eggs were treated 25 min after fertilization. Division number was determined after 2-hour experiment (lower curve). Horizontal lines indicate control values of oxygen consumption and cell division. (From CLOWES and KRAHL, 1936)

The phenolic hydroxyl group is essential to the effect on respiration and cleavage, as shown by the fact that respiration is not stimulated and cleavage is not blocked when a methyl group is substituted for phenolic hydrogen, as in p-nitroanisole, or when hydroxyl is separated from the ring by a -CH_2-group, as in p-nitrobenzyl alcohol. As in mononitro derivatives, when an amino group replaces hydroxyl in DNP, giving 2,4-dinitroaniline, activity is almost completely lost (Fig. 17.11).

Of the dinitrophenol derivatives, 4,6-dinitrocresol (DNC) and 4,6-dinitrocarvacrol are especially active, but 2,4-dinitrothymol in very low concentrations (10^{-6} *M*) blocks division and only slightly stimulates oxidation. It is possible that in some instances oxygen consumption is stimulated while cell division is unaffected, and in others oxygen consumption and cell division are both suppressed (CLOWES and KRAHL, 1936).

After 30-60 minutes, optimal concentrations of DNC (8 x 10^{-6} *M*) increase 6-fold the oxygen consumption of unfertilized *Arbacia* eggs (Fig. 17.12a). Even at high concentrations (2.5 x 10^{-3} *M*), DNC does not inhibit fertilization, and eggs treated with 10^{-4} *M* DNC remain fertilizable longer than controls (CLOWES and KRAHL, 1936).

Optimal DNC concentrations in fertilized and unfertilized eggs are nearly the same. At those levels, respiration in fertilized eggs is about 3 times as fast as normal and in unfertilized eggs about 6 times as fast. Oxygen consumption in normal fertilized *Arbacia* eggs is 4 times as great as in normal unfertilized eggs, and the optimal absolute rate in DNC-treated fertilized eggs is about 2 1/2 times as great as in DNC-treated unfertilized eggs. When DNC concentrations are higher than 2.5 x 10^{-4} *M*, oxygen consumption in fertilized eggs is below normal.

In unfertilized eggs, a DNC concentration of 5 x 10^{-4} *M* causes oxygen consumption almost twice as great as normal, while at higher concentrations oxygen consumption does not fall below normal until the eggs are destroyed (Figs. 17.12b, c). In *Strongylocentrotus purpuratus*, a DNP concentration of 2 x 10^{-4} *M* causes respiration in unfertilized eggs to be 3 times as great as endogenous respiration and in fertilized eggs causes endogenous respiration 2 or 3 times greater than normal (GONSE, 1960).

In eggs treated with DNC at 20°C within 25 minutes after fertilization, cell division is not impaired until DNC concentration reaches the level optimal for respiration. Then cleavage is inhibited (8 x 10^{-6} *M*) and, at slightly higher concentrations, division is inhibited completely and in most cases reversibly (Fig. 17.12c). As temperature increases, DNC stimulation of oxygen consumption declines (CLOWES and KRAHL, 1936).

Halophenols are similar to nitrophenols in their effect on oxygen consumption and their reversible block of cleavage. Exceptions are monohalophenols, which do not produce these effects, or produce them only slightly, and the symmetric trihalophenols, which produce both effects.

In fertilized *Arbacia* eggs, halophenols and nitrophenols induce increased respiration as follows: p-nitrophenol (optimum concentration 10^{-4} *M*), 342%; 4,6-dinitrocresol (8 x 10^{-6} *M*), 300%; 2,4-dinitrophenol (3 x 10^{-5} *M*), 292%; 2,4,5-trichlorophenol (2 x 10^{-5} *M*), 264%; 2,4-dichlorophenol (10^{-4} *M*), 236%; 2,6-dichloro-4-nitrophenol (6 x 10^{-5} *M*), 232%; 2,4,6-tribromophenol (10^{-5} *M*), 227%; 2,6-dibromo-4-nitrophenol (2 x 10^{-5} *M*), 222%; 2,4,6-trichlorophenol (6 x 10^{-5} *M*), 214%; 2,6-dinitro-4-chlorophenol (3 x 10^{-4} *M*), 205%; 2,4,6-triiodophenol (2 x 10^{-6} *M*), 175%; and 2,4,6-trinitrophenol, 100% (KRAHL and CLOWES, 1936).

Substituted phenols penetrate fertilized *Arbacia punctulata* eggs as undissociated molecules. Inhibition of cell division depends on the concentration of phenol molecules (not anions) in the external medium and cell cytoplasm. Respiration stimulation in concentrations lower than required for optimal respiration depends on the cytoplasmic concentrations of phenol anions (not molecules). Oxygen consumption in concentrations greater than required for optimal respiration depends on cytoplasmic concentrations of both phenol molecules and anions (KRAHL and CLOWES, 1936, 1938a, 1938b, 1940; HUTCHENS et al., 1939).

In unfertilized *Paracentrotus* eggs, dimethyl-p-phenylene diamine (DMPD) induces oxygen consumption equal to that of normal fertilized eggs in early cleavage (RUNNSTRÖM, 1930, 1932; ÖRSTRÖM, 1932). DMPD acts as an oxidizable substrate of cytochrome oxidase, increasing oxygen consumption. DMPD also increases oxygen consumption in fertilized eggs. At low concentrations (2 to 14 x 10^{-3}%), oxygen consumption is about 20% higher in fertilized eggs than unfertilized eggs. And at higher concentrations, it is about equal in both or slightly higher in unfertilized eggs (RUNNSTRÖM, 1932, 1956a).

DNP (5×10^{-5} *M*) and DMPD (4×10^{-4} *M*) enhance the respiratory rate of unfertilized *Paracentrotus* eggs (0.77 and 1.13 µl O_2/min/1.8×10^5 eggs) to equal that after fertilization (at the 2-cell stage, 1.58 µl O_2/min). After fertilization, DMPD stimulates the respiratory rate less and, at the late blastula stage, has no effect.

The effect of DNP is greater in fertilized eggs than in unfertilized eggs. The respiratory control ratio decreases and at the late blastula stage is rather constant. DNP and DMPD together stimulate fertilized egg respiration more than DNP alone (RUNNSTRÖM et al., 1970).

In unfertilized eggs it is possible that a break in the middle of the respiratory chain limits the respiratory rate. DMPD might bypass the break, raising oxygen consumption to the level of the normal 2-cell stage:

$$\text{NADH} \rightarrow \underbrace{F_p \xrightarrow{\quad \text{DMPD} \quad} c}_{} \qquad \text{NADH} \rightarrow F_p \rightarrow UQ \rightarrow b \rightarrow c_1 \rightarrow c \rightarrow a \rightarrow a_3 \rightarrow O_2.$$

After fertilization the break in the chain closes and disappears altogether in the late blastula and gastrula stages (RUNNSTRÖM et al., 1970).

17.2.5 Release of Respiratory Control during Development

Release of respiratory control means maximal stimulation of respiration. It is caused by hydrolysis of the primary product of oxidative phosphorylation, X~P, as the result of treatment with DNP and related compounds (LARDY and WELLMAN, 1952).

30 to 90 minutes after fertilization, DNP stimulates respiration of *Paracentrotus lividus* eggs most strongly in concentrations of 5×10^{-5} *M* and *Psammechinus miliaris* eggs in concentrations of 1×10^{-4} *M*. The DNP-released respiratory rate increases during early cleavage parallel with the normal rate. It reaches a maximum before hatching, after which it declines sharply, sinking to its lowest point at the time the normal respiratory curve reaches a plateau. The controlled and released respiratory curves then increase equally until the pluteus stage (Fig. 17.13a, b)(IMMERS and RUNNSTRÖM, 1960).

The respiratory control ratio compares respiration released by optimal concentrations of DNP with the respiration of control eggs. It is high (about 6) in unfertilized eggs and low (about 3) in the early post-fertilization stage. It declines from fertilization until primary mesenchymal cell migration and remains almost constant from the mesenchymal blastula stage to the pluteus stage (Fig. 17.13a′, b′)(CLOWES and KRAHL, 1936).

Respiration suddenly increases at fertilization, and the respiratory rate rises exponentially from fertilization to primary mesenchymal cell migration, which may indicate partial release of respiratory control. Such a release occurs when ADP (the most important acceptor of the energy-rich phosphate of X~P) is added to the cell-free mitochondrial system. It is suggested that the release results from an increased ADP-level due to ~P used for embryo activity. The respiratory rate in embryo development is thought to be controlled by the ADP/ATP ratio (ZOTIN et al., 1967).

The Cartesian diver respirometer shows that the respiratory rate of separated meridional halves is almost exactly half what it is in whole embryos. At the late cleavage stage, the respiratory rate is higher

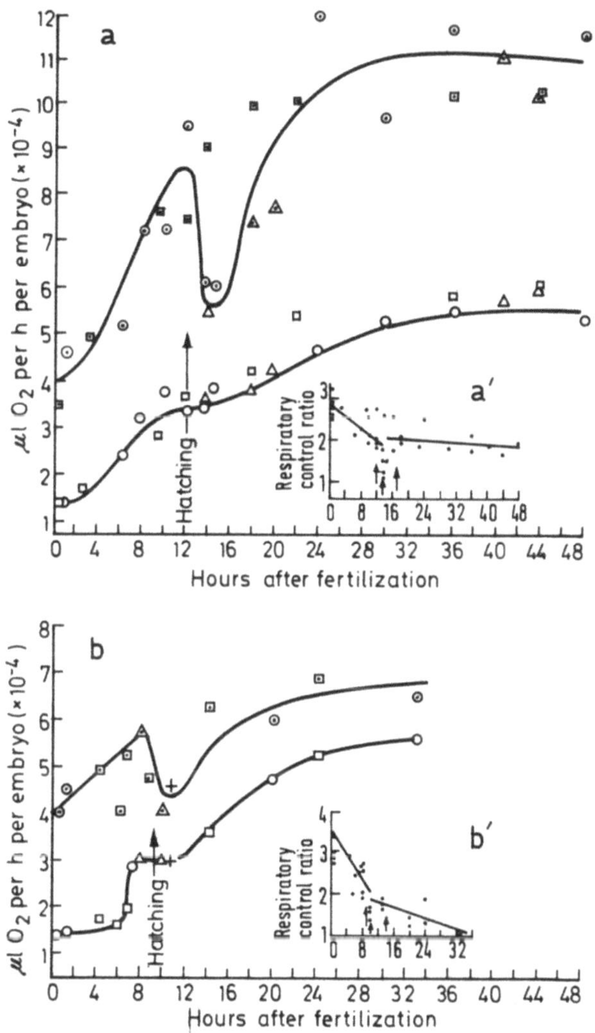

Fig. 17.13 a and b. Respiratory rate released by 2,4-dinitrophenol of *Paracentrotus lividus* at 24°C (a, a'; 5 x 10^{-5} *M*) and *Psammechinus miliaris* at 18°C (b, b'; 10^{-4} *M*) during embryo development. Upper curve: released respiration. Lower curve: controls. Figures a' and b' indicate respiratory control ratios. Arrows in a' and b' indicate time of hatching, onset of primary mesenchymal cell immigration, archenteron invagination. (From IMMERS and RUNNSTRÖM, 1960)

in vegetal halves than in animal halves, but at the mesenchymal blastula stage it is about the same in both (Fig. 17.14).

In animal halves at the late cleavage and mesenchymal blastula stages, 5 x 10^{-5} *M* of DNP causes strong respiratory release and a respiratory control ratio of 2.9, about the same as the maximal value found in whole eggs soon after fertilization. When DNP is added to vegetal halves, the respiratory rate decreases at late cleavage, and a respiratory control ratio of 1.3 occurs in the mesenchymal blastula (Fig.

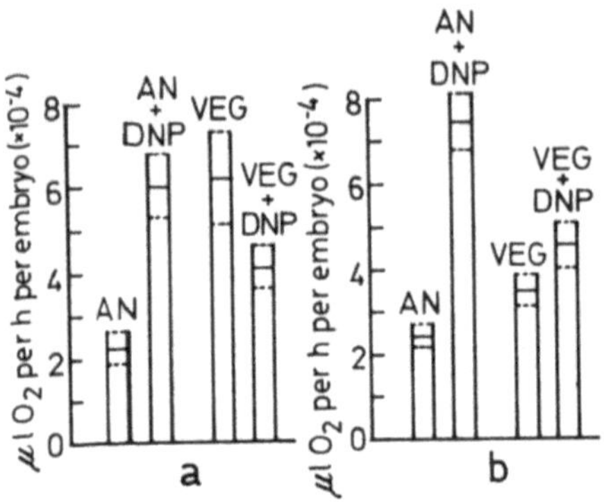

Fig. 17.14 a and b. Controlled and released respiration in animal and vegetal halves of *Paracentrotus lividus* eggs in late cleavage stage (a) and mesenchymal blastula stage (b). Glass-needle technique was used to separate embryos into halves. Controlled, normal respiration was released by adding 5 x 10^{-5} *M* DNP. Respiration was measured by Cartesian diver. (From DE VINCENTIIS et al., 1966)

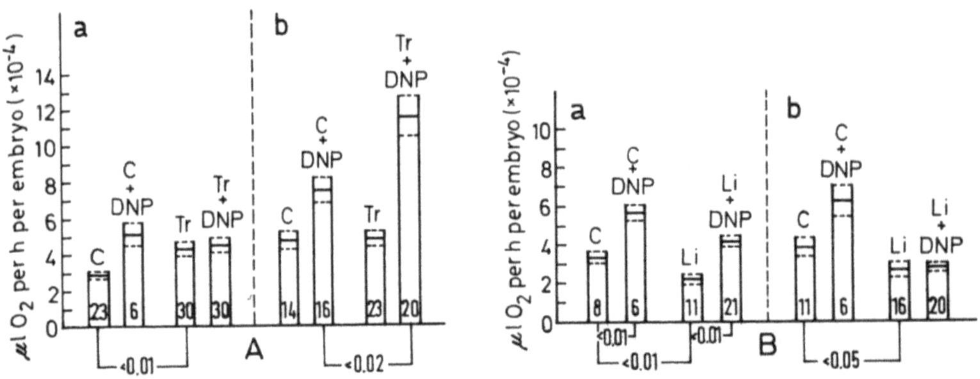

Fig. 17.15 A and B. Controlled and released respiration in animalized (A) and vegetalized (B) *Paracentrotus lividus* embryos. 30 minutes after insemination, embryos were animalized with 0.05% trypsin or vegetalized with 0.074 *M* lithium. Just before respiration was measured, 5 x 10^{-5} *M* DNP was added to control embryos (C) or treated embryos. a, cleavage stage; b, mesenchymal blastula stage. (From DE VINCENTIIS and RUNNSTRÖM, 1967)

17.14). This indicates that the isolated vegetal halves come uncoupled during cleavage. As released respiration is an expression of maximal electron transport capacity, it may also be considered an approximate measure of the respiratory enzyme chain of the embryo. Considerable synthesis of these enzymes seems to take place in isolated animal halves. In isolated vegetal halves at the mesenchymal blastula stage, respiratory enzyme chains are fewer (DE VINCENTIIS et al., 1966).

In embryos animalized with 0.05% trypsin and others vegetalized with 0.074 *M* lithium, the release of respiratory control is similar to separated animal and vegetal halves. In the advanced cleavage stage, exposure to trypsin alone brings about a release which does not increase with DNP. But when DNP is added to trypsin-treated embryos at the mesenchymal blastula stage, the respiratory control released is significantly higher than that of embryos in pure seawater. While vegetalization with lithium reduces oxygen consumption, the addition of DNP causes a normal release of respiratory control at the advanced cleavage stage but no release at the blastula and mesenchymal blastula stages (Fig. 17.15)(DE VINCENTIIS and RUNNSTRÖM, 1967).

17.2.6 Respiratory Activity of Egg Homogenates

The respiratory inertness of unfertilized sea-urchin eggs may be due to physical barriers in the cell cortex. These are usually eliminated by homogenizing the eggs. The following methods have been used for this purpose: freezing and thawing in *M*/50 phosphate buffer (pH 7.16) or $NaHCO_3$ buffer in *Echinocardium cordatum*; shaking in 1.5% (0.26 *M*) NaCl + 1.5% (0.2 *M*) KCl + 0.02 *M* phosphate buffer, pH 7.7, in *Strongylocentrotus droebachiensis*; sucking and blowing through a narrow pipet with a quarter of 0.35 *M* Na-citrate, pH 7.0, in *Psammechinus miliaris* and *Paracentrotus lividus*; drawing in and forcing out of a syringe in 0.4 *M* KCl + 0.01 *M* Na-citrate + 0.05 *M* glycylglycine, pH 7.4, in *Arbacia punctulata*; homogenates prepared in 0.02 *M* phosphate buffer, pH 7.5, with or without 0.3 *M* KCl, in *Echinus esculentus*; homogenates prepared in 0.067 *M* phosphate buffer, pH 6.8, in *Lytechinus pictus* and *Strongylocentrotus purpuratus*; spouting into 50% seawater with a graduated pipet in *Pseudocentrotus depressus*; and homogenates prepared in 0.9 *M* sucrose + 0.02 *M* glycylglycine, pH 7.35, + 2.5 millimol Na_2HPO_4 + 1 millimol EDTA, in *Strongylocentrotus purpuratus* (LINDBERG, 1943; LINDBERG and ERNSTER, 1948; HULTIN, 1949, 1950a, 1950b; KELTCH et al., 1950; CLELAND and ROTHSCHILD, 1952b; YČAS, 1954; ISHIHARA, 1958d; GONSE, 1960).

A homogenization medium must be chosen to minimize difficulties. In 1 instance, citrate was included to stabilize echinochrome granules, a nontoxic buffer was used to maintain the homogenate pH at the desired level (7 - 7.4), and KCl concentration was 0.4 *M* (KELTCH et al., 1950). Despite the microscopically identical appearance of KCl and sucrose homogenates, their respiratory rates were quite different. KCl gave poor and inconsistent results, while respiration was lower at all phases than in sucrose. Oxygen consumption was about the same in sucrose as in seawater (GONSE, 1960).

Complete saponin cytolysis of unfertilized *Strongylocentrotus purpuratus* eggs induces an oxidation rate as high as that at fertilization (LOEB and WASTENEYS, 1913). Oxygen consumption of the homogenate is substantially larger than that of the unfertilized eggs from which it is derived. When α-ketoglutarate is the substrate, oxygen uptake of the cell-free material derived from 1 gm of wet eggs averages 340 mm^3 per hour at 20°C. By comparison, 1 gm wet weight of intact unfertilized eggs consumes about 70 mm^3 and fertilized eggs about 310 mm^3 (KELTCH et al., 1950). The respiration of sea-urchin eggs increases 6-fold at fertilization, while the respiration of homogenates of fertilized and unfertilized eggs is surprisingly similar (Fig. 17.16)(GONSE, 1960).

Egg homogenate respiration declines rapidly, and at 15-20 minutes after preparation a more constant rate begins that lasts for several hours (HULTIN, 1949, 1950a). In *Strongylocentrotus* egg homogenate, the rate of endogenous respiration is not constant. Rate changes are abrupt, indicating 2 different activities: an initial rapid rate which progressively decreases in 20 minutes (phase 1) and a constant rate at a lower level (phase 2)(GONSE, 1960).

Small amounts of $CaCl_2$ (4 millimol or less) increase homogenate respiration very rapidly for 10-20 minutes and cause a strong acid forma-

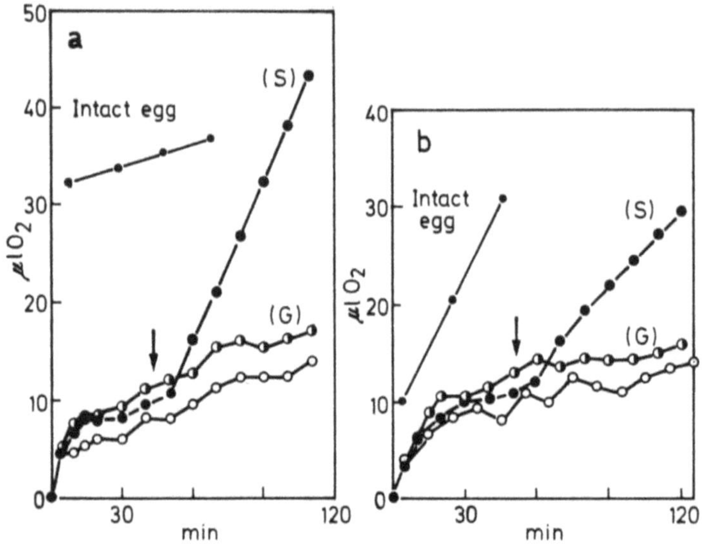

Fig. 17.16 a and b. Respiration in whole homogenate of unfertilized (a) and fertilized (b) *Strongylocentrotus purpuratus* eggs in iso-osmotic sucrose medium. At 16°C respiration was measured by WARBURG manometer. Arrow indicates addition of 0.02 *M* glucose (G), 0.02 *M* succinate (S), or medium (o). Respiratory rate of whole eggs in normal seawater is shown for comparison. (From GONSE, 1960)

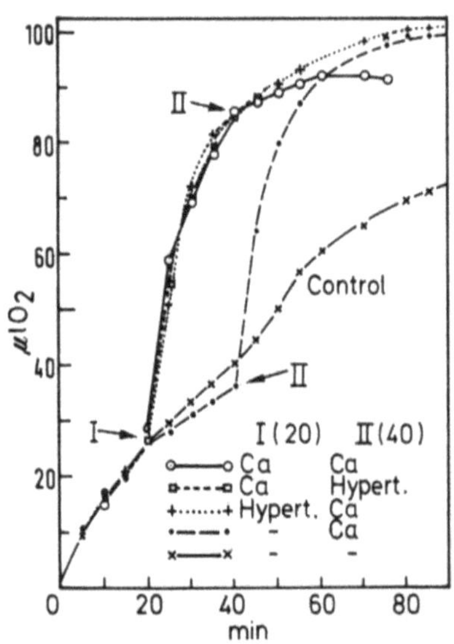

Fig. 17.17. Homogenate respiration of *Paracentrotus lividus* eggs is stimulated by repeated additions of $CaCl_2$ (4 millimol). Hypertonicity is 0.4 *M* NaCl excess. (From HULTIN, 1950a)

tion. Respiration and acid formation then decline, and after an hour or 2 reach to a lower constant level of the calcium-free control. Hypertonicity (0.25 ml of 25% NaCl in a final volume of 3 ml) has the same effect (Fig. 17.17). It declines rapidly about 1 to 1 1/2 hours after homogenization at 18°C and declines even faster at higher temperatures. At low temperature (1°C) it could be observed even after 24 hours (HULTIN, 1949, 1950b).

The rapid respiratory increase caused by addition of calcium ions to a homogenate is not the result of increased glycogen breakdown. Carbohydrate metabolism is not involved, as oxygen uptake is not linked to carbon dioxide production. The oxidation energy cannot be used for phosphate esterification, and as oxygen uptake also occurs in the presence of pyocyanine, it is apparently not caused by NAD oxidation. Monoiodoacetate inhibition of oxygen uptake seems to indicate that the uptake is caused or transmitted by a substance containing SH groups, most likely by a change of protein-bound -SH into S-S (HULTIN, 1950a).

17.3 Respiratory Change after Fertilization

Studies of oxygen consumption just after fertilization, with rapid absorption of released carbon dioxide, show that oxygen consumption and acid formation increase for 5 minutes after fertilization and then fall to levels sometimes lower than before fertilization. Oxygen consumption then increases to the normal post-fertilization level (Fig. 17.18).

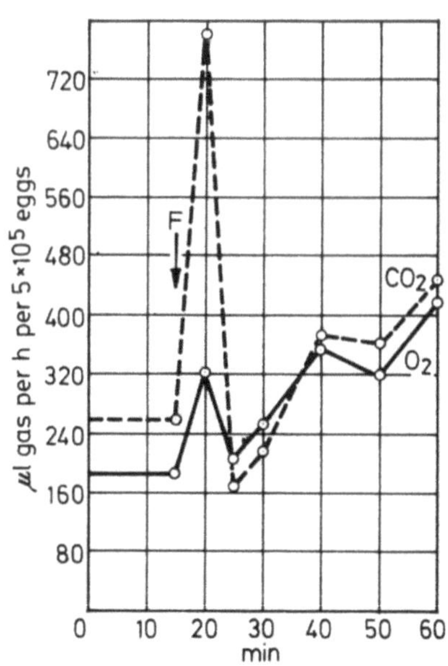

Fig. 17.18. Total CO_2 production and oxygen consumption after fertilization of *Psammechinus miliaris* eggs as measured at 20°C by modified WARBURG differential method. F: fertilization. Gas is 0.3% CO_2 in oxygen. (From LASER and ROTHSCHILD, 1939)

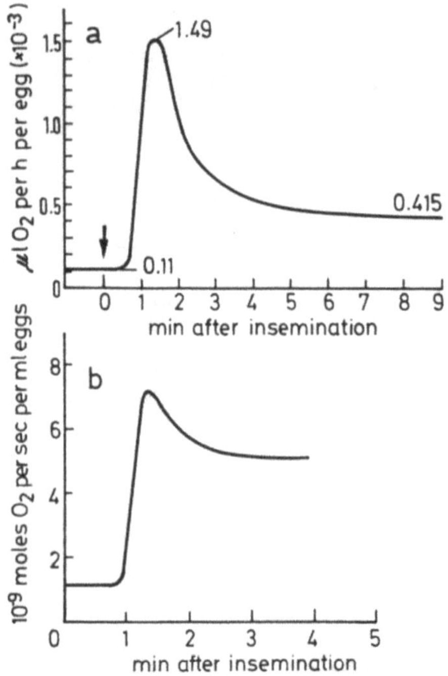

Fig. 17.19 a and b. Polarographic measurement of altered oxygen consumption rate in *Pseudocentrotus depressus* eggs at 20°C (a) and *Lytechinus variegatus* eggs at 30°C (b) a few minutes after fertilization. (From OHNISHI and SUGIYAMA, 1963, and EPEL, 1969)

In *Hemicentrotus*, *Pseudocentrotus*, *Arbacia punctulata*, *Asterias forbesi*, *Strongylocentrotus purpuratus*, and *Lytechinus variegatus*, continuous polarographic measurements have been made from the moment of insemination on, without carbon dioxide interference (OHNISHI and SUGIYAMA, 1963; EPEL, 1965, 1969; HORWITZ, 1965). In less than a minute after fertilization, oxygen uptake increases sharply. It is greatest just after membrane elevation, more than 10 times as great as before fertilization. Next it declines and remains constant (Fig. 17.19). Acid production at fertilization occurs before the increase of oxygen uptake (OHNISHI and SUGIYAMA, 1963).

The first transient increase of respiration occurs at the same time as cortical change; fertilization membrane elevation; change in permeability to water, inorganic cations, and anions; electrical change in membrane potential, resistance, and capacitance; and release of bound calcium in the cytoplasm (LILLIE, 1916; MAZIA, 1937; ISHIKAWA, 1954; TYLER et al., 1956; HIRAMOTO, 1959a, 1959b; TYLER and MONROY, 1959; WHITELEY and CHAMBERS, 1960).

How transient increase in respiration is related to cortical change is not clear. Cortical change in starfish (*Asterias forbesi*) is similar to that of sea-urchin eggs. It accompanies transient increase of oxygen consumption just after fertilization, but oxygen consumption then reverts to the pre-fertilization level (HORWITZ, 1965).

Estimates of respiration due to the oxidative breakdown of lipids, via the citrate cycle, and of carbohydrates, via the glycolysis-KREBS cycle and the pentose-phosphate shunt, have been made on the basis of increased oxygen consumption, which roughly reflects increased overall metabolism, of respiratory quotient (R.Q.) change, which indicates altered respiratory substrates, and of the relative participation of glycolysis-KREBS cycle and pentose-phosphate pathway in carbohydrate oxidation measuring $^{14}CO_2$ evolved from glucose-1-^{14}C and glucose-6-^{14}C (Figs. 17.20, 17.21). Respiratory increase during cleavage seems to be due mainly to increased carbohydrate oxidation via the pentose-phosphate shunt and to a slight but continuous carbohydrate breakdown via the glycolysis-KREBS cycle. The respiratory increase after the mesenchymal blastula stage is due partly to lipid oxidation and partly to the gradual shift of carbohydrate oxidation pathways from the pentose-phosphate shunt to the glycolysis-KREBS cycle (ISONO and YASUMASU, 1966, 1968).

The respiratory rate in sea-urchin eggs seems to depend on the availability of phosphate acceptors, especially ADP, and on the ADP/ATP ratio. In some instances, only ADP increases, and in others, the ADP/ATP ratio. Increased concentration of phosphate acceptors seems to be one of the factors that stimulate respiration during cleavage (Fig. 17.22, Table 17.4)(HULTIN, 1957; IMMERS and RUNNSTRÖM, 1960; NILSSON, 1961; TAGUCHI, 1962; MONROY, 1965; YANAGISAWA and ISONO, 1966; ZOTIN et al., 1967).

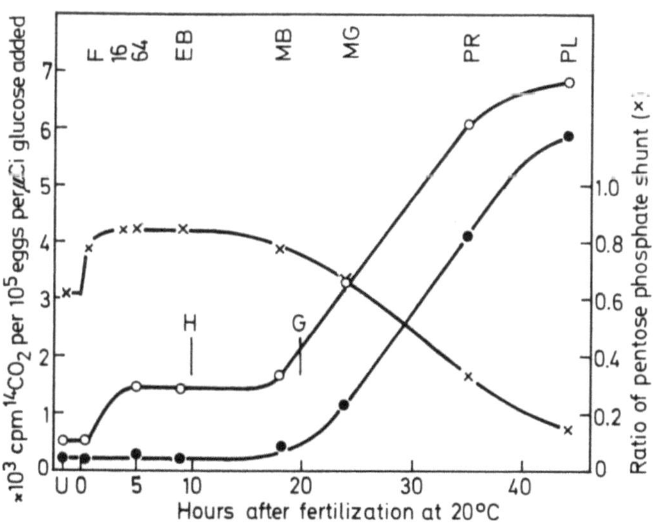

Fig. 17.20. $^{14}CO_2$ produced by ^{14}C-glucose and pentose phosphate shunt participation in glucose oxidation in developing embryos of *Pseudocentrotus depressus*. $^{14}CO_2$ was produced by 7.8 mμmol glucose-1-^{14}C (0.200 μCi)(o) and glucose-6-^{14}C (0.175μCi)(●) in 20 min at 20°C in 2 ml egg suspension (5 x 10^5 eggs). Pentose phosphate shunt participation (x) was calculated from 1 - $^{14}CO_2$ from G-6-^{14}C/$^{14}CO_2$ from G-1-^{14}C. (From ISONO and YASUMASU, 1966, 1968)

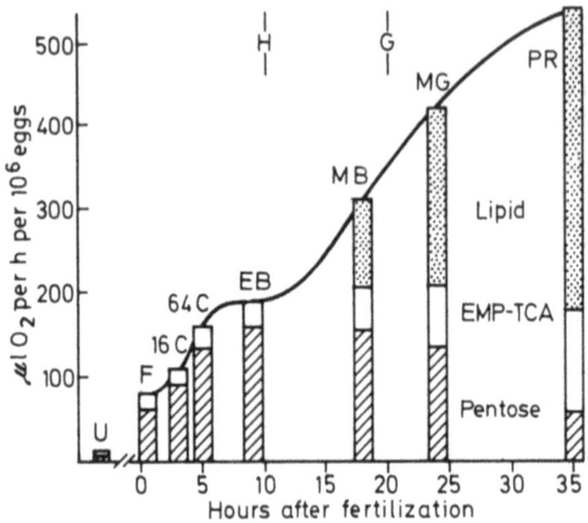

Fig. 17.21. Changes in substrate and respiration pathway during early embryo development of *Pseudocentrotus depressus*. Solid line indicates rate of oxygen uptake; PENTOSE, carbohydrate oxidation by means of the pentose phosphate shunt; EMP-TCA, carbohydrate oxidation by means of glycolysis and the KREBS cycle; LIPID, respiration from lipid oxidation. (From ISONO and YASUMASU, 1968)

If the increased ADP/ATP ratio is strictly related to respiratory change, the ratio should increase 60 seconds after fertilization when respiration increases. While ATP sometimes declines just after fertilization, adenosine nucleotide levels essentially remain unchanged (Fig. 17.23)(CHAMBERS and MENDE, 1953a; ROTHSCHILD, 1956; MONROY, 1965; YANAGISAWA and ISONO, 1966; ZOTIN et al., 1967; YANAGISAWA, 1968; EPEL, 1969; MACKINTOSH and BELL, 1969; YASUMASU et al., 1973). Instead of ADP, arginine (another phosphate acceptor) is phosphorylated just after fertilization in *Strongylocentrotus droebachiensis*, *Anthocidaris crassispina*, *Pseudocentrotus depressus*, and *Hemicentrotus pulcherrimus*, about half a minute after respiration increases (Fig. 17.33)(CHAMBERS and MENDE, 1953b; YANAGISAWA, 1968; NEMOTO and YANAGISAWA, 1969).

There are 3 hypotheses to explain the connection between energy metabolism and mitosis: energy for mitosis is supplied by an energy reservoir synthesized during interphase and used only for mitosis; energy for constructing the mitotic apparatus activates a structure that provides energy for anaphase movement; and any phase of mitosis may be arrested by inhibitors of ATP synthesis or ATP utilization, a continuous energy supply being required during mitosis (SWANN, 1957; MAZIA, 1961; EPEL, 1963).

The energy reservoir hypothesis is based mainly on inhibitory effect of carbon monoxide on sea-urchin cleavage. The carbon monoxide-cytochrome oxidase complex is stable only at wavelengths outside its absorption bands, making it possible to switch inhibition of respiration on and off by alternating green and white lights (SWANN, 1953, 1954).

In *Psammechinus miliaris*, about 35 minutes after fertilization at 19°C, carbon monoxide suppression of oxidation does not inhibit the mitosis in progress but inhibits the next mitosis. This suggests

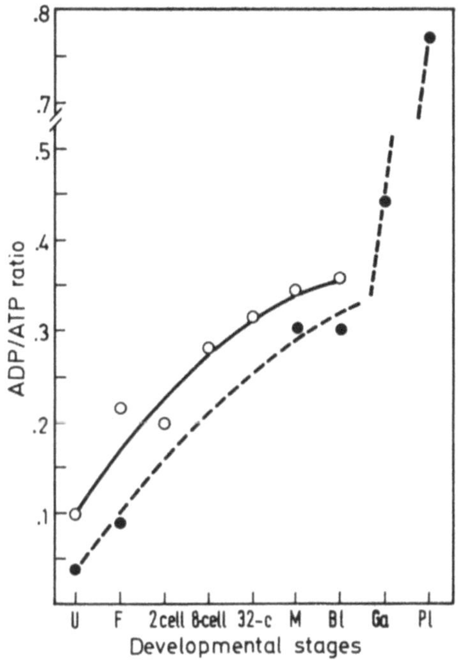

Fig. 17.22. Change in ADP/ATP ratio in *Paracentrotus lividus* eggs (o) and *Pseudocentrotus depressus* eggs (●) after fertilization and during development. (From MONROY, 1965, and YANAGISAWA and ISONO, 1966)

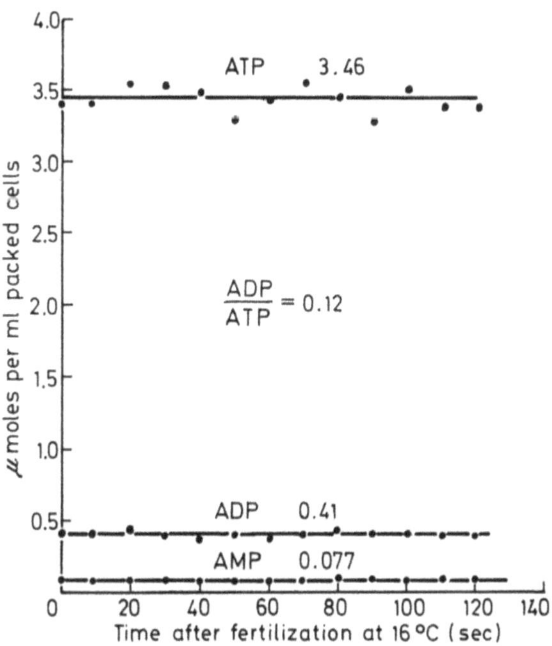

Fig. 17.23. Adenosine nucleotide levels after fertilization of *Strongylocentrotus purpuratus* eggs (1 ml eggs: 4.35 x 10^6). ATP, ADP, and AMP were assayed by NAD-linked enzyme and fluorescence methods. (From ESTABROOK and MAITRA, 1962, and EPEL, 1969)

Table 17.4. Acid-soluble nucleotides in *Pseudocentrotus* eggs (mole percent of total nucleoties). (From YANAGISAWA and ISONO, 1966)

	Unfertilized egg	Fertilized egg	64-cell stage	Early blastula	Mesenchymal blastula	Late gastrula	Pluteus
CMP + unknown	14.2	3.9	6.3	6.0	7.2	19.9	35.7
AMP	0.6	0.9	2.5	2.8	3.6	6.5	11.7
GMP + UMP	1.0	1.5	2.3	2.4	3.4	6.2	9.4
ADP	1.9	4.3	11.1	8.3	10.7	12.0	10.9
UDP-sugar	13.1	11.9	17.9	20.4	15.0	7.9	5.8
CTP + GDP + UDP	2.9	4.5	8.5	6.7	4.3	7.5	8.5
ATP	53.5	54.9	36.3	38.4	35.6	27.3	14.2
GTP	5.3	8.5	6.0	6.2	7.7	6.5	1.9
UTP	7.5	9.6	9.1	8.8	12.5	6.2	1.9
Total nucleotides (µmoles)	26.02	25.23	20.64	18.24	18.01	17.62	19.05

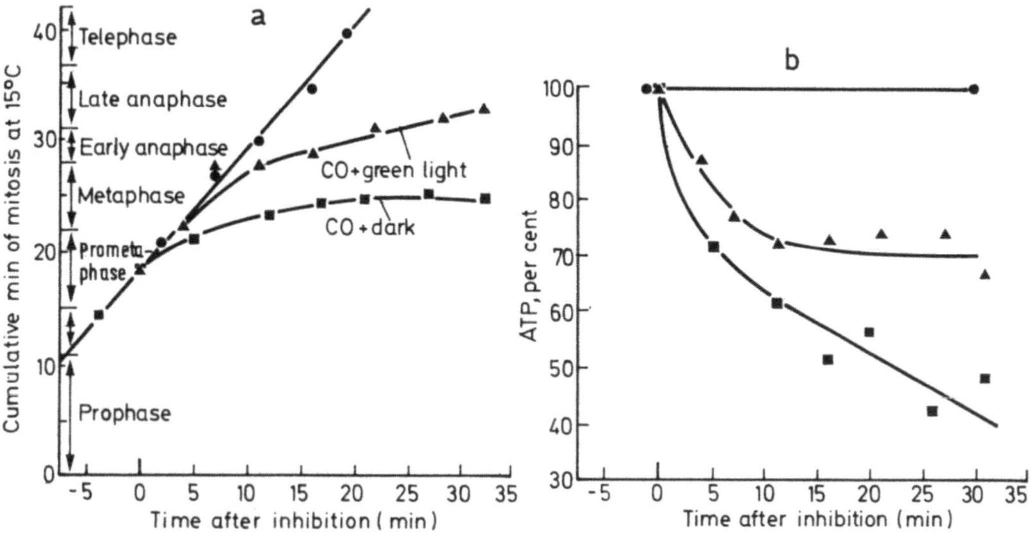

Fig. 17.24 a and b. Relative progress of mitosis (a) and ATP level (b) during 2 types of CO inhibition in *Strongylocentrotus purpuratus*. (●) control; (▲) 97% CO/air in 537 nm green light (330-ft candles); (■) 97% CO/air in total darkness. ATP was determined by firefly-luminescence technique. In CO in the dark mitosis continued for about 20 minutes before it was arrested. Progress during that time was equal to about 10 minutes of normal mitosis. In CO in green light, mitosis was retarded but not arrested. (From EPEL, 1963)

that energy for the second mitosis and cleavage is stored up during the first and that this energy store is a reservoir continually being filled and siphoned out when it is full. The energy once siphoned out is used for mitosis and cleavage, even if carbon monoxide inhibition keeps the reservoir from refilling (SWANN, 1953). As the amounts of ATP and arginine phosphate do not seem to change significantly during cleavage, it is unlikely that they are the energy-carrying compounds (SWANN, 1954). In both the energy reservoir and the activated mitotic apparatus hypotheses, there is a point in mitosis after which inhibitors cannot prevent the completion of cell division.

Specific carbon monoxide inhibition of ATP synthesis in the dark arrests mitosis at any stage, and mitosis is completely inhibited when the ATP content falls to less than half of normal (Fig. 17.24). In green light, carbon monoxide inhibition of ATP is incomplete. The ATP content of cleaving eggs is reduced by about 25%, and mitosis proceeds slowly. This indicates that the energy reservoir and activated mitotic apparatus hypotheses are no longer tenable because their basic premise is that all or certain phases of mitosis are insensitive to inhibition of ATP synthesis (EPEL, 1963; MAZIA, 1963).

17.4.1 Cyclic Change of Respiratory Rate during Cleavage

Anaerobiosis or treatment with metabolic inhibitors retards respiration which delays or arrests the cleavage of sea-urchin eggs (KRAHL, 1950; BARNETT, 1953; KRISZAT, 1954; SWANN, 1957; ZOTIN et al., 1965; YANAGISAWA, 1968). This suggests that cleavage requires the energy from oxidative metabolism and that eggs breathe in rhythm with the cleavage cycle. As shown by the WARBURG manometer, in a large number of *Echinus esculentus* eggs in a relatively small volume of seawater at 11°C, cleavage synchrony is disturbed and no cyclic change in respiration is observed (GRAY, 1925). But in *Paracentrotus lividus* respiration increases at cleavage furrow formation and in *Arbacia punctulata* this effect is more pronounced at 25°C than at 15°C (RUNNSTRÖM, 1933; TANG, 1948). It would seem, therefore, that respiration change is not observable at too low temperatures.

In eggs of *Psammechinus miliaris* at 16°C and *Psammechinus microtuberculatus* and *Dendraster excentricus* at 19°C, the Cartesian diver technique shows that respiration is slowest at anaphase and telophase and that almost 10 respiratory cycles correspond to 10 cleavage cycles (Fig. 17.25) (ZEUTHEN, 1947, 1950a, 1950b). Respiratory rhythm is also evident in single *Psammechinus miliaris* eggs (Fig. 17.26)(ZEUTHEN, 1955; SCHOLANDER et al., 1958). Even when spindle formation and cytokinesis are stopped by colchicine, the rhythm continues, indicating that it may be related to nuclear phenomena or to events preceding cytokinesis, rather than to cytokinesis itself (Fig. 17.25)(ZEUTHEN, 1951, 1953). In *Lytechinus variegatus* eggs at 27°C, respiratory oscillations observed by microrespirometer equipped with automatic electromagnetic diver balance suggest that bursts of energy consumption link to epigenetic events other than cell division (LØVTRUP and IVERSON, 1969).

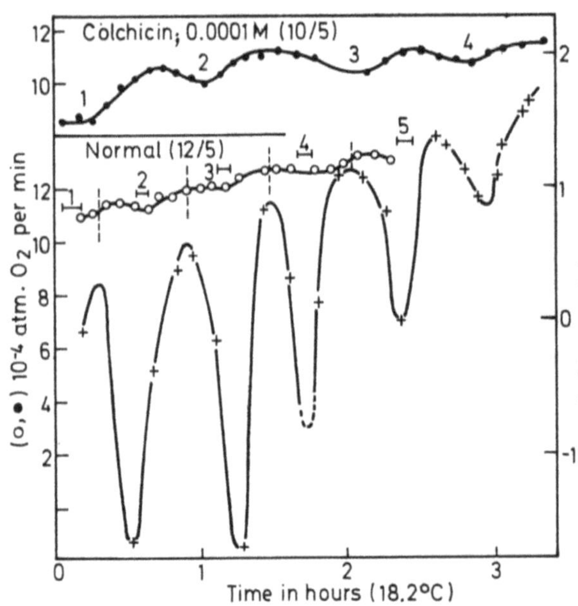

Fig. 17.25. Respiratory rhythms in cleavage of normal (o) and colchicine-treated (•) *Psammechinus microtuberculatus* eggs. Division numbers are indicated and log nuclei per egg (+) determined. Colchicine treatment (10^{-4} *M*) was started 15 min after fertilization, and after 3 hours 10% of the eggs were in a poor 2-cell stage and 90% were uncleaved. (From ZEUTHEN, 1953)

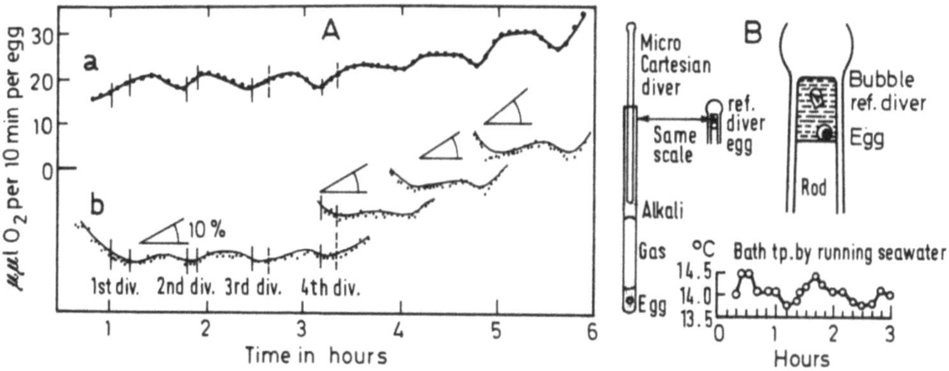

Fig. 17.26 A and B. Respiration in single eggs of *Psammechinus miliaris* at 16.9°C. (A) Rate curve a is constructed from curve b which gives deviation from 4 straight lines corresponding to cumulative readings. Slope of 10% deviation is indicated. (B) Micro-Cartesian diver for single cells (left); reference diver (right); reference diver drawn to scale of micro-Cartesian diver (middle). (From ZEUTHEN, 1955, and SCHOLANDER et al., 1958)

Experiments with substituted phenols show that cell division and respiratory rate are independent and that cell division is more closely related to oxidative phosphorylation (CLOWES and KRAHL, 1936). In concentrations of 10^{-5} to 10^{-4} *M*, ATP reverses mitotic block produced by CN, low oxygen tension, DNP, and malonate, while AMP and inorganic pyrophosphate do not. Block produced by malonate, which inhibits succinic dehydrogenase, is reversed by 5.7×10^{-2} *M* of succinate or fumarate (BARNETT, 1953).

As it is unlikely that a large polar molecule such as ATP penetrates the eggs, it must act only on the cell cortex (KRISZAT, 1954; LANDAU et al., 1955). It follows, therefore, that cleaving eggs would show rhythmic changes in the ~P compounds they contain and in their metabolic activities. During *Dendraster excentricus* cleavage, the rate of ^{32}P uptake decreases slightly during cytokinesis (Fig. 17.27); and during cell division of *Amoeba proteus* the ratio between ^{32}P incorporation at interphase and during mitosis is 1.8. The eggs of *Lytechinus pictus* show no cyclic change (WHITELEY, 1949; ZEUTHEN, 1951; MAZIA and PRESCOT, 1954).

Cyclic change in glucose oxidation of *Strongylocentrotus droebachiensis* is 35% at first cleavage, 16% at second cleavage, and 16% at third. It peaks just before prophase and is lowest at telophase (Fig. 17.28) (ZOTIN, 1967).

At the start of cytokinesis in *Lytechinus*, ~P content per egg increases about 10 μg P/ml and the acid-soluble organic P fraction decreases about 10 μg P, while total acid-soluble P and inorganic P fractions remain unchanged. This suggests that ~P derived from acid-soluble organic P via the KREBS cycle is directly involved in cell division (BARNETT and DOWNEY, 1955).

When the first *Pseudocentrotus* cleavage is delayed by low concentrations of DNP ($2\text{-}10 \times 10^{-6}$ *M*), NaCN ($1\text{-}5 \times 10^{-5}$ *M*), or NaN_3 ($1\text{-}4 \times 10^{-3}$ *M*), the arginine phosphate level falls rapidly while the inorganic phosphate level rises, depending on the length of time of cleavage delay. The ATP level is not affected (Fig. 17.29)(YANAGISAWA, 1968).

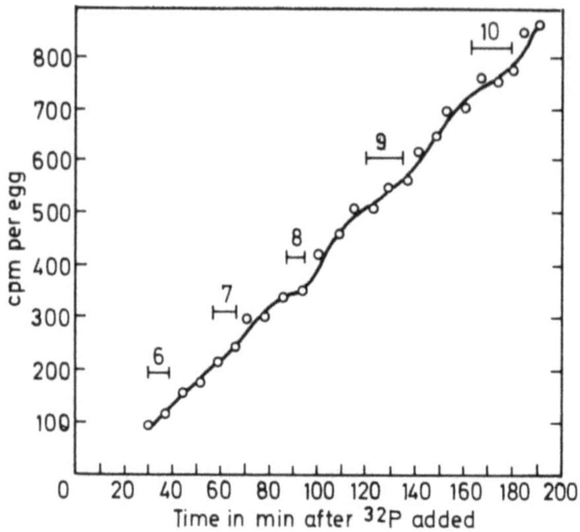

Fig. 17.27. ^{32}P accumulation in *Dendraster excentricus* eggs during cleavage stages. Flow chamber was used for continuous reading of uptake. (From ZEUTHEN, 1951)

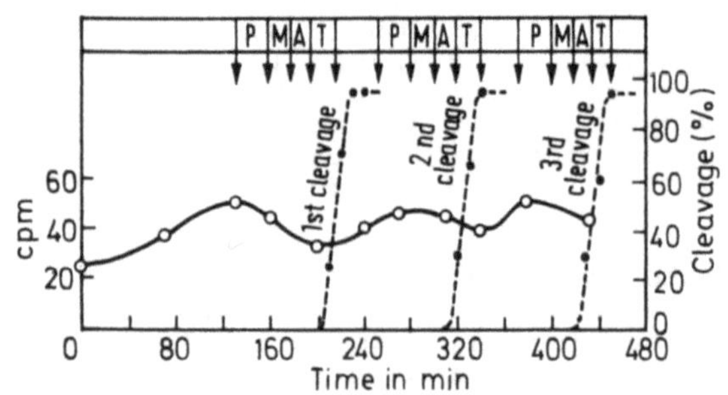

Fig. 17.28. Rate of 40-min ^{14}C-glucose oxidation at different stages during first 3 cleavage cycles of 8,000 to 12,000 *Strongylocentrotus droebachiensis* eggs at 6°-7°C. P, prophase; M, metaphase; A, anaphase; T, telophase. (From ZOTIN, 1967)

In *Anthocidaris* cleavage at 22°C, arginine phosphate content rapidly declines 10% (0.09 μμmole/egg) at late anaphase but promptly recovers. When DNP treatment delays cleavage for 5, 8, and 12 minutes, arginine phosphate decreases 22% (0.12 μμmole/egg), 32% (0.2 μμmole/egg), and 53% (0.24 μμmole/egg), respectively. DNP treatment also lowers the arginine phosphate plateau at interphase (Fig. 17.30) (NEMOTO, 1970).

17.5 High-Energy Phosphate Compounds and Their Metabolic Activity

It is generally believed that adenosine nucleotides and related compounds are essential to energy metabolism during embryo development. Acid-soluble compounds in sea-urchin eggs were first found in the acid extract of *Dendraster excentricus* and *Paracentrotus lividus* (Table 17.5) (NEEDHAM and NEEDHAM, 1930; ZIELIŃSKI, 1939).

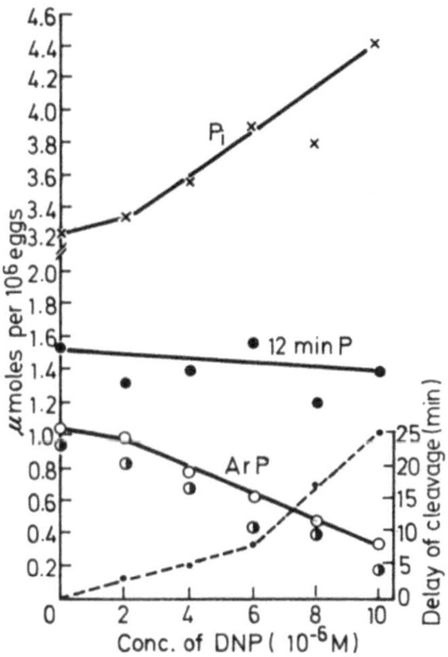

Fig. 17.29. Relationship among inorganic phosphate (x), 12-min phosphorus (ATP) (●), arginine phosphate (labile phosphorus: o; arginine produced after acid hydrolysis: ◑), and cleavage delay (·) of *Pseudocentrotus* eggs treated for 60 min with different DNP concentrations. Control eggs cleaved 84 min after fertilization. (From YANAGISAWA, 1968)

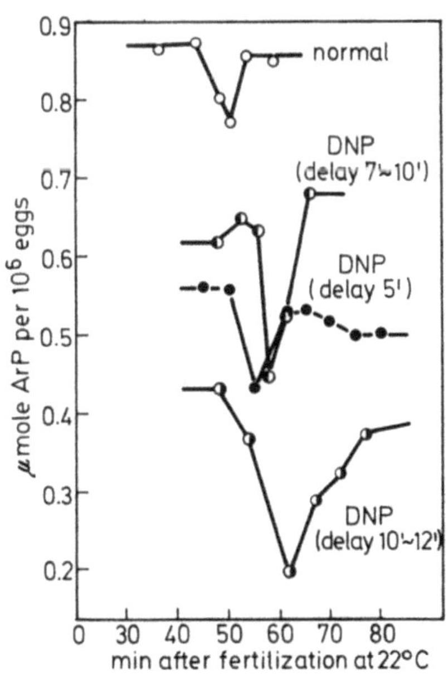

Fig. 17.30. Changes in arginine phosphate content during first cleavage of *Anthocidaris* eggs: normal and treated with 10^{-6} *M* DNP. (From NEMOTO, 1970)

Table 17.5. Phosphorus compounds in unfertilized echinoderm eggs. Data have been recalculated. Some compounds are identified more definitely than they originally were. (From WHITELEY, 1949)

	Patiria miniata	*Dendraster excentricus*	*Paracentrotus lividus*	*Arbacia punctulata*		*Strongylocentrotus purpuratus*	*Lytechinus pictus*
Total phosphorus	6.64 mg/gm dry weight	6.94 mg/gm dry weight	134.7 mg/gm *N*	130 mg/gm *N*	960 μg/10^6 eggs[b]	817 μg/10^6 eggs	2.78 mg/gm wet weight
				3.1 mg/gm wet weight[a]	4.13 mg/gm wet weight[c]	2.94 mg/gm wet weight	
Acid-insoluble phosphorus (%)	48	50	53.0	62.4	62.2	69.1	56.8
RNA-phosphorus	16 (RNA + DNA)	32 (RNA + DNA)	30.5 (RNA + DNA + phosphoprotein)	4.6	22.2	38.0	24.7
DNA-phosphorus				1.1	0.9	1.7	0.5
Phosphoprotein-phosphorus	trace	12		19.1	2.6	2.4	1.3
Phospholipid-phosphorus	32	6	22.5	37.6	36.5	27.0	20.5
Acid-soluble phosphorus (%)	52	47	47.0	37.6	37.8	30.3	38.2
Inorganic phosphorus	20	29	28.7	11.2		7.4	15.1
ATP + ADP phosphorus	13	14	11.4	21.3		8.2	13.8
Propanediol phosphate phosphorus (barium-soluble ethanol-soluble fraction)	19	4	7.6			12.0	7.3
Unidentified phosphorus				5.1		2.7	2.0
Investigators	NEEDHAM and NEEDHAM, 1930		ZIELIŃSKI, 1939	CRANE, 1947	SCHMIDT et. al., 1948	WHITELEY, 1949	

[a]CRANE (1947) reported 0.31 mg percent of wet weight.

[b]SCHMIDT et al. (1948) reported 96 x 10^{-3} μg/egg; this should probably have read 96 x 10^{-5} μg/egg.

[c]Calculated per egg, using conversion data from HARVEY (1932).

[d]9.8 percent lost in fractionating acid-insoluble compounds.

17.5.1 Nucleotides

Phosphorus analysis shows that the average pyrophosphate content of Ba-precipitable P fractions is 15.5%, while the precipitable P that is harder to hydrolyze makes up 8.4% of the acid-soluble P. This ratio of nearly 2:1 indicates that pyrophosphate is in the form of ATP (ZIELIŃSKI, 1939).

When phosphate liberated after 7-minute hydrolysis in normal HCl at 100°C is quantitatively analyzed and total phosphate, ribose, nitrogen, and adenine determined spectroscopically in *Asterias forbesi* and *Strongylocentrotus droebachiensis*, most of the organic phosphate in the Ba-insoluble fraction of the acid extract is found to be adenosine triphosphate (CHAMBERS and MENDE, 1953a). The same is true in *Arbacia punctulata*, *Strongylocentrotus purpuratus*, and *Lytechinus pictus* and in enzymatic dephosphorylation with myosin ATPase and potato adenylpyrophosphatase and deamination with muscle AMP-deaminase after adenylpyrophosphatase hydrolysis (CHAMBERS and WHITE, 1949; WHITELEY, 1949; CHAMBERS and MENDE, 1953a).

No significant amounts of AMP are found in *Strongylocentrotus droebachiensis*, and no change in optical density occurs after AMP-deaminase is added to the neutralized acid extract. When myokinase is added, absorption slowly decreases at 265 nm and increases at 240 nm. When potato pyrophosphatase is added, density falls rapidly at 265 nm. This shows the presence of 2% AMP and 6% ADP (CHAMBERS and MENDE, 1953a).

Anion exchange chromatography has shown several ribose nucleotides in the acid extract of sea-urchin eggs: CMP, (CDP), CTP, GMP, GDP, GTP, UMP, UDP, UTP, and such nucleotide derivatives as UDP-glucose, UDP-acetylglucosamine, and UDP-acetylgalactosamine (Table 17.4, Figs. 17.31, 17.32) (HULTIN, 1957; SUGINO et al., 1957, 1960; NILSSON, 1959, 1961; ISONO and YANAGISAWA, 1966; YANAGISAWA and ISONO, 1966).

In *Hemicentrotus* eggs, anion exchange chromatography has shown small amounts of 3 deoxyribonucleotides distinguishable as separate peaks, one of which was identified as deoxy-CDP-choline (SUGINO et al., 1957, 1960; SUGINO, 1960).

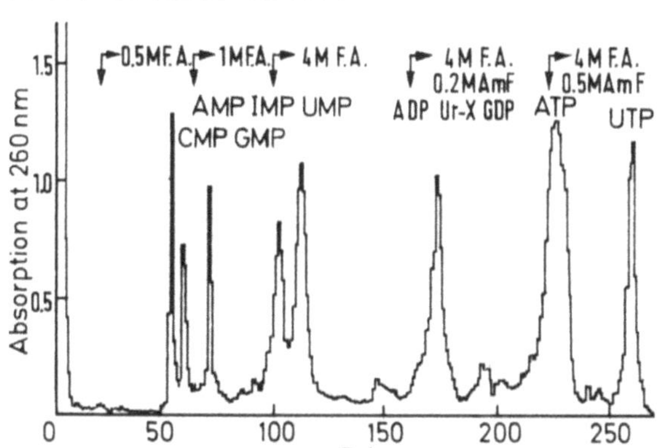

Fig. 17.31. Formic acid chromatogram of acid-soluble fraction (perchloric acid extract) of 10^6 unfertilized *Psammechinus miliaris* eggs. Column: Dowex-1 formate. Rate of elution: 1.4 to 1.8 ml/30 min. E_{260} of 30-min samples was determined after dilution to 2.5 ml. (From HULTIN, 1957)

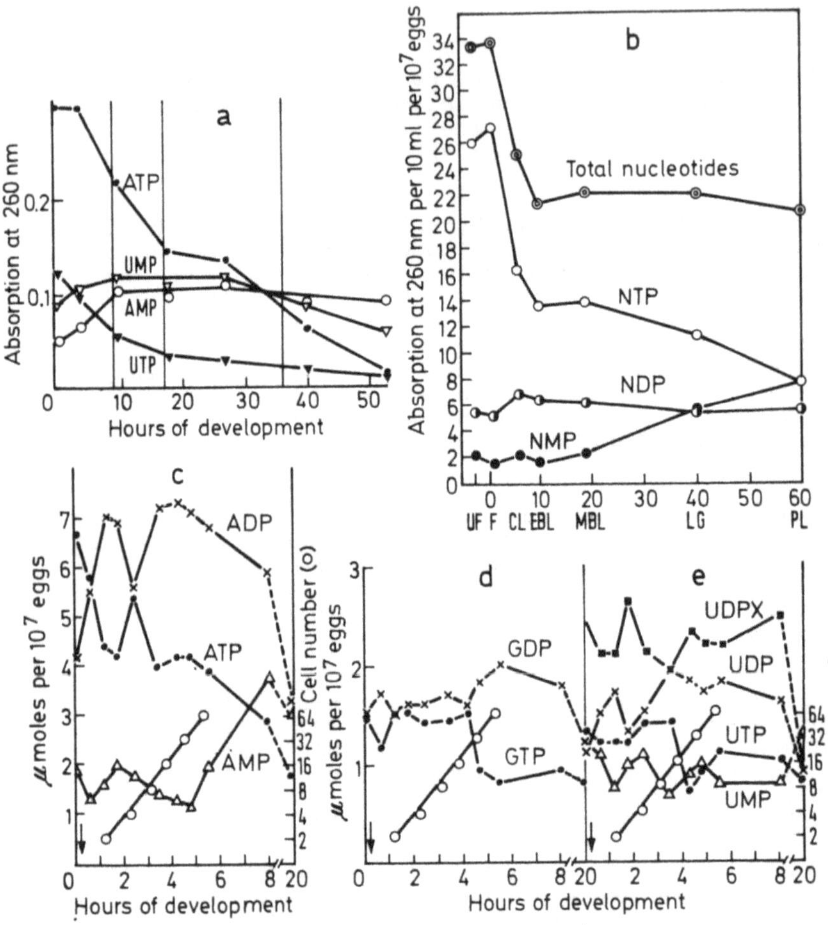

Fig. 17.32a-e. (a) Change in ATP, AMP, UTP, and UMP content of *Psammechinus miliaris* embryos at 18°-19°C. E_{260} of each fraction was recalculated to standard column of 100 ml. (b) Change in total nucleotides (◎), nucleoside triphosphate (NTP)(o), nucleoside diphosphate (NDP)(◑), and nucleoside monophosphate (NMP)(●) in *Pseudocentrotus* embryos at 18°C. (c-e) Adenosine nucleotide, guanosine nucleotide, and uridine nucleotide change in *Paracentrotus lividus* embryos at 16°C. (From HULTIN, 1957; NILSSON, 1961; and YANAGISAWA and ISONO, 1966)

In unfertilized *Psammechinus* eggs, 45 molar percent nucleoside triphosphate (30 molar percent ATP) and in unfertilized *Pseudocentrotus* and *Hemicentrotus* eggs 70 molar percent nucleoside triphosphate (50 molar percent ATP) are found, calculated on the basis of total nucleotide content determined by the absorption at 260 nm (E_{260})(Table 17.4, Fig. 17.32)(HULTIN, 1957; YANAGISAWA and ISONO, 1966). Low nucleoside triphosphate and high nucleoside monophosphate values may be due to pyrophosphate breakdown at time of preparation (SUGINO, 1960; TAGUCHI, 1962).

Psammechinus eggs contain about 9 molar percent IMP of the total nucleotide amount based on E_{260}, while *Paracentrotus*, *Hemicentrotus*, and *Pseudocentrotus* eggs appear to have no IMP at all. During embryo development nucleoside triphosphates and UDP sugars decrease, nucleoside monophosphates increase, and nucleoside diphosphates increase slightly in the early stage and then remain rather constant (HULTIN, 1957; NILSSON, 1959, 1961; YANAGISAWA and ISONO, 1966).

17.5.2 Arginine Phosphate

The gastrula and pluteus of *Paracentrotus lividus* eggs contain a small amount of arginine phosphate or a related compound (NEEDHAM et al., 1932, ZIELIŃSKI, 1939), while the ovaries of *Brissopsis* but not the eggs contain it (LINDBERG, 1943).

Arginine phosphate is recovered as a barium-soluble, ethanol-insoluble fraction. Differential hydrolysis is performed in 0.1 *N* TCA for 1 minute at 100°C or in 0.01 *N* HCl (pH 2) for 180 minutes at 37°C, causing the equimolar liberation of inorganic phosphate and arginine. In eggs of *Pseudocentrotus*, *Hemicentrotus*, *Anthocidaris*, *Toxopneustes pileolus*, and *Diadema setosum*, the enzyme arginine phosphokinase is also present (MENDE and CHAMBERS, 1953; YANAGISAWA, 1959a, 1959b).

In *Strongylocentrotus droebachiensis*, *Pseudocentrotus*, *Anthocidaris*, and *Hemicentrotus*, after fertilization the ATP level remains fairly constant while the amount of arginine phosphate doubles. In 2 or 3 minutes the arginine phosphate reaches a plateau which lasts until the 16-cell stage. Then it declines sharply and drops more gradually as it approaches the gastrula stage (Fig. 17.33)(CHAMBERS and MENDE, 1953b; YANAGISAWA, 1968; NEMOTO and YANAGISAWA, 1969).

17.5.3 Propanediol Phosphate

Sea-urchin eggs contain large amounts of acid-soluble phosphate compounds, the barium salts of which are soluble in alcohol. Unfertilized eggs of *Strongylocentrotus droebachiensis*, for example, contain 9.3% of these compounds and *Strongylocentrotus purpuratus* 58% (MENDE and CHAMBERS, 1953; BOLST and WHITELEY, 1957). Vertebrate tissue contains 1%-8% (SACKS, 1949). Propanediol phosphate was believed to be main component of these compounds in early development of *Brissopsis* eggs, but it seems not to occur in the eggs of *Strongylocentrotus purpuratus* (LINDBERG, 1943, 1946; RUDNEY, 1954; BOLST and WHITELEY, 1957).

The barium- and ethanol-soluble fraction is composed of at least 3 compounds distinguishable by paper chromatography. From fertilization to early pluteus they slowly decrease. As determined by ^{32}P incorporation, metabolic activity of the fraction is very low (Fig. 17.34) (BOLST and WHITELEY, 1957).

17.5.4 Metabolic Activity of Energy-Rich Phosphate Compounds

The acid-labile phosphate of fertilized *Psammechinus miliaris* eggs is quickly marked by exogenous ^{32}P (LINDBERG, 1949). In *Strongylocentrotus purpuratus* eggs at the 2-cell stage exposed to pulses of ^{32}P, most of the ^{32}P incorporated from seawater is in the form of acid-soluble phosphate compounds. Dowex-1 formate column chromatography and formic acid-ammonium formate elution show that short pulses produce more ^{32}P incorporation into inorganic phosphate and less into arginine phosphate and ATP (Table 17.6, Fig. 17.35). The same reactions occur in *Anthocidaris* and *Hemicentrotus* eggs at the 8-cell stage (Fig. 17.36)(YANAGISAWA, 1968). Arginine phosphate and ATP are the phosphate compounds that in-

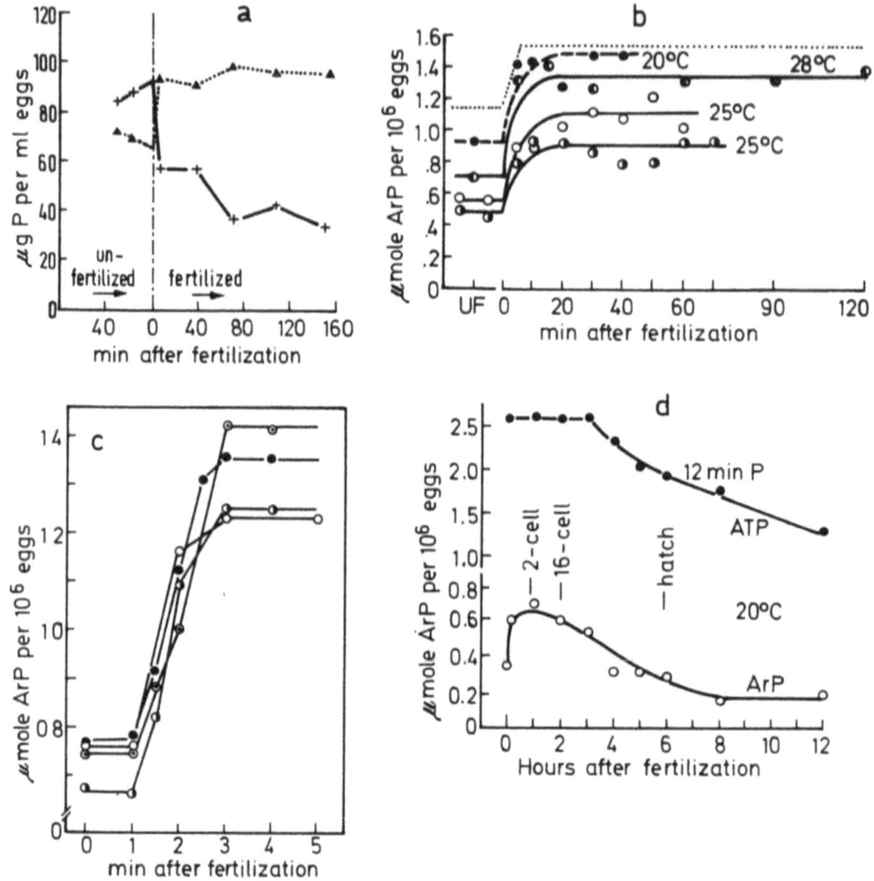

Fig. 17.33a-d. Change in amount of arginine phosphate during fertilization and development of *Strongylocentrotus droebachiensis* eggs at 14.5°C (a)(▲, hydrolyzable phosphorus in 2% TCA at 1 min and 100°C; +, inorganic phosphate), *Pseudocentrotus* (●----●) and *Anthocidaris* (O, ◑, ◐) eggs (b), *Pseudocentrotus* eggs just after fertilization (c), and ArP and ATP of *Anthocidaris* embryos (d). (From CHAMBERS and MENDE, 1953b; YANAGISAWA, 1968; NEMOTO and YANAGISAWA, 1969)

corporate ^{32}P most rapidly. The specific activity of labile phosphorus in arginine phosphate is the same as that of terminal phosphorus in ATP (Table 17.7, Fig. 17.36)(YANAGISAWA, 1969a).

Once the ^{32}P uptake system is established, it is not inhibited by CN^-, N_3^-, monoiodoacetate, anaerobiosis, or DNP. NaN_3 (5×10^{-3} *M*) and KCN (10^{-4} *M*) cause ^{32}P uptake equal to or higher than that of the control but with different distribution, reduce incorporation into arginine phosphate and ATP, and increase incorporation into inorganic phosphate (Table 17.6)(LITCHFIELD and WHITELEY, 1959; GRIFFITHS and WHITELEY, 1964). These inhibitors reduce the ATP content while others, such as NaF and malonate, even at concentrations that inhibit cleavage, do not (Fig. 17.37)(ZOTIN et al., 1965).

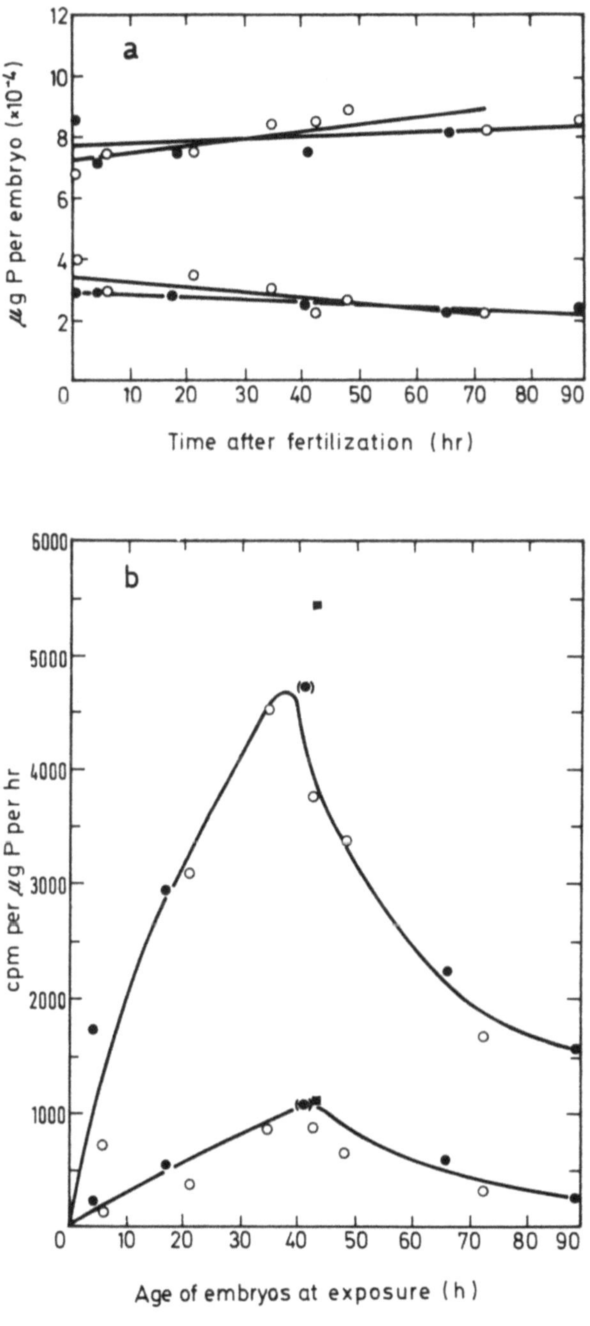

Fig. 17.34 a and b. Total phosphorus (upper curve) and barium-soluble, ethanol-soluble fractions (lower curve) in *Strongylocentrotus purpuratus* embryos at 11°C. (a) 2 experiments. (b) Rate of ^{32}P uptake; embryos incubated with 0.04μCi ^{32}P per μl for 60 min. (From BOLST and WHITELEY, 1957)

Fig. 17.35. Anion-exchange chromatogram of acid-soluble phosphate compounds in fertilized *Strongylocentrotus purpuratus* eggs exposed 1 min to 100μCi $^{32}PO_4$. PCA extract (0.4 *M*) was adsorbed on Dowex-1 chloride 9.0 mm x 100 mm. Eluting solutions: (1) water; (2) 0.025 *M* NH_4Cl + 0.001 *M* $Na_2B_4O_7$; (3) 0.03 *M* NH_4Cl; (4) 0.005 *N* HCl; (5) 0.01 *N* HCl; (6) 0.02 *N* HCl + 0.02 *M* KCl; (7) 0.02 *N* HCl + 0.2 *M* KCl; (8) 1.0 *N* HCl. The percentage of the acid-soluble activity is indicated above each peak. Full scale deflection equals 1000 c.p.m. (not hatched); 10,000 c.p.m. (hatched); or 25,000 c.p.m. (solid black). (From GRIFFITHS and WHITELEY, 1964) ▶

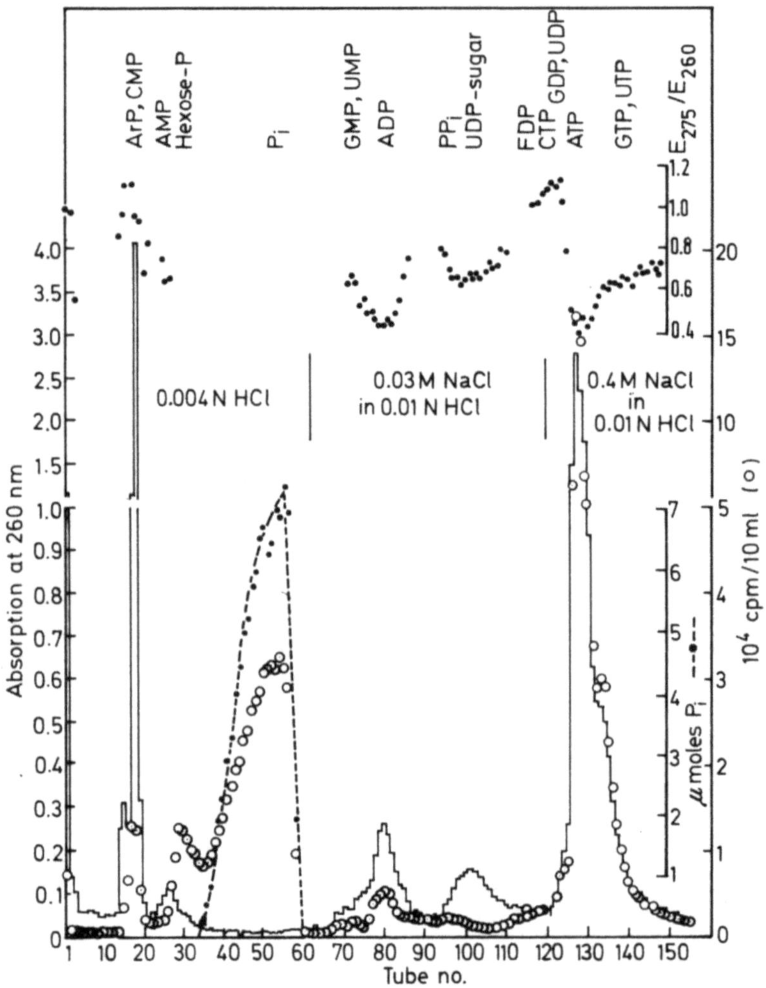

Fig. 17.36. Anion-exchange chromatogram of acid-soluble phosphate compounds of 8-cell stage *Anthocidaris* embryos labeled with 10-min ^{32}P. Continuous line: UV absorbancy at 260 nm; --·--: inorganic phosphate; O, total counts per minute in each tube; ●, UV absorbancy ratio at 275 nm against at 260 nm (E_{275}/E_{260}). Middle: eluting solutions; Upper: anticipated location of phosphate compounds. (From YANAGISAWA, 1968)

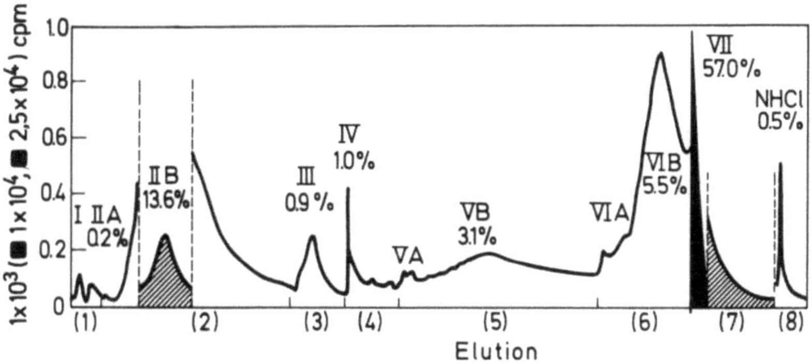

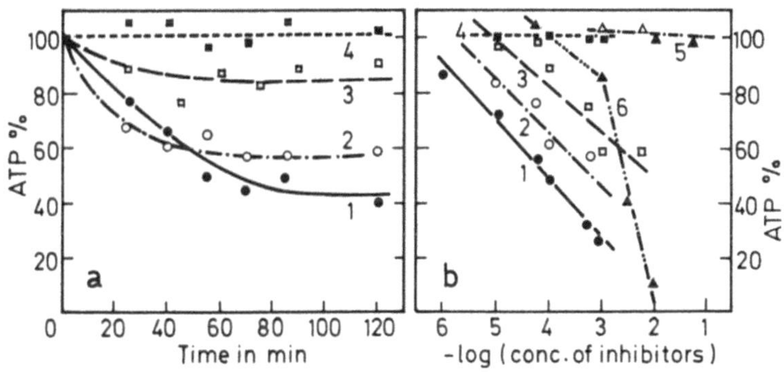

Fig. 17.37 a and b. (a) 10^{-4} *M* NaCN (●), 2,4-DNP (○), NaN_3 (□), and 5 x 10^{-3} *M* malonate (■) reduce ATP level of *Strongylocentrotus droebachiensis* eggs. (b) ATP level after 2-hr treatment with different concentrations of NaCN (●), DNP (○), NaN_3 (□), malonate (■), NaF (△), and monoiodoacetate (▲). (From ZOTIN et al., 1965)

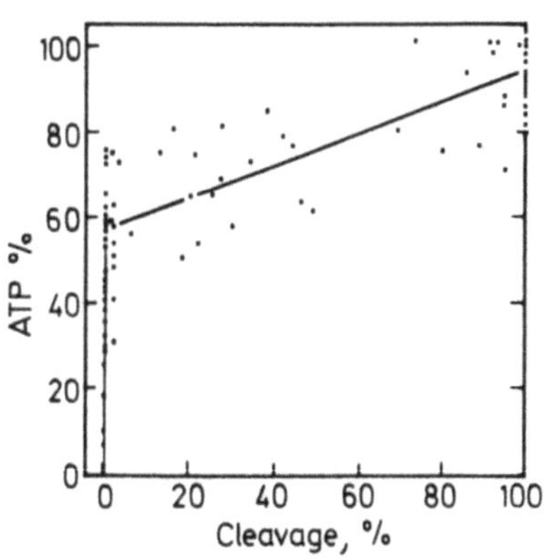

Fig. 17.38. Percent of cleavage rate plotted against ATP level after *Strongylocentrotus droebachiensis* eggs were treated with inhibitors. (From ZOTIN et al., 1965)

Table 17.6. Distribution of ^{32}P in acid-soluble fractions from fertilized eggs; effect of metabolic inhibitors

Strongylocentrotus purpuratus (2-cell stage). (From GRIFFITH and WHITELEY, 1964)

	Duration of ^{32}P pulse (min)	^{32}P uptake by eggs (%)	Eluted from Dowex-50 arginine ^{32}P (%)	Fractions eluted from Dowex-1 column (%)											Recovery (%)
				I	IIA	IIB P_i	III Glucose-6-P	IV	VA	VB ADP	VIA	VIB	VII ATP	1 *N* HCl strip	
Experiment 1, control	10	39.1	11.9	0.1	0.2	11.5	0.9	1.4	0.3	1.3	0.7	3.3	55.3	0.3	87.2
$5x10^{-3}$*M* NaN_3	10	47.8	0.3	0.2	0.5	58.2	1.0	1.7	-	4.3	1.1	1.8	23.0	-	92.1
Experiment 2, control	1	6.4	27.0	0.4	0.9	6.8	0.8	0.8	-	2.0		4.2	53.7	0.3	97.0
$5x10^{-3}$*M* NaN_3	1	9.7	1.2	0	0	76.9	1.9	1.2	0	4.2		3.9	14.3	0.2	104.8
Experiment 3, control	1	29.3	19.1	0.2	-	13.7	0.9	1.0	-	3.1	-	5.5	57.0	0.5	101.0
10^{-4} *M* KCN	1	42.1	0.6	-	-	71.7	1.8	1.2	-	4.0		2.6	16.9	0.2	99.4
Experiment 4, control	1	15.0	9.0	-	2.3	16.1	5.7	0.4	-	2.2	1.4	4.7	47.8	0.3	89.8
0.1 *M* IAA	1	14.9	9.0	-	-	35.0	3.3	0.5	-	1.9	1.2	3.4	38.3	-	92.6
Experiment 5, control	1	24.5	12.5	0.9	0.8	6.1	5.9	1.3	-	4.1	1.7	4.7	61.2	0.7	99.9

Anthocidaris crassispina (8-cell stage), 7.1 x 10^6 eggs. (From YANAGISAWA, 1969a)

	CMP + ArP	AMP + G-6-P	P_i	GMP+ UMP	ADP	UDP-sugar	CTP	ATP	GTP+ UTP
E_{260}	6.45	0.63	(104 µmoles)	0.55	1.83	1.85	1.16	10.20	4.02
^{32}P (%)	2.7	6.1	32.4	0.7	3.3	1.5	2.8	36.8	12.7

Hemicentrotus pulcherrimus (8-cell stage), 8 x 10^6 eggs

	CMP + ArP	AMP + G-6-P	P_i	GMP+ UMP	ADP	UDP-sugar	CTP	ATP	GTP+ UTP
E_{260}	3.14	0.75	(27.8 µmoles)	0.33	1.45	2.91	1.12	12.37	3.09
^{32}P (%)	1.6	0.7	10.6	0.9	3.0	2.9	4.8	62.0	13.4

In *Pseudocentrotus* 2 x 10^{-5} *M* DNP completely inhibits cleavage, destroys arginine phosphate after 60-minute treatment, reduces ATP content, and increases ADP (YANAGISAWA, 1969a). NaCN, DNP, and NaN_3 (10^{-4} *M*) arrest cleavage and reduce ATP content. The first cleavage is completely blocked when the ATP level falls below 58%, but proceeds normally when it is over 93% (Fig. 17.38)(ZOTIN et al., 1965).

Table 17.7. Metabolic activity of arginine phosphate and ATP in 8-cell stage of *Hemicentrotus* and *Anthocidaris* embryos. After 10-minute pulse labeling, arginine phosphate and ATP extracted with TCA were fractionated by Dowex-1 chromatography. Phosphorus of arginine phosphate was liberated by differential hydrolysis in 0.01 *N* HCl at 37°C for 10 hr, while terminal phosphorus of ATP (parentheses) was liberated with myosin ATPase. Values are in cpm/μmole phosphorus. (From YANAGISAWA, 1969a)

	Hemicentrotus	*Pseudocentrotus*
Arginine phosphate	17,200	66,200
ADP	6,480	31,800
UDP-sugar	1,730	18,400
CTP	11,100	34,100 (30,400)
ATP	21,000 (17,300)	91,500 (58,500)
GTP + UTP	18,300	63,700 (39,100)

When inhibitors in low concentration delay but do not arrest cleavage, arginine phosphate content is reduced in proportion to the length of cleavage delay, while ATP level is hardly affected at all (Fig. 17.29) (YANAGISAWA, 1968).

Such nucleic acid and protein biosynthesis inhibitors as actinomycin D, 5-fluorouridine, puromycin, and chloramphenicol do not reduce ATP or arginine phosphate levels even at concentrations that block cleavage (ZOTIN et al., 1965). It is possible that they prevent use of the energy reservoir without affecting the energy source (YANAGISAWA, 1968).

References, see p. 594.

18. Carbohydrate Metabolism and Related Enzymes

T. Yanagisawa

Based on 20th century biochemical research, an explanation has been found for that once-mysterious phenomenon - glycolysis. Yeast fermentation, characterized by "boiling" and spirit formation, is the result of the vigorous evolution of CO_2 bubbles and the production of ethanol as an end product of the anaerobic breakdown of starch. Lactate formation in muscle tissues during their "movement" is also a result of that carbohydrate catabolism via glycolysis (a process almost identical to that of fermentation) that accompanies the production of energy for muscle contraction.

In general, living cells obtain the necessary energy for their activities from either or both of the major catabolic pathways of carbohydrates: glycolysis and the pentose phosphate shunt. Studies of the respiration of the sea-urchin embryo show that carbohydrate metabolism is important mainly during early development. In this chapter, our main problem will be to ascertain in which pathway sea-urchin eggs metabolize carbohydrates, which may contribute to embryo development.

At present, there is little indication that the major products of carbohydrate metabolism have an effect on embryo development, but carbohydrates in general are known to constitute such biologically important cell structures as mucopolysaccharides in membraneous structures (plasma membrane, mitochondrial membrane, endoplasmic reticulum) and the pentose moieties of nucleic acids contained in self-duplicating and protein-synthesizing structures.

18.1 Glycolysis and the Pentose Phosphate Shunt

18.1.1 Glycogen

At first, no glycogen (or reducing sugar) was found in *Paracentrotus lividus* eggs (MEYERHOF, 1911). However, in later investigations hydrolysis of unfertilized *Arbacia punctulata* eggs with acid (*N* HCl, 100°C, 1 hour) yielded readily fermentable reducing sugar from which glucosazone was obtained (PERLZWEIG and BARRON, 1928). In another experiment a similar amount of sugar, after repeated precipitations from water extract, showed typical reactions with iodine and orseillin BB. Upon hydrolysis with *N* HCl, it yielded 97% glucose identified by the formation of the typical phenylhydrazone (HARVEY, 1932; BLANCHARD, 1935).

These results have been confirmed in various sea-urchin species (Table 18.1). In 1 demonstration, of 110 mg acid-hydrolyzable carbohydrates, 46% was isolated as typical glycogen, which tested red-brown in color with an iodine solution, as expected. Further evidence for the identity of *Arbacia* glycogen with that in other animal tissues was provided by experiments in which glycogen, after several alcohol pre-

Table 18.1. Total carbohydrate and glycogen content of sea-urchin eggs

Species	Total carbohydrate (mg glucose/gm eggs)	Glycogen (mg glucose/gm eggs)	Authors
Paracentrotus lividus	13.6	13.1	EPHRUSSI and RAPKINE, 1928; EPHRUSSI, 1933
	13.4	12.1	ZIELIŃSKI, 1939
		10.1-19.5	CHAIGNE, 1934
		8.2[a]	ÖRSTRÖM and LINDBERG, 1940
Echinus esculentus	12-5	6-1	STOTT, 1931
		3.3[b]	CLELAND and ROTHSCHILD, 1952a
Arbacia punctulata	17.6	8	HUTCHENS et al., 1942a
	4		PERLZWEIG and BARRON, 1928
		8-14	BLANCHARD, in HARVEY, 1932
Psammechinus miliaris		10.9[a]	LINDBERG, 1943
Echinocardium cordatum		6.4[a]	LINDBERG, 1943
Hemicentrotus pulcherrimus	19.4	9.0	AKETA, 1957

[a]Recalculated on the assumption that 100 mg *N* = 3.05 gm wet weight.
[b]Recalculated on the assumption that 1 ml = 1.1 gm wet weight.

cipitations, was estimated from its optical rotation (the specific rotation is 196.6°) before acid hydrolysis and later by the ferricyanide method. Results showed the average value obtained by hydrolysis to be 98% of that obtained by rotation (HUTCHENS et al., 1942a).

The partially purified glycogen sample also contained small amounts of galactose, mannose, fucose, arabinose, and xylose (IMMERS, 1952). The nature of the carbohydrates other than glycogen in sea-urchin eggs is still uncertain, although polysaccharide-containing amino sugar has been reported (IMMERS, 1952). However, no free reducing sugar has been found (PERLZWEIG and BARRON, 1928; HUTCHENS et al., 1942a; KRAHL, 1950; AKETA et al., 1964).

18.1.2 Changes in Carbohydrate Content

As sea-urchin eggs contain a considerable amount of carbohydrates in a form presumably used as a metabolic substrate, it is interesting to observe carbohydrate metabolism during embryo development. Various investigations of carbohydrate content at fertilization have yielded contradictory results. In *Paracentrotus lividus*, for example, there was no change in total sugar and only a slight decrease in glycogen during 12 hours of development (very early gastrula stage); total sugar and glycogen accounted for 5.43% and 5.13%, respectively, of the dry egg in unfertilized eggs, and for 5.46% and 4.8% in the 12-hour larva, with total sugar content decreasing rapidly (0.72%) and glycogen disappearing completely by 40 hours of development (pluteus stage)

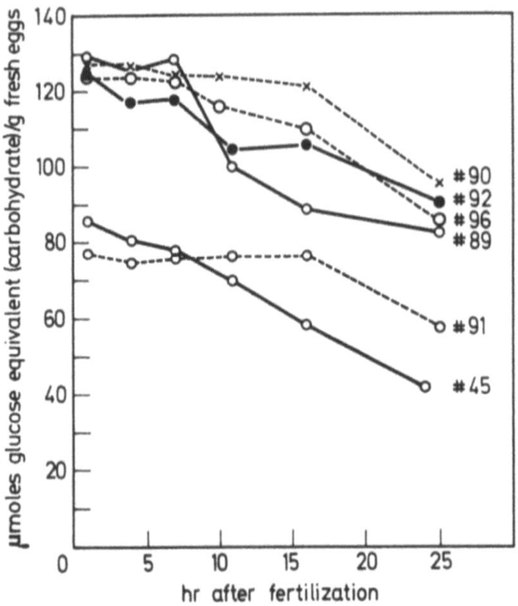

Fig. 18.1. Changes in total carbohydrate content during *Arbacia* egg development. 6 determination series are shown. Carbohydrate is reducing sugars produced after 1-hour hydrolysis at 100°C in *N* HCl from packed eggs or embryos. (From HUTCHENS et al., 1942a)

(EPHRUSSI and RAPKINE, 1928; EPHRUSSI, 1933; ZIELIŃSKI, 1939). Also, no decrease in carbohydrates was observed in *Arbacia* eggs after fertilization. It seems as though little or no carbohydrate is consumed in the 6-hour period, although the amount consumed is quite large from the 15th to 27th hours (Fig. 18.1)(HUTCHENS et al., 1942a).

On the other hand, *Paracentrotus lividus* eggs exhibit marked decrease in total reducing sugars, especially glycogen (from 8.02 to 6.15 mg/gm egg) during the 10 minutes following fertilization (ÖRSTRÖM and LINDBERG, 1940). The same decrease is also observed in *Psammechinus miliaris* (from 33.3 to 29.2 mg/gm) and *Echinocardium cordatum* (LINDBERG, 1943). The decrease in glycogen (from 20.0 to 17.7) is due chiefly to free glycogen (not precipitable when 0.5 ml of extract is treated with 1 ml 3.6% $ZnSO_4$ and 1 ml 0.15 *N* NaOH) reduced from 4.9 to 2.8. Seemingly, the free glycogen is first combined with protein and then metabolized as a fuel (LINDBERG, 1945).

The TCA extract of unfertilized *Paracentrotus lividus* eggs contains a considerable amount of glycoprotein demonstrated as a fluffy precipitate at the interphase between the water and ether phases when the extract is shaken with ether; the glycoprotein disappears shortly after fertilization. As the contents of the total and TCA-soluble carbohydrate are not altered, it seems that at fertilization a change in a glycoprotein fraction present in the probable cortex of unfertilized egg takes place, after which the fraction can no longer be precipitated by ether or alcohol (MONROY and VITTORELLI, 1960).

In *Hemicentrotus* eggs, acid-hydrolyzable carbohydrate is reduced from 592 µg to 402 µg/mg *N* by 15 minutes after fertilization, while the glycogen level (about 270 μg) does not change (AKETA, 1957). A decrease in acid-soluble polysaccharides also has been reported (ISHIHARA, 1957).

There are various opinions as to the pathways of carbohydrate breakdown in sea-urchin eggs. In animal cells, glycogen is introduced into catabolism via G-6-P, either by phosphorylase followed by mutase or by amylase and subsequent phosphorylation by hexokinase. G-6-P is then broken down via the glycolytic pathway followed by the KREBS cycle or via the pentose phosphate shunt.

One explanation is that phosphorylase is the rate-limiting enzyme in sea-urchin eggs, and its activation may take place after fertilization (AKETA et al., 1964; YASUMASU et al., 1973).

An analysis of sea-urchin egg homogenate (0.25 mol sucrose + 0.04 mol glycylglycine buffer, pH 8.2 + 0.10 mol NaF + 0.01 mol Na_4-EDTA) shows the presence of glycogen phosphorylase, by measuring G-1-P formed in the following medium: 0.1-0.2 ml homogenate + 0.3 ml of 0.083 mol phosphate buffer, pH 6.5 + 0.833% glycogen + 0.017 mol NaF + 0.50% bovine serum albumin + 0.017 mol Na_4-EDTA (+ 3.33 millimol 5'-AMP). Total phosphorylase (+ AMP) and active phosphorylase (- AMP) are, respectively, 0.16 and 0.10 unit/gm total *N* in unfertilized *Sphaerechinus* eggs, 0.18 and 0.10 in newly fertilized eggs, 0.15 and 0.09 in blastulae and 0.09 and 0.06 in plutei. (Muscles from adult sea urchins contained 0.50 and 0.44 unit). Phosphorylase activity remains unchanged at fertilization. However, the enzyme in unfertilized *Paracentrotus* eggs is localized in a subcellular structure sedimentable mostly by centrifugation at 3,600 g from homogenate and releases up to 20,000 g of supernatant at fertilization, a fact that may have some bearing on the activation of carbohydrate metabolism at fertilization (BERGAMI et al., 1968).

Adenyl cyclase occurs in sea-urchin eggs especially in the plasma membrane. Catalytic conversion of ATP to cyclic AMP indicates that adenyl cyclase regulates phosphorylase and glycogenolytic activity in the cell. Activity is very slight before fertilization but gradually increases afterwards and reaches a plateau after 30 minutes (Fig. 18.2a). The supernatant is much less active than the pellet. NaF is known to activate adenyl cyclase maximally and greatly increases activity in unfertilized eggs (Fig. 18.2b). In NaF-treated preparations, after fertilization the activity drops to less than half what it is in unfertilized eggs, apparently because enzymes are lost or, less likely, because responsiveness to NaF is reduced. Cyclic phosphodiesterase, which destroys cyclic AMP, may also be present (CASTAÑEDA and TYLER, 1968).

Arbacia egg homogenate (in 0.4 mol KCl, 0.01 mol Na-citrate, 0.05 mol glycylglycine, pH 7.5) consumes hexoses and equimolar amounts of ATP in the presence of 10 millimol $MgCl_2$ (KRAHL et al., 1953, 1954b). *Echinocardium* eggs also show hexokinase activity which is enhanced with 0.02 mol phosphate and weakly inhibited by 0.05 mol NaF. Located mainly in the supernatant fraction (20,000 g, 30 minutes), hexokinase is not associated with particles.

In sea-urchin eggs, hexokinase phosphorylates glucose (relative rate, 1), 2-deoxyglucose (2), fructose (1.8), mannose (1.2), and glucosamine (0.6). Practically no phosphorylation occurs with F-6-P, G-6-P, and galactose. Hexokinase activity increases during development: relative values (μg glucose consumed/hr/mg protein) were 67 in unfertilized eggs, 72 in eggs 1 hour after fertilization, 94 in 24-hour embryos,

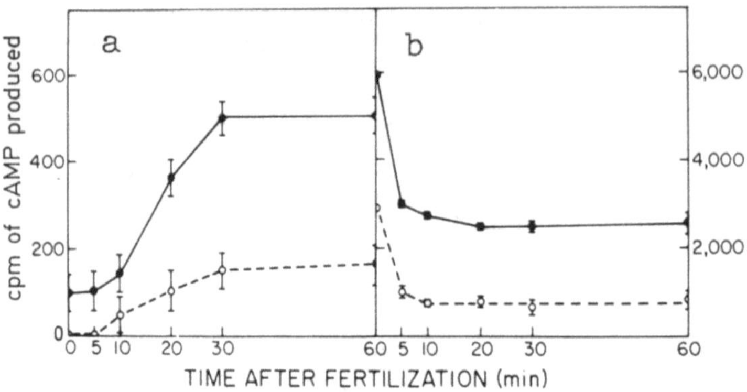

Fig. 18.2 a and b. Adenyl cyclase activity of *Lytechinus pictus* membrane preparations (—●—) and supernatant (--o--) after fertilization without (a) and with (b) 10 millimol NaF. Enzyme activity is measured as rate of cAMP production at 30°C in system containing (in 0.2 ml) 30 mmol tris (pH 7.6), 5 mmol $MgSO_4$, 15 mmol theophylline, 5 mmol β-mercaptoethanol, 1 mmol ATP (1 μCi ^{14}C at carbon atom 8), 10 mmol phosphoenol pyruvate, 5 μg PEP-kinase and 350 μg of enzyme. (From CASTAÑEDA and TYLER, 1968)

and 226 in 48-hour plutei at 27°C. If the turnover number is assumed to be the same as for crystalline hexokinase of yeast (13,000 mol/min/ 10^5 gm protein), 0.005% of the protein extracted from eggs - or about 0.001% of the total egg protein - is hexokinase (KRAHL et al., 1954b).

When glucose used by intact eggs and embryos is measured in seawater containing 0.001 mol glucose, unfertilized eggs consume about one-third as much glucose, and plutei about as much glucose, as would be expected from the hexokinase activity of homogenates. Hexokinase activity in intact fertilized and unfertilized eggs and embryos is supported by rapid phosphorylation of ^{14}C-glucose absorbed from seawater (KELTCH et al., 1956; KRAHL, 1956; MATSUNAGA and YANAGISAWA, 1970).

18.1.3 Pentose Phosphate Shunt

Although a knowledge of resting-egg metabolism is essential for understanding the metabolic changes that occur at fertilization and during development, it is still not definitely known whether the main pathway of carbohydrate breakdown in sea-urchin eggs is by classic glycolysis or by the pentose phosphate shunt.

That monoiodoacetate (0.03 mol), an inhibitor of glycolysis, does not inhibit normal respiratory increase at fertilization suggests that glycolysis does not influence respiration at fertilization. Its respiratory inhibition in early embryo may be prevented by the addition of pyruvate or lactate (RUNNSTRÖM, 1935d). Moreover glycogen decrease after fertilization is not inhibited by 0.02 mol NaF and 1 millimol iodoacetate but is inhibited by 4 millimol phloridizin, indicating that carbohydrates are metabolized by hexose-6-phosphate and phosphogluconate (ÖRSTRÖM and LINDBERG, 1940; LINDBERG, 1943; RUNNSTRÖM, 1935c).

In the homogenate of *Strongylocentrotus droebachiensis* eggs, glucose is oxidized to form an energy-rich phosphate (probably ATP) in the pres-

Table 18.2. The effect of monoiodoacetic acid on respiration and phosphorylation in the presence of glucose, hexosemonophosphate, and hexosediphosphate in homogenized sea-urchin eggs. Each vessel contained 0.1 mg NAD, 0.5 mg AMP, and 0.03 mol NaF. Glucose concentration = 100 μmol, HMP = 50, HDP = 50, and monoiodoacetic acid = 0.005 mol. (From LINDBERG and ERNSTER, 1948)

	μmol	Without IAA				With IAA			
		H_2O	Glucose	HMP	HDP	H_2O	Glucose	HMP	HDP
20-min incubation	Oxygen	0.7	2.5	2.4	0.8	1.1	2.8	2.3	1.6
	P uptake	0.6	3.4	0.7	2.2	0.4	3.0	0.2	1.2
30-min incubation	Oxygen		2.3	2.2	2.1		3.2	2.9	1.6
	P uptake		7.3	3.6	2.0		6.2	2.5	1.1
45-min incubation	Oxygen	0.6	3.5	2.5	0.7	1.4	4.2	2.7	1.5

ence of adenylate. The rate of glucose oxidation is at least as rapid as the oxidation of hexose monophosphate and considerably faster than the oxidation of hexose diphosphate. Monoiodoacetic acid does not inhibit glucose oxidation, and when hexose diphosphate is incubated with homogenized eggs in the presence of monoiodoacetic acid, alkali-labile phosphate (triose phosphate) does not form. This implies glucose oxidation via the pentose phosphate shunt. NAD and pyocyanine stimulate glucose oxidation markedly; it is possible that NAD is phosphorylated to NADP in the homogenate (Table 18.2)(LINDBERG and ERNSTER, 1948).

That sea-urchin egg glycogen is broken down via the pentose phosphate shunt is suggested by the disappearance of carbohydrate; the absence of aerobic accumulation of lactate or pyruvate; the fact that monoiodoacetate does not inhibit glucose oxidation; the absence of arginine and creatine phosphate; the stimulation of glucose oxidation when pyocyanine or cozymase is added; the transient increase in the pentose content of cytolyzed eggs; hexose monophosphate and phosphogluconic acid stimulation of pentose formation; the absence of alkali-labile phosphate (triose phosphate) formation when hexose diphosphate is incubated with homogenized eggs and monoiodoacetate; the fact that phosphogluconic acid is as good a substrate as glucose; and the absence of triose phosphate dehydrogenase (RUNNSTRÖM, 1933, 1949; JANDORFF and KRAHL, 1942; LINDBERG, 1943; LINDBERG and ERNSTER, 1948). But this evidence is inconclusive. It is low monoiodoacetate concentrations that fail to inhibit glucose oxidation (ROTHSCHILD, 1951).

The supernatant of *Arbacia* egg homogenate contains glucose-6-phosphate dehydrogenase (G6PDH) and 6-phosphogluconate dehydrogenase (6PGDH) determined from the rates of NADP reduction (ΔE_{340} nm), but the particulate fraction itself is inactive (KRAHL et al., 1954a, 1955).

With G-6-P as a substrate, the supernatant fraction from 1 gm wet weight of fertilized or unfertilized eggs reduces NADP at a rate of 1.8 - 3 μmol/minute. This is sufficient to support an oxygen consumption rate (1,700 μl/hour/gm egg) 24 times as great as that of unfertilized eggs (70 μl) and 6 times as great as that of fertilized eggs (310 μl). Pentose is formed from G-6-P at a rate of 0.3 to 0.5 times as great as that of NADP reduction. [In cytolysate of *Echinocardium* eggs, pentose forms at the expense of phosphogluconate and hexose-

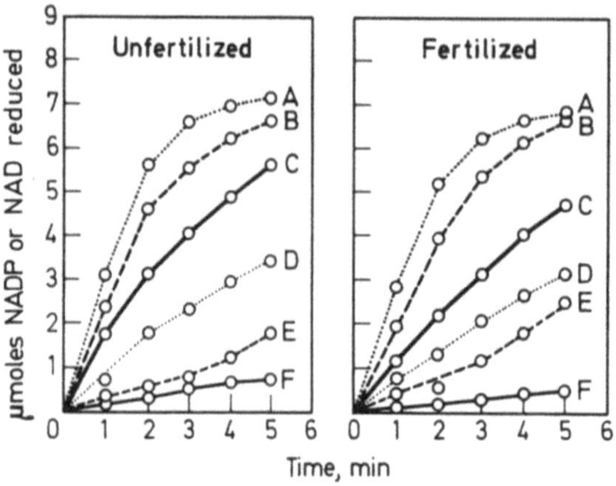

Fig. 18.3. NADP or NAD reduction by supernatant fraction from 1 gm wet weight unfertilized or fertilized *Arbacia* eggs. Fraction is prepared from homogenate in 0.4 mol KCl in 4 millimol EDTA and 0.05 mol glycylglycine (pH 7.5) by centrifugation at 20,000 g for 30 minutes at 5°C. Initial concentrations of reaction system are $MgCl_2$, 0.03 mol; glycylglycine, 0.025 mol; KCl, 0.08 mol; substrate, 0.001 mol; and NADP or NAD, 0.00004 mol. Total volume: 2.17 ml; 0.05 ml egg fraction from 0.01 ml eggs; pH 7.4, 23°C. (A) NADP reduction with glucose-6-phosphate. (B) fructose-6-phosphate. (C) 6-phosphogluconate. (D) fructose-1,6-diphosphate. (E) glucose-1-phosphate. (F) NAD reduction with fructose-1,6-diphosphate. (From KRAHL et al., 1955)

6-phosphate (LINDBERG, 1943).] Concentrations of G-6-P and 6-phosphogluconate for half maximal activity are approximately 4×10^{-5} mol. Maximal NADP reduction toward 6-phosphogluconate is 50% to 60% what it is toward G-6-P, and G6PDH activity is inhibited 50% in the presence of 6×10^{-5} mol 2,4,5-trichlorophenol (KRAHL et al., 1955).

At 1 millimol, other substrates reduce NADP at the following rates: G-1-P, 0.4 µmol/minute; F-6-P, 2.6 µmol/minute; F-1,6-P, 0.8 µmol/minute, indicating in the egg homogenate the presence of phosphoglucomutase, phosphoglucoisomerase, and phosphofructokinase, respectively. Reduced NAD in the presence of F-1,6-P, and ADP is 0.1 to 0.2 µmol/minute/gm wet eggs, showing that the glycolytic pathway metabolizes G-6-P at about 5% of the rate at which G-6-P is oxidized by the NADP system of fertilized or unfertilized *Arbacia* eggs (Fig. 18.3). G6PDH activity does not change significantly at fertilization (KRAHL et al., 1955).

G-6-P, 6-phosphogluconate, and, to a lesser extent, HDP, are converted to pentose during incubation with homogenates in 0.1 mol tris buffer (pH 7.8) of unfertilized *Pseudocentrotus* and *Hemicentrotus* eggs, indicating the pentose phosphate shunt. The conversion rate does not alter after fertilization (ISHIHARA, 1958b).

G6PDH activity in unfertilized *Paracentrotus* eggs is very slight or even nonexistent (BÄCKSTRÖM, 1959). However, the low value may be due to incomplete extraction of the bound enzyme (ISONO, 1963c). After fertilization, G6PDH activity increases rapidly in a few hours, peaking at the cleavage stage and again after hatching. The first peak corresponds to the period of determining the relative size of the ectoderm, and the second peak may be important to endomesoderm development, although

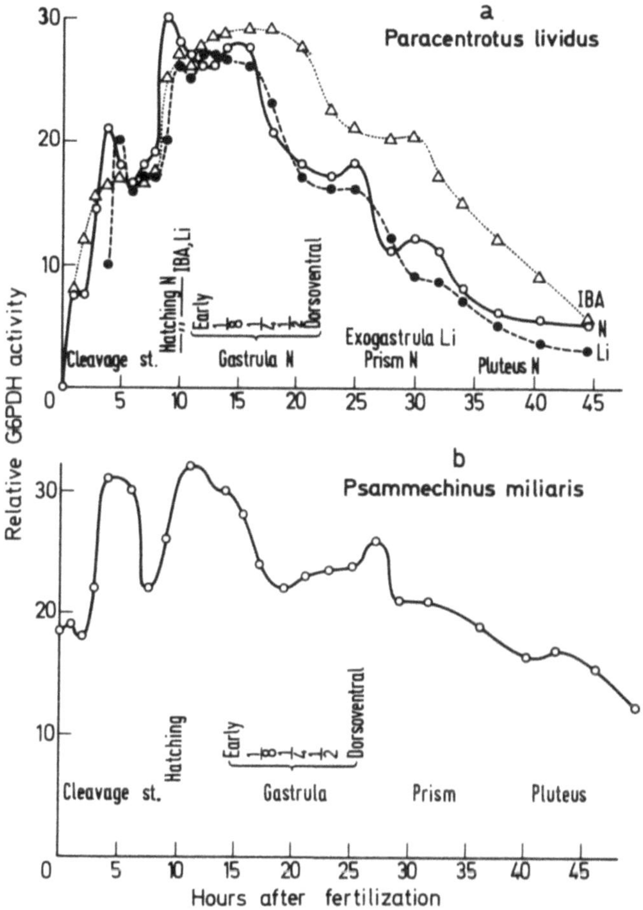

Fig. 18.4 a and b. Changes in G6PDH activity in early development of sea-urchin eggs. (a) *Paracentrotus lividus*: normal (—o—), vegetalized (--●--), and animalized (..△..). (b) *Psammechinus miliaris*. Assays of enzyme activity were carried out in spectrophotometer at 340 nm in 0.3 ml, 0.04 mol glycylglycine buffer (pH 7.5); 0.3 ml, 0.1 mol $MgCl_2$; 0.3 ml egg extract; 1.8 ml H_2O; 0.15 ml, 0.02 mol G-6-P; and 0.15 ml, 1.5×10^{-3} mol NADP. (From BÄCKSTRÖM, 1959)

by this time the relative size of the germ layers is already determined. In later stages, G6PDH activity decreases stepwise (Fig. 18.4).

Before hatching, larvae with hypertrophied ectoderm (cultured in 0.4 - 0.8 mmol of iodosobenzoate) or endoderm (treated for 3 hours, from the 8- to the 16-cell stage, with 10%-15% LiCl) show an enzyme activity pattern similar to that of normal larvae. After hatching, animalized larvae have a higher level of enzyme activity than vegetalized larvae (Fig. 18.4)(BÄCKSTRÖM, 1959).

Although unfertilized egg activity is much higher in *Psammechinus*, the general trend of enzyme activity change during embryo development is the same as that of *Paracentrotus* (Fig. 18.4b). And 6PGDH activity during *Paracentrotus* egg development is almost identical to G6PDH activity (Fig. 18.5)(BÄCKSTRÖM, 1963).

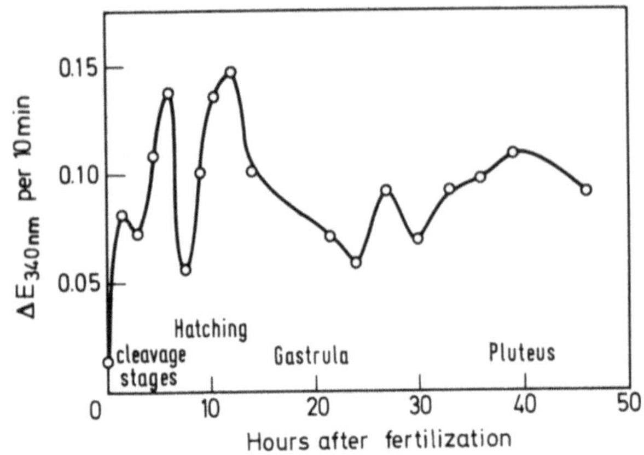

Fig. 18.5. Changes in G6PDH activity during early development of *Paracentrotus lividus*. Same assay system as Fig. 18.4 except that 6-phosphogluconate is used as substrate. (From BÄCKSTRÖM, 1963)

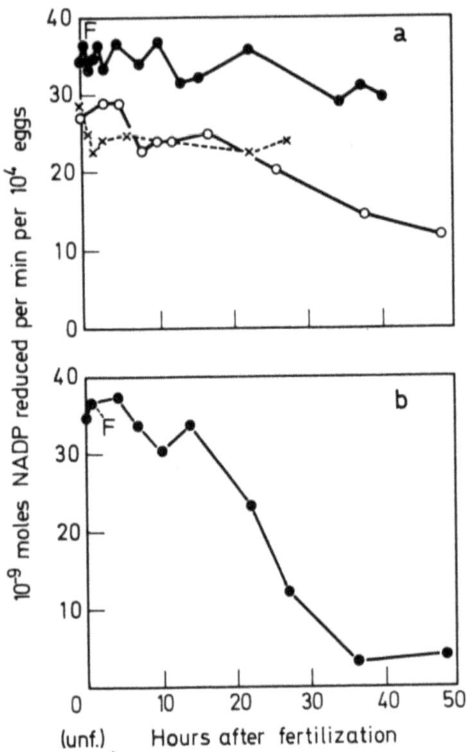

Fig. 18.6 a and b. Changes in G6PDH (solid line) and 6PGDH (broken line) activity during early development of sea-urchin eggs. (a) *Hemicentrotus pulcherrimus* cultured at 20°C; activity determined at 19°C (●), 18°C (x), and 13°C (o). (b) *Pseudocentrotus depressus* cultured at 20°C; activity determined at 18°C. Assay mixture contains 1.0 ml, 0.1 mol tris-HCl buffer (pH 8.75); 0.3 ml, 0.1 mol $MgCl_2$; 0.5 ml, 0.01 mol G-6-P or 6-phosphogluconate; 0.2 ml egg homogenate in water (10^4 eggs); 0.8 ml H_2O; and 0.2 ml, 1 millimol NADP. (From ISONO, 1962)

In the water homogenates of unfertilized *Hemicentrotus* eggs, G6PDH and 6PGDH reduce NADP at a rate of 34 mμmol/minute at 15°C, and in *Anthocidaris* they reduce it at a rate of 9 mμmol/minute (ISONO and ISHIDA, 1962). In *Hemicentrotus* and *Pseudocentrotus*, G6PDH and 6PGDH activity does not change markedly from the unfertilized egg to the hatching stage, after which it declines (Fig. 18.6)(KRAHL et al., 1955; ISHIHARA, 1958b; ISONO, 1962).

Based on these facts, it seems that the G6PDH system is the major pathway of carbohydrate utilization in sea-urchin eggs.

Ribose-5-phosphate added to cell-free extracts of *Arbacia lixula* and *Paracentrotus lividus* eggs (12,000 rpm supernatant of homogenate in 0.01 mol phosphate buffer, pH 7.2) disappears rapidly, possibly because G-6-P, via the G6PDH system, is further degraded. Its disappearance, shown by orcinol reaction measured at 660 nm, is accompanied by a definite increase in light absorption at 580 nm, the absorption peak of heptulose and an indication of the presence of transketolase. Ribulose-5-phosphate formed by phosphopentose isomerase shows the 540-nm peak in the cysteine-carbazole reaction, while ketopentose appears after 15 minutes, becomes more concentrated up to 60 minutes, and is later replaced by fructose, as indicated by a peak shift to 560 nm (GHIRETTI and D'AMELIO, 1956).

In the homogenate of unfertilized *Anthocidaris* and *Hemicentrotus* eggs, at pH 8.5 over 60% of R-5-P (10 μmol) disappears within 5 minutes at 37°C, and all of it disappears after 30 minutes. From unfertilized eggs to plutei the rate of R-5-P elimination from homogenates does not change (ISONO and ISHIDA, 1962).

Sequential alteration of specific absorption peaks in the resorcinol test shows that R-5-P first converts to a ketopentose (peak: 630-640 nm; probable form: ribulose-5-phosphate or xylulose-5-phosphate) and then to a ketoheptose (peak: 610 nm; probable form: sedoheptulose-7-phosphate). After prolonged incubation, ketohexose (peak: 470-480 nm) forms from ketoheptose. After 30 minutes of incubation ketoheptose and ketopentose production in the reaction mixture is demonstrated by absorption peaks at 505 and 415 nm, appearing after 24 hours of the H_2SO_4-cysteine test.

Thus, the complete pentose phosphate cycle seems to occur in sea-urchin eggs (Fig. 18.7).

18.1.4 Glycolysis

Although it has been maintained that monoiodoacetic acid (IAA) cannot inhibit the respiration of sea-urchin eggs (RUNNSTRÖM, 1935c, 1949; LINDBERG and ERNSTER, 1948), IAA (a representative inhibitor of glycolysis at the point of triose phosphate dehydrogenase) not only inhibits the respiration of the egg homogenate but also that of the intact egg, if applied in sufficiently high concentration. In the homogenate of unfertilized *Echinus esculentus* eggs during 90 minutes of incubation, 10^{-3} mol IAA inhibits respiration as much as 22%, while 10^{-2} mol IAA inhibits it as much as 46% (CLELAND and ROTHSCHILD, 1952b). During 3 hours of incubation, 2 x 10^{-3} mol IAA inhibits the respiration of unfertilized *Hemicentrotus* eggs 40%, fertilized *Anthocidaris* eggs 19%, and unfertilized *Anthocidaris* eggs 44% (ISHIHARA, 1958a, 1958b).

Although in oxygenated seawater sea-urchin eggs contain only traces of lactate (CLELAND and ROTHSCHILD, 1952a; YČAS, 1954), whole sea-

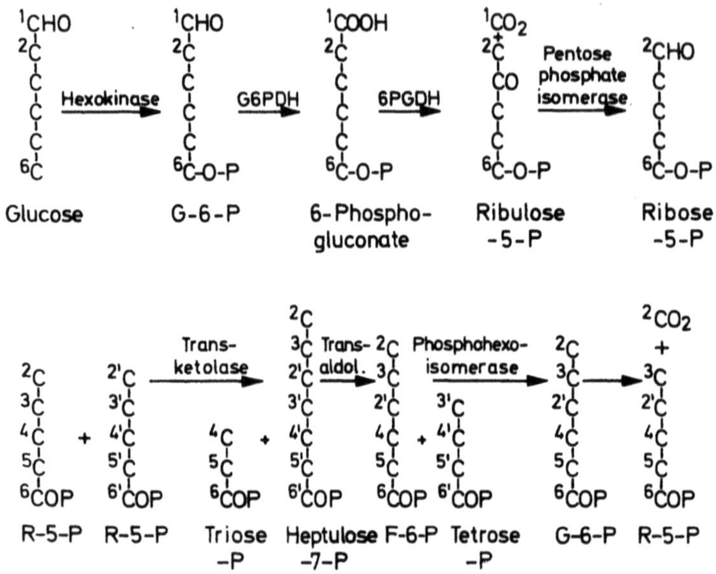

Fig. 18.7. Pathway for glucose oxidation via pentose phosphate shunt. Related enzymes: glucose-6-phosphate dehydrogenase[1-5], 6-phosphogluconate dehydrogenase[1-4], pentose phosphate isomerase[4,6], ketopentose epimerase[4], transketolase[4,6], transaldolase[4]. 1. LINDBERG, 1943; 2. KRAHL et al., 1955; 3. BÄCKSTRÖM, 1959; 4. ISONO and ISHIDA, 1962; 5. ISONO, 1963b; 6. GHIRETTI and D'AMELIO, 1956

urchin eggs are capable of anaerobic lactate production. The lactic acid content of unfertilized *Arbacia punctulata* eggs (3.14 mg/gm of egg protein) increases as much as 81% during 1 to 6 hours of respiratory inhibition with 0.02 mol KCN. Fertilized eggs are only slightly affected (PERLZWEIG and BARRON, 1928).

Echinus esculentus eggs produce 300 μg of lactic acid per milliliter of eggs during 4 hours of anaerobiosis in N_2 gas after 1 hour of induction but produce only 3 μg of pyruvic acid per milliliter of eggs (CLELAND and ROTHSCHILD, 1952a). *Strongylocentrotus purpuratus* eggs accumulate 860 μg of lactate per milliliter of eggs during 3 hours of anaerobiosis at 23°C. Lactate production in intact eggs is suppressed by 10^{-2} mol IAA (YČAS, 1954).

Even in oxygenated seawater, lactic acid accumulates immediately after fertilization in *Hemicentrotus pulcherrimus*, *Anthocidaris crassispina*, *Pseudocentrotus depressus* eggs (Fig. 18.8a). 2-hour egg pretreatment with 5×10^{-3} mol phloridzin inhibits lactic acid production, while pretreatment with 10^{-3} - 5×10^{-4} mol IAA and 10^{-2} mol NaF does not (AKETA, 1957). In *Echinus esculentus* eggs, however, the lactic acid level remains constant throughout fertilization: 15 μg per milliliter of unfertilized eggs and 16 μg per milliliter of fertilized eggs 2 to 10 minutes after insemination (ROTHSCHILD, 1958).

In *Arbacia lixula*, lactic acid accumulates markedly in 2 minutes after fertilization but returns to the initial level in 5 minutes. In *Paracentrotus lividus* it does not change significantly, and in *Psammechinus miliaris* it remains unchanged for 2 minutes after fertilization and then increases considerably for the next 3 minutes (AKETA, 1964).

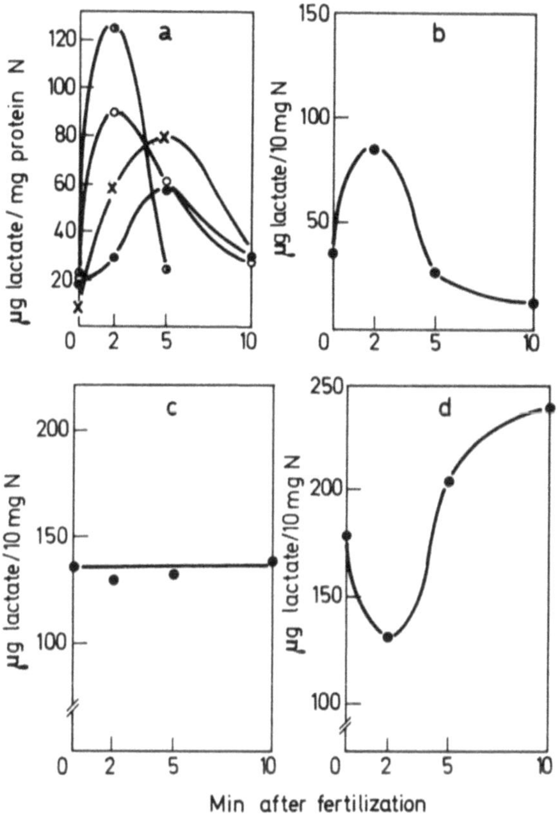

Fig. 18.8a-d. Changes in lactic acid content of sea-urchin eggs at fertilization. (a) *Hemicentrotus pulcherrimus* cultured at 18°-20°C; 4 measurements; 120 µg lactic acid/mg protein-N = 10 mg lactic acid/ 100 mg N. (b) *Arbacia lixula.* (c) *Paracentrotus lividus* (Palermo). (d) *Psammechinus miliaris* (S-form, Kristineberg). (From AKETA, 1957, 1964)

In egg homogenates after carbohydrates are added, NAD and ATP stimulate oxygen uptake and anaerobic lactate and pyruvate production (Tables 18.3 - 18.6), while IAA and fluoride inhibit them (Tables 18.7, 18.8) (CLELAND and ROTHSCHILD, 1952a, 1952b). Acid production from these carbohydrates implies the presence of phosphorylase and hexokinase. Phloridzin inhibition of lactate production from glycogen also indicates the presence of phosphorylase (ROTHSCHILD, 1937).

Table 18.3. The effect of carbohydrates on oxygen uptake in *Echinus esculentus* egg homogenates. (From CLELAND and ROTHSCHILD, 1952b)

Addition	µl O_2/75 min	Percent
None	110	100
Glycogen (10^{-2} mol)[a]	121	110
Glucose (10^{-2} mol)	138	125
Fructose (10^{-2} mol)	153	139

[a]Glucose equivalent.

Table 18.4. Comparison of *Echinus esculentus* acid production under anaerobic conditions, using different substrates. Each THUNBERG tube contained 1 ml of 50% extract (homogenate supernatant), 200 μg NAD, 1 mg ATP, *M*/20 phosphate buffer (pH 7.5). Final volume: 2 ml. Substrate concentration: *M*/40 (or *M*/40 glucose equivalent in the case of starch and glycogen). Incubation: 60 min at 20°C. (From CLELAND and ROTHSCHILD, 1952a)

Substrate (μg)	None	Glycogen	Starch	Glucose	Fructose	G-1-P
Lactic acid	137	239	273	334	337	275
Pyruvic acid	93	149	147	162	176	234
Total	230	384	420	496	513	509
Percent	100	167	183	216	223	221

Table 18.5. The effect of cofactors on *Echinus esculentus* acid production. Each THUNBERG tube contained *M*/20 phosphate buffer, pH 7.5. Cofactors added: glycogen, 12 mg; NAD, 200 μg; ATP, 1 mg; AMP, 1 mg; 1 ml of 50% or 30% extracts. Final volume: 2 ml. Incubation: 60 min at 20°C. (From CLELAND and ROTHSCHILD, 1952a)

	50% Extract			30% Extract		
Extract plus	Lactic acid (μg)	Pyruvic acid (μg)	Total	Lactic acid (μg)	Pyruvic acid (μg)	Total
	-8	55	47	0	5	5
Glycogen	28	119	147	3	8	11
Glycogen + NAD	244	14	258	44	27	71
Glycogen + ATP	66	147	213	7	6	13
Glycogen + ATP + AMP	220	24	244	40	67	107
NAD + ATP	193	17	210	40	45	85
Glycogen + NAD + AMP	-	-	-	40	25	65

In egg homogenates the intermediate products of glycolysis accelerate oxygen consumption and break down rapidly into lactate and pyruvate (Tables 18.9, 18.10). And as might be expected in a complex enzyme system containing labile enzymes, the rate of acid production may vary from one preparation to the next. That the intermediates from G-1-P to HDP produce acids less effectively than phosphoglycerate may indicate that some of the enzymes in the glycolytic pathway - phosphoglucomutase and hexose isomerase in particular - are labile. As phosphogluconate does not affect acid production, acids evidently are not formed via the pentose phosphate shunt (CLELAND and ROTHSCHILD, 1952a, 1952b). In the homogenate of *Arbacia punctulata* eggs, phosphohexoisomerase is 20 times as active as hexokinase (KRAHL et al., 1953).

In the homogenates of *Lytechinus pictus* and *Strongylocentrotus purpuratus* eggs prepared with *M*/15 phosphate buffer (pH 6.8), 9 μmol of G-1-P are consumed during 90 minutes of incubation at 23°C and produce

Table 18.6. The effect of cofactors on endogenous oxygen uptake and on fructose oxidation by egg homogenates in *Echinus esculentus*. (From CLELAND and ROTHSCHILD, 1952b)

Homogenate plus	µl O_2/120 min (20°C)
	122
NAD (200 µg)	157
ATP (1 mg)	139
ATP + NAD	179
Fructose (10^{-2} mol)	184
Fructose + NAD	262
Fructose + NAD + ATP	272
ATP + NAD + IAA (5 x 10^{-3} mol)	115
Fructose + NAD + ATP + IAA	135

Table 18.7. The effect of inhibitors on the reconstructed glycolytic system of *Echinus esculentus*. Each THUNBERG tube contained 1 ml of 40% egg extract, *M*/20 phosphate buffer (pH 7.5), 200 µg NAD, 1 mg ATP, and 12 mg soluble starch. Final volume: 2 ml. Incubation: 60 min at 20°C. (From CLELAND and ROTHSCHILD, 1952a)

Inhibitor	None	Iodoacetate		Fluoride		Phloridzin	Phenylmercuric nitrate
(µg)		10^{-2} mol	10^{-3} mol	10^{-2} mol	10^{-3} mol	10^{-2} mol	10^{-3} mol
Lactic acid	111	18	80	20	82	48	3
Pyruvic acid	78	4	5	-4	-36	62	8
Total	189	22	85	16	46	110	11

Table 18.8. The effect of inhibitors on fructose oxidation by egg homogenates in *Echinus esculentus*. (From CLELAND and ROTHSCHILD, 1952b)

Homogenate plus	µl O_2/120 min	Excess O_2	Inhibition (%)
	183	-	-
Fructose (10^{-2} mol)	234	51	-
Fluoride (5 x 10^{-3} mol)	141	-	23
Fluoride + fructose	154	13	75
IAA (5 x 10^{-3} mol)	125	-	32
IAA + fructose	140	15	71

4.3 µmol of fructose ester, probably F-6-P; 12 µmol of F-6-P disappear in 7 minutes and produce 5 µmol of a stable compound, presumably G-6-P, which resists 180-minute hydrolysis in *N* HCl at 100°C; as determined colormetrically (SIBLEY and LEHNINGER, 1949), FDP is degraded to triose phosphate at pH 8.6; sodium phosphoglycerate (0.02 mol) is converted to 0.30 mg phosphoenolpyruvate in 7 minutes

Table 18.9. Acid production in *Echinus esculentus* using phosphorylated intermediates as glycolytic substrates. Each THUNBERG tube contained 1 ml 45% extract, 200 µg NAD, 1 mg ATP, 1 mg AMP, *M*/20 phosphate buffer (pH 7.5). Final volume: 2 ml. Final substrate concentration: *M*/200 at 20°C. (From CLELAND and ROTHSCHILD, 1952a)

	45 min				60 min (a)				60 min (b)			
	Lactic acid (µg)	Pyruvic acid (µg)	Total (µg)	Greater than control (µg)	Lactic acid (µg)	Pyruvic acid (µg)	Total (µg)	Greater than control (µg)	Lactic acid (µg)	Pyruvic acid (µg)	Total (µg)	Greater than control (µg)
Control	57	60	117	-	21	5	26	-	60	35	95	-
Glucose	121	90	211	94	28	8	36	10	143	55	198	103
G-1-P	101	139	240	123	33	7	40	14	122	68	190	95
G-6-P	95	157	252	135	55	26	81	55	136	108	244	149
P-gluconate	63	68	131	14	4	18	22	-4	61	39	100	5
F-6-P	103	252	355	238	56	117	273	247	176	174	350	255
HDP	136	248	384	267	136	196	332	306	205	232	437	342
P-glycerate	61	670	731	614	29	750	779	753	57	755	812	717

Table 18.10. The effect of phosphorylated intermediates on the oxygen uptake of *Echinus esculentus* homogenates. (From CLELAND and ROTHSCHILD, 1952b)

Substrate	Control (%)
G-1-P, 5 x 10^{-3} mol	115
G-6-P, 5 x 10^{-3} mol	136
Phosphogluconate	122
F-6-P	130
F-1,6-P	146

and is reduced to 0.03 mg by 0.01 mol NaF; endogenous homogenate respiration is inhibited by 0.01 - 0.05 mol NaF. In the methylene blue reduction system, triose phosphate dehydrogenase is sensitive to 10^{-1} or 3 x 10^{-2} mol IAA. As the reaction is very slow, inhibition requires at least 1 hour of preincubation (YČAS, 1954).

These findings indicate the presence of phosphoglucomutase, phosphohexoisomerase, aldolase, triose phosphate dehydrogenase and enolase in egg homogenates (YČAS, 1954).

Aldolase activity occurs in the homogenates of *Echinus esculentus* and *Paracentrotus lividus* eggs and in the latter is constant for about 40 hours of development (GUSTAFSON and HASSELBERG, 1951). In unfertilized eggs, aldolase seems bound to the heavy cytoplasmic fraction (and presumably to the egg cortex) and by centrifugation at 400 g is precipitable from the homogenate in a solution of 0.375 mol sucrose and 0.26 mol KCl. 5 minutes after fertilization it is apparently released from the bound state and becomes soluble and functional. In the next hour, no change in distribution or activity occurs. Aldolase is released when unfertilized eggs are treated for 7 seconds with water or parthenogenetic reagents such as 0.1% choleate, 0.1% saponine, and 1 mol urea. 1-minute treatment with butyric acid-seawater (3 ml 0.1 *N* butyric acid and 50 ml seawater) also induces aldolase release as well as increased oxygen consumption and anaerobic accumulation of lactate; 10-minute treatment cancels these effects. While these results seem to indicate that aldolase release is probably responsible for the sudden respiratory increase of sea-urchin eggs at fertilization, further verification is needed (ISHIHARA, 1957, 1958c).

Chemical analysis shows that *Strongylocentrotus droebachiensis* eggs contain 0.48 μmol R-5-P and about 0.1 μmol G-1-P per milliliter of eggs. But they contain no phosphoenolpyruvate, triose phosphate, or fructose phosphate (MENDE and CHAMBERS, 1953). In *Arbacia* eggs, hexose phosphate is shown by paper chromatograms of the Ba-soluble, ethanol-insoluble fraction, while in *Strongylocentrotus purpuratus* eggs G-6-P and G-1-P are shown by anion exchange chromatography (KRAHL, 1956; GRIFFITHS and WHITELEY, 1964).

Experiments with specific enzyme reaction systems show that concentrations of sugar metabolism intermediates are very low in unfertilized *Arbacia lixula* and *Paracentrotus lividus* eggs (Table 18.11), and the G-6-P level determined with a G6PDH-NADP system increases about 5-fold immediately after fertilization. Almost undetectable in unfertilized eggs when determined with a triose phosphate dehydrogenase-NAD system with or without aldolase, FDP and triose phosphate are detected at fertilization. These results suggest that the low concentration of

Table 18.11. Concentration of glucose-6-phosphate, fructose diphosphate, and triose phosphate in unfertilized and fertilized sea-urchin eggs. Values are expressed as μmol/100 mg of total *N*. Glucose-1-phosphate and pyruvate content is undetectable. (From AKETA et al., 1964; MONROY, 1965)

Species		G-6-P	FDP	Triose phosphate
Arbacia lixula	Unfertilized	0.053		
	Fertilized (5 min)	0.260		
	Unfertilized	0.055		
	Fertilized (10 min)	0.470		
	Unfertilized	0.076	0.003	0.007
	Fertilized (10 min)	0.166	0.014	0.023
Paracentrotus lividus	Unfertilized		0.002	0.002
	Fertilized (5 min)		0.039	0.043
	Unfertilized	0.068		
	Fertilized (8 min)	0.456		
	Unfertilized	0.063		
	Fertilized (12 min)	0.408		

hexose phosphate in unfertilized, slowly respiring eggs results from blockage of a pathway leading from glycogen to G-6-P, and that the sudden rise at fertilization is due to blockage removal. Thus, it seems possible that phosphorylase is also the rate-limited enzyme in sea-urchin eggs and that it is activated after fertilization (AKETA et al., 1964).

NAD and NADP-linked enzyme measurement reveals practically all the glycolytic intermediates in unfertilized *Anthocidaris crassispina* eggs. G6P and F6P increase 4 times in 20 minutes, while the others, from FDP to 2PG and pyruvate, increase 3 to 10 times, 30 minutes after fertilization and then decline. At fertilization, PEP does not show any change, and lactate gradually decreases (Table 18.12)(YASUMASU et al., 1973).

These data show that phosphorylase is activated just after fertilization, and then phosphofructokinase and pyruvate kinase are activated. Evidence for the existence of the glycolytic pathway in sea-urchin eggs is summarized (Table 18.13).

18.1.5 Pathway Contribution to Carbohydrate Metabolism

As the enzymes needed to oxidize G-6-P via the pentose phosphate shunt are in great excess over the enzymes of the glycolytic pathway (KRAHL et al., 1955), it is interesting to measure intact-egg oxidation of glucose-1-^{14}C, glucose-2-^{14}C, and glucose-6-^{14}C.

Glucose molecules oxidized via the glycolytic pathway and KREBS cycle yield the same amount of $^{14}CO_2$ whether labeled in the 1 position or 6 position. But if glucose is metabolized via the pentose-phosphate shunt, carbon atom 1 of the glucose molecule is oxidized as $^{14}CO_2$, and carbon atom 2 is oxidized at the 6PGDH step after a complete cycle (Fig. 18.7). In CO_2 carbon atom 6 appears only if triose phosphate reenters the glycolysis scheme. The ratio of CO_2 from carbon atom 6

Table 18.12. Change in glycolytic intermediate and adenosine nucleotide content of *Anthocidaris crassispina* eggs after fertilization. Values are expressed as mμmol/10^6 eggs. Those marked with an (a) are expressed as μmol/10^6 eggs. All intermediates and adenosine nucleotides are measured with appropriate NAD or NADP-linked enzyme systems. AMP is determined with phosphatase plus adenosine deaminase. (From YASUMASU et al., 1973)

	Unfertilized	Fertilized eggs (time after fertilization)				
	eggs	10 min	20 min	30 min	40 min	60 min
Experiment	6	5	5	5	4	5
G6P	11.6	45.2	40.6	32.4	30.6	21.3
F6P	2.1	8.6	7.4	6.2	6.8	4.0
FDP	3.6	3.9	8.9	9.5	9.0	7.8
DHAP	4.4	4.3	9.8	12.3	14.9	9.4
GA3P	2.1	2.2	3.4	3.6	3.9	2.7
3PG	2.0	2.5	6.6	8.9	7.8	5.0
2PG	1.0	0.7	1.2	1.3	1.2	1.0
PEP	27.3	23.1	24.2	23.9	21.7	12.9
Pyr	9.1	29.4	60.3	50	50.2	79.1
Lac	289.3	280.4	245.6	206.2	189.2	128.3
αGP	0.6		2.4			
6PG	0.7		2.3			
Pi[a]	43	39	37	44	43	47
ATP[a]	1.72	1.74	1.79	1.53	1.47	1.06
ADP	123.5	186.2	255.3	266.4	252.0	236.2
AMP	71.0	98.3	99.8	98.6	95.4	77.5

to CO_2 from carbon atom 1 indicates glycolytic breakdown in relation to total glucose oxidation, and from this the extent of the pentose-phosphate shunt contribution to glucose oxidation can be calculated.

Glucose-1-^{14}C produces $^{14}CO_2$ more rapidly than glucose-2-^{14}C and glucose-6-^{14}C (Table 18.14). The ratio of $^{14}CO_2$ produced by glucose-2-^{14}C or glucose-6-^{14}C to $^{14}CO_2$ produced by glucose-1-^{14}C is 0.31 : 0.07 in unfertilized *Arbacia* eggs, 0.52 : 0.12 25 minutes after fertilization, and 0.42 : 0.28 in 24-hour embryos. This indicates glucose utilization via the pentose phosphate shunt rather than the glycolytic pathway. In later development, $^{14}CO_2$ production by glucose-2-^{14}C, and especially by glucose-6-^{14}C, increases. Dinitrocresol causes glucose-6-^{14}C to produce nearly as much $^{14}CO_2$ as glucose-1-^{14}C, suggesting that glucose utilization is diverted from the pentose phosphate shunt to the glycolytic pathway. Considerable glucose is esterified and even incorporated into the acid-insoluble fraction containing nucleoprotein, a process strongly inhibited by treatment with DNC (KELTCH et al., 1956).

In *Psammechinus miliaris* embryos, the rate of $^{14}CO_2$ production by uniformly labeled glucose (100 μg, 1.08 μCi) and glucose-1-^{14}C (100 μg, 0.18 μCi) increases gradually in the early blastula stage, but falls

Table 18.13. Evidence for the existence of classic glycolytic pathway in sea-urchin eggs. (Based on ROTHSCHILD, 1956)

	GLYCOGEN[1-3,9]	ADENYL CYCLASE [21]
	↓	cyclic AMP
	glucose-1-P	PHOSPHORYLASE[3,20]
	↓	
hexose —HEXOKINASE[6,7] / ATP[a]→	GLUCOSE-6-P[8,13,17]	PHOSPHOGLUCOMUTASE[3-5]
	↓	
	FRUCTOSE-6-P[13]	PHOSPHOHEXO-ISOMERASE[3-5,6,18,19]
	↓	
	DIHYDROXY-ACETONE-P[4,13,17]	PHOSPHOFRUCTO-KINASE[3,5,18]
	D-GLYCERALDE-HYDE-3-P[4,13,17]	ALDOLASE[3-5,14,15,18]
	↓	
	D-1,3-diphosphoglycerate	triose phosphate isomerase
	↓	
	D-3-PHOSPHOGLYCERATE[13]	TRIOSE PHOSPHATE DEHYDROGENASE[3-5]
	↓	
	D-2-PHOSPHOENOLPYRUVATE[13]	transphosphorylase
	↓	
	PYRUVATE[3,4,13]	phosphoglyceromutase
	↓	ENOLASE[3,4]
	LACTATE[3,4,10-13,16]	
		transphosphorylase
		LACTATE DEHYDROGENASE[3,4]

1. ÖRSTRÖM and LINDBERG, 1940
2. HUTCHENS et al., 1942a
3. CLELAND and ROTHSCHILD, 1952a
4. YČAS, 1954
5. KRAHL et al., 1953
6. KRAHL et al., 1954a, 1955
7. KRAHL et al., 1954b
8. KRAHL, 1956
9. BLANCHARD, 1935
10. PERLZWEIG and BARRON, 1928
11. ZIELIŃSKI, 1939
12. RUNNSTRÖM, 1933
13. YASUMASU et al., 1973
14. GUSTAFSON and HASSELBERG, 1951
15. ISHIHARA, 1957
16. AKETA, 1957, 1961a, 1961b, 1964
17. AKETA et al., 1964
18. ISONO and ISHIDA, 1962
19. ISONO, 1963b
20. CASTAÑEDA and TYLER, 1968
21. BERGAMI et al., 1968

[a]Evidence omitted.
Lower case letters: not identified; capital letters: identified, or most probably existing.

sharply a short time before hatching, possibly due to reduced glucose uptake while respiration remains unchanged (LINDAHL, 1939b). Reduced $^{14}CO_2$ output lasts until the mesenchymal blastula stage begins and it again increases. The pattern of $^{14}CO_2$ production by glucose-1-^{14}C, glucose-2-^{14}C, and glucose-6-^{14}C roughly resembles the production pattern of uniformly labeled glucose. Glucose-1-^{14}C produces con-

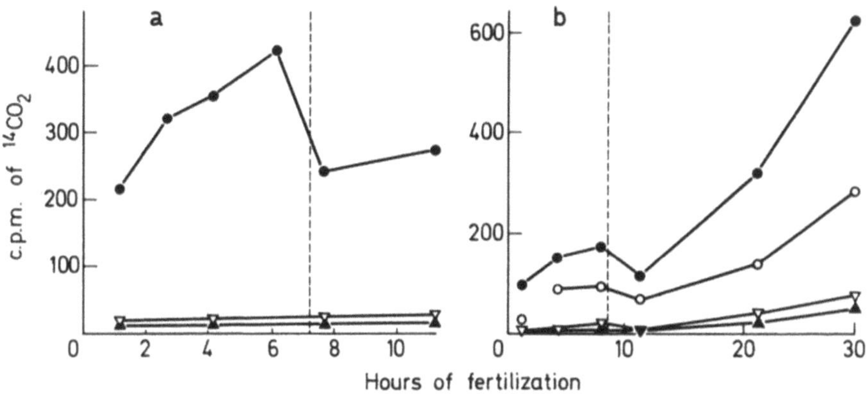

Fig. 18.9 a and b. $^{14}CO_2$ production from *Psammechinus miliaris* eggs at various stages of development with glucose-U-^{14}C (●)(10.8 μCi/mg), glucose-1-^{14}C (o), glucose-2-^{14}C (▽), and glucose-6-^{14}C (▲)(1.8 μCi/mg); incubated with radioactive glucoses at 20°C for 20 minutes; ---, hatching. (From BÄCKSTRÖM et al., 1960)

siderably more $^{14}CO_2$ than do glucose-2-^{14}C and glucose-6-^{14}C. At the 2-cell stage, about 30% of the $^{14}CO_2$ (which expires in the presence of glucose-U-^{14}C) comes from carbon atom 1, and shortly before hatching 45%-65% comes from there. From the late blastula stage on, the predominance of $^{14}CO_2$ from carbon atom 1 declines and by the pluteus stage is almost the same as in the 2-cell stage. Simultaneously, glucose-6-^{14}C produces $^{14}CO_2$ at an increasing rate (Fig. 18.9)(BÄCKSTRÖM et al., 1960).

Glucose-1-^{14}C and glucose-6-^{14}C produce $^{14}CO_2$ in *Pseudocentrotus* embryos at about the same rate as in the embryos of *Arbacia* and *Psammechinus*. The ratio of $^{14}CO_2$ produced by glucose-6-^{14}C to that produced by glucose-1-^{14}C is higher in unfertilized and just fertilized *Pseudocentrotus* eggs than in *Arbacia* eggs and at later stages is higher than in *Psammechinus* eggs. $^{14}CO_2$ production does not slow down at the mesenchymal blastula stage (Fig. 17.20)(ISONO and YASUMASU, 1966, 1968).

There seems to be slightly more glucose oxidation via the pentose phosphate shunt than via the glycolysis-KREBS cycle. Breakdown via the pentose phosphate shunt seems to increase at fertilization and remain on the same level from the 16-cell stage until the early blastula stage, after which it gradually declines. This may indicate that glucose oxidation via the pentose phosphate shunt increases at fertilization and during early cleavage and remains the major route until the gastrula stage. And the glycolytic-KREBS cycle route, low when development begins, gradually increases after hatching and exceeds the pentose phosphate shunt at the prism stage.

Enzymes participating in the pentose phosphate shunt, G6PDH and 6PGDH as well as the R-5-P consuming system, seem to be localized in the supernatant (GHIRETTI and D'AMELIO, 1956; BÄCKSTRÖM, 1959, 1963).

Homogenized in a nonionic medium of 0.75 mol mannitol or sucrose and centrifuged at 24,000 g for 30 minutes, 6PGDH, phosphohexose isomerase, and enzymes involved in R-5-P consumption are found mainly in the supernatant of unfertilized and fertilized *Hemicentrotus* and *Anthocidaris* eggs, and 83% of the G6PDH is associated with the sedimentable frac-

Table 18.14. Oxygen consumption, glucose utilization, and radioactivity recovered in fractions from *Arbacia punctulata* eggs and embryos after 2 hours of incubation with radioactive glucose; the effect of dinitrocresol. Eggs and embryos (88 mg wet weight) are incubated at 20°C for 2 hours in 4 ml seawater containing 644 x 10^3 cpm (2.38 µmol of glucose) of G-1-^{14}C, G-2-^{14}C, or G-6-^{14}C. Oxygen and glucose (µmol) are shown in parentheses. No glucose is liberated into the medium during incubation, and no lactate is formed with or without additional glucose. (From KRAHL, 1956; KELTCH et al., 1956)

	Control			8 x 10^{-6} mol			1.28 x 10^{-4} mol		
DNC	G1C	G2C	G6C	G1C	G2C	G6C	G1C	G2C	G6C
Unfertilized eggs									
O_2 consumed	12 µl (0.54)			76 µl (3.39)			31 µl (1.38)		
Glucose used									
Medium	68 µg (0.38)			118 µg (0.65)			100 µg (0.56)		
Glycogen (egg)	157 µg (0.87)			236 µg (1.31)			162 µg (0.90)		
Total uptake (x 10^3 cpm)	17.2	18.9	15.0	17.2	19.7	16.5	13.3	13.2	10.1
% $^{14}CO_2$ output	34.4	10.0	2.3	65.7	62.5	57.0	49.8	35.7	15.9
% Ba-insoluble	23.3	30.7	35.2	18.6	17.3	24.9	21.1	23.6	27.8
% Ba-soluble, alc-insoluble	34.9	50.3	53.2	13.9	19.3	17.0	28.7	40.2	45.7
% Acid-insoluble	7.6	9.0	9.3	1.8	1.0	1.2	0.5	0.5	0.5
Fertilized eggs (25 min after fertilization)									
O_2 consumed	55 µl (2.46)			139 µl (6.20)			63 µl (2.81)		
Total uptake (x 10^3 cpm)	9.0	12.6	10.6	9.7	10.3	8.3	9.3	-	12.3
% $^{14}CO_2$ output	37.8	15.9	5.4	49.5	42.7	38.6	48.6	44.5	18.0
% Ba-insoluble	14.5	18.3	18.0	9.3	11.7	13.3	13.0	23.4	13.1
% Ba-soluble, alc-insoluble	44.4	63.5	73.7	39.2	44.6	45.8	37.8	-	68.5
% Acid-insoluble	3.3	2.4	2.8	2.1	1.0	2.4	0.5	0.7	0.4
24-hour embryos									
O_2 consumed	143 µl (6.38)			272 µl (12.1)			211 µl (9.42)		
Glucose used									
Medium	91 µg (0.51)			77 µg (0.43)			90 µg (0.50)		
Glycogen (egg)	235 µg (1.30)			287 µg (1.59)			235 µg (1.30)		
Total uptake (x 10^3 cpm)	139.0	148.0	137.0	122.2	145.0	140.0	101.7	115.7	113.6
% $^{14}CO_2$ output	37.4	16.0	13.1	67.8	60.0	47.8	86.5	84.6	72.1
% Ba-insoluble	21.6	23.6	25.6	13.9	16.6	17.2	7.8	6.7	14.1
% Ba-soluble, alc-insoluble	18.0	27.0	30.6	5.9	11.0	22.8	7.1	7.4	12.3
% Acid-insoluble	23.0	33.2	30.6	12.3	12.4	12.2	1.6	1.2	1.4

% means distribution of radioactivity in fractions.
"Ba-soluble, alc-insoluble" means Barium salt soluble in water but insoluble in 80% ethanol.

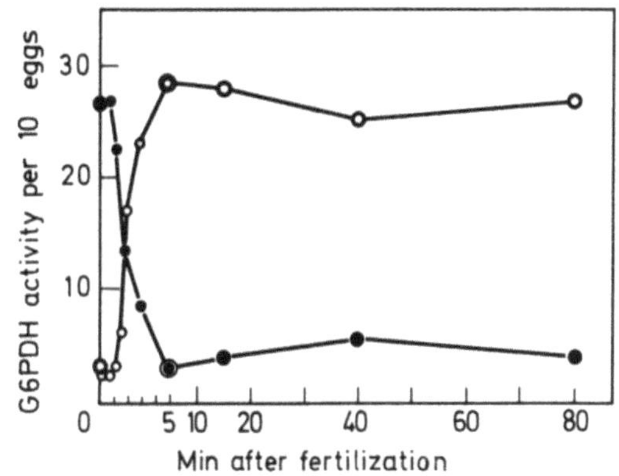

Fig. 18.10. Release of glucose-6-phosphate dehydrogenase after fertilization of *Anthocidaris crassispina* (0-5 minutes; o,•) and *Hemicentrotus pulcherrimus* (0-80 minutes; O,●) eggs. Enzyme in supernatant fraction (O,o); enzyme in particulate fraction (●,•); both obtained by centrifugation at 24,000 g for 20-30 minutes from egg homogenate in 0.75 mol mannitol. (From ISONO, 1963b, 1969)

tion of unfertilized eggs. After fertilization, 81.3% of the G6PDH appears in the supernatant fraction after centrifugation at 24,000 g for 30 minutes or 100,000 g for 2 hours. The release begins within 1 1/2 minutes after fertilization and is completed within 5 minutes (Fig. 18.10)(ISONO, 1963b, 1963c, 1969; ISONO et al., 1963).

Regardless of the kinds of cations and anions, G6PDH is released from precipitate fractions treated for 10 minutes in the cold with a salt solution of ionic strength higher than 0.15 - 0.20. At 1 quarter of this ionic strength, the enzyme released recombines into a precipitable fraction within 10 minutes. Partial release results from fraction treatment with 0.01 mol EDTA, 0.05 mol oxalate, or 0.01% Duponol-PC. Egg jelly, the supernatant fraction of fertilized eggs, water, 0.75 mol mannitol, 0.01 - 0.05 mol cysteine, 0.001% - 0.01% crude trypsin, or 10% acetone do not cause enzyme release.

G6PDH is released at a pH higher than 8, and is released at a faster rate as the pH rises. It reaches 72% (tris buffer) and 56% (veronal buffer) at pH 9.5, but remains in a bound state at pH values lower than 7. Release by alkalinity (for example, pH 8.75) is inhibited by divalent cations such as 0.01 mol $MgCl_2$ or $CaCl_2$. Monovalent cations do not inhibit release but antagonize release with divalent cations.

In 0.01 mol tris buffer and 0.01 mol $MgCl_2$ (pH 8.75), G6PDH is released with 1 millimol of G-6-P, galactose-6-P, 6-phosphogluconate, 0.1 millimol NADP, or $NADPH_2$. Related compounds such as glucose, G-1-P, glucose-6-sulfate, 2-deoxyglucose-6-phosphate, glucosamine-6-phosphate, mannose-6-phosphate, R-5-P, FDP, NAD, $NADH_2$, AMP, ATP, and inorganic phosphate do not affect release. At pH 7, 1 millimol of G-6-P or 0.1 millimol of NADP alone releases only 5%-10% of the enzyme, but in combination they release 90% (ISONO, 1963b, 1963c, 1967a, 1967b; ISONO et al., 1963).

Ionic strength in unfertilized eggs is 0.13 and in fertilized eggs is 0.20 (CHAMBERS et al., 1948; ROTHSCHILD and BARNES, 1953; TYLER et

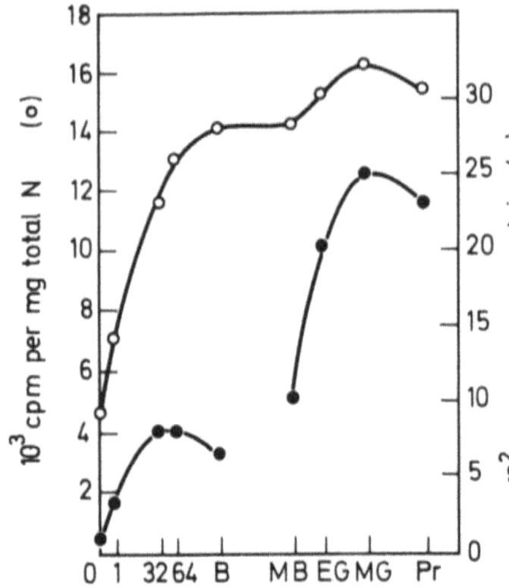

Fig. 18.11. Total ^{14}C uptake from glucose-U-^{14}C in TCA-soluble fraction (o) and incorporation into protein (•) of *Paracentrotus lividus* eggs and embryos cultured in seawater containing 1 μCi of glucose-U-^{14}C (10 μCi/mg glucose) per milliliter at 18°C for 1 hour. (From MONROY and VITTORELLI, 1962)

al., 1956; TYLER and MONROY, 1959). This change in strength may be enough to alter the binding state of G6PDH. As intracellular pH in unfertilized and fertilized sea-urchin eggs is between 6 and 7, G6PDH is probably not released in vivo by the effect of pH unless intracellular pH varies locally (NEEDHAM and NEEDHAM, 1926; PANDIT and CHAMBERS, 1932; WIERCINSKI, 1944; WINTERS, 1962; ISONO, 1963c).

Oxygen consumption and CO_2 production increase markedly after fertilization, but $^{14}CO_2$ production from ^{14}C-glucose does not. This might be explained if exogenous ^{14}C-glucose is diluted about 5 times with nonlabeled glucose or labeled G-6-P is diluted by enhanced production of nonlabeled G-6-P via the phosphorylase system (KRAHL, 1956; ISONO and YASUMASU, 1966).

Carbon atoms 2 and 6 of ^{14}C-glucose are oxidized to CO_2 by as much as 16%, while carbon from carbon atom 1 is oxidized about 40% during 2-hour incubation (Table 18.14). From 45% to almost 90% of the radioactivity from ^{14}C-glucose taken up is distributed in the acid-soluble pool.

Although radioactivity uptake into the acid-soluble fraction of *Arbacia* eggs does not alter significantly after fertilization, in *Paracentrotus lividus* it increases rapidly just after fertilization and continues until the early blastula stage when it reaches a plateau that lasts until the mesenchymal blastula stage (KRAHL, 1956; MONROY and VITTORELLI, 1962). The curve rises again and declines after the mid-gastrula stage (Fig. 18.11). In the *Hemicentrotus* eggs, the increase after fertilization is even more rapid and does not decline noticeably at later stages (Fig. 18.12)(YANAGISAWA, 1969b).

Paper chromatography of the acid-soluble pool shows that the Ba-soluble and ethanol-insoluble fraction is composed mainly of hexose monophosphates derived from labeled glucose (KRAHL, 1956). Analysis of

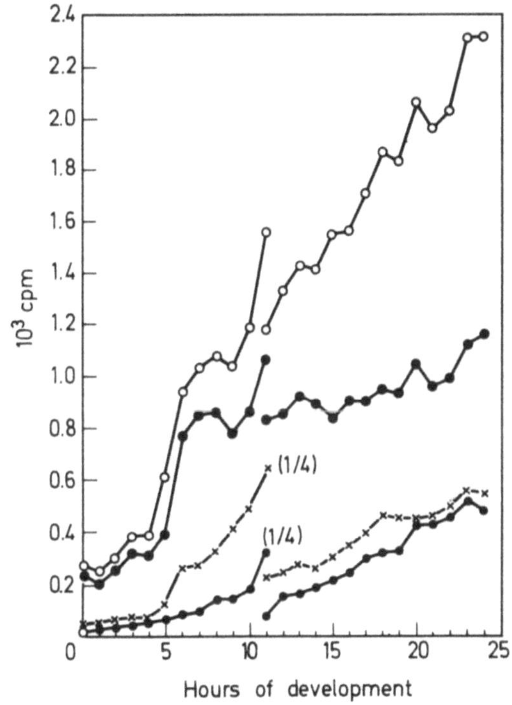

Fig. 18.12. Glucose-U-^{14}C incorporation into total (o), acid-soluble (●), nucleic acid (x), and protein (·) fractions of *Hemicentrotus* embryos (2 x 10^4) pulse-labeled in 1 ml seawater with 25 mμCi glucose-U-^{14}C (65.7 mCi/mmol) for 30 minutes at 20°C. (From YANAGISAWA, 1969b)

the pool by Dowex-1 chromatography shows that glucose-U-^{14}C is rapidly phosphorylated, and in the free-sugar fraction a very small part is recovered that is not adsorbed on Dowex-1 or Dowex-50. 10-minute pulse labeling indicates that although much of the phosphorylated glucose remains in the form of HMP, a considerable amount is converted into amino acids (probably aspartic and glutamic acids)(Table 18.15) (MATSUNAGA and YANAGISAWA, 1970). When the pulsing time is prolonged, the proportion of HMP declines while amino acids increase.

In *Paracentrotus* eggs, glucose-U-^{14}C is converted to amino acids - alanine, glutamic and aspartic acids, serine, and traces of glycine and proline (MONROY and VITTORELLI, 1962). Small amounts of FDP, triose monophosphate and diphosphate, and phosphoenol-pyruvate, as well as some unidentified peaks, also occur (Fig. 18.13).

While oxidation of glucose to CO_2 is mainly via the pentose phosphate shunt, more than half of the ^{14}C taken up remains in the eggs incorporated into amino acids, FDP, triose phosphate, PEP, and macromolecules, which seem to be derived from glucose mainly via the glycolytic pathway. If these compounds are taken into account, the glycolysis contribution to glucose metabolism, from the beginning of embryo development on, is roughly equal to that of the pentose phosphate shunt.

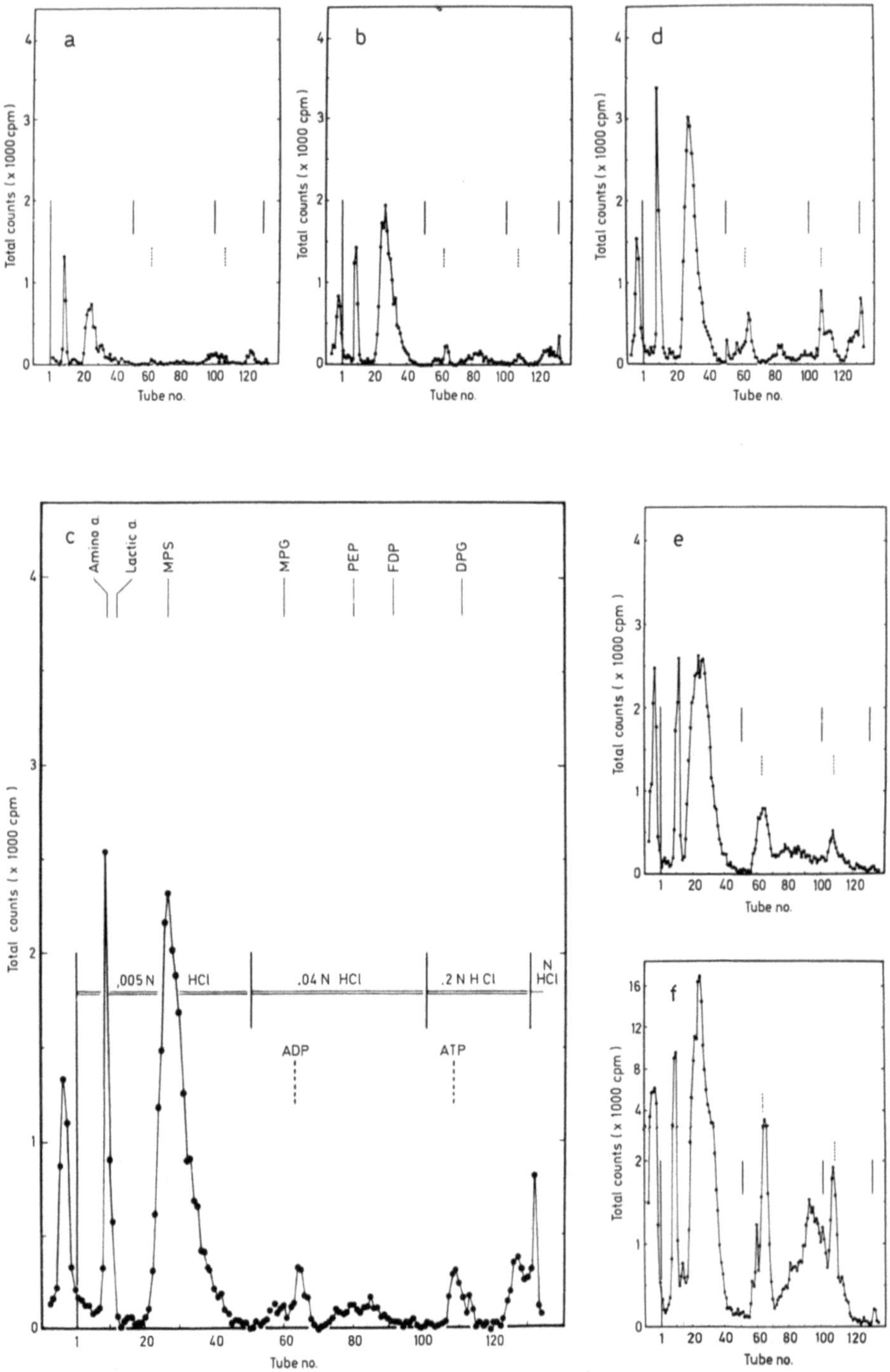
a
b
c
d
e
f
Total counts (x 1000 cpm)
Tube no.
Amino a.
Lactic a.
MPS
MPG
PEP
FDP
DPG
.005 N HCl
.04 N HCl
.2 N H Cl
N HCl
ADP
ATP

Table 18.15. Distribution of radioactivity derived from glucose-U-^{14}C in hexose monophosphate and acidic amino acid fractions of *Hemicentrotus pulcherrimus* eggs and embryos. Unfertilized eggs were pulse-labeled with 2.5 μCi glucose-U-^{14}C for 2 hours; fertilized eggs and embryos, for 10 minutes. (From MASTUNAGA and YANAGISAWA, 1970)

	Un-fertilized	Hours after fertilization							
		1/4	2	4	6	8	10	22.5	42
Total cpm in acid-soluble fraction ($x\ 10^3$)	88.0	13.6	30.7	39.9	56.4	59.5	59.8	75.5	254.9
% in hexose monophosphate fraction	30.6	36.4	56.6	51.8	42.3	45.9	44.0	50.9	53.6
% in Asp + Glu fraction			19.1	22.7	18.9	18.4	20.7	15.5	17.3

Another possible explanation of amino acid formation from glucose-^{14}C is that aspartic and glutamic acids are produced from pyruvate by means of $^{14}CO_2$ fixation by malic enzyme and the KREBS cycle. An appreciable amount of $^{14}CO_2$ is incorporated into the TCA-soluble and protein fractions, and the rate of isotope uptake into the TCA-soluble fraction increases progressively during cleavage, reaches a maximum about the time of hatching, and decreases shortly before the appearance of the primary mesenchymal cells. The specific activity of the protein carboxyl carbon increases shortly after fertilization, peaks at the same time as the TCA-soluble fraction, and increases again when gastrulation begins (Fig. 18.14)(HULTIN and WESSEL, 1952; HULTIN, 1953).

Pathway contributions to egg carbohydrate metabolism are altered in various ways. In eggs treated with dinitrocresol, increased oxygen uptake is sustained mainly via the glycolytic pathway (KELTCH et al., 1956). In unfertilized and fertilized *Pseudocentrotus* eggs and fertilized *Hemicentrotus* eggs treated with 10^{-3} - 10^{-2} mol monoiodoacetate, the pentose content sometimes increases and respiration is not affected. But in unfertilized and fertilized *Anthocidaris* eggs and unfertilized *Hemicentrotus* eggs, the pentose content is reduced and respiration is inhibited. In those instances in which pentose content and respiration are unaffected, it is possible that the glycolytic pathway is converted to a pentose phosphate shunt which compensates for the reduced respiration by means of glycolysis induced by IAA treatment (ISHIHARA, 1958b).

Fig. 18.13a-f. Anion-exchange chromatography of acid-soluble fractions of *Hemicentrotus* eggs and embryos (5 x 10^5) pulse-labeled for 10 minutes with 2.5 μCi glucose-U-^{14}C (65.7 mCi/mmol/10 ml). PCA-soluble fraction is chromatographed on Dowex-1 (Cl) column. Elution system: gradient elutions with 0.005 *N* HCl, 0.04 *N* HCl, and 0.2 *N* HCl. Elution positions of the main intermediates of carbohydrate metabolism are indicated. MPS: hexose monophosphates; MPG, monophosphoglycerate; DPG: diphosphoglycerate; PEP: phosphoenolpyruvate; FDP: fructose diphosphate. Those of ADP and ATP are used as elution position references. Development at 20°C is as follows: (a) 15 min; (b) 2 hr; (c) 4 hr; (d) 10 hr; (e) 23 hr; (f) 42 hr. (From MATSUNAGA and YANAGISAWA, 1970)

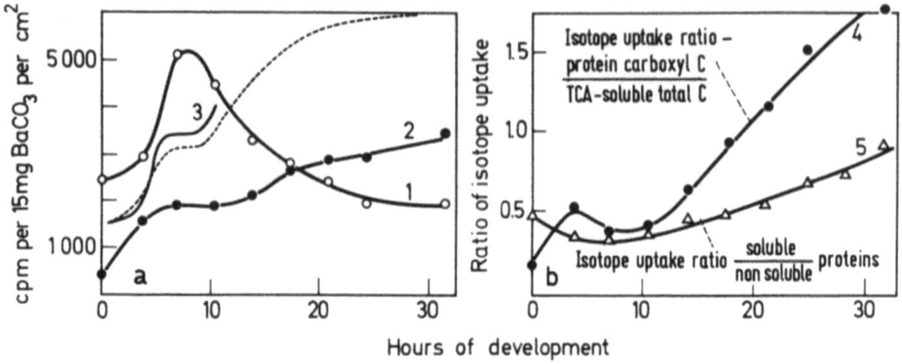

Fig. 18.14 a and b. Rate of $^{14}CO_2$ uptake into TCA-soluble and protein fractions of *Psammechinus miliaris* eggs incubated with $NaH^{14}CO_3$ (2-5 µCi) in 5 ml seawater for 3 1/2 hr. (a)(1) Total incorporation into TCA-soluble fraction (-o-); (2) protein fraction (-●-); (3) relative respiration rate (solid line) *Psammechinus miliaris*; (broken line) *Paracentrotus lividus*. (b) Uptake ratio of protein carboxyl-C/ TCA soluble-C (-●-) and soluble protein/nonsoluble protein (-Δ-). (From HULTIN and WESSEL, 1952)

18.1.6 NAD and NADP Levels

NAD is a coenzyme necessary to dehydrogenation via the glycolytic pathway, while NADP is required in dehydrogenation via the pentose phosphate shunt. Unfertilized *Arbacia punctulata* eggs contain 0.38 - 0.75 µmol of total NAD per gm of eggs, and 25%-40% of that is in reduced form. Total, oxidized and reduced NAD are extracted with hot molar NaCl under neutral (pH 4-8), acid (pH 2), and alkaline (pH 12) conditions (JANDORF and KRAHL, 1942). Although reduced NAD increases slightly after fertilization, total NAD and oxidized NAD decline significantly (Table 18.16). The NAD content of *Hemicentrotus* eggs also decreases after fertilization, possibly as the result of increased NADase activity (Table 18.17)(OHNISHI, 1961; FUJII and OHNISHI, 1962).

Table 18.16. Nicotinamide adenine dinucleotide content of *Arbacia punctulata* eggs. Eggs were developed at 20°C. Total NAD and oxidized and reduced NAD were extracted with hot molar NaCl under neutral (pH 4-8), acid (pH 2), and alkaline (pH 12) conditions and measured manometrically (µmol/gm of eggs). Mean is given in parentheses. (From JANDORF and KRAHL, 1942)

	Unfertilized eggs	Fertilized eggs	
		30 min	10 hr
Total NAD	0.550-0.721(0.623)	0.393-0.650(0.541)	0.252-0.603(0.408)
Reduced NAD	0.154-0.226(0.195)	0.152-0.314(0.219)	0.113-0.291(0.228)
Oxidized NAD	0.518-0.655(0.575)	0.428-0.520(0.461)	0.356-0.448(0.390)

In *Arbacia punctulata* eggs, native fluorescence or fluorescence induced by methyl ethyl ketone combined with dehydrogenases and substrates has demonstrated a 3- to 7-fold increase in the concentration of NADPH within 1 hour after fertilization but no significant change in

Table 18.17. NAD and nicotinamide content and proton release at fertilization. Values are per 10^6 eggs. (From FUJII and OHNISHI, 1962)

Species	NAD (mμmol) Unfertilized	NAD (mμmol) Fertilized (5 min)	Nicotinamide (mμmol) Unfertilized	Nicotinamide (mμmol) Fertilized (10 min)	H ions released (mμmol)
Hemicentrotus	92	47			7.2
Pseudocentrotus	77	50	94	124	3.6
Anthocidaris	157	89			
Hemicentrotus + 10^{-2} mol nicotinamide	92	51			2.7

the concentration of oxidized NADP (KRANE and CRANE, 1958, 1960). NAD kinase occurs in *Arbacia punctulata* and *Strongylocentrotus purpuratus* eggs, and while in vivo results show that it is active only after fertilization, in vitro assays of its activity in homogenates show that it is equally active in both unfertilized and fertilized eggs. NAD kinase seems to be activated in vivo by fertilization and possibly after cell breakage in vitro (KRANE and CRANE, 1958; EPEL, 1965; BLOMQUIST, 1969).

Enzyme analysis of *Strongylocentrotus purpuratus* and *Lytechinus variegatus* eggs shows that after insemination NADPH increases 3- to 5-fold, while NADH decreases initially and then later surpasses its prefertilization level. Within 30-40 seconds after insemination, NAD begins to decline and NADP to increase (Fig. 18.15). Total NAD declines 24%, while total NADP increases 2.6 times (EPEL, 1964a, 1964b, 1965).

18.1.7 Fertilization Acid

When the sea-urchin egg is fertilized, acid (about 20 μmol/100 mg *N*) evolves within 5 minutes or so (Fig. 17.18)(RUNNSTRÖM, 1933). It is assumed to be one of the intermediates of the glycolytic pathway or KREBS cycle, synthesized in large amounts in connection with rapidly increased respiration.

In some sea-urchin eggs the lactic acid content increases at fertilization (Fig. 18.8a,b)(AKETA, 1957). In *Echinus esculentus* eggs, however, the difference between the amounts of lactic acid produced by unfertilized eggs and newly fertilized eggs is too small to explain CO_2 formation in seawater. And there is evidence that the lactic acid does not diffuse into the surrounding seawater (ROTHSCHILD, 1958).

The lactate content of *Paracentrotus lividus* and *Psammechinus miliaris* is highly variable and has no definite relation to acid production (AKETA, 1964). Thus, the fertilization acid is not lactate, pyruvate, phosphoric acid, malate, or any other member of the KREBS cycle (ROTHSCHILD, 1956).

While *Echinus esculentus* and *Psammechinus miliaris* eggs produce 4 and 5.6 μmol of acid per milliliter of eggs, respectively, starting 1 or 2 minutes after fertilization and continuing for 2 or 3 minutes, the pH titration curve shows that no acid with a pK of 3 to 8 is released into the medium in sufficient quantities to account for the fertilization acid. Acid production declines with decreasing pH of the medium

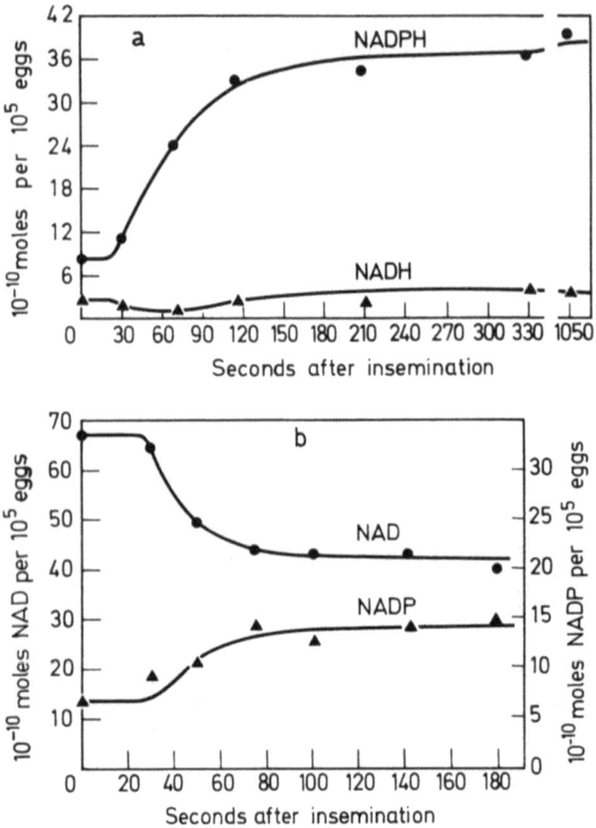

Fig. 18.15 a and b. (a) Changes in NADH and NADPH and (b) changes in NAD and NADP after insemination of *Strongylocentrotus purpuratus* eggs. Reduced and oxidized pyridine nucleotides extracted at room temperature with molar KOH-0.05 mol versene, and at $0^{o}C$ with 5% PCA and measured enzymatically. (From EPEL, 1964a)

and is nullified at pH 5.2 - 5.3. A fertilization base is produced at pH lower than 5.2 (MEHL and SWANN, 1961).

As no acid production is observed when rock-oyster eggs are fertilized, it is likely that most of the acid production is not metabolic (CLELAND, 1950b). Or the reaction ATP + glucose $\longrightarrow$ ADP + G-6-P + H^+ (at pH 7) may suggest a means of producing acid in the form of hydrogen ions (ROTHSCHILD, 1956).

NAD hydrolysis is related to acid production. Measured with yeast alcoholic dehydrogenase, the NAD content declines sharply just after *Hemicentrotus* eggs are fertilized: 101 mμmol NAD/10^6 unfertilized eggs; 43.7 in fertilized eggs 5 minutes after fertilization; 46.7 at the fusion nucleus stage; 55.7 at the streak stage; 60.3 at metaphase; 61.8 at anaphase; and 64.8 at telophase. The nicotinamide content of *Pseudocentrotus* eggs increases from 94.2 to 123.7 mμmol, the molar increase of free nicotinamide equivalent to the molar decrease of egg NAD (Table 18.17). The levels of NADH, NADP, and NADPH do not change appreciably. NADase activity at fertilization increases 80% in *Hemicentrotus* eggs and 64% in *Anthocidaris* eggs (OHNISHI, 1961; FUJII and OHNISHI, 1962).

At fertilization, NADase inhibitors 10^{-2} mol nicotinamide, ethyl nicotinate, 2.5 x 10^{-2} mol isonicotinic acid hydrazide, and 3-acetylpyrimidine inhibit acid production. Acid production is believed to be at least the partial result of what happens when NADase splits NAD. Hydrogen ions are released according to the reaction NAD + $H_2O \longrightarrow$ nicotinamide + ADP-ribose + H^+. Only about one-seventh of the hydrogen ions due to NAD breakdown are liberated into the surrounding medium as fertilization acid (FUJII and OHNISHI, 1962).

Acid polysaccharide changes at fertilization may possibly contribute to acid production. And similar acid production occurs in unfertilized eggs after hypotonic or hypertonic treatment, after freeze-thawing or after cytolyzing with saponin. Acid is also produced in the homogenates of unfertilized eggs after Ca-ions are added or after treatment with papain (RUNNSTRÖM, 1930a, 1930b, 1933, 1935b; ROTHSCHILD, 1939; HULTIN, 1950a, 1950b; RUNNSTRÖM and IMMERS, 1956; RUNNSTRÖM et al., 1959).

In the cortical bodies of *Echinus esculentus*, *Strongylocentrotus droebachiensis*, and *Paracentrotus lividus* eggs, acid mucopolysaccharide, demonstrated by HALE's staining, may play a role in acid production (IMMERS, 1956, 1960).

In *Hemicentrotus*, *Pseudocentrotus*, *Anthocidaris*, and *Temnopleurus* eggs, acid is produced after fertilization or after activation with 0.5% Lipon or Monogen, 0.2% Na choleinate, 0.3 part saturated thymol, or isotonic urea. Pretreatment with such metabolic inhibitors as iodoacetamide, Na-malonate, NaN_3, phloridzin, eserine, and diisopropyl-fluorophosphate does not inhibit acid production (AKETA, 1961a).

10-minute treatment with butyric acid-seawater or 5-minute narcotization with urethane-seawater prevents cortical body breakdown and inhibits acid production. At least part of the fertilization acid is probably liberated from the acid mucopolysaccharide in the cortical bodies (AKETA, 1961b, 1962).

Paper chromatography analysis of the surrounding seawater concentrated in vacuo after removing cations with Dowex 50-H shows only 1 spot of inorganic sulfate. Barium-precipitation shows that the amount of sulfate in the surrounding seawater increases at fertilization from 24.25 mg SO_4/10 ml seawater to 24.81 mg SO_4, which means release of 7.57 mg SO_4/100 mg *N*. It seems that the acid formed at fertilization corresponds to the inorganic sulfate split off from mucopolysaccharide in the cortical bodies. Phenol sulfatase and sulfate-bound phenols are natural tanning substances of the fertilization membrane (MONNE and HÅRDE, 1951; AKETA, 1963).

At fertilization, acid mucopolysaccharide is liberated into seawater by jellyless eggs and by eggs treated with trypsin to remove the vitelline membrane. In *Arbacia*, the acid mucopolysaccharide contains 6% sulfate and in *Hemicentrotus*, 20% sulfate, in addition to fucose, mannose, glucose, galactose, hexosamine, and amino acids. It occurs at the egg surface, expels at fertilization into the perivitelline space, and then into the surrounding seawater, accounting for at least 62% of the fertilization acid (IMMERS, 1956; ISHIHARA, 1968a, 1968b).

An enzyme, exo-β-D-1,3-glucanase, occurs in cortical bodies and is released rapidly at fertilization (EPEL et al., 1969; MUCHMORE et al., 1969).

18.2 Pyruvate Metabolism and the Tricarboxylic Acid (KREBS-) Cycle

18.2.1 Pyruvate Utilization

Pyruvate does not occur in *Paracentrotus lividus* eggs, but in *Arbacia punctulata* it is shown colorimetrically to be present in quite small amounts, compared with total carbohydrate and lactate. During incubation with 5 millimol pyruvate, fertilized *Arbacia* eggs use about 7 times as much pyruvate as unfertilized eggs (RUNNSTRÖM, 1933; ÖRSTRÖM and LINDBERG, 1940; BARRON and GOLDINGER, 1941; HUTCHENS et al., 1942a; GOLDINGER and BARRON, 1946).

The cocarboxylase method shows that cytolysate (1 part egg to 3 parts water) from unfertilized *Arbacia* eggs consumes 160 µg pyruvate/hr/gm dry weight (0.87 µmol or 38 µg/gm wet weight), while eggs 30 minutes after fertilization consume 310 µg/gm dry weight (1.7 µmol or 75 µg/gm wet weight)(KRAHL et al., 1942). As oxygen uptake is generally lower in the presence of pyruvate, without chemical analysis it may not be conclusive that pyruvate is not utilized or does not penetrate the cell (GOLDINGER and BARRON, 1946).

In *Echinus esculentus* egg homogenate, pyruvate (10^{-2} mol) stimulates respiration with considerable variation. In active preparations, pyruvate oxidation proceeds to completion, using 2 1/2 mol of oxygen/mol of pyruvate, while less active preparations use 1/2 to 2 1/2 mol of oxygen and their pyruvate reacts only with CoA to produce acetyl CoA (Fig. 18.16). Even in homogenates in which oxygen consumption

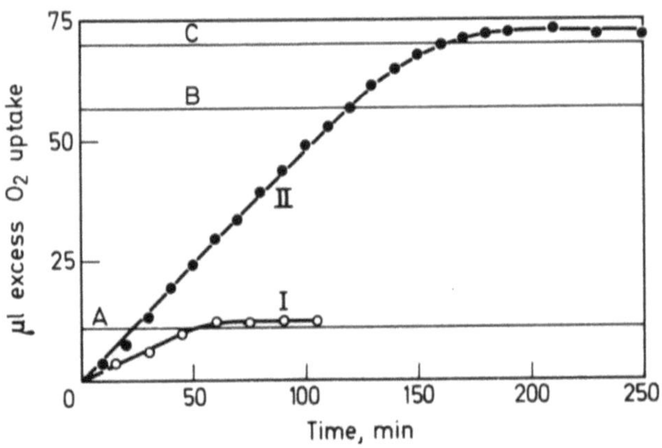

Fig. 18.16. Oxidation of added pyruvate in *Echinus esculentus* egg homogenates. 110 µg Na-pyruvate oxidized by inactive preparation (curve I) and active preparation (curve II). Lines A and B indicate theoretical excess oxygen uptake for complete combustion in 0.5 O_2/mol pyruvate and 2.5 O_2/mol pyruvate, while line C indicates theoretical O_2 uptake for complete combustion of pyruvate plus lactate. In curve II, 20 µg more lactic acid disappears than in the control, and the added pyruvate disappears completely. (From CLELAND and ROTHSCHILD, 1952b)

Table 18.18. Pyruvate used by *Echinus esculentus* egg homogenates. (From CLELAND and ROTHSCHILD, 1952b)

Homogenate plus	O_2 (μl)	Terminal change (μg) Pyruvate	Lactate
	72	+ 2	0
Pyruvate (110 μg)	81	-100	+12
Fumarate (10^{-2} mol)	83	-	-
Fumarate + pyruvate	90	-105	+12
Cyanide (10^{-3} mol)	10	+ 20	+60

is relatively unaffected by the addition of pyruvate, the added pyruvate is used up (Table 18.18). This suggests that pyruvate is produced during endogenous metabolism and is immediately removed by oxidation.

In preparations in which pyruvate stimulates oxygen uptake moderately, respiration stimulation is obvious because of reduction in endogeneous pyruvate production. In a concentration of *M*/200, fluoride inhibits glycolysis by about 85% (CLELAND and ROTHSCHILD, 1952b).

Most pyruvate reactions in cells require thiaminepyrophosphate as a coenzyme. Extracting twice at 100°C for 3 minutes with molar NaCl containing 40 μg of thiamine hydrochloride per milliliter, the thiaminepyrophosphate content was found to be highest in unfertilized *Arbacia punctulata* eggs, somewhat smaller 30 minutes after fertilization, and smaller yet 10 hours after fertilization (KRAHL et al., 1942). Other studies of *Arbacia punctulata* show the same amount of thiaminepyrophosphate in unfertilized and fertilized eggs (BARRON and GOLDINGER, 1941; GOLDINGER and BARRON, 1946).

18.2.2 Tricarboxylic Acid Cycle

As succinate and malate are not metabolized in *Arbacia punctulata* eggs, the KREBS cycle does not appear even 24 hours after fertilization (BALL and MEYERHOF, 1940; GOLDINGER and BARRON, 1946). In the egg homogenate of *Echinocardium cordatum*, fumarate and malate stimulate the decolorization of methylene blue (LINDBERG, 1943).

Further evidence of the operation of the KREBS cycle appears in particulates sedimentable at 2,000 g for 10 minutes of *Arbacia punctulata* egg homogenate (1 ml of egg + 3 ml of 0.17 mol KCl, 0.23 mol Na-citrate, pH 7.7). Echinochrome must be eliminated to obtain active preparations that utilize oxaloacetate, succinate, α-ketoglutarate, glutamate, and citrate in the following medium: 0.36 mol KCl, 14 millimol $MgCl_2$, 3.6 millimol glycine buffer (pH 7.0), 36 millimol glucose with hexokinase, 18 millimol NaF, 0.014 millimol cytochrome c, and 0.7 millimol ATP. The oxygen uptake of this cell-free system (cyclophorase) prepared from unfertilized and fertilized eggs is about twice that of an equivalent quantity of intact unfertilized eggs and half that of intact fertilized eggs. It can be stimulated by 0.64 x 10^{-4} mol of dinitrocresol or 7 x 10^{-4} mol of inorganic phosphate (CRANE and KELTCH, 1949; KELTCH et al., 1949, 1950).

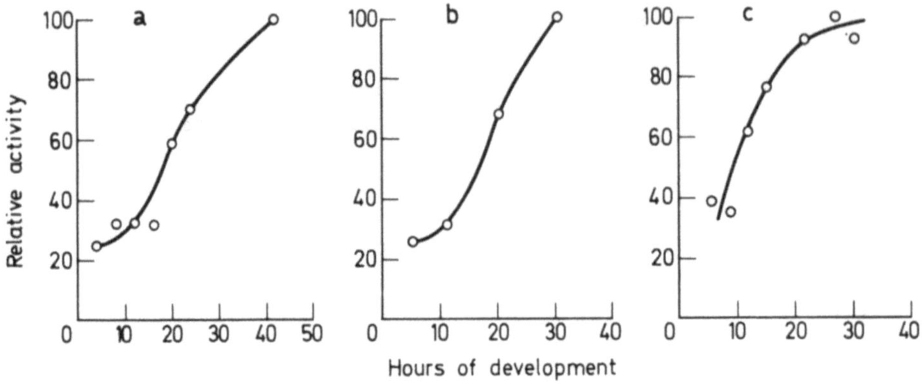

Fig. 18.17a-c. Succinic dehydrogenase (a, b) and malic dehydrogenase activities in various developmental stages of *Psammechinus miliaris* (a, c) and *Echinocardium cordatum*. Succinic dehydrogenase is determined by using triphenyl tetrazolium chloride as a hydrogen acceptor. Malic dehydrogenase is determined using the following reaction mixture: 0.25 ml 0.20 mol Na-L-malate, 0.50 ml 0.13 mol phosphate buffer (pH 8.0), 0.10 ml 0.06% methylene blue, 0.10 ml 2 mol KCN, 0.25 ml 400 μg/ml NAD, 0.10 ml 0.026 mol $MgCl_2$, 0.10 ml 0.018 mol AMP, 0.20 ml 0.60 mol NaF and 0.50 ml homogenate. (From GUSTAFSON and HASSELBERG, 1951)

Coloration of triphenyltetrazolium with succinate as a substrate and discoloration of methylene blue with malate as a substrate show the presence of succinate dehydrogenase in the egg homogenates of *Paracentrotus lividus*, *Sphaerechinus granularis*, *Echinocardium cordatum*, and *Psammechinus miliaris* and malic dehydrogenase in the homogenate of *Psammechinus miliaris*. Both enzymes are constant during cleavage until the mesenchymal blastula stage when they become much more active (Fig. 18.17)(GUSTAFSON and HASSELBERG, 1951). In egg homogenate in 0.01 mol tris buffer (pH 7.4), malic dehydrogenase activity increases 3-fold at the young pluteus stage, mostly after hatching (OZAKI and WHITELEY, 1970).

Respiration of *Echinus esculentus* egg homogenate in *M*/50 phosphate buffer (pH 7.5) significantly increases when KREBS cycle intermediates (10^{-2} mol) are added: 35%, with citrate; 45%, with α-ketoglutarate; 42%, with succinate; 15%, with fumarate; 21%, with malate; and 10%, with glutamate (Table 18.18)(CLELAND and ROTHSCHILD, 1952b). Results are essentially the same in the egg homogenate of *Strongylocentrotus purpuratus* in *M*/15 phosphate buffer (pH 7.1) using citrate, α-ketoglutarate, succinate, and malate as substrates (YČAS, 1954). Egg homogenate respiration in 0.9 mol sucrose + 0.02 mol glycylglycine (pH 7.35) + 2.5 millimol phosphate + 1-5 millimol EDTA is also stimulated by succinate (0.02 mol) but not by pyruvate (0.02 mol) (Fig. 17.16)(GONSE, 1960).

KREBS cycle inhibitors 10^{-2} mol malonate and fluoroacetate reduce the endogenous respiration of egg homogenates 44% and 23%, respectively (CLELAND and ROTHSCHILD, 1952b). Malonate also inhibits the respiration of *Strongylocentrotus purpuratus* egg homogenate (YČAS, 1954).

Complete pyruvate oxidation in the egg homogenate is the most cogent proof of the operation of the KREBS cycle. In animal tissue it is the only mechanism known that completely combusts pyruvate to CO_2 and H_2O (CLELAND and ROTHSCHILD, 1952b).

18.2.3 Malic Dehydrogenase Isozymes

Malic dehydrogenase (MDH) occurs in various forms in sea-urchin eggs. Starch gel electrophoresis (0.025 mol barbital buffer, pH 8.7) and DEAE-cellulose column chromatography show that the supernatant (10,000 g at 20 minutes) of unfertilized *Arbacia punctulata* egg homogenate in 2 parts of the same buffer contains 5 electrophoretically separable peaks of NAD-L-MDH. The major peak (peak II, counting from anode to cathode in starch gel electrophoresis) represents about 60% of the total activity. With NAD analogs as cofactor - 3-acetyl pyridine adenine dinucleotide (APAD) or thionicotinamide adenine dinucleotide (TNAD) - the supernatant contains 6 or 7 peaks of APAD-L-MDH and 3 peaks of TNAD-L-MDH. Only peak II of NAD-L-MDH coincides with the major peak of APAD-L-MDH, and 3 TNAD-L-MDH peaks coincide with the 3 peaks of APAD-L-MDH. 5 other peaks of D-MDH catalyze the dehydrogenation of D-malic acid in the presence of APAD but do not interact with NAD (MOORE and VILLEE, 1962, 1963a).

In *Arbacia punctulata* eggs, more than 10 molecularly distinct proteins with MDH activity are demonstrable by electrophoresis and by different heat stability. TNAD-L-MDH and NAD-L-MDH are the most labile. They also react differently to concentrations of cofactors and enzyme solutions (MOORE and VILLEE, 1963a).

By the 64-cell stage of *Arbacia punctulata* eggs, the 5 molecular forms are reduced to 3 (peaks I, II and IV), which remain through the cleavage stage up to hatching.

After 12 hours at 20°C (at the swimming blastula stage), peak V reappears and the 4 electrophoretically distinct NAD-L-MDH bands remain until 48 hours after fertilization (until the pluteus stage). APAD-L-MDH and TNAD-L-MDH patterns do not change markedly. The ratio of APAD-L-MDH and NAD-L-MDH activity of the unfractionated supernatant is 0.68 in unfertilized eggs, 0.63 in 6-hour embryos, and 2.2 in 48-hour embryos (MOORE and VILLEE, 1962, 1963a).

Disc acrylamide gel electrophoresis in tris buffer at pH 8.3 separates L-MDH of unfertilized *Arbacia punctulata* eggs into 6 bands (Fig. 18.18A). The major bands are 3, 5, and 6, counting from cathode to anode. There are 3 bands in 4-hour and 6-hour embryos and 4 in 12-hour embryos. Embryos grown for 6 to 12 hours in seawater containing 10^{-3} mol D-malate (Fig. 18.18Bb, Cb) have 1 more band of L-MDH activity than embryos grown in normal seawater (Fig. 18.18Ac, d) or seawater containing 10^{-2} mol L-malate (Fig. 18.18Ba, Ca)(BILLIAR et al., 1964).

The specific activities and electrophoretic patterns of L-MDH of embryos grown in the presence of actinomycin D (10-20 µg/ml), which inhibits protein synthesis after the early blastula stage but not in early cleavage, are the same as those of untreated embryos until at least 22 hours after fertilization. Puromycin (20-40 µg/ml) inhibits cell division, embryo development, and protein synthesis but does not alter L-MDH patterns (BILLIAR et al., 1966). The small blastomeres of 64-cell stage embryos have 3 L-MDH bands, and the large ones have 2 (Fig. 18.19)(MOORE and VILLEE, 1963b). The biologic and embryologic significance of multiple forms of MDH is still unclear.

Acrylamide gel electrophoresis of nucleate halves of unfertilized *Arbacia punctulata* eggs reveals 7 bands of MDH activity comparable to those of whole eggs. In the more dense anucleate halves, bands I and VI are omitted. Fertilization reduces the number of isozymes from 7 to 4 in nucleate halves and from 5 to 3 in anucleate halves (Fig. 18.20a).

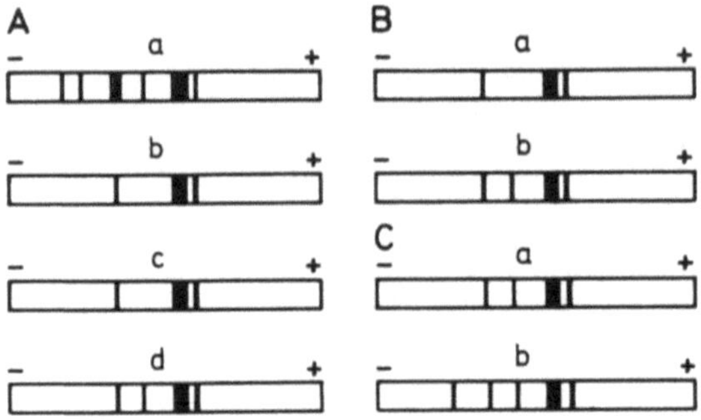

Fig. 18.18A-C. Patterns of L-malate dehydrogenase isozymes in *Arbacia punctulata* separable by disc electrophoresis on acrylamide gel; changes during early development. Supernatant (10,000 g, 20 min) of egg homogenate in 0.025 mol barbital buffer (pH 8.7) is used. Staining is with nitro blue tetrazolium (pH 7.4) reagents: 50 millimol L-malate, 1 millimol NAD, 2 millimol KCN, 0.5 millimol $MgCl_2$, 50 millimol glycylglycine, 0.05 mg/ml phenazine methosulfate, and 0.3 mg/ml nitro blue tetrazolium. (A) Patterns of unfertilized eggs (a, 6 distinct bands and 1 very thin one) and fertilized eggs at 4 hr (b, 3 bands), 6 hr (c, 3 bands), and 12 hr (d, 4 bands) after fertilization. Embryos cultured for 6 hr (B) and 12 hr (C) in seawater containing 10^{-2} mol L-malate (a) or 10^{-3} mol D-malate (b). (From BILLIAR et al., 1964)

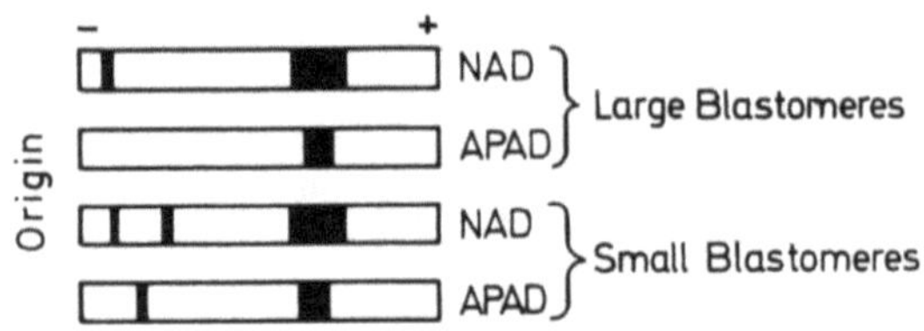

Fig. 18.19. Isozyme (NAD-L-MDH and APAD-L-MDH) patterns of large and small *Arbacia punctulata* blastomeres dissociated from denuded 16-cell-stage embryos in glass homogenizer and separated by centrifugation on 0.27-0.87 mol sucrose gradient (750 g, 25 min). Other procedures same as Fig. 18.18 except that acetyl pyridine adenine dinucleotide (APAD) is used as coenzyme in addition to NAD. (From MOORE and VILLEE, 1963b)

Spermatozoa and the particulate and soluble fractions of unfertilized eggs have different isozyme patterns (Fig. 18.20b,c)(PATTON et al., 1967).

From unfertilized egg to early pluteus, *Strongylocentrotus purpuratus* has only 2 distinct L-MDH fractions separable by disc electrophoresis and DEAE-cellulose column chromatography. One, a supernatant enzyme, occurs only in the cytoplasm, and the other occurs in the mitochondria (Fig. 18.21).

A strong concentration of malate (10^{-2} *M*) inhibits the supernatant enzyme but oxaloacetate (10^{-4} *M*) does not, while of the mitochondrial enzyme, the reverse is true. Total enzyme activity increases during development, but the ratio of supernatant MDH and mitochondrial MDH (about 4:6 at early cleavage) does not change until the pluteus stage.

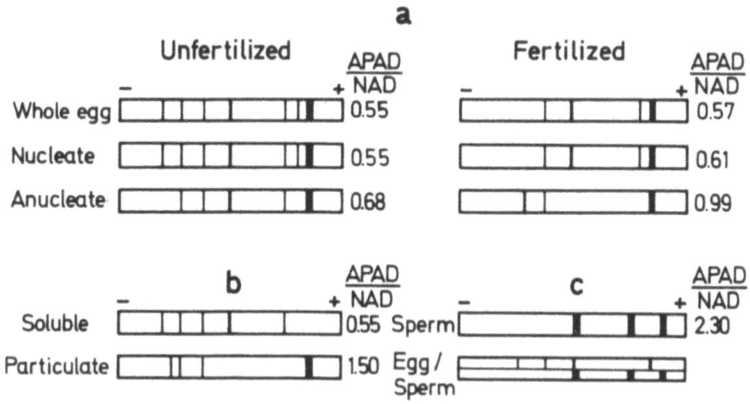

Fig. 18.20a-c. L-MDH isozyme patterns in *Arbacia* egg fractions. (a) Nucleate and anucleate halves of unfertilized and fertilized eggs. (b) Particulate and nonparticulate fractions of unfertilized eggs; both fractions obtained from homogenate in seawater after centrifugation at 22,000 g for 30 min. (c) Comparison of isozyme patterns of fertilized eggs and spermatozoa. (From PATTON et al., 1967)

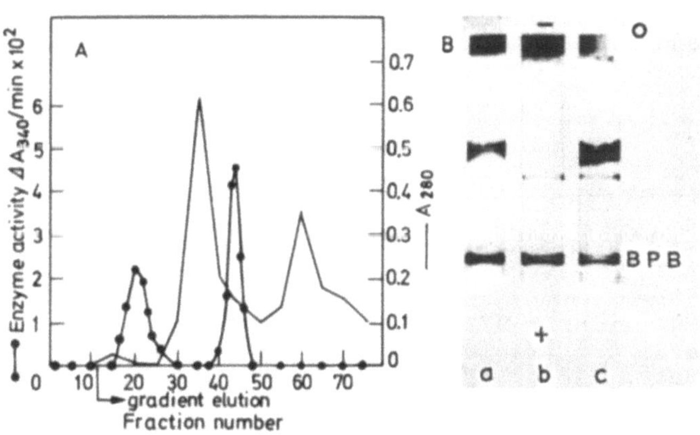

Fig. 18.21 A and B. Chromatographic (A) and disc electrophoretic (B) separation of mitochondrial and supernatant L-MDH of *Strongylocentrotus purpuratus*. Eggs homogenized in 0.44 mol sucrose + 0.12 mol KCl + 1 millimol EDTA + 0.05 mol tris-HCl (pH 7.6); heavy particles removed (750 g, 20 min) and mitochondrial and supernatant fractions obtained by centrifugation at 17,000 g for 20 min; enzyme extracted with 0.01 mol tris-HCl (pH 7.4). (A) DEAE-cellulose column (1.5 x 20 cm) eluted with 0.04 mol NaCl in 0.01 mol tris-HCl (pH 7.4) for first 10 fractions and then eluted with linear gradient elution to 0.4 mol NaCl in 0.01 mol tris-HCl (pH 7.4); 3-ml fractions. (B) Acrylamide gel electrophoresis. (a) First active chromatographic component (supernatant enzyme). (b) Second active chromatographic component (mitochondrial enzyme). (c) Mixture of both preparations. O: origine; BPB: bromophenol blue dye front. (From OZAKI and WHITELEY, 1970)

Table 18.19. Effect of KCN and CO on sea-urchin egg respiration. (From RUNNSTRÖM, 1930)

	1 ml egg suspension (%)	Duration (min)	Temperature	Unfertilized ($\mu l\ O_2$) Control	Unfertilized ($\mu l\ O_2$) Inhibitor	Fertilized ($\mu l\ O_2$) Control	Fertilized ($\mu l\ O_2$) Inhibitor
KCN							
10^{-4} mol in *Paracentrotus lividus*	3.8	120	20.5°C	4.4	2.45	22.6	5.2
2 x 10^{-4} mol in *Arbacia punctulata*	1.8	20	25°C	5.8	5.8	23	6.7
$CO + O_2$							
95:5 in *Paracentrotus lividus*	4.2	75	23°C	6.3	8.6	41.5	17
97:3 in *Arbacia punctulata*	1.8	30	25°C	21	12.2	-	-

By this time a significant increase in total enzyme activity above the level of the blastula stage is observed. Relative total activity is 1.0 in unfertilized eggs, 1.6 in morulae, 2.1 in blastulae, 2.2 in gastrulae, 2.3 in prisms, and 2.8 in young plutei (OZAKI and WHITELEY, 1967, 1970).

18.3 Terminal Electron Transport System

18.3.1 Effect of Respiratory Inhibitors

The respiration of fertilized *Paracentrotus* and *Arbacia* eggs is markedly depressed by 10^{-5} - 10^{-4} mol CN^- and 92%-97% CO, while that of unfertilized eggs is scarcely affected (Table 18.19) (RUNNSTRÖM, 1930). In confirmation of the KCN effect, unfertilized *Arbacia* eggs seem to breathe through a nonferrous carrier, and on fertilization functional connection between the substrate-dehydrogenase system and oxidase is established. Cytochrome is suddenly thrown into circulation, increasing the respiratory rate and introducing a large cyanide-sensitive fraction of respiration (KORR, 1937).

Higher concentrations of KCN (10^{-3} mol) inhibit respiration in unfertilized *Paracentrotus lividus* eggs by 40% and by 88% in fertilized eggs (LINDAHL, 1939a, 1940). The apparent lack of effect of KCN on the respiration of unfertilized eggs has been attributed to its extreme volatility at a pH less than 8 (ROBBIE, 1946a, 1946b). In manometrically determining the CN^- effect on the respiration of unfertilized sea-urchin eggs, the main difficulty is to keep the concentration of this extremely volatile and chemically reactive reagent constant long enough to measure the very low oxygen consumption. To meet this requirement, a balanced $Ca(CN)_2$-10% $Ca(OH)_2$ mixture is used for the center-well alkali.

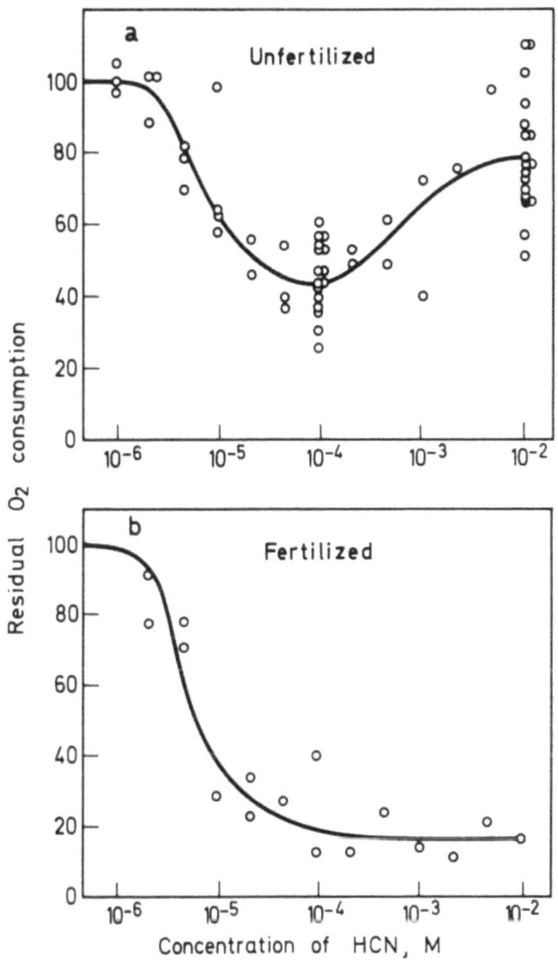

Fig. 18.22 a and b. Relative oxygen consumption of unfertilized (a) and fertilized (b) *Arbacia* eggs during 4-hour exposure to cyanide concentrations. Oxygen consumption measured in WARBURG manometer with balanced $Ca(CN)_2$-$Ca(OH)_2$ solution as center-well mixture and fertilized eggs started 1 hour after fertilization. (From ROBBIE, 1946b)

Even in 10^{-4} mol concentrations, HCN inhibits oxygen consumption of unfertilized *Arbacia punctulata* eggs (43% of the control)(ROBBIE, 1946a, 1946b). Higher concentrations of CN^-reverse inhibition, and the residual oxygen uptake in 10^{-2} mol HCN for a 4-hour exposure is 75% of the control (Fig. 18.22).

A combination of 75% CO and 25% oxygen stimulates respiration of unfertilized *Paracentrotus* eggs in the dark (44%) and on illumination respiration increases 100%. And when oxygen tension decreases to 75% CO + 20% N_2 + 5% O_2, the stimulation decreases. However, freshly fertilized eggs in the dark are slightly stimulated in 75% CO + 25% O_2 and markedly inhibited in 95% CO + 5% O_2. In the light, fertilized eggs show marked stimulation by 75% CO + 25% O_2; this effect declines when the oxygen concentration is reduced to 5% (LINDAHL, 1939a).

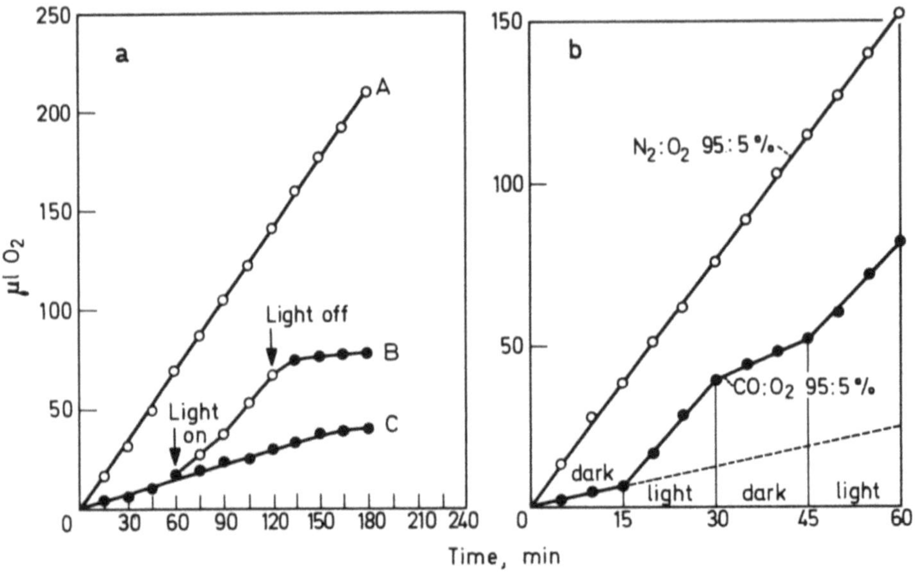

Fig. 18.23 a and b. Effect of CO on oxygen consumption in light and dark. (a) In *Arbacia punctulata*, main compartment of WARBURG flask contains 0.5 ml 0.33 mol glycylglycine (pH 6.9), 0.5 ml oxidase preparation (cytolysate in 0.067 mol Na_2HPO_4), 1.0 ml 4.5 x 10^{-5} mol cytochrome c, and 1.0 ml H_2O. Side arm contains 0.3 ml of substrate (0.22 mol hydroquinone). Center cup is empty. (A) Control in air, (B) in 7.5% O_2-92.5% CO in light for length of time indicated, and (C) in 7.5% O_2-92.5% CO in dark. (b) In *Sphaerechinus granularis*, WARBURG flask contains 0.033 mol phosphate buffer (pH 7.3), 0.01 mol Na-ascorbate, 2.46 x 10^{-5} mol cytochrome c, and 0.15 mg/ml mitochondrial protein at 24°C. (From KRAHL et al., 1941; MAGGIO and GHIRETTI-MAGALDI, 1958)

In studies of the effect of light and CO on the respiration of *Arbacia punctulata* and *Psammechinus miliaris* eggs, particularly unfertilized eggs, of 14 comparisons of the effect of 95% CO/O_2 with 95% N_2/O_2 in the dark, 3 show an average decrease in respiration in 95% CO/O_2 of 11%, 2 are not affected, and the remaining 9 show an average increase in 95% CO/O_2 of 14%. In 24 determinations of respiration in 95% CO/O_2 in dark and light, the average increase in light is 44%. Light alone inhibits the respiration of unfertilized eggs (44 cases) and fertilized eggs (43 cases) in air by 38% and 9%, respectively. With 80% CO/O_2 in the dark, an average respiratory increase of 55% above respiration in 80% N_2/O_2 occurs, and no significant change in light takes place (Fig. 18.23a)(KRAHL et al., 1941; ROTHSCHILD, 1949).

The finding that CO at low oxygen pressure in the dark inhibits the respiration of unfertilized eggs the same as N_2 does, does not necessarily indicate the presence of the cytochrome system, in spite of the positive photoreversal of CO inhibition. However, 95% CO/O_2 in the dark has a marked inhibitory effect even when compared with respiration in 95% N_2/O_2 (Fig. 18.23b)(MAGGIO and GHIRETTI-MAGALDI, 1958).

18.3.2 Oxidation of CO

The stimulating action of CO is probably due to its ultimate oxidation to CO_2. This reaction has been shown to happen in heart muscle. KCN inhibits CO respiration with additional inhibition due to CO itself (RUNNSTRÖM, 1930; LINDAHL, 1939a; ROTHSCHILD, 1949).

In *Strongylocentrotus purpuratus* eggs and embryos, in the light in 80% CO/O_2, gas uptake exceeds the uptake of the air controls at all stages of development (Table 18.20)(BLACK and TYLER, 1959a, 1959b). The excess gas uptake relative to respiration in air is considerably lower in later embryo stages (less than 20%) than in the unfertilized (83%) or freshly fertilized eggs (43%). In the dark, the eggs show greater gas uptake in the CO/O_2 mixture than in air during the early stages of development but considerable inhibition at later stages.

From the data of concurrent measurement of CO_2 and total gas uptake, respiratory quotients (RQ) in air and apparent respiratory quotients (ARQ) in CO/O_2 are calculated (Table 18.21). The ARQ refers to experiments in the presence of CO, as occurrence of CO oxidation means that some of the gas uptake represents CO. Such oxidation has an RQ of 2, but manometrically it is recorded as an RQ of 0.67 (2 parts CO_2 per 2 parts CO + 1 part O_2). It is expected, therefore, that the values for the ARQ in the presence of CO will be lower than the RQ values in air wherever CO oxidation occurs; however, most of the data point to the opposite. Along with inhibiting ordinary respiration in the dark and its own oxidation, CO may be stimulating glycolysis (BLACK and TYLER, 1959b).

By measuring ^{13}CO oxidation to $^{13}CO_2$ in light and dark, the rate of CO oxidation has been determined (Table 18.20). The rate of CO oxidation in the light nearly doubles after fertilization but increases much less during later development. In the dark, the rate of CO oxidation during development declines only slightly (BLACK and TYLER, 1959b).

Table 18.20. Rate of respiration and CO oxidation in light and dark in fertilized and unfertilized *Strongylocentrotus purpuratus* eggs and at later stages of development. (From BLACK and TYLER, 1959b)

Hours after fertilization	Gas uptake ($\mu l/10^6$ eggs/hr) In air; Light	In 80% CO/O_2 Light	In 80% CO/O_2 Dark	Excess gas uptake in 80% CO/O_2 ($\mu l/10^6$ eggs/hr) Light	Dark	CO oxidized from $^{13}C/^{12}C$ ($\mu l/10^6$ eggs/hr) Light	Dark
Unfertilized	16.0	29.9	28.5	+13.5	+12.3	8.9	8.0
Unfertilized	22.5	40.9	37.5	18.4	15.0	13.4	10.6
1 - 5	69.9	97.4	88.3	27.5	18.4	20.5	9.9
1 - 7	81.9	118.4	105	36.5	23.1	24.1	10.8
26 - 30	166	190	132	24.0	-34.0	25.0	3.4
27.5 - 31.5	200	238	203	38.0	+ 3.0	25.0	9.8
26 - 50	211	242	164	31.0	-47.0	24.4	3.3
49 - 53	264	321	216	57.0	-46.0	34.0	6.4

Table 18.21. Manometric measurement of respiratory quotient in *Strongylocentrotus purpuratus* eggs and embryos in air and CO. (From BLACK and TYLER, 1959b)

Hours after fertilization	Respiratory quotient in air	Apparent respiratory quotient in 80% CO/O_2	
		Light	Dark
Unfertilized	0.58	0.63	0.73
Unfertilized	0.85	0.65	0.77
1 - 5	0.81	0.77	0.83
1 - 7	0.77	0.71	0.85
26 - 30	0.85	0.85	0.92
27.5 - 31.5	0.76	0.77	0.81
46 - 50	0.84	0.91	1.0
49 - 53	0.81	0.81	1.04

18.3.3 Flavin-Adenine Dinucleotide

Very few reports of investigations of the flavoproteins in sea-urchin eggs connect NAD to the cytochrome system. Manometric measurement of their coenzyme activity in the enzymic oxidation of an α-amino acid (alanine) shows that unfertilized *Arbacia* eggs contain flavin-adenine dinucleotide (8.5 μg/gm wet egg) and that the content is not altered significantly during development (9.8 in eggs 20 minutes after fertilization, 11.5 in 10-hour embryos, 8.8 in 24-hour embryos). The flavin dinucleotide found in *Arbacia* eggs appears to be similar to or identical with that found in other tissue (KRAHL et al., 1940).

18.3.4 Cytochrome System

CO and CN^- inhibition of respiration suggests that a cytochrome system is present in unfertilized eggs and that their apparent insensitivity to these reagents is due to lack of contact between cytochrome system and substrates (RUNNSTRÖM, 1930). It is possible that cytochrome c occurs in fertilized *Arbacia punctulata* eggs in concentrations disproportionately small (2 μg/gm wet weight), compared to cytochrome oxidase, and too low to be significant in fertilized egg respiration (KRAHL et al. , 1941). And it is possible that cytochrome c occurs in *Echinus esculentus* eggs in concentrations less than 5×10^{-4} μg/mg dry matter, the limit of detection of the micro-method used (BOREI, 1950).

Manometric measurement of cytochrome oxidase activity in dialyzed egg brei (0.1 mol phosphate buffer, pH 6.9) with 1.35×10^{-5} mol reduced cytochrome c and 0.02 *M* substrate (hydroquinone, cysteine, or p-phenylene diamine) confirms that *Arbacia* eggs contain as much cytochrome oxidase as many mammalian tissues and show no succinic dehydrogenase activity (KRAHL et al., 1941). The enzyme occurs in nearly equal amounts in unfertilized eggs (QO_2 = 9.8/mg dry weight) and fertilized eggs (QO_2 = 8.5) and in sufficient concentration to account for egg respiration (unfertilized eggs, QO_2 = 0.4 - 0.5; fertilized eggs, 2; fertilized eggs treated with dinitrocresol, 7-8). The enzyme is completely inhibited by 2.5×10^{-4} mol NaCN, 10^{-3} mol Na_2S, and 10^{-2} mol NaN_3. In 92.5% CO/O_2, it is inhibited in the dark and nearly restored in the light (Fig. 18.23a).

Cytochrome oxidase activity in *Psammechinus miliaris* initially increases 1 1/2 to 2 times and falls markedly by 6 to 8 hours of development, somewhat before the number of mitochondria increases (DEUTSCH and GUSTAFSON, 1952). Cytochrome oxidase occurs mainly in the cytoplasmic matrix residue (67%)(supernatant obtained after centrifugation at 9,000 g for 10 minutes) and in yolk granules (12%) but not in the mitochondrial fraction (about 1%)(HUTCHENS et al., 1942b; MAGGIO and GHIRETTI-MAGALDI, 1958; RAPOPORT et al., 1958; BLACK and TYLER, 1959c; MAGGIO, 1959).

Cytochrome oxidase is found in the particulate fraction of unfertilized *Strongylocentrotus purpuratus* and *Paracentrotus lividus* eggs in which CO is rapidly oxidized in the presence of ascorbic acid, cytochrome c, succinic dehydrogenase, and NADH-cytochrome c reductase. In the light, the rate of CO oxidation is fastest at the cytochrome concentration that gives about half of the maximum oxygen uptake. With maximal oxygen uptake, CO inhibits oxygen uptake as well as CO oxidation. in the level of cytochrome c (Table 18.22). It appears, therefore, In the dark, the rate of CO oxidation is not much affected by changes in the level of cytochrome c (Table 18.22). It appears, therefore, that cytochrome oxidase participates in the CO oxidation pathway (BLACK and TYLER, 1959c).

In a mitochondrial fraction from jelly-free, unfertilized *Paracentrotus lividus* and *Sphaerechinus granularis* eggs homogenized in 0.44 mol sucrose + 0.1 mol citrate, pH 6.3 + 5 x 10^{-4} mol EDTA, or 0.5 mol sucrose + 5 x 10^{-4} mol EDTA, pH 7.3, and sedimented centrifugally at 20,000 g, cytochrome oxidase activity occurs only there. Optimum pH is about 7.1. In *Sphaerechinus* preparations, 95% CO + oxygen inhibits cytochrome oxidase activity 80% in the dark, and in the light this is reversed (Fig. 18.23b). In *Paracentrotus* preparations, inhibition is 90% in the dark and reversed in the light (MAGGIO and GHIRETTI-MAGALDI, 1958).

The mitochondria of unfertilized *Paracentrotus* eggs become saturated at a cytochrome c concentration of 3.28 x 10^{-5} mol, while those of fertilized eggs become saturated at a concentration of 5.74 x 10^{-5} mol. At fertilization, cytochrome oxidase activity increases about 30% and remains constant until the blastula stage (MAGGIO, 1959).

Table 18.22. Rate of gas uptake and CO oxidation in granule preparations in light and dark. Final concentration: 6% egg granules (based on egg volume; sedimentable fraction at 44,000 g for 20 minutes); 0.06 mol phosphate buffer (pH 7.4); 8 x 10^{-4} mol $AlCl_3$; 0.02 mol ascorbate. (From BLACK and TYLER, 1959c)

Experiment	Cytochrome c (mol/liter)	Gas uptake ($\mu l/10^6$ eggs/hr) Light		Dark		CO oxidized ($\mu l/10^6$ eggs/hr) Light	Dark
		Air	80% CO/O_2	Air	80% CO/O_2		
1	2 x 10^{-4}	762	462			16.3	
	4 x 10^{-5}	364	382			26.0	
	2 x 10^{-5}	187	254			21.4	
2	2 x 10^{-4}	779	495	639	61	16.8	6.1
	4 x 10^{-5}	356	408	357	139	26.5	6.1
	2 x 10^{-5}	192	244	217	125	24.0	6.3

The usual spectroscopic techniques do not reveal any specific absorption bands of cytochromes (BRACHET, 1934; LINDAHL, 1936, 1939b; KRAHL et al., 1939; BALL and MEYERHOF, 1940; CLOWES and KRAHL, 1940). And even when echinochrome, which is believed to cause difficulty in spectroscopic examination, is reduced with hyposulfite, cytochrome absorption bands are not observed (KORR, 1939). Pyridine added to the reduced-egg suspension reveals a typical hemochromogen spectrum at 550-560 nm. This suggests that sea-urchin eggs contain an iron-porphyrin catalyst other than cytochromes (BALL and MEYERHOF, 1940).

Cytochromes a and b were first identified spectroscopically in unfertilized *Psammechinus miliaris* eggs frozen in liquid air (ROTHSCHILD, 1949; BOREI, 1950). Spectroscopic identification of cytochrome a_3 was believed to be prevented by egg pigment. After repeated sedimentation at 10,000 g for 1 hour, *Lytechinus pictus* egg homogenate (13 ml *M*/15 phosphate buffer, pH 7.1 + 7 ml jellyless egg) yielded a precipitate that showed strong a and b bands (YČAS, 1954).

A complete and active cytochrome system (flavoprotein→b→c→a→a_3) occurs in a mitochondrial fraction obtained from *Sphaerechinus* and *Paracentrotus* eggs by differential centrifugation in a sucrose density gradient (1.5 ml of mitochondrial suspension in 0.5 mol sucrose/1.5 ml of 1.0 mol sucrose/1.5 ml of 1.5 mol sucrose, 1 hour, 30,000 rpm), as may be shown by difference spectra (Fig. 18.24). When the preparation is reduced with succinate, in the SORET region the spectrum shows 2 distinct peaks at 430 nm and 444 nm corresponding to the γ bands of cytochromes b and a_3. Although the γ band of cytochrome c should appear at 415 nm, it is not visible because of the low concentration of cytochrome c in washed mitochondria.

In the visible region of the spectrum, 3 bands are present: the one at 520 nm corresponds to the β band of cytochromes b and c, while those at 562 nm and 602 nm are the α bands of cytochromes b and c and of cytochrome a. The trough at 470 nm corresponds to the reduced flavins. The addition of dithionite increases the γ band of cytochrome b, and in this spectrum the γ band of cytochrome a_3 appears as a shoulder at 444 nm.

Terminal respiration in sea-urchin eggs is mediated through the cytochrome system. In unfertilized eggs this system is similar to that of mammals, in that the cytochrome bands in the difference spectra have peaks in the same positions (MAGGIO and GHIRETTI-MAGALDI, 1958).

18.3.5 Natural Inhibitors of Cytochrome Oxidase

Although oxygen consumption in sea-urchin eggs increases severalfold at fertilization, cytochrome activity in the mitochondria increases only slightly - 35% at most (MAGGIO, 1959). It is possible that mitochondrial structural and physicochemical rearrangement at fertilization causes the increased respiration (MAGGIO, 1957; MONROY, 1957a, 1957b; NAKANO and MONROY, 1958).

Tested in isolated mitochondria, cytochrome oxidase from unfertilized eggs is fully active and is not modified by aging. This indicates that other factors in the eggs, probably extra-mitochondrial, act on the mitochondria and reduce the respiratory rate (MAGGIO, 1959).

The supernatant (105,000 g, 60 minutes) of unfertilized *Paracentrotus lividus* egg homogenate (in sucrose or 0.01 mol phosphate buffer,

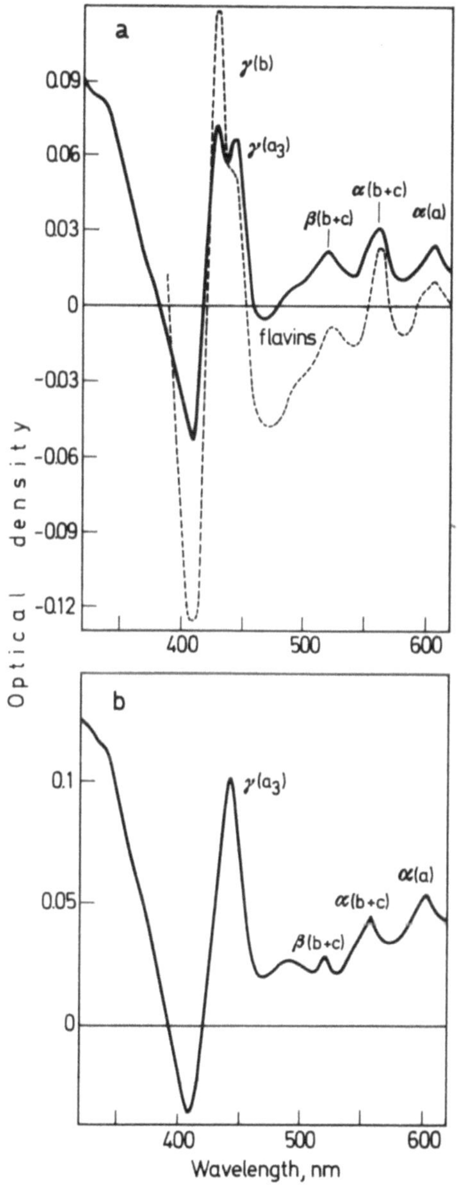

Fig. 18.24 a and b. Differential spectra (reduced spectra minus oxidized spectra) of mitochondrial suspension. (a) *Sphaerechinus granularis* eggs homogenized in 0.5 mol sucrose containing 5 x 10^{-4} mol versene (pH 7.3) sedimented at 20,000 g for 30 min. (b) *Paracentrotus lividus* eggs homogenized in 0.44 *M* sucrose containing 0.1 mol citrate buffer (pH 6.3) and 5 x 10^{-4} mol versene. Solid line indicates reduction with succinate; dotted line, reduction with dithionite. (From MAGGIO and GHIRETTI-MAGALDI, 1958)

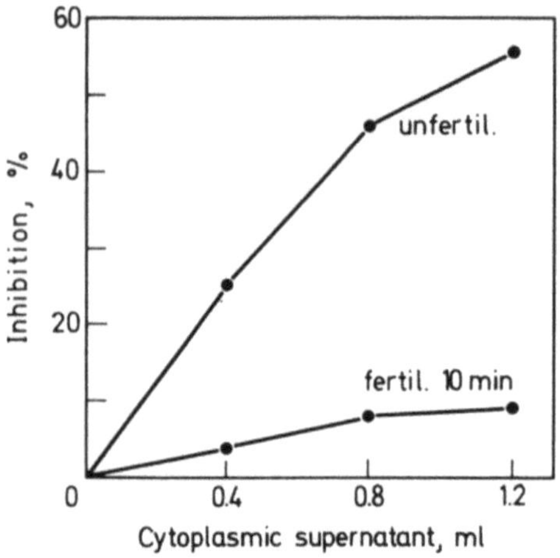

Fig. 18.25. Inhibitory effect of cytoplasmic supernatant of unfertilized and fertilized *Paracentrotus lividus* eggs on cytochrome oxidase activity of mouse liver mitochondria. Each vessel contains in final volume of 3 ml: 0.2 ml mitochondrial suspension, 2.46×10^{-5} mol cytochrome c, 30 μmol Na-ascorbate, 0.033 mol phosphate buffer (pH 7.4), and 0.2 mol sucrose. (From MAGGIO and MONROY, 1959)

pH 5.0) contains an inhibitor of mitochondrial cytochrome oxidase of mouse liver or sea-urchin eggs, while the supernatant from fertilized eggs (10 minutes after fertilization) has much less inhibitory effect. In the supernatant from eggs 1 hour after fertilization, the inhibitory effect is negligible (Fig. 18.25)(MAGGIO and MONROY, 1959). This inhibitor interferes with cytochrome oxidase oxidation of reduced cytochrome c but does not inhibit the reduction of cytochrome c.

The nature of this inhibitory factor is not yet clear. It is dialyzable and quite stable in 5% TCA or PCA but is destroyed by alkaline treatment or by heating at 95°C for 30 minutes. It is adsorbed on Dowex-50 in neutral solution and on Ecteola-cellulose in 0.04 mol phosphate solution (pH 7.2). It is eluted with pH 5 but not with pH 6-6.5. As chromatograms of the pH 5 fraction do not show ninhydrin-positive spots and nitrogen determinations are negative, the inhibitor seems not to contain nitrogen (MAGGIO et al., 1960).

References

AKETA, K., 1957. Quantitative analyses of lactic acid and related compounds in sea urchin eggs at the time of fertilization. Embryologia 3, 267-278.

AKETA, K., 1961a. Studies on the production of the fertilization acid in sea urchin eggs. I. Acid production at fertilization and activation, and the effect of some metabolic inhibitors. Embryologia 5, 397-405.

AKETA, K., 1961b. Studies on the production of the fertilization acid in sea urchin eggs. II. Experimental analysis of the production mechanism. Embryologia 5, 406-412.

AKETA, K., 1962. Studies on the production of the fertilization acid in sea urchin eggs. III. Cytochemical examination on the possible role of acid mucopolysaccharide components in the acid production. Embryologia 7, 223-227.
AKETA, K., 1963. Studies on the acid production at fertilization of sea urchin eggs. Exp. Cell Res. 30, 93-97.
AKETA, K., 1964. Some comparative remarks on the transient change in lactic acid content in sea urchin eggs following fertilization. Exp. Cell Res. 34, 192-194.
AKETA, K., BIANCHETTI, R., MARRE, E., MONROY, A., 1964. Hexose monophosphate level as a limiting factor for respiration in unfertilized sea urchin eggs. Biochim. Biophys. Acta 86, 211-215.
ASHBEL, R., 1930. Sul quoziente respiratorio delle uova fecondate e non fecondate dei ricci di mare (Arbacia pustulosa). Boll. soc. ital. biol. sper. 5, 72-74.
BÄCKSTRÖM, S., 1959. Activity of glucose-6-phosphate dehydrogenase in sea urchin embryos of different developmental trends. Exp. Cell Res. 18, 347-356.
BÄCKSTRÖM, S., 1963. 6-Phosphogluconate dehydrogenase in sea urchin embryos. Exp. Cell Res. 32, 566-569.
BÄCKSTRÖM, S., HULTIN, K., HULTIN, T., 1960. Pathways of glucose metabolism in early sea urchin development. Exp. Cell Res. 19, 634-636.
BALL, E.G., MEYERHOFF, B., 1940. On the occurrence of iron-porphyrin compounds and succinic dehydrogenase in marine organisms possessing the copper blood pigment hemocyanine. J. Biol. Chem. 134, 483-493.
BALLENTINE, R., 1940. Analysis of the changes in respiratory activity accompanying the fertilization of marine eggs. J. Cell. Comp. Physiol. 15, 217-232.
BARNETT, R.C., 1953. Cell division inhibition of Arbacia and Chaetopterus eggs and its reversal by KREBS cycle intermediates and certain phosphate compounds. Biol. Bull. 104, 263-274.
BARNETT, R.C., DOWNEY, M., 1955. Phosphorus content and ^{32}P uptake of marine eggs during cell division. Fed. Proc. 14, 9.
BARRON, E.S.G., 1932. Studies on cell metabolism. I. The oxygen consumption of Nereis eggs before and after fertilization. Biol. Bull. 62, 42-45.
BARRON, E.S.G., GOLDINGER, J.M., 1941. Intermediary carbohydrate metabolism of eggs and sperm of Arbacia punctulata before and after fertilization. Biol. Bull. 81, 289.
BERGAMI, M., MANSOUR, T.E., SCARANO, E., 1968. Properties of glycogen phosphorylase before and after fertilization in the sea urchin eggs. Exp. Cell Res. 49, 650-655.
BILLIAR, R.B., BRUNGARD, J.C., VILLEE, C.A., 1964. D-malate: Effects on activity of L-malate dehydrogenase in developing sea urchin embryos. Science 146, 1464-1465.
BILLIAR, R.B., ZELEWSKI, L., VILLEE, C.A., 1966. L-malate dehydrogenase activity and protein synthesis in sea urchin embryos. Develop. Biol. 13, 282-295.
BLACK, R.E., TYLER, A., 1959a. The oxidation of C^{13}-labelled carbon monoxide by embryos of Urechis caupo and Strongylocentrotus purpuratus. Anat. Rec. 134, 535.
BLACK, R.E., TYLER, A., 1959b. Effects of fertilization and development on the oxidation of carbon monoxide by eggs of Strongylocentrotus and Urechis as determined by use of ^{13}C. Biol. Bull. 117, 443-453.
BLACK, R.E., TYLER, A., 1959c. Cytochrome oxidase and oxidation of CO in eggs of sea urchin Strongylocentrotus purpuratus. Biol. Bull. 117, 454-457.
BLANCHARD, K.C., 1935. The nucleic acid of the eggs of Arbacia punctulata. J. Biol. Chem. 108, 251-256.
BLOMQUIST, C.H., 1969. Reversible inactivation of nicotinamide adenine

dinucleotide kinase in extracts of unfertilized sea urchin eggs. Exp. Cell Res. 56, 172-174.
BOLST, A.L., WHITELEY, A.H., 1957. Studies of the metabolism of phosphorus in the development of the sea urchin, Strongylocentrotus purpuratus. Biol. Bull. 112, 276-287.
BOREI, H., 1933. Beiträge zur Kenntnis der Vorgänge bei der Befruchtung des Echinodermeneies. Z. Physiol. 20, 258-266.
BOREI, H., 1948. Respiration of oocytes, unfertilized eggs and fertilized eggs from Psammechinus and Asterias. Biol. Bull. 95, 124-150.
BOREI, H., 1949. Independence of post-fertilization respiration in the sea urchin egg from the level of respiration before fertilization. Biol. Bull. 96, 117-122.
BOREI, H., 1950. Cytochrome c in sea-urchin eggs. Acta Chem. Scand. 4, 1607-1608.
BOREI, H., LYBING, S., 1949. Temperature coefficients of respiration in Psammechinus eggs. Biol. Bull. 96, 107-116.
BOYD, M., 1928. A comparison of the oxygen consumption of unfertilized and fertilized eggs of Fundulus heterocritus. Biol. Bull. 55, 92-100.
BRACHET, J., 1934. Étude du métabolisme de l'oeuf de Grenouille (Rana fusca) au cours du développement. 1-La respiration et la glycolyse, de la segmentation à l'éclosion. Arch. Biol. 45, 611-727.
BROCK, N., DRUCKREY, H., HERKEN, H., 1938. Der Stoffwechsel des geschädigten Gewebes. III. Arch. exp. Pathol. Pharmark. 188, 451-464.
CASTANEDA, M., TYLER, A., 1968. Adenyl cyclase in plasma membrane preparations of sea urchin eggs and its increase in activity after fertilization. Biochem. Biophys. Res. Comm. 33, 782-787.
CHAIGNE, M., 1934. Sur la surcharge en glycogène des organes reproducteurs de quelques invertébrés au moment de la ponte. Compt. rend. soc. biol. 115, 174-176.
CHAMBERS, E.L., MENDE, T.J., 1953a. The adenosine triphosphate content of the unfertilized and fertilized eggs of Asterias forbesii and Strongylocentrotus droebachiensis. Arch. Biochem. Biophys. 44, 46-56.
CHAMBERS, E.L., MENDE, T.J., 1953b. Alterations of the inorganic phosphate and arginine phosphate content in sea urchin eggs following fertilization. Exp. Cell Res. 5, 508-519.
CHAMBERS, E.L., WHITE, W.E., 1949. The accumulation of phosphate and evidence for synthesis of adenosinetriphosphate in the fertilized sea urchin egg. Biol. Bull. 97, 225-226.
CHAMBERS, E.L., WHITE, W.E., JEUNG, N., BROOKS, S.C., 1948. Penetration and effects of low temperature and cyanide on penetration of radioactive potassium into the eggs of Strongylocentrotus purpuratus and Arbacia punctulata. Biol. Bull. 95, 252-253.
CLELAND, K.W., 1950a. Respiration and cell division in developing oyster eggs. Proc. Linn. Soc. N.S.W. 75, 282-295.
CLELAND, K.W., 1950b. Intermediary metabolism of unfertilized oyster eggs. Proc. Linn. Soc. N.S.W. 75, 296-319.
CLELAND, K.W., ROTHSCHILD, LORD, 1952a. The metabolism of the sea-urchin egg. Anaerobic breakdown of carbohydrate. J. Exp. Biol. 29, 285-294.
CLELAND, K.W., ROTHSCHILD, LORD, 1952b. The metabolism of the sea urchin egg. Oxidation of carbohydrate. J. Exp. Biol. 29, 416-428.
CLOWES, G.H.A., KRAHL, M.E., 1936. Studies on cell metabolism and cell division. I. On the relation between molecular structures, chemical properties, and biological activities of nitrophenols. J. Gen. Physiol. 20, 145-171.
CLOWES, G.H.A., KRAHL, M.E., 1940. Studies on cell metabolism and cell division. III. Oxygen consumption and cell division of fertilized sea urchin eggs in the presence of respiratory inhibitors. J. Gen. Physiol. 23, 401-411.

CONNORS, W.M., SCHEER, B.T., 1947. Adenosine triphosphatase in the sea urchin egg. J. Cell. Comp. Physiol. 30, 271-283.
CRANE, R.K., 1947. The distribution of phosphorus in the unfertilized egg of Arbacia punctulata. Biol. Bull. 93, 192-193.
CRANE, R.K., KELTCH, A.K., 1949. Dinitrocresol and phosphate stimulation of the oxygen consumption of a cell-free oxidative system obtained from sea-urchin eggs. J. Gen. Physiol. 32, 503-509.
DEUTSCH, H.F., GUSTAFSON, T., 1952. The changes in catalase and cytochrome oxidase activity in developing sea-urchin eggs. Arkiv Kemi 4, 221-231.
DE VINCENTIIS, M., HÖRSTADIUS, S., RUNNSTRÖM, J., 1966. Studies on controlled and released respiration in animal and vegetal halves of the embryo of the sea urchin, Paracentrotus lividus. Exp. Cell Res. 41, 535-544.
DE VINCENTIIS, M., RUNNSTRÖM, J., 1967. Studies on controlled and released respiration in animalized and vegetalized embryos of the sea urchin Paracentrotus lividus. Exp. Cell Res. 45, 681-689.
DICKENS, F., SIMER, F., 1930. Carbohydrate metabolism of normal and tumor tissue. I. A method for the measurement of the respiratory quotient. Biochem. J. 24, 905-913.
EPEL, D., 1963. The effects of carbon monoxide inhibition on ATP level and the rate of mitosis in the sea urchin egg. J. Cell Biol. 17, 315-319.
EPEL, D., 1964a. A primary metabolic change of fertilization: Interconversion of pyridine nucleotides. Biochem. Biophys. Res. Comm. 17, 62-68.
EPEL, D., 1964b. Simultaneous measurement of TPNH formation and respiration following fertilization of the sea-urchin egg. Biochem. Biophys. Res. Comm. 17, 69-73.
EPEL, D., 1965. Some aspects of metabolic control in the fertilization transition of sea urchin eggs. In: Control of Energy Metabolism (B. Chance, R.W. Estabrook, J.R. Williamson, eds.), pp. 267-272. New York: Academic Press.
EPEL, D., 1969. Does ADP regulate respiration following fertilization of sea urchin eggs? Exp. Cell Res. 58, 312-319.
EPEL, D., WEAVER, A.M., MUCHIMORE, A., SCHIMKE, R.T., 1969. β-1,3-glucanase of sea urchin eggs: release from particles at fertilization. Science 163, 294-296.
EPHRUSSI, B., 1933. Contribution à l'analyse des premiers stades du développement de l'oeuf. Action de la température. Arch. Biol. 44, 1-147.
EPHRUSSI, B., RAPKINE, L., 1928. Composition chimique de l'oeuf d'Oursin Paracentrotus lividus LK. et ses variations au cours du développement. Ann. Physiol. Physicochem. Biol. 4, 386-398.
ESTABROOK, R.W., MAITRA, P.K., 1962. A fluorimetric method for the quantitative microanalysis of adenine and pyridine nucleotides. Anal. Biochem. 3, 369-382.
FAURE-FREMIET, E., 1922. Echanges respiratoires des oeufs de Sabellaria alveolata L. au cours de la segmentation et de la cytolyse. C.R. soc. Biol. 86, 20-23.
FUJII, T., OHNISHI, T., 1962. Inhibition of acid production by nicotinamide and other inhibitors of DPNase in the sea urchin. J. Fac. Sci., Univ. Tokyo, Sect. IV 9, 333-348.
GERARD, R.W., RUBENSTEIN, B.B., 1934. A note on the respiration of Arbacia eggs. J. Gen. Physiol. 17, 375-381.
GHIRETTI, F., D'AMELIO, V., 1956. The metabolism of pentose phosphate in sea urchin sperm and eggs. Exp. Cell Res. 10, 734-737.
GOLDINGER, J.M., BARRON, E.S.G., 1946. The pyruvate metabolism of sea urchin eggs during the process of cell division. J. Gen. Physiol. 30, 73-82.

GONSE, P.H., 1960. Respiratory levels in mature sea urchin eggs. J. Embryol. Exp. Morphol. 8, 73-93.
GRAY, J., 1925. The mechanism of cell-division. II. Oxygen consumption during cleavage. Proc. Camb. Phil. Soc. (Biol. Sci.) 1, 225-236.
GRAY, J., 1927. The mechanism of cell-division. III. The relationship between cell-division and growth in segmenting eggs. J. Exp. Biol. 4, 313-321.
GRIFFITHS, W.M., WHITELEY, A.H., 1964. A study of the mechanism of phosphate transport in sea urchin eggs by ion exchange analysis of rapidly labeled compounds. Biol. Bull. 126, 69-82.
GUSTAFSON, T., HASSELBERG, I., 1951. Studies on enzymes in the developing sea urchin egg. Exp. Cell Res. 11, 642-672.
HARVEY, E.N., 1932. Physical and chemical constants of the egg of the sea urchin, Arbacia punctulata. Biol. Bull. 74, 267-277.
HERK, A.W.H., VAN, 1933. The metabolism of the eggs of the sea urchin. I. The influence on respiration and lactic acid formation through dyestuffs. Arch. Neerl. Physiol. 18, 578-602.
HIRAMOTO, Y., 1959a. Changes in electric properties upon fertilization in the sea urchin egg. Exp. Cell Res. 16, 421-424.
HIRAMOTO, Y., 1959b. Electric properties of echinoderm eggs. Embryologia 4, 219-235.
HOLTER, H., ZEUTHEN, E., 1944. The respiration of the egg and embryos of the ascidian, Ciona intestinalis L. C.R. Trav. Lab. Carlsberg, Sér. Chim. 25, 33-65.
HORWITZ, B.A., 1965. Rates of oxygen consumption of fertilized and unfertilized Asterias, Arbacia and Spisula eggs. Exp. Cell Res. 38, 620-625.
HULTIN, T., 1949. The effect of calcium on respiration and acid formation in homogenates of sea-urchin eggs. Ark. Kemi, Mineral. Geol. 26A (27), 1-10.
HULTIN, T., 1950a. On the oxygen uptake of Paracentrotus lividus egg homogenates after the addition of calcium. Exp. Cell Res. 1, 159-168.
HULTIN, T., 1950b. On the acid formation, breakdown of cytoplasmic inclusions, and increased viscosity in Paracentrotus egg homogenates after the addition of calcium. Exp. Cell Res. 1, 272-283.
HULTIN, T., 1953. Incorporation of C^{14}-labeled carbonate and acetate into sea urchin embryos. Ark. Kemi 6, 195-200.
HULTIN, T., 1957. Acid-soluble nucleotides in the early development of Psammechinus miliaris. Exp. Cell Res. 12, 413-415.
HULTIN, T., WESSEL, G., 1952. Incorporation of C^{14}-labeled carbon dioxide into the proteins of developing sea urchin eggs. Exp. Cell Res. 3, 613-616.
HUTCHENS, J.O., KELTCH, A.K., KRAHL, M.E., CLOWES, G.H.A., 1942a. Studies on cell metabolism and cell division. VI. Observations on the glycogen content, carbohydrate consumption, lactic acid production, and ammonia production of eggs of Arbacia punctulata. J. Gen. Physiol. 25, 717-731.
HUTCHENS, J.O., KRAHL, M.E., CLOWES, G.H.A., 1939. Physiological effects of nitro- and halo-substituted phenols on Arbacia eggs in the presence of ammonia. J. Cell. Comp. Physiol. 14, 313-325.
HUTCHENS, J.O., KOPAC, M.J., KRAHL, M.E., 1942b. The cytochrome oxidase content of centrifugally separated fractions of unfertilized Arbacia eggs. J. Cell. Comp. Physiol. 20, 113-116.
IMMERS, J., 1952. Carbohydrate components in unfertilized sea urchin eggs. Arkiv Zool. 3, 367-371.
IMMERS, J., 1956. Changes in acid mucopolysaccharides attending the fertilization and development of the sea urchin. Arkiv Zool. 9, 367-375.
IMMERS, J., 1960. Studies on cytoplasmic components of sea urchin eggs stratified by centrifugation. Exp. Cell Res. 19, 499-514.

IMMERS, J., RUNNSTRÖM, J., 1960. Release of respiratory control by 2,4-dinitrophenol in different stages of sea urchin development. Develop. Biol. 2, 90-104.
ISHIHARA, K., 1957. Release and activation of aldolase at fertilization in sea urchin eggs. J. Fac. Sci., Tokyo Univ., Sect. IV 8, 71-93.
ISHIHARA, K., 1958a. Activation of glycolytic process at the time of fertilization in sea urchin eggs. Annot. Zool. Japon. 31, 1-5.
ISHIHARA, K., 1958b. Compensatory respiration and pentose formation in sea urchin eggs by the treatment with monoiodoacetate. Sci. Rep. Saitama Univ., Ser. B 3, 1-10.
ISHIHARA, K., 1958c. Effect of butyric acid on aldolase complex in sea urchin eggs. Sci. Rep. Saitama Univ., Ser. B 3, 11-20.
ISHIHARA, K., 1958d. Enhanced respiration of sea urchin eggs induced by mechanical stimulation. Sci. Rep. Saitama Univ., Ser. B 3, 21-32.
ISHIHARA, K., 1968a. Chemical analysis of glycoproteins in the egg surface of the sea urchin, Arbacia punctulata. Biol. Bull. 134, 425-433.
ISHIHARA, K., 1968b. An analysis of acid polysaccharides produced at fertilization of sea urchin. Exp. Cell Res. 51, 473-484.
ISHIKAWA, M., 1954. Relation between the breakdown of the cortical granules and permeability to water in the sea urchin egg. Embryologia 2, 57-62.
ISONO, N., 1962. Carbohydrate metabolism in sea urchin eggs. II. Pentose phosphate cycle in developing eggs. J. Fac. Sci., Univ. Tokyo, Sect. IV 9, 369-377.
ISONO, N., 1963a. Carbohydrate metabolism in sea urchin eggs. III. Changes in respiratory quotient during early embryonic development. Annot. Zool. Japon. 36, 126-132.
ISONO, N., 1963b. Carbohydrate metabolism in sea urchin eggs. IV. Intracellular localization of enzymes of pentose phosphate cycle in unfertilized and fertilized eggs. J. Fac. Sci., Univ. Tokyo, Sect. IV 10, 37-53.
ISONO, N., 1963c. Studies on glucose-6-phosphate dehydrogenase in sea urchin eggs. II. J. Fac. Sci., Univ. Tokyo, Sect. IV, 10, 67-74.
ISONO, N., 1967a. Release of glucose-6-phosphate dehydrogenase of unfertilized sea urchin eggs, in vitro (preliminary report)(in Japanese with English summary). Zool. Mag. 76, 57-59.
ISONO, N., 1967b. Increase in respiratory rate following fertilization of sea urchin eggs (in Japanese with English summary). Zool. Mag. 76, 207-215.
ISONO, N., 1969. Release of glucose-6-phosphate dehydrogenase following fertilization of sea urchin eggs (preliminary report with English summary). Zool. Mag. 78, 305-306.
ISONO, N., ISHIDA, J., 1962. Carbohydrate metabolism in sea urchin eggs. I. Pentose phosphate cycle in unfertilized sea urchin eggs. J. Fac. Sci., Univ. Tokyo, Sect. IV 9, 357-367.
ISONO, N., YANAGISAWA, T., 1966. Acid-soluble nucleotides in the sea urchin egg. II. Uridine diphosphate sugars. Embryologia 9, 184-195.
ISONO, N., YASUMASU, I., 1966. Carbohydrate metabolism in sea urchin embryos (preliminary report with English summary). Zool. Mag. 75, 276-279.
ISONO, N., YASUMASU, I., 1968. Pathways of carbohydrate breakdown in sea urchin eggs. Exp. Cell Res. 50, 616-626.
ISONO, N., TSUSAKA, A., NAKANO, E., 1963. Studies on glucose-6-phosphate dehydrogenase in sea urchin eggs. I. J. Fac. Sci., Univ. Tokyo, Sect. IV 10, 55-66.
JANDORF, B.J., KRAHL, M.E., 1942. Studies on cell metabolism and cell division. VIII. The diphosphopyridine nucleotide (cozymase) content of eggs of Arbacia punctulata. J. Gen. Physiol. 25, 749-754.

KELTCH, A.K., KRAHL, M.E., CLOWES, G.H.A., 1956. Alteration by dinitrocresol of pathways for glucose oxidation in eggs of Arbacia punctulata. J. Gen. Physiol. 40, 27-35.

KELTCH, A.K., STRITTMATER, C.F., WALTERS, C.P., CLOWES, G.H.A., 1949. Oxidative phosphorylation by a cell-free particulate enzyme system from unfertilized Arbacia eggs. Biol. Bull. 97, 242-243.

KELTCH, A.K., STRITTMATER, C.F., WALTERS, C.P., CLOWES, G.H.A., 1950. Oxidative phosphorylation by a cell-free particulate system from unfertilized Arbacia eggs. J. Gen. Physiol. 33, 547-553.

KORR, I.M., 1937. Respiratory mechanisms in the unfertilized and fertilized sea urchin egg. A temperature analysis. J. Cell. Comp. Physiol. 10, 461-485.

KORR, I.M., 1939. Oxidation-reductions in heterogeneous systems. Cold Spring Harb. Symp. 7, 74-93.

KRAHL, M.E., 1950. Metabolic activities and cleavage of egg of the sea urchin, Arbacia punctulata. A review, 1932-1949. Biol. Bull. 98, 175-217.

KRAHL, M.E., 1956. Oxidative pathways for glucose in eggs of the sea urchin. Biochim. Biophys. Acta 20, 27-32.

KRAHL, M.E., CLOWES, G.H.A., 1936. Studies on cell metabolism and cell division. II. Stimulation of cellular oxidation and reversible inhibition of cell division by dihalo- and trihalophenols. J. Gen. Physiol. 20, 173-184.

KRAHL, M.E., CLOWES, G.H.A., 1938a. Physiological effects of nitro- and halo-substituted phenols in relation to extracellular and intracellular hydrogen ion concentration. I. Dissociation constants and theory. J. Cell Comp. Physiol. 11, 1-20.

KRAHL, M.E., CLOWES, G.H.A., 1938b. Physiological effects of nitro- and halo-substituted phenols in relation to extracellular and intracellular hydrogen ion concentration. II. Experiments with Arbacia eggs. J. Cell. Comp. Physiol. 11, 21-39.

KRAHL, M.E., CLOWES, G.H.A., 1940. Studies on cell metabolism and cell division. IV. Combined action of substituted phenols, cyanide, carbon monoxide, and other respiratory inhibitors on respiration and cell division. J. Gen. Physiol. 23, 413-428.

KRAHL, M.E., JANDORF, B.J., CLOWES, G.H.A., 1942. Studies on cell metabolism and cell division. VII. Observations on the amount and possible function of diphosphothiamine (cocarboxylase) in eggs of Arbacia punctulata. J. Gen. Physiol. 25, 733-747.

KRAHL, M.E., KELTCH, A.K., CLOWES, G.H.A., 1939. Oxygen consumption and cell division of fertilized Arbacia eggs in the presence of respiratory inhibitors. Biol. Bull. 77, 318-319.

KRAHL, M.E., KELTCH, A.K., CLOWES, G.H.A., 1940. Flavin-dinucleotide in eggs of the sea urchin, Arbacia punctulata. Proc. soc. Exp. Biol. Med. 45, 719-721.

KRAHL, M.E., KELTCH, A.K., NEUBECK, C.E., CLOWES, G.H.A., 1941. Studies on cell metabolism and cell division. V. Cytochrome oxidase activity in the eggs of Arbacia punctulata. J. Gen. Physiol. 24, 597-617.

KRAHL, M.E., KELTCH, A.K., WALTERS, C.P., CLOWES, G.H.A., 1953. Hexokinase and isomerase activity in eggs of the sea urchin, Arbacia punctulata, and other marine forms. Biol. Bull. 105, 377.

KRAHL, M.E., KELTCH, A.K., WALTERS, C.P., CLOWES, G.H.A., 1954a. Activity of glucose-6-phosphate and 6-phosphogluconate dehydrogenase in relation to glycolytic enzymes of Arbacia eggs. Biol. Bull. 107, 315-316.

KRAHL, M.E., KELTCH, A.K., WALTERS, C.P., CLOWES, G.H.A., 1954b. Hexokinase activity from eggs of the sea urchin, Arbacia punctulata. J. Gen. Physiol. 38, 31-39.

KRAHL, M.E., KELTCH, A.K., WALTERS, C.P., CLOWES, G.H.A., 1955. Glucose-6-phosphate and 6-phosphogluconate dehydrogenase from eggs of the sea urchin, Arbacia punctulata. J. Gen. Physiol. 38, 431-439.

KRANE, S.M., CRANE, R.K., 1958. Changes in the levels of triphosphopyridine nucleotide in the eggs of Arbacia punctulata subsequent of fertilization: presence of pyridine nucleotide transhydrogenase and diphosphopyridine nucleotide kinase. Biol. Bull. 115, 355.
KRANE, S.M., CRANE, R.K., 1960. Changes in levels of triphosphopyridine nucleotide in marine eggs subsequent to fertilization. Biochim. Biophys. Acta 43, 369-373.
KRISZAT, G., 1954. Die Wirkung von Purinen, Nucleosiden, Nucleotiden und Adenosintriphosphat auf die Teilung und Entwicklung des Seeigeleies bei Anwendung von Dinitrophenol. Exp. Cell Res. 6, 425-439.
KUN, E., ABOOD, L.G., 1949. Colorimetric estimation of succinic dehydrogenase by triphenyltetrazoliumchloride. Science 109, 144-146.
LANDAU, J.V., MARSLAND, D., ZIMMERMAN, A., 1955. The energetics of cell division: Effects of adenosine triphosphate and related substances on the furrowing capacity of marine eggs (Arbacia and Chaetopterus). J. Cell. Comp. Physiol. 45, 309-329.
LARDY, H.A., WELLMAN, H., 1952. Oxidative phosphorylations: role of inorganic phosphate and acceptor systems in control of metabolic rates. J. Biol. Chem. 195, 215-224.
LASER, H., ROTHSCHILD, LORD, 1939. The metabolism of the eggs of Psammechinus miliaris during the fertilization reaction. Proc. Roy. Soc. London B 126, 539-557.
LENTINI, R., 1961. The oxygen uptake of Ciona intestinalis eggs during development in normal and in experimental conditions. Acta Embryol. Morphol. Exp. 4, 209-218.
LILLIE, R.S., 1916. Increase of permeability of water following normal and artificial activation in the sea urchin eggs. Am. J. Physiol. 40, 249-266.
LINDAHL, P.E., 1936. Zur Kenntnis der physiologischen Grundlagen der Determination im Seeigelkeim. Acta Zool. 17, 179-365.
LINDAHL, P.E., 1939a. Über die biologische Sauerstoffaktivierung nach Versuchen mit Kohlenmonoxyd an Seeigeleiern und Keimen. Z. Physiol. 27, 136-168.
LINDAHL, P.E., 1939b. Zur Kenntnis der Entwicklungsphysiologie des Seeigeleies. Z. Physiol. 27, 233-250.
LINDAHL, P.E., 1940. Über die CN-resistente Atmung des Seeigeleies. Arkiv Kemi, Mineral. Geol. 14 A (12), 1-31.
LINDAHL, P.E., HOLTER, H., 1941. Über die Atmung der Ovozyten erster Ordnung von Paracentrotus lividus und ihre Veränderung während der Reifung. Comp. Rend. Trav. Lab. Carlsberg, Ser. Chim. 24, 49-57.
LINDAHL, P.E., ÖHMAN, L.O., 1938. Weitere Studien über Stoffwechsel und Determination im Seeigelkeim. Biol. Zentralblatt 58, 179-218.
LINDBERG, O., 1943. Studien über das Problem des Kohlenhydratabbaus und der Säurebildung bei der Befruchtung des Seeigeleies. Arkiv Kemi, Mineral. Geol. 16 A (15), 1-15.
LINDBERG, O., 1945. On the metabolism of glycogen in the fertilization of the sea urchin egg. Arkiv Kemi, Mineral. Geol. 20 B (1), 1-8.
LINDBERG, O., 1946. On the occurrence of propanediol phosphate and its effect on the carbohydrate metabolism in animal tissues. Arkiv Kemi, Mineral. Geol. 23 A (2), 1-45.
LINDBERG, O., 1949. On the turnover of adenosine triphosphate in the sea urchin egg. Arkiv Kemi, Mineral. Geol. 26 B (13), 1-4.
LINDBERG, O., ERNSTER, L., 1948. On carbohydrate metabolism in homogenized sea urchin eggs. Biochim. Biophys. Acta 2, 471-477.
LITCHFIELD, J.B., WHITELEY, A.H., 1959. Studies on the mechanism of phosphate accumulation by sea urchin embryos. Biol. Bull. 117, 133-149.
LOEB, J., 1896. Untersuchungen über die physiologischen Wirkungen des Sauerstoffmangels. Pflüger's Arch. Physiol. 62, 249-295.

LOEB, J., 1906a. Versuche über den chemischen Charakter des Befruchtungsvorganges. Biochem. Z. 1, 183-206.
LOEB, J., 1906b. Über die Hemmung der toxischen Wirkung hypertonischer Lösungen auf das Seeigelei durch Sauerstoffmangel und Cyankalium. Pflüger's Arch. Physiol. 113, 487-511.
LOEB, J., WASTENEYS, H., 1911. Sind die Oxydationsvorgänge die unabhängige Variable in den Lebenserscheinungen? Biochem. Z. 36, 345-356.
LOEB, J., WASTENEYS, H., 1913. The influence of hypertonic solution upon the rate of oxidations in fertilized and unfertilized eggs. J. Biol. Chem. 14, 469-480.
LØVTRUP, S., IVERSON, R.M., 1969. Respiratory phases during early sea urchin development, measured with the automatic diver balance. Exp. Cell Res. 55, 25-32.
MACKINTOSH, F.R., BELL, E., 1969. Labelling of nucleotide pools in sea urchin eggs. Exp. Cell Res. 57, 71-73.
MAGGIO, R., 1957. Mitochondrial and cytoplasmic protease activity in sea urchin eggs. J. Cell. Comp. Physiol. 50, 135-144.
MAGGIO, R., 1959. Cytochrome oxidase activity in the mitochondria of unfertilized and fertilized sea urchin eggs. Exp. Cell Res. 16, 272-278.
MAGGIO, R., AJELLO, F., MONROY, A., 1960. Inhibitor of the cytochrome oxidase of unfertilized sea urchin eggs. Nature 188, 1195-1196.
MAGGIO, R., GHIRETTI-MAGALDI, A., 1958. The cytochrome system in mitochondria of unfertilized sea urchin eggs. Exp. Cell Res. 15, 95-102.
MAGGIO, R., MONROY, A., 1959. An inhibitor of cytochrome oxidase activity in the sea urchin egg. Nature 184, 68-69.
MAGGIO, R., VITTORELLI, M.L., RINALDI, A.M., MONROY, A., 1964. In vitro incorporation of amino acids into protein stimulated by RNA from unfertilized sea urchin eggs. Biochem. Biophys. Res. Comm. 15, 436-441.
MATSUNAGA, Y., YANAGISAWA, T., 1970. Glucose metabolism during the early development of sea urchin eggs (abstract in Japanese). Zool. Mag. 79, 356-357.
MAZIA, D., 1937. The release of calcium in Arbacia eggs on fertilization. J. Cell. Comp. Physiol. 10, 291-304.
MAZIA, D., 1961. Mitosis and the physiology of cell division. In: The Cell (J. Brachet, A.E. Mirsky, eds.), vol. 3, pp. 77-412. New York: Academic Press.
MAZIA, D., 1963. Synthetic activities leading to mitosis. J. Cell. Comp. Physiol. 62, 123-140.
MAZIA, D., PRESCOTT, D.M., 1954. Nuclear function and mitosis. Science 120, 120-122.
MEHL, J.W., SWANN, M.M., 1961. Acid and base production at fertilization in the sea urchin. Exp. Cell Res. 22, 233-245.
MENDE, J.M., CHAMBERS, E.L., 1953. The occurrence of arginine phosphate in echinoderm eggs. Arch. Biochem. Biophys. 45, 105-116.
MEYERHOF, O., 1911. Untersuchungen über die Wärmetönung der vitalen Oxydationsvorgänge in Eiern. I-III. Biochem. Z. 35, 246-328.
MINGANTI, A., 1957. Experiments on the respiration of Phallusia eggs and embryos (ascidians). Acta Embryol. Morphol. Exp. 1, 150-163.
MONNE, L., HÅRDE, S., 1951. On the cortical granules of the sea urchin egg. Arkiv Zool. 1, 487-498.
MONROY, A., 1957a. Swelling properties of the mitochondria of unfertilized and newly fertilized sea urchin eggs. Experientia 13, 398-399.
MONROY, A., 1957b. Adenosinetriphosphatase in the mitochondria of unfertilized and newly fertilized sea-urchin eggs. J. Cell. Comp. Physiol. 50, 73-82.

MONROY, A., 1965. Chemistry and physiology of fertilization. New York: Holt Rinehart and Winston, Inc.
MONROY, A., VITTORELLI, M.L., 1960. On a glycoprotein of the sea urchin eggs and its changes following fertilization. Experientia 16, 56-59.
MONROY, A., VITTORELLI, M.L., 1962. Utilization of ^{14}C-glucose for amino acids and protein synthesis by the sea urchin embryo. J. Cell. Comp. Physiol. 60, 285-287.
MOORE, R.O., VILLEE, C.A., 1962. Malic dehydrogenases in sea urchin eggs. Science 138, 508-509.
MOORE, R.O., VILLEE, C.A., 1963a. Multiple molecular forms of malate dehydrogenase in echinoderm embryos. Comp. Biochem. Physiol. 9, 81-94.
MOORE, R.O., VILLEE, C.A., 1963b. Malate dehydrogenase: Multiple forms in separated blastomeres of sea urchin embryos. Science 142, 389-390.
MUCHMORE, A.V., EPEL, D., WEAVER, A.M., SCHIMKE, R.T., 1969. Purification and properties of an exo-β-D-1,3-glucanase from sea urchin eggs. Biochim. Biophys. Acta 178, 551-560.
NAKANO, E., 1953. Respiration during maturation and at fertilization of fish eggs. Embryologia 2, 21-30.
NAKANO, E., MONROY, A., 1958. Incorporation of ^{35}S-methionine in the cell fractions of sea urchin eggs and embryos. Exp. Cell Res. 14, 236-244.
NEEDHAM, J., NEEDHAM, D.M., 1926. The hydrogen-ion concentration and oxidation-reduction potential of the cell-interior before and after fertilization and cleavage: A micro-injection study on marine eggs. Proc. Roy. Soc. London, B 99, 173-199.
NEEDHAM, J., NEEDHAM, D.M., 1930. On phosphorus metabolism in embryonic life. I. Invertebrate eggs. J. Exp. Biol. 7, 317-348.
NEEDHAM, D.M., NEEDHAM, J., BALDWIN, E., YUDKIN, J., 1932. A comparative study of the phosphagens, with some remarks on the origin of vertebrates. Proc. Roy. Soc. London, B 110, 260-294.
NEMOTO, S., 1970. Changes in the content of arginine phosphate during cell cycle of sea urchin eggs (abstract in Japanese). Zool. Mag. 79, 342-343.
NEMOTO, S., YANAGISAWA, T., 1969. Arginine phosphokinase and creatine phosphokinase in echinoderms. IX. From when the content of arginine phosphate begins to increase in fertilized eggs? (Abstract in Japanese). Zool. Mag. 78, 378-379.
NILSSON, R., 1959. Acid-soluble nucleotides in the unfertilized eggs of the sea-urchin Paracentrotus lividus. Acta Chem. Scand. 13, 395-408.
NILSSON, R., 1961. Acid-soluble nucleotides during early embryonic development of the sea-urchin Paracentrotus lividus. Acta Chem. Scand. 15, 583-591.
ÖHMAN, L.O., 1940. Über die Veränderung des respiratorischen Quotienten während der Frühentwicklung des Seeigeleies. Arkiv Zool. 32 A (15), 1-9.
OHNISHI, T., 1961. Changes in diphosphopyridine nucleotide contents of sea-urchin eggs after fertilization and during mitotic phases, with special reference to the effect of nicotinamide. J. Fac. Sci., Univ. Tokyo, Sect. IV, 9, 205-211.
OHNISHI, T., SUGIYAMA, 1963. Polarographic studies of oxygen uptake of sea-urchin eggs. Embryologia 8, 79-88.
ÖRSTRÖM, Å., 1932. Zur Analyse der Atmungssteigerung bei der Befruchtung des Seeigeleies auf der Grundlage von Versuchen über Oxydation und Reduktion von Dimethylparaphenylendiamin in der Eizelle. Protoplasma 15, 566-589.
ÖRSTRÖM, Å., LINDBERG, O., 1940. Über den Kohlenhydratstoffwechsel bei der Befruchtung des Seeigeleies. Enzymologia 8, 367-383.
OZAKI, H., WHITELEY, A.H., 1967. L-malate dehydrogenase of the sea-

urchin Strongylocentrotus purpuratus. Biochim. Biophys. Acta 146, 587-590.
OZAKI, H., WHITELEY, A.H., 1970. L-malate dehydrogenase in the development of the sea urchin Strongylocentrotus purpuratus. Develop. Biol. 21, 196-215.
PANDIT, C.G., CHAMBERS, R., 1932. Intracellular hydrion-concentration studies. IX. The pH of the egg of the sea-urchin, Arbacia punctulata. J. Cell. Comp. Physiol. 2, 243-249.
PATTON, G.W., Jr., METS, L., VILLEE, C.A., 1967. Malic dehydrogenase isozymes: distribution in developing nucleate and anucleate halves of sea urchin eggs. Science 156, 400-401.
PERLZWEIG, W.A., BARRON, E.S.G., 1928. Lactic acid and carbohydrate in sea urchin eggs under aerobic and anaerobic conditions. J. Biol. Chem. 79, 19-26.
PHILIPS, F.S., 1940. Oxygen consumption and its inhibition in the development of Fundulus and various pelagic eggs. Biol. Bull. 78, 256-274.
RAPOPORT, S., HOFMANN, E.C.G., GHIRETTI-MAGALDI, A., 1958. Über die Atmungsenzyme des Seeigeleies. Experientia 14, 169-170.
ROBBIE, W.A., 1946a. The quantitative control of cyanide in manometric experimentation. J. Cell. Comp. Physiol. 27, 181-209.
ROBBIE, W.A., 1946b. The effect of cyanide on the oxygen consumption and cleavage of the sea urchin egg. J. Cell. Comp. Physiol. 28, 305-324.
ROTHSCHILD, LORD, 1939. The effect of phlorizin on the metabolism of cytolyzing sea-urchin eggs. J. Exp. Biol. 16, 49-55.
ROTHSCHILD, LORD, 1949. The metabolism of fertilized and unfertilized sea-urchin eggs. The action of light and carbon monoxide. J. Exp. Biol. 26, 100-111.
ROTHSCHILD, LORD, 1951. Sperm-egg interacting substances and metabolic changes associated with fertilization. Biochem. Soc. Symp. 7, 40-51.
ROTHSCHILD, LORD, 1956. Fertilization. London: Methuen.
ROTHSCHILD, LORD, 1958. Acid production after fertilization of sea-urchin eggs. A re-examination of the lactic acid hypothesis. J. Exp. Biol. 35, 843-849.
ROTHSCHILD, LORD, BARNES, H., 1953. The inorganic constituents of the sea-urchin egg. J. Exp. Biol. 30, 534-544.
RUBENSTEIN, B.B., GERARD, R.W., 1934. Fertilization and the temperature coefficients of oxygen consumption in eggs of Arbacia punctulata. J. Gen. Physiol. 17, 677-685.
RUDNEY, H., 1954. The synthesis of dl-propanediol-1-phosphate and C^{14}-labeled propanediol and their isolation from liver tissue. J. Biol. Chem. 210, 353-360.
RUNNSTRÖM, J., 1930a. Atmungsmechanismus und Entwicklungserregung bei dem Seeigelei. Protoplasma 10, 106-173.
RUNNSTRÖM, J., 1930b. Spaltung und Atmung bei der Entwicklungserregung des Seeigeleies. Arkiv Zool. 21B (8), 1-5.
RUNNSTRÖM, J., 1932. Über den Mechanismus der Entwicklungserregung bei dem Seestern- und Seeigelkeim. Protoplasma 15, 448-452.
RUNNSTRÖM, J., 1933. Zur Kenntnis der Stoffwechselvorgänge bei der Entwicklungserregung des Seeigeleies. Biochem. Z. 258, 257-279.
RUNNSTRÖM, J., 1935a. On the influence of pyocyanine on the respiration of the sea-urchin egg. Biol. Bull. 68, 327-334.
RUNNSTRÖM, J., 1935b. Acid formation in frozen and thawed Arbacia punctulata eggs and its possible bearing on the problem of activation. Biol. Bull. 69, 345-350.
RUNNSTRÖM, J., 1935c. Influence of iodoacetate on activation and development of the eggs of Arbacia punctulata. Biol. Bull. 69, 351-355.
RUNNSTRÖM, J., 1949. The mechanism of fertilization in metazoa. Adv. Enzymol. 9, 241-327.

RUNNSTRÖM, J., 1956a. Some considerations on metabolic changes occurring at fertilization and during early development of the sea-urchin egg. Pubbl. Sta. Zool. Napoli 28, 315-340.
RUNNSTRÖM, J., 1956b. Some aspects of the initiating processes in the fertilization of the sea urchin egg. Zool. Anz. 156, 91-101.
RUNNSTRÖM, J., BOTTE, L., DE VINCENTIIS, M., 1970. Experiments and considerations on the state of respiratory chains before and after fertilization and in the early development of the sea urchin egg. Rendiconti di Istituto Lombardo di Scienze e Lettere 104, 20-32.
RUNNSTRÖM, J., HAGSTRÖM, B.E., PERLMAN, P., 1959. Fertilization. In: The Cell (J. Brachet, A.E. Mirsky, eds.), vol. 1, pp. 327-397. New York, London: Academic Press.
RUNNSTRÖM, J., IMMERS, J., 1956. The role of mucopolysaccharides in the fertilization of the sea-urchin eggs. Exp. Cell Res. 10, 354-363.
SACKS, J., 1949. A fractionation procedure for the acid-soluble phosphorus compounds of liver. J. Biol. Chem. 181, 655-666.
SCHMIDT, G., HECHT, L., THANNHAUSER, S.J., 1948. The behavior of the nucleic acids during the early development of the sea urchin egg (Arbacia). J. Gen. Physiol. 31, 203-207.
SCHMIDT, G., THANNHAUSER, S.J., 1945. A method for the determination of deoxyribonucleic acid, ribonucleic acid, and phosphoproteins in animal tissues. J. Biol. Chem. 161, 83-89.
SCHOLANDER, P.F., CLAFF, C.L., SVEINSSON, S.L., 1952a. Respiratory studies of single cells. I. Methods. Biol. Bull. 102, 157-177.
SCHOLANDER, P.F., CLAFF, C.L., SVEINSSON, S.L., SCHOLANDER, S.I., 1952b. Respiratory studies of single cells. III. Oxygen consumption during cell division. Biol. Bull. 102, 185-199.
SCHOLANDER, P.F., LEIVESTAD, H., SUNDNES, G., 1958. Cycling in the oxygen consumption of cleaving eggs. Exp. Cell Res. 15, 505-511.
SHAPIRO, H., 1935. The respiration of fragments obtained by centrifuging the egg of the sea urchin, Arbacia punctulata. J. Cell. Comp. Physiol. 6, 101-116.
SHEARER, C., 1922a. On the oxidation processes of the echinoderm egg during fertilization. Proc. Roy. Soc. London B 93, 213-229.
SHEARER, C., 1922b. On the heat production and oxidation processes of the echinoderm egg during fertilization and early development. Proc. Roy. Soc. London B 93, 410-425.
SIBLEY, J.A., LEHNINGER, A.L., 1949. Determination of aldolase in animal tissues. J. Biol. Chem. 177, 859-872.
STEFANELLI, A., 1938. Il metabolismo dell'uovo e dell'embrione studiato negli Anfibi Anuri. II. L'assunzione di ossigeno. Arch. Sc. Biol. 24, 411-441.
STOTT, F.C., 1931. The spawning of Echinus esculentus and some changes in gonad composition. J. Exp. Biol. 8, 133-150.
SUGINO, Y., 1960. Studies on deoxynucleosidic compounds. II. Deoxycytidine diphosphate choline in sea urchin eggs. Biochim. Biophys. Acta 40, 425-434.
SUGINO, Y., SUGINO, N., OKAZAKI, R., OKAZAKI, T., 1957. Deoxyribosidic compounds of sea urchin eggs. Biochim. Biophys. Acta 26, 453-454.
SUGINO, Y., SUGINO, N., OKAZAKI, R., OKAZAKI, T., 1960. Studies on deoxynucleosidic compounds. I. A modified microbioassay method and its application to sea urchin eggs and several other materials. Biochim. Biophys. Acta 40, 417-424.
SWANN, M.M., 1953. The mechanism of cell division. A study with carbon monoxide on the sea-urchin egg. Quart. J. Micro. Sci. 94, 369-379.
SWANN, M.M., 1954. The mechanism of cell division. Experiments with ether on the sea-urchin egg. Exp. Cell Res. 7, 505-517.
SWANN, M.M., 1957. The control of cell division: A review. I. General mechanisms. Cancer Res. 17, 727-757.

SWANN, M.M., 1958. The control of cell division: A review. II. Special mechanisms. Cancer Res. 18, 1118-1160.
TAGUCHI, S., 1962. Changes in the content of adenosine nucleotides during early development of sea urchins, Pseudocentrotus depressus and Hemicentrotus pulcherrimus. Annot. Zool. Japon. 35, 183-187.
TANG, P.S., 1931a. The oxygen tension-oxygen consumption curve of unfertilized Arbacia eggs. Biol. Bull. 60, 242-244.
TANG, P.S., 1931b. The rate of oxygen consumption of Asterias eggs before and after fertilization. Biol. Bull. 61, 468-471.
TANG, P.S., 1948. Rhythmic respiration in the sea urchin. Nature 162, 189.
TANG, P.S., GERARD, R.W., 1932. The oxygen tension-oxygen consumption curve of fertilized Arbacia eggs. J. Cell. Comp. Physiol. 1, 503-513.
TYLER, A., HUMASON, W.D., 1937. On the energetics of differentiation. VI. Comparison of temperature coefficients of the respiratory rates of unfertilized and of fertilized eggs. Biol. Bull. 73, 261-279.
TYLER, A., MONROY, A., 1959. Changes in rate of transfer of potassium across the membrane upon fertilization of eggs of Arbacia punctulata. J. Exp. Zool. 142, 675-690.
TYLER, A., RICCI, N., HOROWITZ, N.H., 1938. The respiration and fertilizable life of Arbacia eggs under sterile and non-sterile conditions. J. Exp. Zool. 79, 129-143.
TYLER, A., MONROY, A., KAO, C.Y., GRUNDFERT, H., 1956. Membrane potential and resistance of the starfish egg before and after fertilization. Biol. Bull. 111, 153-177.
WARBURG, O., 1908. Beobachtungen über die Oxidationsprozesse im Seeigelei. Z. physiol. Chem. 57, 1-16.
WARBURG, O., 1915. Notizen zur Entwicklungsphysiologie des Seeigeleies. Pflüger's Arch. Physiol. 160, 324-332.
WHITAKER, D.M., 1931a. On the rate of oxygen consumption by fertilized and unfertilized eggs. I. Fucus vesiculosus. J. Gen. Physiol. 15, 167-182.
WHITAKER, D.M., 1931b. On the rate of oxygen consumption by fertilized and unfertilized eggs. II. Cumingia tellinoides. J. Gen. Physiol. 15, 183-190.
WHITAKER, D.M., 1931c. On the rate of oxygen consumption by fertilized and unfertilized eggs. III. Nereis limbata. J. Gen. Physiol. 15, 191-200.
WHITAKER, D.M., 1933a. On the rate of oxygen consumption by fertilized and unfertilized eggs. IV. Chaetopterus and Arbacia punctulata. J. Gen. Physiol. 16, 475-495.
WHITAKER, D.M., 1933b. On the rate of oxygen consumption by fertilized and unfertilized eggs. V. Comparisons and interpretation. J. Gen. Physiol. 16, 497-528.
WHITELEY, A.H., 1949. The phosphorus compounds of sea-urchin eggs and the uptake of radio-phosphate upon fertilization. Amer. Nat. 83, 249-267.
WHITELEY, A.H., BALTZER, F., 1958. Development, respiratory rate and content of desoxyribonucleic acid in the hybrid Paracentrotus ♀ x Arbacia ♂. Pubbl. Sta. Zool. Napoli 30, 402-457.
WHITELEY, A.H., CHAMBERS, E.L., 1966. Phosphate transport in fertilized sea urchin eggs. II. Effects of metabolic inhibitors and studies on differentiation. J. Cell Physiol. 68, 309-324.
WIERCINSKI, F.J., 1944. An experimental study of protoplasmic pH determination. I. Amoebae and Arbacia punctulata. Biol. Bull. 86, 98-112.
WINTERS, R.W., 1962. Intracellular pH in Arbacia eggs. Biol. Bull. 123, 519-520.
YANAGISAWA, T., 1959a. Studies on echinoderm phosphagens. I. Occurrence and nature of phosphagens in sea-urchin eggs and spermatozoa. J. Fac. Sci., Univ. Tokyo, Sect. IV, 8, 473-479.

YANAGISAWA, T., 1959b. Studies on guanidine phosphoryltransferases. I. Occurrence in spermatozoa and eggs of sea-urchins. J. Fac. Sci., Univ. Tokyo, Sect. IV, 8, 481-486.
YANAGISAWA, T., 1968. Studies on echinoderm phosphagens. IV. Changes in the content of arginine phosphate in the sea-urchin egg after fertilization and the effect of some metabolic inhibitors. Exp. Cell Res. 53, 525-536.
YANAGISAWA, T., 1969a. Cell division and energy metabolism (in se). Japanese). Prot. Nuc. Acid Enzyme 14, 677-687.
YANAGISAWA, T., 1969b. Nucleic acid metabolism during the early development of sea-urchin eggs (in Japanese). Jap. J. Develop. Biol. 23, 138-139.
YANAGISAWA, T., ISONO, N., 1966. Acid-soluble nucleotides in the sea-urchin egg. I. Ion-exchange chromatographic separation and characterization. Embryologia 9, 170-183.
YASUMASU, I., NAKANO, E., 1963. Respiratory level of sea-urchin eggs before and after fertilization. Biol. Bull. 125, 182-187.
YASUMASU, I., ASAMI, K., SHOGER, R., FUJIWARA, A., 1973. Glycolysis in sea-urchin eggs. Rate-limiting steps and activation at fertilization. Exp. Cell Res. 80, 361-371.
YČAS, M., 1954. The respiration and glycolytic enzymes of sea-urchin eggs. J. Exp. Biol. 31, 208-217.
ZEUTHEN, E., 1944. Oxygen uptake during mitosis. Experiments on the eggs of the frog (Rana platyrrhina). C.R. Trav. Lab. Carlsberg, Sér. Chim. 25, 191-228.
ZEUTHEN, E., 1947. Respiration and cell division in eggs of the sea urchin, Psammechinus miliaris. Nature 160, 577-578.
ZEUTHEN, E., 1950a. Cartesian diver respirometer. Biol. Bull. 98, 139-143.
ZEUTHEN, E., 1950b. Respiration during cell division in the egg of the sea urchin Psammechinus miliaris. Biol. Bull. 98, 144-151.
ZEUTHEN, E., 1950c. Respiration and cell division in the egg of Urechis caupo. Biol. Bull. 98, 152-160.
ZEUTHEN, E., 1951. Segmentation, nuclear growth and cytoplasmic stage in eggs of echinoderms and amphibia. Pubbl. Sta. Zool. Napoli 13, 47-69.
ZEUTHEN, E., 1953. Biochemistry and metabolism of cleavage in the sea urchin egg, as resolved into its mitotic steps. Arch. Neer. Zool. 10, Suppl., 31-58.
ZEUTHEN, E., 1955. Mitotic respiratory rhythms in single eggs of Psammechinus miliaris and of Ciona intestinalis. Biol. Bull. 108, 366-385.
ZIELIŃSKI, M.A., 1939. Carbohydrate metabolism and phosphorus compounds in the fertilized eggs of the sea urchin (Paracentrotus lividus Lm.). Acta Biol. exp., Varsovie 13, 35-48.
ZOTIN, A.I., 1967. Rate of glucose oxidation in sea urchin eggs. Nature 213, 529-530.
ZOTIN, A.I., FAUSTOV, V.S., RADZINSKAJA, L.I., OZERNYUK, N.D., 1967. ATP level and respiration of embryos. J. Embryol. Exp. Morphol. 18, 1-12.
ZOTIN, A.I., MILMAN, L.S., FAUSTOV, V.S., 1965. ATP level and cleavage of sea-urchin eggs Strongylocentrotus droebachiensis (O.F. MÜLLER). Exp. Cell Res. 39, 567-576.

19. Lipids

Y. Isono and N. Isono

The lipids of sea-urchin eggs have been studied since the beginning of the century, in many cases before the development of modern lipid chemistry. Those results therefore need to be critically reinterpreted from the standpoint of recent lipid research.

The meanings of a number of terms have changed, such as cephalin, which was once used to mean a single phospholipid, phosphatidyl ethanolamine. But the cephalin fraction is now known to consist of several phospholipids. A single substance or group of substances has often been called by different names; for example, phospholipids and phosphatides. To avoid confusion, the terms used in recent lipid chemistry will be used here.

19.1 Lipids in General

19.1.1 Total Lipid Content of Unfertilized Eggs

Total lipid content often lies in the range of 15-35 mg/10^6 eggs, 100-250 mg/100 mg total N, or 10%-30% of the dry weight (Table 19.1). The results for *Paracentrotus* are in general agreement (26-32 mg/10^6 eggs, 170-230 mg/100 mg N, 19%-26% of dry weight), while values for *Arbacia* vary widely.

In studies of lipid distribution in *Arbacia punctulata* half-eggs obtained by centrifugation, free and bound lipids were extracted with ether and ethanol-ether, respectively. Light (pigment-free) halves of 10^6 eggs were found to contain 2.2 mg free lipids and 9.6 mg bound lipids (total, 11.8 mg), while the same number of heavy (pigmented) halves contained 6.6 mg free lipids and 12.2 mg bound lipids (total, 18.8 mg) (HUNTER and PARPART, 1946).

19.1.2 Relative Composition of Lipids

The ether extract from unfertilized *Paracentrotus lividus* eggs is equivalent to 140 mg total N and consists of 0.282 gm of saponifiable substances (86%) and 0.046 gm of unsaponifiable substances (14%) (MEYERHOFF, 1911). The ethanol-ether extract obtained from 183 x 10^6 unfertilized *Arbacia punctulata* eggs consists of 8.3 gm (84.4%) of acetone-soluble materials such as neutral fats, and 1.539 gm (15.6%) of acetone-insoluble materials such as phospholipids (PAGE, 1927a). The ethanol-ether extract of unfertilized *Arbacia punctulata* eggs contains 38% phospholipids and 7.5% cholesterol (HAYES, 1938). Unfertilized *Echinocardium cordatum* eggs contain a total lipid fraction of 29.9% phospholipids and 11.3% cholesterol (ÖHMAN, 1945).

The chloroform-methanol extract of *Paracentrotus lividus* gastrulae consists of 41.5% triglycerides, 27.4% phospholipids, 8.8% cholesterol, and only small amounts of cholesterol esters and free fatty acids (MOHRI, 1964a). In unfertilized *Anthocidaris crassispina* eggs, the chloroform-methanol extract contains 3.50 mg of simple lipids (neutral fats, cholesterol, and pigments) per 10^5 eggs (74.8% of the total lipids), 0.80 mg of phospholipids and sulfolipids (17.1%), and 0.38 mg of mucolipids (8.1%). In unfertilized *Pseudocentrotus depressus* eggs, the comparable values are 3.40 mg/10^5 eggs (65%), 1.63 mg (31.2%), and 0.20 mg (3.8%)(HOSHI and NAGAI, 1970).

19.1.3 Lipoproteins

Most lipid components are bound to proteins and occur in the form of lipoproteins. A water-soluble lipoprotein fraction isolated from *Arbacia punctulata* eggs contains 15.9% protein, 8.7% carbohydrate, and 75.4% lipid. The lipids consist of 2.6% cholesterol (no esters), 9.4% phospholipid, and the rest triglycerides and echinochrome. The lipoprotein contains only even-numbered fatty acids up to $C_{22:1}$, 44% of which are saturated (MARSH, 1965).

19.1.4 Iodine and Saponification Values

Iodine and saponification values of lipids obtained from unfertilized eggs and ovaries are generally comparable to those of lipids from other marine animals in which the iodine value is 70-180 and the saponification value is 120-200 (Table 19.2). An exceptional saponification value of 606 has been reported (PAGE, 1927a).

In *Paracentrotus lividus* the iodine value decreases slightly during embryo development (unfertilized egg, 150; gastrula and pluteus, 140), while in *Arbacia punctulata* it remains almost unchanged (before fertilization, 149-152; after fertilization, 150-153)(EPHRUSSI and RAPKINE, 1928; PAYNE, 1930).

19.1.5 Changes in Total Lipid Content during Development

Studies of lipid component changes during early embryo development and calculations based on wet and dry weight indicate a slight decrease in total lipids in early *Paracentrotus lividus* development (Table 19.3). And nitrogen content suggests that lipids decrease slightly between fertilization and gastrulation and then more noticeably thereafter (EPHRUSSI and RAPKINE, 1928).

The total lipid content of *Arbacia punctulata* drops steadily after fertilization, increases from 9 to 25 hours, and then declines again (Fig. 19.1). Lipids are the main source of energy until the time of hatching (8 hr), again after about 24 hours, and act as building materials from hatching to 24 hours (HAYES, 1938).

Total lipids make up 5.4% of *Arbacia punctulata* eggs, and protein-bound lipids account for 77% of the total. During the first 5 hours of development, neither total nor bound lipids change in amount (PARPART, 1941).

In *Echinocardium cordatum*, free lipids extracted with ether-chloroform (1:1) are much greater at all stages than bound lipids extracted with ethanol-chloroform (2:1)(Table 19.4). The total lipid content (used

Table 19.1. Total lipid content of unfertilized sea-urchin eggs

Species	Author	Method of extraction[a]	Lipids: mg/10^6 eggs	Lipids: mg/100 mg total N	Lipids: % of dry weight	Lipids: % of wet weight
Arbacia punctulata	MC CLENDON, 1909	Ether	7.25*	116*	12.5	2.25
		Ether/EtOH	23.6*	378*	40.5	7.33
	PAGE, 1927a	EtOH-ether/Ether	53.8	917*	43.4*	
	HAYES, 1938	EtOH-ether	5.65[b]	97*	9.8*	
	PARPART, 1941	Ether/EtOH-ether	34.1	251*	26.9	5.4
	HUNTER and PARPART, 1946[c]	Ether/EtOH-ether	30.6	523*	52.8*	
Echinus esculentus	CLELAND and ROTHSCHILD, 1952	EtOH/EtOH-ether			20	
Paracentrotus lividus	WETZEL, 1907	EtOH	28.8*	178*	19.1	4.3
	MEYERHOF, 1911	Ether	26.0*	234	17.2	
	EPHRUSSI and RAPKINE, 1928	Saponification	32.0*	198	21.2	4.81
		EtOH	39.5*	244	26.2	
	MOHRI, 1964a	EtOH/EtOH-ether	26.8	166	17.8*	
Strongylocentrotus purpuratus	LEITCH, 1934	Petroleum ether	14.1	204		
Strongylocentrotus franciscanus	LEITCH, 1934	Petroleum ether	56.3	204		
Pseudocentrotus depressus	HOSHI and NAGAI, 1970	CH-MtOH	52.3			
Anthocidaris crassispina	HOSHI and NAGAI, 1970	CH-MtOH	46.8			
Echinocardium cordatum	ÖHMAN, 1945[d]	Ether-CH/EtOH-CH	30.0*	174		
			35.0*	203		

* Values without asterisk are original data or calculated on the basis of other data appearing in original report. Values with asterik are calculated on the basis of the following data: *Arbacia punctulata*: 10^6 eggs= 5.86 mg N = 58.0 dry weight; dry weight = 24.2% of wet weight (BALLENTINE, 1940); 10^6 eggs = 0.906 mg P (PAGE, 1927b). *Paracentrotus lividus*: 10^6 eggs = 16.1 mg N (MOHRI, 1964a); 100 mg dry weight = 10.7 mg N (EPHRUSSI, 1933); 100 mg N = 13.5 mg P (ZIELINSKI, 1939). *Strongylocentrotus purpuratus*: 10^6 eggs = 817 µg P (WHITELEY, 1949). *Echinocardium cordatum*: 10^6 eggs = 17.24 mg N (CHEN, 1958).

[a]Solvent A/Solvent B; extracted first with A and then with B. CH: chloroform; EtOH: ethanol; MtOH: methanol.

[b]Cited in HUNTER and PARPART, 1946.

[c]Sum of data obtained for half-eggs.

[d]Total lipids are all free lipids in ether-CH fraction plus bound phospholipids in EtOH-CH fraction.

Table 19.2. Iodine and saponification values of lipids of unfertilized eggs and ovaries

Species	Author	Fraction	Iodine value	Saponification value
Arbacia punctulata[a]	PAGE, 1927a	Neutral fats	146-148	606
Arbacia punctulata[a]	PAYNE, 1930	Total lipids	149-152	
Arbacia punctulata[a]	NAVEZ, 1938	Total lipids	180-190	About 200
Paracentrotus lividus[a]	EPHRUSSI and RAPKINE, 1928	Total lipids	150	
Echinus esculentus[b]	MOORE, WHITLEY, and ADAMS, 1913	Total lipids	133-166	188-205
		Phospholipids	73- 84	207-259
Echinometra lucunter[b]	TENNENT, GARDINER, and SMITH, 1931	Total lipids	84	
		Neutral fats	100	
		Lecithin	108	
		Cephalin	47	

[a]Unfertilized eggs.
[b]Ovaries.

Table 19.3. Lipid content of *Paracentrotus lividus* eggs and embryos. (From EPHRUSSI and RAPKINE, 1928)

Fraction		Unit	Unfertilized eggs	Gastrula (12 hr)	Pluteus (40 hr)
Total lipids	Experiment A	% of dry weight	21.2	19.55	17.4
		% of wet weight	4.81	4.43	3.69
		mg/100 mg N	198.0	192.0	179.0
	Experiment B	% of dry weight	26.2	25.2	18.8
		mg/100 mg N	244.0	247.0	194.0
Unsaponifiable fraction	Total	% of dry weight	3.3	3.0	2.7
	Cholesterol	% of dry weight	1.8	1.65	1.4
	Unidentified	% of dry weight	1.5	1.3	1.3
Total nitrogen		% of dry weight	10.7	10.2	9.7
Dry weight		% of wet weight	22.7	22.7	21.2

here to mean all free lipids - neutral fats, phospholipids, sterols - and bound phospholipids) drops slightly up to 7 hours after fertilization. Changes in the amount of free and bound lipids vary according to changes in the amount of free and bound phospholipids (ÖHMAN, 1945).

The total lipid content of *Paracentrotus lividus* eggs hardly changes until the mesenchymal blastula stage but at gastrulation drops considerably (Fig. 19.2)(MOHRI, 1964a). And during the early embryo development of *Anthocidaris crassispina* and *Pseudocentrotus depressus*, the amount of simple lipids (acetone-soluble) gradually declines (Table 19.5)(HOSHI and NAGAI, 1970).

It is generally agreed that total lipid content changes little or decreases only slightly until the mesenchymal blastula stage and then declines significantly after gastrulation. Oxygen uptake greatly increases after hatching, and respiratory quotient determinations suggest that large amounts of lipids are used for respiration after the mesenchymal blastula stage (ISONO, 1963; ISONO and YASUMASU, 1968). In the early stages, respiration probably depends mostly on the oxidation of carbohydrates and after the mesenchymal blastula stage on the oxidation of lipids (MOHRI, 1964a).

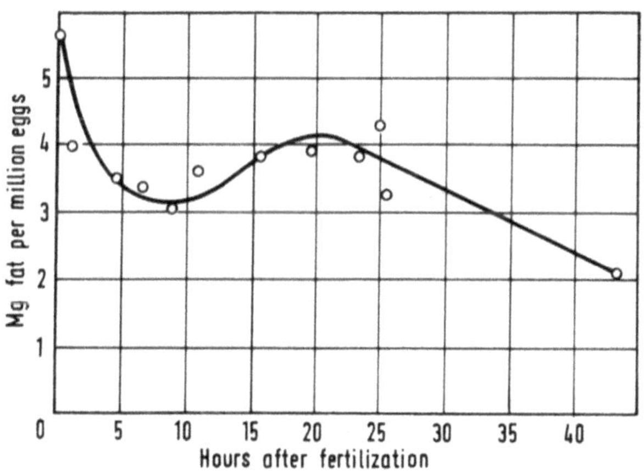

Fig. 19.1

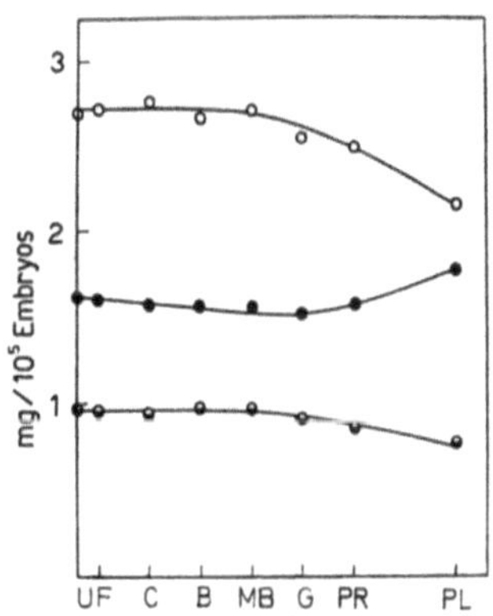

Fig. 19.2

Fig. 19.1. Changes in total lipid content during the early development of *Arbacia punctulata* (HAYES, 1938). The time of hatching (8 hrs after fertilization) corresponds to the initial trough of the curve

Fig. 19.2. Changes in content of total lipids (o), phospholipids (◒) and total nitrogen (•), during the early development of *Paracentrotus lividus* (MOHRI, 1964a). U, unfertilized egg; F, newly fertilized egg; C, cleavage stage (32-64 cells); B, blastula; MB, mesenchyme blastula; G, gastrula; PR, prism larva; PL, pluteus

Table 19.4. Total lipid content of *Echinocardium cordatum* eggs and embryos. Numerical values represent mg/100 mg total N. (From ÖHMAN, 1945)

Experiment	Type of lipid	Unfertilized eggs	Fertilized eggs (10 min)	Fertilized eggs (1.5 hr)	Early blastula (7 hr)	Late blastula (21 hr)
1 (1942)	Free	145.0	138.0	140.0	146.0	-
	Bound	29.2	35.0	33.7	23.0	-
	Total	174.2	173.0	173.7	169.0	-
2 (1943)	Free	174.0	168.0	-	175.0	177.0
	Bound	28.8	34.3	-	23.9	21.8
	Total	202.8	202.3	-	198.9	198.8

Table 19.5. Changes in simple lipid content during early development. Numerical values represent mg/10^5 eggs. (From HOSHI and NAGAI, 1970)

Species	Unfertilized eggs	Fertilized eggs	32-Cell	Early blastula	Hatching blastula	Early gastrula	Late gastrula	Prism larva	Pluteus
Anthocidaris crassispina	3.50	4.18	3.08	3.20	-	2.20	-	1.94	2.23
Pseudocentrotus depressus	3.40	3.34	3.24	-	2.54	-	2.32	2.14	-

19.2 Phospholipids

19.2.1 Phospholipid Content of Unfertilized and Developing Eggs

The phospholipid content of unfertilized sea-urchin eggs is generally in the range of 5-15 mg/10^6 eggs, 50-150 mg/100 mg total N, 20%-40% of the total lipid content, or 20-40 mg lipid-P/100 mg total P (Table 19.6).

The ether-soluble, acetone-insoluble fraction of unfertilized *Arbacia punctulata* eggs contains lecithin (phosphatidyl choline), cephalin (noncholine-containing phospholipid), and sphingomyelin. Much of the acetone-insoluble material is also insoluble in boiling alcohol, probably due to a high cephalin content. A slight white precipitate when ether is added to the alcohol fraction suggests the presence of sphingomyelin (PAGE, 1927a). *Arbacia* eggs have considerably more cephalin and less lecithin than *Asterias forbesi*. Today cephalin is known to be a mixture of phosphatidyl ethanolamine, phosphatidyl serine, and other phospholipids.

Sphingomyelin occurs in the ovaries of *Echinometra lucunter*, while lysolecithin (lysophosphatidyl choline) is found in the eggs of *Hemicentrotus pulcherrimus* (TENNENT, GARDINER, and SMITH, 1931; NUMANOI, 1959). *Anthocidaris crassispina* eggs contain plasmalogens (phospholipids that have an aldehydogenic group)(MOHRI, 1959).

In eggs of *Paracentrotus lividus*, *Hemicentrotus pulcherrimus*, and *Anthocidaris crassispina*, lecithin is the most abundant phospholipid, while phosphatidyl ethanolamine is second. A small amount of lysolecithin is present. Negligible in *Paracentrotus*, inositol phospholipids are absent from *Hemicentrotus* and *Anthocidaris*. Phosphatidyl serine and sphingomyelin occur in small amounts in *Hemicentrotus* and *Anthocidaris* (MOHRI, 1964a; ISONO, 1965).

While the phospholipid content seems not to change between the 2-cell and blastula stages of *Arbacia punctulata*, phospholipid-P declines in *Strongylocentrotus purpuratus*: it is 43% of the total phosphorus in unfertilized eggs, 37.7% at the blastula stage, and 35.2% in the pluteus (SHACKELL, 1911; ROBERTSON and WASTENEY, 1913). In *Dendraster excentricus*, both phospholipid-P and total phosphorus increase considerably between fertilization and gastrulation (Table 19.7)(NEEDHAM and NEEDHAM, 1930).

The total phospholipid content of *Echinocardium cordatum* (described earlier) remains unchanged up to the late blastula stage. Lecithin decreases between fertilization and early blastula, while cephalin increases proportionately. It is possible that part of the lecithin is converted into other phospholipids. Free phospholipids are reduced at fertilization but recover at early blastula to a level slightly higher than in unfertilized eggs. 29% of the cephalin is bound to proteins within 10 minutes after fertilization (Table 19.8). Cephalin is sometimes considered a structural lipid, and its binding may be similar to protein binding at fertilization (MIRSKY, 1936). As it is possible that the insoluble protein fraction forms a network, so is it possible that phospholipids do the same (ÖHMAN, 1945).

In *Arbacia lixula* the amount of free phospholipid extracted with cold ether declines just after fertilization and in 15 minutes returns to the level of unfertilized eggs (Table 19.9)(RICOTTA, 1956).

Table 19.6. Phospholipid content of unfertilized sea-urchin eggs[a]

Species	Author	Method of extraction[b]
Arbacia punctulata	MC CLENDON, 1909	Ether Ether/EtOH
	PAGE, 1927a	EtOH-ether/Ether
	HAYES, 1938	EtOH-ether
	CRANE, 1947	EtOH/EtOH-ether
	SCHMIDT, HECHT, and THANNHAUSER, 1948	EtOH/EtOH-ether
Arbacia lixula	RICOTTA, 1956	Ether/EtOH-ether
Echinus esculentus	CLELAND and ROTHSCHILD, 1952	EtOH/EtOH-ether
Lytechinus pictus	WHITELEY, 1949	EtOH/EtOH-ether
Paracentrotus lividus	ZIELIŃSKI, 1939	EtOH/EtOH-ether
	MOHRI, 1964a	EtOH/EtOH-ether
Strongylocentrotus purpuratus	ROBERTSON and WASTENEYS, 1913	EtOH
	WHITELEY, 1949	EtOH/EtOH-ether
Pseudocentrotus depressus	YAMAGAMI, 1963	EtOH/EtOH-ether
Anthocidaris crassispina	MOHRI, 1959	EtOH/EtOH-ether
Dendraster excentricus	NEEDHAM and NEEDHAM, 1930	EtOH/EtOH-ether
Echinocardium cordatum	ÖHMAN, 1945	Ether-CH/EtOH-CH

[a]Values without asterisk * are original data or calculated on the basis of other data appearing in original report. Values with asterisk are calculated on the basis of data listed in Table 19.1. It is assumed that phospholipids contain 4% P.

[b]Solvent A/Solvent B; extracted first with A and then with B. CH: chloroform; EtOH: ethanol.

[c]PL: phospholipids; PL-P: phospholipid-P; TL: total lipid; TP: total P.

[d]Determined as phospholipid-P.

In eggs of *Anthocidaris crassispina*, the phospholipid-P content in unfertilized eggs is 34.4 (33.8 - 35.5) µg/10^5 eggs and in fertilized eggs after 10 minutes is 31.5 (31.3 - 31.7) µg. About one-fifth of the egg phospholipids are plasmalogens. In unfertilized eggs 6.1 (5.0-6.7) µg plasmalogen-P/10^5 eggs and in fertilized eggs 6.1 (5.1 - 6.6) µg are found (MOHRI, 1959).

Original expression[c]	Phospholipids mg/10^6 eggs	mg/100 mg N	% of total lipid	Phospholipid-P % of total P
PL-P, 0.069% wet weight	5.54*	89 *	76.5	10.4
PL-P, 0.194% wet weight	15.6*	250 *	64.8	29.3
PL, 1.539 gm/183 x 10^6 eggs	8.4	143 *	15.6	37.1*
PL, 2.17 mg/10^6 eggs	2.17	37 *	38.4	9.6*
PL-P, 1.26 mg/gm wet weight	7.6*	122		37.6
PL-P, 0.35 mμg/egg	8.75*	148 *		36.5
PL-P, 22.8 μg/100 μg TP				22.8
PL, 24% TL[d]			24	
PL-P, 20.5% TP				20.5
PL-P, 22.5% TP	12.2*	76		22.5
PL, 0.96 mg/10^5 eggs[d]	9.6	59.6	35.8	17.7*
PL-P, 43.0% TP	8.8*	128 *		43.0
PL-P, 27.0% TP	5.5*	82*		27.0
PL-P, 29.0% TP	8.7			29.0
PL-P, 34.4 μg/10^5 eggs	8.6			
PL-P, 0.422 mg dry weight				6.1
PL-P 28.3 μmol/5 mg TP	10.3*	59.9	34.4	17.5
PL-P 28.1 μmol/5 mg TP	10.4*	60.6	29.9	17.4

In *Pseudocentrotus depressus* the phospholipid-P content of unfertilized eggs is 29% of the total phosphorus; in fertilized eggs still 29%; in the morula, 28%; in the hatching blastula, 27.5%; in early gastrula, 27.5%; late gastrula, 27%; and in the prism larva, 24%. Total phosphorus content remains unchanged during embryo development (YAMAGAMI, 1963).

Similarly, in unfertilized *Paracentrotus lividus* eggs, phospholipids are about one-third of the total lipids. The phospholipid-P content does not change significantly until the mesenchymal blastula stage and then decreases (Fig. 19.2)(MOHRI, 1964a).

Total phospholipid content generally seems to remain almost unchanged until the mesenchymal blastula stage or gastrulation, after which it decreases. Or an earlier slight drop may occur transiently just after fertilization. The metabolism of sea-urchin eggs and embryos thus differs sharply from the metabolism of sea-urchin spermatozoa in which phospholipids are the energy source (MOHRI, 1957, 1964b).

Table 19.8. Phospholipid content of *Echinocardium cordatum* embryos. Embryos developed at 19°C. Numerical values represent µmol phospholipid-P/5.0 mg total P. (From ÖHMAN, 1945)

Experiment	Compounds[a]		Unfertilized eggs	Fertilized eggs (10 min)	Fertilized eggs (1.5 hr)	Early blastula (7 hr)	Late blastula (21 hr)
1 (1942)	Phospholipids	Free	14.4	11.5	12.1	17.0	-
		Bound	13.7	16.4	15.8	10.7	-
		Total	28.1	27.9	27.9	27.7	-
	Phospholipids	Free	14.9	12.0	-	16.9	17.7
		Bound	13.5	16.1	-	11.2	10.2
		Total	28.3	28.1	-	28.1	27.9
2 (1943)	Lecithin	Free	5.6	5.4	-	6.4	6.8
		Bound	13.2	13.0	-	9.3	8.6
		Total	18.8	18.4	-	15.7	15.5
	Cephalin	Free	9.3	6.6	-	10.5	10.9
		Bound	trace	3.1	-	1.9	1.6
		Total	9.5	9.7	-	12.4	12.4

[a]Amount of phospholipid and lecithin (phosphatidyl choline) was determined from respective phosphorus and choline contents. Difference between the 2 was taken as content of noncholine-containing phospholipid called cephalin.

Table 19.7. Phospholipid-P content of *Dendraster excentricus* embryos. (From NEEDHAM and NEEDHAM, 1930)

Stage	Total P	Phospholipid-P	
	mg P/gm dry weight	mg P/gm dry weight	% of total P
Unfertilized eggs	6.94	0.42	6.09
Gastrula	8.80	0.77	8.75
Pluteus	9.50	0.75	7.90

19.2.2 Cortical Changes and Phospholipids

Dark-field illumination shows that immediately after fertilization the interference color of the egg surface in *Psammechinus microtuberculatus*, *Sphaerechinus granularis*, *Paracentrotus lividus*, and *Echinocardium cordatum* changes from yellow-red to silver-white. This may be due to the dispersion of lipids that make up the unfertilized egg cortex (RUNNSTRÖM, 1928). Fertilization or artificial activation weakens or removes the birefringence of the cortical layer in *Psammechinus miliaris*, probably because it rearranges the chemical composition. The lipids and proteins in the cortical layer of unfertilized eggs are strongly connected, and fertilization weakens the connections. In the reaction between sperm and egg cortex, the first event may be the splitting of the lipoprotein complex (MONROY, 1945, 1947, 1957; MONROY and MONTALENTI, 1947).

Live spermatozoa of *Arbacia lixula* release phospholipids from hen's egg lipovitelline and transiently produce hemolytic derivatives (probably lysophospholipids) which are then degraded into nonhemolytic compounds such as glyceryl phosphoryl choline (MAGGIO and MONROY, 1955). And long before, a hypothesis was advanced in which the key event of egg activation was transient cytolysis of the egg surface (LOEB, 1913). Putting the 2 together, it is possible that where the fertilizing

Table 19.9. Free and bound phospholipids of *Arbacia lixula* eggs. Numerical values represent μg phospholipid-P/100 μg total P. (From RICOTTA, 1956)

Experiment	Type[a]	Unfertilized eggs	Minutes after fertilization 3	6	15
1	Free	4.3	1.9	2.9	3.0
	Bound	17.2	17.4	16.2	16.7
	Total	21.5	19.3	19.1	19.7
2	Free	9.2	5.9	7.3	9.3
	Bound	13.7	14.6	15.4	13.7
	Total	22.9	20.5	22.7	23.0
3	Free	6.1	5.0	5.2	5.8
	Bound	17.9	17.9	17.8	17.7
	Total	24.0	22.9	23.0	23.5

[a]Free phospholipids, extracted with cold ether; bound phospholipids, extracted with hot ethanol-ether (3:1).

sperm enters, a reaction begins, involving splitting the lipoprotein cortical complex and releasing free lipids that are subsequently degraded to lysophospholipids. The cytolytic products formed in the reaction may be the basis for the transient cytolysis thought to be the first step of activation (MAGGIO and MONROY, 1955).

19.3 Glycolipids

19.3.1 Sphingoglycolipids

From ethanol-ether extracts of the ovaries of *Echinometra lucunter*, a fraction insoluble in cold ether contained glucose and phosphorus was obtained and thought to consist of cerebroside, sphingomyelin, and other substances (TENNENT, GARDINER, and SMITH, 1931).

Arbacia punctulata eggs may contain lipids that have sialic acid and N-glycolylneuraminic rather than N-acetylneuraminic acid. The sialolipid differs from gangliosides and other mucolipoproteins of mammalian brains (WARREN and HATHAWAY, 1960).

Mucolipids are sphingoglycolipids that contain sialic acid and/or hexosamine. Glycolipids may be extracted with chloroform-methanol and by means of a Folch's partitition fractionated further into components soluble in the organic phase and others soluble in the water phase.

Mucolipids occur in the eggs and spermatozoa of *Pseudocentrotus depressus* and are composed of sphingosine, N-acetylneuraminic acid, and glucose. Thin-layer chromatography shows that they resemble ox-brain monosialogangliosides. Unlike common mammalian sialoglycolipids, they contain only glucose as the carbohydrate moiety (ISONO and NAGAI, 1966).

In *Hemicentrotus pulcherrimus*, similar sialoglycolipids are found and the content seems to decline during early embryo development, and thin-layer chromatography reveals that the number of molecular species of sialoglycolipids increases in later developmental stages. Increasing complexity is suggested by changes in the ratio of glucose to sialic acid to sphingosine in the sialoglycolipid fraction (ISONO, 1967).

In the gonads of *Strongylocentrotus intermedius*, sialoglycolipids are composed of phytosphingosine, a higher aliphatic acid, glucose (no other sugars), and a sialic acid identical to the one in mammalian brain gangliosides (KOCHETKOV et al., 1967).

In studies of mucolipids (sialoglycolipids soluble in the water phase of the Folch's partition), the total mucolipid content of unfertilized *Anthocidaris crassispina* eggs is 0.38 mg/10^5 eggs and of *Pseudocentrotus depressus* eggs is 0.20 mg/10^5 eggs. The quantity of mucolipids remains unchanged until hatching and declines between the blastula and gastrula stages. While the content of the main molecular species of mucolipids does not change markedly during early development, the content of minor species may alter (HOSHI and NAGAI, 1970).

The principal mucolipid in *Anthocidaris* eggs is monosialo-diglucosyl ceramide, and the main molecular species of *Anthocidaris* spermatozoa are disialo-monoglucosyl ceramide (N-acylneuraminyl (2⟶8) N-acylneuraminyl (2⟶6) glucosyl ceramide) and monosialo-monoglucosyl ceramide (N-acylneuraminyl (2⟶6) glucosyl ceramide). The first of these sperm mucolipids is present in larger amounts than the second (HOSHI and NAGAI, 1970; NAGAI and HOSHI, 1975).

19.3.2 Glyceroglycolipids

Sulfur-containing glycolipids occur in the gametes of *Pseudocentrotus depressus*, *Anthocidaris crassispina*, and *Hemicentrotus pulcherrimus*. In *Pseudocentrotus depressus*, their content is 0.14% of the wet weight of the unfertilized egg and 0.05% of the sperm. This lipid lacks phosphorus and nitrogen but contains sulfonic acid (not sulfuric acid) and glycerol. The sugar moiety probably consists only of 6-sulfoquinovose (= 6-sulfo-6-deoxy-D-glucose). Infrared spectroscopy shows that the glycolipid resembles the so-called plant sulfolipid, and all the data suggest that it belongs to the group of sulfolipids that occur only in plants (NAGAI and ISONO, 1965; ISONO and NAGAI, 1966).

The difference between animal and plant sulfolipids is the fatty acid composition. The plant sulfolipid is characterized by a predominance of unsaturated fatty acids, while the egg sulfolipid contains only saturated fatty acids (mainly myristic acid and palmitic acid). The ratio between myristic acid and palmitic acid is approximately 2:3 in *Pseudocentrotus* and approximately 1:2 in *Hemicentrotus*.

The sulfolipid content of unfertilized *Hemicentrotus* eggs is 76.2 µg (as hexose)/10^5 eggs; of fertilized eggs, 84.8; of mesenchymal blastula, 62.5; of late gastrula, 58.7; and of the pluteus, 59.0. The fatty acid composition does not change during development. This egg sulfolipid stimulates the respiration of sea-urchin spermatozoa (ISONO, 1967; ISONO, MOHRI, and NAGAI, 1967).

19.4 Cholesterol

Cholesterol in sea-urchin eggs was first observed in *Arbacia punctulata* (MATHEWS, 1913). The cholesterol content of unfertilized eggs is 5%-10% of the total lipids (Table 19.10). Protein-bound cholesterol does not occur in sea-urchin eggs (ÖHMAN, 1945).

Table 19.10. Cholesterol content of unfertilized sea-urchin eggs

Species	Author	Total cholesterol: % of total lipids	Total cholesterol: mg/100 mg N	Free cholesterol (% of total cholesterol)
Arbacia punctulata	HAYES, 1938	7.5	7.2	
	MONROY and RUFFO, 1945	4.09		41
Paracentrotus lividus	EPHRUSSI and RAPKINE, 1928	8.7	17.2	
	LINDVALL and CARSJÖ, 1948		21.9	65
			20.5	71
(gastrula)	MOHRI, 1964a	8.8	14.6	
Echinometra lucunter (ovary)	TENNENT, GARDINER, and SMITH, 1931	8.3		
Echinocardium cordatum	ÖHMAN, 1945	11.3	23.0	93

Table 19.11. Cholesterol content of *Echinocardium cordatum* eggs and embryos. Numerical values represent μmol/5 mg of total P. (From ÖHMAN, 1945)

Cholesterol	Unfertilized eggs	Fertilized eggs (10 min)	Early blastula	Late blastula
Esterified	1.4	1.2	1.4	1.5
Free	17.6	17.4	17.3	17.6
Total	19.3	18.6	18.7	18.7

The total cholesterol content declines during early development of *Paracentrotus lividus* but remains virtually unchanged in *Arbacia punctulata* (EPHRUSSI and RAPKINE, 1928; HAYES, 1938). In the unfertilized eggs of the latter it is 0.430 mg/10^6 eggs. 1 hour after fertilization it is 0.413; 6 1/2 hours, 0.413; 19 1/2 hours, 0.431; 43 hours and 10 minutes, 0.416. Hatching occurs at 8 hours. Free and total cholesterol content does not change significantly during early development of *Echinocardium cordatum* (Table 19.11)(ÖHMAN, 1945).

Table 19.12. Cholesterol content of *Arbacia lixula* eggs. (From MONROY and RUFFO, 1945)

Cholesterol	Unfertilized eggs		Fertilized eggs (10 min)	
	% of total lipids	% of dry weight	% of total lipids	% of dry weight
Total	4.09	1.43	4.43	1.34
Free	1.66	0.68	3.88	1.20

The total cholesterol content of *Arbacia lixula* eggs does not change at fertilization, but the amount of free cholesterol increases suddenly, possibly due to the activation of cholesterol esterase (Table 19.12) (MONROY and RUFFO, 1945). In *Paracentrotus lividus*, the total amount of cholesterol is not affected until 80 minutes after fertilization (Table 19.13)(LINDVALL and CARSJÖ, 1948). The free cholesterol content remains unchanged for the first 25 minutes and then appears to be esterified.

19.5 Carotenoids

The major components of carotenoids in sea-urchin eggs are echinenone and β-carotene (Table 19.14). Substantial quantities of 2 new carotenoids, paracentrotins A and B, occur in *Paracentrotus lividus*. Estimated from the optical density of the ethanol-petroleum ether extract, the carotenoid content of *Paracentrotus* embryos declines sharply between fertilization and gastrulation. After the gastrula stage, the quantity of pigments increases until, at the pluteus stage, it reaches the level of unfertilized eggs. Metabolism of the carotenoid pigments may well be important to early morphogenesis and to the process underlying gastrulation in particular (MONROY, MONROY-ODDO, and DE NICOLA, 1951).

Table 19.13. Cholesterol content of *Paracentrotus lividus* eggs. Numerical values represent mg cholesterol/ mg total P. (From LINDVALL and CARSJÖ, 1948)

Cholesterol	Unfertilized eggs	Minutes after fertilization													
		15	20	25	30	35	40	45	50	55	60	65	70	75	80
Total	1.60	1.68	1.70	1.63	1.80	1.67	1.56	1.75	1.70	1.76	1.69	1.60	1.57	1.52	1.56
Free	1.13	1.26	1.28	1.23	1.12	1.02	1.02	1.03	1.01	1.04	1.04	0.90	0.90	0.82	0.81
Esters	0.47	0.42	0.42	0.40	0.68	0.65	0.54	0.72	0.69	0.72	0.65	0.70	0.67	0.70	0.75

Table 19.14. Carotenoids in unfertilized sea-urchin eggs. Numerical values represent percentage of total carotenoids

	DE NICOLA, 1954		DE NICOLA and GOODWIN, 1954	GRIFFITHS, 1966
Carotenoid	*Psammechinus microtuberculatus*	*Sphaerechinus granularis*	*Paracentrotus lividus*	*Strongylocentrotus purpuratus*
α-Carotene				0
β-Carotene	11	16	3.7	7.9-36.1[a]
Neo-β-carotene B				1-2
Echinenone[b]	85	70	24.1	57.3-86.4[a]
Isocryptoxanthin[c]				3.5-5.5
Zeaxanthin[d]				2-3
Paracentrotin A	0	0	30.9	0
Paracentrotin B	0	0	11.0	0
Unidentified	4[e]	14[f]	10.2[g]+20.1[h]	trace

[a] β-Carotene + echinenone = 86.8-94.3%

[b] 4-Oxo-β-carotene.

[c] 4-Hydroxy-β-carotene.

[d] 3,3'-Dihydroxy-β-carotene.

[e] Epiphasic carotenoid.

[f] Dihydroxy xanthophylls present.

[g] Polyhydroxy xanthophylls.

[h] Hydroxy derivative of echinenone.

In *Psammechinus microtuberculatus* embryos, cyclic changes in carotenoid content occur: maxima in the swimming blastula and pluteus stages, and minima at the end of cleavage and at the gastrula stage. In *Sphaerechinus granularis* embryos, the total carotenoid content decreases slightly during cleavage and remains constant until gastrulation when it drops suddenly (DE NICOLA, 1954).

The total carotenoid content of unfertilized *Paracentrotus* eggs is 10-20 μg/mg total N. Measured in eggs before fertilization, in swimming blastulae, and in plutei, carotenoid pigment composition does not change significantly (DE NICOLA and GOODWIN, 1954).

The average carotenoid content of unfertilized *Strongylocentrotus purpuratus* eggs (0.27 mg/gm protein) remains almost constant until the pluteus stage. The quantity of pigments β-carotene, neo-β-carotene-B, echinenone, isocryptoxanthin, and zeaxanthin does not change appreciably. During early development, egg carotenoids appear metabolically inert. Conceivably they may serve as structural components or light shields in the embryo (GRIFFITHS, 1966).

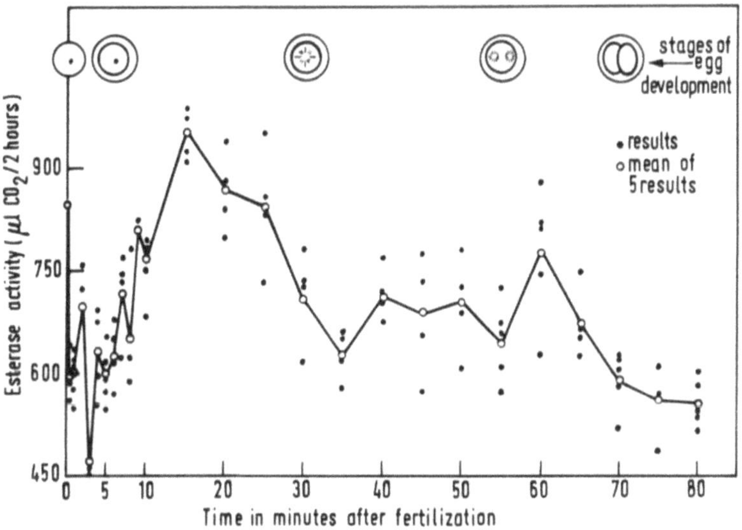

Fig. 19.3. Changes in tributyrin-splitting activity after fertilization of *Paracentrotus lividus* eggs (LINDVALL, 1948). The esterase activity was determined as the amount of CO_2 which was released from a bicarbonate solution by the butyric acid split from tributyrin, and expressed in relation to the number of eggs corresponding to 1.0 mg total P

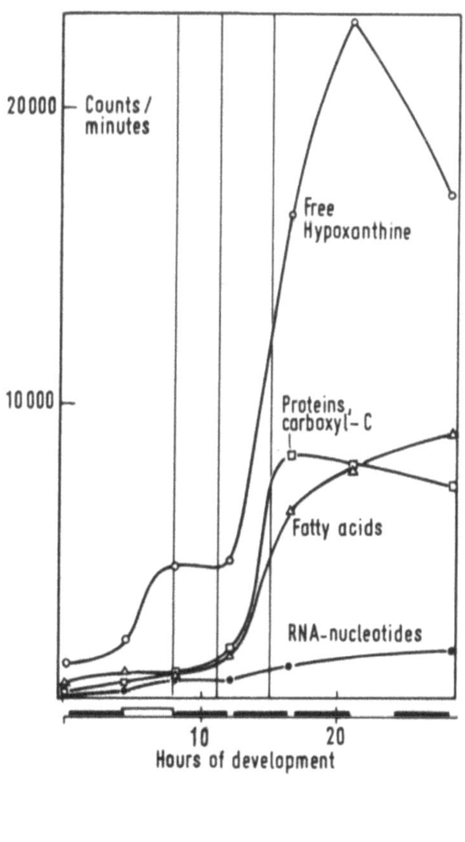

Fig. 19.4. Incorporation of acetate-1-^{14}C into *Psammechinus miliaris* embryos (HULTIN, 1953). Specific activity, cpm per 15 mg $BaCO_3$ per cm^2. Vertical lines indicate hatching, appearance of primary mesenchyme cells and beginning of invagination. The periods of isotope treatment (each 4 hrs) are indicated by horizontal lines at the bottom of the figure

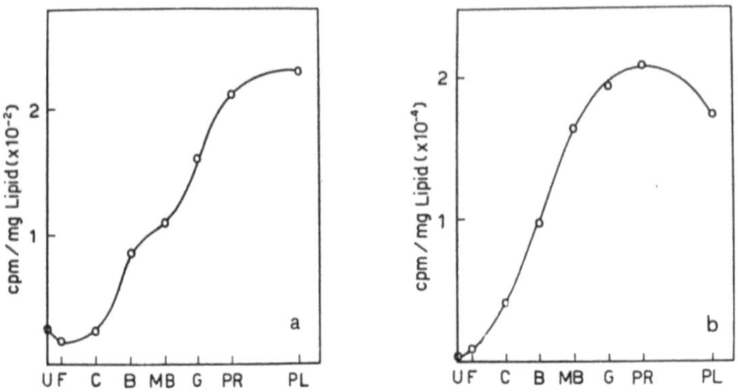

Fig. 19.5 a and b. Incorporation of acetate-1-^{14}C (a) and glycerol-1-^{14}C (b) into the total lipids of *Paracentrotus lividus* embryos (MOHRI, 1964a). Periods of isotope treatment, 2 hrs. U, unfertilized egg; F, newly fertilized egg; C, cleavage stage (32-64 cells); B, blastula; MB, mesenchyme blastula; G, gastrula; PR, prism larva; PL, pluteus

19.6 Lipid Metabolism

19.6.1 Lipase

In assaying lipase in homogenates of *Paracentrotus lividus* eggs, tributyrin splitting activity shows a cyclic change after fertilization (Fig. 19.3). Minima are reached at 3 minutes and at 35-55 minutes after insemination, while maxima occur at 15 minutes and again at 60 minutes when the first cleavage begins (LINDVALL, 1948).

19.6.2 Radioisotope Incorporation into Lipids

Acetate-1-^{14}C is incorporated into fatty acids in *Psammechinus miliaris* embryos at a remarkably low rate until the mesenchymal blastula stage. But the rate of incorporation increases markedly at gastrulation (Fig. 19.4)(HULTIN, 1953).

In *Paracentrotus lividus* eggs, more than 95% of the total radioactivity of acetate-1-^{14}C is incorporated into the fatty acid moiety of triglycerides and phospholipids. The phosphatidyl ethanolamine fraction has a much higher specific activity (cpm/mg lipid) than the phosphatidyl choline fraction, while the specific activity of the triglyceride fraction is fairly low.

Only a slight amount of acetate 1-^{14}C is incorporated into the lipids of unfertilized eggs. After fertilization the rate of incorporation increases very rapidly in an S-shaped curve, peaks at the prism stage, and then declines (Fig. 19.5a).

The distribution pattern of ^{14}C incorporated into lipid fractions changes little during early development. Increased labeling of most fractions parallels the rise in total incorporation (MOHRI, 1964a).

Much less glycerol-1-^{14}C is incorporated than acetate-1-^{14}C (Fig. 19.5b). The rate of incorporation declines slightly after fertilization, then increases, and peaks at the pluteus stage.

In general, lipid metabolism increases during later embryo development. Taken into the phospholipids at the blastula stage, ^{14}C is not later lost in any significant amount, suggesting that in some of the lipid components, especially phospholipids, new synthesis occurs which is essential to later development (MOHRI, 1964a).

References

BALLENTINE, R., 1940. Total nitrogen content of the Arbacia eggs. J. Cell. Comp. Physiol. 15, 121-122.

CHEN, P.S., 1958. Further studies on free amino acids and peptides in eggs and embryos of different sea urchin species and hybrids. Experientia 14, 369-371.

CLELAND, K.W., ROTHSCHILD, LORD, 1952. The metabolism of the sea urchin egg. Anaerobic breakdown of carbohydrate. J. Exp. Biol. 29, 285-294.

CRANE, R.K., 1947. The distribution of phosphorus in the unfertilized egg of Arbacia punctulata. Biol. Bull. 93, 192-193.

DE NICOLA, M., 1954. Further investigations of the change in the pigment during embryonic development of echinoderms. Exp. Cell Res. 7, 368-373.

DE NICOLA, M., GOODWIN, T.W., 1954. Carotenoids in the developing eggs of the sea urchin Paracentrotus lividus. Exp. Cell Res. 7, 23-31.

EPHRUSSI, B., 1933. Contribution à l'analyse des premiers stades du développement de l'oeuf. Action de la température. Arch. Biol. 44, 1-147.

EPHRUSSI, B., RAPKINE, L., 1928. Composition chimique de l'oeuf d'oursin (Paracentrotus lividus Lk.) et ses variations au cours du développement. Ann. Physiol. Physicochim. Biol. 4, 386-399.

GRIFFITHS, M., 1966. The carotenoids of the eggs and embryos of the sea urchin Strongylocentrotus purpuratus. Develop. Biol. 13, 296-309.

HAYES, F.R., 1938. The relation of fat changes to the general chemical embryology of the sea urchin. Biol. Bull. 74, 267-277.

HOSHI, M., NAGAI, Y., 1970. Biochemistry of mucolipids of sea urchin gametes and embryos. III. Mucolipids during early development. Jap. J. Exp. Med. 40, 361-365.

HOSHI, M., NAGAI, Y., 1975. Novel sialosphingolipid from spermatozoa of the sea urchin, Anthocidaris crassispina. Biochim. Biophys. Acta, in press.

HULTIN, T., 1953. Incorporation of C^{14}-labeled carbonate and acetate into sea urchin embryos. Arkiv.Kemi 6, 195-200.

HUNTER, F.R., PARPART, A.K., 1946. The distribution of lipid between the light and heavy halves of the Arbacia egg. Biol. Bull. 91, 222.

ISONO, N., 1963. Carbohydrate metabolism in sea urchin eggs. III. Changes in respiratory quotient during early embryonic development. Annot. Zool. Japon. 36, 126-132.

ISONO, N., YASUMASU, I., 1968. Pathways of carbohydrate breakdown in sea urchin eggs. Exp. Cell Res. 50, 616-626.

ISONO, Y., 1965. Phospholipids of sea urchin eggs. I. Thin-layer and paper chromatographic studies. Sci. Papers Coll. Gen. Educ. Univ. Tokyo 15, 87-94.

ISONO, Y., 1967. Changes of glycolipids during early development of sea urchin embryos. Jap. J. Exp. Med. 37, 87-96.

ISONO, Y., MOHRI, H., NAGAI, Y., 1967. Effect of egg sulpholipid on respiration of sea urchin spermatozoa. Nature 214, 1336-1338.
ISONO, Y., NAGAI, Y., 1966. Biochemistry of glycolipids of sea urchin gametes. I. Separation and characterization of new type of sulfolipid and sialoglycolipid. Jap. J. Exp. Med. 36, 461-476.
KOCHETKOV, N.K., ZHUKOVA, I.G., SMIRNOVA, G.P., VAS'KOVSKII, V.E., 1968. Isolation of sphingoglycolipids containing sialic acid from Strongylocentrotus intermedius gonads. Dokl. Akad. Nauk SSSR 177, 1472-1474. Chem. Abstr. 68, No. 84412a.
LEITCH, J.L., 1934. The water exchanges of living cells. II. Nonsolvent volume determinations from swelling and analytical data. J. Cell. Comp. Physiol. 4, 457-473.
LINDVALL, S., 1948. Changes in activity of tributyrin splitting enzyme during early sea urchin egg development. Arkiv Kemi, Mineral. Geol. 26B, No. 9, 1-3.
LINDVALL, S., CARSJÖ, A., 1948. Cholesterol during early sea-urchin egg development. Arkiv Kemi, Mineral. Geol. 26B, No. 12, 1-3.
LOEB, J., 1913. Artificial parthenogenesis and fertilization. Chicago Univ. Press, Chicago.
MAGGIO, R., MONROY, A., 1955. Some experiments pertaining to the chemical mechanisms of the cortical reaction in fertilization of sea urchin eggs. Exp. Cell Res. 8, 240-244.
MARSH, J.B., 1965. Isolation and composition of a water-soluble lipoprotein from Arbacia eggs. Biol. Bull. 129, 415.
MATHEWS, A.P., 1913. An important chemical difference between the eggs of the sea urchin and those of the starfish. J. Biol. Chem. 14, 465-467.
MCCLENDON, J.F., 1909. Chemical studies on the effects of centrifugal force on the eggs of the sea urchin (Arbacia punctulata). Am. J. Physiol. 23, 460-466.
MEYERHOF, O., 1911. Untersuchungen über die Wärmetönung der vitalen Oxydationsvorgänge in Eiern. II. Versuche an Eiern und Larven von Strongylocentrotus lividus. Biochem. Z. 35, 280-315.
MOHRI, H., 1957. Endogenous substrates of respiration in sea-urchin spermatozoa. J. Fac. Sci. Univ. Tokyo IV, 8, 51-63.
MOHRI, H., 1959. Plasmalogen content in sea-urchin gametes. Sci. Papers Coll. Gen. Educ. Univ. Tokyo 9, 263-267.
MOHRI, H., 1964a. Utilization of C^{14}-labeled acetate and glycerol for lipid synthesis during the early development of sea urchin embryos. Biol. Bull. 126, 440-455.
MOHRI, H., 1964b. Phospholipid utilization in sea-urchin spermatozoa. Pubbl. Sta. Zool. Napoli 34, 53-58.
MONROY, A., 1945. Di alcuni fenomeni corticali che accompagnano la fecondazione e le prime divisioni dell 'uovo di riccio di mare. Experientia 1, 335-336.
MONROY, A., 1947. Further observations on the fine structure of the cortical layer of unfertilized and fertilized sea urchin eggs. J. Cell. Comp. Physiol. 30, 105-109.
MONROY, A., 1957. An analysis of the process of fertilization and activation of the egg. Inter. Rev. Cytol. 6, 107-127.
MONROY, A., MONROY-ODDO, A., DE NICOLA, M., 1951. The carotenoid pigments during early development of the egg of the sea urchin Paracentrotus lividus. Exp. Cell Res. 2, 700-702.
MONROY, A., MONTALENTI, G., 1947. Variations of the submicroscopic structure of the cortical layer of fertilized and parthenogenetic sea urchin eggs. Biol. Bull. 92, 151-161.
MONROY, A., RUFFO, A., 1945. Variazioni del colesterolo durante la fecondazione di uova di riccio di mare. Boll. Soc. Ital. Biol. Sper. 20, 6-7.

MOORE, B., WHITLEY, E., ADAMS, A., 1913. The rôle of glycogen, lecithides, and fats in the reproductive organs of echinoderms. Biochem. J. 7, 127-141.
NAGAI, Y., HOSHI, M., 1975. Sialosphingolipids of sea urchin eggs and spermatozoa showing a characteristic composition for species and gamete. Biochim. Biophys. Acta, in press.
NAGAI, Y., ISONO, Y., 1965. Occurrence of animal sulfolipid in the gametes of sea urchins. Jap. J. Exp. Med. 35, 315-318.
NAVEZ, A.E., 1938. Indophenoloxidase in Arbacia eggs and the Nadi reaction. Biol. Bull. 75, 357-358.
NEEDHAM, J., NEEDHAM, D.M., 1930. On phosphorus metabolism in embryonic life. I. Invertebrate eggs. J. Exp. Biol. 7, 317-348.
NUMANOI, H., 1959. Studies on the fertilization substance. IX. Enzymic degradation of lecithin during development of sea urchin eggs. Sci. Papers Coll. Gen. Educ. Univ. Tokyo 9, 285-296.
ÖHMAN, L.O., 1945. On the lipids of the sea urchin egg. Arkiv Zool. 36A, No. 7, 1-95.
PAGE, I.H., 1927a. The oils of the sea urchin and starfish egg. Biol. Bull. 52, 164-167.
PAGE, I.H., 1927b. The electrolytic content of the sea urchin and starfish egg. Biol. Bull. 52, 168-172.
PARPART, A.K., 1941. Lipo-protein complexes in the egg of Arbacia. Biol. Bull. 81, 296.
PAYNE, N.M., 1930. Some effects of low temperature on internal structure and function in animals. Ecology 11, 500-504.
RICOTTA, C.M.B., 1956. Quantitative changes of the free phospholipids in the sea-urchin egg at fertilization. Experientia 12, 104-106.
ROBERTSON, T.B., WASTENEYS, H., 1913. On the changes in lecithin-content which accompany the development of sea urchin eggs. Arch. f. Entw. Mech. 37, 485-496.
RUNNSTRÖM, J., 1928. Die Veränderungen der Plasmakolloide bei der Entwicklungserregung des Seeigeleies. Protoplasma 4, 388-514.
SCHMIDT, G., HECHT, L., THANNHAUSER, S.J., 1948. The behavior of the nucleic acids during the early development of the sea urchin egg (Arbacia). J. Gen. Physiol. 31, 203-207.
SHACKELL, L.F., 1911. Phosphorus metabolism during early cleavage of the echinoderm egg. Science, N.S. 34, 573-576.
TENNENT, D.H., GARDINER, M.S., SMITH, D.E., 1931. A cytological and biochemical study of the ovaries of the sea urchin Echinometra lucunter. Carnegie Inst. Washington Publ. No. 413, 1-46.
WARREN, L., HATHAWAY, R., 1960. Lipid-soluble sialic acid containing material in Arbacia eggs. Biol. Bull. 119, 354-355.
WETZEL, G., 1907. Die Entwicklung des Ovarialeies und des Embryos, chemisch untersucht mit Berücksichtigung der gleichzeitigen morphologischen Veränderungen. II. Die chemische Zusammensetzung der Eier des Seeigels, der Seespinne, des Tintenfisches und des Hundshaies. Arch. Anat. Physiol., Physiol. Abt., Suppl. 1907, 507-542.
WHITELEY, A.H., 1949. The phosphorus compounds of sea urchin eggs and uptake of radio-phosphate upon fertilization. Am. Naturalist 83, 249-267.
YAMAGAMI, K., 1963. Changes in the amounts of some endogenous phosphates in the development of the sea urchin. Zool. Mag. 72, 252-255. (In Japanese, with Abstract in English).
ZIELIŃSKI, M.A., 1939. Carbohydrate metabolism and phosphorus compounds in the fertilized eggs of the sea urchin (Paracentrotus lividus Lm.). Acta Biol. Exp. Warsaw 13, 35-48.

20. DNA in Gametes and Early Development

M. von Ledebur-Villiger

20.1 DNA in Gametes

Unfertilized sea-urchin eggs which are shed after the second maturation division contain much more DNA than sperm. But earlier reports of very large amounts of DNA in the mature eggs of several species of animals (several hundred times the haploid amount of DNA) are due mainly to failure to distinguish between DNA and other materials that interfere with determination. As methods have improved, values have dropped. In unfertilized *Paracentrotus lividus* eggs, microbiologic assay shows a DNA content of 24-25 x 10^{-12} g (ELSON and CHARGAFF, 1952) and in unfertilized *Arbacia lixula* eggs, microfluorometric methods reveal a DNA content of 25 x 10^{-12} g (BALTUS et al., 1965). Sperm DNA content ranges from 0.67 x 10^{-12} g to 1.0 x 10^{-12} g per sperm, or only one twenty-fifth of the DNA content of unfertilized eggs. Unfertilized *P. lividus* eggs have about 20 times as much DNA as sperm have (HOFF-JØRGENSEN, 1954), unfertilized *Lytechinus pictus* eggs have about 9.5 times as much, and unfertilized *Strongylocentrotus purpuratus* eggs have about 4.3 times as much (PIKÓ et al., 1967).

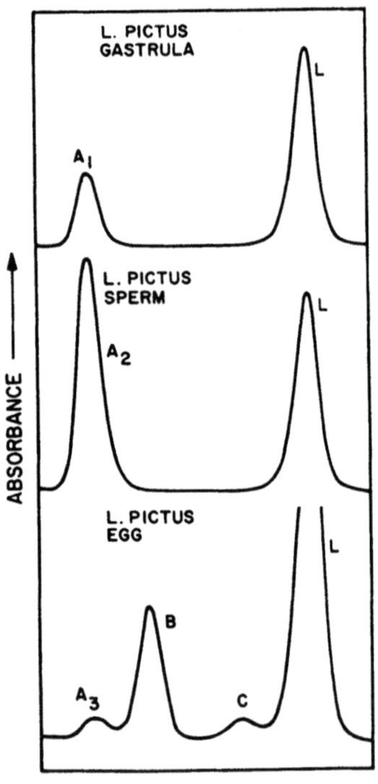

Fig. 20.1. Tracing of direct scans at 265 nm of buoyant density bands (in CsCl) of 3 DNA preparations from *Lytechinus pictus* after 25 hours of centrifugation in the same rotor at 44,770 rpm in the Beckman Model E centrifuge. L, density marker DNA (1.731 gm/cu cm) of *Micrococcus lysodeikticus*. A_1, A_2, A_3, nuclear DNA in the 3 preparations; buoyant density, 1.693. B, DNA derived from mitochondria and yolk; buoyant density, 1.703. C, unidentified nucleic acid band; buoyant density, 1.719. (From PIKÓ et al., 1967)

The amount of DNA in the isolated nuclei of unfertilized sea urchin eggs is $0.89 \pm 0.25 \times 10^{-12}$ g. The nucleus of mature eggs contains essentially the same amount of DNA as the sperm, which suggests that the additional DNA is located in the cytoplasm (HINEGARDNER, 1961).

Recent evidence indicates that cytoplasmic DNA is found in the mitochondria (PIKÓ et al., 1967). In contrast to earlier datas (BIBRING et al., 1965; CARDEN et al., 1965; BALTUS et al., 1965), PIKÓ and co-workers who have intensively studied the cytoplasmic DNA in sea-urchin eggs found that it differed in buoyant density and base composition from the nuclear DNA. In a buoyant density centrifugation of purified whole egg DNA of *Lytechinus pictus* they observed 3 bands at 265 nm with buoyant densities of 1.693 g/cu cm, 1.703 g/cu cm and 1.719 g/cu cm (Fig. 20.1). The relative amount of DNA in these 3 bands is approximately 1:7:1. Sperm DNA and gastrula DNA of the same species band at 1.693, which represents the nuclear DNA, whereas the 1.703 band represents the cytoplasmic DNA. The 1.719 band is unidentified. The average melting temperature of sperm and gastrula DNA is 84°C and of unfertilized egg DNA, 86.8°C. The guanine-cytosine content of whole egg DNA is 42%, of sperm DNA, 35%, and of gastrula DNA, 34% (PIKÓ et al., 1967).

The sedimentation behavior of sea urchin cytoplasmic DNA suggests 2 circular DNA components corresponding to the twisted and relaxed forms found in the mitochondria of bacteria and animal viruses (VINOGRAD and LEBOWITZ, 1966). Electron microscopy shows that sea-urchin cytoplasmic DNA occurs mostly in the form of closed circular filaments, twisted and relaxed (Fig. 20.2). Fairly uniform in size, their average perimeter is 4.45 ± 0.25 µm.

Of catenated DNA molecules in sea-urchin mitochondria, 88% are monomers, 11% dimers, 1% trimers, and 0.3% tetramers (PIKÓ et al., 1968). Based on these values, a mature unfertilized *Lytechinus pictus* egg has about 400,000 circles of mitochondrial DNA, about 350,000 mitochondria or about one DNA molecule per mitochondrion (BERGER, 1968). Mitochondrial DNA is synthesized during oogenesis (MATSUMO and PIKÓ, 1971).

Purified DNA from the egg cytoplasm of *Lytechinus pictus* and *Strongylocentrotus purpuratus* is as effective a primer of RNA synthesis as nuclear DNA prepared from blastulae and plutei (BALTUS et al., 1965; PIKÓ et al., 1967). This activity is insignificant during the first few hours of development and does not account for the activation of protein synthesis at fertilization (GIUDICE and MONROY, 1958; NAKANO and MONROY, 1958).

The function of cytoplasmic DNA is not clear, and whether or not the circular units of all mitochondria are genetically alike is unknown. Nor is it established what morphogenetic effects are due to mitochondrially produced proteins. It is unlikely that cytoplasmic DNA is the precursor of nuclear DNA (BALTUS et al., 1965; LØVTRUP, 1966). While mitochondrial DNA does not contain enough information to code for all mitochondrial components, it is possible that animal mitochondrial DNA codes for the machinery that synthesizes mitochondrial proteins (NASS, 1969; DAWID, 1970). Because the extremely large amount of cytoplasm in the egg is divided among many blastomeres, cytoplasmic DNA is destined for many cells (DAWID, 1966; CZIHAK, 1971). Already in the cell of the sea-urchin blastula, cytoplasmic DNA is only 1% of the total DNA (PIKÓ et al., 1967).

The amount of DNA in sea-urchin spermatozoa varies from species to species. Feulgen microspectrophotometry shows in *P. lividus* sperm a

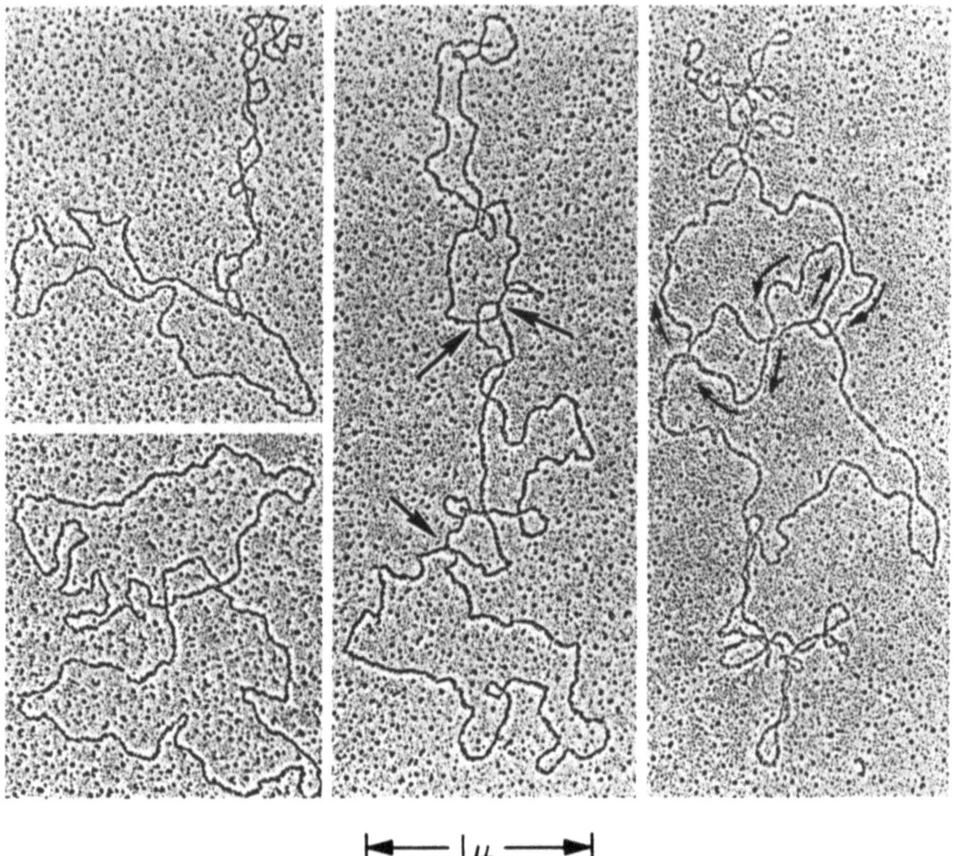

Fig. 20.2. Electron micrographs of catenated molecules from middle-band material of *Lytechinus pictus* egg DNA. Lower left: dimer with 2 relaxed circles. Upper left: dimer with 1 relaxed and 1 twisted circle. Middle: trimer with 3 relaxed circles; arrows indicate cross-overs of interlocking circles. Right: tetramer with 2 relaxed and 2 twisted circles; arrows indicate course of 1 relaxed circle. (From PIKÓ et al., 1968)

relative DNA value of 1; in *Arbacia lixula*, 0.75; in *Sphaerechinus granularis*, 1.61; in *Echinocardium cordatum*, 1.65; and in *Stylocidaris affinis*, 1.71. The hydrolysis curve (extinction values in relation to increasing hydrolysis time) shows 2 peaks possibly caused by 2 classes of DNA complexes with different resistance to acid hydrolysis (VILLIGER et al., 1970).

Circular DNA in sand-dollar sperm is indicated by similar reactions in sperm DNA and circular viral DNA during exposure to sonic vibration and during denaturation-renaturation experiments (ROSENKRANZ et al., 1967).

20.2 DNA Synthesis during Development

20.2.1 Nuclear DNA Synthesis

For many years it was assumed that the DNA present in the unfertilized egg is a reserve which is used up in the course of the first divisions and that DNA synthesis does not begin until this reserve is exhausted (LØVTRUP, 1966), i.e. after 4 or 5 cell divisions. The reason for this assumption was the impossibility to show an increase in the total amount of DNA in the cleaving egg from insemination up to about the 16-cell stage (HOFF-JØRGENSEN, 1954; AGRELL and PERSSON, 1956; OLSSON, 1965). Only thereafter a marked increase in the total DNA content can be observed (BRACHET, 1931). The values reported for the DNA increase between fertilization and the pluteus stage range from 6 to 14 times (BALTZER and CHEN, 1960) up to 2,000 times (FICQ et al., 1963).

Today it is known that marked nuclear DNA synthesis begins shortly after fertilization, but the nuclear DNA synthesized, much less than the cytoplasmic DNA, is undetectable by measuring the total DNA. Nuclear DNA increase may be measured in isolated nuclei, thus excluding cytoplasmic DNA, or by exposing the developing eggs to exogenous labeled DNA precursors followed by autoradiography and/or scintillation counting.

In *Paracentrotus lividus*, nuclear DNA measured by Feulgen microspectrophotometry shows that the S-phase immediately follows the telophase and lasts 5 minutes (LISON and PASTEELS, 1951). Measurements of labeled thymidine incorporated into the acid-insoluble fraction in sand dollar *Echinarachinus parma* show that nuclear DNA synthesis begins in the pronuclei 15-20 minutes after fertilization, before pronuclear fusion (SIMMEL and KARNOFSKY, 1961). In *Paracentrotus lividus* the same finding is supported by electron microscopic autoradiography (ANDERSON, 1969), but not by autoradiography analyzed by light microscopy (VON LEDEBUR, 1972).

In some species, haploid nuclei seem to form diploid values before syngamy, creating a zygote containing a tetraploid complement. After first division, daughter nuclei receive a diploid complement of chromosomes.

In *Strongylocentrotus purpuratus* (BUCHER and MAZIA, 1960) and *Arbacia punctulata* (ZIMMERMAN and SILBERMAN, 1967), DNA synthesis does not begin until after pronuclear fusion. Syngamy, however, is not a pre-requisit for the onset of DNA synthesis. This is confirmed by the observation that in polyspermic zygotes, unfused male pronuclei synthesize DNA (SIMMEL and KARNOFSKY, 1961; LONGO and PLUNKETT, 1973) and that parthenogenetically activated eggs synthesize DNA (VON LEDEBUR, 1972). DNA synthesis also does not depend on centriole duplication (BUCHER and MAZIA, 1960).

In *Paracentrotus lividus* embryos, exogenous cytidine incorporation into DNA possibly indicates an endogenous reservoir of related ribonucleotides of 4×10^{-4} mμmol per unfertilized egg. The large amounts of thymidine and deoxyuridine incorporated into DNA from exogenous sources are possible only with a small endogenous reservoir of thymidine. The available thymidine in the egg seems inadequate to support DNA synthesis beyond the 8-cell stage when 5-fluorodeoxyuridine blocks development (NEMER, 1962). DNA synthesis is thought to occur during early interphase (NEMER, 1962), but autoradiography suggests that it occurs early in telophase (FICQ et al., 1963).

Accumulated acid-insoluble radioactivity in developing *Strongylocentrotus purpuratus* eggs incubated with radioactive thymidine shows that the first major synthesis of nuclear DNA takes place about 30 minutes

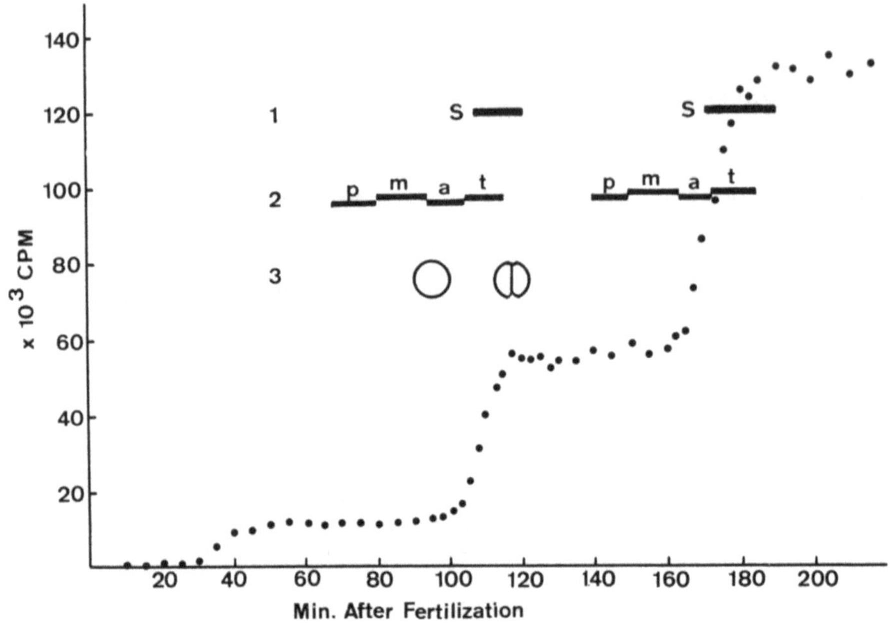

Fig. 20.3. The DNA synthetic period during early development of *Strongylocentrotus purpuratus*. Accumulation of acid insoluble radioactive thymidine (data points) and corresponding cytological data (1-3). 1) time and duration of DNA synthesis (s); 2) time and duration of each mitotic phase: p, prophase; m, metaphase; a, anaphase; t, telophase. 3) appearance of the egg. (After HINEGARDNER et al., 1964)

after fertilization, when nuclear fusion occurs (Fig. 20.3). The second synthesis begins at telophase and lasts about 13 minutes. DNA synthesis begins after furrowing commences and ends soon after cleavage is complete and the nucleus has formed a sphere. Nuclear DNA synthesis amounts to 1.8×10^{-12} g DNA per diploid nucleus at 16°C in 12 minutes (HINEGARDNER et al., 1964).

In early sea-urchin development there is no G_1 period, and the M-phase of one cycle overlaps with the S-phase of the next. In blastula nuclei, however, DNA is replicated during interphase; the G_1 period is relatively long and the G_2 period fairly short (VILLIGER et al., 1970; VON LEDEBUR-VILLIGER, 1972).

^{3}H-thymidine incorporation into the nuclear DNA of *Psammechinus miliaris* and *Paracentrotus lividus* is similar. The incorporation rate of each nucleus is small in the first S-phase and increases until the third S-phase, after which it remains constant (Fig. 20.4)(CZIHAK and POHL, 1970). It increases as the egg pool for thymidine gradually becomes saturated. Radioactivity decreases in ^{3}H-thymidine labeled sea water in which fertilized eggs are developing (Fig. 20.5). Little ^{3}H-TdR is taken up before fertilization, but quite a bit is taken up afterward. The egg pool appears saturated about 50 minutes after fertilization. This suggests that reduced ^{3}H-TdR incorporation at the first S-phase is due to the small amount of labeled thymidine in the thymidine pool and not to reduced DNA synthesis (VON LEDEBUR, 1972).

The uptake mechanism of cytidine may be by specific carrier molecules at the cell surface (MITCHISON and CUMMINS, 1966), and also for thymidine uptake, an active transport is postulated (VON LEDEBUR, in press).

In *Strongylocentrotus purpuratus* embryos, single- and double-stranded nuclear DNA of low molecular weight (8-16 S) is synthesized. Single-

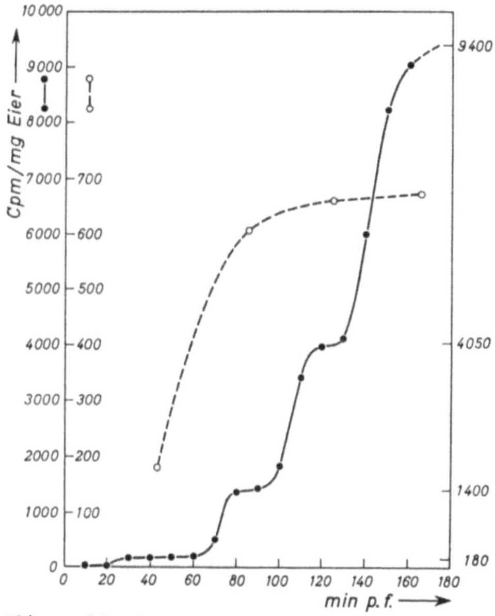

Fig. 20.4. DNA synthesis in early embryogenesis of *Paracentrotus lividus*. Solid line: accumulated acid-insoluble radioactive thymidine in early cleavage stages. Broken line: increased incorporation rate in single nucleus. (From CZIHAK and POHL, 1970)

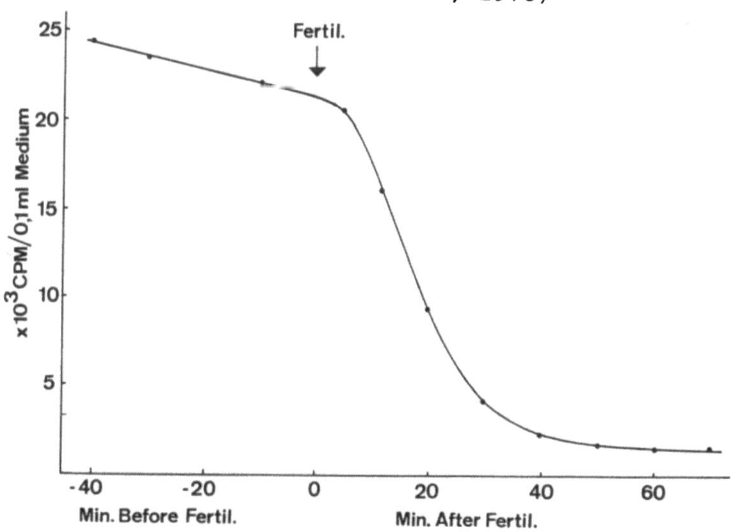

Fig. 20.5. Decreasing radioactivity in ^3H-thymidine-labeled sea water in which *Paracentrotus lividus* eggs are fertilized and developed. Thymidine uptake begins at fertilization and the thymidine pool gradually becomes saturated. (After VON LEDEBUR, 1972)

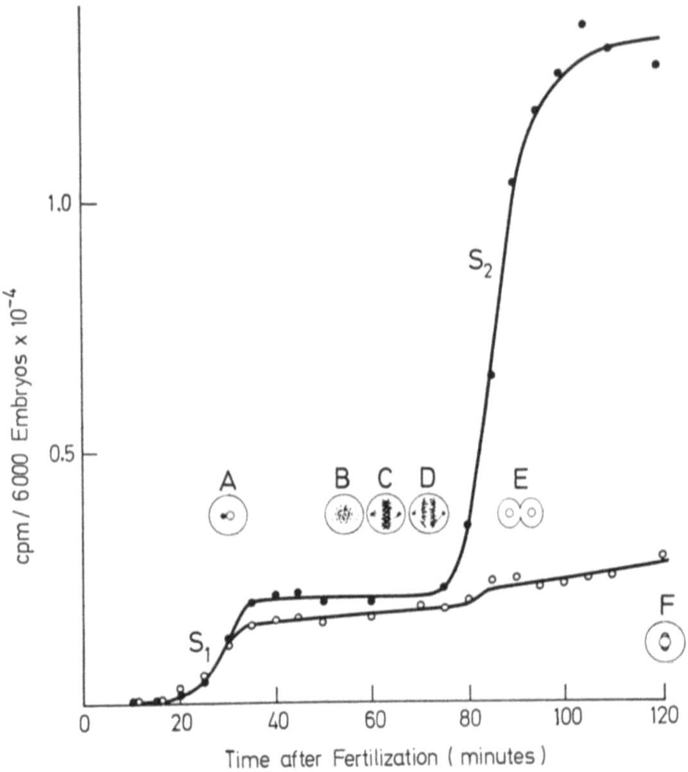

Fig. 20.6. Cumulative incorporation of ^3H-thymidine into DNA of *Echinarachnius parma* embryos; effect of cycloheximide. Abscissa: minutes after fertilization; ordinate: cpm/6,000 embryos x 10^{-4}. Closed circles: controls; open circles: 1 μg/ml cycloheximide from fertilization. Morphological events: A, pronuclear fusion; B, prophase; C, metaphase; D, anaphase; E, cleavage; F, embryo altered by cycloheximide treatment. S_1, first synthetic period; S_2, second synthetic period. (From YOUNG et al., 1969)

stranded pieces made at any stage are transient, while double-stranded molecules of low molecular weight, made in early development, seem to accumulate and become part of DNA of intermediate molecular weight (16-70 S), appearing later in development. Nuclear DNA fractions of low molecular weight are thus able to hybridize to DNA's of higher molecular weight taken from different stages of early embryo development (BAKER, 1971, 1972).

DNA replication seems to begin at the nuclear membrane. In a preparation containing DNA bound to the membrane of the nuclei of developing sea-urchin embryos, DNA synthesis takes place in vivo and in vitro. The complex virtually disappears when DNA is not being synthesized, as in unfertilized eggs or during the G_2 phase of the division cycle (INFANTE et al., 1973).

DNA synthesis in cleaving sea-urchin eggs is blocked by puromycin and cycloheximide, both inhibitors of protein synthesis. Puromycin only slightly affects the first period of thymidine incorporation in fertilized *Arbacia* eggs but inhibits subsequent periods. Embryos seem to synthesize proteins in order to incorporate thymidine into DNA. In early stages, these proteins may be synthesized about 1 duplication cycle before they are used (BLACK et al., 1967). Actinomycin D and puromycin also delay cleavage (CZIHAK, 1968).

In the sand dollar, *Echinarachnius parma*, cycloheximide (1 μg/ml) inhibits protein synthesis by 97%, but does not prevent DNA synthesis during the first period (Fig. 20.6). But in later period, it inhibits DNA synthesis and prevents cleavage. Embryos exposed to cycloheximide at progressively later intervals show that proteins probably involved in DNA synthesis, karyokinesis, and cytocleavage are made at different times, during the S-period or slightly after it. As DNA is still synthesized in newly introduced sperm in animals rendered incapable of replicating their nuclear DNA, the drug-induced defect seems to involve DNA itself rather than the availability of precursors or their polymerisation on a suitable template (YOUNG et al., 1969).

At a concentration of 1 mol^{-3}, hydryxurea greatly inhibits labeled thymidine incorporation into DNA. It also stops cell division after 1 or 2 hours in *Arbacia punctulata* embryos at the 16- to 32-cell stage (KEDES et al., 1969).

20.2.2 Mitochondrial DNA Synthesis

Mitochondrial DNA synthesis seems independent of nuclear DNA synthesis (ANDERSON, 1969). When isolated from *Lytechinus pictus* embryos cultured in the presence of radioactive thymidine, mitochondrial DNA is not labeled up to 2 days after fertilization, while nuclear DNA is labeled greatly, demonstrating that mitochondrial DNA does not replicate appreciably during early development (PIKÓ, 1969). These findings are similar in embryos of *Arbacia punctulata* (MATSUMO and PIKÓ, 1971).

20.3 Enzymes

20.3.1 Enzymes Involved in DNA Synthesis

Nuclei isolated from *Strongylocentrotus purpuratus* embryos during early cleavage stages contain a high level of DNA polymerase. There is no evidence of polymerase outside the nuclei. The enzyme activity is similar to that of DNA polymerases from other sources where pH optimum is about 7.4; the reaction is inhibited in a medium of increased ionic strength; NaCl added to a final concentration of 0.1 *M* reduces the activity by one half; synthesis is reduced by one half in 4×10^{-6} *M* hydroxymercuribenzoate; Mg^{++} or Mn^{++} is required for significant synthetic activity; and calcium does not activate the enzyme. Experiments with tetraploid nuclei show that nuclear polymerase activity is proportional to the number of chromosomal sets (MAZIA and HINEGARDNER, 1963).

In cells of *Strongylocentrotus franciscanus* embryos at the time of hatching, 56%-95% of the total DNA polymerase activity is found in the nuclei. During early development, the overall DNA polymerase activity changes little, and not much DNA polymerase is synthesized. This suggests that the polymerase necessary for DNA synthesis, at least up to gastrulation, is present in the egg before fertilization. The percent of activity associated with the nuclear fraction increases progressively, while

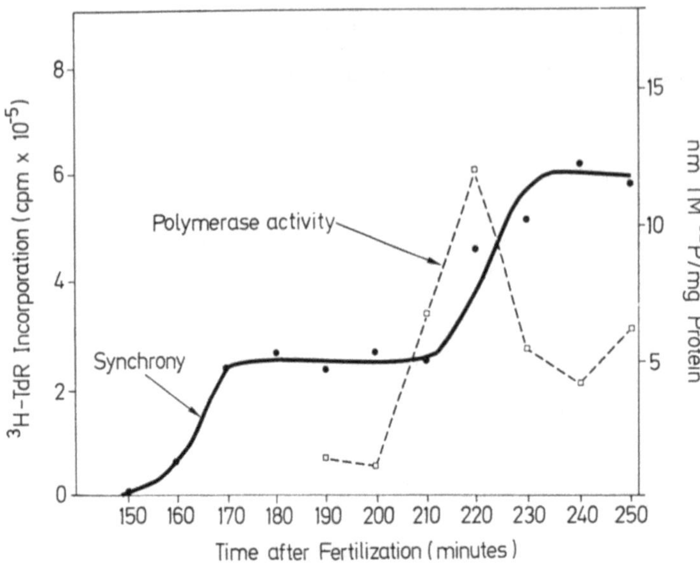

Fig. 20.7. DNA polymerase activity per milligram of protein in nuclei isolated from *Strongylocentrotus purpuratus* embryos during the 4-8-cell cycle. Abscissa: minutes after fertilization; left ordinate (solid line): ^{3}H-TdR incorporation (cpm x 10^{-5}); right ordinate (broken line): nanomoles of TM ^{32}P/mg protein. Synchrony: incorporation of ^{3}H-TdR into DNA by whole embryos. (From FANSLER and LOEB, 1972)

at the same time activity in the soluble cytoplasmic fraction declines. By the hatching stage most of the activity occurs in the nucleus. This migration of DNA polymerase activity represents a molecular transfer of the enzyme from the cytoplasm to the nucleus (FANSLER and LOEB, 1969; LOEB et al., 1969; LOEB and FANSLER, 1970).

DNA polymerases purified from *Strongylocentrotus franciscanus* and *Strongylocentrotus purpuratus* are heat labile and are stabilized in the presence of polyglycols. Their molecular weight is approximately 150,000. DNA polymerase activity depends entirely on the presence of added DNA. Native DNA, degraded to a limited extent with pancreatic DNAse, is the most efficient primer (LOEB, 1969).

Studies of the relationship between DNA polymerase activity and nuclear DNA replication show that polymerase activity in nuclei isolated from S-phase cells is 5 to 10 times greater than in nuclei from mitotic or G_2 embryos. When DNA synthesis is completed, polymerase activity drops to a level slightly higher than before the S-phase (Fig. 20.7). It may be that DNA polymerase attaches to the chromosomes as mitosis is completed and then dissociates from them. DNA polymerase associated with DNA could determine the size of replicating units (FANSLER and LOEB, 1972). If the polymerase is evenly distributed on the chromosomes, one molecule of polymerase is associated with chromosomal units equal in length to 2,000 to 3,000 nucleotide pairs. This corresponds to about 20,000 molecules of polymerase per chromosome. This applies to sea-urchin embryos at the time of hatching, when division is rapid and DNA synthesis is great (LOEB, 1970; LOEB et al., 1971).

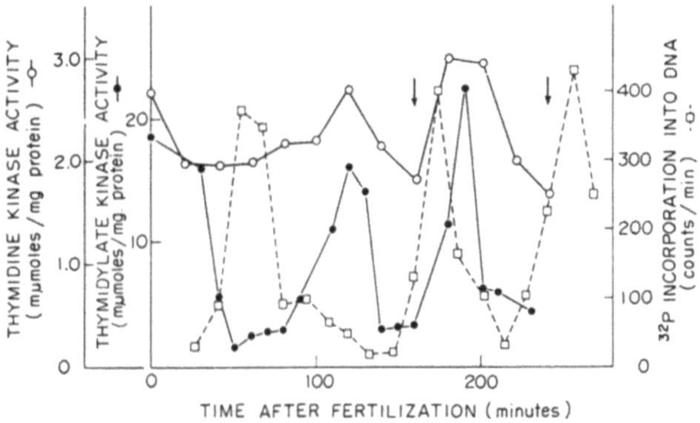

Fig. 20.8. Activity of thymidine kinase (open circles), thymidilate kinase (closed circles), and DNA synthesis (open squares) in early cleavage stages of *Hemicentrotus pulcherrimus*. (From NAGANO and MANO, 1968)

DNA polymerase from *Strongylocentrotus franciscanus* blastulae contains about 4 gm atoms of zinc per mol of enzyme. Bound zinc seems to be essential to DNA-DNA polymerase interaction (SLATER et al., 1971). Partially purified polymerase fractions prepared from 16-cell stages are able to support RNA-directed as well as DNA-directed DNA synthesis, and it is possible that the RNA-dependent DNA synthesis is mediated by polymerases other than those responsible for DNA-dependent DNA synthesis (SLATER and SLATER, 1972). No DNA polymerase activity is detected in spermatozoa.

Deoxyribonucleoside kinases are enzymes that catalyze the phosphorylation of deoxyribonucleosides into monophosphates, diphosphates and triphosphates, supplying thus the substrate for DNA polymerase. In *Psammechinus miliaris*, thymidine kinase activity shows rhythmic variations during mitosis. It increases before DNA synthesis begins (HANSEN-DELKESKAMP and DUSPIVA, 1966). This periodic change in thymidine kinase and thymidylate kinase activity also occurs during early embryogenesis in *Hemicentrotus pulcherrimus* (Fig. 20.8). Maximal kinase activity occurs in unfertilized eggs, demonstrating that kinase is synthesized during oogenesis. Kinase activity decreases after fertilization and changes periodically during the cell cycle. It is greatest before DNA synthesis (NAGANO and MANO, 1968).

Puromycin and ethionine depress thymidine kinase activity in vivo indicating that increased activity may be caused at least partly by enzyme synthesis de novo. Unlike thymidine kinase in such organisms as *Escherichia coli*, that of sea-urchin eggs and embryos is labile: 60%-70% of its activity is lost after heat treatment for 10 minutes at 50°C. Thymidine stabilizes thymidine kinase activity, dTTP is a feedback inhibitor, but dCDP does not activate it (NAGANO and MANO, 1968).

In *Strongylocentrotus franciscanus*, kinase activity is fairly constant up to the 100-cell stage. Specific kinase activity is slightly less in the nuclear fraction than in the cytoplasmic fraction (FANSLER and LOEB, 1969).

Other enzymes are known to be involved in DNA biosynthesis of developing sea-urchin embryos. Deoxycytidylate aminohydrolase hydrolytically deaminates dCMP to dUMP and methyl-dCMP is hydrolytically deaminated to TMP (SCARANO and MAGGIO, 1959). In the early development of *Sphaer-*

echinus granularis the amount of dCMP aminohydrolase decreases strikingly. Actinomycin D changes the dCMP aminohydrolase pattern: after a decrease in the first 20 hours of development similar to the controls, dCMP aminohydrolase remains constant or increases to values higher than those of unfertilized eggs. DNA-dependent RNA synthesis seems to be required for the control of the normal pattern of dCMP aminohydrolase during early embryo development, and it is possible that actinomycin interferes with the synthesis of a specific repressor (SCARANO et al. 1964).

Cell-free extracts of unfertilized *Arbacia punctulata* eggs appear not to contain an enzyme system for reducing ribonucleotides to deoxyribonucleotides, but within 5 hours after fertilization at 23°C, an enzyme system develops, capable of catalyzing the reduction of CDP to dCDP in the presence of Mg^{++}, EDTA, ATP, and a reducing agent, dithiothreitol. The activity begins about 1 hour after fertilization and peaks at about 5 hours. If emetine or puromycin, each an inhibitor of protein synthesis, is added to the cells before fertilization, ribonucleotide reductase does not develop. If, however, actinomycin D is added to the cell suspension at a concentration that inhibits messenger RNA synthesis, ribonucleotide reductase production is not appreciably affected. Pre-existing maternal RNA seems to be used for reductase synthesis (NORONHA et al., 1972).

20.3.2 Enzymes Involved in DNA Degradation

In developing *Arbacia punctulata* embryos, most DNAse occurs in the cytoplasm rather than in the nucleus (MAZIA, 1949). *Paracentrotus* sperm DNAse hydrolyzes calf thymus DNA optimally at pH 5.25 and 3.5 and *Arbacia* sperm DNAse hydrolyzes it optimally at pH 6.5 and 4.5. The DNAse of both species is of the endonucleolytic type. It is possible that DNAse and proteinases are located in the acrosomal region of the sperm (LUNDBLAD and JOHANSSON, 1968).

Alkaline DNAse activity in sea-urchins changes during embryo development (Fig. 20.9). The enzyme level remains almost constant until the hatching blastula stage but suddenly decreases between late blastula and the onset of gastrulation. It increases from the gastrula stage on. This activity pattern parallels DNA synthesis: reduced DNAse activity corresponds to reduced mitotic activity during gastrulation (DE PETROCELLIS and PARISI, 1972).

20.3.3 Enzymes Involved in DNA Modification

The DNA methylases are the most interesting DNA-modifying enzymes known. In bacteria they methylate DNA bases after the primary structure has been synthesized. The methyl group of methionine, as S-adenosylmethionine, is the source of the methyl groups for these reactions. The DNA methylases are species-specific (GOLD et al., 1963).

In *Lytechinus variegatus*, DNA base methylation appears not to occur before gastrulation (COMB, 1965), while in *Paracentrotus lividus* and *Sphaerechinus granularis* DNA methylation occurs at all stages of embryo development (SCARANO et al., 1965). 5-methyl-cytosine is the only methylated base found in sea urchins (GRIPPO et al., 1968). It is possible that highly specific DNA-modifying enzymes are synthesized sequentially during embryogenesis and differentiation. According to this hypothesis each class of genes is turned on or off together, and has an initial specific array of bases making up the binding site

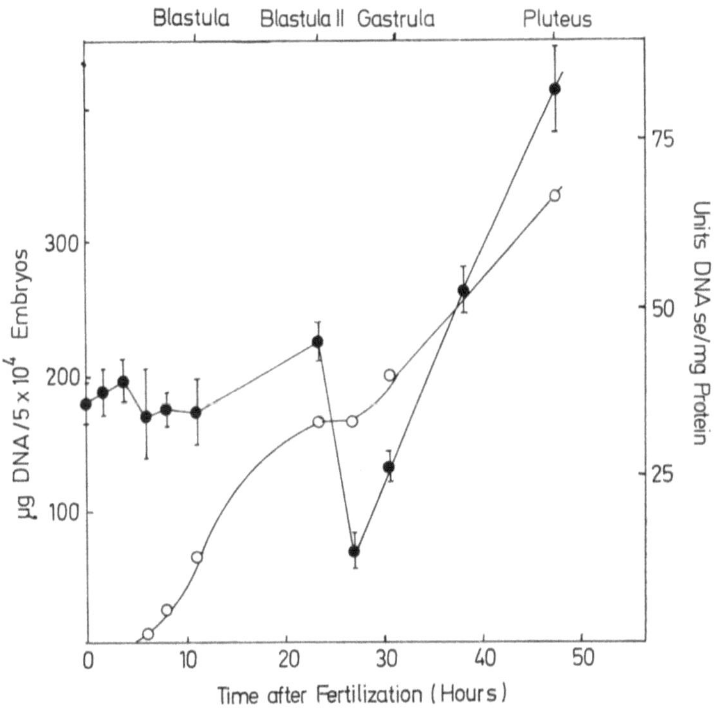

Fig. 20.9. DNA synthesis and DNAse activity during early development of *Paracentrotus lividus* embryos. Abscissa: hours after fertilization. Left ordinate (open circles): micrograms of DNA per 5 x 10^4 embryos; right ordinate (closed circles): units of DNAse per milligram of protein. (From DE PETROCELLIS and PARISI, 1972)

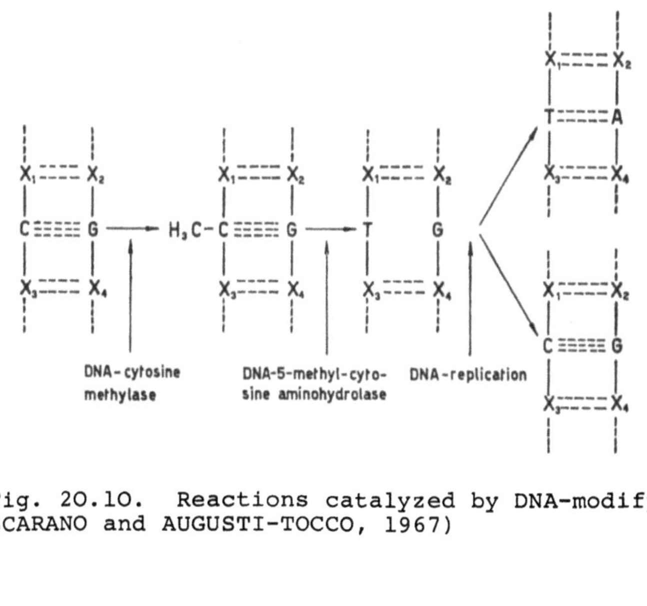

Fig. 20.10. Reactions catalyzed by DNA-modifying enzymes. (From SCARANO and AUGUSTI-TOCCO, 1967)

of highly specific DNA-modifying enzymes. DNA-cytosine methylating enzymes catalyze the methylation of DNA-cytosine to DNA-5-methylcytosine which is transformed into DNA-thymidine by DNA-5-cytosine aminohydrolase (Fig. 20.10)(SCARANO et al., 1967; SCARANO, 1969, 1971; GRIPPO et al., 1970; TOSI et al., 1972). DNA-thymidine basepairs with adenine, and this base pair change is maintained on further cell duplication by the normal mechanism of DNA replication, while the other daughter cell has the same DNA base sequence as the mother cell had. Differentiated cells arise, while the germinal line retains the initial DNA base sequence (SCARANO and AUGUSTI-TOCCO, 1967).

References

AGRELL, I., PERSSON, H., 1956. Changes in the amount of nucleic acids and free nucleotides during early embryonic development of sea urchins. Nature 178, 1398-1399.

ANDERSON, W.A., 1969. Nuclear and cytoplasmic DNA synthesis during early embryogenesis of Paracentrotus lividus. J. Ultrastr. Res. 26, 95-110.

BAKER, R.F., 1971. Changing size pattern of newly synthesized nuclear DNA during early development of the sea urchin. Biochem. Biophys. Res. Com. 43, 1415-1420.

BAKER, R.F., 1972. Isolation of homologous nuclear DNA's from sea urchin embryos. J. Cell Sci. 11, 153-171.

BALTUS, E., QUERTIER, J., FICQ, A., BRACHET, J., 1965. Biochemical studies of nucleate and anucleate fragments isolated from sea urchin eggs. A comparison between fertilization and parthenogenetic activation. Biochim. Biophys. Acta 95, 408-417.

BALTZER, F., CHEN, P.S., 1960. Über das zytologische Verhalten und die Synthese der Nukleinsäuren bei den Seeigelbastarden Paracentrotus o x Arbacia o und Paracentrotus o x Sphaerechinus o. Rev. Suisse Zool. 67, 183-194.

BERGER, E.R., 1968. A quantitative study of sea urchin egg mitochondria in relation to their DNA content. J. Cell Biol. 39, 12a-13a.

BIBRING, T.J., BRACHET, J., GAETA, F.S., GRAZIOSI, F., 1965. Some physical properties of cytoplasmic deoxyribonucleic acid in unfertilized eggs of Arbacia lixula. Biochim. Biophys. Acta 108, 644-665.

BLACK, R.E., BAPTIST, E., PILAND, J., 1967. Puromycin and cycloheximide inhibition of thymidine incorporation into DNA of cleaving sea urchin eggs. Exp. Cell Res. 48, 431-439.

BRACHET, J., 1931. La synthèse de l'acide thymonucléique pendant le développement de l'oeuf d'oursin. Compt. Rend. Soc. Biol. 108, 813-815.

BUCHER, N.L.R., MAZIA, D., 1960. Deoxyribonucleic acid synthesis in relation to duplication of center of dividing eggs of the sea urchin Strongylocentrotus purpuratus. J. Biophys. Biochem. Cytol. 7, 651-655.

CARDEN, G.A., ROSENKRANZ, S., ROSENKRANZ, H.S., 1965. Deoxyribonucleic acids of sperm, eggs and somatic cells of the sea urchin Arbacia punctulata. Nature 205, 1338-1339.

COMB, D.G., 1965. Methylation of nucleic acids during sea urchin embryo development. J. Mol. Biol. 11, 851-855.

CZIHAK, G., 1968. Molekularbiologische Aspekte der frühen Embryonalentwicklung von Seeigeln. Pubbl. Sta. Zool. Napoli 36, 321-345.

CZIHAK, G., 1971. Echinoids. In: Experimental Embryology of Marine and Freshwater Invertebrates (G. Reverberi, ed.), pp. 361-506. North Holland.

CZIHAK, G., POHL, E., 1970. DNS-Synthese in frühen Furchungsstadien von Seeigelembryonen. Z. Naturforsch. 25b, 1047-1052.
DAWID, I.B., 1966. Evidence for the mitochondrial origin of frog egg cytoplasmic DNA. Proc. Nat. Acad. Sci. US 56, 269-276.
DAWID, I.B., 1970. Cytoplasmic DNA in Oogenesis (J.D. Biggers, A.W. Schuetz, eds.). Baltimore: University Park Press.
DE PETROCELLIS, B., PARISI, E., 1972. Changes in alkaline deoxyribonuclease activity in sea urchin during embryonic development. Exp. Cell Res. 73, 496-500.
ELSON, D., CHARGAFF, E., 1952. On the deoxyribonucleic acid content of sea urchin gametes. Experientia 8, 143-145.
FANSLER, B., LOEB, L.A., 1969. Sea urchin nuclear DNA polymerase. II. Changing localization during early development. Exp. Cell Res. 57, 305-310.
FANSLER, B., LOEB, L.A., 1972. Sea urchin nuclear DNA polymerase. IV. Reversible association of DNA polymerase with nuclei during the cell cycle. Exp. Cell Res. 75, 433-441.
FICQ, A., AIELLO, F., SCARANO, E., 1963. Metabolism des acides nucléiques dans l'oeuf d'oursin en développement. Exp. Cell Res. 29, 128-136.
GIUDICE, G., MONROY, A., 1958. Incorporation of S^{35}-methionine in the proteins of the mitochondria of developing and parthenogenetically activated sea urchin eggs. Acta Embryol. Morphol. Exp. 2, 58-65.
GOLD, M., HURWITZ, J., ANDERS, M., 1963. The enzymatic methylation of RNA and DNA. On the species specificity of the methylation enzymes. Proc. Nat. Acad. Sci. US 50, 164-169.
GRIPPO, P., IACCARINO, M., PARISI, E., SCARANO, E., 1968. Methylation of DNA in developing sea urchin embryos. J. Mol. Biol. 36, 195-208.
GRIPPO, P., PARISI, E., CARESTIA, C., SCARANO, E., 1970. A novel origin of some deoxyribonucleic acid thymine and its nonrandom distribution. Biochemistry 9, 2605-2609.
HANSEN-DELKESKAMP, E., DUSPIVA, F., 1966. Aktivitätsverlauf der enzymatischen Phosphorylierung von Thymidin während der Entwicklung des Seeigels Psammechinus miliaris von der Befruchtung bis zum Zweizeller. Experientia 22, 381-382.
HINEGARDNER, R.T., 1961. The DNA content of isolated sea urchin nuclei. Exp. Cell Res. 25, 341-347.
HINEGARDNER, R.T., RAO, B., FELDMAN, D.E., 1964. The DNA synthetic period during early development of the sea urchin egg. Exp. Cell Res. 36, 53-61.
HOFF-JØRGENSEN, E., 1954. Deoxynucleic acid in some gametes and embryos. Proc. of the 7th Symposium Colston Res., 79-88.
INFANTE, A.A., NAUTA, R., GILBERT, S., HOBART, P., FIRSHEIN, W., 1973. DNA synthesis in developing sea urchins. Role of a DNA nuclear membrane complex. Nature, New Biology 242, 5-8.
KEDES, L.H., GROSS, P.R., COGNETTI, G., HUNTER, A.L., 1969. Synthesis of nuclear and chromosomal proteins on light polyribosomes during cleavage in the sea urchin embryo. J. Mol. Biol. 45, 337-351.
LISON, L., PASTEELS, J., 1951. Etudes histophotometriques sur la teneur en acide desoxyribonucleique des noyaux au cours du développement embryonaire chez l'oursin Paracentrotus lividus. Arch. Biol. 62, 1-64.
LOEB, L.A., 1969. Purification and properties of deoxyribonucleic acid polymerase from nuclei of sea urchin embryos. J. Biol. Chem. 244, 1672-1681.
LOEB, L.A., 1970. Molecular association of DNA polymerase with chromatin in sea urchin embryos. Nature 226, 448-449.
LOEB, L.A., FANSLER, B., WILLIAMS, R., MAZIA, D., 1969. Sea urchin DNA polymerase. I. Localization in nuclei during rapid DNA synthesis. Exp. Cell Res. 57, 298-304.

LOEB, L.A., FANSLER, B., 1970. III. Intracellular migration of DNA polymerase in early developing sea urchin embryos. Biochim. Biophys. Acta 217, 50-55.
LOEB, L.A., FANSLER, B.S., SLATER, J.P., 1971. The control of DNA replication in early developing sea urchin embryos. From: Informative molecules in biological systems (L.G.H. Ledoux, ed.). North Holland Publishing Co.
LONGO, F.J., PLUNKETT, W., 1973. The onset of DNA synthesis and its relation to morphogenetic events of the pronuclei in activated eggs of the sea urchin Arbacia punctulata. Develop. Biol. 30, 56-67.
LØVTRUP, S., 1966. Chemical basis of sea urchin embryogenesis. Bull. Schweiz. Akad. med. Wiss. 22, 201-276.
LUNDBLAD, G., JOHANSSON, B., 1968. Proteinase and deoxyribonuclease activity in sea urchin sperm. Enzymologia 35, 345-367.
MATSUMOTO, L., PIKO, L., 1971. In vivo radioactive labeling of mitochondrial DNA in Arbacia punctulata oocytes. Biol. Bull. 141, 397.
MAZIA, D., 1949. The distribution of deoxyribonuclease in the developing embryo (Arbacia punctulata). J. Cell. Comp. Physiol. 34, 17-32.
MAZIA, D., HINEGARDNER, R.T., 1963. Enzymes of DNA synthesis in nuclei of sea urchin embryos. Proc. Nat. Acad. Sci. US 50, 148-156.
MITCHISON, J.M., CUMMINS, J.E., 1966. The uptake of valine and cytidine by sea urchin embryos and its relation to the cell surface. J. Cell Sci. 1, 35-47.
NAGANO, H., MANO, Y., 1968. Thymidine kinase, thymidilate kinase and ^{32}P and (^{14}C)thymidine incorporation into DNA during early embryogenesis of the sea urchin. Biochim. Biophys. Acta 157, 546-557.
NAKANO, E., MONROY, A., 1958. Incorporation of ^{35}S-methionine in the cell fractions of sea urchin eggs and embryos. Exp. Cell Res. 14, 236-243.
NASS, M.M.K., 1969. Mitochondrial DNA: Advances, problems, goals. Science 165, 25-35.
NEMER, M., 1962. Characteristics of the utilization of nucleosides by embryos of Paracentrotus lividus. J. Biol. Chem. 237, 143-149.
NORONHA, J.M., SHEYS, G.H., BUCHANAN, J.M., 1972. Induction of a reductive pathway for deoxyribonucleotide synthesis during early embryogenesis of the sea urchin. Proc. Nat. Acad. Sci. US 69, 2006-2010.
OLSSON, T., 1965. Changes in metabolism of ribonucleic acid during the early embryonic development of the sea urchin. Nature 206, 843-844.
PIKÓ, L., 1969. Absence of synthesis of mitochondrial DNA during early development in sea urchins. Amer. Zool. 9, 1118.
PIKÓ, L., BLAIR, D.G., TYLER, A., VINOGRAD, J., 1968. Cytoplasmic DNA in the unfertilized sea urchin egg: Physical properties of circular mitochondrial DNA and the occurrence of catenated forms. Proc. Nat. Acad. Sci. US 59, No. 3, 838-845.
PIKÓ, L., TYLER, A., VINOGRAD, J., 1967. Amount, localization, priming capacity, circularity and other properties of cytoplasmic DNA in sea urchin eggs. Biol. Bull. 132, 68-90.
ROSENKRANZ, H.S., CARDEN, III, G.A., ROSENKRANZ, S., 1967. Circular DNA from sand dollar sperm. Naturwiss. 54, 46.
SCARANO, E., 1969. Enzymatic modification of DNA and embryonic differentiation. Ann. embryol. morphogénèse suppl. 1, 51-61.
SCARANO, E., 1971. The control of gene function in cell differentiation and embryogenesis. In: Advances in Cytopharmacology (F. Clementi, B. Ceccarelli, eds.), pp. 13-24
SCARANO, E., AUGUSTI-TOCCO, G., 1967. Biochemical pathways in embryos. Comprehensive Biochemistry (M. Florkin, E.H. Stotz, eds.), pp. 55-111
SCARANO, E., DE PETROCELLIS, B., AUGUSTI-TOCCO, G., 1964. Studies on the control of enzyme synthesis during the early embryonic development of the sea urchins. Biochim. Biophys. Acta 87, 174-176.

SCARANO, E., IACCARINO, M., GRIPPO, P., PARISI, E., 1967. The heterogeneity of thymine methylgroup origin in DNA pyrimidine isostichs of developing sea urchin embryos. Proc. Nat. Acad. Sci. US 57, 1394-1400.
SCARANO, E., IACCARINO, M., GRIPPO, P., WINCKELMANS, D., 1965. On methylation of DNA during development of the sea urchin embryo. J. Mol. Biol. 14, 603-607.
SCARANO, E., MAGGIO, R., 1959. The enzymatic deamination of 5'-deoxycytidylic acid and of 5-methyl-5-deoxycytidylic acid in the developing sea urchin embryo. Exp. Cell Res. 18, 333-346.
SIMMEL, E.B., KARNOFSKY, D.A., 1961. Observation on the up-take of tritiated thymidine in the pronuclei of fertilized sand dollar embryos. J. Biophys. Biochem. Cytol. 10, 59-65.
SLATER, I., SLATER, D.W., 1972. DNA polymerase potential of sea urchin embryos. Nature, New Biol. 237, 81-85.
SLATER, J.P., MILDVAN, A.S., LOEB, L.A., 1971. Zinc in DNA polymerase. Biochem. Biophys. Res. Com. 44, 37-43.
TOSI, L., GRANIERI, A., SCARANO, E., 1972. Enzymatic DNA modifications in isolated nuclei from developing sea urchin embryos. Exp. Cell Res. 72, 257-264.
VON LEDEBUR-VILLIGER, M., 1972. Cytology and nucleic acid synthesis of parthenogenetically activated sea urchin eggs. Exp. Cell Res. 72, 285-308.
VON LEDEBUR-VILLIGER, M., in preparation.
VILLIGER, M., CZIHAK, G., TARDENT, P., BALTZER, F., 1970. Feulgen microspectrophotometry of spermatozoa and blastula nuclei of different sea urchin species. Exp. Cell Res. 60, 119-126.
VINOGRAD, J., LEBOWITZ, J., 1966. Physical and topological properties of circular DNA. J. Gen. Physiol. 49, 103-125.
YOUNG, C.W., HENDLER, F.J., KARNOFSKY, D.A., 1969. Synthesis of protein for DNA replication and cleavage events in the sand dollar embryo. Exp. Cell Res. 58, 15-26.
ZIMMERMAN, A.M., SILBERMAN, L., 1967. Studies on incorporation of ^{3}H-thymidine in Arbacia eggs under hydrostatic pressure. Exp. Cell Res. 46, 469-476.

21. Integrating Factors

J. Runnström

21.1 Organization of Egg Cytoplasm

The sea urchin egg has already in the ovary an animal-vegetal polarity that is maintained in embryo development even when the axis through the animal and vegetal pole is bent after gastrulation. The cytoplasm around the animal and vegetal poles shows structural differences (RUNNSTRÖM, 1928c, 1955b).

Enlarged polar bodies in *Psammechinus miliaris* eggs may be fertilized and develop into embryos with enlarged ciliary tufts but without archenterons (GUSTAFSON, 1946). This means that development goes in the animal direction and does not depend on the nucleus. The female pronucleus in the enlarged polar body is equivalent to the egg nucleus, and the male nucleus, even in the haploid state, is able to support normal embryo development (BOVERI, 1889). The animal character of the developing polar body therefore depends on the cytoplasm surrounding the animal pole.

Whole-egg cytoplasm may become animalized when thiocyanate or iodate is added to calcium-free seawater (LINDAHL, 1936). But this method sometimes fails completely or gives only a few animalized eggs, probably because the cytoplasmic structure changes rapidly after fertilization (RUNNSTRÖM, 1967a). If the change is inhibited by added detergent (20μg/ml laurylsulfate) immediately after fertilization, a high percentage of animalized embryos appears. Differences along the egg axis thus seem to depend on stratification of chemically different macromolecules and on the colloidal state of the cytoplasmic matrix containing macromolecular complexes or phases that may separate or intermingle (RUNNSTRÖM, 1962, 1963).

A concavity develops at sperm attachment, particularly in the subequatorial region (RUNNSTRÖM and MANELLI, 1964). Opposite it may appear a lesser concavity. Visible for 20-50 seconds after *Paracentrotus lividus* eggs are fertilized, the concavities reappear if the eggs are kept between slide and coverslip in semi-anaerobic conditions. But as soon as the zygote nucleus dissolves prior to the first cleavage, they disappear.

Nuclear influence on the cytoplasm is also apparent at maturation division. As long as the spindle is visible, the cytoplasm appears homogenous. But as soon as maturation division is complete, the cytoplasm is no longer homogenous and is likely to form birefringent fibers under the influence of low hypertonicity. The position of the first and second cleavage planes is due to properties of the cortical gel at a level slightly above an equatorial plane (RUNNSTRÖM, 1928c). Thus, as early as before fertilization and during first division, regions are structurally - and possibly chemically - different. At the third cleavage, polar movement during certain phases may be the reason why conditions in the cytoplasm cause the spindle to be parallel to the animal-vegetal axis.

Micromere formation is due to structural changes in the vegetal region. In the two-cell stage and particularly in the four-cell stage, pigment granules of *Arbacia* eggs are driven out of the cytoplasmic region where micromeres appear at a set time after fertilization (in *Paracentrotus* it is three hours), even if cleavage is delayed or distorted (MORGAN, 1893; BOVERI, 1901; HÖRSTADIUS, 1928; HULTIN, 1961). Micromeres normally appear even in the presence of actinomycin D, indicating that micromere formation does not immediately depend on genetic information (GIUDICE and HÖRSTADIUS, 1965).

The third cleavage furrow is not always in an equatorial position. In subequatorial eggs it is displaced in a vegetal direction but does not influence differentiation levels in developing embryos (HÖRSTADIUS, 1935). Thus it appears that later differentiation, which depends more directly on genes, is not closely linked to primary egg organization. This organization is heteropolar and is a holdover from the oocyte. At maturation, displacement may accentuate this heteropolarity as when cortical pigment granules accumulate in *Paracentrotus* eggs. Before accumulating, the granules are spread over the whole cortex, and afterward they are apparently driven out of the animal half of the egg, and out of a smaller vegetal region, and form the pigment girdle (BOVERI, 1901).

Pigment girdle formation is stimulated under semi-anaerobic conditions or by the addition of 1/20,000 mol KCN solution (LINDAHL and ÖRSTRÖM, 1932). The pigment girdle then becomes more narrow and more sharply delineated than in nontreated eggs. Presupposing a compensatory flow in the opposite direction, pigment girdle concentration may possibly be due to cytoplasmic flow in the vegetal direction (SCHLEIP, 1929). Or separation may occur in a more condensed phase and a more fluid phase, and the granules may join in the fluid phase. (In the vegetal region, similar phase separation may also occur.) This process may be the first step in preparing for micromere formation (RUNNSTRÖM, 1928a, 1928c).

In unfertilized eggs, treatment with NaSCN or NaJ in calcium-free seawater counteracts the pigment concentration and induces animalization (LINDAHL and ÖRSTRÖM, 1932; LINDAHL, 1936). Phase separation in the egg cytoplasm thus seems to be related to morphogenesis.

In a number of phase systems, pairs of chemically related substances in water solution may separate as phases. Added macromolecules or cell particles settle in a phase with which they are compatible, greatly enriching it. Changes in ion concentration may alter these phases systems, and distribution of substances or particles between phases may change and phase boundaries disappear. Egg exposure to hypertonic seawater causes remixing after fertilization, or at least an intermingling of phases (ALBERTSSON, 1960). For other pertinent model systems reference may be made to BOOIJ and BUNGEBERG DE JONG (1956).

In fertilized *Paracentrotus lividus* eggs, phase separation occurs after pre-fertilization treatment with low concentrations of trypsin (RUNNSTRÖM, 1961a). In unfertilized *Psammechinus miliaris* eggs kept in 30% concentrated seawater, a considerable phase separation occurred where mitochondria seem compatible with one phase of the cytoplasmic matrix and endoplasmic vesicles with another (RUNNSTRÖM, 1928c, 1968).

21.2 Differentiation along the Egg Axis

21.2.1 The Double Gradient Concept

The two polar cytoplasmic regions are believed to be dynamic centers that form animal and vegetal morphogenetic gradients by diffusing animalizing (*an*) and vegetalizing (*veg*) substances (RUNNSTRÖM, 1928a, 1928b, 1933, 1935). The region between the poles is divided into levels indicated by *an/veg* ratios (RUNNSTRÖM, 1961b, 1967b). These substances are also called morphogenes (CRICK, 1970).

The double gradient concept is founded partly on results obtained by dividing the 16-cell stage of sea-urchin embryos into animal and vegetal halves. Each animal half develops the enlarged tuft of stereocilia developed by the whole embryo (HÖRSTADIUS, 1935). This indicates that in the animal half of the whole embryo inhibition normalizes acron and stereocilia growth.

The double gradient concept is also founded on the fact that in embryos treated with 13 - 45.5 millimols of lithium in seawater and transferred to a normal medium after 10-14 hours, differentiation at the animal pole is suppressed and the animal gradient weakened, while the vegetal gradient is stimulated. This is due to competition or antagonism between the two gradients (HERBST, 1892; MACARTHUR, 1924; RUNNSTRÖM, 1928a, 1928b, 1961b, 1967b; LINDAHL, 1941).

Differentiation depends on the range of *an/veg* ratios along the normal, elongated, or shortened egg axis (Fig. 21.1). Where the gradients cross each other, *an/veg* is unity. Above this level the ratios approach infinity, and below it they approach zero. Displacement of *an/veg* = 1 in an anterior direction causes vegetalization, while displacement in a posterior direction causes animalization. Greater displacement in the animal direction causes the opposite part of the embryo to lose control and to become dependent on the more diffuse vegetal gradient, while displacement in the vegetal direction causes the reverse.

Lithium treatment of sectioned embryos fixed in CARNOY's fluid makes cytoplasm more rigid and proteins of the inner layer of columnar cells more resistant to treatment with trypsin (30μg/ml). The sections are then stained by HALE's method and the cytoplasm counterstained with naphthol yellow. This treatment breaks down completely the same cell regions of control embryos (RUNNSTRÖM, 1928b, 1935; RANZI et al., 1962; LALLIER, 1964; HAGSTRÖM and LÖNNING, 1967; IMMERS, MARKMAN and RUNNSTRÖM, 1967; RUNNSTRÖM and IMMERS, 1971).

Gradient centrifugation of embryo homogenates shows that the ratio between free and membrane-bound ribosomes may be higher after lithium treatment than in the controls (CITRO, NUZZOLO and RUNNSTRÖM, unpublished).

Properties that indicate greater rigidity (the presence of open intercellular spaces, for instance) last longer in the animal region than in the vegetal region. After treatment with low doses of lithium (13-19 millimols for 8 hours), the ectoderm is more abnormal than the entomesoderm. It is possible that initially the effect of lithium is reversible and that the enhancement of the vegetal gradient suppresses regulation. The increased rigidity is gradually reversed after transfer to normal seawater, but by then regulation is disturbed and is either no longer possible or takes an abnormal course (RUNNSTRÖM and IMMERS, 1970, 1971).

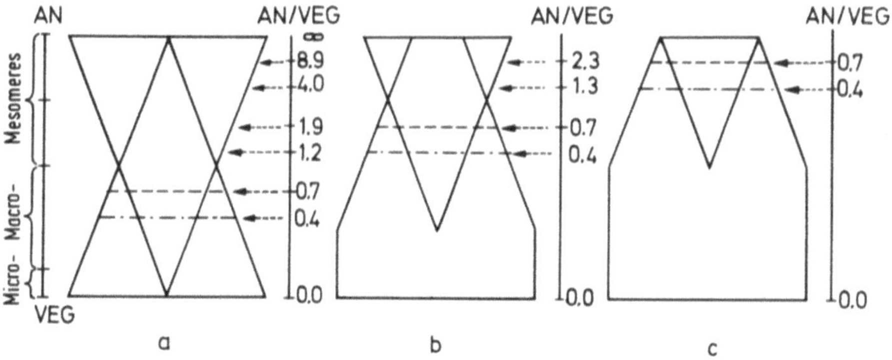

Fig. 21.1a-c. Model of animal-vegetal gradient systems in sea-urchin embryos. (a) Normal; (b, c) degrees of vegetalization. Gradients are caused by decreased concentration of animalizing (*an*) and vegetalizing (*veg*) substances from animal and vegetal centers. Vegetalization causes animal gradient to become narrow and shorter in range, whereas vegetal gradient broadens and becomes less steep. The *an* and *veg* substances are determined by measuring distance between two points at same level of animal and vegetal gradients, respectively, giving *an/veg* ratios for those levels. Cell boundaries in 16-cell stage are indicated on line left of (a). Upper line indicates attachment zone; lower line, ectoderm-entoderm boundary, both determined by *an/veg* levels 0.7 and 0.4 obtained from (a). (b, c) Positions of corresponding *an/veg* ratios are found by trials. The *an/veg* ratios are logarithmic functions of their levels at animal-vegetal axis. Anterior ectoderm is suppressed above *an/veg* values of 2.5 and 1, respectively. Suppressed ectoderm corresponds to acron region and part of oral region derived from more animal 60% of mesomeres (b) and to acron region and nearly entire oral region derived from whole mesomere and 5% of adjacent macromere material (c)

In cytoplasm trypsin causes the animal cells to join and form an enlarged acron zone made up of cells having stereocilia. But the vegetal cells may become repressed or damaged as the result of the repression of the vegetal center, while the animal center is probably increased and the animal gradient broadened. Prospective entoderm cells that have been animalized are resistant to trypsin-induced cytolysis. Animalization protects the cells as it does in embryos partly animalized in seawater without SO_4 (RUNNSTRÖM et al., 1964; RUNNSTRÖM and IMMERS, 1970, 1971).

Treatment with trypsin (2.5μg/ml) for 45 hours causes the greatly enlarged acron to become smaller, and even a small amount of remaining entoderm may grow out and attach itself to a stomodeum that forms only after the beginning of acron region retrogression. Primary and secondary mesenchyme cells appear, the primary cells arranged in a ring with 2 symmetric accumulations of cells, in each of which forms a skeletal piece that grows symmetrically. That part opposite the animal pole separates from the anterior part. A new, small acron may appear in the posterior part of the embryo (RUNNSTRÖM and IMMERS, 1966).

Experiments with trypsin, as those with lithium, emphasize the competitive nature of animal-vegetal interaction (RUNNSTRÖM, 1933; LINDAHL, 1941; RUNNSTRÖM, 1966a, 1966b, 1967). As the result of treat-

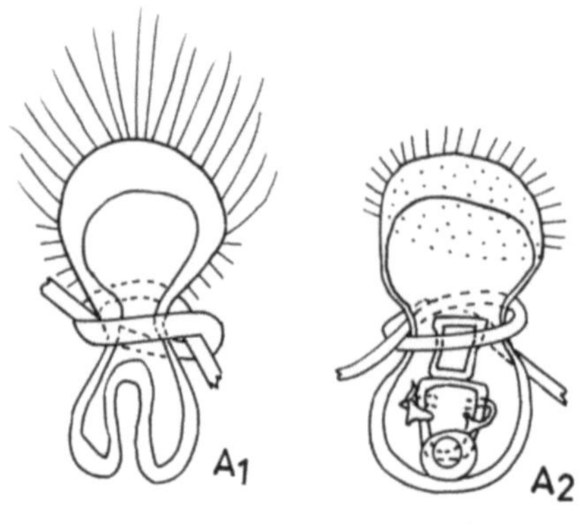

Fig. 21.2. Constriction experiment with silk thread; 16-cell stage. (From HÖRSTADIUS, 1938)

ment with trypsin concentrations of up to 20μg/ml, the enlarged acron remains and the vegetal region may be animalized.

To test the hypothesis that *an* and *veg* are distributed by diffusion, constriction experiments were carried out (HÖRSTADIUS, 1938). Constriction from a silk thread in the 16-cell stage induces animalization of the anterior part of the embryo; whereas the posterior part resembles an isolated vegetal half, the primary mesenchyme cells do not migrate to form the skeleton. The halves of the embryo are not entirely separate, but diffusion is delayed by cell flattening inside the constriction and by overall embryo elongation. A looser constriction induces a uniform, somewhat elongated embryo, while a tighter constriction divides the embryo in two, as if it were separated by glass needles. The diffusion of *an* and *veg* probably takes place in the cytoplasm rather than in the blastocoel.

When a labeled animal half is transplanted onto a nonlabeled vegetal half, and vice versa, and when labeled amino acids are used, autoradiography shows label diffusion from one half to the other. But when labeled uridin or orotic acid is used, label diffusion is not observed (BÄCKSTRÖM, 1969).

21.2.2 The Demonstration of Morphogenes

In an attempt to purify *an* and *veg* substances, lyophilized material is extracted from unfertilized *Paracentrotus lividus* eggs. Acid treatment yields a strong precipitate which is discarded, and the supernatant is brought onto a column of Dowex 50 W-X2 resin and eluted with buffers in three stages of increasing pH. The chromatogram obtained is shown (Fig. 21.3). The peaks are tested on animal and vegetal half embryos separated at the 16-cell stage. Half embryos are generally more sensitive to extracted substances than whole embryos are. Peak IV animalizes animal and vegetal halves, while peak V vegetalizes them. Together, peaks IV and V counteract each other (HÖRSTADIUS, JOSEFSSON, and RUNNSTRÖM, 1967).

On Sephadex G-25, peak IV splits into A_1, A_2, A_3, and A_4. A_1 and A_4 are animalizing, while A_2 and A_3 are inactive. Peak V splits into V_1, V_2, V_3, V_4, V_5, and V_6. V_1 and V_2 are vegetalizing, V_5 is weakly animalizing, and V_6 is strongly animalizing. The absorption curve of V_5 indicates the presence of tryptophane, while that of V_6, like fractions A_1 and A_4, at about 260 nm indicates the presence of a nucleotide. A_4 and V_6 are the most strongly animalizing and may be

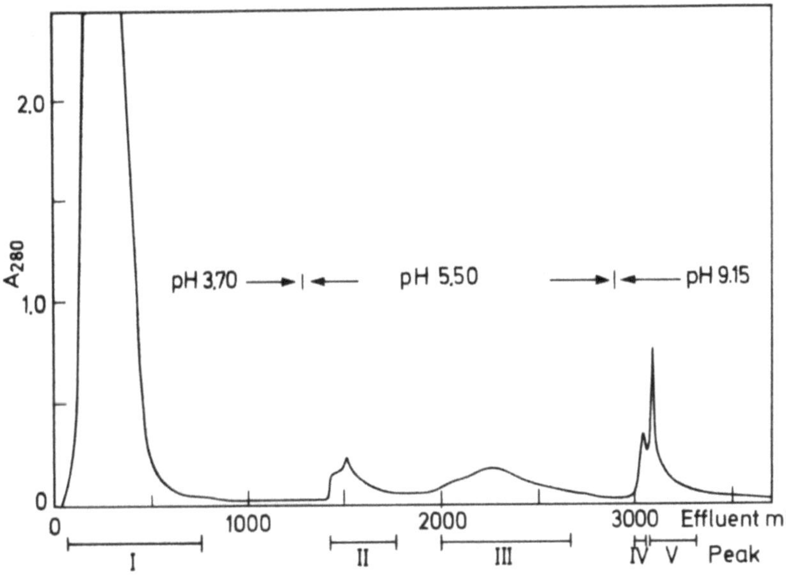

Fig. 21.3. Chromatography of unfertilized *Paracentrotus lividus* homogenate. IV, animalizing fraction; V, vegetalizing fraction. (From HÖRSTADIUS, JOSEFSSON, and RUNNSTRÖM, 1967)

the same fraction. V_5 animalizes only animal half embryos, and its absorption indicates it is protein in nature. The absorption curves of the other animalizing fractions indicate a nucleotide content (JOSEFSSON and HÖRSTADIUS, 1970).

In concentrations of 5 x 10^{-4} millimols, tyrosine has a strong vegetalizing effect on whole embryos and animal half embryos 3 to 5 hours after fertilization (MASTRANGELO, 1965, 1966). It is synergistic with lithium. Phenylalanine is not vegetalizing when used alone, but it enhances the effect of lithium. The addition of other egg amino acids reduces or blocks the vegetalizing effect of tyrosine. In general, the effect of morphogenes may be modified by substances that have no morphogenetic effect of their own.

21.2.3 Morphogenes as Genetic Effectors

In a test solution of highly active *an*, with and without actinomycin D (7.5 - 15μg/ml), vegetal halves separated at the 16-cell stage are treated 8 to 15 hours after fertilization, returned to normal seawater, and observed 20-60 hours after fertilization. The abscissa shows the kind of embryo; the ordinate, the frequency of each kind (Fig. 21.4). The two types of controls have large entoderms, some larger than others, divided into 3 compartments. Treatment with *an* or with actinomycin D reduces or represses the entoderm in more than one-third of the cases, while treatment with both has almost no effect (MARKMAN and RUNNSTRÖM, 1970).

In the control and in actinomycin D, in the animal region stereocilia and acron are absent (Fig. 21.5). The addition of *an* causes stereocilia formation, an indication that animalization is taking place. But

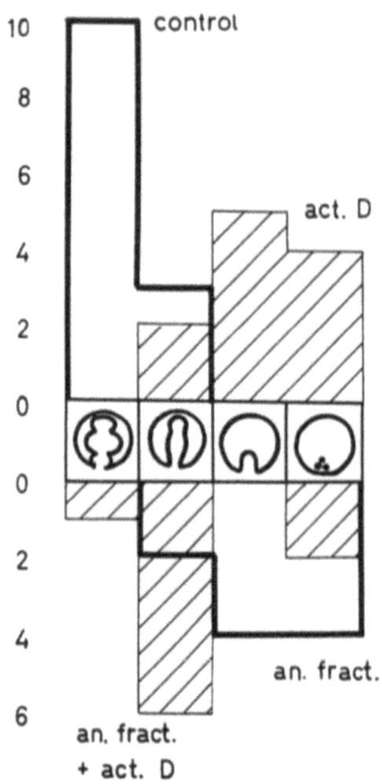

Fig. 21.4

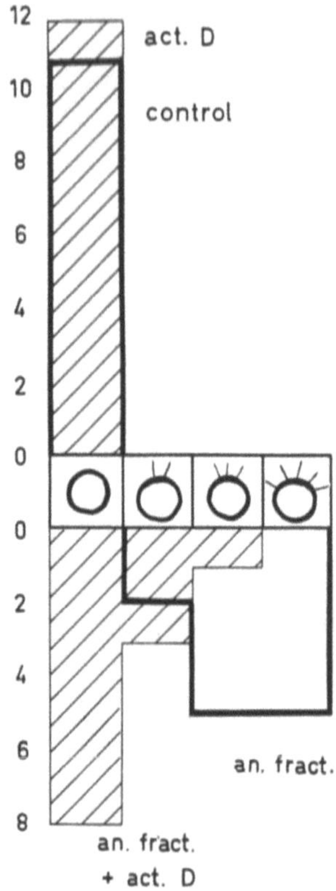

Fig. 21.5

Fig. 21.4. Histogram showing, on abscissa, 4 types of vegetal half embryo differentiation. Upward directed: control and actinomycin D; downward directed: animal fraction (animal morphogene + actinomycin D, and animal morphogene alone). (From MARKMAN and RUNNSTRÖM, 1970)

Fig. 21.5. Four degrees of animal ciliation; conditions as in Fig. 21.4. (From MARKMAN and RUNNSTRÖM, 1970)

an and actinomycin D combined repress stereocilia formation in most instances.

The animalizing effect of trypsin is very similar (Fig. 21.6). Treatment with trypsin alone represses intestine formation, while trypsin and actinomycin D combined almost counteract trypsin's animalizing effect.

In vegetal halves, red pigment in mesenchyme cells is repressed by the animalizing effect of endogenous substances or trypsin, while the result of treatment with animalizing agents and actinomycin D is similar to the condition of nontreated vegetal halves. In animalized halves, the secondary mesenchyme cells that produce pigment cells become elongated and form characteristic strands parallel to the egg axis.

Actinomycin D specifically reacts with DNA to inhibit nuclear transcription processes from DNA to template RNA (REICH and GOLDBERG, 1964).

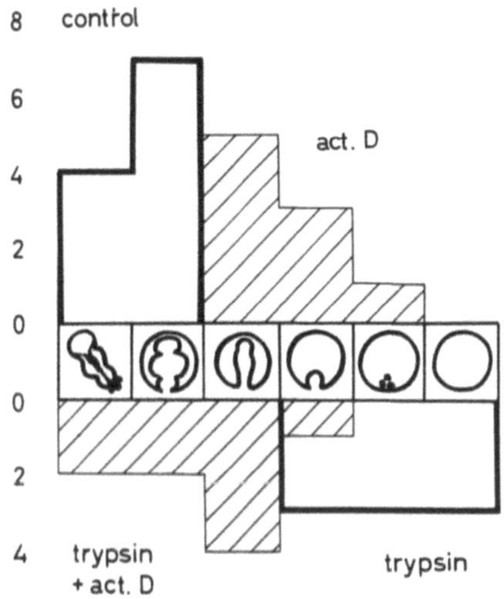

Fig. 21.6. Same as Fig. 21.4 except that endogenous morphogene is replaced by trypsin. (From MARKMAN and RUNNSTRÖM, 1970)

This implies that the animalizing agents described above are active, directly or indirectly, at the gene level. With trypsin particularly, an endogenous substance is released.

Other actinomycins having little or no specific affinity for DNA include desaminoactinomycin C3, an analog of actinomycin D that allows sea-urchin development from embryo to pluteus has the same lethal effect on rats as actinomycin D. Evidently it has no specific effect on DNA (MÜLLER, 1962; DE VINCENTIIS and LANCIERI, 1970).

In animal halves separated at the 16-cell stage, the acron develops and extends, and the stereocilia are larger than in whole embryos (Fig. 21.7). Adding fraction V of the egg extract to isolated animal halves moderates acron extension, while adding outside vegetalizing morphogenes changes the *an/veg* ratio. The level *an/veg* = 1 is thought to move in the animal direction, indicating that vegetalization is taking place (HÖRSTADIUS, JOSEFSSON, and RUNNSTRÖM, 1967).

Lithium added to a suspension of animal half embryos causes development to be more normal and small plutei to form (VON UBISCH, 1925). Animal half embryos separated about 3 1/2 hours after fertilization and placed in 43 millimols of lithium in seawater, about 20 hours past fertilization are rather evenly covered by short, motile cilia. The epithelium is fairly uniform, and mesenchyme cells and a small archenteron appear. Larvae develop, sometimes closely resembling pluteus larvae (Fig. 21.7, column Li_O, III and IV). A ciliary tuft does not form, but a stomodeum develops. They are reduced by lithium treatment but not enough to suppress the stomodeum completely. In whole embryos treated with lithium for 6 hours during late cleavage, stomodeum formation is almost completely, if not completely, repressed.

Unlike lithium alone, lithium and actonimycin D combined do not vegetalize animal half embryos (Fig. 21.7, column 4). The *an/veg* ratios are lower than in nontreated animal halves but higher than after lithium treatment because actinomycin D inhibits the transcription processes that lithium begins. In whole embryos also, actinomycin D blocks the vegetalizing effect of lithium (RUNNSTRÖM and MARKMAN, 1966).

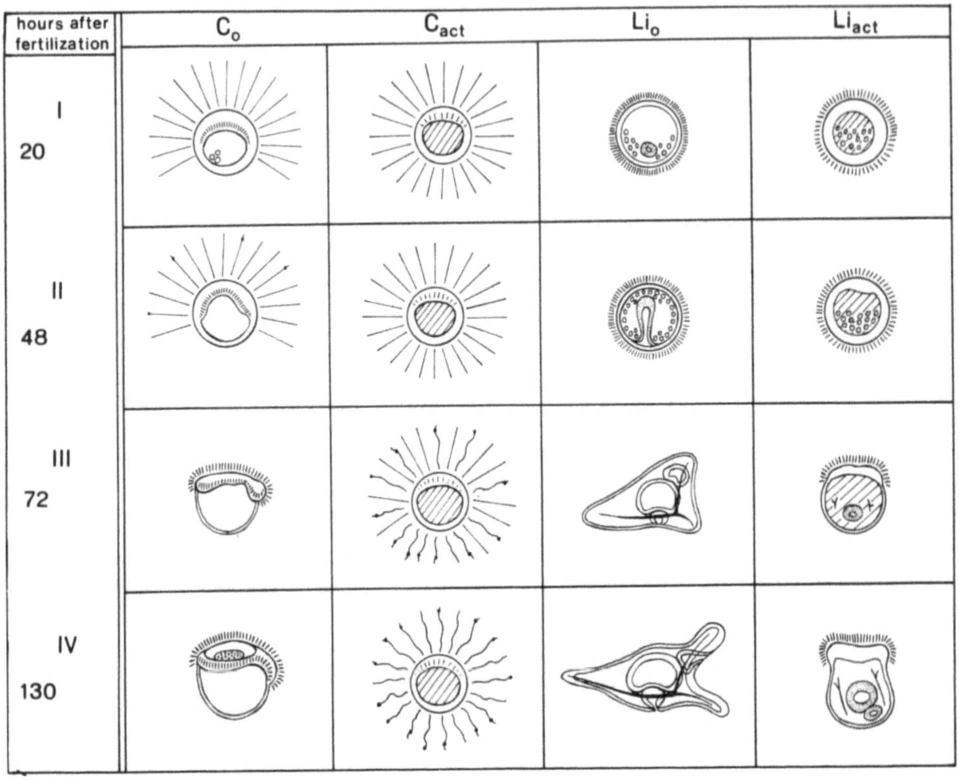

Fig. 21.7. Animal half embryos (C_O) with added actinomycin D (C_{act}), lithium (Li_O), and lithium + actinomycin D (Li_{act}). (From DE ANGELIS and RUNNSTRÖM, 1970)

Animal half embryos treated with actinomycin D do not lose the ciliary tuft 2 or 3 days after fertilization (Fig. 21.7, column C_O). And 130 hours after fertilization the stereocilia are attached to the epithelial cells. On the third day of development, stereocilia become weaker and semi-motile. Reduction rather than formation of stereocilia thus seems to depend on a transcription process.

Similarly, in embryos the enzyme deoxycytidilate amino-hydrolase begins to decrease as early as the blastula stage, so that by the pluteus stage larvae contain about one-fourth as much of the enzyme as the egg after fertilization (SCARANO and MAGGIO, 1959; SCARANO, DE PETROCELLIS, and AUGUSTO-TOCCO, 1964). But if actinomycin D (40μg/ml) is added when development begins, enzyme activity increases. It is proposed that an inhibitor is built up which prevents this hydrolase synthesis and actinomycin D represses the transcription process in the inhibitor's pathway of synthesis, allowing enzyme synthesis to proceed (SCARANO and AUGUSTO-TOCCO, 1964, 1967).

It is also possible that a protein inhibitor interferes with acron and stereocilia development and maintenance in animal halves. If the inhibitor formation is impaired by actinomycin D, stereocilia persist in spite of signs of regression. It may be that in animal halves the protein inhibitor is *veg*.

The polarity of C_{act} animal halves is much less pronounced than that of the control C_O. Seemingly important to the differentiation of animal halves, polarization is due to increased intracellular adhesion that decreases in the vegetal direction. Junctions allow rather large molecules to cross cell boundaries (LOEWENSTEIN, 1967). These phenomena seen in the relatively simple animal half embryo occur also in normal development (GUSTAFSON and WOLPERT, 1963, 1967).

It seems that degree of adhesion is related to RNA labeling with ^3H-labeled precursors (adenine and uridin): increased adhesion favors transcription processes which in turn favor adhesion and junction. At a certain point, reduced cell contact breaks the cycle (IMMERS and RUNNSTRÖM, 1965; MARKMAN, 1965).

21.2.4 Regulation Processes in Animal and Vegetal Directions

In animal halves, animalization is indicated by an acron region and stereocilia more extensive than in whole embryos, and vegetal differentiation is suppressed or rudimentary. Added exogenous or endogenous animalizing or vegetalizing substances stimulate or suppress differentiation in one direction or the other.

Some variation in animal and vegetal half embryo development can be ascribed to asymmetry in the double gradient system (for example, Fig. 21.8). If the animal gradient has a linear slope and the vegetal gradient has a tapering slope, the animal character is predominant in both animal and vegetal halves. It is likely that the animal center becomes functional first, as the vegetal center is incomplete until fertilization (RUNNSTRÖM, 1967a). It may be that the vegetal gradient tapers more than the animal gradient only because the vegetal gradient has not yet become symmetrical.

Animal and vegetal halves were separated at the 8- and 32-cell stage. The animal halves of the 8-cell stage become almost completely covered by stereocilia, while those separated at the 32-cell stage are covered less. This indicates that the predominant animal state of animal halves becomes more balanced (see Fig. 20.1a). Half embryos separated from the 16-cell stage to the 32-cell stage are of every possible type. The animal tendency of vegetal halves depends on the vegetal gradient flow rate (HÖRSTADIUS, 1965a).

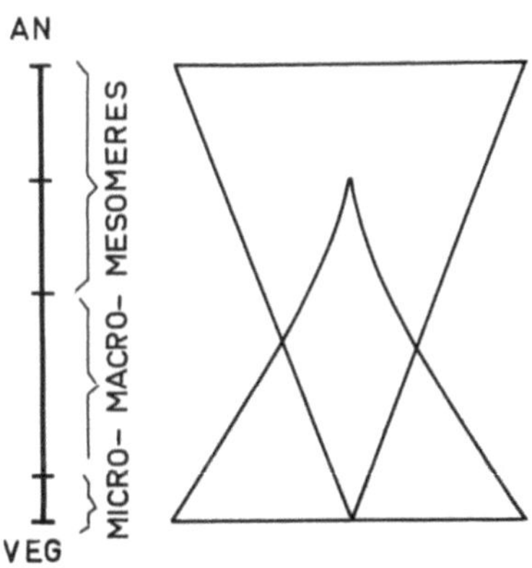

Fig. 21.8. Asymmetric gradient system caused by delayed formation of vegetal gradient

In isolated vegetal halves in which the vegetal gradient is fully formed, differentiation may correspond to much higher *an/veg* ratio than those of the whole vegetal region (0 at the vegetal pole to 1 at the equator). Vegetal-half stereocilia correspond to very high *an/veg* ratios and appear, if they appear at all, reduced in size and after considerable delay.

Stomodeum appearance in vegetal halves is a complicated differentiation that corresponds to *an/veg* ratios substantially higher than the original ratios. This implies a regulation process in which a normal gradient system tries to form with the help of the genome, from which the morphogenes develop. While the animalizing center may form in the oocyte, the vegetalizing center develops completely only after fertilization. The components of both depend on the genome. If a ciliary tuft develops in vegetal halves, it is rudimentary and appears 1 or 2 days later, demonstrating time-dependent regulation (HÖRSTADIUS, 1935).

The reason may be that when the halves are separated, the descendants of the veg_1 ring are rearranged, and adhesion and interaction among them increases, with adhesion being strongest between cells with the highest *an/veg* ratios. A-group genes that form *an* may receive delayed signals. Additional *an* raises the *an/veg* ratio, while genome and cytoplasm interact to raise the *an/veg* level. In many vegetal halves, resistance blocks animalization. The acron with ciliary tuft is stable only when it is surrounded by animal cells, which is why it shifts when micromeres are implanted into the animal region of animal halves (HÖRSTADIUS, 1935).

In some studies, the number of vegetal halves with stomodea is very low, but in others it may be as high as 50%. Of the latter, about one-third have stomodea that are very narrow, due to a low *an/veg* ratio, while the remaining plutei forms with axially elongated stomodea resemble those of whole larvae. Asymmetric slopes of animal and vegetal gradients account for this, while regulation explains the nearly normal state (HÖRSTADIUS, 1935).

In less differentiated material, a small stomodeum appears even in animal halves with weak animal gradients. The great number of vegetal halves indicates that resistance against regulation is overcome in order for much-reduced stomodea to develop (HÖRSTADIUS, 1935).

Whole embryos treated with lithium resemble isolated vegetal halves. The animal gradient in the former is possibly suppressed and the vegetal gradient enhanced. The vegetal gradient is predominant in vegetal halves after isolation, but regulation may supplement the animal gradient system. With the help of the genome, compensation makes the embryo more proportionate and regulation makes it less so (HÖRSTADIUS, 1935; LINDAHL, 1941; RUNNSTRÖM and IMMERS, 1971).

21.2.5 Torsion Regulation in the Embryo

Half embryos separated at the 2-cell stage develop into open half blastulae that eventually form a closed blastula, thereby joining the most animal and the most vegetal regions of the embryo (DRIESCH, 1891). Gastrulae with archenterons develop. The archenteron top is derived from material that, in the whole embryo, forms a lateral region of the entoderm. In the whole embryo, the most animal region forms part of the ectoderm at approximately the *an/veg* 2 level. Regulation occurs in the ectoderm and entomesoderm, making the half

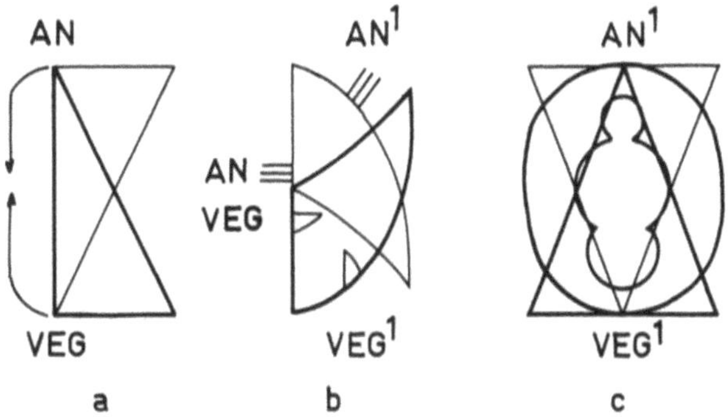

Fig. 21.9a-c. Separation of first two cleavage furrows; normal axial condition after disorder. (From HÖRSTADIUS, 1928, 1935)

embryo heteropolar. After the half embryo closes, the animal and vegetal gradients become curved and their poles move closer together; then normal heteropolarity returns (Fig. 21.9). The gradients are unable to remain curved. When their poles approach, it is possible that they counteract each other (HÖRSTADIUS, 1928, 1935). In a similar way, endogenous *an* and *veg* counteract each other when they are added to animal halves (HÖRSTADIUS, JOSEFSSON, and RUNNSTRÖM, 1967). When 4 micromeres are implanted along the egg axis of animal halves, polarity is reversed (HÖRSTADIUS, 1935).

When the genome intervenes during regulation, it probably replenishes the supply of *an* and *veg* (RUNNSTRÖM, 1967). Actinomycin D suppresses animal and vegetal centers, but transfer to natural seawater restores normal activity (RUNNSTRÖM and MARKMAN, 1966). As actinomycin D prevents normal development (Fig. 21.7, columns L_{act} and Li_{O}), it probably also inhibits regulation.

In Fig. 21.10, arrows indicate the polarity of developing embryo regions. Arrow bases indicate animal direction; arrowheads, vegetal direction. The mesomeres become normal ectoderm (arrow pointing down). The next region, its polarity inverted and continuous with that ot the ectoderm, develops into the intestine (arrow pointing down). The prospective anterior part becomes hindgut, and the prospective posterior part, the esophagus. The prospective anterior part of the micromeres (veg_1) forms an ectodermic vesicle, its polarity (arrow pointing up) the opposite of that of the anterior regions.

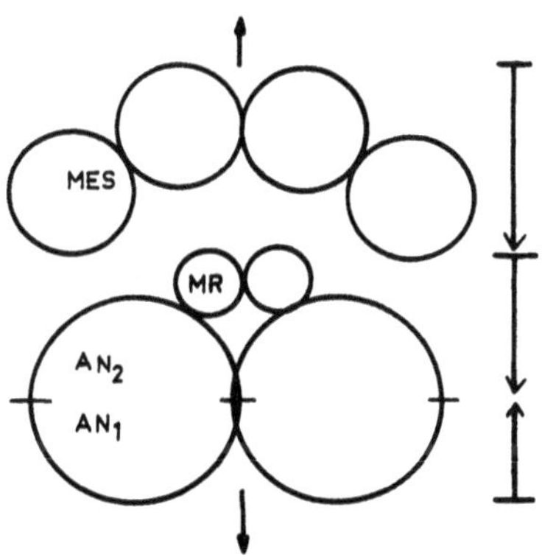

Fig. 21.10 Reversed vegetal half embryo transplanted into inner side of mesomere material. Middle arrow indicates middle-region polarity shifted 180°; in upper region (mesomere material), original polarity prevails (as in an_1 material). (From HÖRSTADIUS, 1928, 1935)

The site of origin of the primary mesenchyme cells is probably dislocated in the direction of the middle arrow. They migrate into the ectoderm that develops from the mesomere ring and form a normal skeleton. Another part of the primary mesenchyme enters the ectodermic vesicle and forms a simplified skeleton.

Believed to function as organizers (SPEMANN, 1936), micromeres transplanted into more animal regions seem to do just that. And as we have seen, *an* also has the same effect in the animal region. These opposing organizational forces are the basis of the double gradient system.

21.3 Integration of DNA and Nuclear Protein

Animalizing and vegetalizing morphogenes may signal the nuclei that activate the groups of genes according to *an/veg* ratios. The double gradient organization of the primary *an/veg* system induces similar organization in the genome. In addition to signals from the cytoplasm, gene activation depends on the receptivity of the regulatory genes.

In studies of DNA-protein interaction in interphase nuclei, the HALE reaction is used, which is dependent on Fe^{3+} binding to DNA and is made on material fixed in Carnoy's fluid and sectioned in histowax. During mitosis, chromosomal DNA reacts with Fe^{3+} at pH 3 and above (Fig. 21.11). In interphase nuclei in which template activity occurs, HALE staining is not demonstrable for the first 5 or 6 hours of development (Fig. 21.11, first period of reaction). But during the first 7 to 12 hours (Fig. 21.11, second period of reaction), strong, uniform staining is observed after section pretreatment with 25μg/ml trypsin in buffer of 7.5 but not after buffer treatment alone. In the advanced blastula stage, after 12 or 13 hours (Fig. 21.11, third period of reaction), the nuclei react to HALE staining after section pretreatment with a 0.1 mol buffer (IMMERS, MARKMAN, and RUNNSTRÖM, 1967).

Curves pH 4-8 (Fig. 21.11) show that staining intensity increases as development proceeds. After pretreatment with buffer of pH 8 at intervals between 12 and 17 hours after fertilization, the curve rises rather steeply. The 50% value of staining intensity of this curve is about pH 2 and occurs about 13 hours after fertilization, beginning at the mesenchyme blastula stage. Pretreatment at lower pH causes it to occur at a more advanced stage. Staining intensity decreases as pH is reduced. The 3 upper curves approach mitotic chromatin values (Fig. 21.11) at which nuclear proteins hinder the Fe^{3+}-DNA reaction very little. From the late blastula stage on, section pretreatment with DNase removes HALE reagent staining. In the earlier stages, DNA is less affected by DNase, as it is in the condensed state as shown by a negative HALE reaction (IMMERS, MARKMAN, and RUNNSTRÖM, 1967).

HALE staining is usually more intensive in the ectoderm than in the entoderm and greater in the entoderm than in mesenchyme cells. Pigmented mesenchyme cells may penetrate between ectoderm cells. The difference in nuclei staining is evident, possibly because in the ectoderm more DNA is released from close association with protein than in the mesenchyme (IMMERS, MARKMAN, and RUNNSTRÖM, 1967).

At the mesenchyme blastula stage, nuclei in all parts of the embryo may be HALE stained after pretreatment with buffer of pH 7.95. Nuclei in the ectodermal zone where the primary mesenchyme cells are attached are particularly stainable after pretreatment with buffer at pH 5.1 -

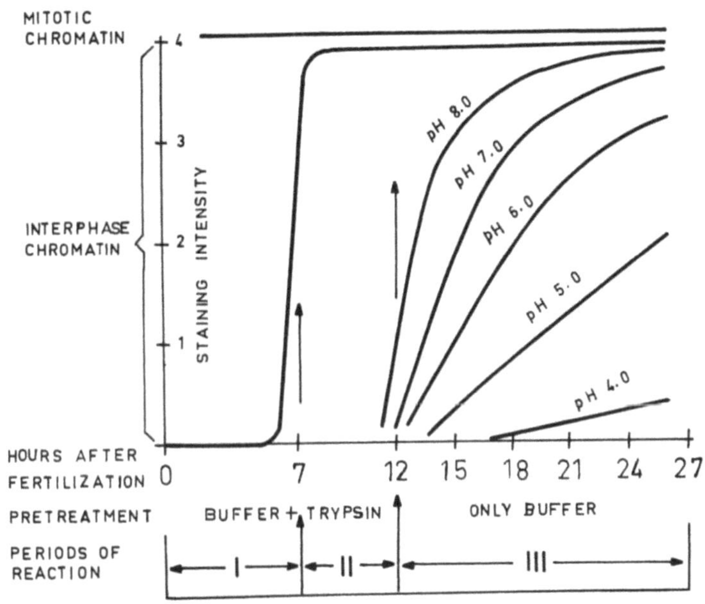

Fig. 21.11. Mitotic chromatin (top curve) shows strong HALE staining before and after pretreatment with buffer; pH values from 3. Intensity of HALE staining of interphase nuclei is given as function of stage of development (hours after fertilizing divided into periods I-III). During first period of reaction (from fertilization to early blastula), HALE staining is negative, regardless of pretreatment. During second and third periods, pretreatment with trypsin at pH 7.5 causes strong staining with HALE reagent. During third period (from late blastula to prism), staining occurs after pretreatment with buffer of different pH. Staining intensity depends on buffer pH and on stage of transfer to buffer. The later the pretreatment, the stronger the staining. Pretreatment with trypsin or buffer lasts for 10 minutes at 25°C

6.2. Even at reduced pH, nuclei in the attachment zone do not have enough positively charged nuclear proteins to prevent HALE reagent Fe^{3+} from reacting with DNA. These nuclei probably become active at the late blastula stage. Treatment with actinomycin D or lithium rapidly affects the function of the attachment zone (IMMERS and RUNNSTRÖM, 1967a).

From the prism stage on, the HALE reagent stains nuclei directly (particularly those of the ventral ectoderm), without pretreatment with enzyme or buffer. Weak at first, this staining is rather strong at the pluteus stage. Thus it seems that HALE staining indicates the degree of nuclear transcription or template activity (IMMERS and RUNNSTRÖM, 1971a).

Also in other studies, template activity was found to increase during early development, reaching a maximum at the pluteus stage when nuclei are most accessible to HALE staining. In earlier stages there are parallels, as well as differences, between template activity and HALE staining. Template activity is lower at the gastrula stage than at the mesenchyme blastula stage, although the reverse is found after HALE staining. Transition between the cleavage stage and the mesen-

chyme blastula stage is less abrupt in template activity than HALE staining indicates (HNILICA and JOHNSON, 1970; JOHNSON and HNILICA, 1970).
HALE staining after treatment with trypsin occurs at the early blastula stage when nuclear protein character changes and the increasing nuclear responsiveness to HALE staining may be the result of nuclear protein changes. The first histones are rich in arginine, while those at later stages are rich in lysine (ORENGO and HNILICA, 1970). During development, basic proteins change so that DNA becomes increasingly responsive to Fe^{3+}, reaching maximum responsiveness at the pluteus stage.

The ability of HALE reagent Fe^{3+} to react with DNA directly, without pretreatment, means that protein-DNA association is so weak that Fe^{3+} is able to break it, while HALE staining of the attachment zone after buffer pretreatment indicates that morphogenetic and nuclear protein changes may be related.

21.4 Differentiation along the Dorsoventral Axis

In *Paracentrotus lividus* eggs elongated by shaking, the pigment girdle makes it possible to determine the angle of elongation in relation to the egg axis. This suggests that elongation gives the eggs bilateral symmetry (BOVERI, 1901a, 1901b).

Eggs elongated by being forced through a glass tube capillary smaller than the egg diameter and placed in a dish of seawater develop bilateral symmetry as well as a dorsoventral axis. The part of the egg that leaves the capillary first, has the highest tension and forms the ventral pole. In an isotonic solution of glucose, cytolysis is likely to occur at the ventral pole (LINDAHL, 1932).

In unfertilized *Arbacia* eggs, centrifugation causes a red-pigment accumulation from which the ventral side of the embryo develops (MORGAN and LYON, 1907; MORGAN, 1927).

When early blastulae of *Paracentrotus lividus* are placed in normal seawater after 10 hours in 10μg/ml actinomycin D in seawater, developing gastrulae are elongated dorsoventrally. The ventral region consists of a uniform, sometimes folded, columnar epithelium, and the dorsal region has a lower epithelium. After 10-15 hours in normal seawater, a few low acron cells with stereocilia develop. First scattered in the ventral region, the primary mesenchyme cells form two groups next to the ventrolateral ectoderm. None enters the dorsal region where the secondary mesenchyme cells gather. Thus it seems that dorsoventral polarity is determined early without direct coding of genes. Transfer to normal seawater restores the transcription processes inhibited or delayed by actinomycin D (RUNNSTRÖM, 1966).

In meridional halves separated in early cleavage and color marked on the side of separation, dorsoventral separation is accompanied by inversion of the dorsoventral axis: the prospective dorsal region is the prospective ventral region. Results are similar in unfertilized eggs separated into meridional halves, fertilized, and allowed to develop. This indicates definitely that the dorsoventral axis is determined before fertilization (HÖRSTADIUS and WOLSKY, 1937).

In fertilized *Paracentrotus* eggs, concavities appear, some larger than others, under anaerobic conditions or after the addition of NaCN. The larger concavities are the temporary ones that reappear after sperm attachment. Local staining suggests that the larger concavities indicate the future ventral side (FOERSTER and ÖRSTRÖM, 1933; RUNNSTRÖM and MANELLI, 1964).

The two sizes of concavities also appear in unfertilized *Echinocardium cordatum* eggs. Upon fertilization, the egg becomes almost spherical (GUSTAFSON, 1952).

Compounds containing a free SH group or histidine and albumin induce concavity formation, while a serum fraction containing globulins does not. Detergents in low concentrations prevent concavity formation, probably by stopping the movement of the cytoplasm. Detergent-treated eggs to not develop bilateral symmetry (GUSTAFSON, 1952).

Local staining shows that the ventral side of the gastrula contains about two-thirds of the original egg material and by the pluteus stage contains three-fourth of it. The ventral side gives rise to the oral field bounded by the ciliary band which includes the acron, by this stage having lost its stereocilia. Continuing on the arms, the ciliary band consists of cells that are very columnar, in contrast to the squamous cells of the apical region, at the terminal end of which the cells are even more columnar. Development of a dorsoventral axis seems to be connected with cell migration in the ventral direction. This, in turn, leads to egg-axis curvature between gastrula and prism stages and between gastrula and pluteus stages (FUKUSHI, 1959).

Animal and vegetal halves separated at the furrow level between an_2 and veg_1 (at the embryo equator), and turned parallel to each other, exhibit uniform bilateral symmetry after growing together, indicating that regulation has taken place. Imperfect healing results in bilaterally symmetrical animal halves in which the direction of invagination is determined by the entoderm and brought into contact with a stomodeum (HÖRSTADIUS and WOLSKY, 1937). It seems unlikely that bilateral symmetry in the vegetal region is determined by the ectoderm.

When the animal regions of whole *Paracentrotus* blastulae are exposed to localized ultraviolet radiation, the bilateral pattern of the attachment zone is suppressed, and when they are exposed to lithium, a homogenous ring of primary mesenchyme cells appears at a more animal level than in normal development. In the region of the esophagus, lithium treatment changes the ordinarily radially symmetrical entoderm to bilaterally symmetrical. This possibly indicates an inherent tendency toward bilateral symmetry, even in the most vegetal region of the intestine. Intestinal symmetry is normally determined by interaction with the ventral ectoderm (CZIHAK, 1962; GUSTAFSON and WOLPERT, 1963, 1967).

When embryos are pretreated with potassium-free seawater, the entomesoderm develops a dorsoventral axis during early invagination, and a small, extra intestine and an extra bilateral skeleton develop (RUNNSTRÖM, 1925a).

Micromere implantation along the egg axis induces intestine and leads to extra skeleton formation. In the most animal region a skeletal system that is bilateral in relation to the extra intestine and which may mirror the larger main skeleton is formed. At a lower level (between an_1 and an_2) it induces a new skeletal piece that may form a bilateral pair with one of the prospective skeletal pieces (Fig. 21.12) (HÖRSTADIUS, 1935).

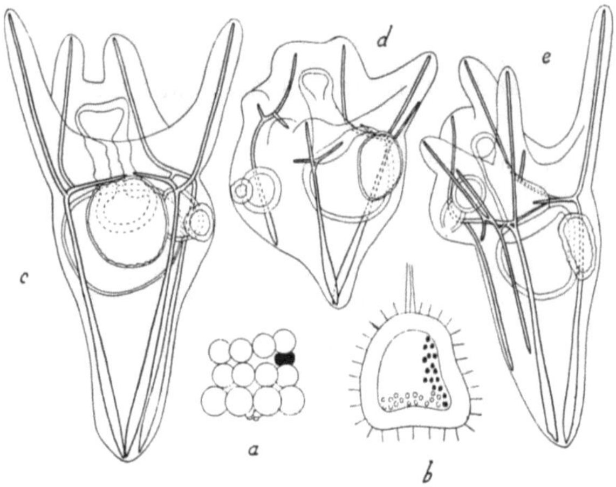

Fig. 21.12. (d) Pluteus from embryo in which 4 micromeres were implanted in one place between mes_1 and mes_2. New ventral side has appeared, in addition to original ventral side. They have the same skeleton. (From HÖRSTADIUS, 1935)

It is clear that micromeres, as well as animal halves, induce bilateral symmetry - usually a new vegetal gradient system that interacts with the existing animal gradient system to form at least an accessory bilateral skeleton. Isolated micromeres do not exhibit bilateral symmetry (HÖRSTADIUS, 1935; HAGSTRÖM and LÖNNING, 1965).

Even before fertilization, it is possible that a dorsoventral gradient is present and that substances diffusing or migrating along it add to the properties of the animal-vegetal gradient. Two oral fields may develop, one better differentiated than the other, although under normal conditions and in weakly elongated embryos, differentiation of more than one oral field is suppressed. The dorsal-side ectoderm becomes more uniform, its epithelium made up mainly of squamous cells.

Dorsal-side animal-vegetal gradients are indicated by the level on which the skeleton is formed. Gradual thickening in the dorsal direction and an increasing number of spines are evidence of a dorsoventral gradient. The slope of the animal-vegetal gradient in relation to the egg axis is greater on the ventral side than on the dorsal side, possibly because diffusion is faster, signals to the genome are more frequent, and template activity is stronger (RUNNSTRÖM, 1925b, 1931; LINDAHL, 1932a, 1936, 1941).

When blastomeres are separated at the 2-cell stage, into approximate prospective ventral side and approximate prospective dorsal side, polarity in the latter is reversed (HÖRSTADIUS and WOLSKY, 1937).

Synthetic metabolism may have a higher rate in the ventral region than the dorsal region. The tuft of stereocilia disappears when acron cells become more motile and migrate in a ventral direction. Anterior to the stomodeum they form protrusions called vela which induce completion of the anterior part of the stomodeum (RUNNSTRÖM, 1958; FUKUSHI, 1959).

The beginning of the asymmetry that is so important in later sea-urchin development is still not understood. But it is first noticeable when the left lobe of the incompletely divided coelom anlage becomes larger than the right (RUNNSTRÖM, 1917). When part of the hydrocoel anlage is irradiated with ultraviolet irradiation, it regenerates from the stone canal. Irradiation of the left lobe tends to make the hydropore grow and the right lobe try to compensate. When regeneration on the left quickens, growth on the right is arrested (CZIHAK, 1964).

The vestibulum (or echinoid invagination) begins in the left oral field of the ectoderm without help from the hydrocoel or the coelom. Then the stone canal and the hydrocoel grow, and the hydrocoel becomes attached to the bottom of the vestibulum and induce formation of the 5 primary ambulacral feet. The ectoderm then covers the mesodermal parts of the feet (RUNNSTRÖM, 1917, 1918).

When the left hydrocoel and the left coelom are removed, the vestibulum still grows normally (CZIHAK, 1964). Premature asymmetry in starfishes indicate early primary gradient-like organization of cytoplasm (RUNNSTRÖM, 1920; HÖRSTADIUS, 1925, 1928).

21.5 Summary

Any theoretical explanation of sea-urchin embryo development must take into account the evident competition between animal and vegetal differentiation processes. Morphogenes *an* and *veg* diffuse and form the double gradient system that divides the egg and embryo into levels indicated by *an/veg* ratios. Extracted and purified, *an* morphogenes, as well as exogenic morphogenes like trypsin and lithium, and vegetal morphogenes, prove to be effectors. Morphogenes probably originate in the oocyte but after a certain stage of development come from producer genes (Fig. 21.13a).

In older embryos treated with lithium and transferred to normal seawater, a new ectoderm may develop in the enlarged entoderm. A cell may begin to develop into ectoderm, under the effect of lithium change into entoderm, and then change back into an ectodermic bud. When the bud begins to form, HALE staining in nuclei is stronger than in the surrounding entoderm. At the level *an/veg* >2 (Fig. 21.13a), gene group A is more often hit by animal morphogenes than gene group V by vegetal morphogenes. In the vegetal direction, this ratio decreases and is later reversed.

The gene content of nuclei throughout the embryo is assumed to be the same: gene groups having one or more regulatory genes (A_1, A_2, for instance; Fig. 21.13) that react with *an* and *veg* and then activate the structural or "producer" genes.

The *an* and *veg* morphogenes genetically formed during development tend to keep the *an/veg* ratios constant. In regulation and in bud processes, it is possible that morphogenes are replenished genetically. If the nucleus containing the A and V gene groups were moved in the vegetal direction, the surrounding *an/veg* ratio and the activity of the gene groups would change, as in embryos treated with lithium. Because *an* diffusion is inhibited, the *an/veg* ratio declines (Fig. 21.1).

Gene groups Z' and Z^n are activated by morphogenes z' and z^n. Z' contains information for oral proteins and represses the activity of gene group A. This facilitates acron cell change, cell migration in the ventral direction, and loss of stereocilia. Z^n is involved in the for-

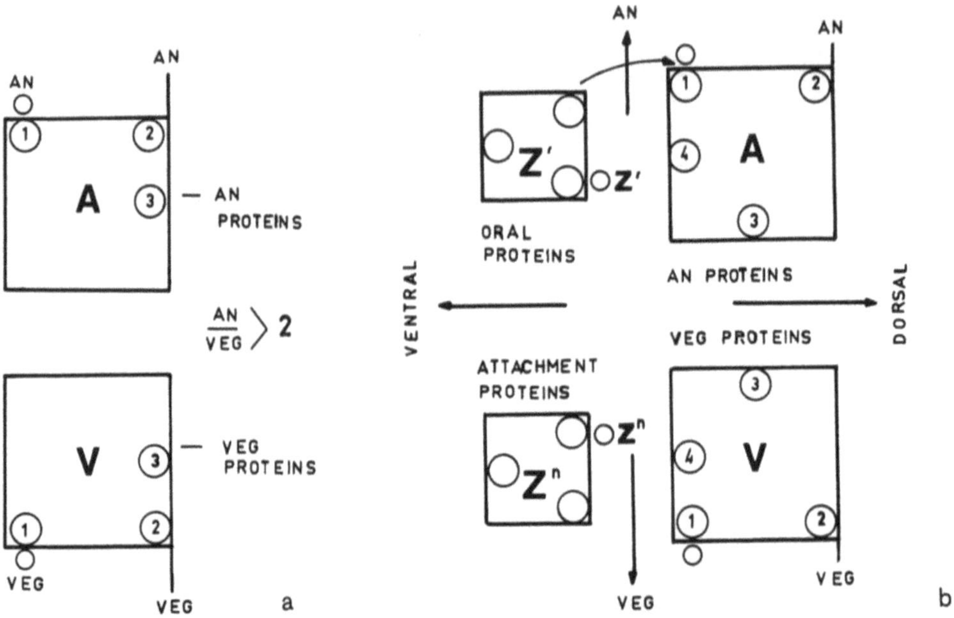

Fig. 21.13a-b. (a) Early blastula stage; 2 gene groups (A and V) and 3 regulatory genes (1, 2, and 3); 1 and 2 produce vegetal morphogenes; 3, animal. When morphogenes begin to form is not indicated. (b) Late blastula or gastrula stage; gene groups Z' and Z^n are activated

mation of the attachment zone and, like Z' codes for proteins that favor the accumulation of ectoderm cells in the ventral direction.

In the early formation of the attachment zone, its nuclei, like other nuclei of the ectoderm, are stained with HALE reagent after pretreatment with a buffer of pH 7.6. Pretreatment at pH 5.1, however, stains most of the ectoderm nuclei weakly and the attachment zone nuclei clearly. This suggests that producer genes in the Z^n group are repressed by nuclear proteins and that increased activity in the attachment zone loosens the binding of nuclear proteins and DNA.

The oral field gradient (decreased cytochrome oxidase content in the dorsoventral direction) probably indicates fewer active mitochondria in that direction (CZIHAK, 1961). In the dorsoventral organization of the embryo, Z' and Z^n groups are on the left (Fig. 21.13b) and A and V cylinders, the first active parts of the genome, radically symmetrical, on the right. The more ventral Z' and Z^n are, the more fully they are activated, and the more dorsal the nuclei are, the more Z' and Z^n are repressed.

The direction of the dorsoventral gradient seems to be determined before fertilization, but it is less stable than the animal-vegetal gradient. The dorsoventral axis may be determined by mechanically elongating the unfertilized egg almost perpendicular to the egg axis (BOVERI, 1901a, LINDAHL, 1932a).

It is possible that metabolism is higher in the ventral direction than in the dorsal direction (LINDAHL, 1941). Z'-Z^n repression may occur earlier in the ventral region of the embryo than in the dorsal region.

References

ALBERTSSON, P.Å., 1960. Partition of cell particles and macromolecules. Uppsala: Almqvist and Wiksell, p. 1-231.

BÄCKSTRÖM, S., 1966. A complex between basic proteins and acid polysaccharides in sea urchin oocytes and eggs. Acta embryol. morph. exp. 9, 37-43.

BÄCKSTRÖM, S., 1966. Basic proteins in parthenogenetically activated sea urchin eggs. Acta embryol. morph. exp. 9, 83-86.

BÄCKSTRÖM, S., 1969. Passage de substance radioactive entre un greffon marqué et son hôte non marque. Etude autoradiographique de l'embryon d'oursin soumis a la transplantation. C.R. Acad. Sci. Paris 269, 1684-1685.

BONNER, J., HUANG, R.C., 1964. Role of histone in chromosomal RNA synthesis. In: The nucleohistones (J.Bonner, P. Ts'O, eds.), pp. 251-261. San Francisco: Holden-Day, Inc.

BONNER, J., HUANG, R.C. Histones as specific repressors of chromosomal RNA synthesis. In: Histones (A.V.S. de Reuck, J. Knight, eds.), pp. 18-33. London: Churchill, Ltd.

BONNER, J., DAHMUS, M.E., FAMBROUGH, D., HUANG, R.C., MARUSHIGE, K., TUAN, D.Y., 1968. The Biology of Isolated Chromatin Chromosomes, biologically active in the test tube, provide a powerful tool for the study of gene action. Science 159, 47-56.

BOOIJ, H.L., BUNGENBERG de JONG, H.G., 1956. Biocolloids and their Interactions. In: Protoplasmatologia (L.V. Heilbrunn, F. Weber, eds.), pp. 1-162. Wien: Springer Verlag.

BOVERI, Th., 1889. Ein geschlechtlich erzeugter Organismus ohne mütterliche Eigenschaften. Sitz.-Ber. Ges. Morphl. Phys. München 5, 78-80.

BOVERI, Th., 1901. Über die Polarität des Seeigels. Verh. phys.-med. Ges. Würzburg 34, 145-176.

BRITTEN, R.J., DAVIDSON, E.H., 1969. Gene regulation for higher cell. A theory. Science 165, 349-357.

CRICK, F., 1970. Diffusion in embryogenesis. Nature 225, 420-422.

CZIHAK, G., 1960. Untersuchungen über die Coelomanlagen und die Metamorphose des Pluteus von Psammechinus miliaris (Gmelin). Zool. Jahrb. Abt. Anat. Ontog. d. Tiere 78, 235-266.

CZIHAK, G., 1961. Ein neuer Gradient in der Pluteusentwicklung. Wilhelm Roux Arch. Entwicklungs-mech. Organ. 153, 353-356.

CZIHAK, G., 1962. Entwicklungsphysiologische Untersuchungen an Echiniden (Topochemie der Blastula und Gastrula, Entwicklung der Bilateral und Radiär-Symmetrie und der Coelomdivertikel). Wilhelm Roux Arch. Entwicklungs-mech. Organ. 154, 29-55.

CZIHAK, G., MEYER, G.F., 1964. Differentiation of the oral field of the sea urchin embryo (Psammechinus miliaris). Nature 201, 315.

CZIHAK, G., 1965. Entwicklungsphysiologische Untersuchungen an Echiniden. Wilhelm Roux Arch. Entwicklungs-mech. Organ. 155, 709-729.

DE ANGELIS, E., RUNNSTRÖM, J., 1970. The effect of temporary treatment of animal half embryos with lithium and the modification of this effect by simultaneous exposure to actinomycin D. Wihelm Roux Arch. Entwicklungs-mech. Organ. 164, 236-246.

DE VINCENTIIS, M., LANCIERI, M., 1970. Observations on the development of the sea urchin embryo in the presence of actinomycin. Exp. Cell Res. 59, 479-481.

DRIESCH, H., 1891. Entwicklungsmechanische Studien. I. Der Wert der beiden ersten Furchungszellen in der Echinodermenentwicklung. Z. wiss. Zool. 53, 160-184.

DRIESCH, H., 1894. Analytische Theorie der organischen Entwicklung. Leipzig: Engelmann-Verlag.

FOERSTER, M., ÖRSTRÖM, Å., 1933. Observations sur la prédétermination de la partie ventrale dans l'oeuf d'oursin. Travaux Stat. Biol. Roscoff 11, 63-83.

FUKUSHI, T., 1959. The presumptive position of the dorsal and ventral areas in the sea urchin egg, studied with local vital staining. Bull. Mar. biol. Sta. Asamushi, Tohoky Univ. 9, 127-132.
GIUDICE, G., HÖRSTADIUS, S., 1965. Effect of actinomycin D on the segregation of animal and vegetal potentialities in the sea urchin egg. Exp. Cell Res. 39, 117-120.
GUSTAFSON, T., 1946. Observations on enlarged polar bodies and oocytary twins in Psammechinus miliaris (Gmelin). Arkiv f. Zool. (Stockholm) 38A (4), 1-10.
GUSTAFSON, T., 1952. Studies on the determination of the oral side of the sea urchin egg. II. The dorsoventral structure of the unfertilized egg. Arkiv f. Zool. (Stockholm) 3 (19), 273-282.
GUSTAFSON, T., 1963. Cellular mechanisms in the morphogenesis of the sea urchin embryos. Cell contacts within the entoderm and between mesenchyme and ectoderm cells. Exp. Cell Res. 32, 570-589.
GUSTAFSON, T., WOLPERT, L., 1962. Cellular mechanism in the morphogenesis of the sea urchin larva. Changes in shape of the cell sheets. Exp. Cell Res. 27, 260-279.
GUSTAFSON, T., WOLPERT, L., 1963. The cellular basis of morphogenesis and sea urchin development. Inter. Rev. Cytol. 15, 139-214.
GUSTAFSON, T., WOLPERT, L., 1967. Cellular movement and contact in sea urchin morphogenesis. Biol. Revs. 42, 442-498.
HAGSTRÖM, B.E., LÖNNING, S., 1965. Studies of cleavage and development of isolated sea urchin blastomeres. Sarsia 18, 1-9.
HAGSTRÖM, B.E., LÖNNING, S., 1967. Cytological and morphological studies on the action of lithium on the development of the sea urchin embryo. Wilhelm Roux Arch. Entwicklungs-mech. Organ. 158, 1-12.
HERBST, C., 1892. Experimentelle Untersuchungen über den Einfluß der veränderten chemischen Zusammensetzung des umgebenden Mediums. Z. wiss. Zool. 55, 446-518.
HERBST, C., 1893. Experimentelle Untersuchungen über den Einfluß der veränderten chemischen Zusammensetzung des umgebenden Mediums auf die Entwicklung der Tiere, II. Mitt. Zool. Stn. Neapel 11, 136-220.
HNILICA, L.S., 1967. Proteins in the cell nucleus. Prog. Nucleic Acid Res. Mol. Biol. 7, 25-106.
HNILICA, L.S., JOHNSON, A.W., 1970. Fractionation and analysis of nuclear proteins in the sea urchin embryos. Exp. Cell Res. 63, 261-270.
HÖRSTADIUS, S., 1925. Entwicklungsmechanische Studien an Asterina gibbosa. Arkif f. Zool. (Stockholm) 17B (6), 1-6.
HÖRSTADIUS, S., 1927. Studien über die Determination bei Paracentrotus lividus, L[k]. Wilhelm Roux Arch. Entwicklungs-mech. Organ. 112, 239-246.
HÖRSTADIUS, S., 1928. Über die Determination des Keimes bei Echinodermen. Acta Zool. (Stockholm) 9, 1-191.
HÖRSTADIUS, S., 1935. Über die Determination im Verlaufe der Eiachse bei Seeigeln. Pubbl. Staz. Zool. Napoli 14, 251-429.
HÖRSTADIUS, S., 1938. Schnürungsversuche an Seeigelkeimen. Wilhelm Roux Arch. Entwicklungs-mech. Organ. 138, 197-259.
HÖRSTADIUS, S., 1965a. Über fortschreitende Determination in Furchinsstadien von Seeigeleiern. Z. Naturforsch. 20, 331-333.
HÖRSTADIUS, S., 1965b. Über die animalisierende Wirkung von Trypsin auf Seeigelkeime. Zool. Jahrb. Abt. Allgem. Zool. Physiol. 71, 241-244.
HÖRSTADIUS, S., JOSEFSSON, L., RUNNSTRÖM, J., 1967. Morphogenetic agents from unfertilized eggs of the sea urchin Paracentrotus lividus. Develop. Biol. 16, 189-202.

HÖRSTADIUS, S., WOLSKY, A., 1936. Studien über die Determination der Bilateralsymmetrie des jungen Seeigelkeimes. Wilhelm Roux Arch. Entwicklungs-mech. Organ. 135, 69-113.
HUANG, R.C., BONNER, J., 1965. Histon-bound RNA, a component of nature. Nucleohistone. Proc. Nat. Acad. Sci. USA 54, 960-967.
HULTIN, T., 1961a. Activation of ribosomes in sea urchin eggs in response to fertilization. Exp. Cell Res. 25, 405-417.
HULTIN, T., 1961b. The effect of puromycin on protein metabolism and cell division in fertilized sea urchin eggs. Experientia 17, 410-413.
IMMERS, J., RUNNSTRÖM, J., 1965. Further studies of the effects of deprivation of sulfate on the early development of the sea urchin Paracentrotus lividus. J. Embryol. Exp. Morph. 14, 289-305.
IMMERS, J., RUNNSTRÖM, J., in preparation. Nuclear and cytoplasmic changes in early development of lithium treated sea urchin embryos.
IMMERS, J., RUNNSTRÖM, J., in preparation. Nuclear and cytoplasmic changes during the advanced development of lithium treated sea urchin larvae.
IMMERS, J., MARKMAN, B., RUNNSTRÖM, J., 1967. Nuclear changes in the course of development of the sea urchin studied by means of Hale staining. Exp. Cell Res. 47, 425-442.
JACOB, F., MONOD, J., 1963. Elements of regulatory circuits in bacteria. In: Biological Organization at the Cellular and Supercellular Level (R.J.C. Harris, ed.), pp. 1-24. New York: Academic Press.
JOHNSON, A.W., HNILICA, L.S., 1970. In vitro synthesis and nuclear proteins of isolated sea urchin embryo nuclei. Biochim. Biophys. Acta 224, 518-530.
JOSEFSSON, L., HÖRSTADIUS, S., 1970. Isolation of animalizing and vegetalizing substances from unfertilized eggs of Paracentrotus lividus. Develop. Biol. 20, 481-500.
LALLIER, R., 1955. Recherches sur le problème de la détermination embryonnaire chez les amphibiens et chez les echinodermes. Arch. biol. (Liège) 66, 223-402.
LALLIER, R., 1956. Recherches sur la détermination chez les echinodermes. L'action des oxydants et de l'acide p. Chloromercuribenzioque sur le développement de l'oeuf de Paracentrotus lividus. Arch. biol. (Liège) 67, 181-209.
LALLIER, R., 1958. Analyse éxpérimentale de la différenciation embryonnaire chez les echinodermes. Experientia 14, 309-315.
LALLIER, R., 1964. Biochemical aspects of animalization and vegetalization of the sea urchin embryos. Advan. Morphogenesis 3, 148-196.
LINDAHL, P.E., 1932a. Zur experimentellen Analyse der Determination der Dorsoventralachse beim Seeigelkeim. I. Versuche mit gestreckten Eiern. Wilhelm Roux Arch. Entwicklungs-mech. Organ. 127, 300-322.
LINDAHL, P.E., 1932b. Zur experimentellen Analyse der Determination der Dorsoventralachse beim Seeigelkeim. II. Versuche mit zentrifugierten Eiern. Wilhelm Roux Arch. Entwicklungs-mech. Organ. 127, 323-338.
LINDAHL, P.E., 1936. Zur Kenntnis der physiologischen Grundlagen der Determination im Seeigelkeim. Acta Zool. (Stockholm) 17, 179-365.
LINDAHL, P.E., 1941. Physiologische Probleme der Entwicklung und Formbildung des Seeigelkeimes. Naturwiss. 29, 673-685.
LINDAHL, P.E., 1942. Contribution to the physiology of form generation in the sea urchin development. Quart. Rev. Biol. 17, 213-217.
LINDAHL, P.E., ÖRSTRÖM, Å., 1932. Beiträge zur Kenntnis des Pigmentringes in dem Ei von Paracentrotus lividus. Protoplasma 17, 25-31.
LOEWENSTEIN, W.R., 1967. Cell surface membranes in close contact. Role of calcium and magnesium ions. J. Colloid interphase Sci. 25, 34-46.

MACARTHUR, J.W., 1924. An experimental study and a physiological interpretation of exogastrulation and related modifications in echinoderm embryos. Biol. Bull. 46, 60-87.
MASTRANGELO, FUDGE, M., 1959. Vegetalization of sea urchin embryos by treatment with tyrosine. Exp. Cell Res. 18, 401-414.
MASTRANGELO, FUDGE, M., 1965. A study of the vegetalizing action of tyrosine on the sea urchin embryo. Dissertation for the Degree of Ph.D., Yale University.
MASTRANGELO, FUDGE, M., 1966. Analysis of the vegetalizing action of tyrosine on the sea urchin embryo. J. Exp. Zool. 161, 109-128.
MARKMAN, B., 1963. Morphogenetic effects of some nucleotide metabolites and antibiotics on early sea urchin development. Arkiv f. Zool. (Stockholm) 16, 207-217.
MARKMAN, B., 1963-65. Histochemical and autoradiographical studies on the role of the nucleus in the early development of sea urchin. Swedish Cancer Soc. Yearbook 4, 494-499.
MARKMAN, B., RUNNSTRÖM, J., 1963. Animal and vegetal halves of sea urchin larvae subjected to temporary treatment with actinomycin D and mitomycin D. Exp. Cell Res. 31, 615-618.
MARKMAN, B., RUNNSTRÖM, J., 1970. The removal by actinomycin D of the effect on endogenous and exogenous agents in sea urchin development. Wilhelm Roux Arch. Entwicklungs-mech. Organ. 165, 1-7.
MORGAN, T.H., 1894. Experimental studies on echinoderm eggs. Anat. Anz. 9, 141-152.
MORGAN, T.H., SPOONER, G.B., 1909. The polarity of the centrifuged eggs. Wilhelm Roux Arch. Entwicklungs-mech. Organ. 28, 104-117.
MOTOMURA, I., 1931. Notes on the effect of centrifugal force on the frog eggs. Sci. Rep. Tohoku Imp. Univ. Serie 6, 4.
MOTOMURA, I., 1949. Artificial alteration of the embryonic axis in the centrifuged eggs of sea urchins. Sci. Rep. Tohoky Univ. Biol. 18, 117-125.
MÜLLER, W., 1962. Bindung von Actinomycinen und Actinomycin-Derivaten an Desoxyribonucleinsäure. Naturwiss. 49, 156-157.
ORENGO, A., HNILICA, L.S., 1970. In vivo incorporation of labelled amino acids into nuclear proteins of the sea urchin embryo. Exp. Cell Res. 62, 331-337.
PEASE, D., 1939. An analysis of the factors of bilateral determination in centrifuged echinoderm embryos. J. exp. Zool. 80, 117-125.
RANZI, S., 1962. The proteins in embryonic and larval development. Adv. Morphogenesis 2, 211-257.
REICH, E., GOLDBERG, I.H., 1964. Actinomycin and nucleic acid function. Prog. Nuc. Acid Res. Mol. Biol. 3, 84-324.
RUNNSTRÖM, J., 1914. Analytische Studien über die Seeigelentwicklung. Wilhelm Roux Arch. Entwicklungs-mech. Organ. 40, 526-564.
RUNNSTRÖM, J., 1917. Analytische Studien über die Seeigelentwicklung. III Mitt. Wilhelm Roux Arch. Entwicklungs-mech. Organ. 43, 223-328.
RUNNSTRÖM, J., 1917-1918. Zur Entwicklungsmechanik der Larve von Parechinus miliaris. Bergens Museums Aarbok. Nat. Sci. 14, 1-23.
RUNNSTRÖM, J., 1918. Analytische Studien über die Seeigelentwicklung. IV Mitt. Wilhelm Roux Arch. Entwicklungs-mech. Organ. 43, 409-431.
RUNNSTRÖM, J., 1920. Entwicklungsmechanische Studien an Henricia sanguinolenta Forbes und Solaster spec. Wilhelm Roux Arch. Entwicklungs-mech. Organ. 46, 459-484.
RUNNSTRÖM, J., 1925a. Über den Einfluß des Kaliummangels auf das Seeigelei. Pubbl. Staz. Zool. Napoli 6, 2-199.
RUNNSTRÖM, J., 1925b. Experimentelle Bestimmung der Dorso-Ventralachse bei dem Seeigelkeim. Arkiv f. Zool. (Stockholm) 18A (4), 1-6.
RUNNSTRÖM, J., 1925c. Zur experimentellen Analyse der Entwicklung von Antedon. Wilhelm Roux Arch. Entwicklungs-mech. Organ. 105, 63-113.
RUNNSTRÖM, J., 1926. Über die Verteilung der Potenzen der Urdarmbildung bei dem Seeigelkeim. Acta Zool. (Stockholm) 7, 117-121.

RUNNSTRÖM, J., 1928a. Plasmabau und Determination bei dem Ei von Paracentrotus lividus Lk. Wilhelm Roux Arch. Entwicklungs-mech. Organ. 113, 556-581.
RUNNSTRÖM, J., 1928b. Zur experimentellen Analyse der Wirkung des Lithium auf dem Seeigelkeim. Acta Zool. (Stockholm) 9, 365-424.
RUNNSTRÖM, J., 1928c. Über die Veränderung der Plasmakolloide bei der Entwicklung des Seeigelkeimes. Protoplasma 5, 201-310.
RUNNSTRÖM, J., 1931. Zur Entwicklungsmechanik des Skelettmusters bei dem Seeigelkeim. Wilhelm Roux Arch Entwicklungs-mech. Organ. 124, 273-297.
RUNNSTRÖM, J., 1933. Kurze Mitteilung zur Physiologie der Determination des Seeigelkeimes. Wilhelm Roux Arch. Entwicklungs-mech. Organ. 129, 442-444.
RUNNSTRÖM, J., 1935. An analysis of the action of lithium on sea urchin development. Biol. Bull. 68, 378-384.
RUNNSTRÖM, J., 1955a. Die Analyse der primären Differenzierungsvorgänge im Seeigelkeim. In: Verh. Deutsch. Zool. Ges. Tübingen, pp. 32-68.
RUNNSTRÖM, J., 1955b. Changes in the submicroscopic structure of the cytoplasm attending maturation and activation of the sea urchin egg. Exp. Cell Res. 8, 49-61.
RUNNSTRÖM, J., 1957. Cellular structure and behavior under influence of animal and vegetal factors in sea urchin development. Arkiv f. Zool. (Stockholm) 10, 523-537.
RUNNSTRÖM, J., 1961a. Effect of pretreatment of the sea urchin egg with trypsin of different doses with respect to cortical changes, cleavage and further development. Exp. Cell Res. 22, 576-608.
RUNNSTRÖM, J., 1961b. The role of nuclear metabolism in the determination of the sea urchin egg. Path. Biol. 9, 781-785.
RUNNSTRÖM, J., 1962. Differential effects of pretreatment of sea urchin eggs (Paracentrotus lividus and Psammechinus miliaris) with low doses of trypsin. Zool. Bidrag, Uppsala, 35, 385-395.
RUNNSTRÖM, J., 1963. Sperm-induced protrusions in sea urchin oocytes: A study of phase separation and mixing in living cytoplasm. Develop. Biol. 7, 38-50.
RUNNSTRÖM, J., 1964. Genetic and epigenic factors involved in the early differentiation of the sea urchin egg (Paracentrotus lividus, Psammechinus miliaris). In: Acidi Nucleici e loro funzione biologica, Ist. Lombarde e Fond. Baselli, pp. 342-351. Milano.
RUNNSTRÖM, J., 1966. Considerations on the control of differentiation in the early sea urchin development. Arch. Zool. Ital. 51, 239-272.
RUNNSTRÖM, J., 1967a. The animalizing action of pretreatment of sea urchin eggs with thiocyanate in calcium-free sea water and its stabilization after fertilization. Arkiv f. Zool. (Stockholm) 19, 251-263.
RUNNSTRÖM, J., 1967b. The mechanism of control of differentiation in early development of the sea urchin. A tentative discussion. Exp. Biol. Med. 1, 52-62.
RUNNSTRÖM, J., 1968. The initiation of the development of the egg with special reference to sea urchin. Atti Accad. naz. Lincei 104, 165-178.
RUNNSTRÖM, J., IMMERS, J., 1966. On the animalizing action of trypsin on the embryos of the sea urchins Psammechinus miliaris and Paracentrotus lividus. Arch. biol. (Liège) 77, 365-410.
RUNNSTRÖM, J., IMMERS, J., 1970. Heteromorphic budding in lithium treated sea urchin embryos. A study of gene expression. Exp. Cell Res. 62, 228-238.
RUNNSTRÖM, J., IMMERS, J., in press 1971. Treatment with lithium as a tool for the study of animal-vegetal interactions in sea urchin embryos. Wilhelm Roux Arch. Entwicklungs-mech. Organ.

RUNNSTRÖM, J., MANELLI, H., 1964. Induction of polyspermy by treatment of sea urchin eggs with mercurials. Exp. Cell Res. 35, 157-193.
RUNNSTRÖM, J., MANELLI, H., 1967. The stereocilia of the sea urchin embryo, the conditions of their formation and disappearance. Atti Accad. naz. Lincei, fasc. 6, Ser. VIII, 12, 1-8.
RUNNSTRÖM, J., MARKMAN, B., 1966. Gene dependency of vegetalization in sea urchin embryos treated with lithium. Biol. Bull. 130, 402-414.
RUNNSTRÖM, J., RUNNSTRÖM, S., 1920. Über die Entwicklung von Cucumaria frondosa Gunnerus und Psolus phantapus Strussenfelt. Bergens Museums Aarbok, Nat. Sci. 5, 5-99.
RUNNSTRÖM, J., HÖRSTADIUS, S., IMMERS, J., FUDGE-MASTRANGELO, M., 1964. An analysis of the role of sulfate in the embryonic differentiation of the sea urchin (Paracentrotus lividus). Rev. Suisse Zool. 71, 21-54.
SCARANO, E., 1969. Enzymatic modifications of DNA and embryonic differentiation. Ann. d'Embryol. Morphogen. Suppl. 1, 51-61.
SCARANO, E., AUGUSTI-TOCCO, G., 1967. Biochemical pathways in embryos. In: Comprehensive Biochemistry, Vol. 28 (M. Florkin, E.H. Stotz, eds.), pp. 55-110. Amsterdam: Elsevier.
SCARANO, E., DE PETROCELLIS, B., AUGUSTI-TOCCO, G., 1964a. Deoxycytidylate aminohydrolase content in disaggregated cells from sea urchin embryos. Exp. Cell Res. 36, 211-213.
SCARANO, E., DE PETROCELLIS, B., AUGUSTI-TOCCO, G., 1964b. Studies on the control of enzyme synthesis during the early embryonic development of the sea urchins. Biochim. Biophys. Acta 87, 174-176.
SCARANO, E., MAGGIO, R., 1969. The enzymatic deamination of 5'-deoxycytidylic acid and of 5-methyl-5'-deoxycytidylic acid in the developing sea urchin embryo. Exp. Cell Res. 18, 333-346.
SCHLEIP, W., 1929. Determination der Primitivenentwicklung. VIII. Echinodermen, pp. 379-553. Leipzig: Akad. Verlagsgesellschaft.
SCHMIDT, H., 1904. Zur Kenntnis der Larvenentwicklung von Echinus microtuberculatus. Verh. Phys. Med. Ges. Würzburg N.F. 36, 297-336.
SEELIGER, O., 1892. Studien zur Entwicklungsgeschichte der Crinoiden (Antedon rosaea). Zool. Jahrb. Anat. 6, 161-444.
SPEMANN, H., 1936. Experimentelle Beiträge zu einer Theorie der Entwicklung. Berlin: Springer Verlag.
TOMKINS, G.M., GELEHRTER, T.D., GRANNER, D., MARTIN, jr., D., SAMUELS, H., THOMPSON, B., 1968. Control of specific gene expression in higher organisms. First Ann. Midec Lecture, Nat. Inst. Health.
VON UBISCH, L., 1925. Entwicklungsphysiologische Studien an Seeigelkeimen. III. Die normale und durch Lithium beeinflußte Anlage der Primitivorgane bei animalen und vegetativen Halbkeimen von Echinocyamus pusillus. Z. wiss. Zool. 124, 449-486.

Subject Index

MIX
Papier aus verantwortungsvollen Quellen
Paper from responsible sources
FSC® C105338

If you have any concerns about our products,
you can contact us on
ProductSafety@springernature.com

In case Publisher is established outside the EU,
the EU authorized representative is:
Springer Nature Customer Service Center GmbH
Europaplatz 3, 69115 Heidelberg, Germany

Printed by Libri Plureos GmbH
in Hamburg, Germany